Efficient Petrochemical Processes

Efficient Petrochemical Processes

Technology, Design and Operation

Frank (Xin X.) Zhu, James A. Johnson, David W. Ablin, and Gregory A. Ernst

This edition first published 2020
© 2020 John Wiley & Sons, Inc.

All rights reserved. No part of this publication may be reproduced, stored in a retrieval system, or transmitted, in any form or by any means, electronic, mechanical, photocopying, recording or otherwise, except as permitted by law. Advice on how to obtain permission to reuse material from this title is available at http://www.wiley.com/go/permissions.

The right of Frank (Xin X.) Zhu, James A. Johnson, David W. Ablin, and Gregory A. Ernst to be identified as the authors of this work has been asserted in accordance with law.

Registered Office
John Wiley & Sons, Inc., 111 River Street, Hoboken, NJ 07030, USA

Editorial Office
111 River Street, Hoboken, NJ 07030, USA

For details of our global editorial offices, customer services, and more information about Wiley products visit us at www.wiley.com.

Wiley also publishes its books in a variety of electronic formats and by print-on-demand. Some content that appears in standard print versions of this book may not be available in other formats.

Limit of Liability/Disclaimer of Warranty
In view of ongoing research, equipment modifications, changes in governmental regulations, and the constant flow of information relating to the use of experimental reagents, equipment, and devices, the reader is urged to review and evaluate the information provided in the package insert or instructions for each chemical, piece of equipment, reagent, or device for, among other things, any changes in the instructions or indication of usage and for added warnings and precautions. While the publisher and authors have used their best efforts in preparing this work, they make no representations or warranties with respect to the accuracy or completeness of the contents of this work and specifically disclaim all warranties, including without limitation any implied warranties of merchantability or fitness for a particular purpose. No warranty may be created or extended by sales representatives, written sales materials or promotional statements for this work. The fact that an organization, website, or product is referred to in this work as a citation and/or potential source of further information does not mean that the publisher and authors endorse the information or services the organization, website, or product may provide or recommendations it may make. This work is sold with the understanding that the publisher is not engaged in rendering professional services. The advice and strategies contained herein may not be suitable for your situation. You should consult with a specialist where appropriate. Further, readers should be aware that websites listed in this work may have changed or disappeared between when this work was written and when it is read. Neither the publisher nor authors shall be liable for any loss of profit or any other commercial damages, including but not limited to special, incidental, consequential, or other damages.

Library of Congress Cataloging-in-Publication data

Names: Zhu, Frank, 1957– author. | Johnson, James Albert, 1946– author. |
  Ablin, David William, author. | Ernst, Gregory Allen, 1982– author.
Title: Efficient Petrochemical Processes : Technology, Design and Operation /
  Frank (Xin X.) Zhu, James Albert Johnson, David William Ablin, Gregory Allen Ernst.
Description: First edition. | Hoboken, NJ, USA : Wiley-AIChE, 2020. |
  Includes bibliographical references and index.
Identifiers: LCCN 2019024987 (print) | LCCN 2019024988 (ebook) | ISBN 9781119487869 (hardback) |
  ISBN 9781119487876 (adobe pdf) | ISBN 9781119487883 (epub)
Subjects: LCSH: Aromatic compounds. | Extraction (Chemistry) | Petroleum products.
Classification: LCC TP248.A7 Z45 2020 (print) | LCC TP248.A7 (ebook) | DDC 661/.8–dc23
LC record available at https://lccn.loc.gov/2019024987
LC ebook record available at https://lccn.loc.gov/2019024988

Cover Design: Wiley
Cover Image: UOP Aromatics Technology at the TPPI Aromatics Complex Plant, Indonesia.
© PT Trans-Pacific Petrochemical Indotama, Indonesia. Used with permission.

Set in 10/12pt Warnock by SPi Global, Pondicherry, India

Printed in the United States of America

V10014474_100719

Contents

Preface *xix*
Acknowledgments *xxi*

Part I Market, Design and Technology Overview *1*

1 Overview of This Book *3*
1.1 Why Petrochemical Products Are Important for the Economy *3*
1.1.1 Polyethylene *3*
1.1.2 Polypropylene *3*
1.1.3 Styrene and Polystyrene *4*
1.1.4 Polyester *4*
1.1.5 Polycarbonate and Phenolic Resins *4*
1.1.6 Economic Significance of Polymers *4*
1.1.7 Petrochemicals and Petroleum Utilization *7*
1.2 Overall Petrochemical Configurations *8*
1.3 Context of Process Designs and Operation for Petrochemical Production *11*
1.4 Who Is This Book Written For? *11*

2 Market and Technology Overview *13*
2.1 Overview of Aromatic Petrochemicals *13*
2.2 Introduction and Market Information *13*
2.2.1 Benzene *13*
2.2.2 Benzene Production Technologies *14*
2.2.3 Toluene *16*
2.2.4 Toluene Production Technologies *17*
2.2.5 Ethylbenzene/Styrene *17*
2.2.6 Ethylbenzene/Styrene Production Technologies *17*
2.2.7 *para*-Xylene *17*
2.2.8 *para*-Xylene Production Technologies *18*
2.2.9 *meta*-Xylene *20*
2.2.10 *meta*-Xylene Production Technologies *20*
2.2.11 *ortho*-Xylene *20*
2.2.12 *ortho*-Xylene Production Technologies *20*
2.2.13 Cumene/Phenol *21*
2.2.14 Cumene/Phenol Production Technologies *21*
2.3 Technologies in Aromatics Synthesis *21*
2.4 Alternative Feeds for Aromatics *27*
2.5 Technologies in Aromatic Transformation *28*
2.5.1 Transalkylation *28*
2.5.2 Selective Toluene Disproportionation *30*

2.5.3	Thermal Hydro-Dealkylation	*31*
2.5.4	Xylene Isomerization	*32*
2.6	Technologies in Aromatic Separations	*35*
2.6.1	Liquid–Liquid Extraction and Extractive Distillation	*35*
2.6.2	Liquid–Liquid Extraction	*37*
2.6.3	Extractive Distillation (ED)	*38*
2.7	Separations by Molecular Weight	*39*
2.8	Separations by Isomer Type: *para*-Xylene	*39*
2.8.1	Crystallization of *para*-Xylene	*40*
2.8.2	Adsorptive Separation of *para*-Xylene	*41*
2.9	Separations by Isomer Type: *meta*-Xylene	*44*
2.10	Separations by Isomer Type: *ortho*-Xylene and Ethylbenzene	*45*
2.11	Other Related Aromatics Technologies	*46*
2.11.1	Cyclohexane	*46*
2.11.2	Ethylbenzene/Styrene	*46*
2.11.3	Cumene/Phenol/Bisphenol-A	*48*
2.11.4	Linear Alkyl Benzene Sulfonate for Detergents	*49*
2.11.5	Oxidation of *para*- and *meta*-Xylene	*50*
2.11.6	Melt-Phase Polymerization of PTA to PET	*53*
2.11.7	Melt-Phase Polymerization and Solid State Polycondensation of PET Resin	*55*
2.11.8	Oxidation of *ortho*-Xylene	*57*
2.12	Integrated Refining and Petrochemicals	*57*
	References	*61*
3	**Aromatics Process Description**	***63***
3.1	Overall Aromatics Flow Scheme	*63*
3.2	Adsorptive Separations for *para*-Xylene	*64*
3.3	Technologies for Treating Feeds for Aromatics Production	*68*
3.4	*para*-Xylene Purification and Recovery by Crystallization	*68*
3.5	Transalkylation Processes	*71*
3.6	Xylene Isomerization	*72*
3.7	Adsorptive Separation of Pure *meta*-Xylene	*76*
3.8	*para*-Selective Catalytic Technologies for *para*-Xylene	*78*
3.8.1	*para*-Selective Toluene Disproportionation	*78*
3.8.2	*para*-Selective Toluene Methylation	*79*
	References	*81*
	Part II Process Design	***83***
4	**Aromatics Process Unit Design**	***85***
4.1	Introduction	*85*
4.2	Aromatics Fractionation	*85*
4.2.1	Reformate Splitter	*85*
4.2.2	Xylene Fractionation	*86*
4.2.3	Heavy Aromatics Fractionation	*87*
4.3	Aromatics Extraction	*88*
4.3.1	Liquid–Liquid Extraction	*89*
4.3.1.1	Operating Variables	*91*
4.3.2	Extractive Distillation	*92*
4.3.2.1	Operating Variables	*94*
4.4	Transalkylation	*96*
4.4.1	Process Flow Description	*96*
4.4.1.1	Combined Feed Exchanger	*97*

4.4.1.2	Charge Heater	*98*
4.4.1.3	Reactor Design	*99*
4.4.1.4	Catalyst Volume	*99*
4.4.1.5	Bed Pressure Drop	*99*
4.4.1.6	Reactor Bed Dimensions	*100*
4.4.1.7	Products Condenser	*100*
4.4.1.8	Separator	*100*
4.4.1.9	Recycle Gas Purity	*100*
4.4.1.10	Recycle Gas Compressor	*100*
4.5	Xylene Isomerization	*101*
4.5.1	Combined Feed Exchanger	*102*
4.5.2	Charge Heater	*102*
4.5.3	Reactor Design	*103*
4.5.4	Catalyst Volume	*104*
4.5.5	Radial Flow Reactor Sizing	*104*
4.5.6	Products Condenser	*104*
4.5.7	Separator	*104*
4.5.8	Recycle Gas Purity	*104*
4.5.9	Recycle Gas Compressor	*105*
4.6	*para*-Xylene Separation	*105*
4.7	Process Design Considerations: Design Margin Philosophy	*106*
4.7.1	Equipment Design Margins	*107*
4.7.1.1	Fired Heaters	*107*
4.7.1.2	Process–Process Heat Exchangers and Water-Cooled Heat Exchangers	*108*
4.7.1.3	Air-Cooled Heat Exchangers	*108*
4.7.1.4	Pumps	*108*
4.7.1.5	Compressors	*108*
4.7.1.6	Fractionation Columns	*108*
4.7.1.7	Reactors	*108*
4.8	Process Design Considerations: Operational Flexibility	*108*
4.9	Process Design Considerations: Fractionation Optimization	*109*
4.10	Safety Considerations	*110*
4.10.1	Reducing Exposure to Hazardous Materials	*110*
4.10.2	Process Hazard Analysis (PHA)	*110*
4.10.3	Hazard and Operability (HAZOP) Study	*110*
	Further Reading	*111*
5	**Aromatics Process Revamp Design**	*113*
5.1	Introduction	*113*
5.2	Stages of Revamp Assessment and Types of Revamp Studies	*113*
5.3	Revamp Project Approach	*115*
5.3.1	Specified Target Capacity	*115*
5.3.2	Target Production with Constraints	*115*
5.3.3	Maximize Throughput at Minimum Cost	*115*
5.3.4	Identify Successive Bottlenecks	*116*
5.4	Revamp Study Methodology and Strategies	*116*
5.5	Setting the Design Basis for Revamp Projects	*118*
5.5.1	Agreement	*118*
5.5.2	Processing Objectives	*119*
5.5.3	Define the Approach of the Study	*119*
5.5.4	Feedstock and Make-Up Gas	*119*
5.5.5	Product Specifications	*119*
5.5.6	Getting the Right Equipment Information	*119*
5.5.7	Operating Data or Test Run Data	*120*

5.5.8	Constraints	*120*
5.5.9	Utilities	*120*
5.5.10	Replacement Equipment Options	*121*
5.5.11	Guarantees	*121*
5.5.12	Economic Evaluation Criteria	*121*
5.6	Process Design for Revamp Projects	*121*
5.6.1	Adjusting Operating Conditions	*121*
5.6.2	Design Margin	*122*
5.7	Revamp Impact on Utilities	*123*
5.8	Equipment Evaluation for Revamps	*124*
5.8.1	Fired Heater Evaluation	*124*
5.8.1.1	Data Required	*124*
5.8.1.2	Fired Heater Evaluation	*124*
5.8.1.3	Heater Design Limitations	*125*
5.8.1.4	Radiant Flux Limits	*125*
5.8.1.5	TWT Limits	*126*
5.8.1.6	Metallurgy	*126*
5.8.1.7	Tube Thickness	*126*
5.8.1.8	Coil Pressure Drop	*127*
5.8.1.9	Burners	*127*
5.8.1.10	Stack	*128*
5.8.2	Vessels: Separators, Receivers, and Drums	*128*
5.8.2.1	Data Required	*128*
5.8.2.2	Separator, Receiver, and Drum Evaluation	*128*
5.8.2.3	Process and Other Modifications	*129*
5.8.2.4	Test Run Data	*130*
5.8.2.5	Possible Recommendations	*130*
5.8.3	Reactors	*130*
5.8.3.1	Data Required	*131*
5.8.3.2	Reactor Process Evaluation	*131*
5.8.3.3	Process and Other Modifications	*132*
5.8.3.4	Test Run Data	*133*
5.8.3.5	Possible Recommendations	*133*
5.8.4	Fractionator Evaluation	*133*
5.8.4.1	Data Required	*133*
5.8.4.2	Fractionator Evaluation	*133*
5.8.4.3	Retraying and Other Modifications	*134*
5.8.4.4	High-Capacity Trays	*135*
5.8.4.5	Test Run Data	*135*
5.8.4.6	Possible Recommendations	*135*
5.8.5	Heat Exchangers	*136*
5.8.5.1	Data Required	*136*
5.8.5.2	Overall Exchanger Evaluation	*136*
5.8.5.3	Thermal Rating Methods	*136*
5.8.5.4	Rating Procedures	*137*
5.8.5.5	Pressure Drop Estimation	*139*
5.8.5.6	Use of Operating Data	*139*
5.8.5.7	Possible Recommendations	*139*
5.8.5.8	Special Exchanger Services	*140*
5.8.5.9	Overpressure Protection	*141*
5.8.6	Pumps	*141*
5.8.6.1	Data Required	*141*
5.8.6.2	Centrifugal Pump Evaluation	*141*
5.8.6.3	Proportioning Pumps	*142*

5.8.6.4	Use of Operating Data	*142*
5.8.6.5	Possible Recommendations	*142*
5.8.6.6	Tools	*143*
5.8.6.7	Special Pump Services	*143*
5.8.7	Compressors	*143*
5.8.7.1	Data Required	*144*
5.8.7.2	Centrifugal Compressor Evaluation	*144*
5.8.7.3	Reciprocating Compressor Evaluation	*145*
5.8.7.4	Driver Power	*145*
5.8.7.5	Materials of Construction	*145*
5.8.7.6	Use of Operating Data	*145*
5.8.7.7	Potential Remedies	*146*
5.8.8	Hydraulics/Piping	*146*
5.8.8.1	New Unit Line Sizing Criteria Are Generally *Not* Applicable	*146*
5.8.8.2	Pressure Drop Requires Replacement of Other Equipment	*146*
5.8.8.3	Approaching Sonic Velocity	*147*
5.8.8.4	Erosion Concerns	*147*
5.8.8.5	Pressure Drop Affects Yields	*147*
5.8.8.6	Pressure Drop Affects Fractionator Operation or Utilities	*147*
5.9	Economic Evaluation	*147*
5.9.1	Costs	*147*
5.9.1.1	Capital Costs	*147*
5.9.1.2	Operating Costs	*148*
5.9.1.3	Downtime	*148*
5.9.1.4	ISBL Vs. OSBL	*148*
5.9.1.5	Other Costs	*148*
5.9.2	Benefits	*149*
5.9.2.1	Increased Product	*149*
5.9.2.2	Lower Cost Feed	*149*
5.9.2.3	Higher Value Product	*149*
5.9.2.4	Lower Operating Cost	*149*
5.9.3	Data Requirements	*149*
5.9.3.1	Feed/Product Pricing	*149*
5.9.3.2	Utility Pricing	*149*
5.9.3.3	Catalyst/Adsorbent	*150*
5.9.3.4	Other Info	*150*
5.9.4	Types of Economic Analyses	*150*
5.9.4.1	Basic Comparison of Alternatives	*150*
5.9.4.2	Simple Payback	*150*
5.9.4.3	Net Present Value (NPV)	*150*
5.9.4.4	Internal Rate of Return (IRR)	*151*
5.9.4.5	Issues	*152*
5.10	Example Revamp Cases	*152*
5.10.1	Aromatics Complex Revamp with Adsorbent Reload	*152*
5.10.2	Aromatics Complex Revamp with Xylene Isomerization Catalyst Change	*153*
5.10.3	Transalkylation Unit Revamp	*153*
	Further Reading	*154*

Part III Process Equipment Assessment *155*

6 Distillation Column Assessment *157*
6.1 Introduction *157*
6.2 Define a Base Case *157*

x | Contents

6.3	Calculations for Missing and Incomplete Data	*159*
6.4	Building Process Simulation	*161*
6.5	Heat and Material Balance Assessment	*162*
6.5.1	Material Balance Assessment	*162*
6.5.2	Heat Balance Assessment	*164*
6.6	Tower Efficiency Assessment	*164*
6.7	Operating Profile Assessment	*166*
6.8	Tower Rating Assessment	*168*
6.9	Guidelines for Existing Columns	*169*
	Nomenclature	*170*
	Greek Letters	*170*
	References	*170*

7 Heat Exchanger Assessment *171*
7.1 Introduction *171*
7.2 Basic Calculations *171*
7.3 Understand Performance Criterion: U-Values *173*
7.3.1 Required U-Value (U_R) *174*
7.3.2 Clean U-Value (U_C) *174*
7.3.3 Actual U-Value (U_A) *175*
7.3.4 Overdesign (OD_A) *176*
7.3.5 Controlling Resistance *176*
7.4 Understand Fouling *176*
7.4.1 Root Causes of Fouling *176*
7.4.2 Estimate Fouling Factor R_f *177*
7.4.3 Determine Additional Pressure Drop Due to Fouling *177*
7.5 Understand Pressure Drop *178*
7.5.1 Tube-Side Pressure Drop *178*
7.5.2 Shell-Side Pressure Drop *178*
7.6 Effects of Velocity on Heat Transfer, Pressure Drop, and Fouling *178*
7.6.1 Heat Exchanger Rating Assessment *179*
7.6.2 Assess the Suitability of an Existing Exchanger for Changing Conditions *179*
7.6.3 Determine Arrangement of Heat Exchangers in Series or Parallel *181*
7.6.4 Assess Heat Exchanger Fouling *183*
7.7 Improving Heat Exchanger Performance *185*
7.7.1 How to Identify Deteriorating Performance *185*
7.7.1.1 Fouling Resistances *185*
7.7.1.2 U-Value Monitoring *185*
7.7.1.3 Pressure Drop Monitoring *185*
7.7.1.4 Avoid Poor Design *186*
7.A TEMA Types of Heat Exchangers *186*
 References *188*

8 Fired Heater Assessment *189*
8.1 Introduction *189*
8.2 Fired Heater Design for High Reliability *189*
8.2.1 Heat Flux rate *189*
8.2.2 Burner to Tube Clearance *192*
8.2.3 Burner Selection *192*
8.2.3.1 NO_x Emission *192*
8.2.3.2 Objective of Burner Selection *192*
8.2.3.3 Flame Envelope *192*
8.2.3.4 Physical Dimension of Firebox *193*
8.2.3.5 Process-Related Parameters *193*

8.2.4	Fuel Conditioning System	*193*
8.3	Fired Heater Operation for High Reliability	*194*
8.3.1	Draft	*194*
8.3.2	Bridge Wall Temperature	*194*
8.3.3	Tube Wall Temperature	*194*
8.3.4	Flame Impingement	*196*
8.3.5	Tube Life	*196*
8.3.6	Excess Air or O_2 Content	*196*
8.3.7	Flame Pattern	*197*
8.4	Efficient Fired Heater Operation	*197*
8.4.1	O_2 Analyzer	*198*
8.4.2	Why Need to Optimize Excess Air	*198*
8.4.3	Draft Effects	*199*
8.4.4	Air Preheat Effects	*200*
8.4.5	Too Little Excess Air and Reliability	*200*
8.4.6	Too Much Excess Air	*200*
8.4.7	Availability and Efficiency	*200*
8.4.8	Guidelines for Fired Heater Reliable and Efficient Operation	*200*
8.5	Fired Heater Revamp	*201*
	References	*202*
9	**Compressor Assessment**	*203*
9.1	Introduction	*203*
9.2	Types of Compressors	*203*
9.2.1	Multistage Beam-Type Compressor	*203*
9.2.2	Multistage Integral Geared Compressors	*204*
9.3	Impeller Configurations	*205*
9.3.1	Between-Bearing Configuration	*205*
9.3.2	Integrally Geared Configuration	*206*
9.4	Type of Blades	*207*
9.5	How a Compressor Works	*207*
9.6	Fundamentals of Centrifugal Compressors	*208*
9.7	Performance Curves	*209*
9.7.1	Design Point	*209*
9.7.2	Surge	*209*
9.7.3	Choking	*209*
9.8	Partial Load Control	*210*
9.8.1	Recycle or Surge Control Valve	*210*
9.8.2	Variable Speed Control	*210*
9.8.3	Inlet Guide Vane (Prerotation Vane)	*211*
9.9	Inlet Throttle Valve	*212*
9.10	Process Context for a Centrifugal Compressor	*212*
9.11	Compressor Selection	*213*
	References	*213*
10	**Pump Assessment**	*215*
10.1	Introduction	*215*
10.2	Understanding Pump Head	*215*
10.3	Define Pump Head: Bernoulli Equation	*216*
10.4	Calculate Pump Head	*218*
10.5	Total Head Calculation Examples	*219*
10.6	Pump System Characteristics: System Curve	*221*
10.6.1	Examples of System Curves	*221*

10.7	Pump Characteristics: Pump Curve *222*	
10.8	Best Efficiency Point (BEP) *224*	
10.9	Pump Curves for Different Pump Arrangement *225*	
10.9.1	Series Arrangement *225*	
10.9.2	Parallel Arrangement *225*	
10.10	NPSH *226*	
10.10.1	Calculation of $NPSH_A$ *226*	
10.10.2	NPSH Margin *227*	
10.10.3	Measuring $NPSH_A$ for Existing Pumps *228*	
10.10.4	Potential Causes and Mitigation *229*	
10.10.4.1	Lower P_S Due to Drop in Pressure at the Suction Nozzle *229*	
10.10.4.2	Lower P_S Due to Low Density of the Liquid *229*	
10.10.4.3	Lower P_S Due to Low Liquid Level *229*	
10.10.4.4	Lower P_S Due to Increase in the Fluid Velocity at Pump Suction *229*	
10.10.4.5	Lower P_S Due to Plugged Suction Line *229*	
10.10.4.6	Higher P_v Due to Increase in the Pumping Temperature *229*	
10.10.4.7	Reduction of the Flow at Pump Suction *229*	
10.10.4.8	The Pump Is Not Selected Correctly *229*	
10.11	Spillback *229*	
10.12	Reliability Operating Envelope (ROE) *230*	
10.13	Pump Control *230*	
10.14	Pump Selection and Sizing *231*	
	Nomenclature *233*	
	Greek Letters *233*	
	References *233*	

Part IV Energy and Process Integration *235*

11	**Process Integration for Higher Efficiency and Low Cost** *237*	
11.1	Introduction *237*	
11.2	Definition of Process Integration *237*	
11.3	Composite Curves and Heat Integration *238*	
11.3.1	Composite Curves *239*	
11.3.2	Basic Pinch Concepts *239*	
11.3.3	Energy Use Targeting *240*	
11.3.4	Pinch Design Rules *240*	
11.3.5	Cost Targeting: Determine Optimal ΔT_{min} *240*	
11.4	Grand Composite Curves (GCC) *244*	
11.5	Appropriate Placement Principle for Process Changes *244*	
11.5.1	General Principle for Appropriate Placement *244*	
11.5.2	Appropriate Placement for Utility *245*	
11.5.3	Appropriate Placement for Reaction Process *245*	
11.5.4	Appropriate Placement for Distillation Column *246*	
11.5.4.1	The Column Grand Composite Curve (CGCC) *246*	
11.5.4.2	Column Integration Against Background Process *246*	
11.5.4.3	Design Procedure for Column Integration *248*	
11.6	Systematic Approach for Process Integration *249*	
11.7	Applications of the Process Integration Methodology *251*	
11.7.1	Column Split for Xylene Column with Thermal Coupling *252*	
11.7.2	Column Split for Extract Column with Thermal Coupling *252*	
11.7.3	Use of Dividing Wall Columns (DWC) *253*	
11.7.3.1	Benzene–Toluene Fractionation Dividing Wall Column *253*	
11.7.3.2	Reformate Splitter Dividing Wall Column *254*	
11.7.4	Use of Light Desorbent *255*	

11.7.5	Heat Pump for Paraxylene Column	*255*
11.7.6	Indirect Column Heat Integration	*256*
11.7.7	Benefit of Column Integration	*256*
11.7.8	Process–Process Stream Heat Integration	*256*
11.7.9	Power Recovery	*257*
11.7.9.1	Organic Rankine Cycle for Low-Temperature Heat Recovery	*257*
11.7.9.2	Variable Frequency Driver on Adsorbent Chamber Circulation Pumps	*259*
11.7.10	Process Integration Summary	*259*
	References	*261*

12 Energy Benchmarking *263*

12.1	Introduction	*263*
12.2	Definition of Energy Intensity for a Process	*263*
12.3	The Concept of Fuel Equivalent (FE) for Steam and Power	*264*
12.3.1	FE Factors for Fuel	*264*
12.3.2	FE Factors for Steam	*264*
12.3.3	FE Factors for Power	*265*
12.3.4	Energy Intensity Based on FE	*265*
12.4	Calculate Energy Intensity for a Process	*265*
12.5	Fuel Equivalent for Steam and Power	*267*
12.5.1	FE Factors for Power (FE_{power})	*267*
12.5.2	FE Factors for Steam, Condensate, and Water	*268*
12.6	Energy Performance Index (EPI) Method for Energy Benchmarking	*271*
12.6.1	Benchmarking: based on the Best-in-Operation Energy Performance (OEP)	*271*
12.6.2	Benchmarking: based on Industrial Peers' Energy Performance (PEP)	*271*
12.6.3	Benchmarking: based on the Best Technology Energy Performance (TEP)	*272*
12.7	Concluding Remarks	*272*
12.7.1	Criteria for Data Extraction	*272*
12.7.2	Calculations Accuracy for Energy Benchmarking	*272*
	References	*273*

13 Key Indicators and Targets *275*

13.1	Introduction	*275*
13.2	Key Indicators Represent Operation Opportunities	*275*
13.2.1	Reaction and Separation Optimization	*275*
13.2.2	Heat Exchanger Fouling Mitigation	*276*
13.2.3	Furnace Operation Optimization	*276*
13.2.4	Rotating Equipment Operation	*277*
13.2.5	Minimizing Steam Letdown Flows	*277*
13.2.6	Turndown Operation	*277*
13.3	Defining Key Indicators	*277*
13.3.1	Simplifying the Problem	*278*
13.3.2	Developing Key Indicators for the Reaction Section	*278*
13.3.2.1	Understanding the Process	*278*
13.3.2.2	Understanding the Energy Needs	*278*
13.3.2.3	Effective Measures for the Energy Needs	*279*
13.3.2.4	Developing Key Indicators for the Energy Needs	*279*
13.3.3	Developing Key Indicators for the Product Fractionation Section	*279*
13.3.3.1	Understand Process Characteristics	*279*
13.3.3.2	Understand the Energy Needs	*280*
13.3.3.3	Effective Measure for the Energy Needs	*280*
13.3.3.4	Developing Key Indicators for the Energy Needs in the Main Fractionation System	*280*
13.3.4	Remarks for the Key Indicators Developed	*280*
13.4	Set Up Targets for Key Indicators	*280*

13.4.1	Problem	*281*
13.4.2	Rationale	*281*
13.4.3	Solution	*281*
13.5	Economic Evaluation for Key Indicators	*283*
13.6	Application 1: Implementing Key Indicators into an "Energy Dashboard"	*285*
13.7	Application 2: Implementing Key Indicators to Controllers	*287*
13.8	It Is Worth the Effort	*287*
	References	*288*

14 Distillation System Optimization *289*

14.1	Introduction	*289*
14.2	Tower Optimization Basics	*289*
14.2.1	What to Watch: Key Operating Parameters	*289*
14.2.1.1	Reflux Ratio	*290*
14.2.1.2	Overflash	*290*
14.2.1.3	Pressure	*290*
14.2.1.4	Feed Temperature	*290*
14.2.1.5	Stripping Steam	*290*
14.2.1.6	Pump Around	*291*
14.2.1.7	Overhead Temperature	*291*
14.2.2	What Effects to Know: Parameter Relationship	*291*
14.2.3	What to Change: Parameter Optimization	*291*
14.2.4	Relax Soft Constraints to Improve Margin	*292*
14.3	Energy Optimization for Distillation System	*293*
14.3.1	Develop Economic Value Function	*293*
14.3.2	Setting Operating Targets with Column Bottom Temperature	*294*
14.3.3	Setting Operating Targets with Column Reflux Ratio	*294*
14.3.4	Setting Operating Pressure	*295*
14.4	Overall Process Optimization	*296*
14.4.1	Basis	*296*
14.4.2	Current Operation Assessment	*297*
14.4.3	Simulation	*298*
14.4.4	Define the Objective Function	*298*
14.4.5	Off-Line Optimization Results	*299*
14.4.6	Optimization Implementation	*300*
14.4.7	Online Optimization Results	*300*
14.4.8	Sustaining Benefits	*301*
14.5	Concluding Remarks	*302*
	References	*302*

15 Fractionation and Separation Theory and Practices *303*

15.1	Introduction	*303*
15.2	Separation Technology Overview	*303*
15.2.1	Single-Stage Separation	*304*
15.2.2	Distillation	*304*
15.2.3	Liquid–Liquid Extraction	*304*
15.2.4	Adsorption	*304*
15.2.5	Simulated Moving Bed Chromatography	*305*
15.2.6	Crystallization	*305*
15.2.7	Membrane	*305*
15.3	Distillation Basics	*305*
15.3.1	Difficulty of Separation	*305*
15.3.2	Selection of Operating Pressure	*306*
15.3.3	Types of Reboiler Configurations	*307*

15.3.4	Optimization of Design	308
15.3.5	Side Products	310
15.4	Advanced Distillation Topics	311
15.4.1	Heavy Oil Distillation	311
15.4.2	Dividing Wall Column	312
15.4.2.1	DWC Fundamentals	312
15.4.2.2	Guidelines for Using DWC Technology	313
15.4.2.3	Application of Dividing Wall Column	313
15.4.3	Choice of Column Internals	315
15.4.3.1	Crossflow Trays	315
15.4.3.2	Packing	316
15.4.3.3	Super System Limit Devices	316
15.4.4	Limitations with Distillation	316
15.4.4.1	Separation Methods for Low Relative Volatility	316
15.4.4.2	Separation Methods for Low Concentration	316
15.5	Adsorption	316
15.6	Simulated Moving Bed (SMB)	317
15.6.1	The Concept of Moving Bed	318
15.6.2	The Concept of Simulated Moving Bed	319
15.6.3	Rotary Valve	319
15.7	Crystallization	320
15.8	Liquid–Liquid Extraction	320
15.9	Extractive Distillation	321
15.10	Membranes	322
15.11	Selecting a Separation Method	323
15.11.1	Feed and Product Conditions	323
15.11.2	Operation Feasibility	323
15.11.3	Design Reliability	323
15.11.4	Selection Heuristics	324
	References	324
16	**Reaction Engineering Overview**	**325**
16.1	Introduction	325
16.2	Reaction Basics	325
16.2.1	Reaction Rate Law	325
16.2.2	Arrhenius Activation Energy (E_a)	325
16.2.3	Reaction Catalyst	325
16.2.4	Order of Reaction	326
16.2.5	Reactor Design	326
16.3	Reaction Kinetic Modeling Basics	326
16.3.1	Elementary Reaction Rate Law	327
16.3.2	Reversible Reaction	327
16.3.3	Nonelementary Reaction Rate Law	327
16.3.4	Steady-State Approximation	327
16.3.5	Reaction Mechanism	327
16.4	Rate Equation Based on Surface Kinetics	328
16.5	Limitations in Catalytic Reaction	330
16.5.1	External Diffusion Limitation	331
16.5.2	Surface Reaction Limitation	331
16.5.3	Internal Pore Diffusion Limitation	331
16.5.4	Mitigating Limitations	331
16.5.5	Important Parameters of Limiting Reaction	332
16.5.5.1	Nature of Reactions	332
16.5.5.2	Physical state	332

16.5.5.3	Concentration	*332*
16.5.5.4	Temperature	*332*
16.5.5.5	Mass Transfer	*332*
16.5.5.6	Catalysts	*333*
16.6	Reactor Types	*333*
16.6.1	General Classification	*333*
16.6.1.1	Homogeneous and Heterogeneous	*333*
16.6.1.2	Continuous Flow Reactors	*333*
16.6.1.3	Semi-Batch Reactors	*333*
16.6.2	Practical Types of Reactors	*334*
16.6.2.1	Fixed Bed Reactors	*334*
16.6.2.2	Trickle Bed Reactors	*334*
16.6.2.3	Moving Bed Reactors	*334*
16.6.2.4	Fluidized Bed Reactors	*334*
16.6.2.5	Slurry Reactors	*334*
16.7	Reactor Design	*335*
16.7.1	Objective	*335*
16.7.2	Temperature and Equilibrium Constant	*335*
16.7.3	Pressure, Reaction Conversion, and Selectivity	*336*
16.7.4	Reaction Time and Reactor Size	*336*
16.7.5	Determine the Rate-Limiting Step	*337*
16.7.6	Reactor Design Considerations	*338*
16.7.6.1	The Overall Size of the Reactor	*338*
16.7.6.2	Products of the Reactor	*338*
16.7.6.3	By-products	*338*
16.7.6.4	The Physical Condition of the Reactor	*338*
16.7.7	General Guidelines	*338*
16.8	Hybrid Reaction and Separation	*340*
16.9	Catalyst Deactivation Root Causes and Modeling	*341*
	References	*343*

Part V	**Operational Guidelines and Troubleshooting**	*345*
17	**Common Operating Issues**	*347*
17.1	Introduction	*347*
17.2	Start-up Considerations	*348*
17.2.1	Catalyst Reduction	*348*
17.2.2	Catalyst Sulfiding	*348*
17.2.3	Catalyst Attenuation	*349*
17.3	Methyl Group and Phenyl Ring Losses	*349*
17.4	Limiting Aromatics Losses	*350*
17.4.1	Olefin Removal in an Aromatics Complex	*350*
17.4.2	Fractionation and Separation Losses	*351*
17.4.2.1	Vent Losses	*351*
17.4.2.2	Losses to Distillate Liquid Product	*351*
17.4.2.3	Losses to Bottoms Liquid Product	*352*
17.4.3	Extraction Losses	*352*
17.4.3.1	Common Variables Affecting Aromatic Recovery	*352*
17.4.3.2	Feed Composition	*352*
17.4.3.3	Foaming	*353*
17.4.4	Reaction Losses	*353*
17.4.4.1	Xylene Isomerization Unit Losses	*354*

17.4.4.2	Transalkylation Unit Losses	*355*
17.4.5	Methyl Group Losses	*355*
17.4.5.1	Fractionation and Separation Losses	*356*
17.4.5.2	Reaction Losses	*356*
17.5	Fouling	*356*
17.5.1	Combined Feed Exchanger Fouling	*356*
17.5.1.1	Chemical Foulants	*356*
17.5.1.2	Particulate Foulants	*356*
17.5.2	Process Heat Exchanger Fouling	*358*
17.5.3	Heater Fouling	*358*
17.5.4	Specialty Reboiler Tube Fouling	*358*
17.5.5	Line Fouling	*359*
17.5.6	Extraction Unit Column Fouling	*360*
17.6	Aromatics Extraction Unit Solvent Degradation	*360*
17.6.1	Oxygen and Oxygenates	*361*
17.6.2	Temperature	*362*
17.6.3	Chloride	*362*
17.6.4	Other Measurements	*362*
17.7	Selective Adsorption of *para*-Xylene by Simulated Moving Bed	*363*
17.7.1	Purity and Recovery Relationship	*365*
17.7.2	*meta*-Xylene Contamination	*366*
17.7.3	Common Poisons	*366*
17.7.3.1	Olefins	*366*
17.7.3.2	Oxygenates	*366*
17.7.3.3	Heavy Aromatics	*366*
17.7.3.4	Water	*367*
17.7.4	Rotary Valve™ Monitoring	*367*
17.7.4.1	Dome Pressure	*367*
17.7.4.2	Alignment	*368*
17.7.4.3	Maintenance	*368*
17.7.5	Flow Meter Monitoring	*368*
17.7.6	Hydration Monitoring	*369*
17.7.7	Shutdown and Restart Considerations	*371*
17.7.7.1	Severe Start-up or Shutdown Conditions	*371*
17.7.7.2	Oxygenate Ingress	*371*
17.7.7.3	Leaking of Adsorption Section Isolation Valves	*371*
17.8	Common Issues with Sampling and Laboratory Analysis	*371*
17.8.1	Bromine Index Analysis for Olefin Measurement	*371*
17.8.2	Atmospheric Contamination of Samples	*372*
17.8.3	Analysis of Unstabilized Liquid Samples	*372*
17.8.4	Gas Chromatography	*372*
17.8.4.1	Nitrogen vs Hydrogen or Helium Carrier Gas	*373*
17.8.4.2	Resolution of *meta*-Xylene and *para*-Xylene Peaks	*373*
17.8.4.3	Wash Solvent Interference	*374*
17.8.4.4	Over-Reliance on a Particular Analytical Method	*374*
17.8.4.5	Impact of Unidentified Components	*374*
17.9	Measures of Operating Efficiency in Aromatics Complex Process Units	*374*
17.9.1	Selective Adsorption *para*-Xylene Separation Unit	*374*
17.9.2	Xylene Isomerization Unit	*375*
17.9.3	Transalkylation Unit	*376*
17.9.4	Aromatics Extraction Unit	*377*
17.10	The Future of Plant Troubleshooting and Optimization	*377*
	References	*377*

18	**Troubleshooting Case Studies** *379*
18.1	Introduction *379*
18.2	Transalkylation Unit: Low Catalyst Activity During Normal Operation *379*
18.2.1	Summary of Symptoms *379*
18.2.2	Root Cause and Solution *379*
18.2.3	Lesson Learned *381*
18.3	Xylene Isomerization Unit: Low Catalyst Activity Following Start-up *381*
18.3.1	Summary of Symptoms *381*
18.3.2	Root Cause and Solution *382*
18.3.3	Lesson Learned *384*
18.4	*para*-Xylene Selective Adsorption Unit: Low Recovery After Turnaround *384*
18.4.1	Summary of Symptoms *384*
18.4.2	Root Cause and Solution *384*
18.4.3	Lesson Learned *385*
18.5	Aromatics Extraction Unit: Low Extract Purity/Recovery *385*
18.5.1	Summary of Symptoms *385*
18.5.2	Root Cause and Solution *386*
18.5.3	Lesson Learned *386*
18.6	Aromatics Complex: Low *para*-Xylene Production *386*
18.6.1	Summary of Symptoms *387*
18.6.2	Root Cause and Solution *387*
18.6.3	Lesson Learned *388*
18.7	Closing Remarks *388*
	Reference *389*

Index *391*

Preface

This book presents technology, chemistry, selection, process design, and troubleshooting for aromatics production processes which are the source of important petrochemical building blocks. The selection of this topic came from realizing the importance of petrochemicals, and specifically aromatics in today's world. The demand for petrochemical products has increased dramatically in recent years due to rapid growth in emerging markets such as China, Southeast Asia, India, and other regions. Economic growth in these regions has driven the ever-increasing need for petrochemicals and the downstream products they create. Petrochemicals have enabled the creation of advanced materials and products in other related industrial sectors including apparel, packaging, consumer goods, automotive products, electronics, detergents, medical devices, agriculture, communication, and transportation. For example, in the new Boeing 787 Dreamliner, the latest modern aircraft to be launched, modern synthetic materials comprise about half of its primary structure. In addition, most of the tools on which we depend for daily existence – such as cars, computers, cell phones, children's toys, fertilizers, pesticides, household cleaning products, and pharmaceutical drugs – are derived from petrochemicals. Therefore, petrochemical products, which are chemical products made from fossil fuels such as petroleum (crude oil), coal, and natural gas, are the foundation of nearly every sector of the world's economy.

Although there has been an increased emphasis on petrochemical technology and products, there are no dedicated books available to discuss technology selection, efficient process design, and reliable operation for petrochemical processes. Thus, this book is written to fill this gap with the focus on aromatics technology, which are arguably the most important building blocks in the area of petrochemicals.

Brief Overview of the Book

The book contains five parts with 18 chapters in total are provided. The first part provides an overview of the petrochemical processes including the typical feeds, products, and technology. The book starts by introducing the topic of petrochemicals in Chapter 1, giving a market and technology overview in Chapter 2, and providing an aromatic process description in Chapter 3. The lens is then zoomed onto more specific aspects such as process design in Part II, equipment assessment in Part III, and process integration and system optimization in Part IV. Last but not the least, Part V deals with operation in which operation guidelines are provided and troubleshooting cases are discussed.

The Audience

This book is written with the following people in mind: plant managers, process engineers, and operators working in process industries, and engineering firms who face challenges and wish to find opportunities for improved process operation and design. The book will provide practical methods and tools to industrial practitioners with the focus on improving aromatics plant energy efficiency, reducing capital investment, and optimizing yields via better design, operation, and optimization. These methods have been proven with successful applications in many aromatics plants over many years, which contributes to the improvements of environmental performance and reduction of foot print from petrochemical production.

In addition, this book may be valuable to engineering students with the design projects, chemistry students for the context and application of chemistry theory, and to those who want to equip themselves with real-world applications and practical methods which will allow them to become more employable after graduation. The contents of this book are based on the knowledge and experience the authors gained over many years in research, engineering, consulting, and service support for petrochemical processes. The authors hope that this book is able to convey concepts, theories, and methods in a straightforward and practical manner.

Acknowledgments

Clearly, it was no small effort to write this book; but it was the authors' desire to provide practical methods for helping people understand the issues involved in improving design and operation for better energy efficiency and lower capital cost which motivated us. In this endeavor, we owe an enormous debt of gratitude to many of our colleagues at Honeywell UOP and Honeywell for their generous support in this effort. First of all, we would like to thank Kelly Seibert, vice president and general manager of Honeywell UOP, who provided strong encouragement at the beginning of this journey for writing this book. We are very grateful to many colleagues for their constructive suggestions and comments on the materials contained in this book. We would especially like to thank UOP process technology specialist and engineering Fellow Jason Corradi for his valuable comments for Chapters 11 and 12, former UOP fractionation specialist Paul Steacy for his writing contribution to Chapter 15, Dr. Seok Hong of Illinois Institute of Technology for his review with constructive comments for Chapter 16, and Simmi Sood for her assistance in collecting data for some of the calculations.

We also like to thank the following people who made critical review: Phil Daily for Chapter 7, Darren Le Geyt for Chapter 8, Bruce Lieberthal for Chapters 9 and 10, Rajesh Rajappan for Chapter 13, and Joe Haas for Chapter 15. Our sincere gratitude also goes to Kevin O'Neil, Mark James, and Ashley Romano for their reviews. Specially, Kevin O'Neil provided a thorough review of the full text with constructive comments. However, any deficiencies in this book are the authors' responsibility and we would like to point out that this book reflects our own opinions but not those of Honeywell UOP.

Finally, we would like to thank the co-publishers, AIChE and John Wiley, for their help. Special thanks go to Cindy (Cynthia) Mascone and Kate McKay at AIChE and Michael Leventhal at John Wiley for guidance as well as Vishnu Priya at John Wiley for managing the copyediting and typesetting.

12 August 2019

Frank Zhu, Jim Johnson,
Dave Ablin, Greg Ernst
Des Plaines, Illinois, USA

Part I

Market, Design and Technology Overview

1

Overview of This Book

1.1 Why Petrochemical Products Are Important for the Economy

To understand the economic significance of petrochemical products, it is important to first look at some of the history of the petrochemical industry. Aromatics such as benzene, toluene, and the C_8 aromatic isomers have long been of interest to chemists and chemical engineers. In the earliest stages of the chemical industry, these aromatics were recovered by distillation from coal tar, which was a by-product of steel production. In addition to being used as solvents, these basic aromatics could be fairly easily converted into functional derivatives by chlorination, nitration, and similar chemistries, usually in batch operation. Basic intermediates, such as di-chloro benzene isomers, nitro benzene, and chloro-nitro benzene, found their way into agricultural chemicals and pharmaceuticals, among other applications.

In the mid-1800s, styrene (aka styrol) had been found by Erlenmeyer to be capable of producing a dimer, which explained how, when styrene was exposed to air, heat, and sunlight, it would form a hard, rubbery substance. This may have been the first recognition of the potential for aromatics to be useful for polymers. Ethylene followed a similar, but much later pathway with the accidental discovery of polymerization in 1898, followed by industrial polyethylene synthesis in 1933. Ethylene became relatively abundant via steam cracking, and benzene by recovery from naphtha fractions. This led to the use of polyethylene and polystyrene for military purposes during WW II.

Propylene was converted via dimers and trimers into a gasoline component that could be blended with FCC gasoline and naphtha. Cumene, the reaction product of benzene and propylene, was produced in large quantities during WW II as a high-octane component for aviation gasoline, but it was not until near the end of WW II that the Hoch process for oxidation of cumene to phenol and acetone was discovered. Phenol and acetone each have many end uses as solvents or derivatives in pharmaceuticals, agricultural chemicals, and other applications. But more than 65% of the world's phenol is converted into polymers.

After WW II, companies began to focus on ways to sustain economic growth beyond supplying wartime materials, and they focused on emerging consumer goods. With increased automobile production came the need for large amounts of higher octane gasoline. UOP discovered and commercialized catalytic reforming of naphtha to aromatics using a supported platinum catalyst. This process was called Platforming™ and it became a mainstay of refining for its capability to produce significant aromatics along with valuable hydrogen. This development, along with the confluence of polymerization knowledge, and the capacity for production of ethylene and propylene, led to the origins of today's petrochemicals industry. Over the years, polymer chemists have created numerous grades of polyethylene, polypropylene, polystyrene, polycarbonate, and polyethylene terephthalate (PET) for use in the apparel, packaging, automotive, electronics, medical, and numerous other industries. Biodegradable detergents also are derived from hydrocarbon feeds, using benzene and the linear high-molecular weight olefins that are derived from n-paraffins. All of these products make use of the abundance of hydrocarbons being produced from crude oil, LPG, and natural gas liquids.

1.1.1 Polyethylene

Today, polyethylene is the most prominent polymer worldwide, with more than 140 million tons per year being produced. It is extremely durable and is used for production of containers, insulators, coatings, pipe, liners, and films. Industries such as packaging, electronics, power transmission, consumer, and household goods produce components or end products made from polyethylene, with significant job creation in finished goods.

1.1.2 Polypropylene

Polypropylene is a thermoplastic polymer with numerous uses, including textiles, packaging, consumer goods, appliances, electrical and manufacturing industries, and automotive and construction industries. More than

Efficient Petrochemical Processes: Technology, Design and Operation, First Edition.
Frank (Xin X.) Zhu, James A. Johnson, David W. Ablin, and Gregory A. Ernst.
© 2020 John Wiley & Sons, Inc. Published 2020 by John Wiley & Sons, Inc.

90 million tons per year of polypropylene are produced worldwide. The job creation and added value impact is very high in both developed and emerging regions.

1.1.3 Styrene and Polystyrene

Styrene is used in the production of latex, synthetic rubber, and polystyrene resins. Primary categories of styrene are: films, foams, composites, ABS plastic, SAN plastic, SB rubber, and SB latex. It is used in plastic packaging; building insulation; cups and containers; composite products such as tubs, showers, auto body parts, boats, and wind turbines; medical devices; optical fibers; tires; and backing for carpets. Worldwide demand is more than 27 million tons per year.

1.1.4 Polyester

para-Xylene oxidized to terephthalic acid is the primary constituent of polyester fibers, used in fine clothing, carpet staple, and industrial yarn applications such as tire cord. For resin applications such as water or carbonated soft drink bottles, comonomer isophthalic acid, derived from oxidation of *meta*-xylene, is added, in the range of 2–3%. An ethylene derivative, ethylene glycol, is used for cross-linking in the continuous polymerization process. Worldwide PET consumption for fiber, resin, and films is nearly 80 million tons per year.

1.1.5 Polycarbonate and Phenolic Resins

Phenol and acetone are combined to form bisphenol-A, which can then be converted into polycarbonate and epoxide resins. Phenol can also be reacted with formaldehyde to produce phenolic resins, such as bakelite. Polycarbonates are widely used in automotive applications since they are lightweight, durable, and their thermoplastic properties allow them to be easily molded. Worldwide demand for polycarbonate is around 6 million tons per year.

1.1.6 Economic Significance of Polymers

Polymer products permeate nearly every sector of the world's economy. It is very easy to see numerous applications in everyday life throughout the world. Here are a few examples (the list could be much longer):

- Construction
- Furniture
- Transportation
- Clothing
- Sports and leisure
- Food and beverage packaging
- Communications
- Health care
- Electronics

Significant value chains are created that start with the production of aromatics or olefins – the building blocks – in very large production plants located throughout the world. The next step downstream is to convert these petrochemical building blocks into transportable monomers and polymers. For example, *para*-xylene can be converted into purified terephthalic acid (PTA), which can then be transported to facilities where PET fiber or PET resin chips can be produced. The fiber can be shipped to textile mills where it is converted into fabrics, dyed, and then sold to companies that convert it into clothing. The PET resin chips can be shipped to bottle converters, where injection molding is used to make preforms for PET bottles. The preforms can then be blow-molded into bottles that can be used for water, juices, carbonated soft drinks, or other food packaging. Each of these steps requires capital, adds value, and usually takes place on a much smaller scale tonnage-wise than the centralized production of the building blocks.

The move from a large centralized production facility for the building blocks all the way to smaller enterprises that are widely dispersed throughout the world creates significant job opportunities and value added at each step. Many countries that are rich in petroleum resources start by selling crude oil, then build refineries to produce fuels, and then build and operate petrochemical plants that utilize some of the refined hydrocarbons for producing the olefinic and aromatic building blocks. Ultimately, they will create the downstream fabrication industries to gain additional foreign exchange as well as create full employment for the population.

Similarly, countries with large populations that are not rich in petroleum resources, such as China and India, will build significant capacities for refining and petrochemicals based on imported crude oil, and then create the downstream industries that feed the changing consumption patterns of the growing middle class population. There are many factors, which taken together, account for the high importance of petrochemicals to the world's economy. As shown in Figure 1.1, world population continues to grow from just over 7 billion people in 2018 and projected to reach nearly 9 billion people in less than 20 years. Most of this population growth will take place outside OECD countries. World GDP is expected to more than double during this time from $90 trillion to more than $200 trillion. The growth in world GDP will be driven primarily by increases in productivity. This translates to growth in the middle classes of these expanding societies, and more disposable income. The Brookings Institute projects that the world's middle class

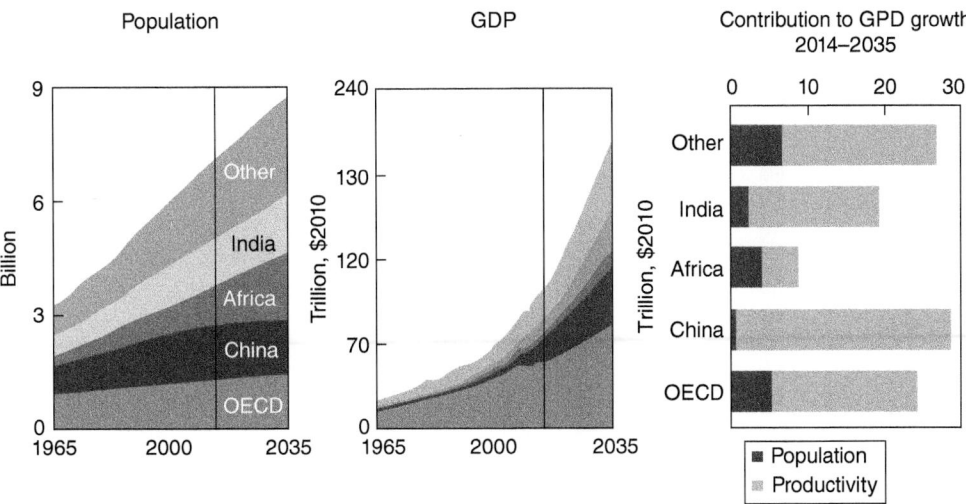

Figure 1.1 World population and GDP growth. *Source:* 2016 Energy Outlook © BP p.l.c 2016.

Figure 1.2 Natural and synthetic fiber demand.

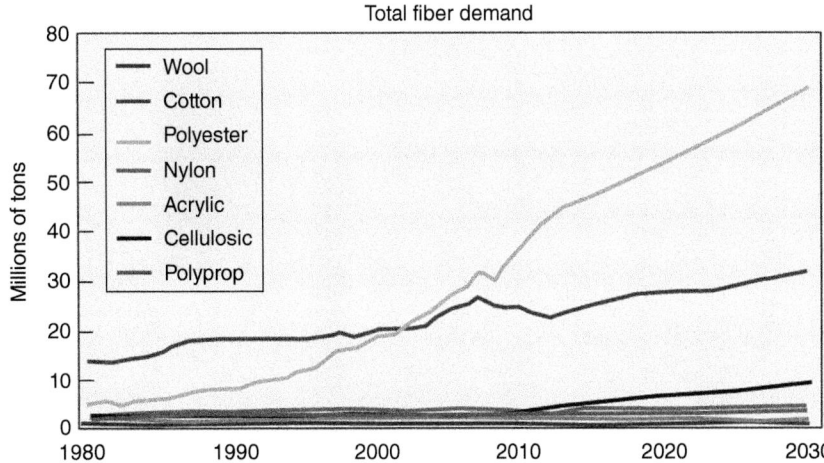

will nearly double from 3 billion to more than 5.5 billion people by 2030.

The polyester industries, both fiber and resins, have been the beneficiaries of these changes, and provide a convenient example to illustrate the global economic impact of aromatic petrochemicals. For example, Figure 1.2 shows the world production of various sources of fiber for clothing. Up until about the year 2000, cotton was the dominant fiber, with an annual demand of about 20 million tons. Starting in 2000, the demand for polyester overtook the demand for cotton, and by 2020, is expected to be double that of cotton. Cotton demand has continued to grow at a rate of about 3–4% per year (and is somewhat climate-dependent), whereas polyester fiber demand has grown at 6–7% per year. By 2025, the worldwide demand for polyester fiber will be about 60 million tons per year, whereas the demand for cotton will be only 30 million tons per year. Figure 1.2 also shows the other man-made fibers that compete with cotton.

The per-capita use of man-made vs. natural fibers by region is shown in Figure 1.3. In North America, man-made fiber use was about 22 kg/capita, while China accounted for approximately 13 kg/capita. Polyester accounted for more than 60% of man-made fiber use in North America and nearly 90% of man-made fiber in China.

Nearly 30% of all polyester is converted into resin using solid-state polymerization. Most of this PET resin is used for beverage packaging – that is, bottles for water, carbonated soft drinks, juices, alcoholic beverages, and other food items. These applications, particularly portable water, have enabled significantly greater mobility, where it is possible to travel some distances to work, school, or commerce while still having a clean and reliable supply of water. This

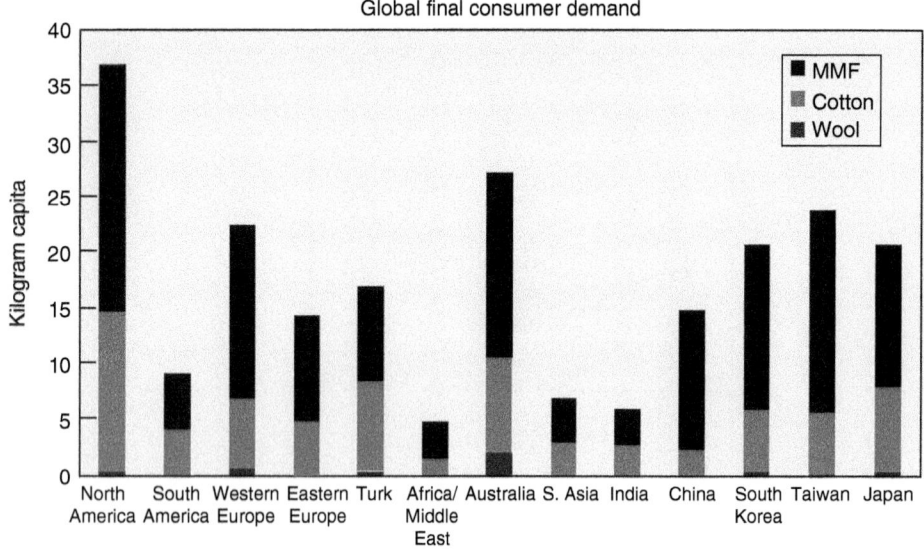

Figure 1.3 Per-capita consumption of fibers by region.

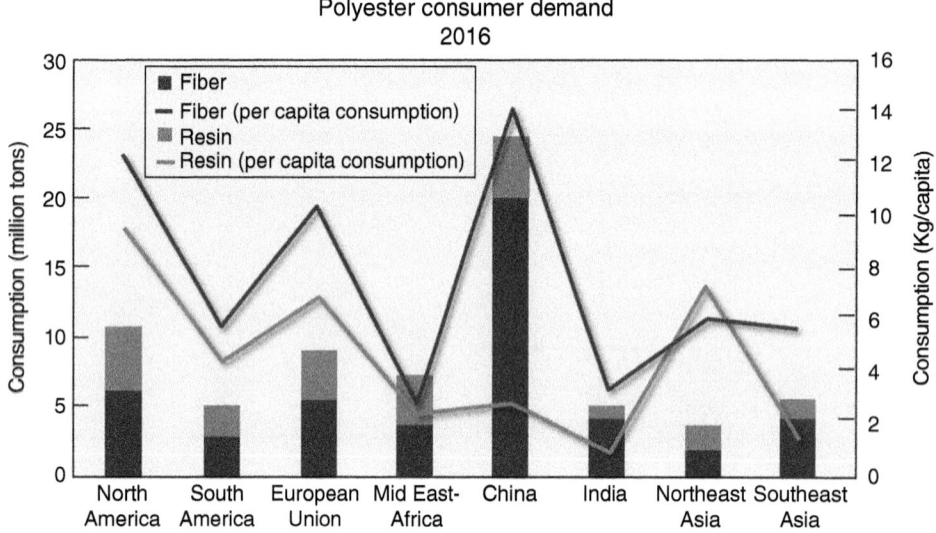

Figure 1.4 PET consumption and per-capita consumption by region.

convenience, in turn, has allowed job growth and expansion of the middle class in many countries. Figure 1.4 provides a breakdown of polyester fiber and resin consumption as well as per-capita usage by region in 2016.

North America, Northeast Asia, and OECD had the highest per-capita usage of resin, in the range of 7–9 kg/capita. China and India, with very large populations had lower per-capita consumption at 1–2 kg/capita, but China's large population still drove the total resin consumption to be very comparable to that of North America at about 5 MM tons. As the middle class grows in China, India, and elsewhere, these per-capita consumption levels are expected to approach those seen in North America, Northeast Asia, and OECD.

Figure 1.4 also shows how important China is to the global consumption of polyester, accounting for nearly 25 million tons in 2016, very close to the combined total of consumption in North America, Northeast Asia, and OECD.

In addition to allowing increased mobility for the population, polyester resin offers significant advantages over glass and aluminum in container weight for beverages and foods. Essentially, the same volume and mechanical strength for retail display purposes can be possible with only a fraction of the container weight compared to glass and aluminum. In addition, distribution and delivery costs from the bottling plants will allow much lower consumption of fuel in delivery vehicles.

Similar detailed examples could be made for styrenics, polycarbonates, urethanes, and polyolefins, all of which have ubiquitous positions in consumer products, transportation, electronics, communications, and construction.

1.1.7 Petrochemicals and Petroleum Utilization

Worldwide, the vast majority of oil production is utilized to produce transportation fuel. Even if gas liquids such as ethane, propane, and butanes are included on a volumetric basis, less than 15% of the world's hydrocarbon resources ended up as petrochemicals in 2016 according to the US Energy Information Agency. During the last 10 or so years, oil prices have been as low as $30/barrel and as high as $100/barrel. Even these higher prices for oil, which translated into costs for naphtha, *para*-xylene (*p*X), and ultimately PET, did not slow down the replacement of natural fiber such as cotton by polyester fiber. Shale oil sources, such as in the United States, can be economically recovered at prices in the $50–60/barrel range and, at this writing, the United States has emerged as a key oil producer at more than 9 MM barrels per day. A large amount of gas liquids, ethane, propane, and butanes are coproduced with shale oil, and have allowed a disconnect to develop between traditional naphtha petrochemical feedstock and these lighter feeds for on-purpose olefin production. As a result, the United States now exports significant amounts of these light hydrocarbons to China, Europe, and other locations, and is able to compete with hydrocarbons produced in the Middle East.

As shown in Figure 1.5, the world consumption of hydrocarbons grew by 36% in the 25 years between 1990 and 2015. Over the next 25 years (2015–2040), world consumption is expected to grow by only about 15%. The highest growth rates are expected in China, India, and the rest of the non-OECD world. Hydrocarbon usage is expected to stay flat or even decline in the United States, OECD, and other highly developed economies. This will be the result of energy efficiency improvements in vehicles, as well as alternatives to internal combustion engines, such as hybrid and electric vehicles (EV's).

Already, many oil refiners in highly developed regions are diverting some of their transportation fuels into petrochemicals, seeking the higher margins that are available from petrochemical intermediates. This is driven by the realization that global demand growth for chemicals will outpace global GDP by 40% over the next several years, as shown in Figure 1.6.

The opportunity to capitalize on the growth rates of petrochemicals is causing those who are investing in new refining capacity to seriously consider adding some processing capability to produce chemicals as well as fuels. While these capabilities add to the capital cost of the project, the high value obtained from the chemical products can significantly improve the net present value and return on investment for a given project. In some cases, such a "pre-investment" may make a great deal of sense.

The prospect of using renewable feed sources has been of interest in transportation fuels, and to some degree in petrochemicals. Renewable fuels such as ethanol, green diesel, or green jet fuel are intended to reduce greenhouse gases, and typically rely on government incentives in order to be economically viable. These can take the form of subsidies, taxes, or renewable fuel credits such as Renewable Identification Numbers (RINs). RINs are a means of accounting for renewable sourced fuels in the US refining industry. Petrochemicals have much lower carbon footprints than fuels, so there have been minimal government incentives to encourage renewable feed sources. There have been a few examples, such as ethylene glycol derived from plant-sourced feeds, but these are largely driven by companies wishing to make a "label claim" that their container incorporates some

Figure 1.5 World hydrocarbon consumption 1990–2040.

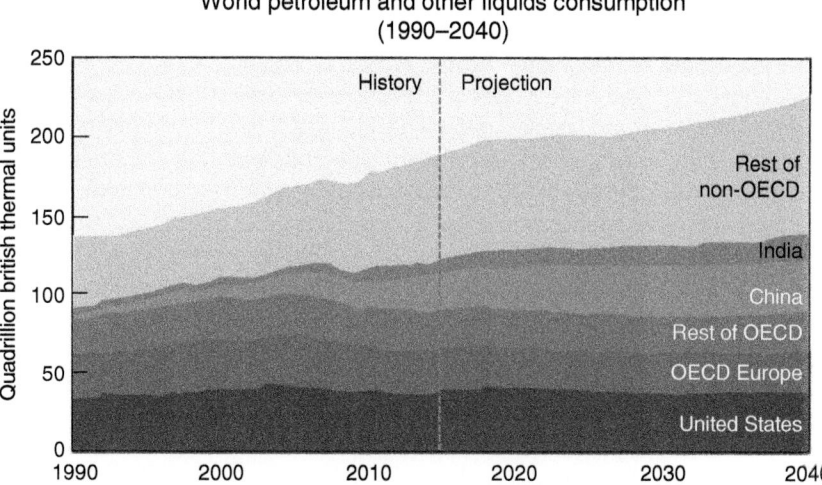

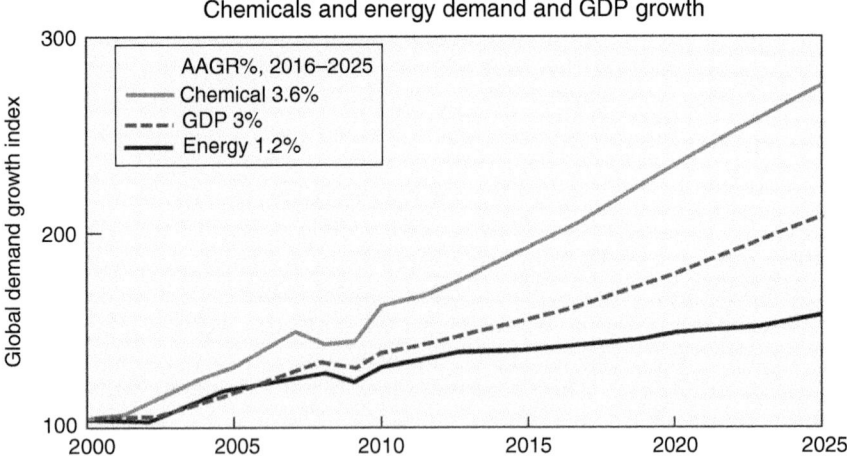

Figure 1.6 Chemicals and energy demand and GDP growth. *Source:* IHS, ExxonMobil 2018 Outlook for Energy, ExxonMobil estimates Birdsall ExMo HP.

renewable-sourced components. The common renewable feed sources, such as plant oils and sugars, are simply too high in price and too uneconomical to convert in high yields. Until there are significant improvements in bioconversion technology and associated polymer performance, hydrocarbon feeds are expected to dominate as sources for petrochemicals.

In summary, the feedstocks for petrochemical products are expected to be abundant, reasonably priced, and will continue to be a small fraction of the total hydrocarbons that are used for transportation fuels. Companies that offer technologies for petrochemical as well as refining will continue to focus on obtaining the highest yields, best use of capital and lowest operating costs for the end products.

1.2 Overall Petrochemical Configurations

In the previous section we concentrated on the economic importance of petrochemicals and how these represent a small fraction of the total amount of hydrocarbons that are processed in industry today. While naphtha fractions from crude oil refining have been dominant, several other sources of feed have gained significance. In order to establish a common frame of reference, Figure 1.7 is a simplified flow diagram that starts with crude oil, and illustrates the link between some common refining products and the associated production of petrochemicals. This also shows how a refiner might find ways to enhance profitability by proceeding along the petrochemical value chain. For example, it may be beneficial to remove benzene from the gasoline pool to be in compliance with environmental regulations. By sending the C_6 stream to an extractive distillation (ED) unit, a high-purity benzene stream can be produced, which could then be combined with propylene produced by the fluid catalytic cracking unit (FCCU) to produce cumene for the petrochemical merchant market. Similarly, from kerosene, the n-paraffins could be recovered in high purity and sold into the petrochemicals merchant market to produce biodegradable detergent. The return stream from the n-paraffin separation could be sent back into the jet fuel pool.

Some companies focus almost exclusively on petrochemicals from the start and will strive to achieve close integration between refining units and the downstream petrochemical units, as well as between those units that produce olefins and those that produce aromatics. This close integration is shown in Figure 1.8.

Modern steam crackers will be designed with flexibility to use ethane, propane, butanes, and naphtha as feed, with the feed cost and availability as the determining factor. Higher oil prices will favor lighter feeds and vice versa. Pyrolysis gasoline by-product from the steam cracker will have high aromatic content, and can supplement the reformate obtained from the straight-run naphtha for production of the key aromatic products. Although not shown here, the straight-run naphtha can be supplemented with natural gas condensate to further enhance aromatics production. Likewise, a vacuum gas oil (VGO) feedstock hydrocracker can be used to produce supplemental heavy naphtha to further enhance aromatics production. Heavy by-product streams like light cycle oil (LCO) from the FCCU can be converted into additional heavy naphtha or even into a supplemental aromatics stream, depending on the technology.

Production costs of petrochemicals, whether aromatic or olefin, are dominated by feedstock costs. Producers who have integrated refining and petrochemical operations often have the most cost-advantaged positions due to the transfer prices they can set for their key feedstocks. When oil prices are high, producers will place great emphasis on production technologies with the highest yields and lowest utility costs. They will also

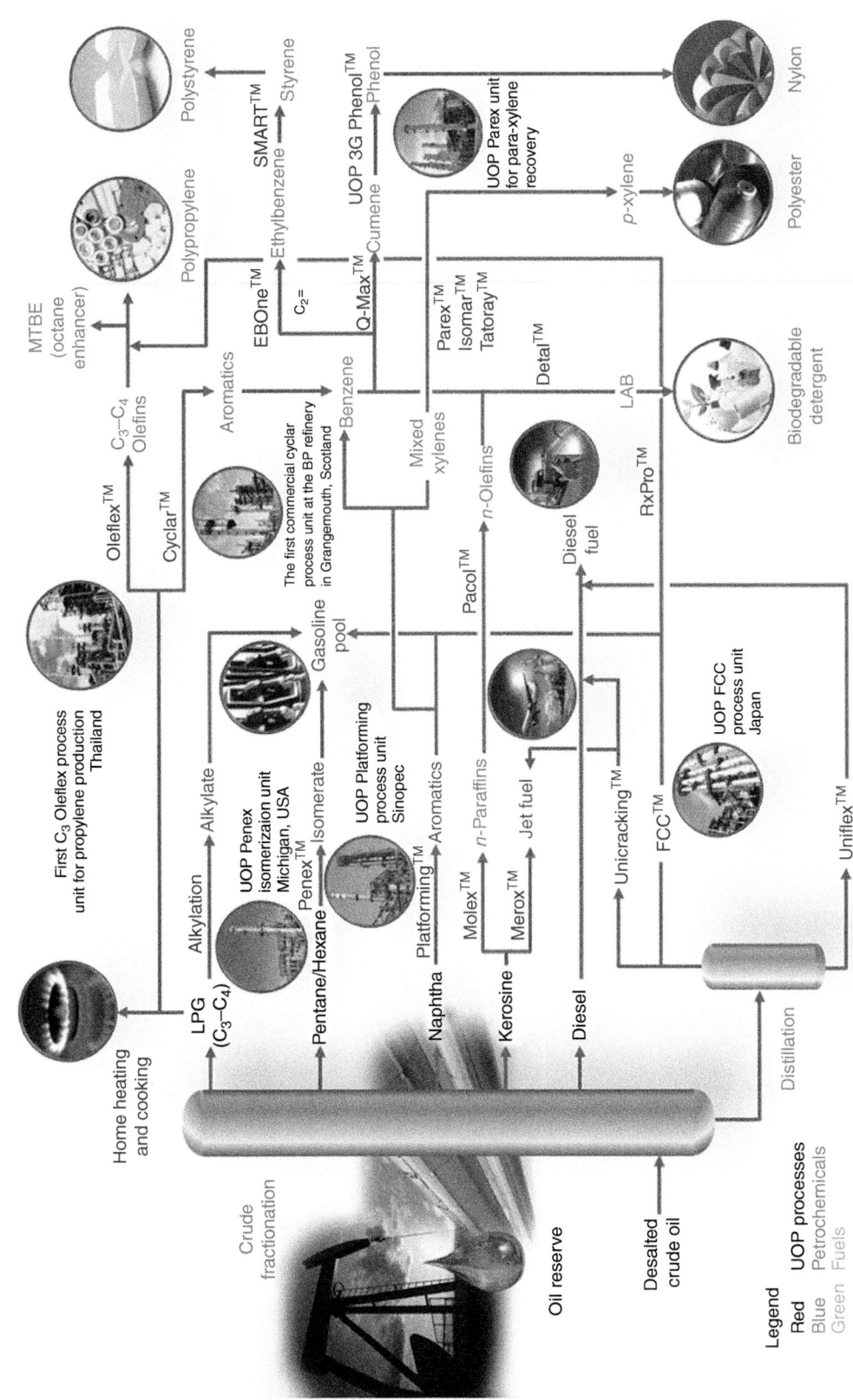

Figure 1.7 Feedstock sources and technologies for petrochemicals and fuels.

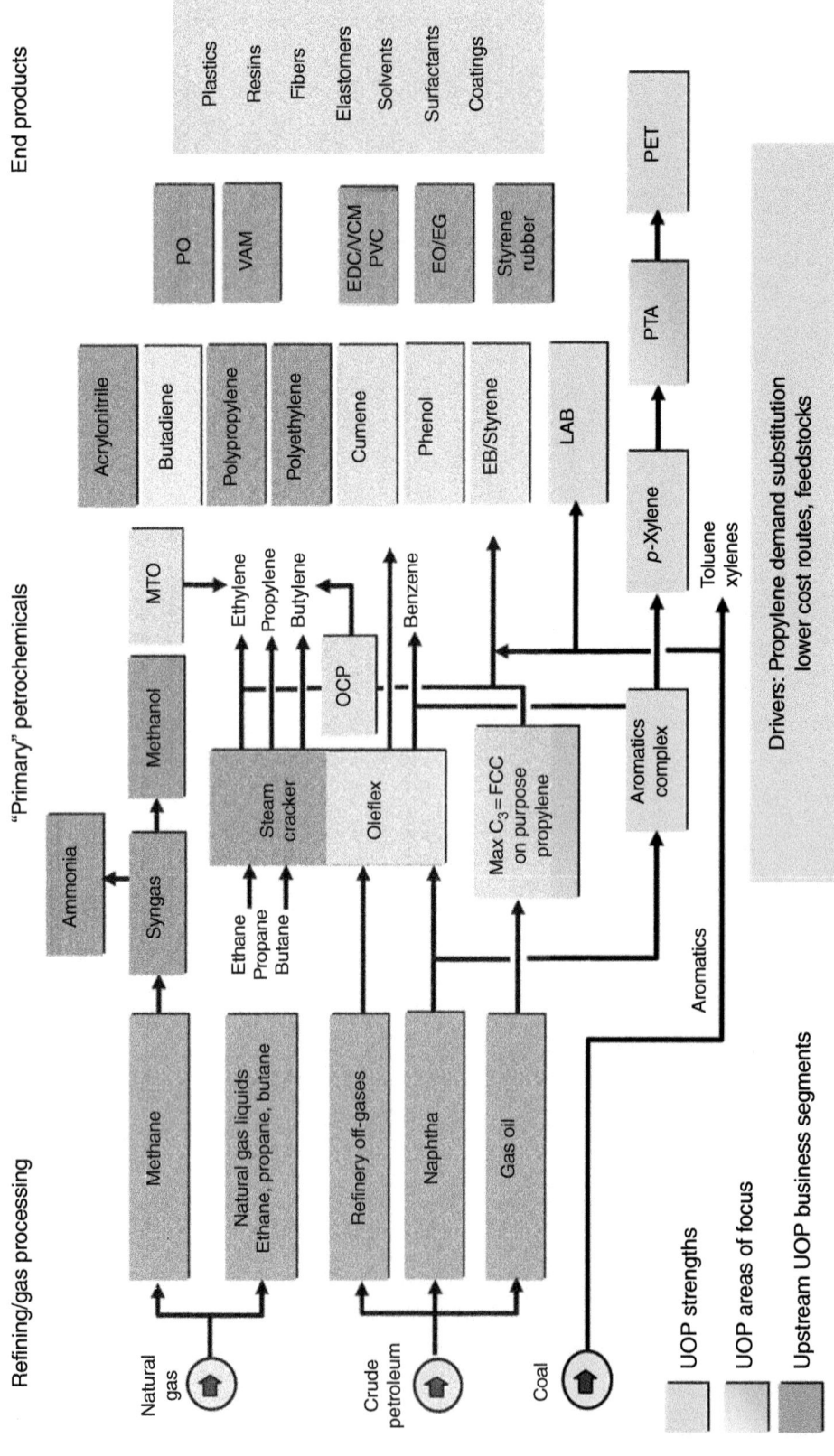

Figure 1.8 Producing petrochemicals from refinery integration.

Figure 1.9 Economies of scale in *para*-xylene production.

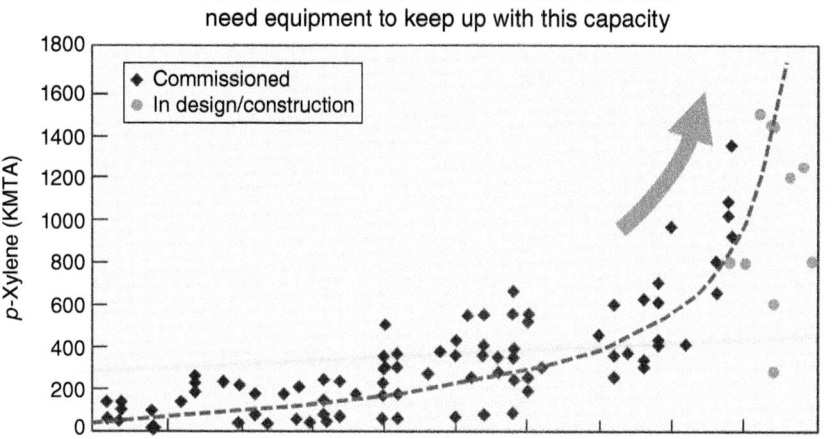

seek technologies that can economically upgrade low-value by-product streams into petrochemical feeds. Some technologies, such as dehydro-cyclodimerization (DHCD) or toluene methylation (TM) can reduce or eliminate the need for conventional refinery feed streams such as naphtha. Advantageous pricing for LPG or methanol relative to naphtha can make these alternatives very attractive, provided the DHCD or TM technology has good yields. More will be said about this in later chapters of this book.

1.3 Context of Process Designs and Operation for Petrochemical Production

The preceding sections have discussed the macroeconomic drivers behind the rapid growth of the petrochemical industry, its impact in everyday life, some of the basics of how the supply chains of fuels and petrochemicals are linked, as well as the value chains that are created from aromatic petrochemicals. It should be clear that a few molecules – we will call them "monomers" for sake of simplicity – are the backbone of the petrochemical industry. In the next chapters, we will explore in great detail how the quantities and purities of these key monomers are maximized from hydrocarbon streams that contain hundreds of other components. This task of efficiently producing these monomers requires high-performing catalysts and process technologies to transform the hydrocarbon feedstocks into a simpler mixture that can then be efficiently separated into the purified monomers. This is not a trivial task, and companies such as UOP, Axens, ExxonMobil, and BP have spent considerable effort to perfect these technologies over the years. There have been steady advancements in catalysis that have been enabled significant improvements in conversion process technologies. Similarly, advances in separation technologies have enabled greater efficiencies in purification and recovery of these monomers.

On top of these advances in catalysis, separations, and process technologies, the industry has rapidly moved to large-scale "mega-plants" to capture the value in economy of scale, as shown in Figure 1.9. As an example, in 1970, when the UOP Parex™ process was introduced for pX, a plant with 100 000 metric tons per year of pX capacity was considered to be very large. In 1999, a 400 000 metric ton per year plant was typical. By 2010, the typical pX plant size was 1 000 000 metric tons per year. Today, it is not unusual for producers to seek pX capacities of 2–2.5 million tons per year for grassroots plants. This evolution toward larger and larger scale plants has driven innovations in equipment such as reactors, adsorption units, fractionators, and other process technologies, complementing the advances in catalysis and separation technologies described earlier.

1.4 Who Is This Book Written For?

This book is intended as a reference on aromatic process technology for students of chemistry and chemical engineering, as well as for industrial business leaders and technologists who are interested in participating in or strengthening their competitive positions in the aromatics industry.

2

Market and Technology Overview

2.1 Overview of Aromatic Petrochemicals

The focus of this section will be the primary aromatic building blocks for the petrochemical industry:

- Benzene
- Toluene
- Ethylbenzene/styrene
- *para*-Xylene
- *meta*-Xylene
- *ortho*-Xylene
- Cumene/phenol

There are other aromatics such as 1,2,4-tri-methyl benzene and 1,2,4,5-tetra-methyl benzene which can be converted, respectively, to tri-mellitic anhydride and pyro-mellitic dianhydride. While these oxidation products have high value as plasticizers, powder coatings, and precursors for heat-resistant polyimide resins, they are much lower volume, considered as niche markets, and will not be discussed in detail. We will first discuss the market dynamics for each of these aromatics, and then focus on the technologies that produce them.

2.2 Introduction and Market Information

2.2.1 Benzene

Benzene can be alkylated with ethylene to produce high-purity ethylbenzene (EB) which is then dehydrogenated to give styrene. Ethylbenzene has also been used to coproduce propylene oxide and styrene monomer (POSM technology commercialized by LyondellBassell and others). Compound annual growth rates (CAGR) for styrene are in the range of 1.7–2% worldwide, whereas they are around 3.5% for propylene oxide. As a result, very few POSM plants have been built, and propylene oxide producers have focused their R&D efforts on direct oxidation technologies for PO.

In the early years of the styrene industry, aluminum chloride catalyst was used for alkylation of benzene with ethylene. This technology operated at very low temperature and low benzene : ethylene ratios, producing a very small amount of di- and tri-ethyl benzenes. However, corrosion and equipment maintenance were problematic. Mobil introduced vapor-phase alkylation for ethylbenzene production using ZSM-5 zeolite (scientifically termed MFI) catalysts (Keown et al. 1973) and this became the industry standard for several years, displacing the aluminum chloride units as the styrene market grew. One drawback of vapor-phase alkylation technology was the coproduction of xylene impurities – up to 0.5 wt.%. This required a purge in the dehydrogenation plant in order to prevent buildup of the xylene impurities in the recycle loop. This resulted in increased feed costs relative to the aluminum chloride catalyst system.

In 1984, Unocal introduced liquid-phase alkylation for ethylbenzene using Y-zeolite catalysts (Inwood et al. 1984). The advantage of this technology was a significant reduction in the amount of xylene impurities, down to about 300 ppm. Like vapor-phase alkylation, some amount of di- and tri-ethylbenzenes was formed. Unocal used liquid-phase transalkylation of these polyalkyl compounds with benzene to maximize ethylbenzene yields. Further advances in zeolitic catalysis led to the use of Beta and UZM-8™ zeolites for alkylation in UOP's EB*One*™ process, and MCM-49-based catalysts in ExxonMobil's Alkymax technology. These liquid-phase ethylbenzene alkylation plants operate at low benzene:ethylene ratios and low temperatures, with significant amounts of heat integration between the ethylbenzene and the styrene units. The typical ethylbenzene alkylation plant consists of several catalyst beds with intermediate ethylene injection to control the exotherm. Reactants typically are introduced upflow to avoid localized accumulation of ethylene. There are also a few ethylbenzene plants that utilize catalytic distillation, where the catalyst bed is located inside the fractionator. One such technology is CDTech, offered by McDermott.

Efficient Petrochemical Processes: Technology, Design and Operation, First Edition.
Frank (Xin X.) Zhu, James A. Johnson, David W. Ablin, and Gregory A. Ernst.
© 2020 John Wiley & Sons, Inc. Published 2020 by John Wiley & Sons, Inc.

Another important use of benzene is for production of phenol. In this case, benzene is alkylated with propylene to form cumene (iso-propyl benzene). Cumene is then oxidized to give phenol and acetone, along with alpha-methyl styrene, if desired. Cumene and acetone can be converted into bisphenol-A, and ultimately into polycarbonate (PC) resin. The CAGR for polycarbonate is similar to that for polystyrene, around 2% worldwide, although localized shortages can create added demand.

The earliest cumene plants employed solid phosphoric acid (SPA) catalysts in either fixed bed or multi-tube reactors. SPA catalyst consists of pelletized crystalline silicon phosphate. In cumene service, it was necessary to have a low temperature rise across the catalyst bed as the alkylation proceeded, in order to avoid release of phosphoric acid. This was done by operating at high benzene:propylene ratios, around 10 or higher. While this would lengthen the catalyst life and minimize operational upsets, the utility consumption was very high due to the amount of benzene recycle. In the mid-1990s, zeolitic catalysts were developed by Chevron Research and Mobil Oil Corp to produce cumene in liquid phase, and these could operate at benzene : propylene ratios in the range of 3–6, depending on the catalyst being used (Innes et al. 1992; Kushnerick et al. 1991). Examples were the UOP's Q-Max™ process and the Badger Cumene process. Zeolites were Beta and later UZM-8 in the case of UOP technology and ExxonMobil's MCM-22 in the case of Badger technology. Similar to ethylbenzene, with polyalkylates, these zeolitic cumene technologies required the use of transalkylation of polyisopropyl benzenes with benzene to maximize cumene yield.

Other uses of benzene include cyclohexane for nylon and various derivatives for agricultural or pharmaceutical uses. It is also worth noting that there are several grades of benzene commonly used by the petrochemical industry. Detergent purity can be as low as 99.5%. Nitration-grade benzene is comparable to EB and cumene-grade benzene at 99.9%. The most stringent purity is that used for cyclohexane, where the trace impurities such as methylcyclopentane and hexanes must be less than 300 ppm.

The aggregate demand for benzene at this time is comparable to that for *para*-xylene (pX), at about 45 million tons per year. But, as shown in Figure 2.1, the demand for benzene in the future will lag that of pX because of the lower CAGR associated with the downstream benzene applications relative to polyester.

2.2.2 Benzene Production Technologies

Benzene is typically a coproduct of many catalytic technologies. For example, in catalytic reforming of naphtha, such as with the UOP CCR Platforming™ process or Axens' Aromizing process, a bifunctional catalyst (acid and metal functionalities) operates at about 50 psig and 550 °C to dehydrogenate C_6 naphthenes to aromatics. This catalyst also isomerizes and cyclizes C_6 paraffins to C_6 naphthenes which go on to be dehydrogenated to benzene. For petrochemical use, most naphthas will contain C_6–C_{10} hydrocarbons, with analogous dehydrogenation and cyclization reactions occurring to produce benzene, toluene, and C_8, C_9, and C_{10} aromatics. A small amount of dealkylation of higher aromatics can occur to further increase the benzene yield. As shown in Figure 2.2, selectivities are more favorable as the paraffin carbon number increases.

In addition, equilibrium for aromatics is favored by lower hydrogen partial pressure and higher reactor temperature. Aromatics yields from naphtha will be a function of the cyclic content, which in turn is dependent on the crude oil source. Mideastern naphthas are typically 20–25% cyclic, whereas shale naphthas can be 30–40% cyclic. The reformate product will typically contain between 75 and 85 wt.% aromatics, with the balance being unconverted C_6 and C_7 paraffins. Aromatics will

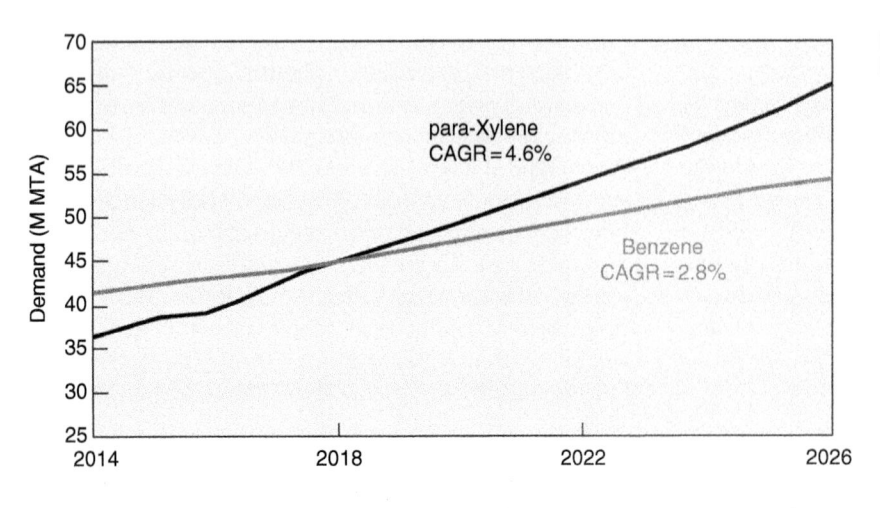

Figure 2.1 Demand for *para*-xylene and benzene. *Source*: IHS.

typically be recovered and purified by downstream fractionation and extractive distillation (ED).

Naphtha reforming of C_6 and C_7 components can also be efficiently accomplished using zeolitic reforming. This technology employs a nonacidic catalyst, where the Pt atoms are deposited in the channels of L-zeolite. There is no acid function in the zeolite or catalyst binder to accomplish isomerization, but the selectivity advantage can be significant, since cracking of C_6 and C_7 paraffins is minimized. There are a few examples of this technology in commercial operation: Chevron's Aromax and UOP's RZ Platforming, but this route accounts for less than 1% of the world's naphtha reforming capacity. There is no selectivity advantage for C_8+ hydrocarbons in RZ Platforming compared to conventional CCR Platforming, so the feed to a zeolitic unit will typically be C_6 and C_7 hydrocarbons. Unconverted paraffins from both CCR Platforming and RZ Platforming can be recovered in ED or liquid–liquid extraction (LLE), and sent back for another pass through the Platformer. We will describe naphtha reforming technology in greater detail in Section 2.3.

A third source of petrochemical benzene comes as a by-product of naphtha steam cracking for olefins production. This by-product stream is called pyrolysis gasoline, or pygas in short. The pygas results from thermal reforming reactions that occur in the high-temperature low hydrogen partial pressure tubes of the cracker furnace. Pygas is an extremely complex mixture, consisting of benzene–toluene–C_8 aromatics as well as paraffins, naphthenes, olefins, di-olefins, and *cyclo*-olefins. Pygas will typically contain some amount of sulfur and even nitrogen compounds. The di-olefins and mono-olefins will be removed by the first-stage pygas hydrotreating catalyst and the polar compounds will be removed by the second-stage pygas hydrotreating catalyst. The hydrotreated pygas can then be sent to ED or LLE. The aromatics yield from a naphtha steam cracker will typically be 20–25% of the feed. As a result, the pygas aromatics are often viewed as supplemental to the on-purpose aromatics produced through naphtha reforming.

Once toluene and A_9+ aromatics are available from naphtha reforming and pygas, on-purpose production of benzene can take place via transalkylation technologies such as the UOP Tatoray™ process or the ExxonMobil TransPlus process. These technologies use efficient zeolitic catalysts to maximize the yields of benzene and xylenes from the aromatic feed. The yield of benzene is predominantly a function of the methyl/phenyl ratio of the feed, governed by equilibrium. Benzene yield increases with the amount of toluene in the feed. This is illustrated in Figure 2.3. For the highest levels of toluene in the feed, the benzene yield can approach 50%. Many operators choose to produce xylenes preferentially to benzene due to the differences in CAGR for end products as described earlier. In this case, the feed will be much higher in C_9+ aromatics than toluene. This choice

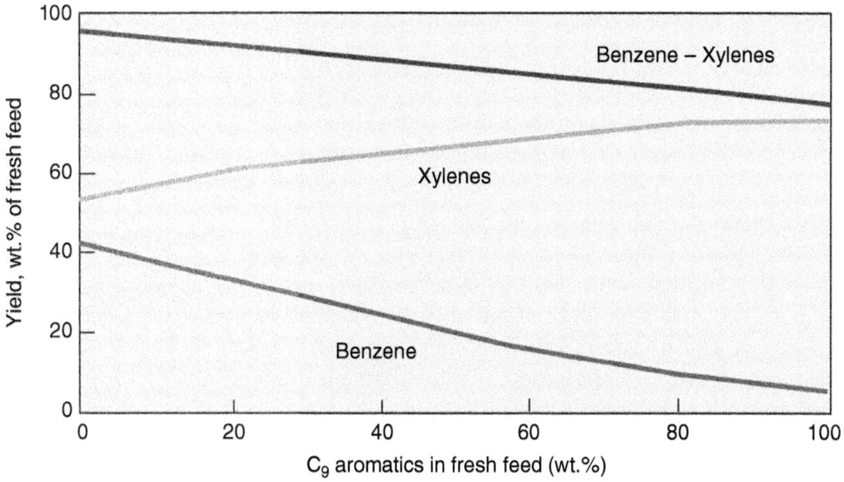

Reaction		Relative Rates	
Step 1	Step 2	Step 1	Step 2
$P_6 \rightarrow N_6$	$\rightarrow A_6$	0	4.01
$P_7 \rightarrow N_7$	$\rightarrow A_7$	0.58	9.03
$P_8 \rightarrow N_8$	$\rightarrow A_8$	1.33	21.5
$P_9 \rightarrow N_9$	$\rightarrow A_9$	1.81	24.5
$P_{10} \rightarrow N_{10}$	$\rightarrow A_{10}$	2.54	24.5

- Rates increase with carbon number
- The rates for competing cracking conversion also increase with C number

Figure 2.2 Reactivity and selectivity of naphtha components.

Figure 2.3 Transalkylation yields vs. feed C_9 content.

will ultimately affect decisions on the naphtha cut point in the feed to the reformer, where C_6 and even C_7 precursors can be reduced in order to minimize benzene and toluene. The details of transalkylation technology will be covered in Section 2.5.1.

Selective toluene disproportionation (STDP) is a technology that feeds only toluene to a catalyst whose pore structure has been engineered to preferentially allow benzene and pX to diffuse out of the selective pores after being formed from the bimolecular toluene disproportionation reaction that occurs inside the pores. This technology can produce up to a 45% yield of benzene, in addition to a xylene product that is 90% or higher pX. Examples of this technology are UOP's PX-Plus™ process and ExxonMobil's MSTDP or PXMax process. These *para*-STDP technologies will be employed when there are local shortages of benzene coupled with the need to further increase pX production. Some operators will install such a technology and then run it intermittently depending on the relative prices of benzene and pX. This technology will be covered in more detail in Section 3.8.

Hydro-dealkylation is the final example of technology for benzene production. This technology is rarely used, as it is commonly regarded as the most expensive means for producing benzene. Examples are UOP's THDA™ process (thermal hydro-dealkylation) and Axens' HDA process. These technologies are strictly thermal, operated at temperatures approaching 1300 °F, and they essentially dealkylate toluene and higher aromatics down to benzene with the by-product being methane. Because of the significant hydrogen consumption and low yield of benzene relative to the aromatic hydrocarbon feed, dealkylation technology has long been abandoned in favor of transalkylation and even STDP. Perhaps the only positive feature of dealkylation technology is that it produces extremely high-purity benzene, but at the expense of generating massive amounts of fuel gas by-product. In order to reduce yield losses to heavies, the recycle of bi-phenyl is typically implemented in THDA designs. This technology will be briefly discussed in Sections 2.2.5 and 2.2.6.

Some final observations about benzene relate to how it is viewed by the transportation fuel industry vs the petrochemical industry. In most locations, there are regulations imposed that cap the maximum amount of benzene in gasoline, and even the amount of total aromatics in gasoline. As of 2019, the United States mandates that on average gasoline can contain no more than 1% benzene, and today in most cases the level is below 0.6%. In addition, toluene has been backed out of some gasoline formulations due to vapor pressure concerns when large amounts of ethanol are added. Rather than producing benzene only to then have to remove it from the gasoline blending stocks, refiners have adjusted to these regulations by avoiding production of benzene in the first place.

This has been done, for example, by removing benzene precursors via naphtha cut point, and then subjecting the light naphtha by-product to processing that further minimizes benzene potential. The petrochemical industry needs a particular amount of benzene to meet today's demand of 45 million tons per year, and will produce it by the most economical means possible to meet demand. Returning to Figure 2.1, in the long term, the lower CAGR for benzene relative to pX may mean that other technologies like toluene methylation may help manage skewed demand between benzene and pX. More about that in Section 3.8.2.

2.2.3 Toluene

Toluene is primarily used as a solvent for paints, adhesives, glues, and coatings. It is difficult to find the industrial volumes of toluene in commerce due to the diversity of applications and the fact that much of the chemical toluene is converted to benzene and xylenes. In Figure 2.4, IHS provides the world chemical toluene consumption by region. Korea consumed about 2 million tons of toluene in 2017, so the world chemical consumption of toluene is approximately 20 million tons per year.

According to IHS Markit (2018a), the largest derivative use of toluene is toluene di-isocyanates (TDI's) for production of polyurethanes, accounting 5% of toluene usage. Polyurethanes have numerous applications as shown in Figure 2.5.

A wide variety of foam textures and densities can be produced, depending on how the blending components are assembled. World production of all polyurethanes is around 20 million tons per year, expected to increase to 26 million tons per year by 2021. TDI's are particular

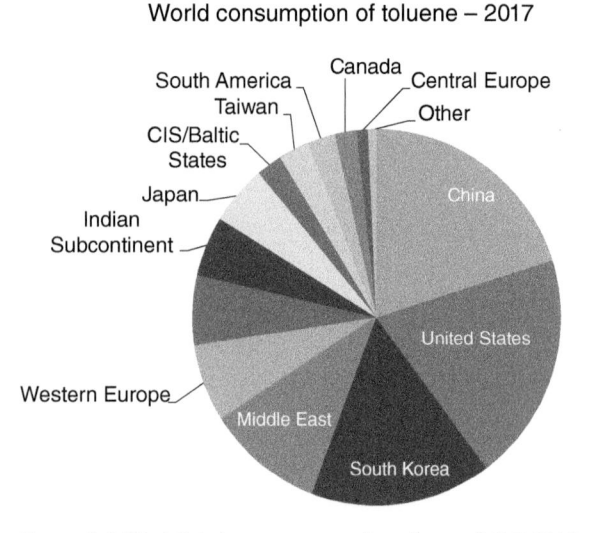

Figure 2.4 World's toluene consumption. *Source:* IHS © 2018 IHS Markit.

2.2 Introduction and Market Information

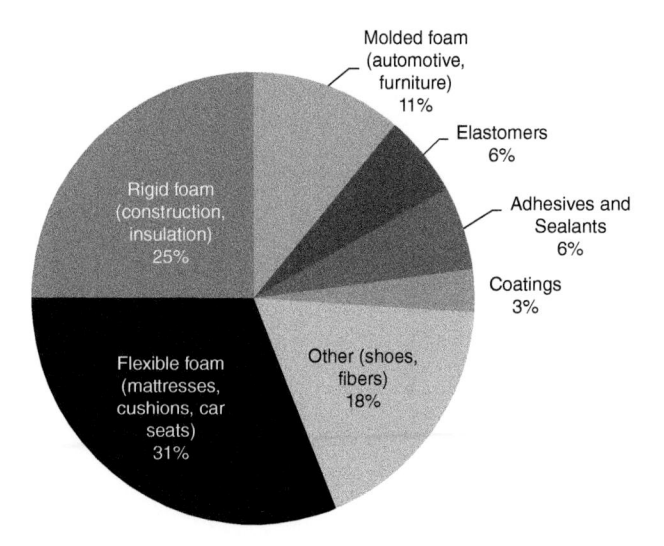

Figure 2.5 Polyurethane applications. *Source:* The Essential Chemical Industry (2017).

components of urethane, and are derived from nitration of toluene, followed by reduction to the corresponding diamines, and then phosgenation to the corresponding di-isocyantes. The TDI is reacted with polyols to form the polyurethane polymer.

2.2.4 Toluene Production Technologies

Naphtha reforming and pygas are the primary sources of toluene. As with benzene, toluene contained in the reformate is admixed with unconverted paraffins and some olefins that are in equilibrium with the paraffins at the reformer last reactor outlet temperature and hydrogen partial pressure. For most chemical uses, such as nitration, it is necessary to purify the toluene to meet downstream specifications. A common practice is to send the C_6–C_7 reformate cut (reformate splitter overhead) to ED, where the extract will contain less than 100 ppm of nonaromatics. It is then a simple matter of fractionation to recover purified benzene overhead and purified toluene at the bottoms. If the toluene is to be used as part of transalkylation feed to produce additional benzene and xylenes, it may be possible to avoid extracting that portion of the toluene, at least for the very stable and robust UOP Tatoray catalysts such as TA-30™ and TA-32™. This will be covered in more detail in Sections 2.5.1 and 3.5.

2.2.5 Ethylbenzene/Styrene

As noted earlier, ethylbenzene is typically synthesized via benzene alkylation with ethylene to produce 99.97% purity ethylbenzene for styrene production. This high purity avoids the need for significant purges of xylene impurities from the EB/styrene loop, saving feed consumption. It is tempting to consider recovering the ethylbenzene contained in the reformate (17% of the C_8 aromatics) or pygas (40% of the C_8 aromatics). One could consider super-fractionation, adsorption, or solvent extraction as a means to go from these purity levels to the 99.97% required for styrene production. There is a 2.2 °C difference in normal boiling point between ethylbenzene and pX. This has led some companies to try super-fractionation of EB (Carson and Haufe 1968), with over 300 trays and reflux : EB ratio of around 80.

2.2.6 Ethylbenzene/Styrene Production Technologies

In the mid-1960s, the high CAPEX and OPEX of super-fractionation were abandoned in favor of direct synthesis via alkylation of benzene with ethylene. Vapor-phase alkylation technology was used extensively, but resulted in xylene impurities. Unocal developed liquid-phase alkylation technology with significantly lower xylene contaminants, and that has been the dominant route ever since. UOP acquired liquid-phase EB technology from Unocal and went on to develop the EB*One* process, which includes alkylation as well as liquid-phase transalkylation of polyethyl benzenes with benzene. ExxonMobil offers EB Max with similar flow scheme and performance. Nearly all EB units are closely heat-integrated with styrene units where the steam from the styrene unit is used as a heat source for the low-temperature liquid-phase alkylation. UOP and McDermott (CBI Lummus) offer EB*One* along with either classic styrene or an oxidative reheat styrene process called SMART™. There are also a few examples of liquid-phase EB units that produce feed for POSM plants. The drawback for this technology is the difference in growth rates between styrene (1.7%) and propylene oxide (4%). See Figure 2.6 for yearly styrene demand and Figure 2.7 by IHS Markit (2018b) for styrene demand by region.

2.2.7 *para*-Xylene

As noted earlier, *para*-xylene is the primary component of PET, used in fiber, resin, and films. Dow, ICI, and DuPont were the first companies to explore oxidation to purified terephthalic acid (PTA) and polymerization of PTA with MEG. A 31 March 1951 article in *Chemical Industries Week* described the prospects for xylene isomers to be purified and recovered from reformate and the use of *para*-xylene for PTA production. In 1941, terylene fiber had been invented by J.R. Whinfield in England. He sold the patent rights to ICI. In the United States, DuPont began exploring PET, and planned construction of a plant in North Carolina to produce its version called Fiber V to be used in Dacron fabrics. DuPont hoped to grow the market to 50 million lb, or about

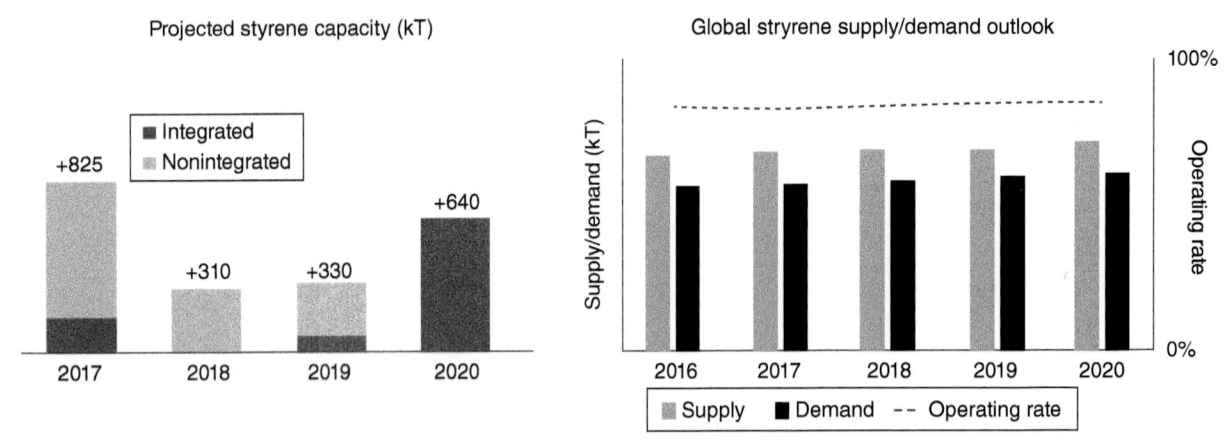

Styrene supply/demand outlook

Projected styrene capacity (kT) — Integrated / Nonintegrated: 2017 +825, 2018 +310, 2019 +330, 2020 +640

Global styrene supply/demand outlook — Supply, Demand, Operating rate (2016–2020)

- Global capacity of about 33 000 kT
- 2016 – 2020 average annual supply growth of 1.6%
  - Over 85% of new capacity over three years is nonintegrated in China
- Estimated annual styrene demand growth is estimated at 2.0%

Supply/demand projections support continued healthy operating rates through 2020

Figure 2.6 Styrene demand outlook. *Source:* Trinseo.

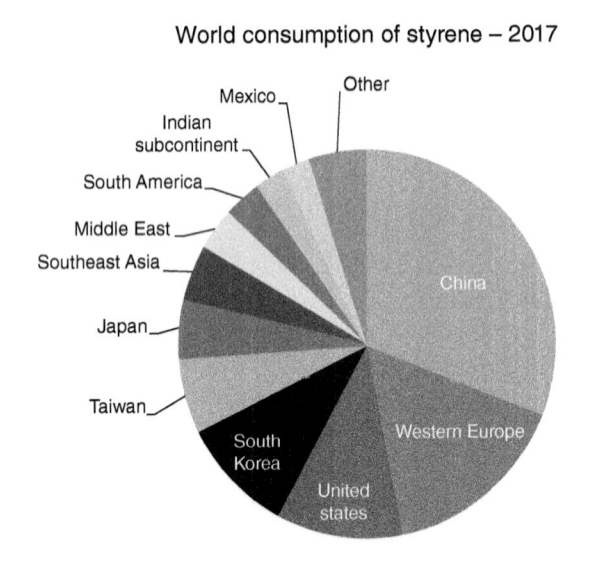

Figure 2.7 Styrene demand by region. *Source:* IHS Markit © 2018 IHS Markit.

World consumption of styrene – 2017: China, Western Europe, United States, South Korea, Taiwan, Japan, Southeast Asia, Middle East, South America, Indian subcontinent, Mexico, Other.

22 600 metric tons per year, and reported buying 95% pure *para*-xylene from Oronite for $0.25–0.30/lb, which would be about $550/MT, at a time when oil prices were only $2.50/barrel.

As shown in Figure 2.8, it took some time for the PET industry and the demand for *para*-xylene to grow, beginning with the introduction of synthetic polyester fabrics. Gradually, polyester became more widely accepted by consumers in clothing and other home goods, and that created the need for larger scale and more efficient means of producing it from the hydrocarbons available in typical gasoline refining operations. The *para*-xylene concentration in a typical reformate is actually fairly low – around 10%. A C_8 aromatic heartcut of reformate might contain as much as 20% *para*-xylene, so there are challenges to achieving high purity and recovery of *para*-xylene from this type of feed.

2.2.8 *para*-Xylene Production Technologies

Fractional crystallization was the earliest means of purifying and recovering *para*-xylene from a mixture of C_8 aromatics. The separation was limited by the eutectic point between *para*-xylene and *meta*-xylene. This allowed recoveries of *para*-xylene to be 60–65%, with purity limited by the ability to exclude *meta*-xylene from the crystals. ICI, Amoco, and Standard Oil of New Jersey were among the first to produce *para*-xylene via crystallization. Through technology improvements, the attainable purity increased from 95% to more than 99%, but recovery remained low due to the eutectic limitation. In 1970, UOP introduced the Parex™ process, which used selective adsorption in simulated moving bed (SMB) to purify and recover *para*-xylene from a mixture of C_8 aromatics and nonaromatics. This technology allowed recovery of *para*-xylene above 90% and purities above 99%. After the first plants proved the technology, the world primarily used adsorption for producing *para*-xylene. Over about 20 years, the purity requirement for *para*-xylene increased in steps from 99.0 to 99.2, then

Figure 2.8 *para*-Xylene capacity growth by technology.

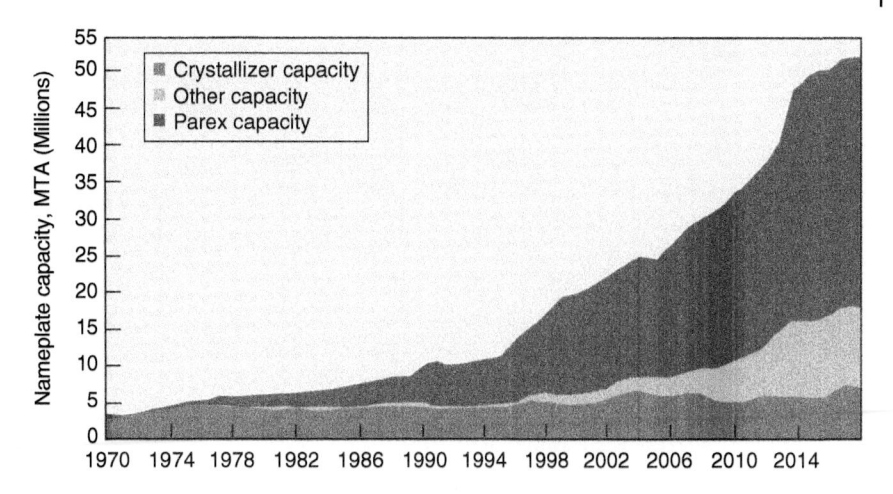

Figure 2.9 PET value chain and growth rates. *Source:* 2018 Demand (WoodMac); Demand CAGR (2018–2028) (WoodMac); RPET Rates (SBA-CCI).

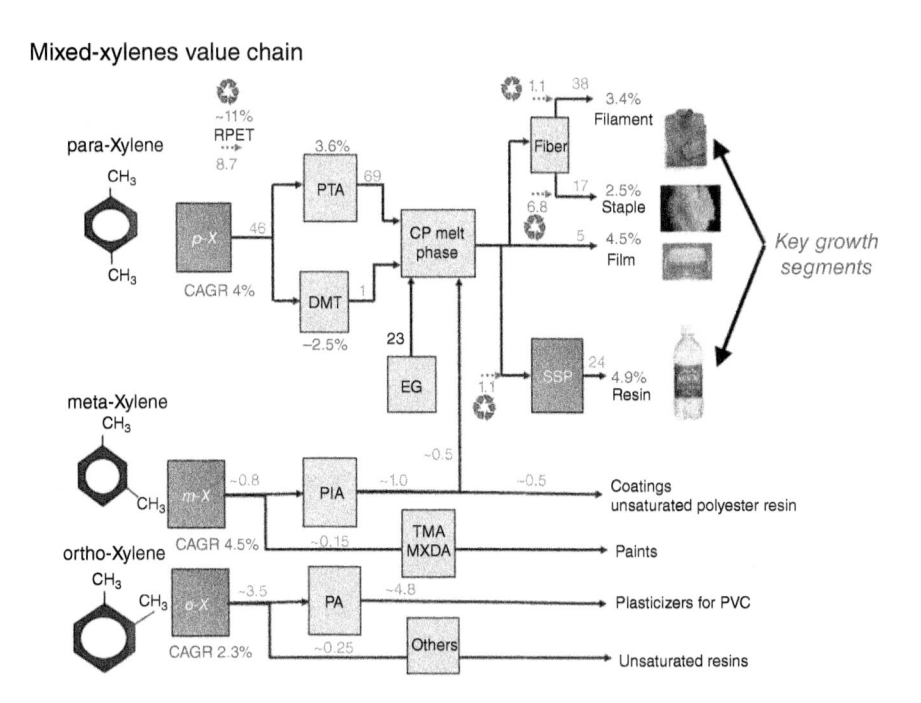

99.5, and 99.8. Several *para*-xylene technologies are able to easily achieve purities above 99.9%, but there is no premium value for achieving such high purity. Most PTA producers can easily handle *para*-xylene purities between 99.7 and 99.8%. In 1994, Axens developed a similar adsorptive separation technology to the Parex process called Eluxyl, and around 2010, Sinopec imitated Axens with their own adsorption technology. More details of these technologies will be provided in Section 3.7.

While fiber applications consume nearly 75% of the *p*X, PET resins represent the fastest growing segment of the PET market, as shown by Wood Mackenzie (2018) and UOP in Figure 2.9.

This growth in PET resins was enabled by increasingly abundant *meta*-xylene (see Section 2.2.9) as a result of technical innovations at UOP. PET resins typically incorporate 2–3% of purified isophthalic acid (PIA) as a comonomer. The inclusion of PIA allows higher molecular weights and crystalline structures within the polymer. These properties allow strong and rigid bottles to be made via injection molding for preforms and blow-molding of bottles. PET resin bottles are used worldwide for water, carbonated soft drinks, juices, and even alcoholic beverages. These bottles are much lower weight than glass or aluminum and have disrupted the container industry over the past 20 years. They allow increased mobility as well as reduced shipping costs and losses due to breakage. Between 2000 and 2010, this market grew at a rate of more than 10% per year. Numerous PET fiber facilities added production lines for solid state polymerization of PET resins, in order to take advantage of this growth.

2.2.9 meta-Xylene

meta-Xylene is the dominant xylene isomer in a typical gasoline refining stream, with more than double the concentration of para-xylene. However, its industrial uses – and subsequently market demand – are significantly smaller than for para-xylene. World capacity for purified meta-xylene has grown to around 1 MM MTA, as shown in Figure 2.10. As noted above, the predominant end use of meta-xylene is for oxidation to PIA, and use as a comonomer in PET resin formulations. A smaller but still important end-use of meta-xylene is for production of meta-xylene diamine (MXDA). The MXDA can then be polymerized with adipic acid to produce coatings and barrier layers for containers that make them impervious to various gases.

2.2.10 meta-Xylene Production Technologies

As shown in Figure 2.10, purified meta-xylene was originally produced using the Mitsubishi HF/BF$_3$ adduct technology. Between mid-1980 and mid-1990, the supply of high-purity meta-xylene was tightly controlled by Amoco and Mitsubishi, with only three plants worldwide using the HF/BF$_3$ technology. UOP developed an alternative production technology – the MX Sorbex™ process – based on adsorption using a SMB, and quickly licensed several units in the United States and Asia. The availability of abundant high-purity meta-xylene accelerated the development of PET resin applications that led to the rapid growth of the PET bottle market. Prior to that, there was strong interest in polyethylene naphthalate (PEN) as a potential polymer for bottles. PEN is derived from 2,6-dimethyl naphthalene (2,6-DMN) oxidation to 2,6-naphthalene dicarboxylic acid (NDCA). UOP participated in the early development and commercial production of high-purity 2,6-DMN using Sorbex technology but the high cost of 2,6-DMN synthesis and purification, along with polymer properties, limited the development of the PEN market for bottles. Today, there is only one plant in the world making 2,6-NDCA with a capacity of around 27 000 MTA. Today, many para-xylene producers take a slipstream from their facilities and coproduce a small but commercially significant amount of high-purity meta-xylene using the MX Sorbex process. Prices of para-xylene and meta-xylene vary with the price of oil, but meta-xylene is generally 1.5 times more valuable than para-xylene.

2.2.11 ortho-Xylene

The concentration of ortho-xylene in typical gasoline refinery streams is very similar to that of para-xylene, around 10%. In the C$_8$ heartcut of gasoline, the ortho-xylene content can be as much as 20%. Since there is a 6 °C boiling point differential between ortho-xylene and the next closest isomer, meta-xylene, it is possible to purify and recover ortho-xylene by fractionation. A few para-xylene production facilities will coproduce ortho-xylene, often in a para : ortho ratio of 4 : 1 or 5 : 1.

2.2.12 ortho-Xylene Production Technologies

There are a few plants that primarily produce mixed xylenes for the merchant market and coproduce a small amount of ortho-xylene. ortho-xylene in commerce is typically 98.5–99% purity. When producing ortho-xylene by fractionation, it is important to have a catalytic reformer that operates at relatively high conversion of C$_9$ paraffins in order to minimize the amount in the product that would co-boil with the fractionated ortho-xylene. For older catalytic reformers that operate at lower C$_9$ paraffin conversion, it would be necessary to send the xylenes' fraction through LLE along with benzene and

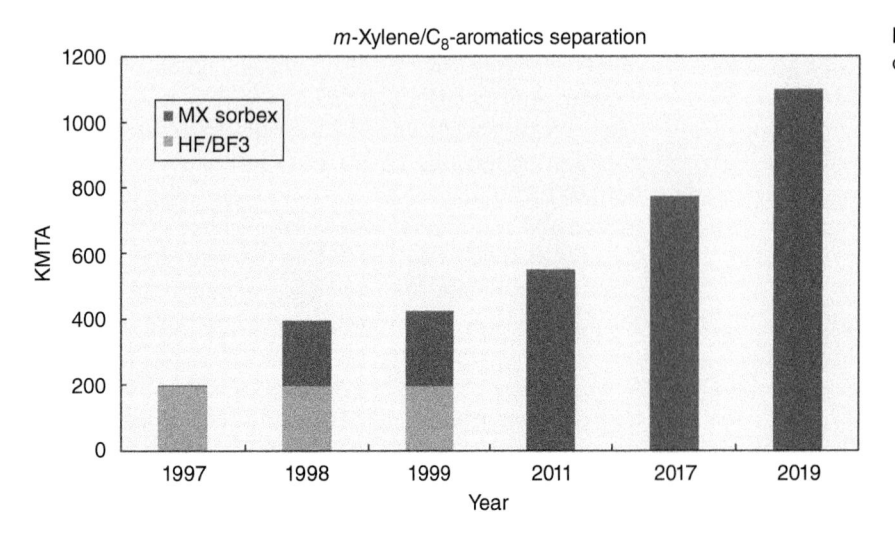

Figure 2.10 meta-Xylene production capacity.

Figure 2.11 Products from *ortho*-xylene.

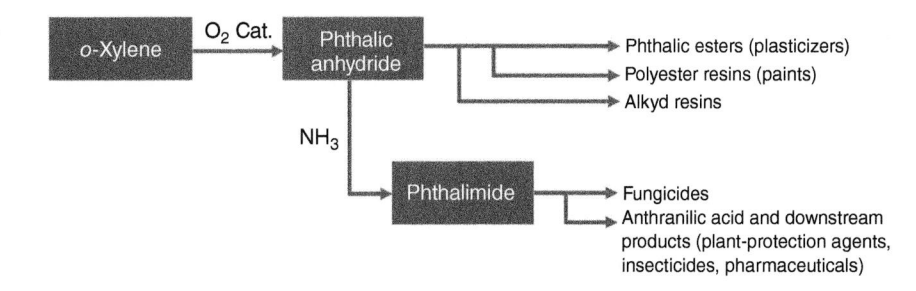

Figure 2.12 *ortho*-Xylene consumption by region. *Source:* IHS Markit © 2018 IHS Markit.

toluene in order to allow high-purity *ortho*-xylene to be achieved by fractionation. More will be said about *ortho*-xylene production technology in Chapter 3.

As shown in Figure 2.11, the primary derivative of *ortho*-xylene is phthalic anhydride (PA), which requires oxidation of the *ortho*-xylene. PA is converted into plasticizer resins for production of poly vinyl chloride (PVC). The growth rates for PA and PVC are approximately in GDP, around 2.5%, in contrast to the growth rates for PET. According to IHS (2018c), the world consumption of *ortho*-xylene was about 10% that of *para*-xylene. Figure 2.12 shows *ortho*-xylene consumption by region.

2.2.13 Cumene/Phenol

As noted earlier, cumene is typically produced by alkylation of high-purity benzene with propylene. Most typically, cumene will be oxidized to produce phenol and acetone, both of which have great importance to the chemical industry. For polymer production, phenol and acetone can be reacted to form bisphenol-A, which in turn can be converted into polycarbonate resin. PC is widely used in consumer goods, automotive parts, appliances, and electronics. Figures 2.13 from ICIS (2016) and 2.14 from IHS (2018d) show the many ways phenol and acetone are used in chemical synthesis, as well as the world consumption growth for phenol. It is worth noting that most of the growth for phenol demand is occurring in Asia. This is tied to the high population and growth in middle class and spending power that were mentioned earlier. In contrast to ethylbenzene, there is a significant merchant market for cumene, and it can be easily transported to phenol producers. However, many producers choose to integrate their cumene and phenol production at the same site.

2.2.14 Cumene/Phenol Production Technologies

As noted earlier, cumene was originally produced as a high-octane additive for aviation gasoline, using SPA catalyst. As the petrochemical industry emerged after WW II, many cumene producers found they could sell this product into the phenolics value chain, with most continuing to use SPA technology, either in adiabatic two-bed reactors or isothermal multi-tube reactors. In the 1990s, UOP introduced the Q-Max process, which utilizes a zeolitic catalyst to alkylate benzene with propylene. Poly-isopropyl benzene by-products are fractionated and transalkylated with benzene to produce additional cumene. Several improvements to Q-Max process technology further improved yields, reduced OPEX as well as CAPEX. Most often, the cumene from the UOP Q-Max process is then oxidized to phenol and acetone using the UOP 3G™ Phenol process. The Badger Cumene process, based on MCM-22 catalyst from ExxonMobil, has also been an important part of the cumene industry, as has been the KBR Phenol process. More will be said about these technologies later in this chapter.

2.3 Technologies in Aromatics Synthesis

This section will discuss the technologies that result in the formation of aromatics. Referring back to Figures 1.7 and 1.8, we see that naphtha provides the most direct route to aromatics. The naphtha cut from a crude oil

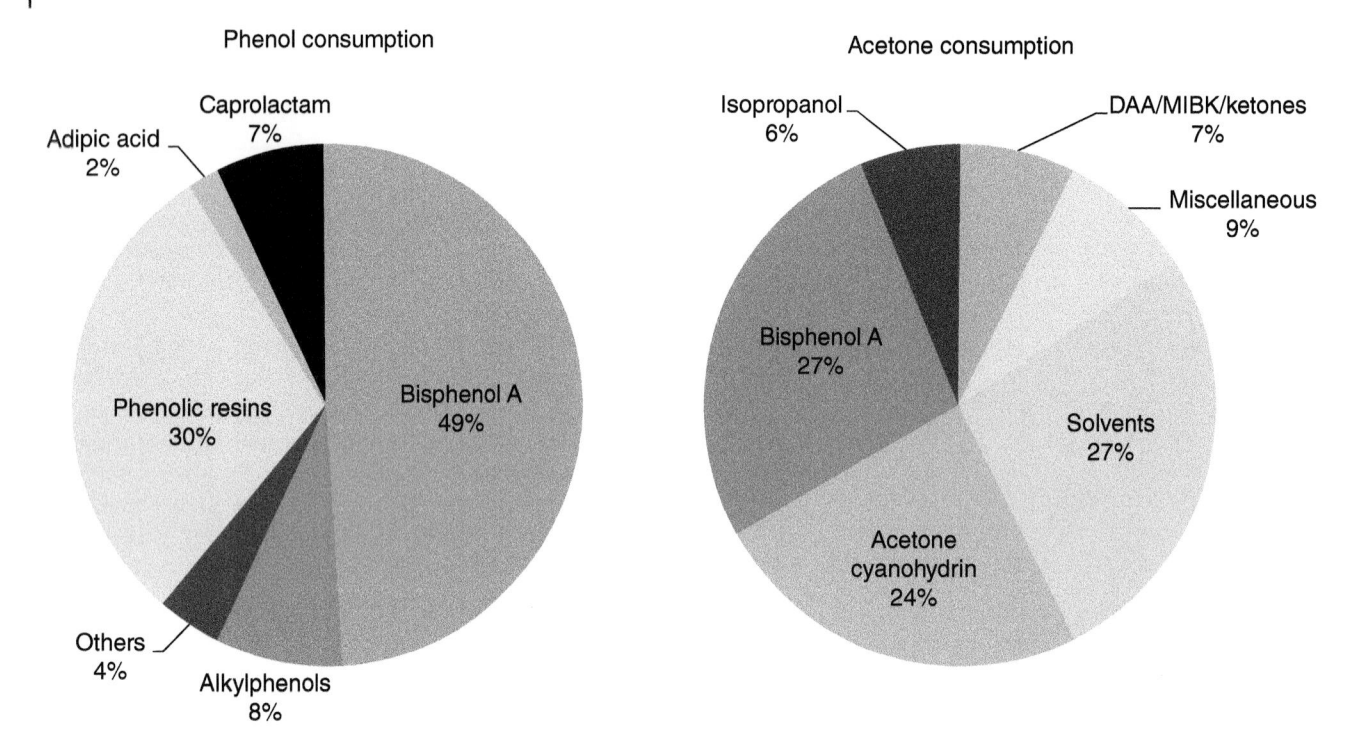

Figure 2.13 Global phenol and acetone consumption – 2015. *Source:* ICIS Analysis and Consulting.

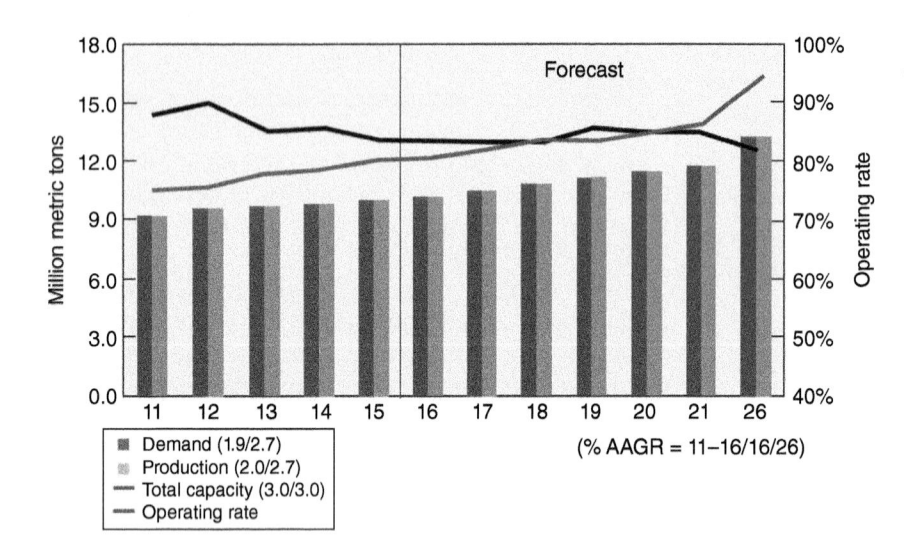

Figure 2.14 World phenol supply and demand. *Source:* IHS © 2016 IHS.

distillation column – called straight-run naphtha (SRN) – will typically contain a carbon number range from C_5 to C_{11}, with a boiling range from about 85 to 175 °C. This naphtha will consist of paraffins (normal and iso) as well as naphthenes (cyclopentyl and cyclohexyl) and aromatics. A typical composition is shown in Table 2.1. Typically, the C_5 and some of the C_6 hydrocarbons would be pre-fractionated out as light naphtha before being fed to the naphtha reformer for aromatics production.

In this example, the total paraffin content is about 79%, the total naphthene content is about 13%, and the total aromatics is around 8%. The total cyclic content, around 21%, is typical of a middle-east paraffinic crude oil source. Crude oils from other locations, such as US Shale, can be much less paraffinic and more cyclic. Crude oil selection can have a significant impact on the attainable aromatics yields from naphtha, as well as the processing severity of the aromatics producing technology, with more cyclic feedstocks being highly preferred.

Feeds such as the one described above will also typically contain some small amount of organic sulfur, in the range of 500–1000 ppm, and possibly some amount of organic nitrogen, in the range of 1–2 ppm. These polar

Table 2.1 Typical detailed straight-run naphtha composition.

C#	Composition in wt.%, ASTM D 6730					
	n–P	i–P	O	N_5	N_6	A
5	10.65	5.26	0.00	0.57	0.00	0.00
6	9.99	8.27	0.00	1.16	1.21	0.94
7	9.08	7.31	0.00	1.85	1.88	2.90
8	5.91	7.79	0.00	1.80	1.89	2.93
9	7.02	4.61	0.00	0.97	0.96	0.66
10+	0.00	3.12	0.00	0.41	0.41	0.47
Total	**42.64**	**36.35**	**0.00**	**6.76**	**6.34**	**7.91**

contaminants will be harmful to downstream catalysts. As a result, it is necessary to send such naphtha feeds to a naphtha hydrotreater for desulfurization and denitrification down to levels below 1 ppm This can usually be done at hydrotreating pressures of about 500 psig with Co/Mo hydrotreating catalysts such as UOP's Unity™ HYT-1119™ catalyst. Many other companies, including Albemarle, Axens, and Haldor Topsoe, offer similar hydrotreating catalysts.

The hydrotreated naphtha is fed to a catalytic reforming unit such as the UOP Platforming process to produce reformate. In the reforming unit, a bifunctional catalyst (acid and noble metal functions) converts the naphthenes and paraffins to aromatics. This is a complex set of reactions, involving isomerization of paraffins and naphthenes as well as ring closure of the paraffins to naphthenes and dehydrogenation of the resultant naphthenes to aromatics. A typical conceptual reaction network is shown in Figure 2.15 for UOP's Platforming process. This figure shows how closely the acid and metal functions of the catalyst must work together to minimize cracked by-products and maximize aromatics.

As shown in Figure 2.16 for C_6 hydrocarbons, the paraffin cyclization and naphthene dehydrogenation reactions produce a significant amount of hydrogen, which is recovered and used in many downstream petrochemical processes, as well as makeup gas to the naphtha hydrotreating unit. It is not unusual to have as much as 4 wt.% hydrogen yield from the naphtha. The catalytic reforming process is therefore highly endothermic and typically requires four reactors with intermediate reheat to achieve target conversion and aromatics production. Similarly, the reactions are equilibrium limited

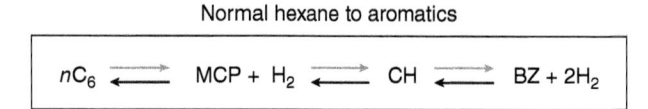

Normal hexane to aromatics

$$nC_6 \rightleftarrows MCP + H_2 \rightleftarrows CH \rightleftarrows BZ + 2H_2$$

Lower H_2 partial pressure to improve equilibrium to aromatics and yield
• Operate at lower pressure

Figure 2.16 Catalytic reforming reactions and hydrogen production.

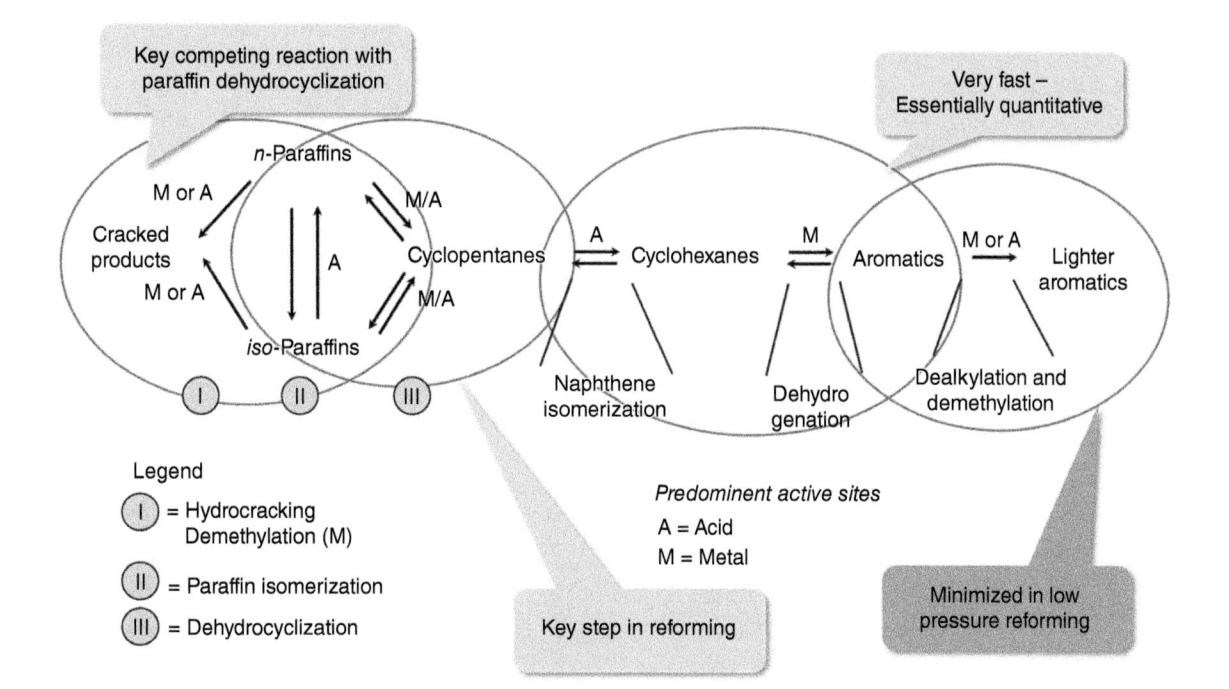

Figure 2.15 Typical catalytic reforming reaction network.

by hydrogen partial pressure. As a result, higher conversion, selectivity, and yields are favored by operating at reduced pressure. Thus, it is advantageous to be able to operate the catalytic reformer at the lowest economical pressure.

When catalytic reforming was first introduced, and for more than 20 years afterward, fixed-bed multi-reactor plants were designed, where the catalyst would operate for between 6 and 12 months before requiring regeneration. As the operating cycle proceeded, the reactor inlet temperatures would need to be raised to compensate for catalyst deactivation. Typically, the yield of aromatics would decrease as a result of poorer and poorer selectivity. Production would stop at the end of the cycle and 1–2 weeks would be required to conduct carbon burning, oxy-chlorination of the platinum, drying, and reduction of the platinum before production could be resumed. These early semi-regenerative reformers need to operate at pressures of 200 psig or higher to obtain reasonable process cycle lengths.

By 1970, UOP led the way to reap the advantages of lower pressure reforming operation by developing and commercializing the first continuously regenerative catalytic reforming technology. This process was called the UOP CCR Platforming process. This breakthrough allowed operation at much lower hydrogen partial pressures, and therefore higher conversion, selectivity, and yields (more than 10 wt.% higher yield) as shown in Figure 2.17.

The more rapid catalyst deactivation at this lower pressure was accommodated by continuously circulating catalyst through the reactors, into a continuous regenerator that performed carbon burning, oxy-chlorination and drying, and returning the catalyst to the first reactor where it could be reduced and then flow through that reactor and the remaining ones. A typical CCR Platforming flow scheme is shown in Figure 2.18.

The naphtha feed is combined with the recycle hydrogen and enters the feed/effluent exchanger where vaporization occurs. From there, the feed stream enters the charge heater which is raised to the desired inlet temperature for reactor #1. In the first reactor, naphthene dehydrogenation is the primary reaction, and the endothermic heat of reaction lowers the temperature at the reactor outlet. This requires reheat using inter-heater #1, from which the reactants then enter the second reactor where still more naphthene dehydrogenation takes place along with some paraffin dehydrocyclization. Again, the endothermic heat of reaction lowers the outlet temperature and reheat is needed. This repeats until the last reactor outlet is reached, after which the reactor effluent is directed through the feed/effluent exchanger. From there, it is condensed and enters the separator. The off gas from the separator is sent to the recontacting section where additional hydrocarbon recovery and gas enrichment take place. Net gas is taken off and the remainder is fed to the recycle hydrogen compressor. The liquid product is then sent to the stabilizer, where the light hydrocarbons are removed overhead. Because the Platforming process operates at relatively low pressures and the flow rates of hydrocarbons and recycle gas are comparatively high, radial flow reactors are used. This minimizes pressure drop and utilities required to operate the recycle compressor. The minimum pressure drop also provides the most favorable pressure for aromatic selectivity while allowing a controlled amount of coke to accumulate on the catalyst. The catalyst flows down through the reactors, with fresh catalyst entering the top of the first reactor and coked catalyst exiting the outlet of the last reactor. The catalyst is then transferred to the continuous catalyst regeneration (CCR) tower where the coke is burned off, the metals are re-dispersed, and the catalyst dried before it is sent to the top of the first reactor for reduction.

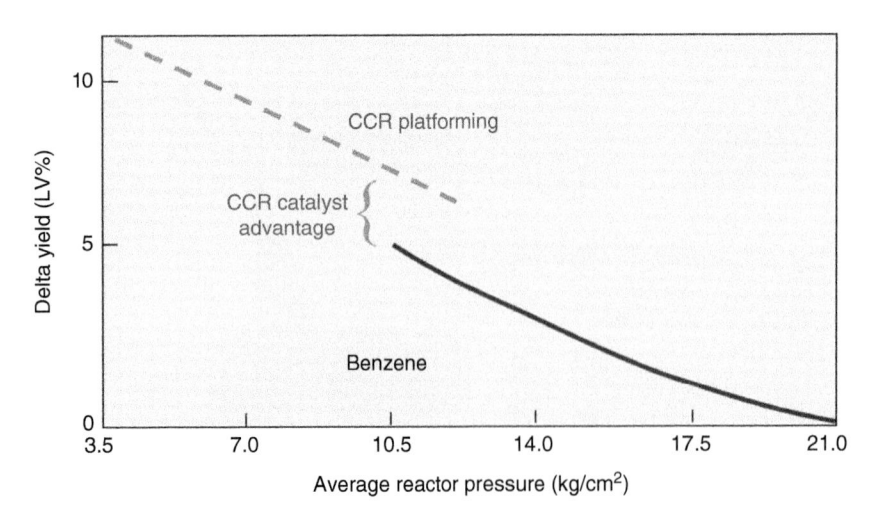

Figure 2.17 Influence of pressure on reforming yields.

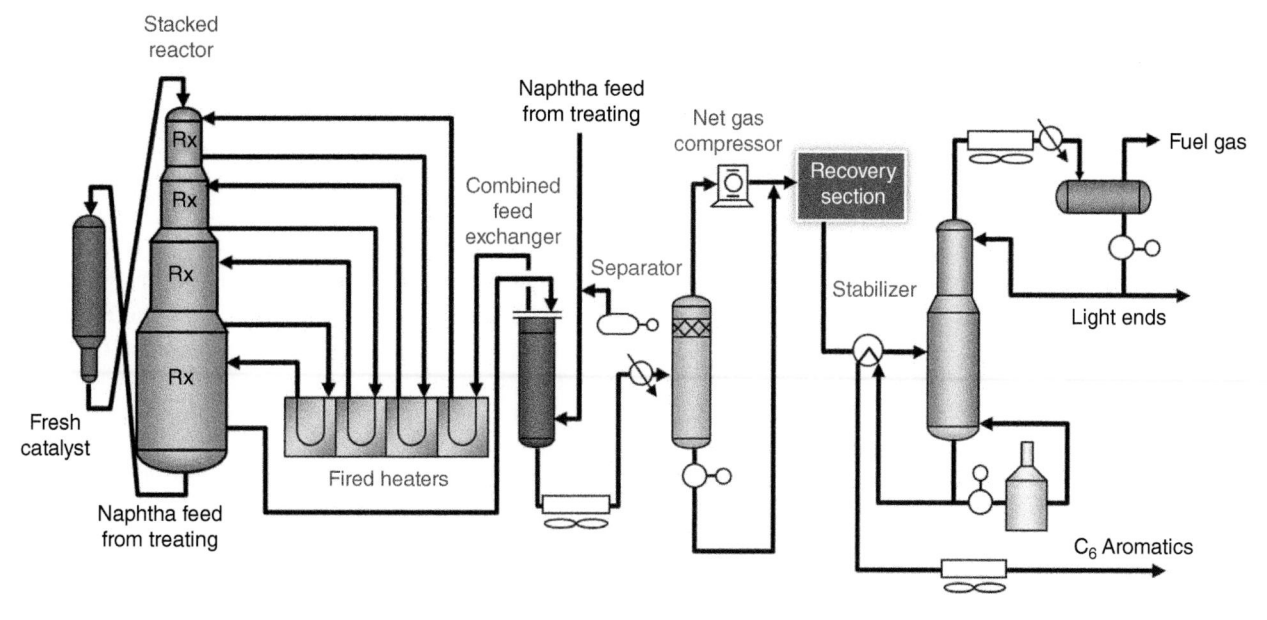

Figure 2.18 UOP CCR Platforming process flow scheme.

Since it was first introduced in 1970, UOP has made numerous improvements – in catalyst, process, and equipment designs – aimed at increased yields, lower CAPEX and OPEX, and improved reliability. At this writing there are more than 200 CCR Platforming units in operation throughout the world, producing aromatics or high-octane gasoline, and providing a continuous supply of hydrogen to associated refining and petrochemical plants. Axens has also offered a CCR technology called Aromizing, with about 30 units in operation. In recent years, Sinopec has developed its own CCR reforming technology and catalyst, primarily for indigenous production facilities of the state-owned enterprise.

Note that the C_5 and some C_6 hydrocarbons have been taken out of the SRN in Table 2.1 to result in the reformer feed shown in Table 2.2. Depending on the feed and operating severity, the C_6–$C_{10}+$ aromatics yield from naphtha can be 70 wt.% or higher. The C_5+ product will be about 85 wt.% aromatic, with the balance being unconverted or cracked paraffins and typically less than 0.5% naphthenes as shown in Table 2.2.

The C_8 and C_9 paraffins and naphthenes will be more than 99% converted at high selectivity to aromatics. The C_7 paraffins will be nearly 90% converted, and the C_6 paraffins about 15% converted, although some amount of cracked C_9 paraffins could end up as C_6 paraffins. This reformate composition shows that it should be fairly easy to fractionate the C_8+ aromatics with very little saturates for downstream processing to high-purity para-xylene, ortho-xylene, or meta-xylene. The C_7-aromatics in the overhead of the reformate splitter will contain significant amounts of saturates which must be dealt with downstream in order to ensure high-purity benzene

Table 2.2 Typical reformer naphtha feed and reformate composition for aromatics production.

	Feed		Product yields	
	LV%	wt.%	LV% FF	wt.% FF
C_5	0.00	0.00	3.10	2.67
P_6	10.77	9.83	9.36	8.52
P_7	24.01	22.75	3.11	2.94
P_8	19.74	19.02	0.12	0.12
P_9	16.26	16.15	0.01	0.01
P_{10}^+	4.21	4.33	0.00	0.00
MCP	1.35	1.40	0.12	0.13
CH	1.55	1.67	0.01	0.01
N_7	4.89	5.18	0.15	0.16
N_8	4.82	5.11	0.01	0.01
N_9	2.43	2.67	0.00	0.00
N_{10}^+	1.03	1.13	0.00	0.00
A_6	0.89	1.08	4.15	5.04
A_7	3.36	4.03	17.99	21.57
A_8	3.39	4.07	20.81	24.99
A_9	0.76	0.92	12.17	14.71
A_{10}^+	0.55	0.65	3.62	4.32

product. Some aromatics producers who are focused on maximizing para-xylene will favor elimination of the benzene precursors from the naphtha. This can easily be done by adjusting the cut point of the naphtha feed to the reformer. There will still be some amount of benzene made as coproduct from isomerization and transalkylation technologies, but that amount usually can be managed.

Referring back to Figures 1.7 and 1.8, there are other potential sources of aromatics besides SRN that can be pursued as fuel demands change. For example, the bottoms of the atmospheric pressure crude column will be fed to another fractionator that operates under vacuum, called the vacuum column. The overhead from this column is referred to as vacuum gas oil and has a boiling range that is much higher than that of SRN – 425–564 °C. Vacuum gas oil can be fed to a fluid catalytic cracking (FCC) unit, or to a hydrocracking unit, usually with the objective of producing fuel such as gasoline, jet, or diesel. However, if there is reduced demand for diesel – for example, due to the effect of pollution abatement regulations – the hydrocracker can be repurposed to produce hydrocracked naphtha, which can in turn be fed to the CCR Platforming unit to produce additional aromatics beyond those achievable with SRN alone. Similarly, if there is reduced gasoline demand, the naphtha from the FCC unit is typically high in aromatics. With suitable pretreatment, FCC gasoline can be fed to the CCR Platforming unit to produce additional aromatics. Going forward, there will likely be many more plants that coproduce aromatics or naphthas designated for aromatics, as fuel demands change. More will be said about refinery–petrochemical integration in a later section. This will involve process technologies like UOP's LCO-X™ process, the FCC process, Steam Cracking, and the UOP Maxene™ process.

Another important source of aromatics comes from pygas, which is a by-product of steam cracking for ethylene production. Feedstock choices for steam cracking range from naphtha (dominant in Asia) to LPG to ethane (dominant in the United States). When naphtha is used as feed to a steam cracker, the pygas by-product can be as much as 20% of the overall yield, with about 50% of the pygas being aromatics. There will also be a large amount of olefins, *di*- and *cyclo*-olefins, as well as saturates in the pygas, along with some sulfur and nitrogen. Pygas hydrotreaters are typically two-stage with intermediate fractionation of heavies. The first stage typically converts the *di*-olefins to *mono*-olefins, while the second stage will remove trace sulfur, nitrogen, and the remaining olefins. The hydrotreated pygas can then be fed along with reformate to produce the desired aromatics. Typically, the first-stage catalyst will be Pd, with shell-impregnated Pd the most widely used. The second-stage catalyst can be Co/Mo, Ni/Mo, or a combination, depending on the amount of nitrogen, and the tolerance for aromatic saturation. Many companies – including UOP, Axens, Shell, and Petrochina – offer pygas hydrotreating technologies.

One final source of aromatics deserves mention because it has been in commercial operation: de-hydro-cyclo-dimerization (DHCD) of LPG – propane and butane – with technologies such as the UOP Cyclar™ process. This technology was developed in the late 1980s (Johnson and Hilder 1984) with a small demonstration unit built and operated in the 1990s at British Petroleum's Grangemouth Scotland facility. A large Cyclar plant was built and operated for more than a decade at SABIC's Ibn Rushd site in Saudi Arabia, producing a concentrated aromatic product that was fed to a fully integrated plant for benzene and *p*X production. Like CCR Platforming, the Cyclar process employs CCR, because the operating conditions that favor DHCD reactions are relatively severe and rapidly deactivate the catalyst. In contrast to CCR Platforming, the Cyclar aromatic product is nearly devoid of any nonaromatics, which avoids the need for ED or LLE. The Cyclar process also produces a significant amount of hydrogen, around 6 wt% of feed. However, the economics of using LPG as feed to the Cyclar process are challenged by the relatively low (at present) price of naphtha and the value of propylene that can be made by on-purpose dehydrogenation of propane. Figure 2.19 provides an overview of Cyclar reaction chemistry, with these reactions catalyzed by a high activity acidic catalyst.

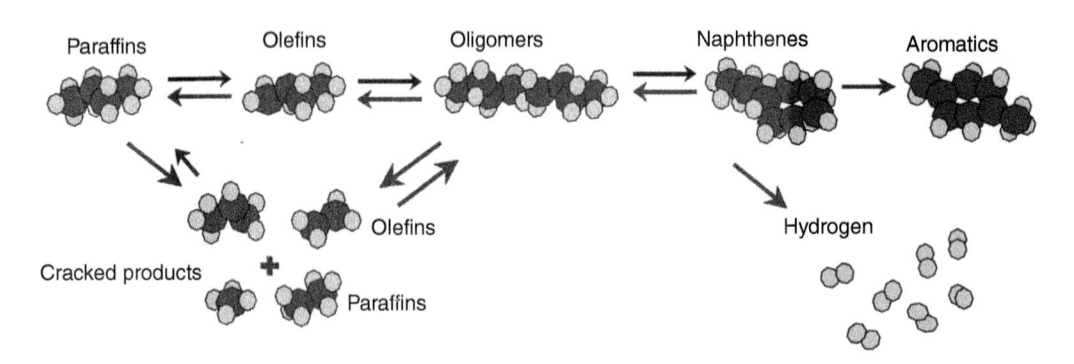

Figure 2.19 UOP Cyclar process chemistry.

Figure 2.20 shows the flow-scheme for the UOP Cyclar process.

The fresh feed and recycled unconverted LPG are sent through the feed/effluent exchanger, where they are vaporized and heated. From the feed/effluent exchanger, the reactants enter the charge heater where they reach the desired temperature for the inlet of the first reactor. The DHCD reactions that take place in the first reactor – primarily C_4 conversion – are highly endothermic and cause the temperature in that reactor to drop. The effluent stream is then reheated in inter-heater #1 and fed to the second reactor. The reaction proceeds through the second reactor, with reheat of the outlet stream, and so on through all four reaction stages. Side-by-side reactors are used for the Cyclar process to minimize the amount of thermal cracking, thus helping to preserve the yield of aromatics. Radial flow reactors are used because of the low pressure and relatively high flow rates. Fresh catalyst is introduced at the top of the first reactor, flows down through that reactor, and is collected and then transferred to the top of the second reactor. This sequence is repeated until the coked catalyst exits the bottom of the last reactor. From there, it is transferred to the CCR where carbon burning and drying take place. Since the Cyclar catalyst does not have noble metals, there is no need for an oxychlorination step to re-disperse the metals. However, the metal on the Cyclar catalyst needs to be reduced, and that step is accomplished at the top of the first reactor. The effluent from the last reactor is sent to the absorber/stripper and gas recovery sections where unreacted LPG is recovered and recycled back to the first reactor. In contrast to the CCR Platformer, there is no bulk recycle of hydrogen in the Cyclar process, only enough to maintain the required sulfur levels on the reactor and heater metallurgy. The stripper bottoms will contain the aromatic stream that can then be fed to the aromatics production processes. One interesting characteristic of the Cyclar C_8 aromatics is that they have very low ethylbenzene content. This allows the Parex unit to be much smaller.

2.4 Alternative Feeds for Aromatics

The majority of the world's aromatics are produced either directly or indirectly from naphtha that would otherwise be used for gasoline production. However, there are some alternative sources to consider beyond naphtha and the LPG example above. Many of these are technically feasible, but fail to provide an economically competitive alternative to conventional feeds. For example, methanol can be converted to a mixture of aromatics

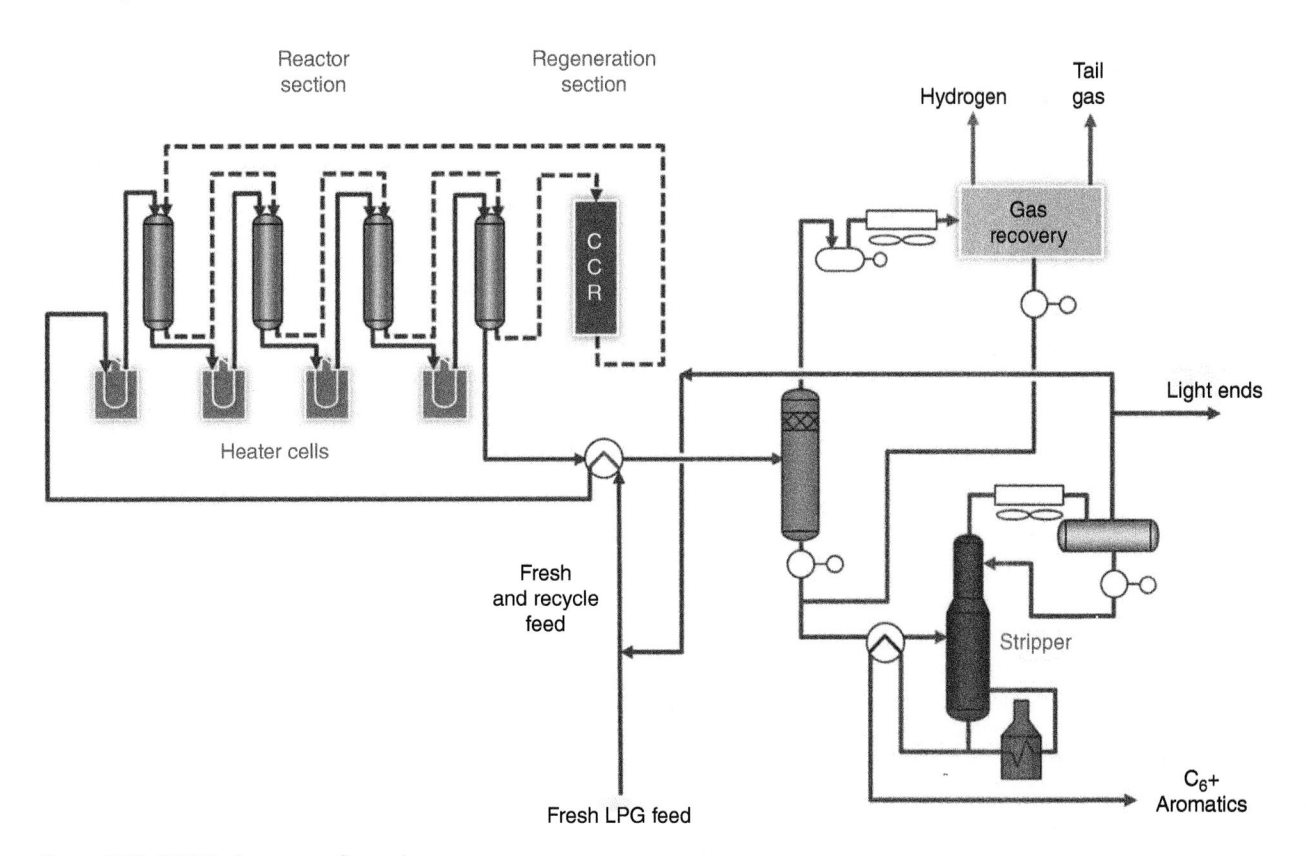

Figure 2.20 UOP Cyclar process flow scheme.

and olefins. The earliest example was the Mobil methanol to gasoline (MTG) process that used an MFI catalyst to produce a mixture of paraffins, olefins, and aromatics from methanol. A demonstration plant was built in New Zealand, but the amount of water by-product dictated a maximum yield of less than 50%. Methanol to propylene (MTP process) technology licensed by Lurgi is designed for on-purpose propylene production from methanol, using an MFI-type catalyst. Methanol is converted to dimethyl ether and then fed to the olefin synthesis section. However, there is still a significant amount of by-product water, similar to the MTG process. This technology produces about 25% gasoline by-product that is mostly aromatic. Two or three of these plants have been licensed, primarily in China.

A significant amount of work has been done in recent years with renewable-sourced feeds for aromatics, such as sugars, lignin, cellulose, and dried biomass (Deischter et al. 2019; Eriksson 2013; Virent 2016). For example, Gevo (Ryan 2016) and Butamax have technology that can ferment sugars to produce isobutanol. Although initially destined for fuel uses, Gevo was able to convert isobutanol to *p*X via a multistep conversion process that included dehydration, oligomerization, ring closure, and dehydrogenation. They worked with Toray in Japan to prove this *p*X source for PET. However, the production costs are extremely high due to the low yields across many of the conversion steps. Virent uses aqueous-based reforming of alcohols and other oxygenates derived from fermentation of sugars to produce an aromatic stream using an MFI catalyst. Virent received support from Coca Cola to build and operate a 10 000 gal per year demonstration plant for *p*X production that could be used for PET resin and renewable-sourced PET bottles under the brand name BioFormPX. Anellotech uses thermo-catalytic pyrolysis of biomass to produce aromatics, but this route requires extensive decontamination of the feed to protect the MFI catalyst. The aromatic product is primarily benzene and toluene (Mazanec and Whiting 2017). Nearly 75% of the feed carbon is converted to char, tar, coke, CO, and CO_2. Thus far, the chemical efficiencies (carbon utilization) of these routes have been too low, and the CAPEX too high compared to conventional routes. As a result, the production costs appear to be 5–10 times those of conventional technology, even with some amount of carbon-tax penalties levied against conventional technology. The growing middle class populations of the future will be seeking the most cost-effective routes to their desired end products. Significant improvements in chemical efficiency will be needed to make renewable sourced feeds competitive, and even then, the economy of scale seen for today's conventional production facilities will be very hard to achieve.

2.5 Technologies in Aromatic Transformation

The aromatics made by CCR Platforming, Cyclar, and those contained in pygas will consist of benzene, toluene, all four C_8 aromatic isomers, all eight C_9 aromatic isomers, and 22 C_{10} aromatic isomers. The amount and type can be influenced by such things as the naphtha cut point and its cyclic content. But in most cases, the reformate will only contain 50% or less of the ultimate amount of C_8 aromatics that will ultimately be needed to produce the required *para*-xylene. It is therefore crucial to have highly selective and robust technologies that can work together to convert these 36 aromatic components into those few that are of primary interest – namely benzene, *para*-xylene, *meta*-xylene, and *ortho*-xylene.

2.5.1 Transalkylation

Transalkylation is the most widely used and economical route to increase the amount of C_8 aromatics – indeed xylenes – coming from the reformate. The UOP Tatoray process (jointly developed by UOP and the Japanese chemical company Toray) transalkylates toluene with C_9 and C_{10} aromatics to produce xylenes and benzene. Uncovered toluene and heavy aromatics are recycled to extinction. Depending on the reformate C_9–C_{10} aromatic content, the Tatoray process can easily double the amount of C_8 aromatics for subsequent recovery into the individual xylene isomers. The Tatoray process also produces benzene coproduct. Other technology suppliers also have transalkylation technology – for example, ExxonMobil has TransPlus, Criterion has ATA, Sinopec has S-TDT technology, and Toray had the TAC-9 process for handling large amounts of C_9 aromatics. Englehard had licensed a technology called Xylenes Plus, with a moving catalyst bed, but that process suffered from converting too much of the feed to coke, since it operated hydrogen-free. The UOP Tatoray process is the most widely used worldwide, and will serve as an example to illustrate the features of aromatic transalkylation.

Aromatic transalkylation is an equilibrium-limited reaction, in which the feed methyl : phenyl ratio governs the extent of conversion as well as the proportion of benzene and xylenes in the net product. The methyl:phenyl ratio in the Tatoray feed is primarily set by the naphtha cut point going to the catalytic reformer, and thus the proportion of toluene vs C_9 and C_{10} aromatics that are in the reformate. Low endpoint naphthas favor more toluene, which will in turn favor increased benzene and decreased xylenes across the Tatoray unit. A higher naphtha endpoint provides more C_9 and C_{10} aromatic precursors which favor increased C_9 and C_{10} aromatics in the reformate, and in turn, more xylenes in the net product from the Tatoray unit. This is illustrated in Figure 2.21. By increasing the naphtha cut point to the CCR Platformer by 20 °C, the amount of attainable *para*-xylene increases by about 50%.

The influence of fresh feed C_9 content on ultimate yields is shown in Figure 2.22 over a wide range of C_9 content. This shows that the amount of xylenes relative to benzene can be very significantly influenced by large concentrations of C_9 aromatics in the feed. The same will be true if C_{10}'s are included. Of course, it is important to understand that very high C_9+ feed content requires a highly active and stable catalyst in order to prevent rapid deactivation and shortened process cycles.

An important and useful characteristic of transalkylation technology is that there is only a small amount of ethylbenzene in the product xylenes. When the transalkylation unit is producing as much as 50% of the xylenes that get converted into *p*X, the low ethylbenzene content of the resultant C_8 aromatic product can be very beneficial to the downstream *para*-xylene recovery operation, particularly when adsorptive separation is being used. This is because the selectivity of *para*-xylene over ethylbenzene has a strong influence on adsorbent efficiency. Figure 2.23 provides a summary of the reaction network for the Tatoray process, and this scheme is applicable to most modern transalkylation catalysts.

There are three primary reactions:

- Transmethylation, where a methyl group is moved from one aromatic ring to another. This is catalyzed by an acid site.
- Disproportionation, where two identical molecules react, moving a methyl group from one to the other via the paring reaction. Again, an acid site is required.
- Hydrodealkylation of C_2+ alkyl groups from a C_9 or C_{10} aromatic such as methyl-ethyl benzene or di-methyl ethyl benzene. This reaction is catalyst by an acid function, but it is important that a metal function be present to saturate the cleaved olefin.

Figure 2.21 Influence of transalkylation and feed on attainable *para*-xylene. Basis: 985 KMTA (25 000 BPD) © Light Arabian Naphtha.

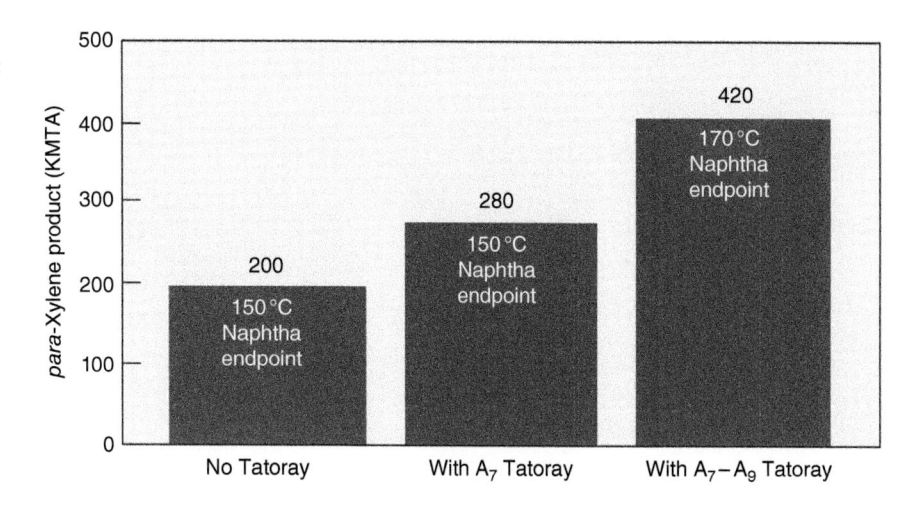

Figure 2.22 Effect of Tatoray process feed composition on yields.

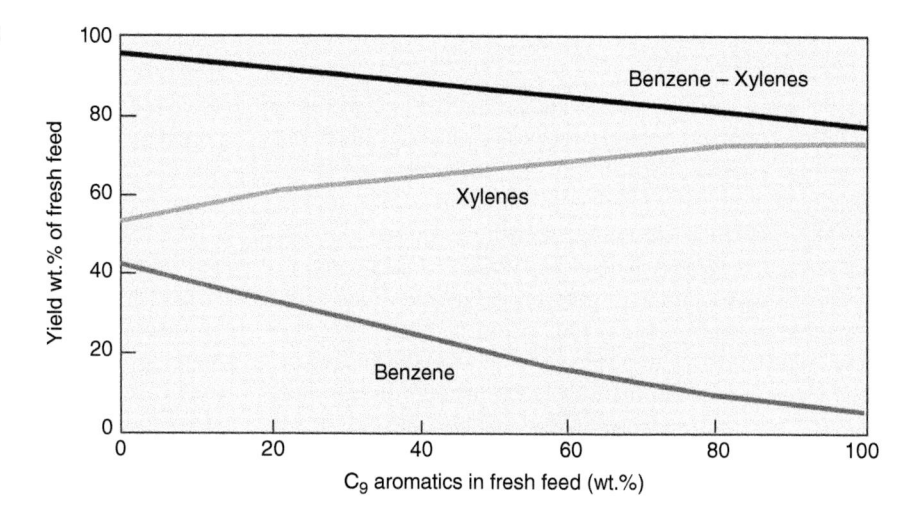

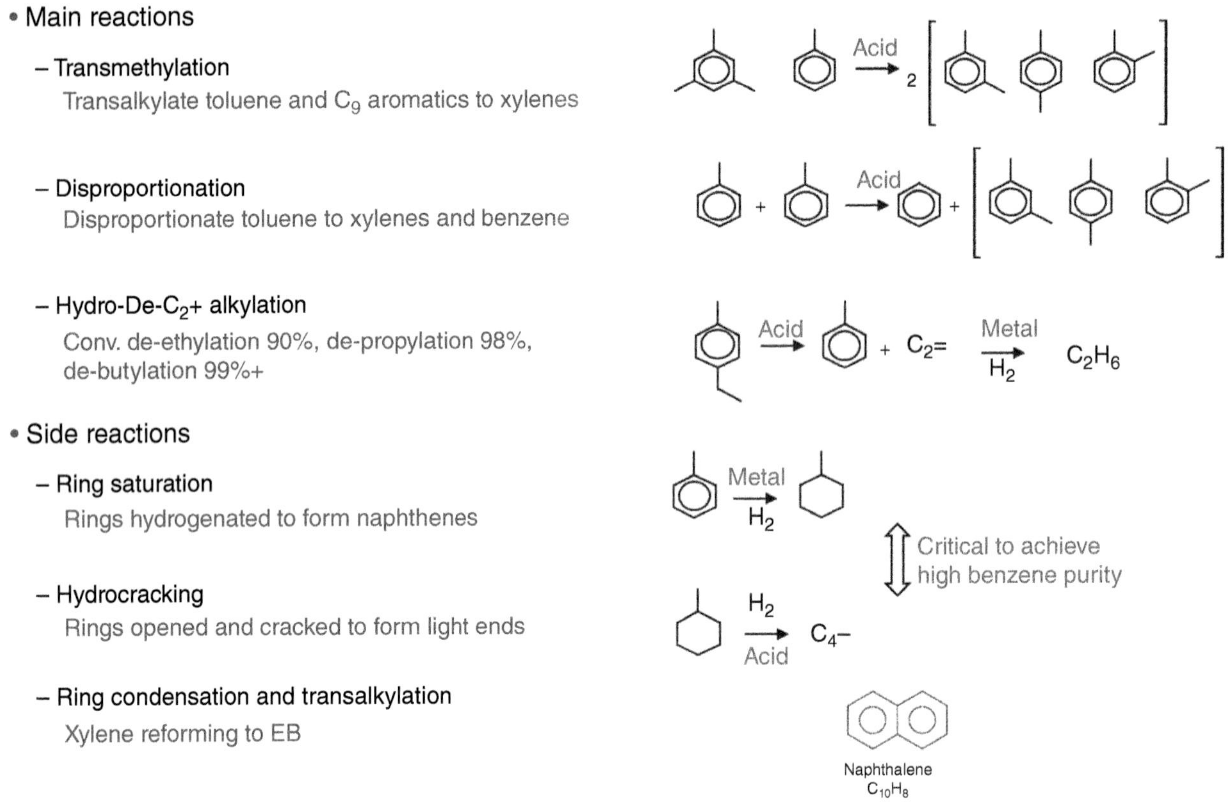

- Main reactions
  - Transmethylation
    Transalkylate toluene and C_9 aromatics to xylenes

  - Disproportionation
    Disproportionate toluene to xylenes and benzene

  - Hydro-De-C_2+ alkylation
    Conv. de-ethylation 90%, de-propylation 98%, de-butylation 99%+

- Side reactions
  - Ring saturation
    Rings hydrogenated to form naphthenes

  - Hydrocracking
    Rings opened and cracked to form light ends

  - Ring condensation and transalkylation
    Xylene reforming to EB

Figure 2.23 Transalkylation reaction network.

The three primary side reactions are:

- Ring saturation, converting an aromatic to a naphthene. This is undesirable because it consumes hydrogen and can reduce benzene purity.
- Naphthene ring opening and cracking to light ends. Hydrogen is consumed, there is an impact on yield, and potentially benzene purity could be affected unless the cracked fragments are lower MW that do not co-boil with benzene.
- Aromatic ring condensation and potential EB formation. This could create additional heavy by-products and represent a yield loss.

UOP has developed and introduced several generations of Tatoray catalysts since the technology was first deployed in the early 1980s, with the current ones being very high activity, stability, and selectivity. The more stable catalysts can be beneficial in situations where lower pressure equipment such a semi-regenerative reformer can be repurposed for petrochemicals production instead of gasoline. For most modern transalkylation technologies, the conversion per pass will range from 40 to 50%, depending on feed C_9+ content, utilities, and feed/product/by-product values. Catalyst design is critical to maximizing the desired primary reactions and minimizing the side reaction. More will be said about transalkylation catalysts in Chapter 3.

2.5.2 Selective Toluene Disproportionation

STDP is related to transalkylation in that toluene is the reactant. In this technology, toluene is disproportionated to benzene and xylenes in a diffusion-controlled pore structure that favors *para*-xylene product over *meta*- and *ortho*-xylene. The resultant xylenes can be upward of 85% pX/X. This reaction is illustrated in Figure 2.24.

The UOP PX Plus™ process is an example of such a technology. ExxonMobil has similar technologies named MSTDP and PXMax. In the early version of PX Plus and MSTDP, the catalyst was intentionally coked in-situ to generate restricted pore mouths that favored *para*-xylene over the other xylenes in the product. More recently, ex-situ selectivation is performed in catalyst manufacturing for both PX Plus and PX Max, allowing much faster start-up and attainment of expected yields.

The high concentration of *pX* in the xylene product can be helpful for both adsorptive separation and fractional crystallization *pX* recovery operations. In the case of adsorptive separation technology such as Parex, an increased *para*-xylene concentration in the feed reduces the degree of competitive adsorption that needs to take place at the active sites of the adsorbent. A step change in *para*-xylene production can be obtained, even when the STDP xylenes are blended with other feed streams.

Figure 2.24 *para*-Selective toluene disproportionation reaction.

Like transalkylation, the ethylbenzene content of the STDP C_8 aromatics is extremely low, which also is beneficial to the adsorptive separation technology.

Similarly, the increased *para*-xylene content from STDP benefits crystallization by allowing higher recovery before the eutectic point with mX is reached. It is important to note that STDP technology produces a significant amount of benzene coproduct, so it is generally most attractive to operate this technology when benzene prices are strong. If benzene prices are weak, it is usually better to route the toluene back through transalkylation. For this reason, some operators will run their STDP units on a campaign basis, alternating with transalkylation, depending on market conditions.

The flow scheme for STDP is very similar to that for transalkylation, with post-fractionation of benzene, toluene, A_8's, and A_9's, except that fresh and recycle toluene are the only aromatic feeds to the reactor. Conversion per pass will range from 25 to 32%, depending on the catalyst and process technology, so the toluene fractionation requirements will be higher than for transalkylation.

2.5.3 Thermal Hydro-Dealkylation

Thermal hydro-dealkylation (UOP THDA process) is rarely used for benzene production in today's aromatics industry. The reason for this is primarily that fuel gas is the main by-product. All the alkyl groups that come off C_7, C_8, or C_9 aromatics will be thermally cracked to methane. Chemical hydrogen consumption is extremely high for this technology, even though the flow scheme is relatively simple, and high-purity benzene is obtained outright. THDA technology is regarded as the highest cost option for producing benzene. The reason for this is highlighted in Table 2.3 which shows the theoretical benzene yields from various potential THDA feed components. For toluene, the maximum yield is less than 85 wt.%, and for C_9 aromatics, it is only 65 wt.%. The primary by-product is methane, and the consumption of hydrogen is very high, usually requiring a pressure swing adsorption (PSA) unit to help supply sufficient recycle hydrogen purity. The basic chemical reactions of THDA are shown in Figure 2.25. Most of the reactions are

Table 2.3 Weight yields of benzene for the THDA process.

Benzene	78.1	1.000	1.000
Toluene	92.1	0.836	0.848
Ethylbenzene	106.2	0.724	0.735
Mixed xylenes	106.2	0.724	0.735
C_9 Aromatics	120.2	0.641	0.650

steeply exothermic and occur at very high reactor inlet temperatures. One reaction that does not consume hydrogen is the formation of bi-phenyl, which is a condensation reaction of benzene that actually produces hydrogen. In the design of the THDA unit, provision is made for the bi-phenyl to be recycled back to the reactor to suppress formation of additional bi-phenyl.

The highly exothermic character of the primary THDA reactions brings significant and critical safety considerations for design and operation of a THDA unit. Typically, the reactors are cold wall to protect the metallurgy. Toluene conversion is usually about 90%, with provision to fractionate benzene, unconverted toluene, and bi-phenyl. The unconverted toluene and bi-phenyl are recycled back to the reactor inlet. Care must be taken to maintain minimum hydrogen content at the reactor outlet to avoid carburization. Reactor outlet temperatures are in the range of 700–740 °C, with the reactor pressure around 35 barg and residence times in the range of 25–35 seconds. There is usually a provision for a quench to be added at the hot combined feed exchanger. A change in feed composition that would add saturates or C_8+ aromatics could significantly increase the temperature rise across the reactor and lower the hydrogen content at the outlet. A simplified block flow diagram for the THDA process is shown in Figure 2.26.

In the early days of the chemical industry, hydrodealkylation was widely used. UOP's initial process was called HYDEAL™ and these early units licensed in the 1960s had a catalyst to accelerate the dealkylation reactions. However, after several years of operation, it soon became apparent that the catalyst was unnecessary and that the process was purely a thermal one. So, in the early

Aromatics conversion:

$$Ar\text{-}R + H_2 \longrightarrow Ar\text{-}H + CH_4$$

Aromatic ring loss:

$$\text{Ar} + 3H_2 \longrightarrow \text{cyclohexane} + 4H_2 \longrightarrow 2CH_2 + 2C_2H_6$$

Saturate cracking:

$$\text{Paraffin or naphthene} + H_2 \longrightarrow CH_4 + C_2H_6$$

Aromatic condensation:

$$2\ \text{Ar} \longrightarrow \text{biphenyl} + H_2$$

Figure 2.25 THDA reaction networks.

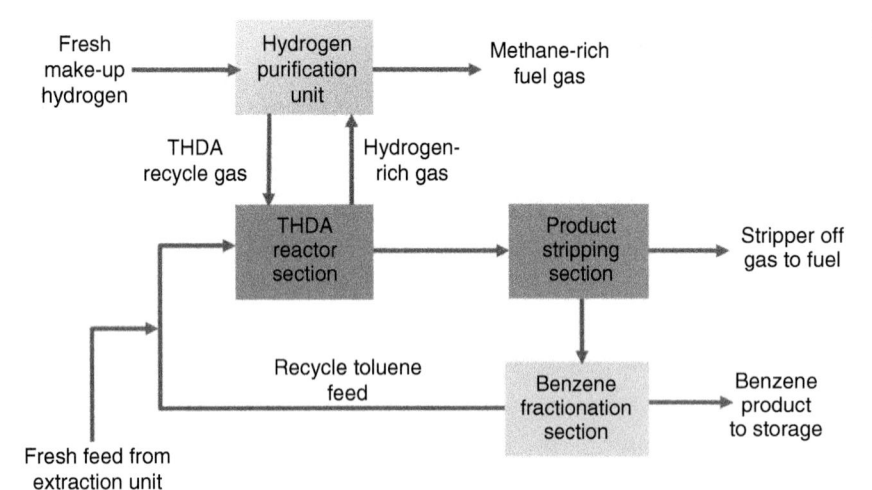

Figure 2.26 THDA process block flow diagram.

1970s, UOP changed the process name to THDA and offered a back-mixed reactor for plants from that point onward. Over the years, UOP designed plants based on a wide range of hydrocarbon feeds. There was even a plant that operated of a very heavy stream to produce naphthalene by dealkylation. Other companies have offered similar thermal processes under the names of HDA (Axens) and DETOL (McDermott).

2.5.4 Xylene Isomerization

The preceding sections have covered technologies that perform molecular weight conversions – higher and lower than the feeds – of aromatic compounds. This section will discuss the conversion of C_8 aromatic isomers to equilibrium mixtures using xylene isomerization. It should be recognized that the C_8 aromatics produced in the catalytic reformer will have a substantial amount of ethylbenzene, as shown in Table 2.4. The ethylbenzene content of the C_8 aromatics is 17.7%, almost as high as that of the *para*-xylene. The *meta*- and *ortho*-xylenes total more than three times that of the *para*-xylene. Yet, the highest demand and volume product of the four isomers is *para*-xylene, which means there needs to be substantial work to drive these other isomers toward *para*-xylene. A similar picture emerges from the distribution of C_8 aromatic isomers in the transalkylation product, shown in Table 2.5. In this case the ethylbenzene content is extremely low, which is helpful for adsorptive separation. But the combined *meta*- plus *ortho*-xylenes are again more than three times the concentration of the *para*-xylene. This helps establish an understanding of the critical role that xylene isomerization technology will have in *para*-xylene production.

Table 2.4 Composition of reformate C$_8$ aromatics.

Reformate component	wt.% of total
Ethylbenzene	17.1
para-Xylene	18.6
meta-Xylene	39.4
ortho-Xylene	24.3
Total A8s	100.0

Table 2.5 Typical C$_8$ aromatic composition of transalkylated product.

Transalkylated component	wt.% of total
Ethylbenzene	1.1
para-Xylene	23.9
meta-Xylene	53.1
ortho-Xylene	21.9
Total A8s	100.0

Xylene isomerization technology has existed nearly since the advent of the *para*-xylene production industry. The design of the isomerization system has been governed to some extent by the characteristics of the *para*-xylene recovery technology that is associated with. For example, adsorptive separation technology typically benefits from lower ethylbenzene content in the feed, whereas fractional crystallization with its low per-pass recovery and high combined feed ratio will benefit from a xylene isomerization technology that produces lower by-products per pass, with some reasonable amount of ethylbenzene conversion that would prevent buildup. The isomerization process can operate in liquid or vapor phase to equilibrate the xylenes in the feed, but the highest ethylbenzene conversion levels are favored by vapor-phase operation, since hydrogen will play a key role in converting the ethylbenzene in the feed.

The UOP Isomar™ process has vapor-phase configurations that can convert ethylbenzene to xylenes in an equilibrium-controlled reaction, as well as configurations that can dealkylate the ethylbenzene to benzene in a kinetically controlled reaction. There is also a liquid-phase version that has a lower EB conversion level, but which may be valuable for de-bottlenecking or saving utilities. ExxonMobil offers its XyMax technology for vapor-phase conversion of ethylbenzene to benzene while driving the xylenes to equilibrium. Previous versions of ExxonMobil technology were named MHAI and AMHAI. ExxonMobil also offers liquid-phase xylene isomerization. Axens offers Oparis, which is a vapor-phase technology that converts ethylbenzene into xylenes. For simplicity, we will refer to the technologies that convert EB to non-C$_8$ aromatics as "EB-destruction" and those that convert EB to xylenes as "EB Isomerization."

The chemistry for most EB-destruction processes is shown in Figure 2.27. The acidic function of the catalyst will isomerize the xylenes to an equilibrium or near-equilibrium composition, in an equilibrium-controlled reaction. This can usually be accomplished in vapor phase at relatively high WHSV at a pressure of 10–20 barg and at temperatures in the range of 350–400 °C. The same reactions can be performed in liquid phase at temperatures around 250 °C. More will be said about the design of these catalysts in Section 3.6, but it is important that they be designed to minimize transalkylation by-products such as toluene and C$_9$ aromatics. These by-products can be recovered and processed through transalkylation, in order to minimize yield losses in the integrated aromatics complex, but it is best to minimize their formation via careful catalyst design.

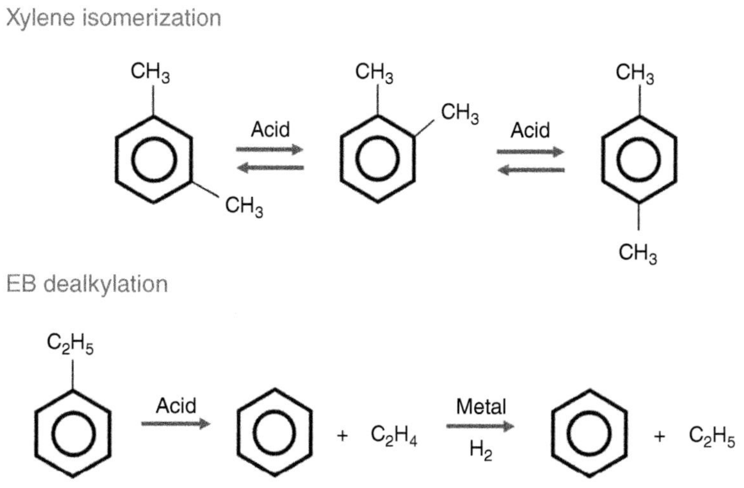

Figure 2.27 EB-destruction xylene isomerization reactions.

For *para*-xylene production, low by-product formation is critical because the equilibrium *para*-xylene content of the xylenes is around 24% pX/X, and the remaining xylenes will make 3–5 more passes through the reactor before they are converted to *para*-xylene, depending on whether adsorption or crystallization is used as the separation technology. For EB-destruction catalysts, the EB reactions require both an acid and a metal function. The acid function cleaves off the ethyl group from the aromatic ring to form benzene plus ethylene, and the metal function of the catalyst then saturates the ethylene to ethane, preventing it from reattaching to a xylene molecule. This reaction is kinetically controlled, not equilibrium controlled. The extent of EB conversion is usually governed by the desired amount of transalkylated by-products from the xylene isomerization steps. Hydrogen is consumed in this reaction, and it is necessary for both stoichiometry as well as catalyst stability to have an excess of hydrogen in the reactor circuit.

Since benzene is a significant by-product of this technology, it is important to minimize aromatic saturation reactions, since those could result in xylenes' yield loss as well as reduce the yield and purity of the benzene by-product. Some technologies can accomplish both the isomerization and the EB-destruction reactions with a single catalyst particle (Sachtler and Lawson 1996; Sharma et al. 1990), whereas others will accomplish these reactions using two different catalyst beds – one for EB conversion and the second for xylene isomerization (Abichandani and Venkat 1996).

The chemistry of the EB-isomerization catalysts is more complicated, as shown in Figure 2.28. In this technology, reactions take place in the vapor phase, with both the xylene isomerization and the EB conversion reactions being equilibrium-limited. This reaction network requires acid and metal functions of the catalysts to work very closely together because there needs to be an appropriate concentration of C_8 naphthenes in balance across the reactor for the EB conversion to favor production of xylenes. Depending on the feed being processed, this concentration might range between 4 and 8% of the C_8 hydrocarbons. The EB conversion to xylenes must proceed through naphthene intermediates, a so-called "naphthene bridge" that creates a flow of EB to ethyl-cyclohexane, then ethyl-cyclohexane conversion to a di-methyl cyclohexane structure, followed by dehydrogenation to xylene. Once the xylene is formed, the acid function of the catalyst will drive the xylenes toward equilibrium. In many cases, *ortho*- and *meta*-xylene are the dominant xylenes formed after dehydrogenation of the naphthenes. Catalysts need to be designed to minimize the formation of excessive amounts of transalkylation by-products as the catalyst drives the xylenes toward equilibrium. The early xylene isomerization catalysts used chloride as the acid function, which caused some corrosion issues in fractionator overheads in the early plants. Since about 1980, zeolites have been widely used in these catalysts, with steady improvements in selectivity and yields. As with the ethylbenzene destruction catalysts, the primary by-products are toluene and C_9 aromatics, which can be recovered and processed in transalkylation to minimize overall yield loss. However, it is best to minimize the formation of these by-products through design of the catalyst. Some examples can be found Whitchurch et al. (2010) and Merien et al. (2000). In US patent 7 745 677 Inventors: Patrick C. Whitchurch (Bossier City, LA), Paula L. Bogdan (Mount Prospect, IL), John E. Bauer (LaGrange Park, IL) 2010 assigned to UOP, and US patent 6 057 486 Inventors: Merien, Elisabeth, Alario, Fabio, Lacombe, Sylvie, Benazzi, Eric, Joly, Jean Fran Assigned to IFP, 2000.

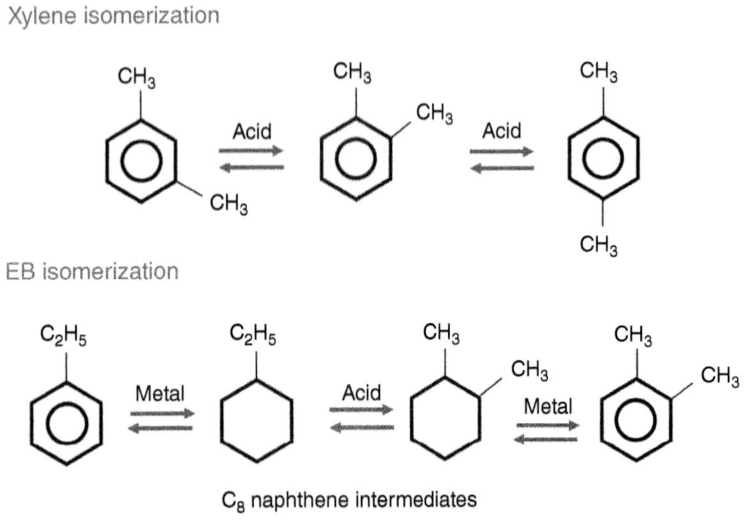

Figure 2.28 EB isomerization to xylenes' reaction network.

The "naphthene bridge" very likely involves C_7 as well as C_5 member naphthene rings to facilitate forming the di-methyl cyclohexane. It is important to minimize significant changes – up or down – in naphthene concentration across the reactor, by carefully matching hydrogen partial pressure with the reactor outlet temperature. If a large net increase in naphthenes takes place across the reactor, the extra naphthenes could potentially be lost in downstream fractionation. The observable EB conversion will be higher, but the extra conversion will not be to xylenes. Likewise, if there is a net loss of naphthenes across the reactor, the "naphthene bridge" will have reduced functionality, and EB conversion to xylenes will suffer. With extended time on stream, the isomerization catalyst will deactivate, requiring an increase in reactor inlet and outlet temperatures. The best operating practice for such catalysts is to gradually increase the hydrogen partial pressure as the reactor inlet temperature is raised, so as to maintain a nearly constant C_8 naphthene concentration. Reactor pressures between 6 and 14 barg are typical, along with temperatures in the range of 350–400 °C.

It is highly desirable that the C_8 naphthenes that leave the reactor be retained through downstream steps of fractionation and recovery of the particular xylene isomer. In fact, there will likely be some C_8 paraffins in equilibrium with the C_8 naphthenes that should also be recovered for maximum chemical efficiency. Both adsorption and fractional crystallization will allow naphthenes to pass through to the raffinate or mother liquor, respectively. In some cases, "local" recycle of the C_8 naphthenes around the isomerization unit may be practiced via fractionation from the C_8 aromatics. The added CAPEX and OPEX of this local recycle must be weighed against the costs of allowing the C_8 naphthenes to circulate throughout the xylene isomer recovery loop.

In today's *para*-xylene industry, the use of EB-destruction catalysts is prevalent, for a couple of reasons.

1) Benzene coproduct typically has a value close to that of *para*-xylene. With modern xylene isomerization catalysts such as those supplied by UOP, the by-product benzene can be sold directly to the merchant market without requiring extraction.
2) The EB-destruction catalysts usually allow much lower CAPEX and OPEX costs due to smaller size fractionation and compressor equipment. Occasionally, facilities that have been designed for EB-isomerization catalysts can be significantly de-bottlenecked for capacity expansions by replacing the EB-isomerization catalyst with and EB-destruction catalyst.

It should be noted that xylene isomerization technology is useful in conjunction with purification and recovery of all three xylene isomers. Of course, *para*-xylene dominates the applications due to its significant market and end uses. However, there are facilities that recover only *ortho*-xylene by fractionation, and many of them will have a xylene isomerization unit associated with the *ortho*-xylene recovery section. Likewise, *meta*-xylene recovery units may have a xylene isomerization operation close by. For these unique applications, there will be important differences in how the xylene isomerization unit approaches equilibrium, as well as the number of recycle passes back through the reactor, and optimal operating severity.

2.6 Technologies in Aromatic Separations

This section discusses the many process technologies that are employed in aromatics separations. The separations can cover three categories:

- Class separations – By hydrocarbon types such as aromatics from nonaromatics.
- Molecular weight – Separating by carbon number, such as between C_7- and C_8-aromatics.
- Separating by isomer – purification and recovery of specific isomers from others such as *para*-, *meta*-, and *ortho*-xylene.

The dominant technologies are fractionation, LLE, ED, selective adsorption, and fractional crystallization.

2.6.1 Liquid–Liquid Extraction and Extractive Distillation

As shown in Table 2.6, there is a significant amount of nonaromatic hydrocarbons in the reformate from a high severity reformer. Even more nonaromatics would be present in the older-style semi-regenerative reformer, where the paraffin conversions are typically much lower. In the early days of the aromatics industry, several approaches were tried for recovering and purifying the aromatics. At UOP, experimental work was done using adsorption, where silica gel was the adsorbent, selective for aromatics over nonaromatics. However, that approach was limited by having no economically effective means to recover the adsorbed aromatics.

In 1952, UOP learned that Dow Chemical was looking at applications for di-ethylene glycol (DEG) and tri-ethylene glycol (Tri-EG) and that such solvents had an affinity for aromatics over nonaromatics. This led to joint development of a process technology called Udex™ (UOP-Dow Extraction). The Udex process was deployed worldwide by UOP to aromatics producers. About 10 years later, Shell Oil Co. in the Netherlands invented and commercialized an organic sulfur compound named

Sulfolane™. This solvent offered significant selectivity and capacity improvements over DEG and Tri-EG. Physically, this meant that the Sulfolane solvent had superior solubility for aromatics, which translated to improved capacity. It also had a much higher specific gravity than the hydrocarbon feed, which helped minimize equipment sizes while promoting countercurrent flow. Finally, it had low specific heat and high boiling point, which minimized utility consumption in solvent recovery and stripping. Shell started by introducing Sulfolane at two Shell refineries in Europe. Recognizing the superior properties and performance of the Sulfolane system, UOP worked with Shell to further develop and broadly commercialize the Sulfolane process, using this solvent for LLE. There was also an early commercial plant designed with ED using the Sulfolane solvent to produce a toluene-xylene feed. In 1999, UOP developed the ED Sulfolane process for benzene–toluene feeds, and that has been UOP's prevalent design for most modern aromatics recovery plants.

Union Carbide had facilities to manufacture tetraethylene glycol (TEG), and in 1973, they commercialized the Union Carbide Tetra process for aromatics purification and recovery. Only a few of these plants were licensed. A few years later, the Carom™ process was developed by Union Carbide, where a cosolvent ether was used with the TEG solvent (Forte 1991). Very few of these plants were licensed, and when UOP merged with Union Carbide's CAPS Division, the Sulfolane process was judged to be most attractive. Thyssenkrupp licenses the Morpholane process using N-formylmorpholine (NFM) as the solvent. GTC licenses a Sulfolane-based ED process with a cosolvent that can further enhance relative volatilities in some situations (Gentry et al. 1995). Figure 2.29 compares the selectivity and capacity of several solvents, and it is clear that Sulfolane is superior to the others. This discussion will focus on Sulfolane solvent to provide some understanding of the technical basis for LLE and ED.

The decision on whether to use LLE or ED is normally based on consideration of the feed composition. LLE is generally favored when there are significant amounts of benzene, toluene, and C_8 aromatics to be recovered. This is generally true of pygas feeds, where these three molecular weight classes are present. LLE is also favored economically when the total aromatics content is less than 50%, again typical of hydrotreated pygas. For most modern high-conversion catalytic reformers, such as the

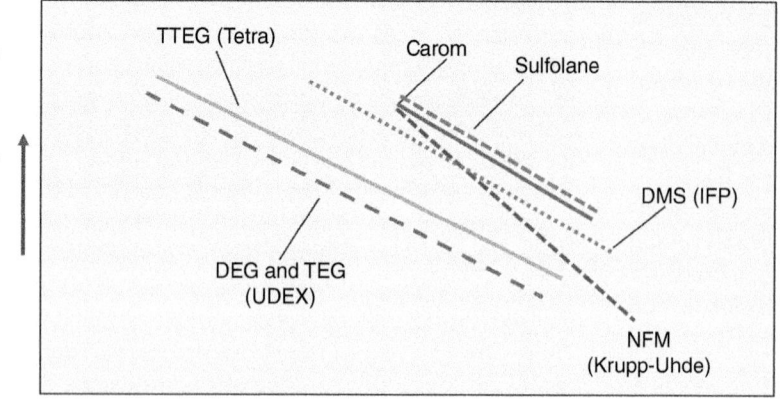

Table 2.6 Typical naphtha composition and reformate C_5+ yields.

	Feed		Product yields	
	LV%	wt.%	LV% FF	wt.% FF
C_5	0.00	0.00	3.10	2.67
P_6	10.77	9.83	9.36	8.52
P_7	24.01	22.75	3.11	2.94
P_8	19.74	19.02	0.12	0.12
P_9	16.26	16.15	0.01	0.01
P_{10}^+	4.21	4.33	0.00	0.00
MCP	1.35	1.40	0.12	0.13
CH	1.55	1.67	0.01	0.01
N_7	4.89	5.18	0.15	0.16
N_8	4.82	5.11	0.01	0.01
N_9	2.43	2.67	0.00	0.00
N_{10}^+	1.03	1.13	0.00	0.00
A_6	0.89	1.08	4.15	5.04
A_7	3.36	4.03	17.99	21.57
A_8	3.39	4.07	20.81	24.99
A_9	0.76	0.92	12.17	14.71
A_{10}^+	0.55	0.65	3.62	4.32

Figure 2.29 Solvent comparison for aromatics extraction.

UOP CCR Platforming process, the C_8 and C_9 paraffins and naphthenes are nearly completely converted. As a result, the C_8 aromatics can be economically recovered by fractionation in high enough purity, so there is very little incentive to extract them. Likewise, the percentage of benzene + toluene in this light reformate fraction will be greater than 60%, which again favors ED for purification and recovery of these aromatics. All other things being equal, LLE technology has about 20% higher CAPEX than ED. This is because LLE has three major columns: the extractor, the stripper column, and the recovery column. The ED unit only has two major columns: the ED column and the solvent stripper column. Both technologies can generate the same purity and recovery of the aromatics being processed.

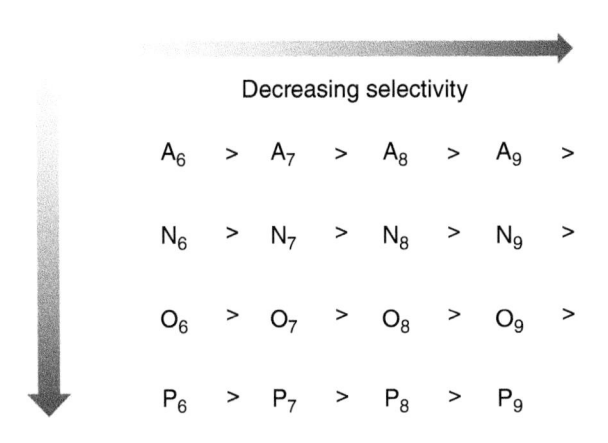

Figure 2.30 Sulfolane solvent selectivity vs. HC type and carbon number.

2.6.2 Liquid–Liquid Extraction

As qualitatively shown in Figure 2.30, Sulfolane solvent is more selective for light aromatics than for heavy aromatics.

The selectivity order for hydrocarbon types is: aromatics > naphthenes > olefins > paraffins. For LLE, the extractor is divided into three zones, as shown schematically in Figure 2.31. A simplified flow diagram of a Sulfolane LLE unit is shown in Figure 2.32. The solvent phase flows downward, countercurrent to the hydrocarbon phase, which flows upward in the main extractor. With the solvent dispersed into small droplets in contact with the hydrocarbon phase, the aromatics are selectively absorbed into the solvent, along with a small amount of heavy nonaromatic impurities.

Most of the light nonaromatic components pass through to the overhead. The extracted aromatics and the solvent move down into the backwash extractor. In this section, the light nonaromatic stream from the top of the extractor is used to displace any entrained heavy nonaromatics from the solvent and return them to the hydrocarbon phase and ultimately to the raffinate. The aromatic-rich solvent phase, along with some small amount of light nonaromatics, then moves down to the extractive stripper section where any remaining light nonaromatics are stripped and recycled into the backwash extractor. The solvent phase with the required aromatics exits the bottom of the extractor, and is sent to the stripper column where any light non-aromatics are removed overhead and then to the solvent recovery column, where the purified aromatic extract is recovered by fractionation from the Sulfolane solvent.

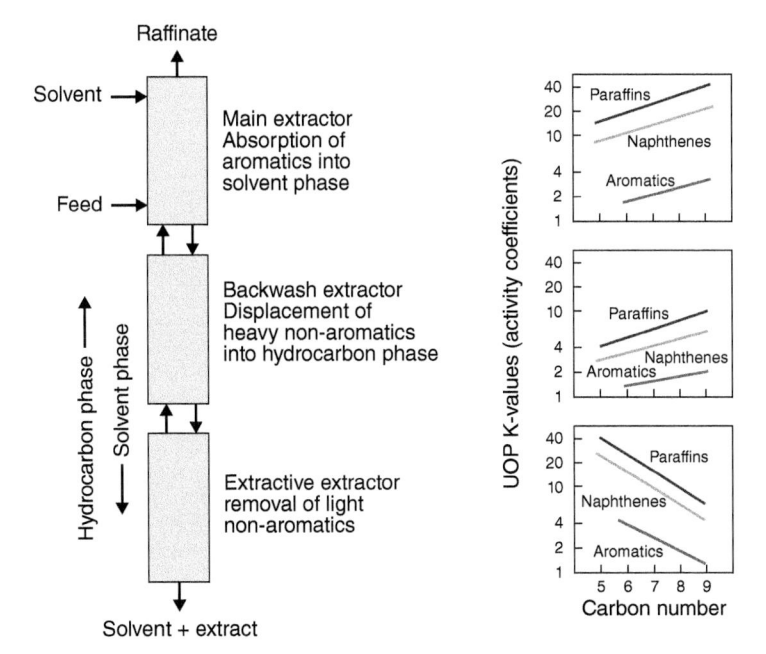

Figure 2.31 Extractor zones in LLE.

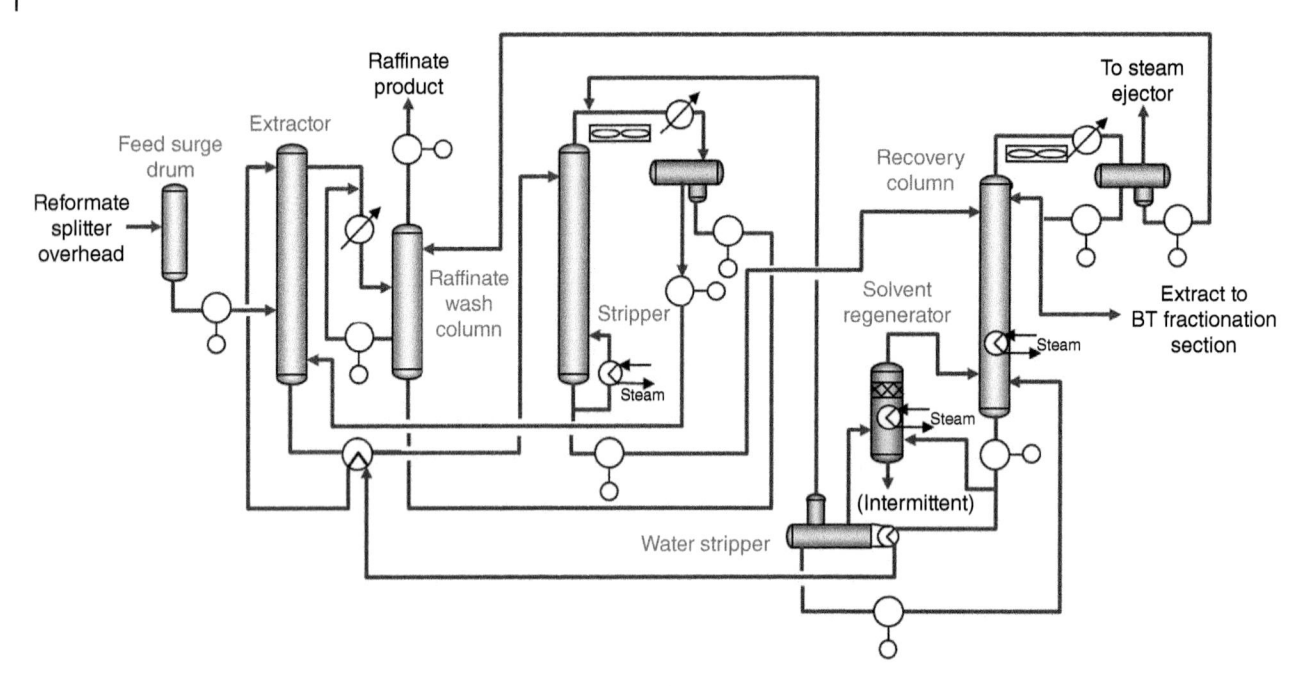

Figure 2.32 Schematic flow diagram for liquid–liquid extraction.

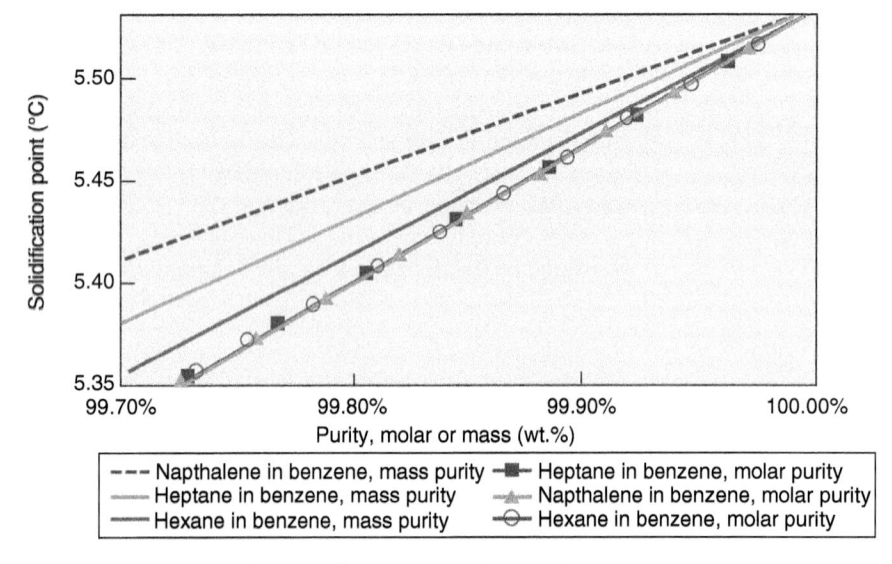

Figure 2.33 Correlation between freeze point and composition.

The recovery column is typically operated under vacuum to minimize thermal exposure to the solvent. The lean solvent is then returned to the extractor, with a small slip stream to the solvent regenerator for removal of oxygenates. The Sulfolane-extracted aromatics are usually sent through a clay-treater to ensure they are free of any olefinic compounds, and then recovered as pure aromatics by simple fractionation. Benzene purity is commonly determined using freeze point measurements, with typical grades being 5.5, 5.45, and 5.35 °C. UOP's Sulfolane LLE and ED units typically produce benzene with 5.5 °C freeze point or higher. There are some useful correlations for relating chemical composition to freeze point, as shown in Figure 2.33.

2.6.3 Extractive Distillation (ED)

As noted earlier, for a variety of reasons, ED has become the dominant technology for recovering purified aromatics from the reformate. The flow scheme for ED is considerably simpler than for LLE, as shown in Figure 2.34, consisting primarily of ED and a solvent recovery column. This technology uses the Sulfolane solvent to enhance the non-idealities of the hydrocarbons.

Figure 2.34 Schematic flow diagram for extractive distillation.

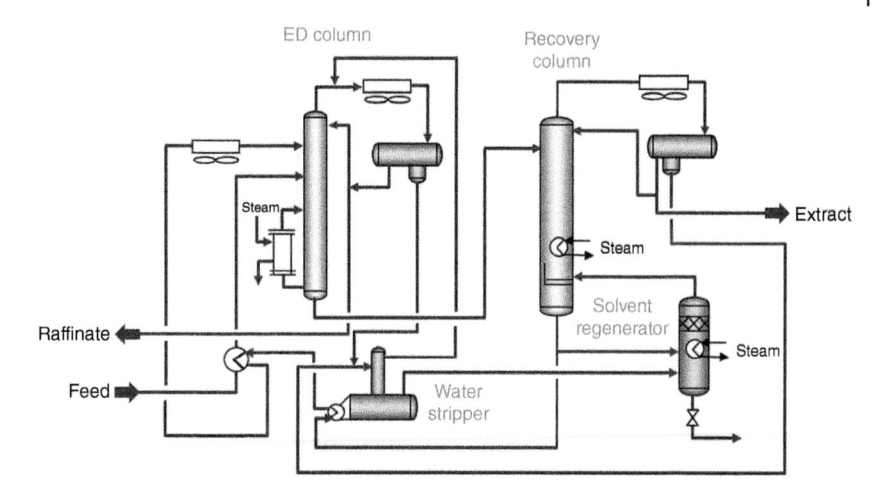

In the ED column, the feed is introduced at a tray halfway up the column, while the solvent is introduced at a tray that is near the top.

This basically increases the relative volatility differences between the lightest aromatics and the heaviest nonaromatics. Selective absorption and recovery of the aromatics into the solvent occurs as the hydrocarbon vapor travels up the column. The nonaromatic hydrocarbons pass upward, are condensed, end up as raffinate, and sent to storage. In the lower section of the ED column, the aromatic-rich solvent is stripped of any remaining nonaromatics, thereby purifying the aromatics in the solvent. The stripped nonaromatics then travel to the upper section of the ED column. The bottoms of the ED column are sent to the solvent recovery column, which uses steam stripping under vacuum to separate the purified aromatics overhead from the Sulfolane solvent. As with LLE, these aromatics are typically sent to a clay-treater, then to fractionation and storage. The stripping steam in the bottom of the solvent recovery column removes any residual hydrocarbons from the Sulfolane solvent, and the lean solvent is then returned to the ED column. As with LLE, a slip stream of solvent is sent to the solvent rerun column to remove any oxygenates.

ED and LLE technologies rely very heavily on highly accurate phase equilibria data, as well as the separation mechanisms of the solvents and the non-idealities of these compositions. These must also include temperature dependence since a significant temperature gradient will exist across the process steps.

2.7 Separations by Molecular Weight

In aromatics processing there are numerous fractionation steps that purify and recover aromatics of different carbon numbers. Some of the most prevalent are listed below.

- Benzene/Toluene (B–T splitter)
- Toluene/Xylenes (T–X splitter)
- Xylenes/C_9+ Aromatics (Xylene rerun column)
- C_7–/C_8+ (Reformate splitter)
- C_{10} and C_{11} Aromatics/Naphthalenes (Heavy aromatics column)
- Benzene/EB/DEB/Tri-EB (used in ethylbenzene synthesis)
- Benzene/Cumene/Di IPB/Tri-IPB (used in cumene synthesis)

Fractionator design pressures can cover a range from pressurized to atmospheric to vacuum. Often this will depend on the heat integration strategies between fractionators or with other process units. There is a wide variety of fractionator internals – trays, packings, and tubes – that may be selected depending on the trade-off between energy costs and capital. Some applications lend themselves to dividing wall columns (DWC). Many of these considerations will be discussed in Section 3.1.

2.8 Separations by Isomer Type: *para*-Xylene

As noted in earlier sections, *para*-xylene is the most widely used isomer of the C_8 aromatics. It is generally oxidized to produce PTA which can then be incorporated into PET fibers, resins, and films. Modern oxidation technology generally requires a minimum *para*-xylene purity of 99.7%, with many producers feeding a purity of 99.8% or slightly higher. As noted earlier, there are several streams containing *para*-xylene in a fully integrated naphtha-to-*para*-xylene facility that may be considered as potential feed sources for *para*-xylene purification and recovery.

- C_8 heartcut of reformate – contains about 19 wt.% *para*-xylene among other aromatics and nonaromatics – considered a fresh feed source.

- C_8 cut from transalkylation – contains about 24 wt.% *para*-xylene among other aromatics – also considered a fresh feed source, since it is produced from reformate A_7 and A_9+.
- C_8 cut from xylene isomerization – contains about 22 wt.% *para*-xylene among other aromatics – considered a recycle stream as a re-equilibrated reject stream originating from the *para*-xylene recovery step.

As noted earlier, depending on feed composition to transalkylation, this source can account for more than 50% of the fresh feed xylenes. Blending the reformate, transalkyation, and isomerization C_8's, the *para*-xylene content of the feed to the *para*-xylene separation step will be about 23% of *p*X. Figures 2.35 and 2.36 show that it is not feasible to recover high-purity *para*-xylene from this mixture by simple fractionation. Either a chemical or a physical property of *para*-xylene needs to be exploited in order to achieve the target purity at high recovery.

2.8.1 Crystallization of *para*-Xylene

In Table 2.7, the pure component freeze points of the pure C_8 aromatic isomers show that *para*-xylene has nearly 40 °C higher freeze point than the closest isomer, meaning that fractional crystallization could be used for the purification and recovery of *para*-xylene.

However, there is a eutectic temperature and composition of a binary mixture of *para*- and *meta*-xylene as shown by Egan and Luthy of California Research

Table 2.7 Freeze points and boiling points of C_8 aromatics.

	*p*X	*m*X	*o*X	EB
Boiling point (K)	411.5	412.3	417.8	409.3
Freezing point (K)	286.4	222.5	248.0	178.2
Density ($g\,cm^{-3}$)	0.858	0.861	0.876	0.867

Ind. Eng. Chem. Res. 2017, 56, 14725–14753.

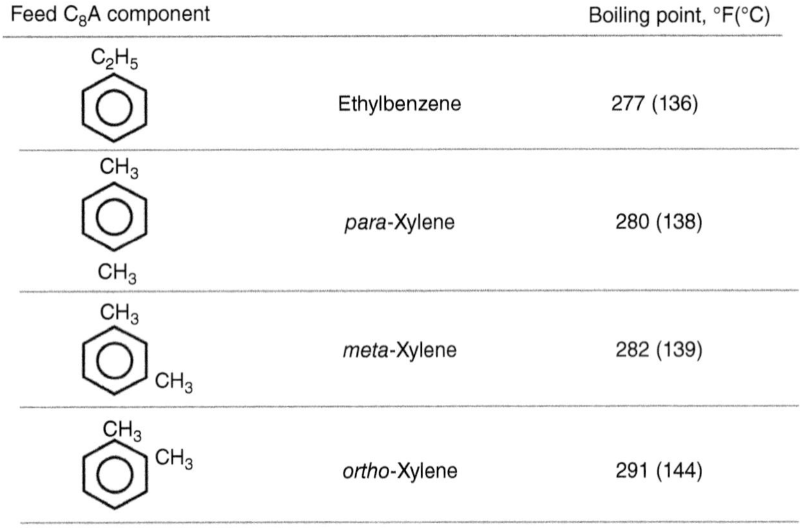
Feed C8A component
Boiling point, °F(°C)
C2H5
Ethylbenzene
277 (136)
CH3
para-Xylene
280 (138)
CH3
CH3
meta-Xylene
282 (139)
CH3
CH3
CH3
ortho-Xylene
291 (144)

Figure 2.35 Normal boiling points of C_8 aromatics.

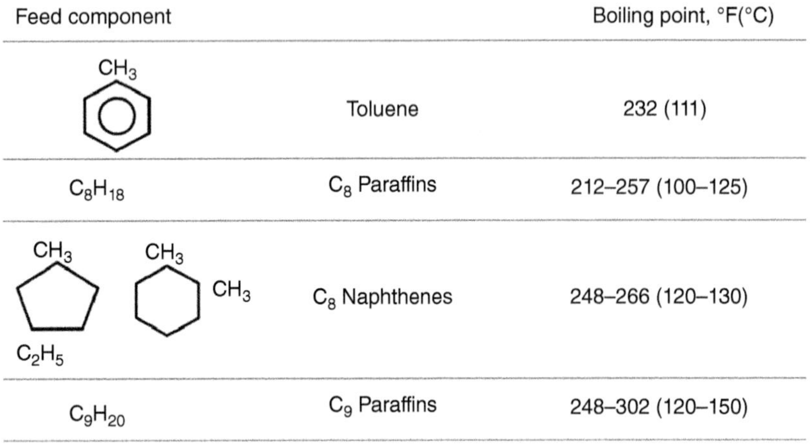
Feed component
Boiling point, °F(°C)
CH3
Toluene
232 (111)
C8H18
C8 Paraffins
212–257 (100–125)
CH3
CH3
CH3
C2H5
C8 Naphthenes
248–266 (120–130)
C9H20
C9 Paraffins
248–302 (120–150)

Figure 2.36 Normal boiling points of toluene and C_8 nonaromatics.

Company (1955). Even with a toluene diluent and further chilling, they state "The effect (of Toluene diluent) is not pronounced; even at filtration temperatures of −120 °F, the diluent-free mother liquor contains about 10% *para*-xylene, which represents an appreciable loss of this isomer." This phenomenon basically limits the recovery of purified *para*-xylene from the above feeds to less than 65%. As was shown in Figure 2.8, in the early years of the *para*-xylene industry, many producers were willing to live with this limitation, including Shell, Exxon, Amoco, and Standard Oil of California (now Chevron). This figure also shows that up to the present time, fractional crystallization remains a minor contributor to the world's supply of *para*-xylene. A later section will discuss the principles of fractional crystallization for *para*-xylene production and some of the innovations for this technology – train capacity, solid–liquid separations, and energy integration – that have been developed. However, the eutectic composition has proven to be an insurmountable obstacle for step-change disruptive improvements.

It should be noted that some catalytic technologies can produce a xylene stream that is highly enriched in *para*-xylene – for example, the UOP PX Plus process and the ExMo PXMax process. In addition, toluene methylation processes offered by UOP, ExMo, and others can achieve *p*X/X that is greater than 90%. These feeds can enable high recoveries in crystallization because they are farther away from the eutectic point with *meta*-xylene. Some companies have implemented STDP technologies and run them with spare crystallization capacity when the market conditions for *para*-xylene and benzene dictate. Of course, these *para*-selective technologies are also highly beneficial for adsorptive separations of *para*-xylene.

2.8.2 Adsorptive Separation of *para*-Xylene

In the 1950s and 1960s, UOP had tested numerous alternatives for efficiently recovering high-purity *para*-xylene from a complex mixture. Selective adsorption from a liquid, using a high-purity synthetic zeolite, had been successfully discovered, developed, and scaled up into a continuous process for production of high-purity *n*-paraffins – the UOP Molex™ process. However, the 5-Å pores of that particular adsorbent would not allow *para*-xylene to be adsorbed. In the meantime, Union Carbide had successfully synthesized larger pore faujasite zeolites – X and Y – and in 1967, samples of these new zeolites were acquired by UOP to determine their usefulness in purification and recovery of *para*-xylene, using selective adsorption from the liquid phase. In 1968, UOP filed its first patent (Neuzil 1968) covering use of X- and Y-zeolites for *para*-xylene purification and recovery. This seminal discovery led to the UOP Parex process.

This technology exploits the small differences in adsorption energies between *para*-xylene and other molecules in the feed, as they interact with the cation sites of the adsorbent, a phenomenon known as competitive adsorption. In contrast, the Molex process uses molecular sieving with selective pore diameters that admit only *n*-paraffins to achieve high purity and recovery.

The Parex process was a truly disruptive technology that avoided the eutectic composition barrier that limits the recovery in crystallization. The first company to adopt the Parex process was URBK in Wesseling Germany in 1970. URBK had long used crystallization for *para*-xylene production, but chose the Parex process when they wanted to expand their production. The first Parex unit had a product capacity of about 70,000 tons per year and contained about 240 tons of adsorbent. Advances in adsorbent and isomerization catalyst technology have been deployed over the years at URBK. Today, the plant is owned by Shell and produces nearly four times as much *para*-xylene from the same adsorbent chambers compared to the original unit. This illustrates the potential for significant advancements in adsorbent technology, and the benefits of not being limited by a physical problem like eutectic compositions. In fact, for the Parex process, UOP has successfully deployed the fifth generation of Parex adsorbents in numerous commercial units as a result of creating step-change improvements in capacity and mass transfer.

The UOP Parex process is one of UOP's Sorbex process technologies. It uses a SMB to provide for countercurrent contact between the bulk liquid phase and the solid adsorbent phase. This equilibrium stagewise countercurrent contacting is analogous to the vapor/liquid equilibrium contacting that takes place in a distillation column. Like a distillation column, there are specific zones that provide particular functions. The principle functions are:

Zone 1   Adsorption of *para*-xylene and some impurities from the liquid phase.
Zone 2   Purification by removing impurities from the *para*-xylene in the solid phase.
Zone 3   Desorption and recovery of the purified *para*-xylene into the extract.
Zone 4   Buffer zone prevents raffinate from the bottom of Zone 1 reaching Zone 3.

These functions must be achieved at high efficiency in the liquid phase, and this requires that the adsorbent must have high mass transfer rates, component selectivities, and capacity for adsorbed components. The efficient mass transfer is aided by setting the adsorbent particle diameter at an average of about 500 μm. The selectivity and capacity of the adsorbent are governed by chemical compositions, such as the Si/Al of the zeolite, the cation types, and proportions. The desorbent is an aromatic hydrocarbon that has comparable affinity for the adsorbent as the aromatic feed components, and which can be easily separated from the feed components by distillation. For the Parex process, two desorbents have been used: *para*-diethyl benzene (PDEB) and toluene. These two desorbents have particularly favorable adsorption affinities compared to the feed aromatics, and they can be recovered and reused via simple distillation. There are numerous references that provide detailed explanations of SMB theory (Broughton 1984; Broughton and Gerhold 1961; Rodrigues 2015; Ruthven and Ching 1989; Yueying 2015).

As a starting point, the principles behind the Parex process can be understood in the context of a chromatography column, as shown in Figure 2.37.

In this example, the chromatography column is filled with the Parex adsorbent. There are two fluids – the feed aromatics with a nonaromatic tracer, and the desorbent. Initially, the desorbent is flowing through the column. At some point a small pulse of the feed enters the column, followed by the flow of desorbent. As the feed components traverse the column, they begin to separate, based on their relative affinities for the adsorbent and the strength of the desorbent to elute them off of the adsorbent. The time-dependent composition profile shows that the nonaromatic tracer is first to elute, because it is essentially non-adsorbed. After that, the *meta*- and *ortho*-xylene peaks appear, because they are the least selectively adsorbed aromatics. Next, the ethylbenzene appears, and finally the most selectively adsorbed, *para*-xylene appears. Note that only a portion of the *para*-xylene peak would have high purity, and indeed a cocurrent chromatographic process such as this depiction would not be economically attractive.

In order to make this technique economically attractive and allow a high-purity product to be withdrawn at high recovery, countercurrent contact of the adsorbent with the liquid would be desirable as shown in Figure 2.38. Here, the adsorbent plus desorbent D circulates upflow, while the liquid feed plus desorbent is pumped downflow.

In this example, the feed stream contains components A (say *para*-xylene) and B (all the other feed components). The feed enters at a fixed point of the adsorbent chamber. The desorbent, D, enters at a fixed point in the chamber well above the feed point. Extract (consisting of A and D) is withdrawn at a fixed point between feed and desorbent. Raffinate (consisting of B and D) is withdrawn at a fixed point below the feed point. If this actual countercurrent contacting were possible, the composition profile shown in Figure 2.37 would be obtained. While this would work in theory, there would be practical difficulties in physically moving the adsorbent countercurrent through the liquid – attrition of the adsorbent, maintaining plug flow, and being able to get ideal addition and withdrawal of feed, desorbent, extract, and raffinate – among others.

These problems are solved by simulating the countercurrent movement of the solid adsorbent and the liquid phases. This is done by dividing the adsorbent chamber into multiple beds, each with a flow distributor above and below. The feed and desorbent inlets, as well as the extract and raffinate outlets, are stepped along the adsorbent chamber over constant time intervals while

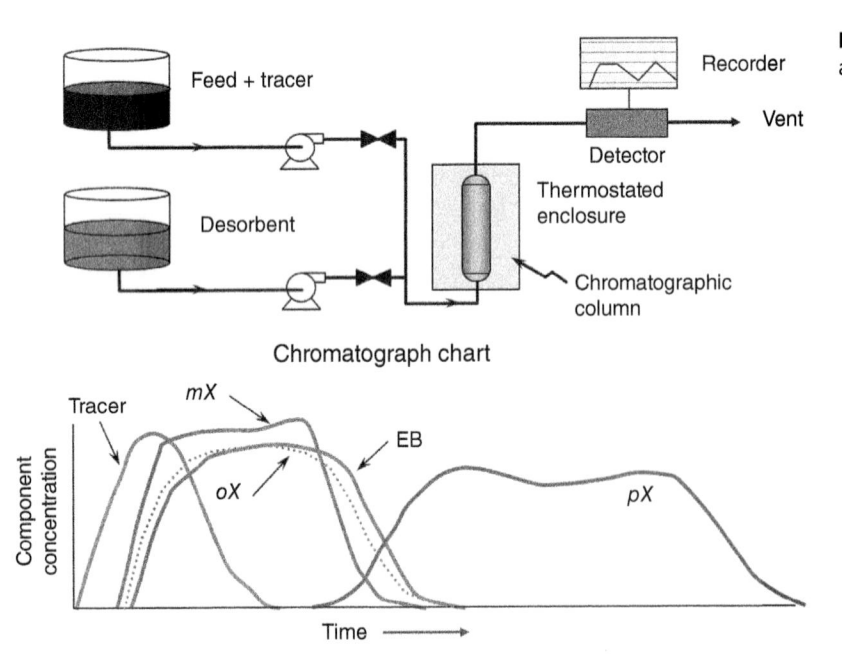

Figure 2.37 Cocurrent chromatography for C_8 aromatic separation.

2.8 Separations by Isomer Type: para-Xylene

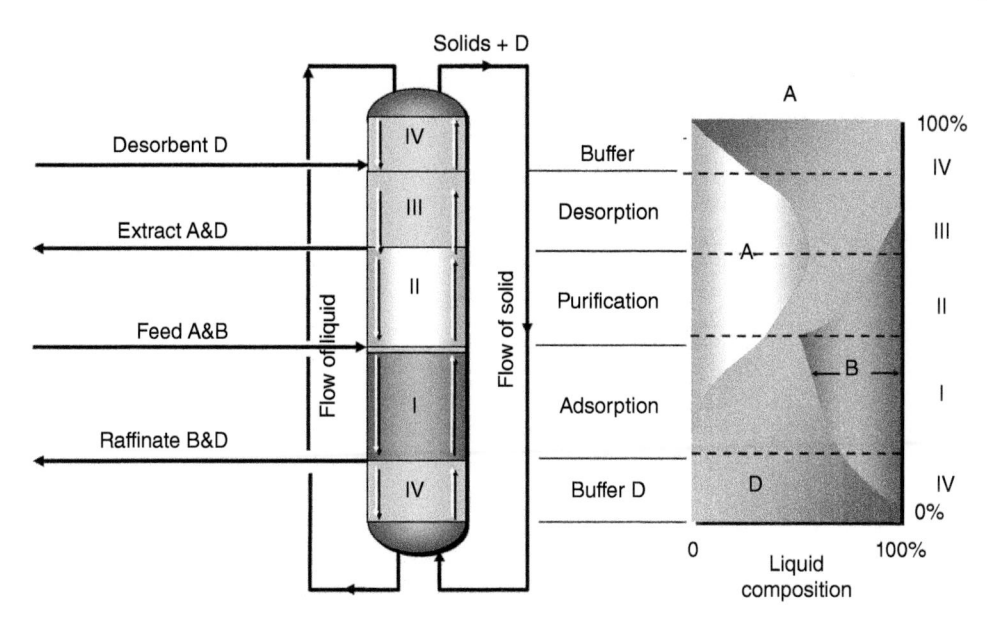

Figure 2.38 Countercurrent contact of adsorbent and liquid.

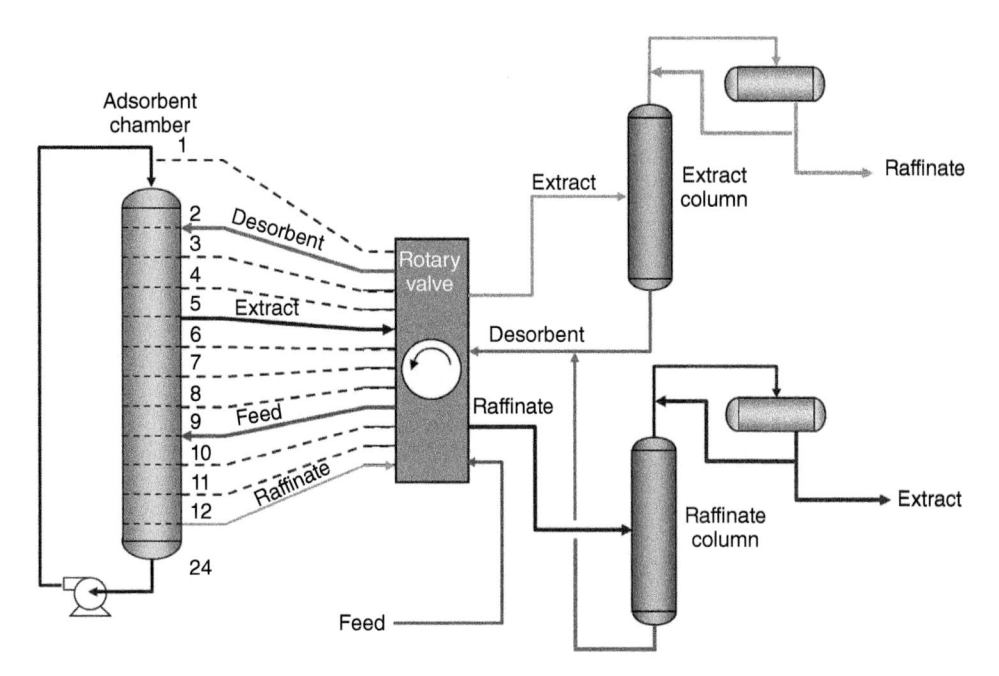

Figure 2.39 Schematic diagram of Sorbex.

the bulk liquid is pumped at particular rates and the flow rates of feed, desorbent, extract, and raffinate are carefully controlled. When this is done, the ideal composition profile can still be obtained, only this time without having to move the adsorbent. This stepping of addition and withdrawal points is accomplished in the Sorbex process through a device called the rotary valve. The rotary valve directs all the external streams to the precise location at the exact time they are needed. A control system links the rotary valve to the liquid pump as well as the external stream flow and pressure controllers to maximize the efficiency of this process. This is shown in Figure 2.39 for a single chamber with 12 adsorbent beds.

In this flow diagram, the use of a "heavy" desorbent – PDEB – is depicted, where the desorbent is recovered from the bottoms of the extract and raffinate fractionators and then recycled back to the adsorption chamber. If toluene were used as the "light" desorbent, it would be recovered overhead in these fractionators. In reality, the

Parex process uses two adsorbent chambers, each with 12 adsorbent beds. Depending on the capacity, the adsorbent chamber diameter can range from 3 m to nearly 12 m. Correspondingly, there are several sizes of rotary valve. The largest single train plants may have two rotary valves that are synchronized to distribute these external streams. The Parex process can produce 3 million tons per year of *p*X from a single train plant with two rotary valves.

The UOP rotary valve is a unique and precisely machined piece of equipment that provides greater than 99.9% onstream efficiency, with only one moving part. It is unique to all UOP Sorbex technology with more than 200 rotary valves deployed across many industries and applications since the invention of the UOP Sorbex process. As noted, UOP invented SMB technology in 1959 using the rotary valve. Some 35 years later, Axens developed and commercialized a SMB technology called the Eluxyl process for *para*-xylene production (Hotier et al. 1995). This technology is similar to Parex, but uses multiple on–off valves, around 144 or more, rather than a single rotary valve, to direct external streams. For high-capacity plants, these valves can be very large. They require a preventive maintenance program to ensure leak-free operation, and the large number of these valves can lead to excessive fugitive emissions. Still, the Eluxyl process has been moderately successful as was shown in Figure 2.8. In 2012, Sinopec developed and commercialized their *para*-xylene adsorptive separation technology using multiple on–off valves, similar to the Eluxyl designs. However, most producers in China have elected to continue use UOP and occasionally Axens technologies. One other SMB technology called Aromax was developed by Toray in 1973 (Yasauki et al. 1974), but was only implemented in a couple of locations. Toray used horizontal adsorption chambers, which tended to have adsorbent settling problems that led to back-mixing and difficulty in maintaining purity and recovery.

One final category of technology for *para*-xylene purification and recovery involves hybrid processing using adsorption and crystallization. In this case, a simplified adsorptive separation unit is used to increase the *para*-xylene purity to 90 + % using a toluene desorbent. This semi-pure extract will have very little *meta*-xylene impurity, and can easily be raised to 99.9% purity with high recovery by being fed to a single-stage crystallizer. UOP developed such a technology called the Hysorb XP™ process. Axens developed a similar technology based on its Eluxyl flow scheme as a means of demonstrating its Eluxyl technology. Axens built a demonstration unit in collaboration with Chevron Chemical in the United States. It is believed that this is the only hybrid unit in operation, although there are a few companies who run their crystallizers when the *para*-xylene market is strong and send the mother liquor to their Parex unit.

With the number of older multistage crystallizers that were built in the early days of the *para*-xylene industry, there were evaluations and even design packages developed to use the hybrid concept and revamp these into single-stage crystallizers to obtain increased capacity. However, the cost of downtime to implement such a revamp prevented such projects from going ahead.

2.9 Separations by Isomer Type: *meta*-Xylene

As noted in earlier sections, *meta*-xylene has a much smaller scale of commercial use than *para*-xylene, with the primary application being PET resin comonomer. In 1970, Japan Gas Chemical Company (later Mitsubishi Gas Chemical Co. and now owned by JX) developed technology for extracting *meta*-xylene using $HF-BF_3$ (Nakano 1971). The technology works on the principal that, when xylenes are contacted with HF/BF_3, two phases form, and the *meta*-xylene preferentially concentrates in the HF phase. The complex of *meta*-xylene with HF/BF_3 is very stable, and thermal decomposition of the complex will release purified *meta*-xylene. Mitsubishi built a plant at their Mizushima facility and later licensed the technology to Amoco, Enichem, and reportedly to a German chemical company. This technology was developed to help remove the eutectic limit on recovery of *para*-xylene by selectively extracting *meta*-xylene using $HF-BF_3$ as the countercurrent extraction agent to form a complex enriched in *meta*-xylene. A flow scheme is shown in Figure 2.40.

The raffinate at the top of the extractor contains less than 0.1% *meta*-xylene and nearly all of the *para*- and *ortho*-xylene as well as ethylbenzene. This allows recovery of the *ortho*-xylene by simple fractionation, since there is minimal *meta*-xylene present. It also allows a high recovery of *para*-xylene by crystallization. The *meta*-xylene in the extract is 99% pure. A portion of this is sent to isomerization using $HF-BF_3$ as the catalyst. The remaining *meta*-xylene complex is sent to the decomposer, where the *meta*-xylene is released back into the hydrocarbon phase, which then proceeds to downstream stripping and rerunning of the light and heavy by-products formed as a result of contact with the $HF-BF_3$.

Interestingly, in 1956, Amoco had one of the first patents on the use of $HF-BF_3$ to overcome the eutectic limitation (Kalfadelis 1956).

UOP had worked steadily to find an economical route to *meta*-xylene purification and recovery using Sorbex technology to avoid the environmental risks and maintenance issues associated with $HF-BF_3$. In 1997, the UOP

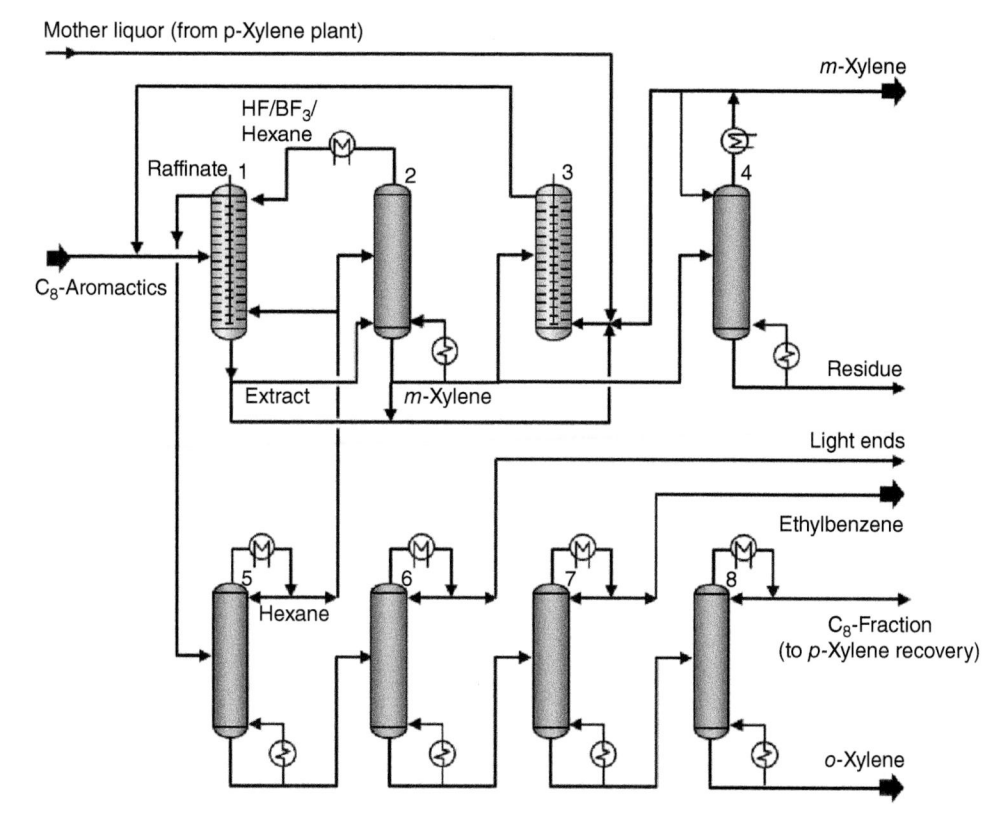

Figure 2.40 Mitsubishi Gas Chemical HF/BF$_3$ process for *meta*-xylene.

MX Sorbex process was introduced and found rapid market acceptance. This technology is based on exactly the same principles of SMB that have been employed in the Parex process described in Section 2.8.2. This technology exploits the basicity differences among the C$_8$ aromatic isomers, with *meta*-xylene being 50 times more basic than *ortho*-xylene, 100 times more basic than *para*-xylene, and several hundred times more basic than ethylbenzene. The MX Sorbex process uses toluene as the desorbent, and the adsorbent composition is different from that of the Parex process, in terms of Si/Al and cation type. The feed can be mixed xylenes from reformate, xylene isomerate, or the raffinate stream from the Parex process. Some operators recover *ortho*-xylene prior to the MX Sorbex unit to help drive up the *meta*-xylene concentration and to increase the production capacity of the MX Sorbex plant. To date, more than 10 MX Sorbex plants have been licensed, and as was shown in Figure 2.10, this technology accounts for 100% of the world's *meta*-xylene production, now up to 1 million tons per year. Mitsubishi Gas Chemical Co. and BP (previously Amoco) were among the MX Sorbex process licensees. The economics of the MX Sorbex have been further improved by advances in adsorbent technology, similar to what has been accomplished with Parex as described in Section 3.2.

2.10 Separations by Isomer Type: *ortho*-Xylene and Ethylbenzene

As noted earlier, there is a small but useful boiling point difference of 5 °C between *ortho*-xylene and *meta*-xylene. As a result, it is possible to recover high-purity *ortho*-xylene by fractionation. This is done using a xylene splitter, which fractionates some percentage of the *ortho*-xylene overhead with ethylbenzene, *para*-xylene, and *meta*-xylene. The remaining *ortho*-xylene exits the bottom of the xylene splitter and then goes to the *ortho*-xylene rerun column, for removal of any C$_9$ aromatics such as *n*-propyl benzene, cumene, etc. Typically, *ortho*-xylene producers will only recover 25–30% of the *ortho*-xylene available due to the limited market for *ortho*-xylene derivatives. Most modern CCR reforming plants have such high conversion of C$_8$ and C$_9$ aliphatics that they no longer pose a threat to *ortho*-xylene purity. Older reformers using semi-regenerative operations will cause more difficulty with C$_9$ aliphatic impurities in the recovered *ortho*-xylene. There is virtually no merchant market for ethylbenzene, and as a result, only a limited number of companies have installed super-fractionators for ethylbenzene recovery. These super-fractionators require about 300 trays, due to the narrow 2.2 °C boiling point differential between ethylbenzene and *para*-xylene.

A recent patent application (Thirasak et al. 2017) uses HYSYS™ simulations to show a minor improvement in ethylbenzene recovery through the addition of water to an ED scheme using nitrobenzene, methyl salicylate, or NMP as the solvents, but it is not clear that the overall economics of ethylbenzene production would improve significantly, nor was there any experimental verification of these simulations.

2.11 Other Related Aromatics Technologies

This section provides a discussion of technologies that are related to derivatives of aromatics, organized by derivatives of benzene, then derivatives of the xylene isomers.

2.11.1 Cyclohexane

The first to consider is hydrogenation of benzene to cyclohexane. This is primarily for production of Nylon 6 and Nylon 66. The cyclohexane is converted to a mixture of cyclohexanone and cyclohexanol via catalytic oxidation. This ketone–alcohol mixture is referred to as KA Oil and is the precursor for adipic acid and caprolactam. The hydrogenation of benzene to cyclohexane is extremely exothermic – with heat of reaction on the order of 216 kJ/mol. The reaction can take place in the vapor or liquid phase but conversion of benzene must be complete and methyl cyclo-pentane (MCP) impurity negligible in order to have nylon-grade cyclohexane. UOP licensed numerous vapor-phase units under the process names Hydrar™ and HB Unibon™. There were two catalysts – one containing Ni, the other containing Pt. The Ni catalyst produced the least amount of MCP, but could be irreversibly poisoned by sulfur such as trace Sulfolane solvent. The Pt catalyst was very tolerant of sulfur, but produced a slightly higher amount of MCP by-product. This technology employed multistage fixed bed reactors with intermediate quench to control the overall heat of reaction. There was usually a small polishing reactor as the last step to ensure complete benzene conversion.

Axens developed a liquid-phase process based on soluble Ni catalyst in a circulating liquid, with hydrogen introduced below the feed point. The cyclohexane is vaporized by the heat of reaction and leaves the reactor, followed by a vapor-phase heterogeneous catalyst reactor for polishing. This technology has had strong market acceptance, with 35 units licensed worldwide. The primary advantage is that most of the heat of reaction is managed by vaporization, rather than by quench or cooling coils, saving both capital and utilities.

2.11.2 Ethylbenzene/Styrene

As noted in earlier sections, benzene is used for polystyrene manufacture. The first step is alkylation with ethylene over a selective zeolitic catalyst, in an excess of benzene. The EBOne process, licensed by UOP, conducts the benzene alkylation reaction in an upflow multistage liquid-phase reactor. The chemistry is fairly simple and shown in Figure 2.41.

In the EBOne process, ethylene is injected between stages to ensure high selectivity and control the temperature. A typical flow scheme is shown in Figure 2.42.

Steam generation is also used for temperature control. Most EBOne plants are closely heat integrated with the adjoining styrene plant, where the ethylbenzene dehydrogenation takes place. The global (that is, overall) benzene to ethylene ratio is in the range of 2–3 for modern EB catalysts that can operate at low temperatures. With the stagewise ethylene injection, the local benzene to ethylene ratio starts much higher, but reaches the low "global" value at the last reactor. This ratio, along with the type of catalyst, will ultimately govern the amount of mono-alkylate that is produced across the alkylation reactor.

The effluent from the alkylation reactor goes to the benzene column, where unconverted benzene is recovered and recycled back to the alkylation reactor, with a small amount going to the polyethyl benzene reactor. The benzene column bottoms are then directed to the EB recovery column, where the product EB is recovered overhead at 99.97 + % purity. The bottoms of the ethylbenzene column consist of di- and tri-ethylbenzene isomers, along with a small amount of di-phenyl ethane (DPE) which is known as flux oil. The poly-ethylbenzene isomers flow to the transalkylation reactor where they are converted to additional ethylbenzene. Yields from the EBOne process approach 99.7%. Badger, in collaboration with ExMo, offers EBMax, with a similar flow scheme.

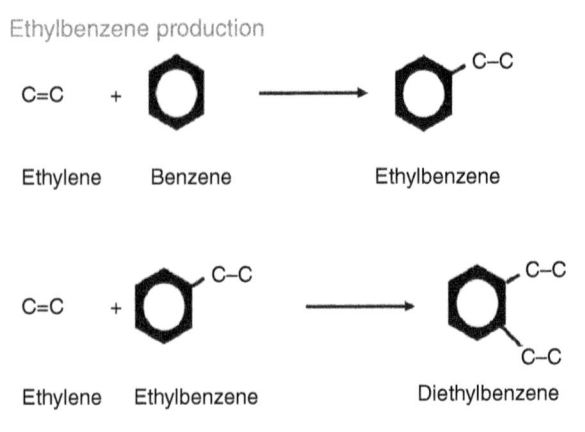

Figure 2.41 Ethylbenzene synthesis reactions.

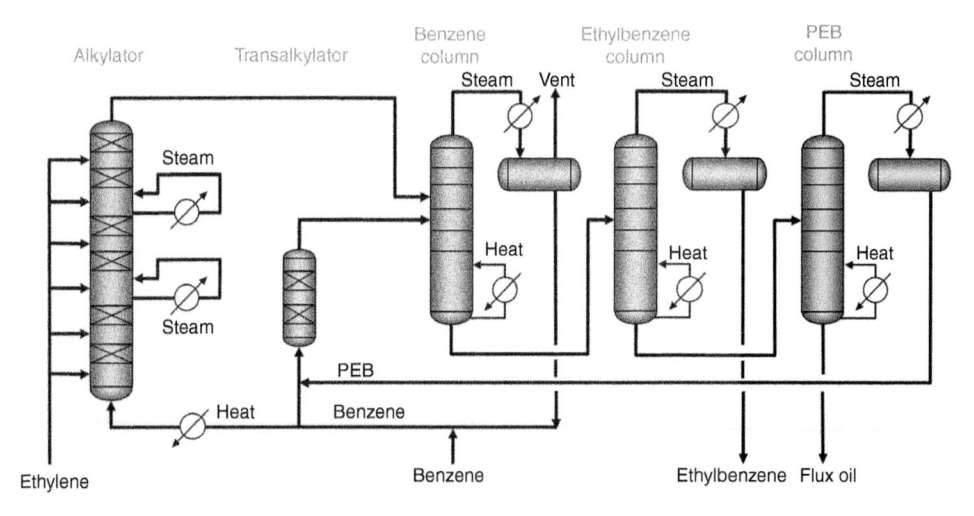

Figure 2.42 Schematic flow diagram for the EB*One* process.

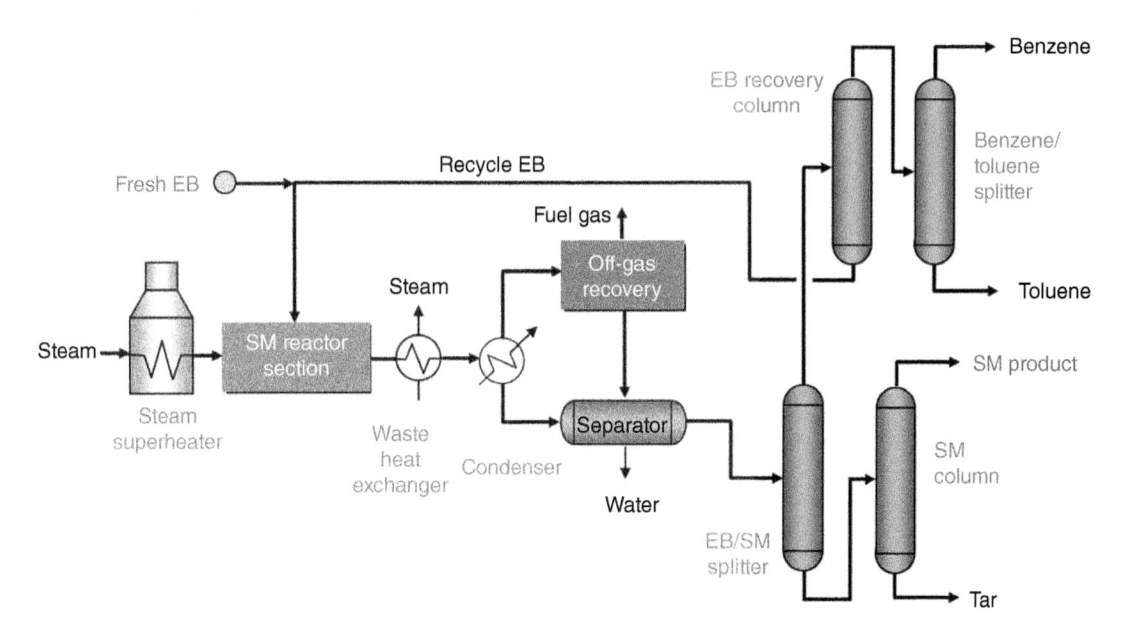

Figure 2.43 Block flow diagram for styrene production.

The ethylbenzene is sent to the styrene section where it is dehydrogenated to styrene monomer. The styrene technology typically requires significant superheated steam dilution, with modern catalysts operating at a 0.5–0.6 M steam : oil ratio. The steam creates favorable equilibrium, provides a heat sink for the endothermic heat of reaction, and helps minimize the rate of carbon deposition on the catalyst. The catalysts are typically iron based with potassium carbonate and other metals, and are reasonably selective with minimal CO and CO_2 generation. There are usually two radial flow reactors in series, operating at very low pressure and inlet temperatures of around 600 °C. A typical flow scheme is shown in Figure 2.43.

In conventional styrene technology, superheated steam is used to achieve the desired inlet temperature in each of the dehydrogenation reactors. This is the so-called "Classic" styrene technology that is offered by McDermott in partnership with UOP's EBOne process. UOP also has the SMART (Styrene Monomer Advanced Reheat Technology) process that selectively combusts the hydrogen formed in the second reactor to provide reheat to a third reactor and to create a favorable equilibrium for further conversion. The SMART technology has often been used to get incremental production capacity by increasing per-pass conversion of the ethylbenzene.

Returning to Figure 2.43, the dehydrogenation effluent, after going through the waste heat exchanger and condenser, enters the phase separator for water removal, then to the EB/SM fractionator, where EB, benzene, and toluene are taken overhead and sent to the EB recovery column.

The EB at the bottom of the EB recovery column is recycled back to dehydrogenation, while the benzene and toluene recovered from the overhead will be sent to the B/T splitter, and the benzene then sent back to the alkylation reactor. The bottoms from the EB/SM column go to the styrene finishing column, where styrene is taken overhead and heavies are rejected. Badger offers a similar styrene technology in collaboration with ExMo.

2.11.3 Cumene/Phenol/Bisphenol-A

Polycarbonates represent the third major outlet for benzene, and this value chain requires high-purity cumene, the product of alkylation of benzene with propylene. As with ethylbenzene, this reaction is best performed in the liquid phase with a zeolitic catalyst. The UOP QMax™ process serves as an illustrative example. This technology uses an extremely efficient and robust zeolitic catalyst with a multiple bed downflow reactor and inter-bed propylene injection under liquid-phase conditions. The benzene:propylene ratio is optimally around 2.0 with today's modern catalysts. UOP technology produces very little n-propyl benzene by-product. Figures 2.44 and 2.45 show the typical chemistry, while Figure 2.46 shows a typical flow scheme.

A certain amount of reactor effluent recycle is used as a means of optimizing reactor temperatures. Effluent from the alkylation reactor goes to the benzene column, where unreacted benzene is taken as a side cut and recycled to the alkylation as well as transalkylation reactors. The bottoms of the benzene column contain cumene and polyisopropyl benzene isomers, and is sent to the cumene

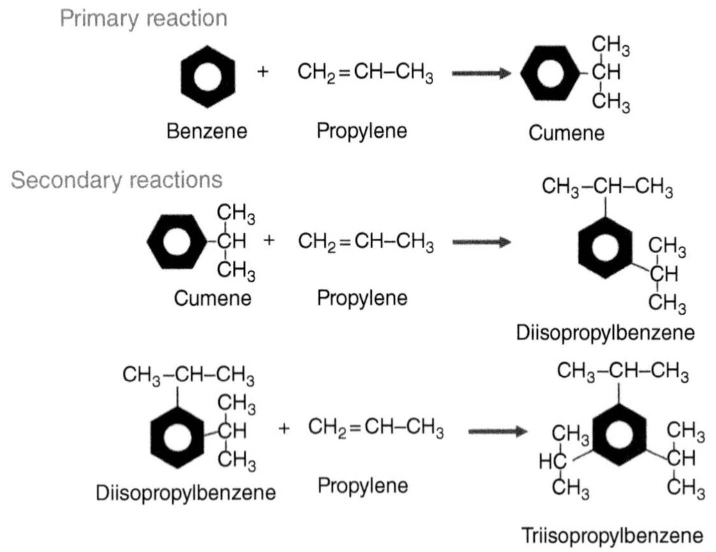

Figure 2.44 Cumene reaction network.

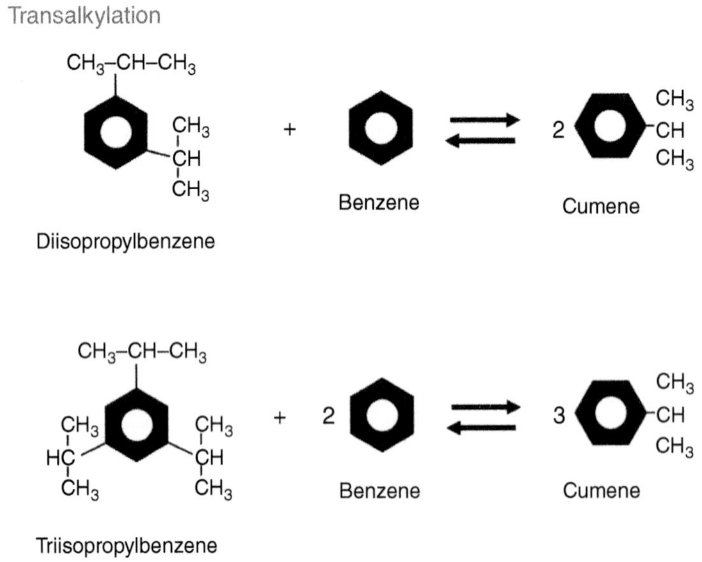

Figure 2.45 Cumene transalkylation reactions.

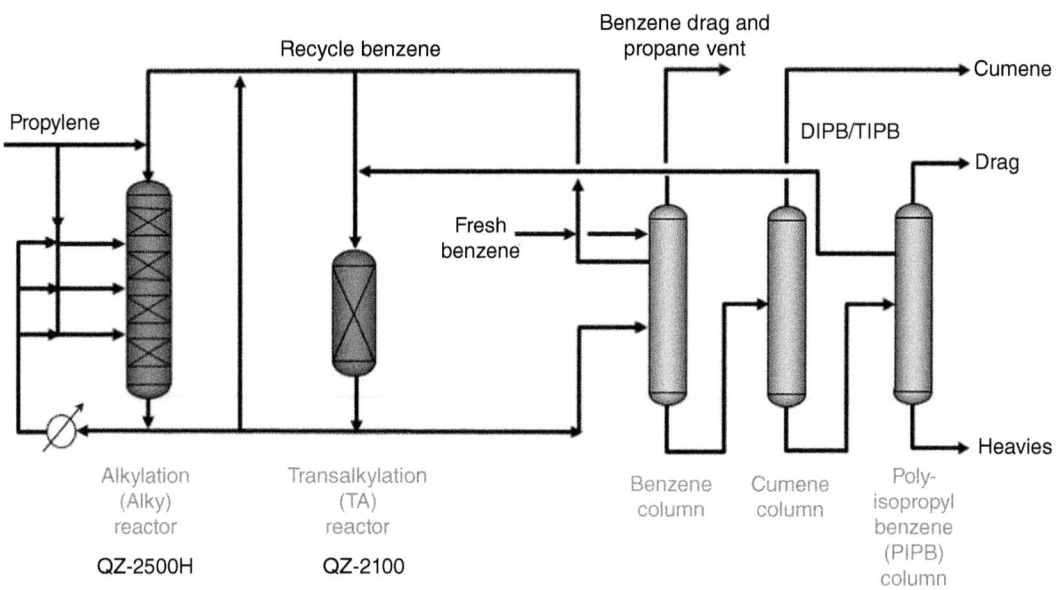

Figure 2.46 Process flow diagram for the UOP QMax process.

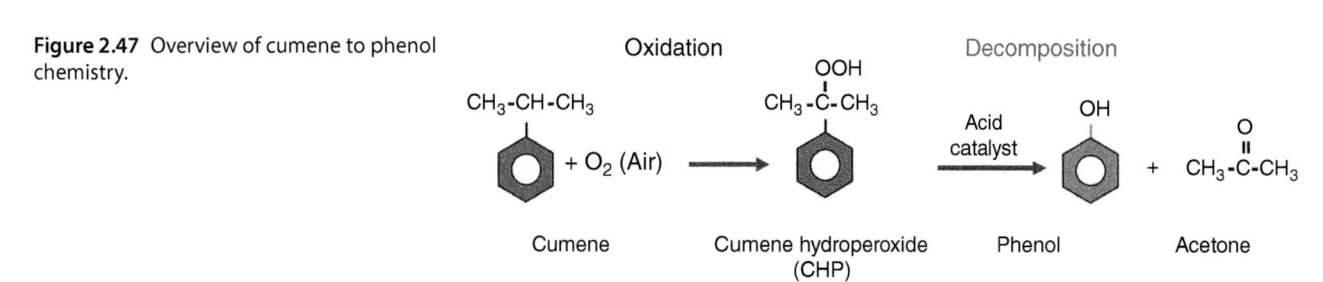

Figure 2.47 Overview of cumene to phenol chemistry.

recovery column where cumene product is recovered overhead. The bottom stream from this column is sent to the poly isopropyl benzene column where the PIB's are separated as a side cut and sent to the transalkylation reactor, where they are converted back to more cumene. A bottom stream from the poly-isopropyl benzene column contains heavies such as di-phenyl propane, and sent to disposal. The cumene product is greater than 99.9% pure with near stoichiometric yields. Badger offers Alkymax technology that has a similar flow scheme.

Most cumene is converted into phenol and acetone using autothermal oxidation technology such as the UOP 3G Phenol™ process. The basic chemistry of phenol production is summarized in Figures 2.47 and 2.48.

A block flow diagram of the 3G Phenol process is given in Figure 2.49.

Cumene is fed to a medium pressure oxidizer where it reacts with air to produce cumene hydroperoxide (CHP). The CHP is then concentrated in pre-flash and flash columns, with the unconverted cumene being returned to the oxidizer. The CHP is then sent to a decomposition step where dilute acid is used to control the decomposition to yield primarily phenol and acetone. A dehydration step is included to further augment the phenol yield. As shown in Figure 2.50, there are also some side reactions in the decomposition step that may form alpha-methyl styrene, which can be a desirable by-product, or alternatively can be selectively hydrogenated back to cumene and then recycled to oxidation.

The dehydrator effluent is then sent to neutralization, and finally to the acetone and phenol purification and recovery section. UOP has made significant improvements to phenol technology in the 3G process, saving 20% CAPEX, and 25% in utilities while minimizing effluents. Phenol technology is also supplied by KBR. Most phenol is converted along with some of the acetone into bisphenol-A. Bisphenol-A is polymerized with phosgene to produce polycarbonate. Chemistry for these routes is summarized in Figures 2.51 and 2.52.

2.11.4 Linear Alkyl Benzene Sulfonate for Detergents

A fourth significant use of benzene is in the production of linear alkyl benzene (LAB), which is a precursor for

Oxidation reactions

Main reaction

Cumene + O₂(Air) → Cumene hydroperoxide (CHP) — UOP phenol selectivity (mol.%): 95.7

By-product reactions

- Dimethylphenylcarbinol (DMPC) — 3.5
- Acetophenone (ACP) + H–C–OH + H₂O — 0.3
- Other by-products — 0.5

Figure 2.48 Cumene oxidation reaction network.

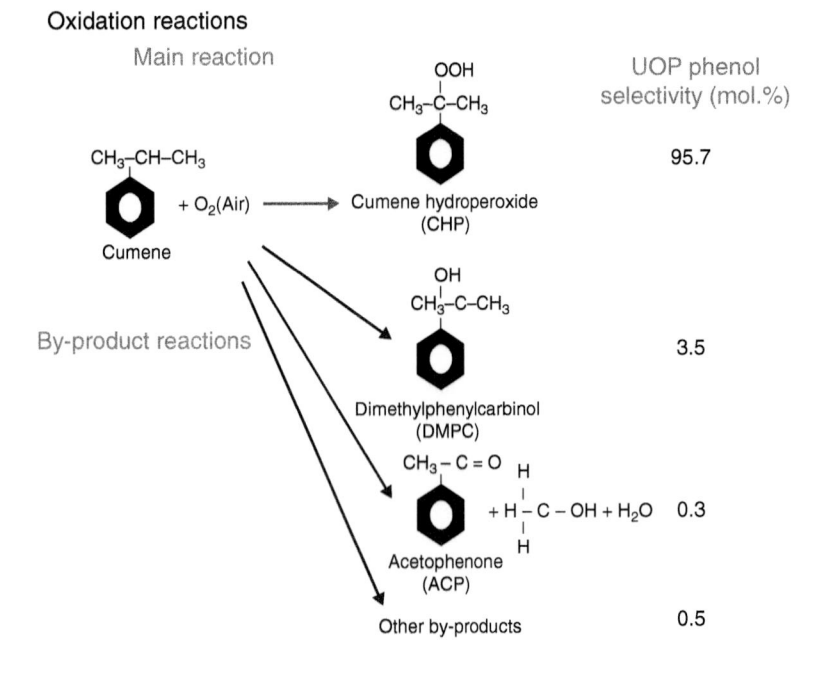

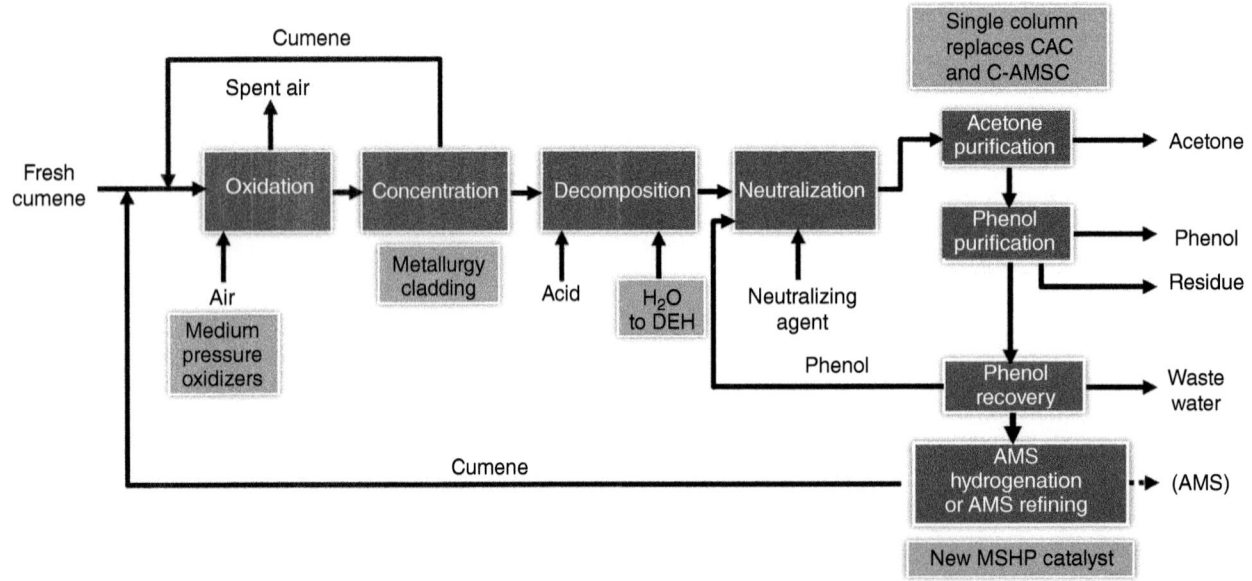

Figure 2.49 Block flow diagram of UOP 3G Phenol process.

biodegradable detergents. In this application, benzene is alkylated with linear olefins in the C_{10}–C_{13} range to form LAB. This use accounts for more than 1 million tons per year of benzene and more than 3 million tons per year of LAB. This technology was first developed and commercialized by UOP, using HF as the alkylation catalyst. Modern LAB is produced using high-performance zeolite catalysts which avoid the environmental and maintenance issues associated with HF. The LAB is sulfonated and then used in biodegradable detergent formulations. Over 90% of the world's biodegradable detergents are produced using UOP technology that starts with kerosene and benzene as the raw materials, as shown in Figures 2.53 and 2.54.

2.11.5 Oxidation of *para*- and *meta*-Xylene

The derivatives of the xylene isomers involve oxidation to more reactive species while preserving the desirable chemical structures of the individual isomers. As noted earlier, the most prevalent is oxidation of *para*-xylene to produce PTA. Several companies

Figure 2.50 Cumene oxidation by-product reactions.

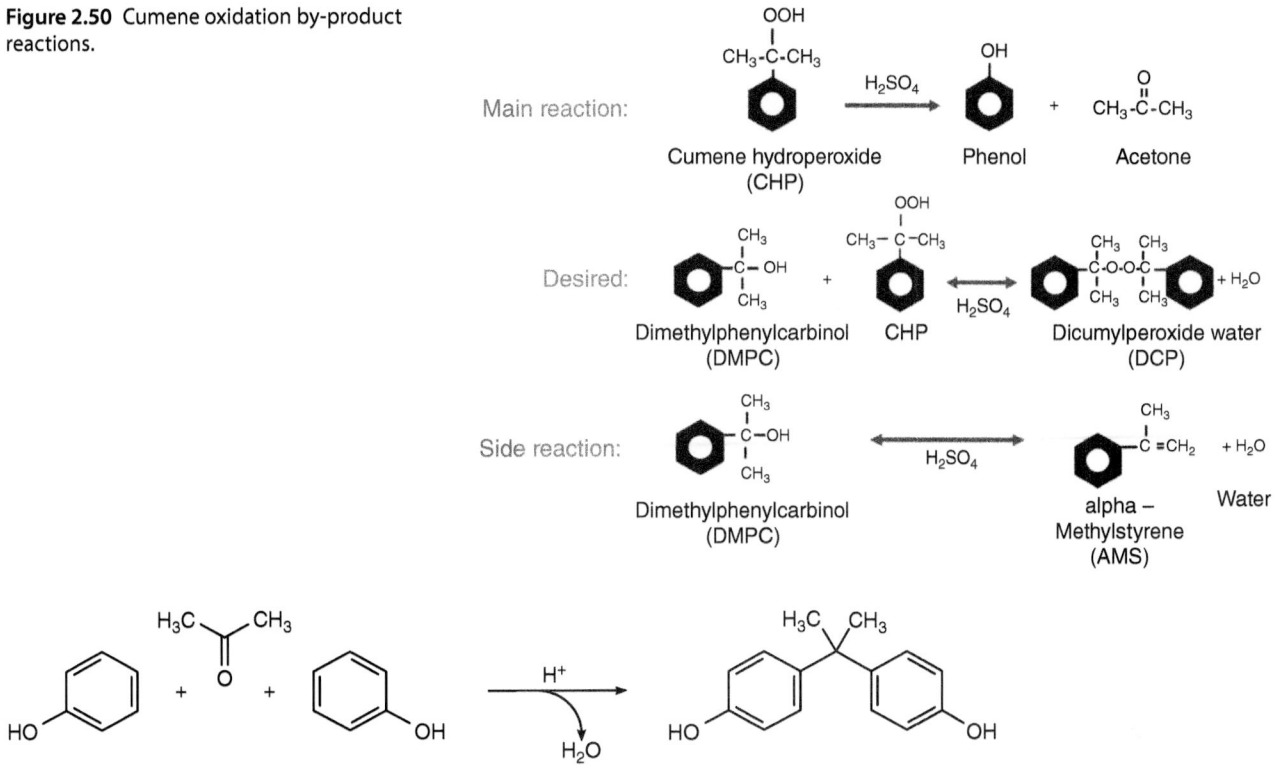

Figure 2.51 Bisphenol-A reactions.

Figure 2.52 Reactions for bisphenol-A to polycarbonate.

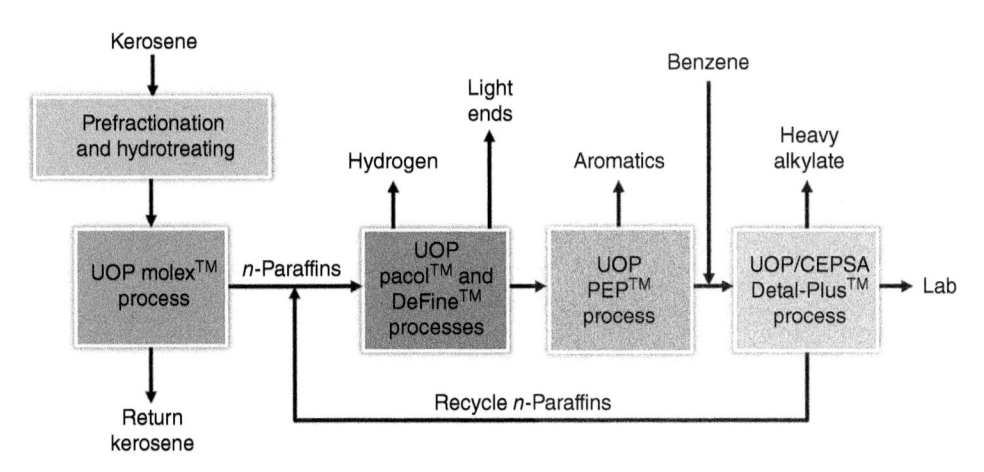

Figure 2.53 Block flow diagram for linear alkyl benzene production.

supply this technology to the industry, including Dow, Davy, Invista, BP, Chemtex, Mitsubishi, Hitachi, and many others. The key chemical steps are shown in Figure 2.55.

para-Xylene is oxidized using air to para-tolualdehyde, then to para-toluic acid, then 4-carboxy benzaldehyde, and finally to crude terephthalic acid (CTA). This happens stagewise in titanium clad stirred tank reactors,

Figure 2.54 Sulfonation chemistry for LAB to LAS.

Figure 2.55 *para*-Xylene oxidation reactions. *Source:* Wikipedia.

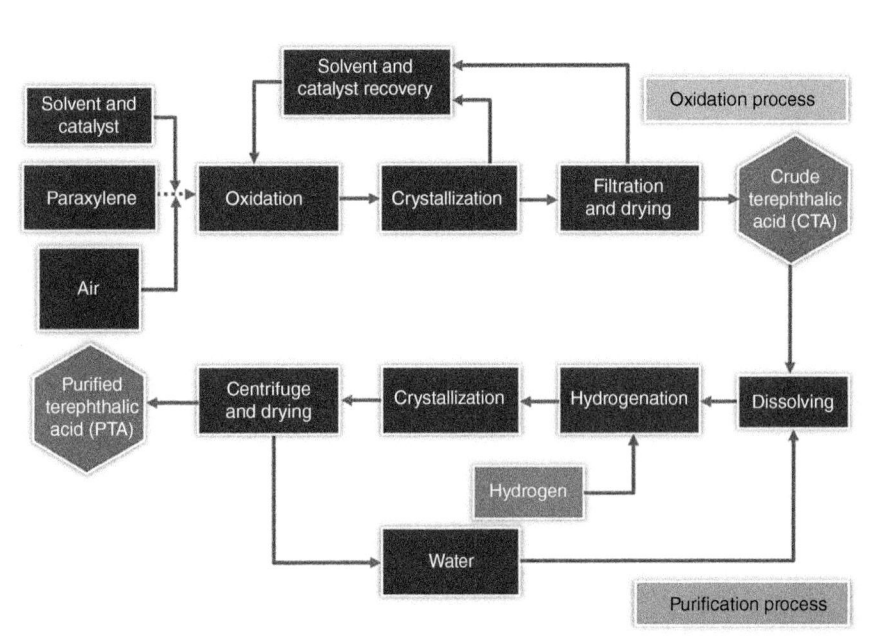

Figure 2.56 PTA process block flow diagram.

using a homogeneous catalyst composed of cobalt, manganese, and bromide in an acetic acid mixture. There can be additional metals in the oxidation catalyst, depending on the technology provider. A simplified block flow diagram is shown in Figures 2.56 and 2.57.

Multistage oxidation cascades the reactants from one stirred tank to another, with the heat of reaction governed by controlled evaporation and recovery of the acetic acid. The catalysts and operating conditions are chosen to minimize the "burn" of acetic acid to CO_2. As the terephthalic acid forms, it begins to crystallize out of solution, creating three-phase conditions in the last stirred tank reactor. As the CTA crystals form, some inclusion of impurities takes place, primarily residual 4-CBA, which must be dealt with by further processing. Ultimately, the CTA crystals must be separated from the mother liquor, usually by filtration or centrifugation and washing, then drying. This allows the mother liquor containing the homogeneous catalyst and any unconverted intermediates in solution to be recovered and recycled back to the oxidation section, while the CTA is dried and sent to the purification section. In the purification section, the CTA is redissolved in water and then subjected to selective hydrogenation over a Pd-carbon catalyst that converts the 4-CBA to *p*-toluic acid, which remains soluble and can be recycled back to oxidation. A series of crystallization, filtration, centrifugation, washing, and drying steps are then performed, after

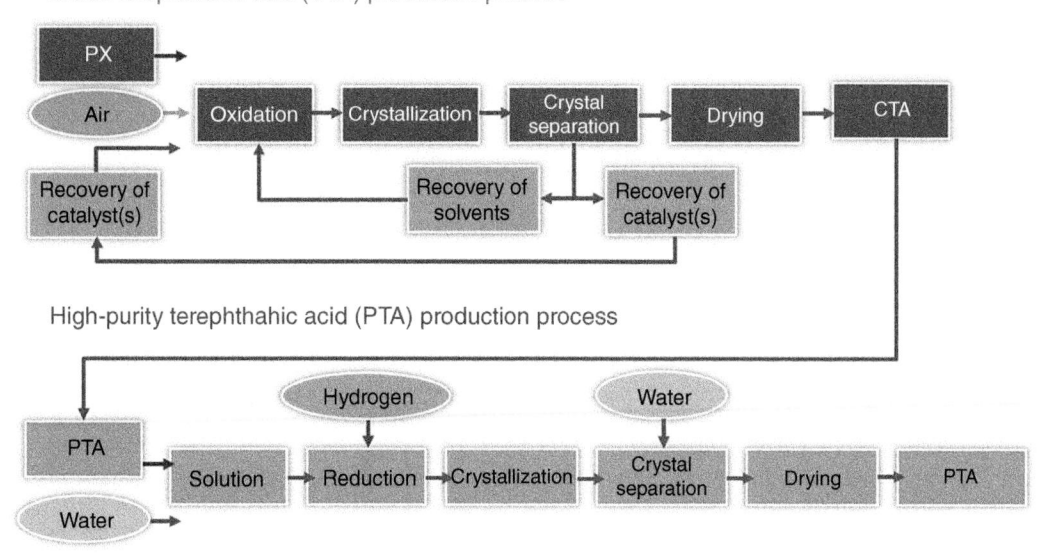

Figure 2.57 Hitachi PTA process.

Figure 2.58 *meta*-Xylene oxidation to iso-phthalic acid. *Source:* De Gruyter.

which PTA is obtained. There are numerous specifications that need to be met by the PTA – chemical impurities as well as particle size. PTA will be 99.99% pure and contains less than 25 ppm 4-CBA, whereas medium quality terephthalic acid (MTA) purity will be 99.90% and contains up to 400 ppm 4-CBA. Since 4-CBA is usually associated with providing a yellow tint to the PET, most PET processors prefer PTA as the starting material.

As noted in an earlier section, *meta*-xylene is predominantly converted to PIA for use as a comonomer in PET resin. The chemical steps in PIA production are summarized in Figure 2.58.

There are many similarities between the oxidation/purification technologies that produce PTA and PIA, but in general, the production of PIA is simpler. This is primarily due to the fact that the crude iso-phthalic acid (CIA) after *meta*-xylene oxidation remains soluble until the crystallization step. In fact, PTA producers, such as Lotte in Korea, have diverted one or more trains of their PTA plants to production of PIA. Companies such as Lonza and Eastman have attempted to improve the PIA process (Buford and Tennant 2010). Once the CIA is formed and crystallized, it is dried, re-slurried in water, and subjected to selective hydrogenation to remove any 3-carboxy benzaldehyde impurity that may have coprecipitated with the isophthalic acid crystals.

Recall that *para*-xylene production has benefitted from economy of scale – going from single-train capacities of 80 000 MTA in the early years to 500 000 MTA in the 1990s and now 2.5–3 MM MTA single-train plants with some recent advances by UOP. Similarly, PTA technology has tracked along this trend with the most recent plants licensed by Invista at 2.5 MM MTA (Invista PTA Reference List 2018).

2.11.6 Melt-Phase Polymerization of PTA to PET

For PET fibers the PET is a homopolymer, meaning PTA is the only di-carboxylic acid that is used. The PTA is esterified with ethylene glycol and fed to a continuous polymerization plant, where esterification and transesterification take place at 3–6 barg pressure and 230–260 °C. The basic chemistry is shown in Figure 2.59.

Several companies provide this technology, including Invista, Technip-Zimmer, Uhde Inventa-Inventa, Hitachi, and China Kunlun. A typical continuous polymerization flow scheme is shown in Figures 2.60 and 2.61.

Ethylene glycol and PTA are continuously fed to a slurry mix tank. This slurry is then combined with

2 Market and Technology Overview

Figure 2.59 PET polymerization chemistry.

Terephthalic acid + Ethylene glycol → PET + Water

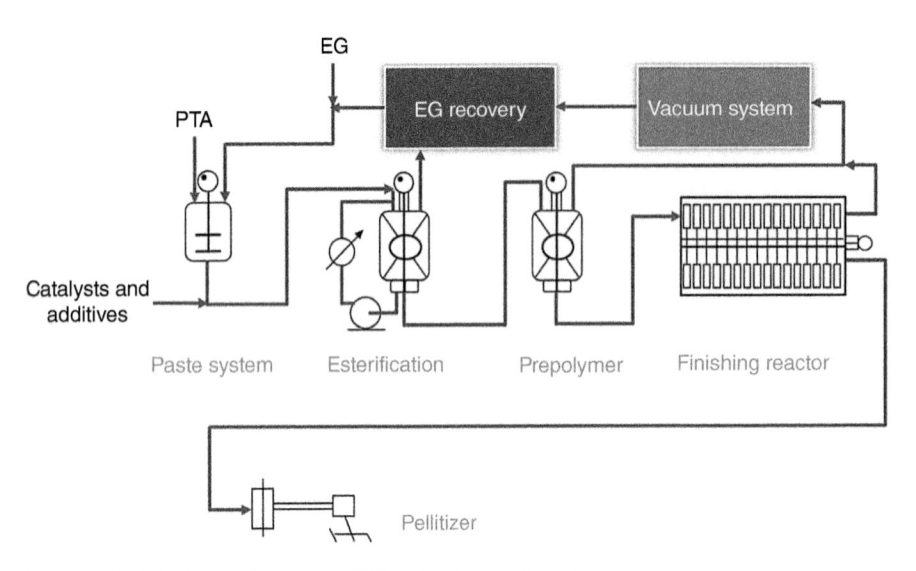

Figure 2.60 Block flow diagram for PET melt-phase polymerization.

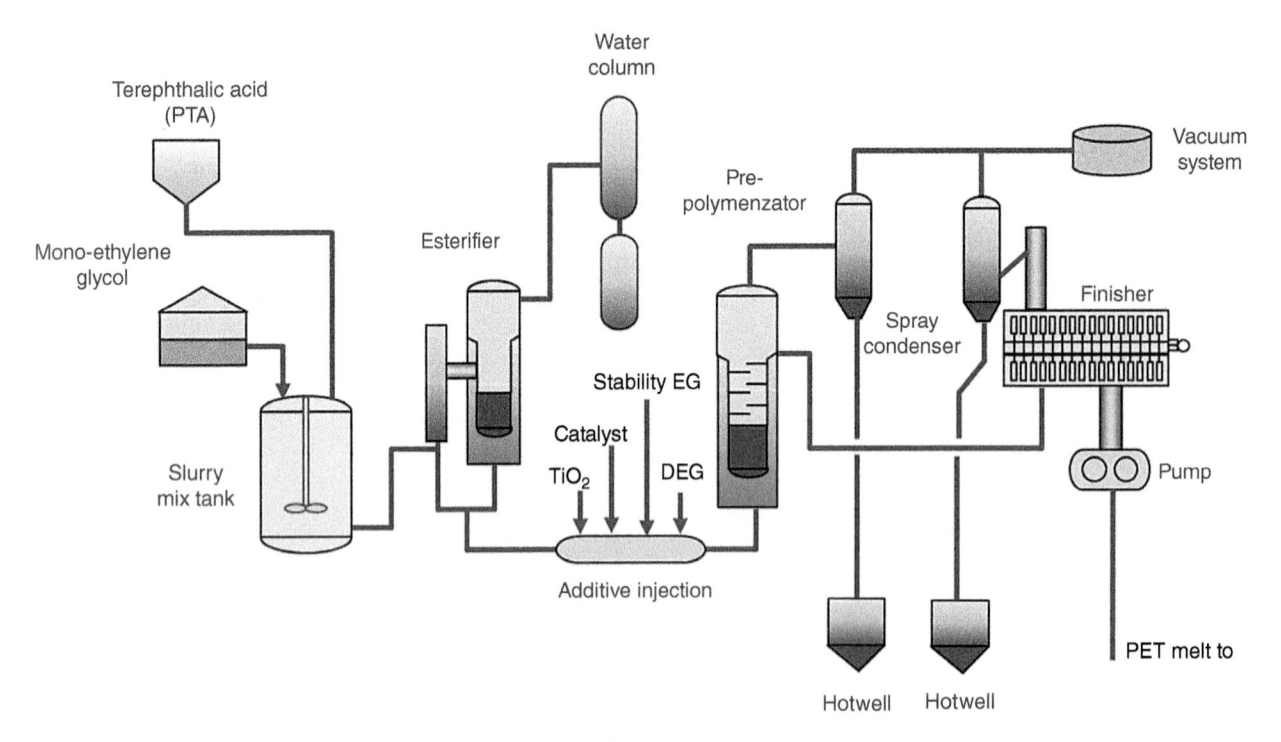

Figure 2.61 Catalyst addition for melt-phase polymerization of PET. *Source:* CPMA India.

a precise amount of catalyst, usually an antimony compound, such as an oxide, acetate, or glycolate. This mixture then is fed to the esterifier, where water byproduct is removed by fractionation. The product of the esterification step goes to a pre-polymerizer where low MW PET is made, and then to the finisher, which operates under vacuum and is designed to handle increasingly viscous liquid as polymerization proceeds, continuously removing water and ethylene glycol, so as to drive the polymerization reactions forward. These plants typically use rotating discs or similar equipment, that become coated with polymer from the bulk melt phase that is moving toward the outlet of the finisher. The discs achieve a thin coating of polymer while allowing mass transfer of excess EG to escape to the vapor phase. The polymer remaining on the disc then becomes remixed with the bulk polymer melt and continues to move toward the finisher outlet. This is intended to prevent dead zones or back-mixing that could cause side reactions to undesirable by-products, such as diethylene glycol or acetaldehyde, while growing to the target MW. As the PET approaches the outlet of the finisher, its MW and viscosity significantly increase. Some PET plants will have two finishers, operating at different pressures and temperatures to maximize the availability of end-groups to complete the polymerization. The outlet typically will have a degree of polymerization between 100 and 140, depending on the end-use. Product quality is usually expressed in terms of intrinsic viscosity (IV). The IV for fiber-grade PET is 0.57–0.65, corresponding to a MW between 17 000 and 21 000. Exiting the finisher, the fiber-grade PET will be extruded and cut into fiber-grade PET chips. These chips can be sent to fiber plants where they would be extruded and drawn into fibers where the amorphous and crystalline phases of the PET are reorganized, to impart the high tensile strength needed for fiber applications.

2.11.7 Melt-Phase Polymerization and Solid State Polycondensation of PET Resin

For PET resins, the PET is a copolymer. The continuous polymerization process is very similar to that described above for PET fiber, except that a comonomer is used, namely PIA. This is usually in the range of 2–3 wt.%, and is done in order to reduce the crystallinity of the PET in order to achieve the greater optical clarity that is needed for bottle applications, while enabling rapid-cycle injection molding and blow-molding operations. The typical IV for bottle-grade resin from continuous polymerization is in the range of 0.4–0.6. Rather than being drawn into fibers, the PET resin will be extruded and then either strand-cut or cut using underwater melt cutting (UMC) to produce PET chips that can then be shipped and processed in solid state polycondensation (SSP) plants to increase the molecular weight to that which would be suitable for bottles.

SSP, as the name implies, achieves the increase in MW by processing the chips at temperature and long residence time in an inert environment. Several changes take place during

- SSP operations
- MW increase
- Crystallinity increase
- Removal of acetaldehyde and light cyclic oligomers
- The MW growth occurs via continued esterification with elimination of ethylene glycol, as well as transesterification with removal of water. These chemical reactions are shown in Figure 2.62.

Since many of these reactions are equilibrium controlled, continuous removal of water and organics is needed. UOP-Sinco, Polymetrix AG, and Uhde Inventa Fisher are the principal suppliers of SSP technology. Flow schemes and details between these technologies will vary, but for purposes of this discussion we will focus on UOP's SSP technology. Two configurations are shown in

Figure 2.62 Solid-state polymerization reactions.

$$PET - COO - CH_2 - CH_2 - OH + HO - CH_2 - CH_2 - OOC - PET$$
$$K_{1,reverse} \uparrow\downarrow K_{1,forward} \quad (1)$$
$$PET - COO - CH_2 - CH_2 - OOC - PET + HO - CH_2 - CH_2 - OH \uparrow$$

$$PET - COOH + HO - CH_2 - CH_2 - OOC - PET$$
$$K_{2,reverse} \uparrow\downarrow K_{2,forward} \quad (2)$$
$$PET - COO - CH_2 - CH_2 - OOC - PET + H_2O \uparrow$$

$$PET - COO - CH_2 - CH_2OOC - PET$$
$$\downarrow K_3 \quad (3)$$
$$PET - COOH + CH_2 = CH - OOC - PET$$

$$PET - COO - CH = CH_2 + HO - CH_2 - CH_2 - OOC - PET$$
$$\text{Vinylesterend}$$
$$\downarrow K_4 \quad (4)$$
$$PET - COO - CH_2 - CH_2 - OOC - PET + CH_3CHO \uparrow$$
$$\text{Acetaldehyde}$$

$$PET - COO - CH = CH_2 + H_2O$$
$$\text{Vinylesterend}$$
$$\downarrow K_5 \quad (5)$$
$$PET - COOH + CH_3CHO$$

Figure 2.63 Flow schemes for UOP Sinco solid state polymerization (SSP).

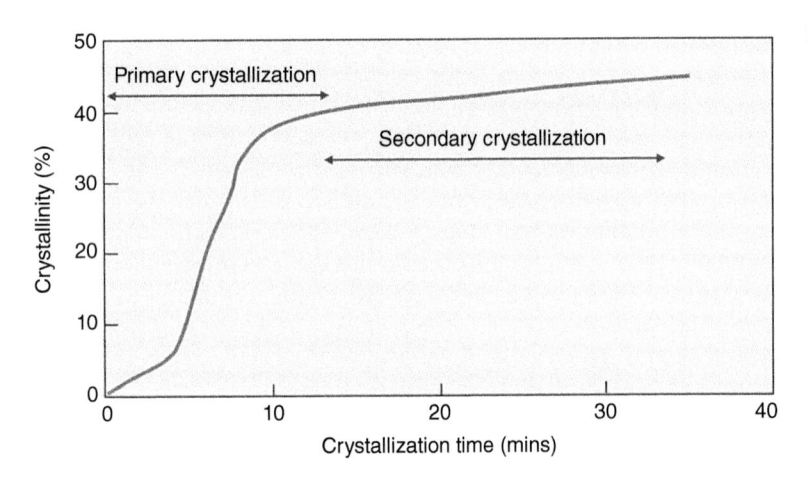

Figure 2.64 SSP crystallinity development.

Figure 2.63 – one based on strand cutting, the other based on UMC.

The UMC configuration reduces CAPEX and OPEX due to elimination of redundant steps and the use of energy integration. Referring to the UMC configuration, the chips are first crystallized, then preheated and hot-lifted to the top of the SSP reactor. Here, the chips flow downward by gravity, while hot nitrogen gas flows upward. Process temperatures are in the range of 200–220 °C, with residence time varied to meet target MW and IV. Ethylene glycol, acetaldehyde, oligomers, and water continually diffuse out of the chips and into the bulk gas phase. The gas exiting the reactor is sent through catalytic oxidation and drying to remove these by-products and drive the reactions to completion. Crystallinity develops rapidly, as shown in Figures 2.64 and 2.65, while the MW increase proceeds much more slowly over several hours as shown in Figure 2.66.

The product properties reach an IV between 0.72 and 0.85 (24 000–31 000 MW), with the higher range suitable for carbonated soft drink bottles, and the lower range for non-carbonated beverages. Crystallinity is typically greater than 50% and the acetaldehyde level below 1 wt. ppm The acetaldehyde specification is particularly important to ensuring no after-taste in water bottles made from PET. These chips will have an opaque appearance due to the increased crystallinity. When the chips are melted and sent to injection molding, they will produce preforms that are very clear. The preforms can then be sent to blow-molding machines where the finished bottles are made (SPE-ANTEC 1997).

2.11.8 Oxidation of *ortho*-Xylene

ortho-Xylene is primarily used in the production of phthalic anhydride (PA) via selective oxidation. This oxidation is significantly more difficult than what is practiced in the production of PTA and PIA. A simplified depiction of the chemistry is provided in Figure 2.67.

It is important to note that the catalysts, reactor designs, and operating conditions need to be very carefully considered due to the potential for explosive compositions and hot spots. This oxidation is done in the vapor phase using a vanadium-containing catalyst, often with other metal promoters such as antimony and cesium. Multi-tube reactors are preferred by several companies, with the heat of reaction being managed by an external molten salt heat transfer medium. Extreme care is taken in the loading of the thousands of individual reactor tubes in these plants, as well as in detection and avoidance of hot spots to prolong catalyst life (Altwasser et al. 2014; Cavani et al. 2009).

2.12 Integrated Refining and Petrochemicals

Most refiners participate solely in the fuels market, producing gasoline, diesel, jet fuel, and maritime oil, trying to maximize profits in the differentials between crude oil and these fuels. Regulations for cleaner fuels, seasonal changes in demand, as well as long-term shifts in demand can make the problem of maximizing profitability more difficult. A number of companies serve the fuels and petrochemical markets, but treat them as separate entities, using transfer prices of petrochemical feedstocks and by-products to establish their economics. As shown in Figure 2.68, by 2030, the number of vehicles is expected to double from 750 million to more than 1.5 billion, with the vast majority being internal combustion engines that will consume gasoline.

However, this graph also shows steady improvements in fuel economy over this period. Figure 2.69 shows that the worldwide demand for refined products has a CAGR of about 1.3%, reducing to less than 0.5% by 2030. In contrast, the demand growth for olefins is very robust, with a CAGR of 3–6%, depending on region.

Aromatics demand is similarly strong. With this scenario, it is not surprising to see fuel producers seek opportunities

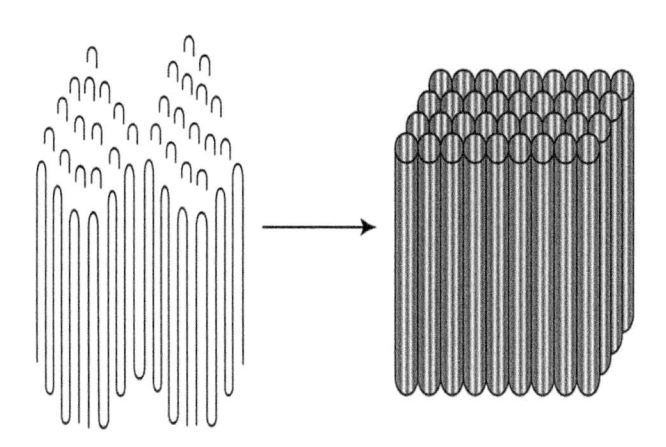

Figure 2.65 SSP PET crystallinity development – molecular view.

Figure 2.66 Pet intrinsic viscosity (IV) increase in SSP reactor.

Figure 2.67 Oxidation chemistry of *ortho*-xylene.

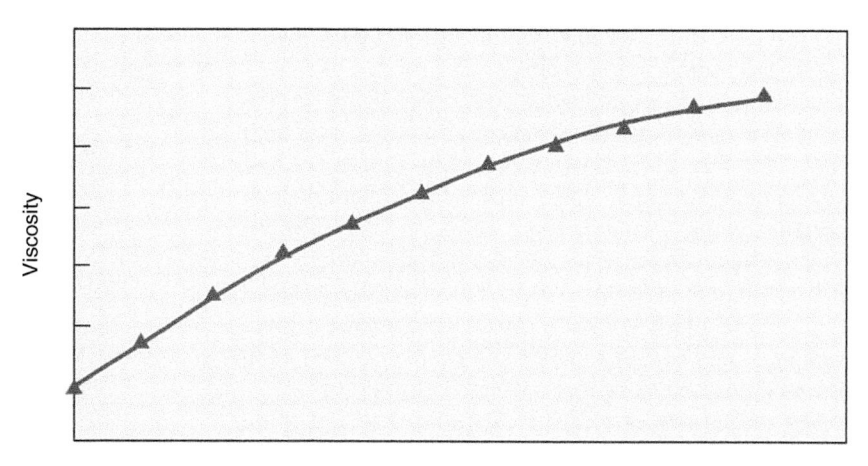

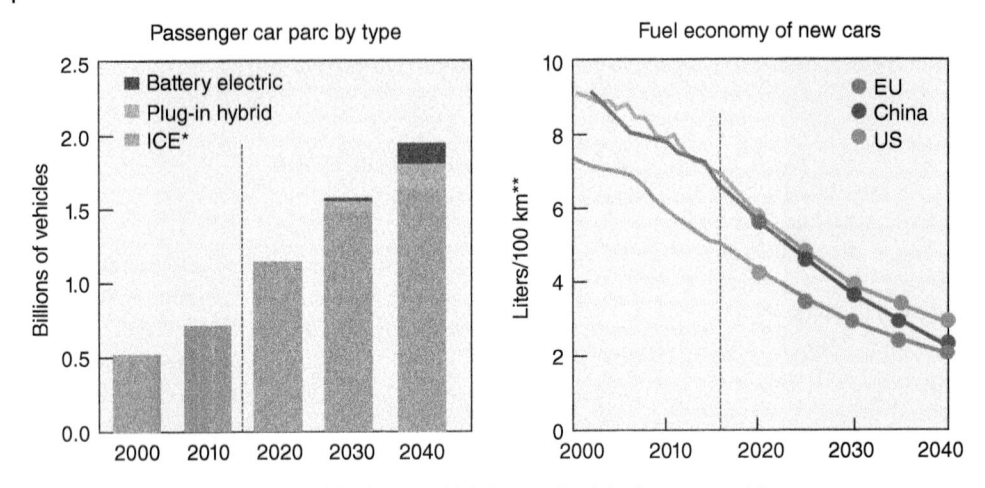

Figure 2.68 Passenger car usage worldwide – 2000–2040. *Source:* 2018 BP Energy Outlook.

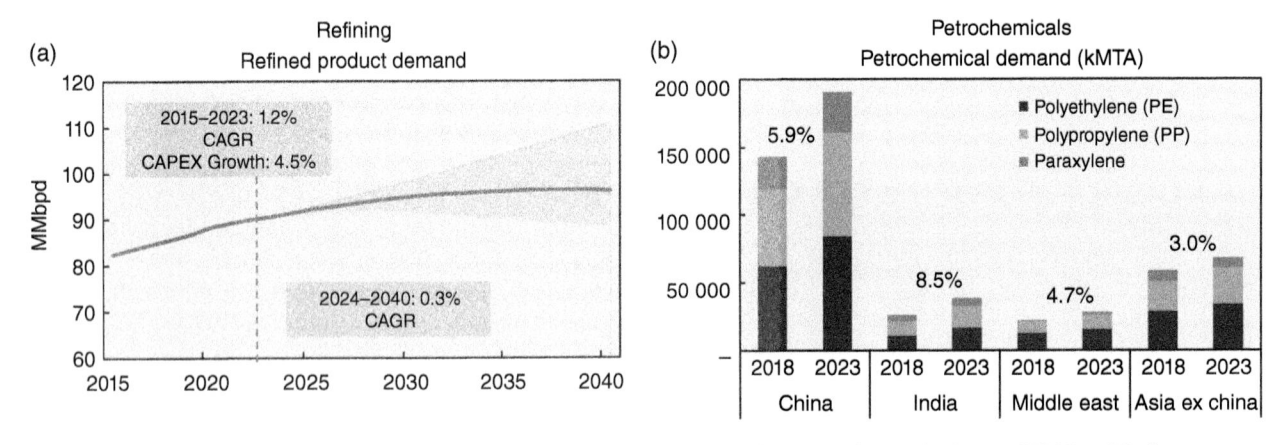

*ICE vehicles includes hybrid vehicles which do not plug into the power grid
**Based on the NEDC (New European Drive Cycle), gasoline fuel

Figure 2.69 Capacity growth rates for (a) refined products (*Source:* IHS, FGE) and (b) petrochemicals (*Source:* IHS, WoodMac).

to increase their margins by shifting toward petrochemical products. Figure 2.70 shows that this can increase the margin per barrel of crude oil by about 40% by shifting some focus to producing petrochemicals.

This shift from refining only to refining plus petrochemicals is depicted in Figure 2.71.

Looking to the future, some companies aspire to produce petrochemicals directly from crude oil, with zero fuels, but this could prove to be capital intensive and impractical, without some significant technological breakthroughs. Certainly there has been and will continue to be much emphasis on understanding the molecular structures of the heavier fractions. This knowledge requires advanced analytical capabilities, and it can then lead to more intelligent design of catalysts, as well as to the process conditions under which they are run.

Most refineries have FCC plants whose main purpose has been to produce gasoline from vacuum gas oil. Light olefins are often sent to alkylation to produce high-octane blending components. Many FCC operators

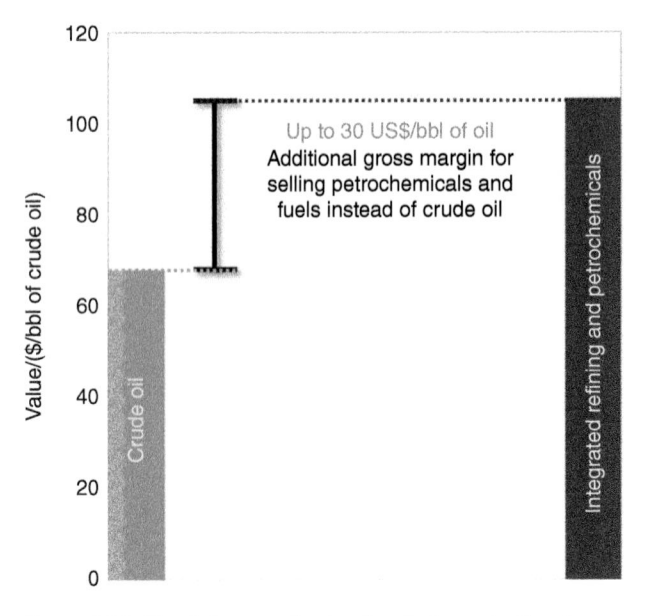

Figure 2.70 Margin impact of petrochemicals production for refinery operations.

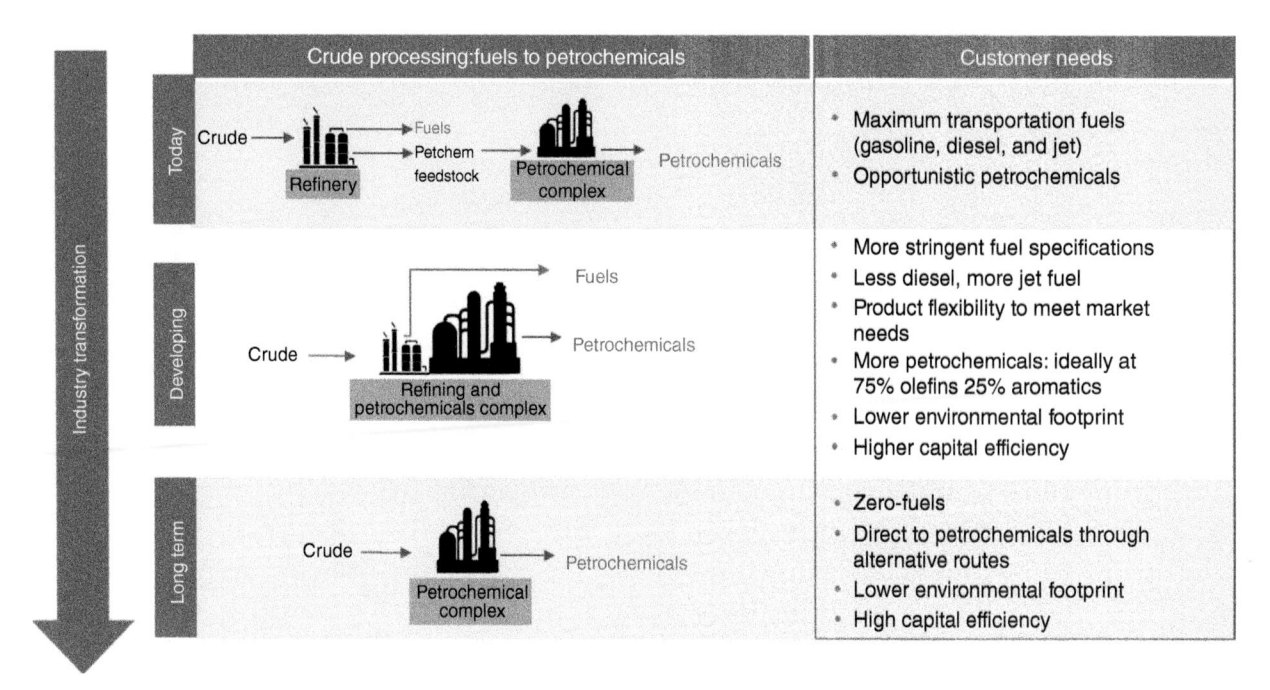

Figure 2.71 Shift in product slate for refinery of the future.

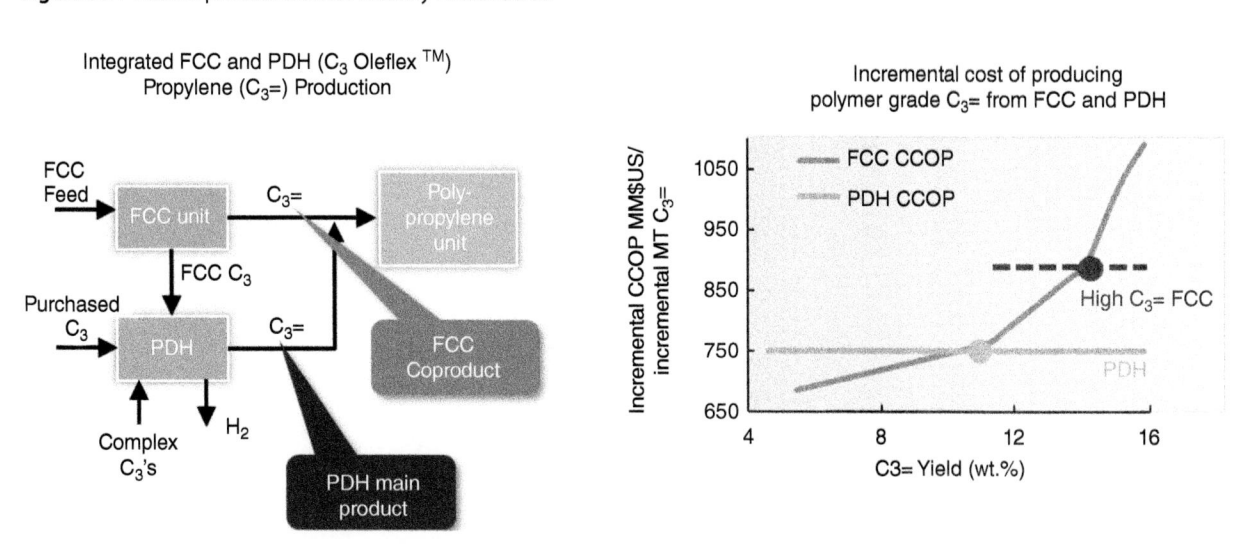

Figure 2.72 Deriving increased petrochemical value from refinery streams.

have chosen to produce propylene from the FCC unit, cutting into the production of gasoline by adding MFI-based catalysts. The propylene yield from FCC can vary from below 10% to more than 15%, dependent on catalyst and operating conditions. With installation of a propane/propylene splitter and appropriate cleanup, polymer-grade propylene can be produced. Propylene production from the FCC unit can increase the margins of the overall plant, but at propylene yields above about 12%, the cost of production becomes much higher than propane dehydrogenation due to the balance of product and by-product values. A fully integrated refinery could collect propane by-product from the units in the refinery, supplement with external propane, and use propane dehydrogenation to produce even more propylene at lower cost, in addition to the propylene made from FCC, as shown in Figure 2.72.

This brings us to the term "molecule management" which has become increasingly important to those who wish to achieve the greatest synergy between refining and petrochemical operations. Figure 2.73 gives an example of how the C_3–C_{10}+ molecules might be managed to maximize the synergy between refining and petrochemicals.

In this case, normal paraffins are segregated and fed to a steam cracker to produce ethylene and propylene. This can be done using UOP's Maxene process, which employs a simplified Sorbex flow scheme. There can even be allowance to isomerize the *iso*-paraffins back to normal to

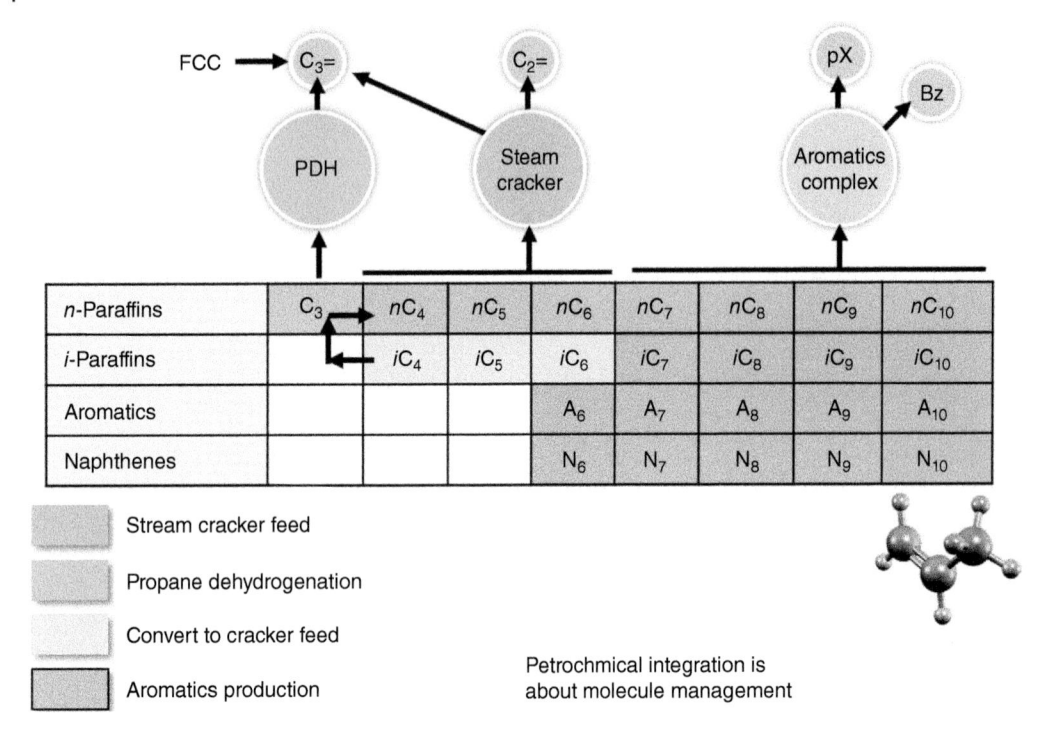

Figure 2.73 Molecule management for maximizing petrochemical yields.

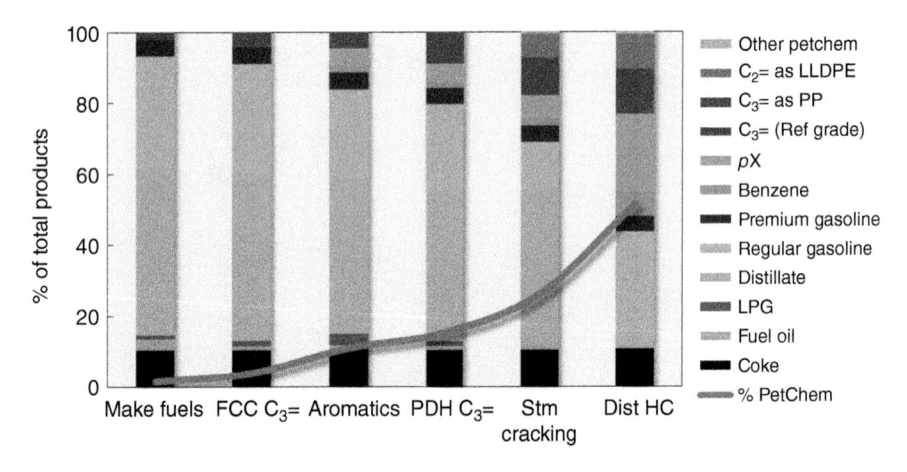

Figure 2.74 Transition from refining to refining plus petrochemicals.

maximize this benefit. The cyclic C_6 hydrocarbons, along with the C_7+, are sent to catalytic reforming to maximize aromatics production. The yields of aromatics in catalytic reforming are enhanced by the increased cyclic content of the naphtha that results from the light normal (and isomerized *iso-*) paraffins being sent to the steam cracker. By-product propane along with supplemental external propane may be fed to a propane dehydrogenation unit, such as the UOP Oleflex™ process, to produce propylene that can be blended with the propylene from the FCC unit.

Configuration studies for new facilities are becoming increasingly easier to perform with modern simulation tools such as HYSYS. In the early stages of project planning, it is possible to evaluate numerous scenarios that can start from a product slate that is 100% fuels, and stepwise add processing capacity to increase the proportion of petrochemical products, such as ethylene, propylene, benzene, and *para-*xylene. This transformation in product mix from 5% to more than 50% petrochemicals is depicted in Figure 2.74.

These configurations can then be evaluated against pricing and supply–demand scenarios for the various fuel and petrochemical products to find the most attractive and beneficial investment. In some cases, a staged investment or pre-investment in particular enabling technologies might be the best option.

References

Abichandani, J. and Venkat, C. (1996). Dual bed xylene isomerization. US Patent 5,516,956.

Altwasser, W., Zihlke, J., Dobner, K., and Rosokski, F. (2014). Process for controlling a gas phase reactor for preparation of phthalic anhydride. US Patent 8,901,320.

Broughton, D. (1984). Production scale adsorptive separations of liquid mixtures by simulated moving bed technology. *Separation Science and Technology* 19: 723–736.

Broughton, D. and Gerhold, C. (1961). Continuous sorption process employing fixed bed of sorbent end moving inlets and outlets. US Patent 2,985,589.

Buford, R. and Tennant, B. (2010). Simplified isophthlic acid process. US Patent 7,714,094.

Carson, D. and Haufe, T. (1968). Process of separating ethylbenzene from C_8 aromatic hydrocarbons by super-distillation with vapor-recompression heat exchange. US Patent 3,414,484.

Cavani, F., Fumagalli, C., Leanza, R., Mazzoni, G., and Panzacchi, B. (2009). Titanium-vanadium-tin comprising catalyst and process for the preparation of phthalic anhydride. US Patent 7,557,223

Deischter, J., Schute, K., Neves, D. et al. (2019). Aromatization of bioderivable isobutyraldehyde over HZSM-5 catalysts. *Journal of Green Chemistry* 21: 1710–1717.

Egan, C. and Luthy, R. (1955). Separation of xylenes. *Ind. Eng. Chem.* 47 (92): 250–253.

Eriksson, N. (2013). *Production of Four Selected Renewable Aromatic Chemicals*. Gothenburg, Sweden: Department of Chemical Engineering, Division of Forest Products and Chemical Engineering, Chalmers University of Technology.

Forte, P. (1991). Method for aromatic hydrocarbon recovery. US Patent 5,073,669.

Gentry, J., Berg, L., McIntyre, J., and Wytcherly, R. (1995). Process to recover benzene from mixed hydrocarbons by extractive distillation. US Patent 5,488,284.

Hotier, G., Guerraz, C., and Thanh, T. (1995). Apparatus for the separation of *p*-xylene in C_8 aromatic hydrocarbons with a simulated moving bed adsorption and a crystallization. US Patent 5,401,476.

ICIS (2016). *Market Outlook: Phenol/Acetone Markets Are under Pressure*. ICIS. https://www.icis.com/resources/news/2016/06/09/10006764/market-outlook-phenol-acetone-markets-are-under-ressure-icis-consulting (accessed 11 April 2019).

IHS Markit (2018a). Toluene – Chemical Economics Handbook (CEH). IHS Markit. https://ihsmarkit.com/products/toluene-chemical-economics-handbook.html (accessed 11 April 2019).

IHS Markit (2018b). Styrene – Chemical Economics Handbook (CEH). IHS Markit. https://ihsmarkit.com/products/styrene-chemical-economics-handbook.html (accessed 11 April 2019).

IHS Markit (2018c). Ortho-Xylene – Chemical Economics Handbook (CEH). IHS Markit. https://ihsmarkit.com/products/ortho-xylene-chemical-economics-handbook.html (accessed 11 April 2019).

IHS Markit (2018d). Phenol – Chemical Economics Handbook (CEH). IHS Markit. https://ihsmarkit.com/products/phenol-chemical-economics-handbook.html (accessed 11 April 2019).

Innes, R., Zones, S., and Nacamuli, G. (1992). Liquid phase alkylation or transalkylation using zeolite beta. US Patent 5,081,323.

Invista PTA Reference List (2018). file:///C:/Users/E340947/Downloads/2018_10_05-PTA%2520Web%2520site%2520reference%2520list%2520-%2520October%25202018.pdf.

Inwood, T., Wright, C., and Ward, J. (1984). Liquid phase alkylation and transalkylation process. US Patent 4,459,426.

Johnson, J. and Hilder, G. (1984). Cyclar process for LPG to aromatics. *NPRA* (March 1984).

Kalfadelis, C. (1956). Recovery of aromatic hydrocarbons from HF-BF3 agent. US Patent 2,773,107.

Keown, P., Meyers, C., and Wetherold, R. (1973). Vapor phase alkylation in the presence of crystalline aluminosilicate catalyst with separate transalkylation. US Patent 3,751,504.

Kushnerick, J., Marler, D., McWilliams, J., and Smith, C. (1991). Process for preparing short chain alkylromatics. US Patent 4,992,606.

Mazanec, T. and Whiting, J. (2017). Fast catalytic pyrolysis with recycle of side products. US Patent 9,534,174.

Merien, E., Alario, F., Lacombe, S., Benazzi, E., and Jean, F. (2000). Catalyst containing a zeolite EUO and the use of the catalyst in a process for isomerizing aromatic compounds containing 8 carbon atoms per molecule. US Patent 6,057,486.

Nakano, T. (1971). Xylene separation and isomerization by HFBF3. Proceedings of the 8th World Petroleum Congress, Moscow (13–18 June).

Neuzil, R. (1968). Aromatic hydrocarbon separation by adsorption. US Patent 3,558,730.

Rodrigues, A. (2015). *Simulated Moving Bed Technology – Principles, Design and Technology Application*. Elsevier Science.

Ruthven, D. and Ching, C. (1989). Countercurrent and simulated countercurrent adsorption separation processes. *Chemical Engineering Science* 44 (5): 1011–1038.

Ryan, C. (2016). *An Overview of Gevo's Bio-Based Isobutanol Production Process*. Gevo Inc, https://gevo.com/wp-content/uploads/2018/02/isobutanol-process.pdf (accessed 11 April 2019).

Sachtler, A. and Lawson, R. (1996). Catalyst for isomerization of aromatics. US Patent 4,899,012.

Sharma, S., Gurevich, S., Riley, B., and Rosinski, G. (1990). Selective xylenes isomerization and ethylbenzene conversions. US Patent 6,143,941.

SPE (1997). *SPE-ANTEC Proceedings*, vol. 1–3. CRC Press.

The Essential Chemical Industry (2017). Polyurethanes – The Essential Chemical Industry. http://www.essentialchemicalindustry.org/polymers/polyurethane.html (accessed 11 April 2019).

Thirasak, A., Kamafoo, A., Tanthapanikachoon, W. et al. (2017). Process for the enhanced separation of ethylbenzene. US Patent Application 2017/0247303 A1.

VIRENT (2016). *Virent BIOFORMPX Paraxylene Used to Produce the world's First Plant Based Polyester Shirts.* VIRENT. http://www.virent.com/news/virent-bioformpx-paraxylene-used-to-produce-worlds-first-100-plant-based-polyester-shirts (accessed 11 April 2019).

Weeks Editor (1951). What's New More Xylenes Loom. Chemistry & Industry, 31 March, p. 17–19.

Whitchurch, P., Bogdan, P., and Bauer, J. (2010). Aromatics isomerization catalyst and isomerization process. US Patent 7,745,677.

Wood Mackenzie (2018). *Polyester and Raw Materials*. Wood Mackenzie, Edinburgh. https://www.woodmac.com/research/products/chemicals-polymers-fibres/polyester-pet-rpet/polyester-raw-materials (accessed 11 April 2019).

Yasauki, T., Ogawa, D., Kanaoka, M. et al. (1974). New *p*-xylene recovery and xylene isomerization processes. *Bulletin of the Japan Petroleum Institute* 16 (1): 60–66.

Yueying, Y. (2015). Simulation and comparison of operational modes in simulated moving bed chromatography and gas phase adsorptive separations. Dissertation. Virginia Polytechnic Institute.

3

Aromatics Process Description

3.1 Overall Aromatics Flow Scheme

Previous sections have described individual technologies in broad categories:

- Aromatic ring production
- Separations of aromatics from nonaromatics
- Aromatic rearrangement
- Individual isomer separations

In this section more detail is offered concerning individual aromatics technologies, as well as how these technologies can be typically integrated to produce the desired products with the most economical designs, while making best use of the available feed stocks. For purposes of discussion, UOP technologies will be referenced, but other technologies can be considered as well. Figure 3.1 shows how the above technologies are used for converting naphtha to pX and benzene. This example represents the most basic configuration for aromatics production.

Several variations – some of which will be discussed later – have been used for significantly reducing capital expense (CAPEX) and energy input. The naphtha will typically be C_6–C_{10} so that significant amounts of C_9 and C_{10} aromatics can be produced. This naphtha is first hydrotreated to remove sulfur and nitrogen contaminants. It is then fed to the catalytic reformer, in this case a UOP CCR Platformer unit that would be run at appropriately high conversion of C_7 paraffins (typically above 85%) to produce aromatics (benzene through C_{10} aromatics). After removal of the C_4 minus hydrocarbons by fractionation, the reformate is typically sent to the reformate splitter that takes the C_5–C_7 fraction overhead. This fraction is typically sent to the extractive distillation (ED) unit – in this case a UOP Sulfolane ED unit that purifies and recovers the benzene and toluene in the extract, while rejecting the nonaromatics into the raffinate. The ED extract will then be fed to a benzene/toluene splitter, where the benzene product is taken overhead, clay-treated, and then sent to benzene product tankage. The clay treating of the benzene product is a precaution to ensure that all benzene product specifications, particularly color, can be met.

The bottoms of the reformate splitter column consist of C_8-plus aromatics. This stream is fed to the xylene rerun column (the start of the xylenes' loop) where the C_8 aromatics are taken overhead and then fed to the pX recovery unit, in this case a UOP Parex unit. The Parex unit will recover the pX in high purity as an extract stream, diluted with desorbent. The desorbent is removed via fractionation, and the pX is sent to a finishing column where any co-extracted toluene is removed. The purified pX is then sent to product tankage. The raffinate stream in the Parex unit will consist of the small amount of unrecovered pX, along with ethylbenzene, *meta*-xylene, *ortho*-xylene, and any nonaromatics that might have entered the xylenes' loop, all diluted by desorbent. The desorbent is removed via fractionation, and the C_8 raffinate is sent to the xylene isomerization unit, in this example a UOP Isomar unit. Here, the xylenes are re-equilibrated and ethylbenzene is either isomerized to produce additional xylenes or converted to benzene and ethane. The isomerate is fractionated to remove light hydrocarbons overhead, as well as any benzene and toluene. With modern isomerization catalysts such as UOP's I-500^{TM} and I-550^{TM} catalysts, the benzene by-product can be blended with the benzene from ED and sold. The toluene by-product from isomerization will combine with the toluene from the benzene–toluene fractionation and become part of the feed to the transalkylation unit – in this example a UOP Tatoray unit, where additional benzene and xylenes will be produced. The C_8 isomerate, along with any C_9-plus aromatic by-products, will combine with the reformate C_8-plus and be sent to the xylene rerun column. It should be noted that the xylene recycle stream will be 4–6 times the size of the fresh C_8 aromatics fed to the xylene rerun column, depending on whether adsorption or crystallization is used for pX recovery.

The bottoms stream from the xylene rerun column will contain the reformate C_9-plus aromatics as well as those from the xylene isomerization unit. This stream is sent to the heavy aromatics column, which will take all the useful C_9 and C_{10} aromatics overhead, while rejecting the heaviest to a drag stream. The overhead of the heavy aromatics column is combined with the fresh and by-product

---

Efficient Petrochemical Processes: Technology, Design and Operation, First Edition.
Frank (Xin X.) Zhu, James A. Johnson, David W. Ablin, and Gregory A. Ernst.
© 2020 John Wiley & Sons, Inc. Published 2020 by John Wiley & Sons, Inc.

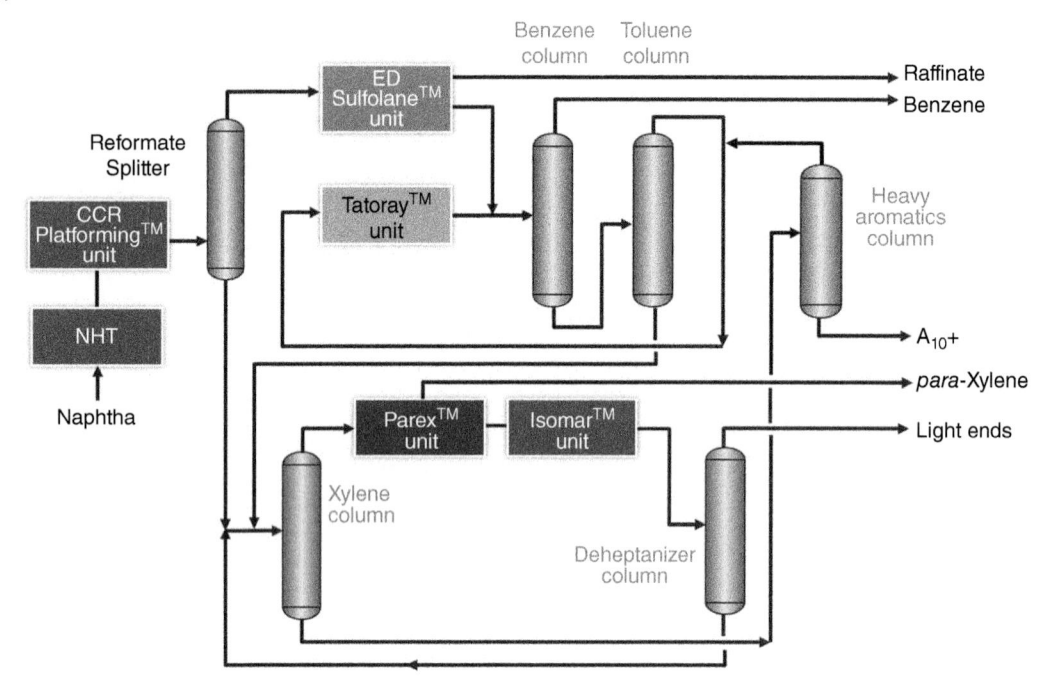

Figure 3.1 UOP aromatics flow scheme.

toluene and sent to transalkylation – in this example the UOP Tatoray unit – where additional xylenes and benzene are made. After removing light hydrocarbons from the Tatoray unit product, the aromatics are sent to the benzene–toluene fractionators. Here, the benzene produced by the Tatoray unit is recovered and blended with the benzene from the ED unit, then into the benzene product tankage. With modern Tatoray catalysts such as UOP's TA-32^{TM} and TA-42^{TM} catalysts, the benzene will be pure enough so as not to require ED. The unconverted toluene is recycled to the Tatoray unit. The xylenes and unconverted C_9-plus aromatics exit the bottoms of the toluene column and are mixed with the fresh C_8-plus and isomerate C_8-plus aromatics and fed to the xylene rerun column.

It should be apparent that a detailed knowledge of the feed and product components, in the form of kinetic models, thermodynamic properties, and associated simulation and control tools, is critical to ensure that the highest yields and efficiencies are obtained. Catalyst performance, along with the process conditions that are employed in the catalytic units, will have a significant effect on the amount of downstream purification and recovery operations that need to take place. These, in turn, can have a profound effect on the capital and operating costs for producing aromatics. For this reason, companies like UOP and others continuously invest in improving these critical technologies. Oftentimes, advances in catalysts and adsorbents can enable operating conditions that provide step-change improvements to individual processes, and which can be used to achieve significant synergies that can lower CAPEX and reduce utilities consumption. In the remaining parts of this section we will discuss underlying theory and technical bases for these processes, as well as highlight some of the significant advances that have been made in separations and catalytic technologies for aromatics.

3.2 Adsorptive Separations for *para*-Xylene

A good starting point is adsorptive separation, since it accounts for the vast majority or pX and mX production. When the UOP Parex process was first developed, it was found that the adsorbent formulation could influence the degree of separation as well as the suitability of particular desorbent compounds. The original Neuzil patent showed that either di-ethylbenzene isomers or toluene could be used as the desorbent, depending on the composition of the adsorbent. At that time, both X- and Y-faujasites were capable of being synthesized in high purity. Neuzil's work showed that by varying the cation types, cation concentrations, and moisture levels of these faujasites, the relative strength of desorbents vs the feed components could be regulated. Figures 3.2 and 3.3 depict a zeolite crystal, as well as cation position and how pX molecules would be attracted to the cations inside the cages of the zeolite crystal. Figure 3.4 depicts an adsorbent particle (bead) that is comprised of numerous crystals that contain the selective pore volume, along with the macropore volume that is nonselective and the

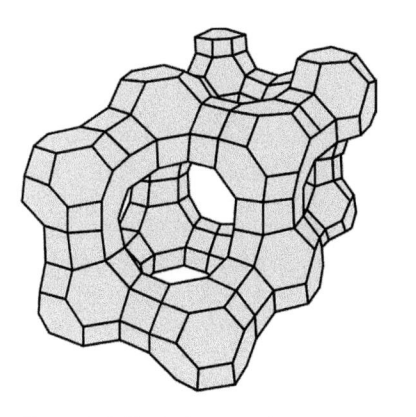

Figure 3.2 Faujasite crystal.

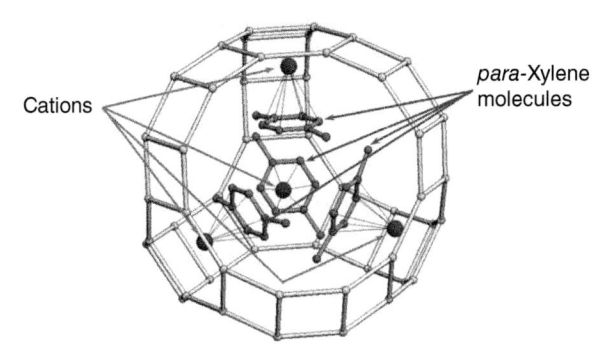

Figure 3.3 Orientation of *para*-xylene with cations in lattice.

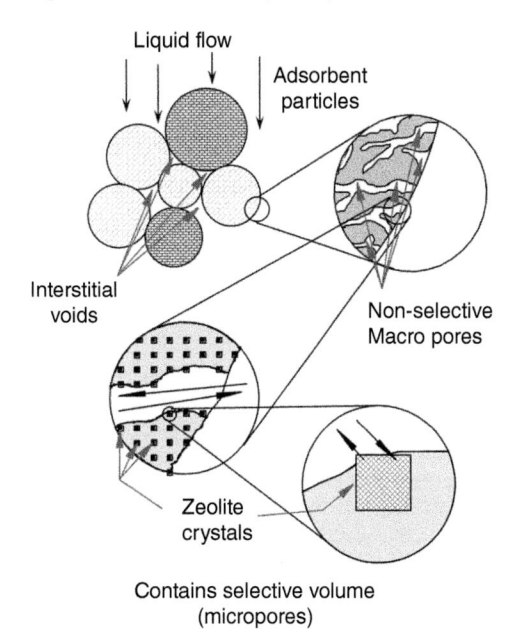

Figure 3.4 Diagram of adsorbent bead.

Figure 3.5 Parex adsorption affinities.

interstitial volume between adsorbent beads that is also nonselective volume.

Any molecule in the feed or desorbent can access the selective pore volume and the cations contained in the crystal lattice, but pX and the desorbent have got the greatest affinities and drive the separation. Adsorption affinities are shown in Figure 3.5.

These can be influenced by the hydration level on the adsorbent, and it is important that the adsorbent be maintained at the hydration level that gives the best combination of capacity, component selectivities, and mass transfer. It is also important to note that the high mass transfer efficiency that is needed for an economically small height equivalent theoretical plate (HETP) comes from judicious selection of the average particle size of the adsorbent. Dr Don Broughton et al. (1970) of UOP did the fundamental analysis that pointed to an average particle diameter of 500 µm, and that dimension has pervaded all of UOP's – and the industry's – adsorbent selection for all varieties of simulated moving bed processes.

While either di-ethylbenzene or toluene would be viable, UOP believed that the most attractive option would be to use di-ethylbenzene isomers, since that choice would allow a 2-carbon number difference between the desorbent and feed components and thus make desorbent recovery by fractionation much easier with 40–45 °C boiling point differentials. This is shown in Tables 3.1–3.3 which compare the normal boiling points of the C_8 aromatic isomers, toluene, several C_9 aromatics, and the di-ethylbenzene isomers.

Toluene, on the other hand, would have only a single carbon number difference and would need to be fractionated overhead, with a 25–30 °C boiling point differential. As a result, the first few Parex units were designed with mixed di-ethylbenzene isomers as the desorbent. These were called "heavy desorbent" Parex, because the desorbent is higher molecular weight than the feed. Two of those first Parex units are still in operation today, and have vastly expanded their capacities dues to advances in adsorbent and desorbent technology. UOP has commercialized five generations of "heavy desorbent" Parex adsorbent, with the latest one having nearly 3X the production capacity of the original adsorbent. Likewise, the desorbent has improved and is now high-purity *p*-diethyl benzene instead of mixed diethylbenzene isomers.

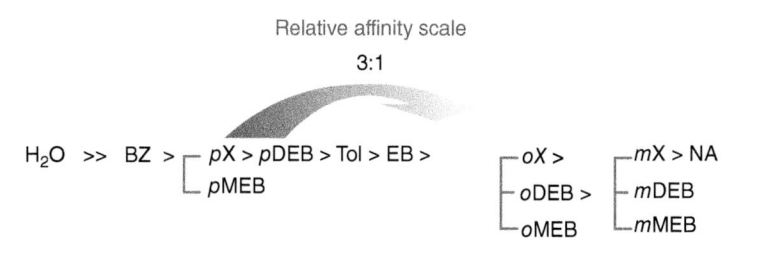

Relative affinity scale
3:1

H_2O >> BZ > ⌈ pX > $pDEB$ > Tol > EB > ⌈ oX > ⌈ mX > NA
                 ⌊ $pMEB$ ⌊ $oDEB$ > ⌊ $mDEB$
                                                  ⌊ $oMEB$ ⌊ $mMEB$

Table 3.1 Normal boiling point of C_8 aromatic isomers.

Feed C_8A component		Boiling point, °F(°C)
	Ethylbenzene	277 (136)
	para-Xylene	280 (138)
	meta-Xylene	282 (139)
	ortho-Xylene	291 (144)

Table 3.2 Normal boiling points of C_9 aromatics.

Feed C_9A component		Boiling point, °F(°C)
	1,4-Methylethylbenzene	324 (162)
	1,3-Methylethylbenzene	322 (161)
	1,2-Methylethylbenzene	329 (165)
	1,3,5-Trimethylbenzene	329 (165)

Table 3.3 Normal boiling points of desorbent components.

Desorbent		Boiling point, °F(°C)
	Toluene	232 (111)
	D-1000 (mostly para-Diethylbenzene)	363 (184)
	meta-Diethylbenzene	358 (181)
	ortho-Diethylbenzene	361 (183)

Figures 3.6 and 3.7 show how advances in adsorbent and desorbent technology can be used to reduce plant size and utility load.

para-Diethyl benzene can be produced via adsorptive separation of DEB isomers using the di-alkylate by-product of EB synthesis, or by direct *para*-selective alkylation of EB with ethylene. Plants that use *p*-diethyl benzene as the desorbent will also have a small fractionator that takes a slipstream of desorbent to remove heavies that might build up via thermal reactions in the bottoms of the extract and raffinate columns.

Despite the success of the first Parex units with mixed di-ethylbenzene desorbent, several companies at that time were concerned about the availability and quality of mixed di-ethylbenzene isomers as a desorbent. For this group of customers, UOP offered a design for Parex that was based on toluene as the desorbent, since toluene was readily available in most facilities. This version of Parex was called "light desorbent," because the desorbent is lower MW than the feed. A total of four of these plants were licensed. Since that time, two of these plants underwent revamps to "heavy desorbent" configurations in order to save utilities and increase capacities. The other two plants remain on stream with toluene as the desorbent, one of them with the original load of adsorbent running for more than four decades. As was shown in Figure 2.8, heavy desorbent Parex was well accepted by the *p*X industry and has accounted for about 70% of world *p*X capacity. The design of the integrated flow scheme evolved substantially over this time, as did the average capacity of a single-train plant. In recent years, however, UOP has returned to the concept of "light desorbent" Parex due to significant advances in adsorbent and catalyst technology, as well as process intensification and advanced simulation tools. The drivers for this resurgence have been reduced CAPEX and utilities input.

Because of this resurgence of interest in "light desorbent" Parex units, there are several million tons per year of *p*X committed to this technology called LDPX^{TM} (light desorbent Parex). Without going too deeply into proprietary details, one key element of the new approach is to allow some amount of feed C_9 aromatics to enter the Parex as well as Isomar units. Rather than have a stringent overhead specification limit of less than 500 ppm of C_9 aromatics, up to 2% can be allowed. This greatly simplifies the design and operation of the xylene rerun column. Another advancement has been to find ways to significantly decrease the desorbent:feed ratio using advanced adsorbent designs and process flow improvements. Where practical, dividing wall columns are used in the overall flow scheme, with pressures, overhead, and bottoms temperatures set to minimize overall fuel fired as well as cooling duties. These changes alone bring about a 25% reduction in energy consumption. The producers using this technology can take

Figure 3.6 Step changes in Parex adsorbent capacity.

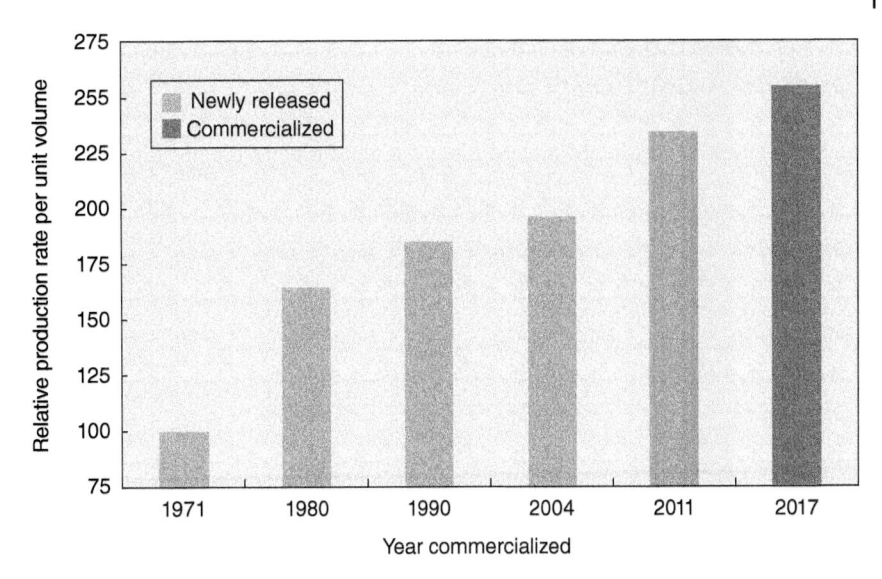

Figure 3.7 Impact on adsorbent chamber size and desorbent circulation.

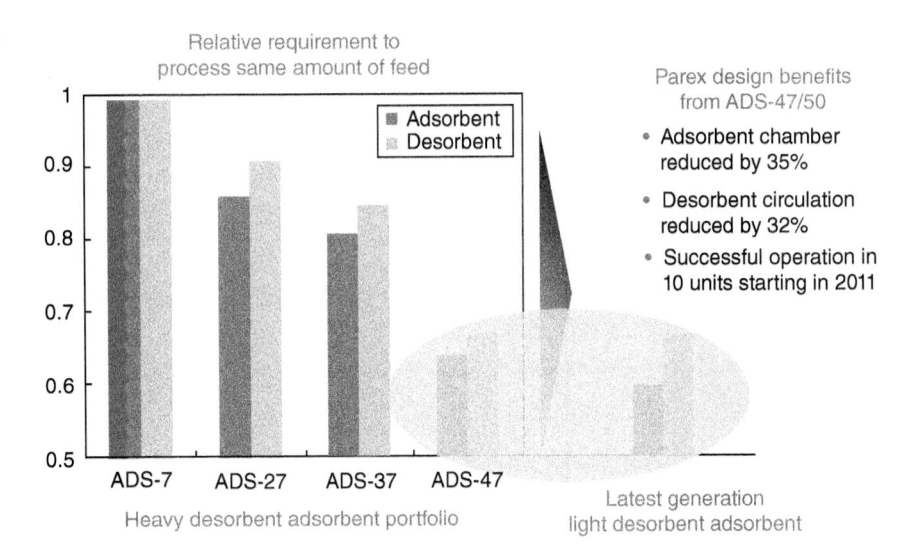

Parex design benefits from ADS-47/50
- Adsorbent chamber reduced by 35%
- Desorbent circulation reduced by 32%
- Successful operation in 10 units starting in 2011

advantage of substantially reduced overall energy input per ton of pX, with as much as a \$24/MT utility advantage over crystallization. Further advances in energy reduction are under active development at UOP.

As noted earlier, Axens developed the Eluxyl process for pX purification and recovery in the early 1990s. The first Eluxyl unit was at Chevron-Phillips Chemical in Pascagoula, Mississippi in the United States, where the adsorptive separation unit increased the pX purity to crystallization. The plant capacity was about 500 KMTA of pX. According to recent reports, this plant was permanently shut down at the end of 2018 to avoid a major maintenance expenditure (Woodmac.com 2018). Worldwide, Axens has licensed about 12 MMT/A of Eluxyl pX capacity with a two-chamber design having 12 adsorbent beds in each chamber, using p-diethyl benzene as the desorbent. The simulated moving bed is accomplished in Eluxyl through the use of a network of multiple on/off valves that set up the adsorption, purification, recovery, and buffer zones. The basic adsorptive separation mechanism for Eluxyl is very similar to that described above for the Parex process, with Axens having the adsorbent produced by Arkema. In an attempt to reduce CAPEX, Axens developed a single-chamber design for Eluxyl with 15 adsorbent beds. This version of the technology is called Eluxyl 1.15. Reportedly, two such plants have been licensed, one in China and another in Turkey. The Chinese unit is expected to start up in 2019. The total amount of adsorbent is very similar between the 24-bed two-chamber version and Eluxyl 1.15, but the adsorbent bed heights are likely much greater for the Eluxyl 1.15 version. It remains to be seen how robust this design will be to valve maintenance operations and process upsets. Since Axens continues to use p-DEB as a desorbent, the fractionation requirements in the xylene rerun column remain very stringent, thus driving up utility consumption.

3.3 Technologies for Treating Feeds for Aromatics Production

As noted earlier, the naphtha feed for *para*-xylene production is hydrotreated to remove polar contaminants such as organic sulfur and nitrogen before being fed to the catalytic reformer. While the reformate will not contain any polar contaminants, there will typically be 1 or 2 wt.% olefins in the reformate that are the result of the reactor effluent temperature and hydrogen partial pressure. Essentially, these olefins are in equilibrium with the corresponding unconverted paraffins in the reformate. There can also be styrenics which are in equilibrium with the ethylbenzene and methyl-ethyl benzene. It is important to have a way to remove these olefins and styrenics so that they do not cause adverse effects in the catalytic or adsorptive separation units that are used in production of *p*X. The conventional method for handling these olefins and styrenics has been with the use of clay treating. Clay treaters operate at low temperature (~100 °C) and use an acidic clay that will alkylate the olefin or styrenic with an aromatic molecule, creating a C_{16} heavy component. These heavy by-products are usually purged out with the heavy aromatic column bottoms, but they can drive up the xylene rerun column reboiler temperatures as well as consume valuable aromatic feed components that get turned into high molecular weight and low value by-products. Another drawback of this technology is the relatively short life of the clay. Depending on where the clay treater is located, the clay life can be as short as a few months, requiring reload and hazardous disposal of the spent clay. ExMo offers a technology called Olgone that is based on an MCM zeolite that is similar to what is used in their cumene technology (Businesswire 2007). While this catalyst system has a very long life and minimizes the need to hazardous disposal of clay, it still consumes aromatic feed components as they alkylate with the olefins and styrenics. Only a limited number of Olgone units have been deployed. To solve both the disposal and the feed consumption problems, UOP developed the Olefin Reduction Process™ or ORP™ (Maher and Hamm 2004). This technology operates at low temperature and in liquid phase to selectively hydrogenate the styrenics and olefins contained in the reformate. A typical reaction network for the ORP process is shown in Figure 3.8.

Figure 3.9 shows an example where the ORP could be inserted into the flow scheme or an aromatics complex, but it can effectively handle full-range reformate found in the debutanizer bottoms, or the light reformate of the reformate splitter overhead.

Many producers will send the raffinate from the ED unit to a steam cracker for production of ethylene and propylene. In such cases, it may be necessary to use the ORP process for reduction of the olefin content of the raffinate stream, in order to minimize coking of the tubes in the steam cracker. UOP has licensed numerous ORP units to producers who wish to minimize hazardous waste disposal costs and maximize feedstock conservation.

3.4 *para*-Xylene Purification and Recovery by Crystallization

As previously noted, crystallization was the technology that was first used in the *para*-xylene industry. Several companies still use crystallization for *p*X production – notably BP, ExMo, and Chevron-Phillips. BP has partnered with McDermott to license their *p*X crystallization and xylene isomerization technologies. A typical crystallizer train has a capacity of about 500 KMTA of *p*X. Referring to Figure 3.10, the train will typically have three crystallizers in series, each one progressively colder as a result of refrigeration, with the final one being at about −39 °C.

As shown in Figure 3.11, each crystallizer is a vessel with jacketed sections where ethylene refrigerant flows through the jacket and is partially vaporized.

The vessel walls are typically scraped using nylon blades. This staged cooling will result in about 72% of the *p*X being solidified. At the outlet of the third crystallizer,

- Selectively hydrogenate olefins with minimal aromatics saturation
  - Stoichiomatric hydrogen consumption
  - No heavies production
  - <0.5% aromatics loss

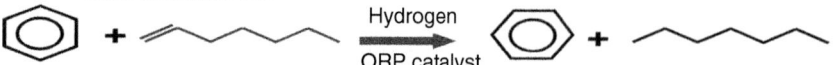

- ORP uses a nickel catalyst in a liquid phase reactor at moderate temperature
- ORP saturates various olefins in the reformate
  - Alkenyl aromatics
  - Primary, secondary and tertiary Olefins
  - C_{10}+ Unsaturates

Figure 3.8 Reactions for selective hydrogenation of aromatics – UOP ORP process.

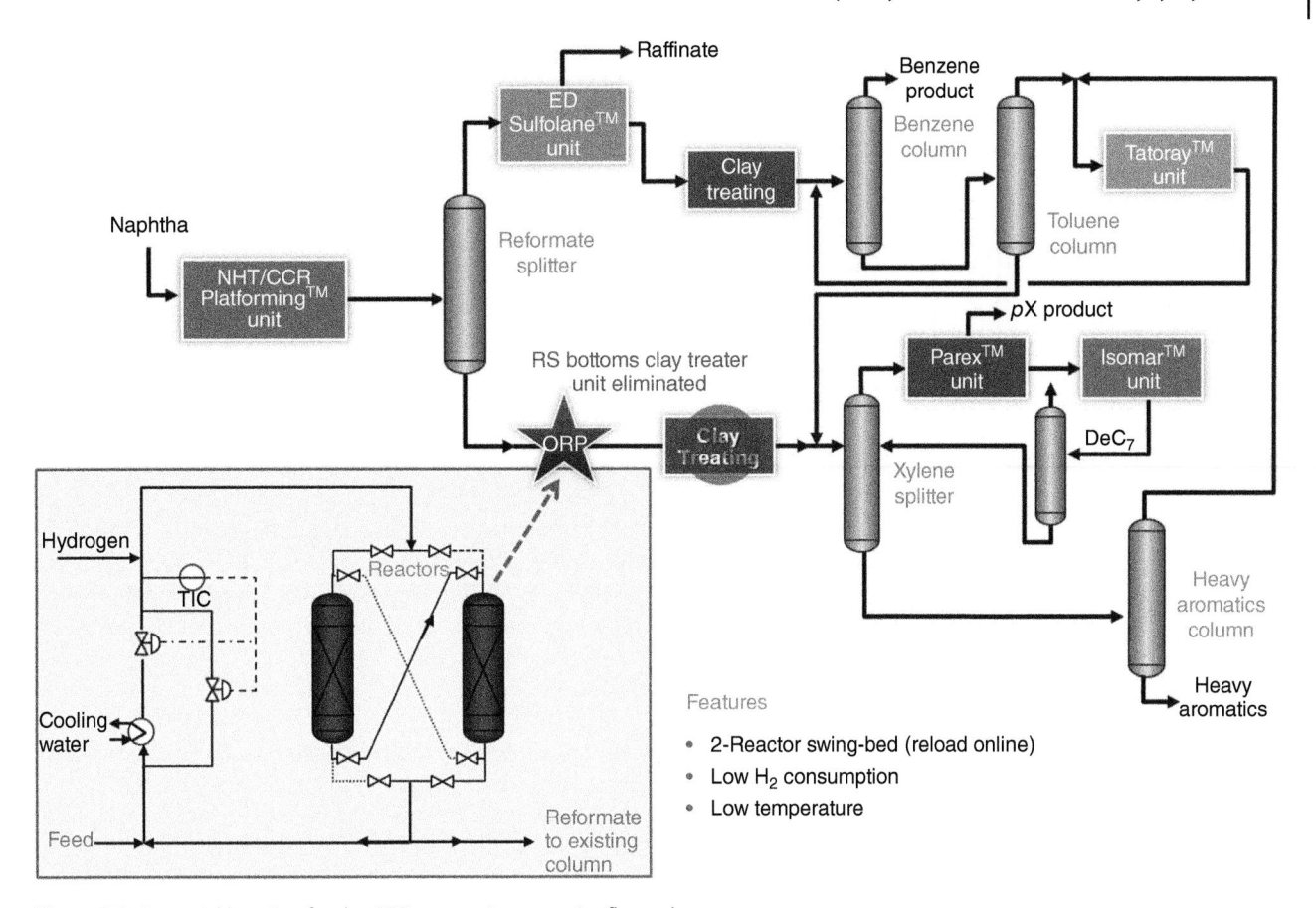

Figure 3.9 Potential location for the ORP process in aromatics flow scheme.

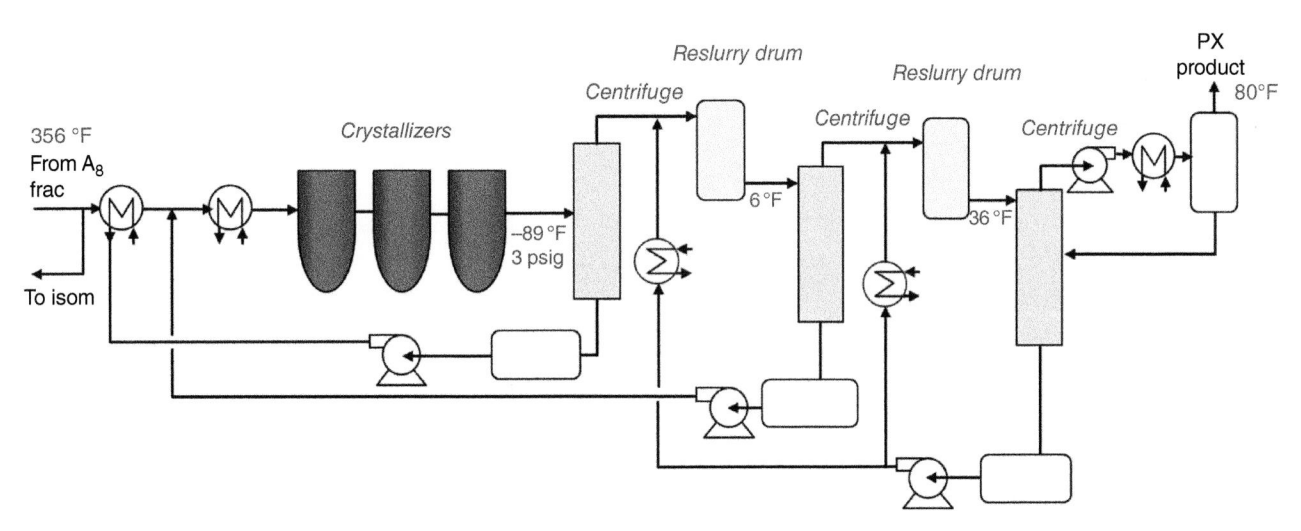

Figure 3.10 Crystallizer flow scheme for *para*-xylene.

a screen bowl centrifuge separates about 97% of the *p*X solids. This cake will have a *p*X purity of about 94%, with residual liquid of about 6%. The *p*X solids are then partially remelted and purified via two re-slurry drums, each with a pusher centrifuge for separations. The final product is 99.85% pure *p*X, and overall *p*X recovery is around 62%. The refrigeration is typically an ethylene–propylene cascade system. It will have three stages of cooling for the ethylene coming from the crystallizer's loop. There will be provision for additional heat exchange

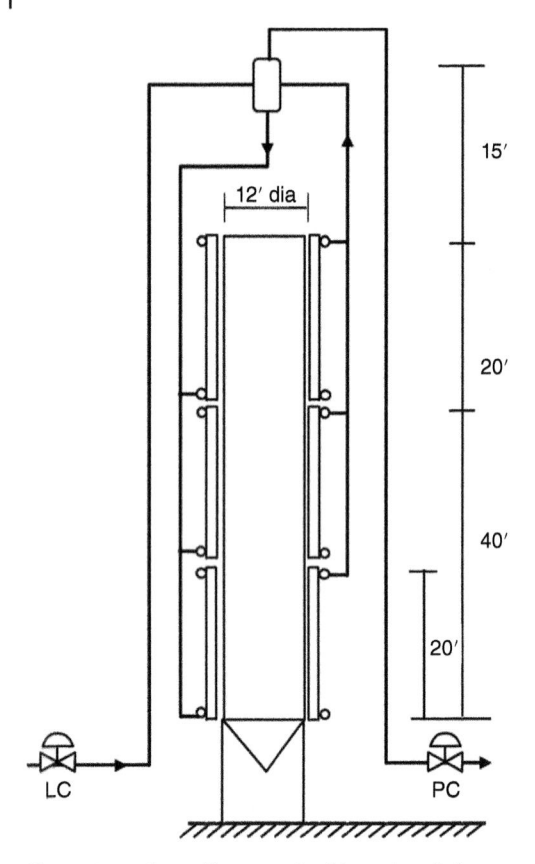

Figure 3.11 Crystallizer vessel refrigeration delivery.

with the crystallization section from the propylene loop. This is shown schematically in Figure 3.12.

Crystallization technology consumes a significant amount of electricity for driving the refrigeration compressors, as well as HP and MP steam. However, a crystallizer–xylene isomerization loop has low fuel consumption since there is minimal associated fractionation, even though the amount of recycle xylenes is comparatively large due to the low pX recovery. There is a substantial amount of rotating equipment in a crystallization-based facility, which can require significant preparations and spare parts inventories for maintenance.

For naphtha- or reformate-based pX facilities, there are numerous opportunities for energy integration when using adsorptive separation technology such as the "light desorbent" Parex offered by UOP. For this reason, the overall energy costs and CAPEX are typically much lower than for a comparable crystallization-based plant. The margins between pX and naphtha can range between $300 and 500/MT, and under these circumstances a grassroots facility can be very attractive in an expanding market. The differentials between pX and mixed xylenes are considerably smaller, often in the range of $100–200/MT. As a result, very few grassroots projects will be built based on mixed xylenes feed. That said, the utility consumption for a mixed-xylenes-based crystallization

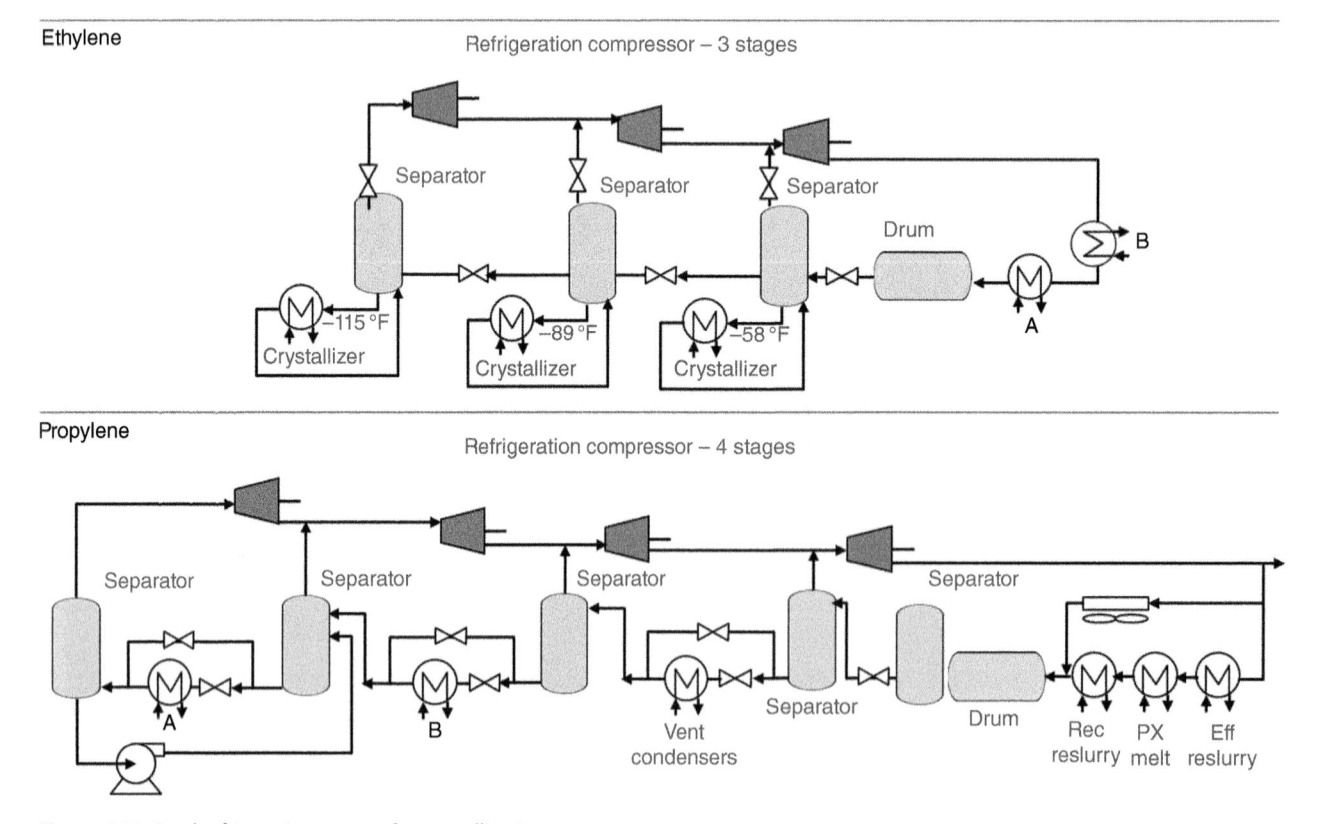

Figure 3.12 Dual refrigeration system for crystallization.

complex will typically be lower than for an adsorptive separation-based complex. In 2012, McDermott and BP were successful in licensing a 2 MM MTA reformate-based pX crystallization plant to Reliance at Jamnagar, India. This large project took several years to complete, but was finally at nameplate capacity in 2018.

3.5 Transalkylation Processes

Referring back to Figure 3.1, the transalkylation unit takes fresh and by-product toluene and C_9 plus aromatics to produce additional xylenes and benzene, beyond that which comes in with the reformate. Depending on the feed carbon number range and distribution, the transalkylation plant can contribute 50% or more of the xylenes to be recovered as pX. Using the example of the UOP Tatoray process, the flow scheme for this unit is very simple, consisting of a charge heater, a fixed-bed reactor, combined feed exchanger, recycle gas compressor, and make-up hydrogen, as shown in Figure 3.13.

As explained earlier, the primary transalkylation reactions are driven by high-activity Bronsted acidic sites that accomplish the following:

- Disproportionation
- Transalkylation
- Dealkylation
- Saturate cracking

The performance of the acidic catalyst is typically enhanced by inclusion of a metal function of appropriate strength. The metal function will saturate any olefins that are formed via dealkylation of C_9 or C_{10} aromatics. However, it is important that the metal function not be so strong as to saturate the aromatic rings, since that would cause lower yields as well as potentially create benzene co-boilers that would need to be removed via ED. As shown in Figure 3.14, the transalkylation and disproportionation reactions are governed by equilibrium, whereas the dealkylation and cracking reactions are kinetically controlled.

Since the transalkylation and disproportionation reactions constitute the majority of the reactions, a typical transalkylation unit will operate at an overall conversion rate of 46–50%. The reactor will typically operate with a mild exotherm, around 10 °C as a result of the dealkylation reactions and any saturate cracking that might take place. Modern transalkylation catalysts like the UOP TA-30™ series are able to achieve high yields of xylenes and benzene, high benzene purity, very long cycles between regenerations, and even accommodate very heavy feed components such as xylene column bottoms. In some countries where gasoline demand is decreasing, refiners are repositioning their assets to produce petrochemicals. The high stability of Tatoray catalysts has allowed simple revamps of medium-pressure semi-regenerative reformers to transalkylation service for production of petrochemical xylenes. UOP has developed and commercialized several generations of Tatoray catalysts. The most recent TA-30 and TA-42 series have particularly attractive selectivity and stability, and are some of the many results of UOP's zeolite discovery process.

The TA-30 catalysts make use of a novel zeolite called UZM-14™ (for UOP zeolitic material) (Broach et al. 2013; Moscoso et al. 2010) in a catalyst that includes a

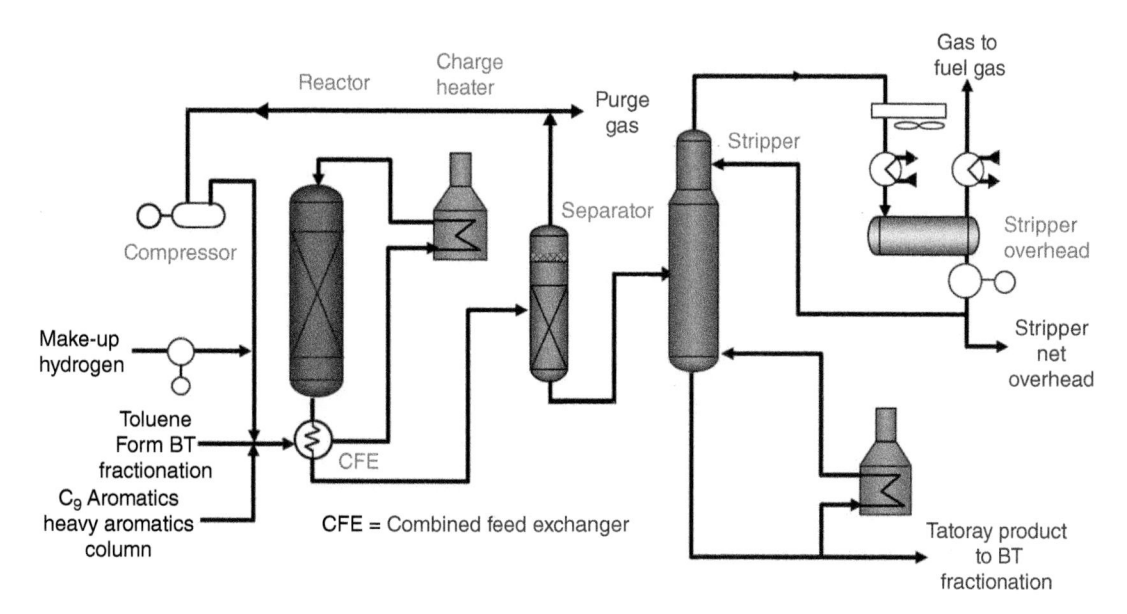

Figure 3.13 Tatoray process flow diagram.

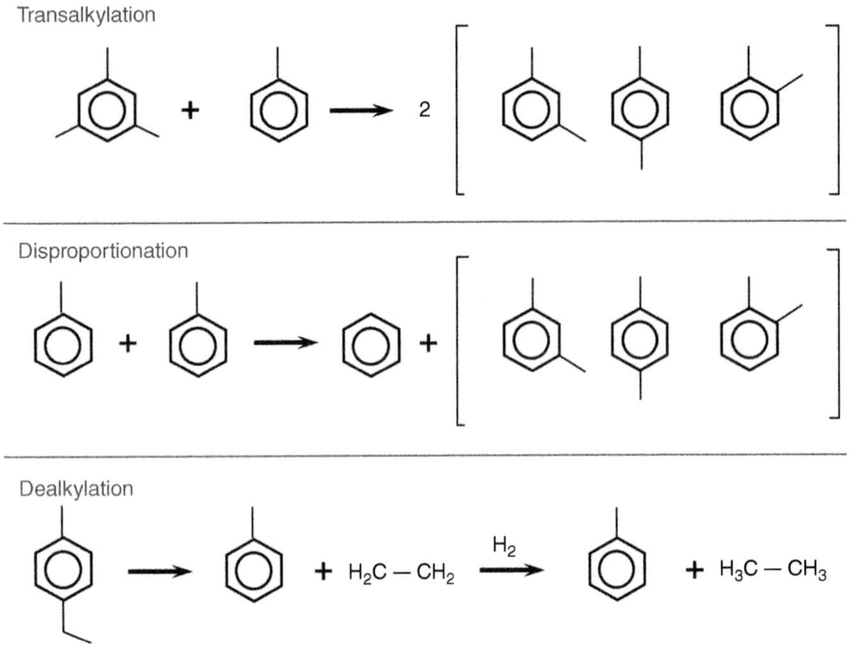

Figure 3.14 Equilibrium control of transalkylation reactions.

well-balanced acid–base metal interaction as well as highly efficient mass transfer. The UZM-14 is a nanocrystalline mordenite-like structure, where the very small dimensions of the crystals allow highly enhanced transport of reactants and products between the bulk vapor phase and the active sites of the catalyst. An example of the UZM-14 morphology and crystal dimensions is shown in Figure 3.15.

The critical dimension is the crystal length along the direction of the pore openings, and this in turn leads to the larger number of pore openings per unit volume. The influence of extremely small crystal structures is shown in the conversion vs crystal dimension plot where the UZM-14 zeolite is compared with the performance of two other commercial mordenite samples that have much larger crystal dimensions in the direction of the pores. The implication is that the reactions can be conducted at lower temperature with UZM-14, resulting in higher selectivity, improved stability, and longer run lengths. The excellent stability of these catalysts can allow un-extracted toluene to be processed in the Tatoray unit, which can de-bottleneck the ED unit.

Other metals have been used in transalkylation catalysts, such as rhenium, platinum, and tin, with tin being the attenuator for platinum. Conventional mordenite has been used (Oh et al. 2005) as well as ZSM-12 (Buchanan et al. 1998). Some technology suppliers have used multiple beds of catalyst to ensure benzene purity (Levin 2011), but generally it is preferable to have all functions built into one well-designed catalyst to avoid complicated differences in deactivation rates.

3.6 Xylene Isomerization

Referring back to Figure 3.1, xylene isomerization is critical to the economics of *para*-xylene production. With the *para*-xylene equilibrium concentration of about 24% of the xylenes, the isomerization unit will be required to handle relatively large flow rates compared to the fresh feed xylenes and the *para*-xylene product. Depending on whether adsorption or crystallization is used for *para*-xylene recovery, the xylenes may make between 4 and 6 passes through the isomerization unit. This magnifies the per-pass performance of the isomerization section, and in particular the performance of the catalyst. Key considerations are:

- *para*-Xylene approach to equilibrium.
- Ethylbenzene conversion (dealkylation to benzene or isomerization to xylenes).
- Xylene or C_8 ring retention across the reactor.
- Overall aromatic retention.
- The types of by-products generated.
- Hydrogen recycle requirements and compressor utility consumption.

The isomerization reactions can be conducted in the vapor or liquid phase. With liquid phase isomerization, there is very limited capability to employ hydrogen to drive ethyl–benzene reactions, so liquid phase operations tend to be used when ethylbenzene buildup is not a serious concern. That may especially be the case with flow schemes where a high-concentration *para*-xylene and low-concentration ethylbenzene stream is present,

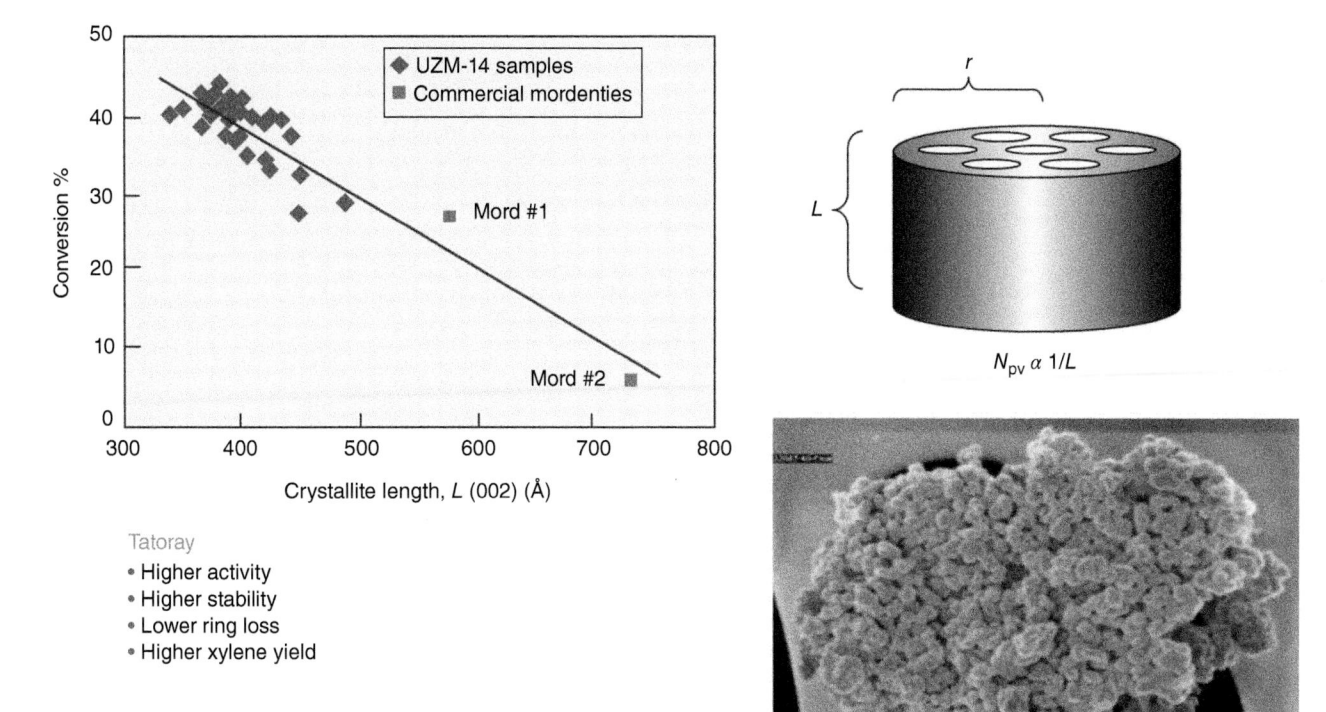

Figure 3.15 Crystal dimensions and reactivity of UZM-14 zeolite.

such as with toluene methylation or *para*-selective toluene disproportionation (TDP) units. The vast majority of xylene isomerization units operate in the vapor phase.

A key consideration in the early stages of designing a facility for *para*-xylene production is whether to use isomerization or dealkylation as the means of ethylbenzene conversion. Ethylbenzene isomerization to xylenes is equilibrium-limited, whereas ethylbenzene dealkylation is kinetically controlled. The equilibrium control for the ethylbenzene isomerization reaction means that the conversion per pass will be a strong function of the ethylbenzene level in the feed. When the fresh feed sources are reformate and/or pyrolysis gasoline C_8 aromatics, there can be a stronger driving force for ethylbenzene isomerization, with per-pass conversions in the range of 30–35%. However, when a transalkylation process like Tatoray is used, the C_8 aromatics from that unit will have very low EB content. If those C_8 aromatics represent 50% of the total fresh feed C_8 aromatics, the overall ethylbenzene content will be lower and there will be significantly less benefit to using an ethylbenzene isomerization catalyst.

Another factor is, of course, the degree to which the producer wants to participate in the benzene market. It can be very challenging – but not impossible – to have a plant configuration that produces zero benzene. The incremental benzene from ethylbenzene dealkylation might add 15–20% to the amount of benzene generated by catalytic reforming and transalkylation. Some producers may start out with a plant that is designed to use an ethylbenzene isomerization catalyst and, after a few years, switch to an ethylbenzene dealkylation catalyst as part of a de-bottlenecking project. Often the *para*-xylene production can be increased by 25–30% when this is done.

The majority of xylene isomerization plants will use an ethylbenzene dealkylation-type isomerization catalyst. This has the benefit of minimizing the size of the *para*-xylene recovery plant as well as the associated fractionation, such as the xylene rerun column. Typical reactions for this type of catalyst are shown in Figure 3.16.

The conversion of ethylbenzene via dealkylation requires a strong and selective acid function to cleave off the ethyl group to make ethylene, along with a selective metal function that will quickly hydrogenate the ethylene to ethane so that the ethylene cannot alkylate onto another aromatic to form a C_{10} by-product such as di-ethyl benzene or di-methyl ethyl benzene. It is also important that this metal function not hydrogenates either the benzene by-product or the C_8 aromatics. The xylene isomerization reactions are also driven by strong but selective acid sites. Here, it is important to minimize disproportionation and transalkylation reactions that

3 Aromatics Process Description

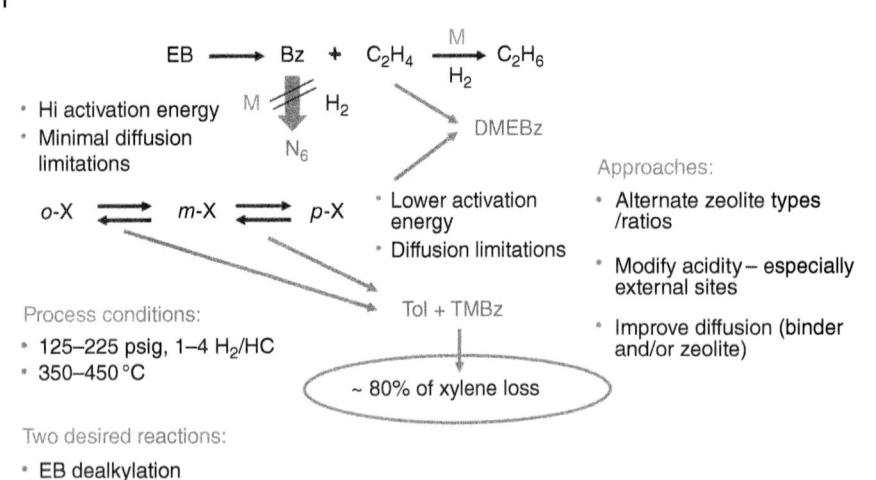

Figure 3.16 Reaction network for ethylbenzene dealkylation xylene isomerization catalyst.

EB ⟶ Bz + C$_2$H$_4$ —M/H$_2$→ C$_2$H$_6$

- Hi activation energy
- Minimal diffusion limitations

M, H$_2$, N$_6$ → DMEBz

o-X ⇌ m-X ⇌ p-X

- Lower activation energy
- Diffusion limitations

→ Tol + TMBz (~80% of xylene loss)

Approaches:
- Alternate zeolite types /ratios
- Modify acidity – especially external sites
- Improve diffusion (binder and/or zeolite)

Process conditions:
- 125–225 psig, 1–4 H$_2$/HC
- 350–450 °C

Two desired reactions:
- EB dealkylation
- Xylene Isomerization

---

could form toluene, tri-methyl benzene, or di-ethyl benzene. (The tri-methyl benzene and toluene by-products can be easily accommodated in the feed to transalkylation to produce more xylenes, but it is much better to minimize their formation in the first place.) This requires careful control of the acid site strength and distribution as well as heavy reliance on catalyst micropore geometry. The vast majority of ethylbenzene dealkylation catalysts are based on MFI zeolite structures (Abichandani and Venkat 1994; Sharma et al. 1999), or MFI derivative structures such as boro-silicates (Amelse and Reichmann 1990).

The MFI structure intrinsically discourages transalkylation reactions. The catalysts can have both the dealkylation and the isomerization functions in a single catalyst, or can have separate catalyst beds where dealkylation occurs in the first bed and isomerization occurs in the second bed. The catalysts associated with crystallization, such as the AMSAC catalyst from BP, are not overly concerned with ethylbenzene conversion efficiency because *meta*-xylene, rather than ethylbenzene, limits the recovery in crystallization. For that reason, and the higher combined feed ratio associated with the low recovery of crystallization, a modest ethylbenzene conversion level is tolerable. The isomerization units associated with adsorptive separation technologies such as the UOP Parex process will be designed to operate anywhere between 50 and 80% ethylbenzene conversion per pass. The *para*-xylene will typically be very close to equilibrium at about 24% *p*X/X. Since these catalysts are very active and stable, typical WHSV will be between 10 and 15 and H$_2$/HC between 0.8 and 1.5. When *ortho*-xylene is being coproduced with *para*-xylene, the isomerization catalyst will be able to drive both *ortho*- and *para*-xylene to equilibrium. In these cases, the combined feed ratio will be lower, since there is more product being taken out of the plant.

Figure 3.17 shows a typical flow diagram for a xylene isomerization unit, in this case for purposes of discussion, a UOP Isomar™ process flow scheme is shown.

The C$_8$ aromatic feed, make-up and recycle hydrogen are combined and sent through the feed/effluent exchanger where vaporization takes place, then through the charge heater which takes the reactants up to the desired reactor inlet temperature. Most UOP Isomar reactors are radial flow in order to minimize pressure drop and utilities, since the hydrocarbon and recycle hydrogen flow rates through these plants can become very large, even at a H$_2$/HC of 1.0. (Other technologies such as the ExMo XyMax isomerization technologies typically employ a downflow reactor with separate catalyst beds for dealkylation and isomerization.) The product steam exits the reactor, goes through the feed/effluent exchanger where it preheats the feed, into a condenser where it is liquefied, and then into the separator where the recycle hydrogen and a small purge gas stream are taken. The liquid from the separator is fed to the deheptanizer, where benzene and toluene are taken off in overhead streams, with the bottoms being returned to the xylene rerun column. If needed, the reactor inlet temperature can be raised incrementally to compensate for catalyst deactivation, but modern xylene isomerization catalysts are typically so stable that they can go for many years of operation before requiring regeneration. By that time, a next-generation catalyst is usually available. Nevertheless, most UOP Isomar process plants will be designed for the possibility of conducting in-situ catalyst regeneration, which consists primarily of a carbon burn, oxidation, dry-out, and reduction.

For ethylbenzene isomerization catalysts, most of the same process principles that were discussed above are still applicable. The most significant difference is that the ethylbenzene conversion to xylenes must proceed through a C$_8$ naphthene intermediate. This means that a

Figure 3.17 UOP Isomar process flow diagram.

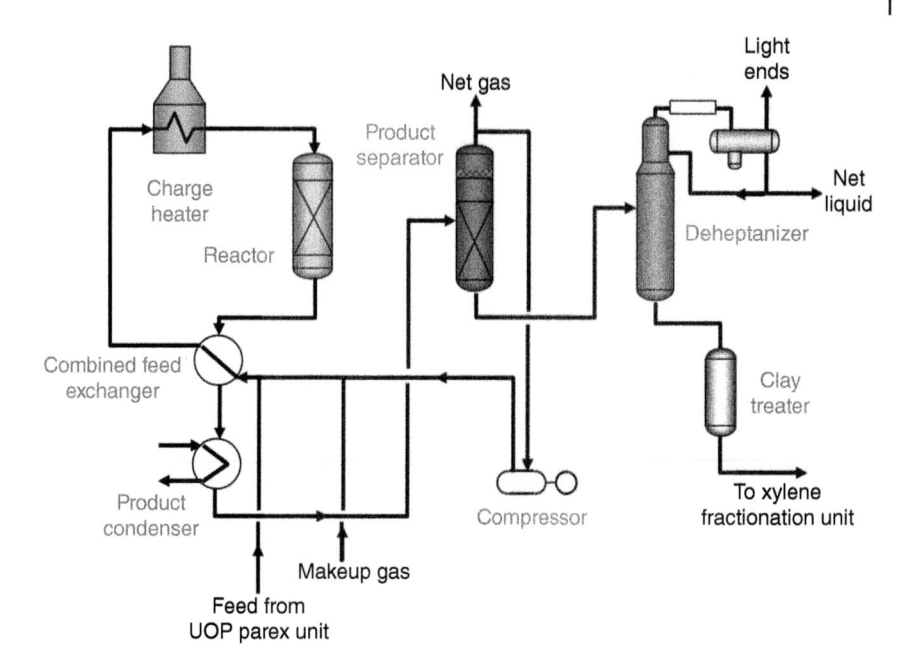

prescribed concentration of C_8 naphthenes needs to be present at the reactor inlet in order to be able to achieve the targeted equilibrium-governed ethylbenzene conversion. Ideally, the amount of C_8 naphthenes should not increase or decrease across the reactor but instead should stay "balanced." This balance is typically achieved by careful control of the hydrogen partial pressure, balanced with the reactor outlet temperature. As a reminder, a simplified reaction network for ethylbenzene isomerization catalysts is shown in Figure 3.18.

Since the equilibrium relationship between C_8 aromatics and naphthenes depends on the third power of the hydrogen partial pressure, most ethylbenzene isomerization catalysts will operate at relatively low pressures – in the range of 6–8 barg, and at modest reactor temperatures, around 350 °C. At such low pressures, a H_2/HC of about 3.0 and relatively high flow rates of the combined stream in the reactor, the radial flow geometry for the reactor is the best way to minimize high pressure drops and excessive compressor utilities. Referring again to the process flow diagram in Figure 3.18, these C_8 naphthenes enter and leave the reactor, proceed through the separator, and then to the deheptanizer. However, the design of the deheptanizer for an ethylbenzene isomerization Isomar unit will need to minimize losses of C_8 naphthenes to the overhead, so that they can be recycled back through the xylene rerun column and the Parex unit, where they then be recovered in the raffinate with the unextracted xylenes.

Figure 3.18 Reaction network for ethylbenzene and xylene isomerization catalyst.

Catalyst includes two types of sites to facilitate reactions:

Xylene isomerization:

EB isomerization:

Reactions are not perfect with losses from aromatic transalkylation and naphthenic cracking both present

→ Goal is to maximize pX formation and EB conversion while minimizing these side reactions

Axens offers an ethylbenzene isomerization catalyst system caller Oparis, which was combined with the design philosophy of Englehard's Octanizer technology. That technology practices local C_8 naphthene recycle around the isomerization unit, which requires very stringent fractionation between the xylenes, the naphthenes, and the by-product toluene. This design philosophy avoids the recycle of the naphthenes back through the xylene rerun column and the *para*-xylene separation unit. But it does require very high reflux/feed and a large number of trays, sometimes even a separate column.

The earliest ethylbenzene isomerization catalysts were non-zeolitic, essentially gamma alumina with platinum and chloride to drive the ethylbenzene xylene isomerization reactions. Both UOP and Englehard offered such catalysts in the early years of the *para*-xylene industry. The biggest issue with these catalysts was corrosion due to the chloride and moisture levels in the overheads of fractionators.

As a result, both companies moved to zeolitic catalysts to replace the chloride acid function. With the significant role of naphthenes in the ethylbenzene isomerization chemistry, a small-pore zeolite like MFI was not a good choice. A larger pore geometry was needed with reasonable acidity and the capability to balance the acid and metal functions. The most suitable medium pore zeolite at the time was mordenite, but special precautions needed to be taken to control the transalkylation reactions. These catalysts were well accepted by the growing *para*-xylene industry, and were widely deployed for the vast majority of producers because they avoided the maintenance problems associated with the chloride catalysts. However, these mordenite-based catalysts still produced 4–5% transalkylated by-products per pass. When the substantial combined feed ratio was taken into account, it was not unusual to see an overall *para*-xylene yield of 83–84% from the fresh feed xylenes. This low overall yield was close to that which would be seen from a well-performing ethylbenzene dealkylation catalyst.

With further advances in zeotype technology (including ALPO, SAPO, and MeAPSO structures), catalyst scientists were able to fine-tune the pore geometry, acidity profile, and metal dispersion to achieve significant reductions in transalkylation by-products while maintaining acceptable ethylbenzene isomerization and xylene isomerization activity. The Axens Oparis catalyst system was based on EU-1 zeolite (Guillon et al. 2011; Merien et al. 2000) whereas UOP's Isomar catalyst was based on MTW zeolite (Whitchurch et al. 2010). Further advances in this type of catalysis have been made in recent years by both companies. When product requirements dictate the use of ethylbenzene isomerization catalysts – which is about 25% of the time – there is a reasonable solution available. However, the significant role of transalkylation technology in the modern-day *para*-xylene facility has led to feed compositions that very strongly favor ethylbenzene dealkylation catalysts.

Earlier in this section there was a discussion about on-purpose *ortho*-xylene production, where the *ortho*-xylene is recovered from the mixed xylenes by fractionation, and the ethylbenzene, *para*- and *meta*-xylene as well as unrecovered *ortho*-xylene are sent to the isomerization reactor. In most cases, the catalyst will be an ethylbenzene isomerization catalyst, since there will be an incentive for converting the ethylbenzene into *ortho*-xylene. In fact, *ortho*-xylene is viewed as the first xylene isomer that comes from ethylbenzene conversion. There are a number of these plants throughout the world, albeit their capacities are rather small. Many of them are associated with a semi-regenerative reformer that might be operating at 98–100 research octane number (RON) severity. The feeds to these fractionators will likely contain some amount of C_9 paraffins that did not get converted in the reformer. While a certain amount of these C_9 paraffins can be tolerated in the *ortho*-xylene product, it is important for the associated isomerization catalyst to be able to crack them to smaller fragments to prevent their buildup in the *ortho*-xylene loop.

The job of the isomerization catalyst in this mode of operation is thus:

- Isomerize the feed to close approach to *ortho*-xylene equilibrium (about 23% oX/X).
- Convert ethylbenzene to xylenes for more *ortho*-xylene production.
- Convert co-boiling saturates to prevent their buildup in the *ortho*-xylene loop.

3.7 Adsorptive Separation of Pure *meta*-Xylene

As noted above, *meta*-xylene is part of a small but growing market for PET resin comonomers. The dominant technology for producing high-purity *meta*-xylene is the UOP MX Sorbex process. This technology replaced the HF-BF3 adduct route and it now accounts for nearly 1 million tons per year of *meta*-xylene capacity as was shown in Figure 2.10. The MX Sorbex process is a "light desorbent" technology, with toluene as the desorbent. A typical MX Sorbex flow scheme is shown in Figure 3.19.

Similar to the "light desorbent" Parex process, the product is recovered from the bottoms of the extract column. Alternative desorbents have been evaluated (Kulprathipanja 1999; Kulprathipanja et al. 2010), but have not provided the level of performance seen with toluene. As with the Parex process, the adsorbent is a faujasite, and in this case it is the *meta*-xylene molecule what is most strongly attracted to the cations that are

3.7 Adsorptive Separation of Pure meta-Xylene

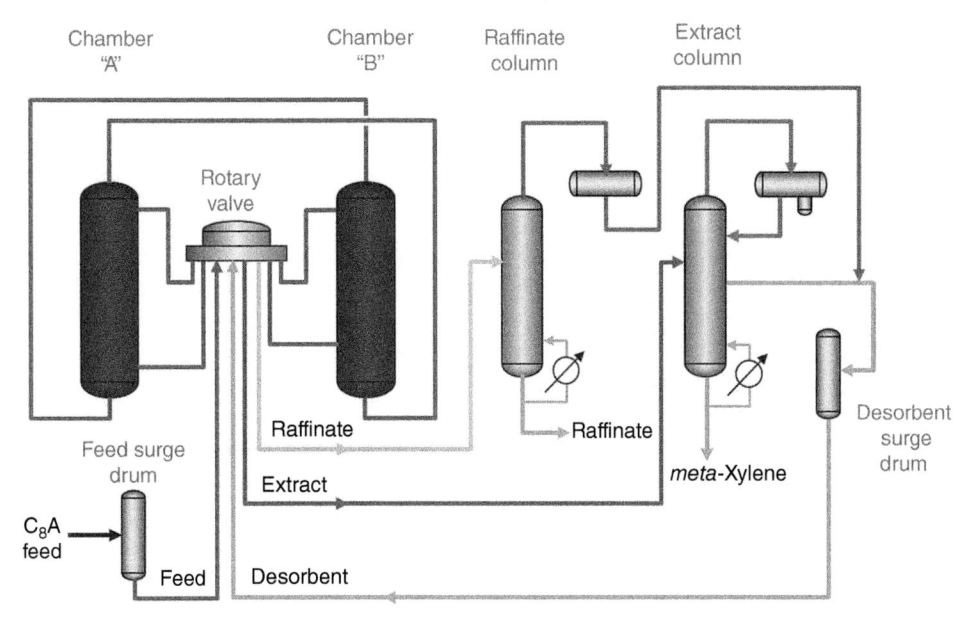

Figure 3.19 UOP MX Sorbex process flow diagram.

Table 3.4 Relative basicity of aromatic hydrocarbons.

Compound	Relative basicity at 0.1 M in HF
Benzene	0.09
Toluene	0.63
p-Xylene	1.0
o-Xylene	1.1
m-Xylene	26

positioned inside the zeolite super cage. There are significant differences among various C_6, C_7, and C_8 aromatics and *meta*-xylene shows the most basicity, as shown in Table 3.4.

This means that the acidity of the zeolitic adsorbent can be tuned via cation type and level, to create appropriate adsorption affinities. It is this difference, rather than steric effects, that provide the required selectivity for *meta*-xylene over the other C_8 aromatics and nonaromatics. Of the aromatic feed components, ethylbenzene is the easiest to reject, followed by *para*-xylene and then *ortho*-xylene. Saturates, if present, will go to the raffinate, since they are not adsorbed. The MX Sorbex unit can process reformate-derived xylenes directly. Alternatively, the MX Sorbex feed can be a slip stream from the Parex raffinate, since it would be nearly depleted in *para*-xylene, giving an increase in *meta*-xylene content. This is shown in Figure 3.20. This could reduce the size and utility load of the MX Sorbex unit.

Some producers have found it useful to fractionate out some of the *ortho*-xylene from the feed to the MX Sorbex unit to further increase the *meta*-xylene content. In nearly all cases, the capacity of the MX Sorbex unit is much smaller than the raffinate stream of the associated Parex

Figure 3.20 UOP MX Sorbex integration with Parex raffinate stream.

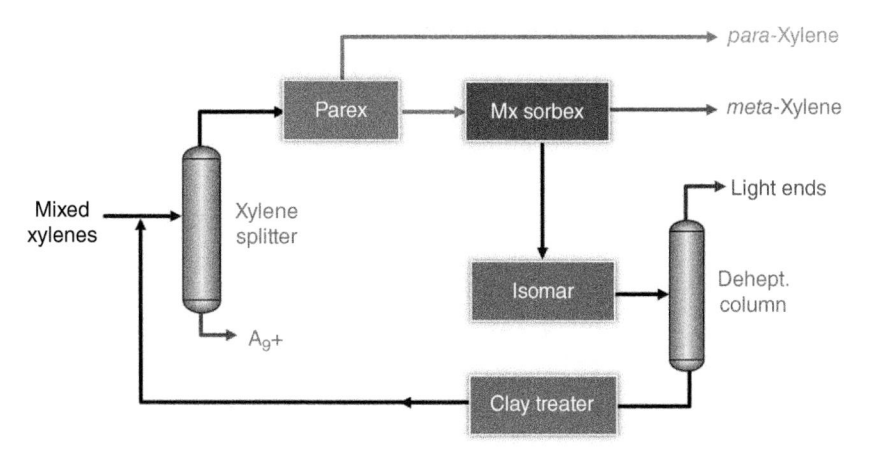

unit, so there is only a small impact on the composition of the feed to the xylene isomerization unit. However, in the case of a stand-alone MX Sorbex unit that is combined with xylene isomerization, it is important to understand that the isomerization reactions will proceed along a different path toward equilibrium, and that the combined feed ratio will be much lower than seen with a plant that produces *para*-xylene.

3.8 *para*-Selective Catalytic Technologies for *para*-Xylene

para-Selective catalytic processes for xylenes have been available to the industry for more than 20 years. They are broadly classified into two categories:

1) *para*-Selective toluene disproportionation
2) *para*-Selective toluene methylation

3.8.1 *para*-Selective Toluene Disproportionation

The most prominent technologies are for selective TDP, which converts toluene to benzene and a xylene product that can contains 80–90% or more *para*-xylene. The benzene coproduct is typically high enough purity to directly meet sales specifications of 99.9%. For many *para*-xylene producers, the use of this type of technology can provide a significant boost to overall *para*-xylene production from adsorptive separation processes by increasing the concentration of the feed to the separation unit. The performance of a crystallizer could also be enhanced by a higher concentration of *para*-xylene, primarily through incrementally higher recovery due to less eutectic limitations.

ExMo invented the Mobil Selective Toluene Disproportionation (MSTD) process when it was realized that controlled carbon deposition on an MFI catalyst could shrink the pore mouth openings of the zeolite to allow predominantly *para*-molecules to enter and exit. The intentional coking could also passivate the acid sites on the external surface of the zeolite, to prevent back-isomerization reactions from occurring. UOP also offered this type of technology in the PXPlus^{TM} process. This mode of catalyst selectivation is depicted in Figure 3.21.

The flow scheme for a PXPlus plant is very similar to that for a Tatoray plant, and is shown in Figure 3.22.

However, in contrast to the UOP Tatoray process, toluene is the sole feed, since the catalyst is not designed to handle C$_9$-plus aromatics. One other significant difference is that the conversion of toluene is much lower in selective TDP technologies – on the order of 30% – due to the diffusion restrictions imposed on the catalyst. Even though thermodynamic equilibrium would allow toluene conversions above 50% in a conventional transalkylation unit, the higher conversions would drive up isomerization reactions as well as by-product formation in a selective TDP unit. Operating conditions are typically 30–35 barg, 3–5 WHSV, and 1–2 H$_2$/HC, with a reactor inlet temperature around 400–420 °C.

The coke-selectivated catalysts proved to be difficult to operate, with principle problems being to establish exactly the right amount of coke at start-up and to maintain the soft-coke content of the catalyst throughout the process cycle. Ex-situ selectivated catalysts began to be offered with the ExMo PXMAX technology and UOP's second-generation PXPlus catalyst. These new catalysts avoided the problems associated with coke-selectivation

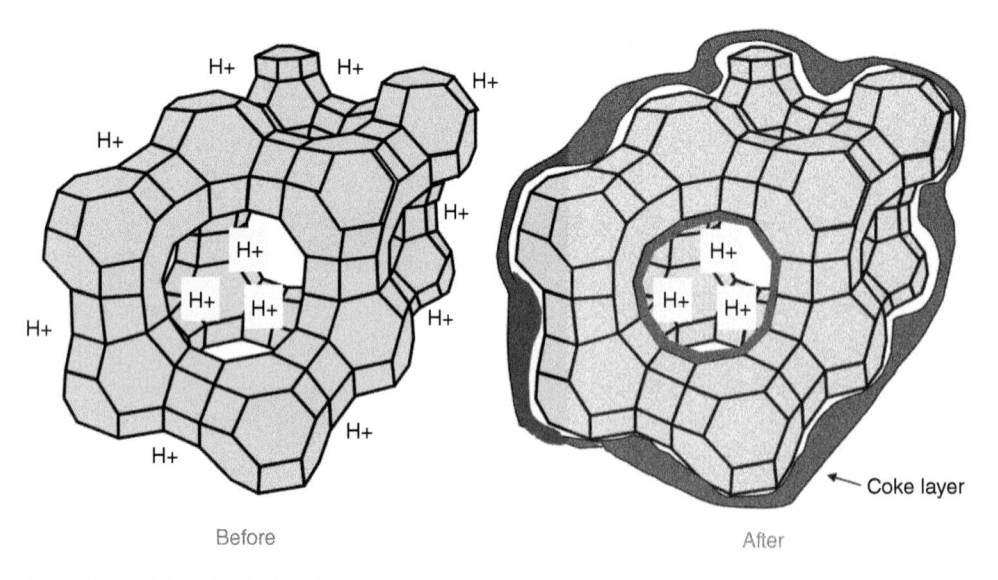

Figure 3.21 Coke selectivation for *para*-xylene production.

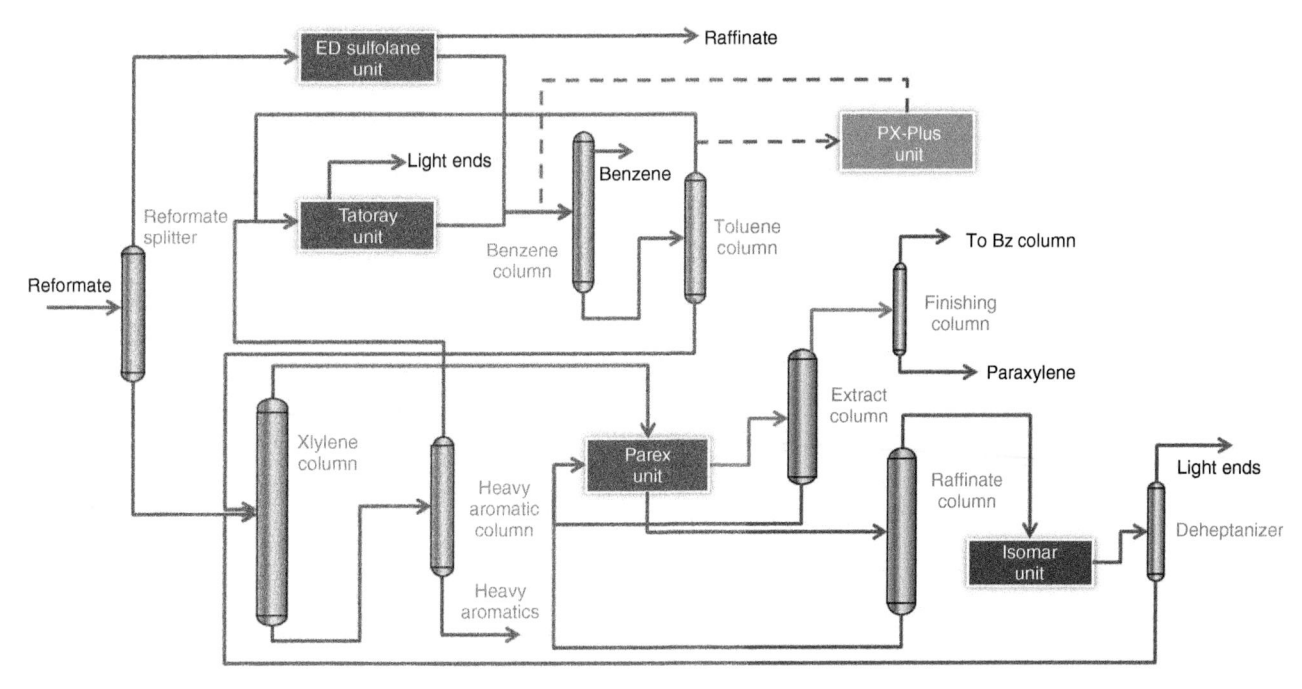

Figure 3.22 Process flow diagram for UOP PXPlus process.

Figure 3.23 Ex-situ selectivation of para-xylene selective catalyst.

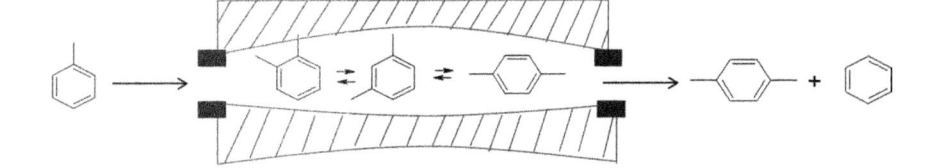

and coke maintenance during the processing cycle. They also offered very stable operations and enabled the use of minimal recycle hydrogen. An example of this type of ex-situ selectivation is shown in Figure 3.23.

Several plants have been built based on this technology, with most of them blending the high concentration xylenes with the fresh feed and isomerized xylenes. This would typically raise the para-xylene concentration in the feed to the recovery unit from around 22 to 27–30%, depending on the size of the TDP unit. A few producers installed single-stage crystallizers that could operate at moderate temperatures and still achieve high para-xylene recoveries due to a lower eutectic limitation. The producers who have built PXPlus or MTPX units will keep a close eye on the para-xylene and benzene prices and their differential. When benzene prices are significantly lower than para-xylene, many of these plants will be idled and the toluene sent to transalkylation or other destinations to produce more xylenes and less benzene.

3.8.2 para-Selective Toluene Methylation

Toluene methylation to produce a high concentration para-xylene product is an emerging technology that is likely to become more prominent in the para-xylene industry in coming years. Methanol for chemical use has gained widespread acceptance in China, as part of the "Coal to Chemicals" initiative that was strongly promoted by the Chinese government. Numerous plants have been built to convert methanol to light olefins such as propylene and ethylene. These include UOP's Advanced MTO^{TM} process, Dalian's DMTO, and Sinopec's SMTO technologies. Although there is mounting environmental pressure on coal gasification in China, there are still opportunities for new methanol to olefins plants in other locations, particularly where methanol can be produced at relatively low cost from stranded gas. Even in the United States, a significant amount of methanol capacity has been installed on the gulf coast, with attractive methane and energy prices. Some of this methanol is destined to be shipped to China for olefin production, but some could be allocated to toluene methylation.

UOP worked on toluene methylation in the 1980s along with DuPont (Herkes 1981, 1984), to develop a technology and examine its product's compatibility with the Parex process. Even though the technology offered attractive economics, the major players in the industry viewed it as too risky vs the alternatives that were available. However, with the success UOP has had and knowledge it has gained in deploying advanced MTO technology

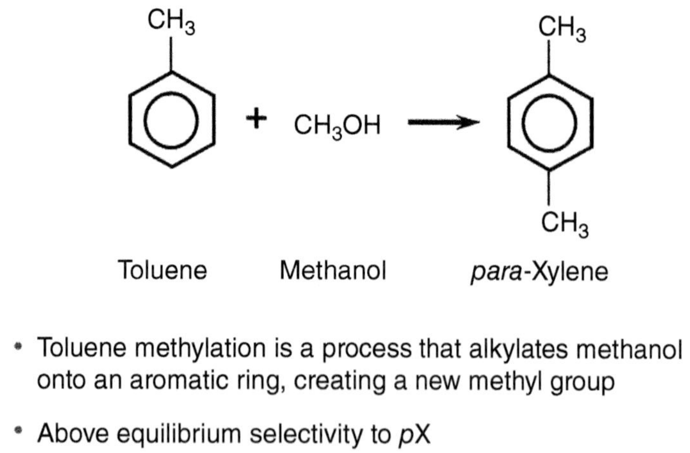

Figure 3.24 Fundamental toluene methylation reactions.

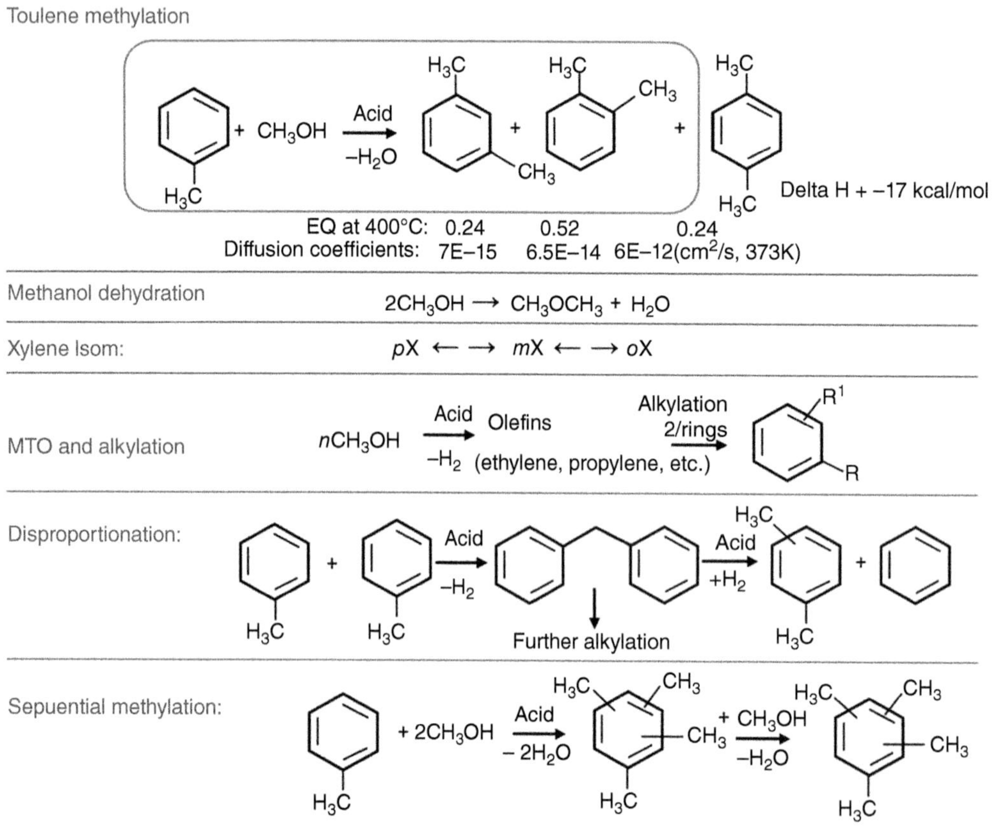

Toluene + CH$_3$OH → para-Xylene

- Toluene methylation is a process that alkylates methanol onto an aromatic ring, creating a new methyl group
- Above equilibrium selectivity to pX

Toluene methylation:

Toluene + CH$_3$OH $\xrightarrow[-H_2O]{Acid}$ mX + oX + pX   Delta H + −17 kcal/mol

EQ at 400°C: 0.24, 0.52, 0.24
Diffusion coefficients: 7E−15, 6.5E−14, 6E−12 (cm^2/s, 373K)

Methanol dehydration: $2CH_3OH \rightarrow CH_3OCH_3 + H_2O$

Xylene Isom: pX ↔ mX ↔ oX

MTO and alkylation: $nCH_3OH \xrightarrow[-H_2]{Acid}$ Olefins (ethylene, propylene, etc.) $\xrightarrow{Alkylation\ 2/rings}$ alkylated aromatic

Disproportionation: 2 toluene $\xrightarrow[-H_2]{Acid}$ diphenylmethane $\xrightarrow[+H_2]{Acid}$ xylene + benzene / Further alkylation

Sepuential methylation: toluene + 2CH$_3$OH $\xrightarrow[-2H_2O]{Acid}$ trimethylbenzene + CH$_3$OH $\xrightarrow{-H_2O}$ tetramethylbenzene

Figure 3.25 Detailed reaction network for toluene methylation.

worldwide, and with its intimate knowledge of FCC and para-xylene technology, UOP committed to developing toluene methylation technology and is now offering it for license (Larson and Krimsky 2019) as a fluidized bed process. This technology can produce a xylene product stream that is nearly 95% pX/X, with very high methanol utilization and toluene conversion. When coupled with other technologies, such as the Tatoray process, it can be used to bring about a significant increase in the available methyl/phenyl ratio in the aromatics complex, close to 2.0 that maximizes para-xylene. Typical toluene methylation chemistry is shown in Figures 3.24 and 3.25.

If the market conditions dictate, it is possible to completely eliminate benzene as a coproduct. Toluene methylation technology can be attractive as a revamp option for significantly boosting a plant's para-xylene capacity. This technology can also be very important in the planning of a grassroots facility when a refinery must be built to supply naphtha feedstock for *para*-xylene production. In essence, by using toluene methylation technology, the size of the associated refinery for a given *para*-xylene capacity can be considerably smaller. In addition to UOP's offering, ExMo also offers its EMTAM technology for toluene methylation.

References

Abichandani, J. and Venkat, C. (1994). Dual bed xylene isomerization. US Patent 5,516,956.

Amelse, J. and Reichmann, M. (1990). Process for isomerization of unextracted, ethylbenzene-containing xylene feeds. US Patent 4,899,010.

Broach, R., Boldingh, E., Jan, D. et al. (2013). Tailoring zeolite morphology by charge density mismatch for aromatics processing. *Journal of Catalysis*, 50th Anniversary Special Issue 308: 142–153.

Broughton, D., Neuzil, R., Pharis, J., and Brearly, C. (1970). The separation of *p*-xylene from C_8-hydrocarbon mixtures by the Parex process. Presentation at Third Joint Annual Meeting, American Institute of Chemical Engineers and Puerto Rican Institute of Chemical Engineers, May 1970.

Buchanan, J., Chester, A., Fung, A. et al. (1998). Transalkyation process for producing aromatic product using a treated zeolite catalyst. US Patent 5,763,720.

Businesswire (2007). ExxonMobil's new Olgone process proves successful at NPRC's Muroran refinery, https://www.businesswire.com/news/home/20070403005960/en (accessed 11 April 2019).

Guillon, E., Sanchez, E., and Lacombe, S. (2011). Catalyst comprising an EUO zeolite, a 10 MR zeolite and a 12 MR zeolite, and its use in a process for isomerizing aromatic C8 compounds. US Patent 7,982,083.

Herkes, F. (1981). Crystalline silica and use in alkylation of aromatics. US Patent 4,283,306.

Herkes, F. (1984). Methylation of toluene to *para*-xylene catalyzed by crystalline silica. US Patent 4,444,989.

Kulprathipanja, S. (1999). Process for adsorptive separation of *meta*-xylene from xylene mixtures. US Patent 5,900,523.

Kulprathipanja, S., Frey, S., Willis, R., and Knight, L. (2010). Adsorbent and process for the separation of *meta*-xylene from C_8 aromatic hydrocarbons. US Patent 7,728,187.

Larson, R. and Krimsky, D. (2019). Processes and apparatuses for toluene methylation in an aromatics complex. US Patent 10,239,802.

Levin, D. (2011). Heavy aromatics process catalyst and process of using the same. US Patent 7,897,825.

Maher, G. and Hamm, D. (2004). Integrated process for aromatics production. US Patent 6,740,788.

Merien, E., Alario, F., Lacombe, S. et al. (2000). Catalyst containing a zeolite EUO and the use of the catalyst in a process for isomerizing aromatic compounds containing 8 carbon atoms per molecule. US Patent 6,057,486.

Moscoso, J., Boldingh, E., Gatter, M., and Koster, S. (2010). Selective catalyst for aromatics conversion. US Patent 7,687,423.

Oh, S., Lee, S., Seong, K., and Park, S. (2005). Disproportionation/transalkylation or aromatic hydrocarbons. US Patent 6,867,340.

Sharma, S., Gurevich, S., Riley, B., and Rosinski, G. (1999). Selective xylenes isomerization and ethylbenzene conversion. US Patent 6,143,941.

Whitchurch, P., Bogdan, P., and Bauer, J. (2010). Aromatics isomerization catalyst and isomerization process. US Patent 7,745,677.

Woodmac (2018). Chevron–Phillips Chemical to cease US *para*-xylene production, 6 December 2018. https://www.woodmac.com/press-releases/chevron-phillips-chemical-to-cease-u.s.-paraxylene-production (accessed 11 April 2019).

Part II

Process Design

4

Aromatics Process Unit Design

4.1 Introduction

The objective of process design is to achieve safe, reliable, and economic performance of the plant and its equipment throughout its life. This is not a trivial task as the designer will encounter trade-offs between capital and operating costs as well as safety, reliability, and environmental issues and measures. The designer must also be aware of the latest technology developments and identify and champion new and improved process concepts. This chapter covers topics related to the process design and optimization of a typical aromatics complex including aromatics fractionation, aromatics extraction, transalkylation, xylene isomerization, and *para*-xylene separation units. In addition, important process design considerations including design margin philosophy, operational flexibility, fractionation optimization, and safety considerations are discussed.

4.2 Aromatics Fractionation

An aromatics complex producing *para*-xylene and benzene products, as well as by-products including nonaromatic raffinate and heavy aromatics streams, contains up to 14 different fractionation services not including those associated with the upstream naphtha hydrotreating and reforming units. Additional columns are required if other xylene products such as *ortho*-xylene are produced. Some of these columns are included in the aromatics extraction, transalkylation, xylene isomerization, and *para*-xylene separation units. Depending on the particular flow scheme, the aromatics fractionation unit includes five or six columns including those to fractionate the reformate feed and separate the aromatics products and recycle streams including benzene, toluene, mixed xylenes, and A_9+. The benzene and toluene columns may be included in the aromatics fractionation unit, or alternatively, they may be contained in a separate benzene–toluene fractionation unit. If there is a separate benzene–toluene fractionation unit, the other columns to fractionate mixed xylenes and A_9+ are included in a unit typically called the xylene fractionation unit.

4.2.1 Reformate Splitter

As discussed earlier, the C_6+ fraction of the product from the naphtha reforming unit (also known as reformate) is a common feed to an aromatics complex. A reformate splitter column is typically the first column in the aromatics complex to which the reformate is directed. The reformate splitter may have several feed streams. In addition to the feed from the reforming unit, recycle streams from downstream columns in the aromatics complex such as the xylene isomerization unit, transalkylation unit, and benzene–toluene fractionation unit may also be directed to the reformate splitter.

The design of the reformate splitter depends on the feeds to the aromatics complex, the complex flow scheme, and the desired products from the complex. The reformate splitter can be designed to make several different splits. Also, the column can be designed to make from two to as many as four different products.

The most common reformate splitter design is a simple column making a split between toluene and C_8 aromatics with toluene and lighter material going to the overhead and C_8 aromatics and heavier going to the bottom. With this design, the overhead product generally goes to an aromatics extraction unit to produce relatively pure benzene and toluene. The bottoms go to a xylene column, or similar column, to separate the C_8 aromatics from heavier aromatics (C_9+ aromatics). This reformate splitter service can make a very good split between toluene and C_8 aromatics recovering up to 99.9% of the toluene to the overhead product and 98–99% of the C_8 aromatics to the bottoms. The column will typically have about 50 real trays and require a reflux : feed ratio of about 0.5. The design split is an economic decision based on a trade-off between the capital and operating cost of the reformate splitter and the cost of the downstream processing of the toluene remaining in the bottoms and C_8 aromatics

---

Efficient Petrochemical Processes: Technology, Design and Operation, First Edition.
Frank (Xin X.) Zhu, James A. Johnson, David W. Ablin, and Gregory A. Ernst.
© 2020 John Wiley & Sons, Inc. Published 2020 by John Wiley & Sons, Inc.

remaining in the overhead. Like many of the economic evaluations required for a rigorous design of an aromatics complex, this evaluation can be quite complex.

If the toluene does not need to have nonaromatics components removed and therefore does not need to be sent to the aromatics extraction unit, the reformate splitter design will be significantly different from that described above. In this case, the benzene cut and toluene cuts are taken from the column separately. This can be done by simply taking a sidecut toluene product from a conventional column, or if a better split is required, a dividing wall column can be used. In either case the benzene and lighter cut is taken overhead, the toluene cut is taken as a side product, and the C_8 aromatics and heavier material is taken as the bottom product. As will be discussed in Chapter 12, a dividing wall column can make a much more defined three-way split than can a conventional column with a sidecut.

A third possible split that can be made in a reformate splitter is a three-way split in which a nonaromatic material lighter than benzene is taken as the overhead product, a benzene-rich cut is taken as the side product, and toluene and heavier material are taken as the bottoms product. The net overhead product is primarily C_5–C_6 nonaromatics that came to the reformate splitter in the reformate or possibly one of the recycle streams. The benzene-rich side stream is sent to the aromatics extraction unit. The design and operation of the reformate splitter in this manner would be done for two reasons: (i) to minimize the load on the aromatics extraction unit by removing some of the nonaromatics by simple fractionation and (ii) the toluene product does not need to be sent to the aromatics extraction unit. Clean splits between the C_5–C_6 nonaromatics, benzene, and toluene cannot be made with a sidecut in a conventional column. In order to make relatively sharp cuts between these products requires the use of a dividing wall column. Even with a dividing wall column, some of the C_6 nonaromatics and most of the C_7 nonaromatics boil with the benzene and therefore cannot be separated by simple fractionation. Thus, the reason an aromatics extraction unit is required.

The reformate splitter is normally operated only slightly above atmospheric pressure. For purposes of this discussion we will refer to columns such as this as "atmospheric" even though it operates with a receiver pressure of about 0.07–0.2 kg/cm^{2}g (1–3 psig). It is desirable to operate the column at as low a pressure as practical, without operating under vacuum conditions. Operating at low pressure allows for the use of lower temperature heating medium which is generally lower cost than higher temperature heating medium. For example, we know that medium pressure steam is lower cost than high pressure steam in most cases. If using a process stream in heat integrated systems it is desirable to use a process stream at the lowest possible temperature as the reboiler heating medium. This will be discussed at length in Chapters 12 and 13 with the discussion of process heat integration. The choice of overhead condensing system and column pressure control is critical to achieve the desired low operating pressure. Operation of the receiver at 0.07–0.2 kg/cm^{2}g (1–3 psig) is achieved by floating the column's receiver directly on the flare knockout drum using a separate dedicated line, or header if more than one column is connected to the flare knockout drum. This provides for not only the low operating pressure but a steady pressure as the receiver is not connected to the relief header but to the knockout drum at the base of the flare, downstream of the relief header, and at a lower pressure than the relief header.

A somewhat higher operating pressure is needed if the overhead system is closed. A closed overhead system is one in which there is no continuous vent of vapor or non-condensibles. If there is a sufficient flow of net vapor product, a closed overhead system with a simple control valve in the net vapor line may be employed. The control valve controls the pressure of the overhead system by regulating the flow of net vapor. This type of overhead system must operate with a receiver pressure at least 0.7 kg/cm^{2}g (10 psig) above the destination pressure of the net vapor for purposes of the control valve operability. The use of this type of overhead pressure control system will always result in a significantly higher operating pressure than when the overhead receiver "floats" on the flare knockout drum.

Another possible type of overhead system that operates at a higher pressure than the "floating" system, but lower than that with a net vapor control valve, is that which utilizes a "hot vapor bypass" system. With a hot vapor bypass a small portion of the hot overhead vapor, upstream of the condenser, is bypassed around the condenser directly to the receiver. The amount of vapor bypassed is controlled by a pressure differential indicating controller (PDIC) which regulates the pressure differential between the column overhead and the receiver. Due to the needed pressure drop for the control valve and the hydraulics of the system, the lowest practical receiver pressure is approximately 0.35 kg/cm^{2}g (5 psig). The hot vapor bypass overhead system is also a closed system as there is not a continuous net vapor product.

4.2.2 Xylene Fractionation

The separation of C_8 aromatics from heavier components is commonly done in a xylene column. While we refer to "xylene fractionation" and "xylene column," these would more accurately be called "C_8 aromatics fractionation" and "C_8 aromatics column." As discussed earlier in this book, traditionally, and for convenience, we use the term "xylenes" to refer to all four of the C_8 aromatic components including ethylbenzene (along with *para*-xylene, *meta*-xylene, and *ortho*-xylene).

The requirements for xylene fractionation vary greatly depending on the choice of downstream processing to separate and produce *para*-xylene, *meta*-xylene, *ortho*-xylene, and/or ethylbenzene. The various processes used to produce *para*-xylene have differing requirements for the allowable contaminants in the "mixed xylenes" or C_8 aromatics feedstock. The predominant technology used to produce *para*-xylene in the world as of the writing of this book is an adsorptive process using a selective, solid zeolitic adsorbent in a simulated moving bed system, using a heavy desorbent (*para*-diethylbenzene). This process requires that the mixed xylenes feed contain not more than 500 wt-ppm C_9 aromatics and not more than 100 wt-ppm methylethylbenzenes. These low levels of C_9 aromatics are achieved in the xylene column and require a large number of fractionation trays and relatively high reflux : feed ratio and thus reboiler duty. The number of trays in the xylene column for this service is typically in the range of 80–110 and the molar reflux : feed ratio is in the range of 1.5–3.0, depending on many factors including the concentration of C_9 aromatics in the feed and the amounts of specific C_9 aromatic components. The high reflux results in a column with large reboiler and condenser duties. As will be discussed in Chapter 12, the overhead heat (condensing duty) from the xylene column can be and is used as a major source of heat for the complex.

Other *para*-xylene separation technologies, such as adsorption using a simulated moving bed but with a light desorbent (toluene), and crystallization, do not have as stringent contaminant requirements for the mixed xylenes feed. Much larger concentrations of C_9 aromatics can be tolerated in the feed. Therefore, the size and energy requirements of the xylene fractionation system are less than that in the foregoing discussion.

When *ortho*-xylene is to be produced as a coproduct along with *para*-xylene, the most common way of separating the *ortho*-xylene is by fractionation. *ortho*-Xylene is the heaviest (highest boiling point) of the four C_8 aromatic isomers and the difference in boiling points between it and the nearest boiling C_8 aromatic, *meta*-xylene, is relatively large (5 °C, mX BP = 139 °C, oX BP = 144 °C). To produce a given quantity of *ortho*-xylene, slightly more than the required fraction of the total *ortho*-xylene that is contained in the feed is taken to the bottom of the xylene fractionator. So, if 100 thousand metric tons per annum (KMTA) of *ortho*-xylene is to be produced and the feed contains 500 KMTA of *ortho*-xylene, about 101 KMTA of oX, or 20.2% of the oX in the feed, is recovered to the bottom of the column. The extra 1 KMTA is to allow for a recovery of 99% in the oX column. In this case, the xylene fractionator is called a xylene splitter since some of the C_8 aromatics go overhead and some go to the bottom of the column. A fairly sharp split is needed in this column to reject most of the *meta*-xylene, *para*-xylene, and ethylbenzene from the bottoms. A typical *ortho*-xylene specification limits the amount of $mX + pX$ to 0.5–1 wt.%. See the typical specifications for *ortho*-xylene in Table 4.1.

Table 4.1 Typical *ortho*-xylene product specifications.

oX purity	98.0–98.5 wt.%
Cumene or A_9+	0.3–0.5 wt.%
Nonaromatics	1.0 wt.%
$mX + pX$ + Ethylbenzene	0.5–1.0 wt.%

The more A_9+ and nonaromatics in the oX product, the less $mX + pX$ + EB is allowed to be able to meet the oX purity requirement; thus, requiring a sharp split in the xylene splitter column.

The number of trays, reflux, and reboiler duty are most reasonable for a xylene splitter in which not more than 30–35% of the *ortho*-xylene is taken from the bottom of the column. Larger percentages of *ortho*-xylene recovery to the bottom can be designed for, but with increasing cost as the split becomes increasingly more difficult. While quite uncommon, *ortho*-xylene is sometimes the only C_8 aromatic product. In this case, in order to minimize the recycle stream to/from the xylenes' isomerization unit, and the resulting cost of that unit, the xylene splitter may be designed to recover 80–90% of the *ortho*-xylene to the bottoms. This high recovery results in a column with many trays (300 or more), and high reflux : feed (10 or greater, mol basis). The economics of such a design and operation are questionable.

As mentioned above, when the xylene column (or xylene splitter) has large reboiler and condenser duties, the column becomes an excellent source of heat for other column reboilers and heating services in the aromatics complex. However, in order to make the overhead vapor useful as a heating medium, the pressure of the column needs to be raised to raise the temperature of the overhead vapor and its condensed liquid. Typically, the column pressure is elevated to 5.5–8.5 kg/cm^{2}g (80–120 psig).

4.2.3 Heavy Aromatics Fractionation

The bottoms of the xylene column consist of that portion of the reformate, and any other feeds to the complex, that are heavier than C_8 aromatics. This includes C_9 and C_{10} aromatics as well as $C_{11}+$ aromatics. The transalkylation unit can use the C_9 and C_{10} aromatics as feed as they contribute a lot of methyl groups to the reaction which is beneficial in producing xylenes in that unit. However, the $C_{11}+$ aromatics can deactivate the transalkylation catalyst reducing catalyst life. It is therefore desirable to separate the C_9/C_{10} aromatics from the $C_{11}+$ aromatics. This is the job of the heavy aromatics column. The term "heavy aromatics" is used in different ways in the industry and can refer to different cuts. The C_9/C_{10} cut is sometimes referred to as "heavy aromatics" and the $C_{11}+$ cut is also referred to as "heavy aromatics," so it can be confusing as the meaning is not clear. The authors prefer to refer to the $C_{11}+$ aromatics as heavy aromatics and the C_9/C_{10} aromatics as C_9/C_{10} aromatics or A_9's and A_{10}'s.

The heavy aromatics column is typically operated at low pressure due to the high boiling point of the bottoms material and its tendency to thermally degrade and cause coking and fouling of equipment at high temperatures. Most heavy aromatics columns are operated just above atmospheric pressure using a pressure control system that "floats" on the flare knockout drum as discussed for the reformate splitter column above. It is possible to operate the heavy aromatics column at a somewhat higher pressure of 1.0–3.5 kg/cm^{2}g in order to raise the overhead temperature high enough to generate low pressure steam or serve as heating medium for some process heat sink such as the reformate splitter reboiler, benzene column reboiler, or column preheat.

4.3 Aromatics Extraction

An aromatics extraction process unit is usually incorporated within an aromatics complex to recover high-purity benzene and toluene products from the reformate and possibly other feedstocks. Typically, the aromatics extraction unit is located downstream of the reformate splitter column. The C_6–C_7 fraction from the overhead of the reformate splitter is fed to the aromatics extraction unit. The aromatics extraction unit separates aromatics (the extract product) from nonaromatics (the raffinate product). The aromatic extract from the unit contains trace olefins that need to be removed to produce a saleable benzene product. There are several technologies available to remove trace olefins such as clay treating and selective hydrogenation. These technologies will not be discussed in this chapter. Individual benzene and toluene products are recovered by fractionation. The toluene may or may not be a final product from the complex. In most cases the toluene is not sold as a product but is recycled in the complex to extinction via fractionation and processing in the transalkylation unit. The nonaromatic raffinate from the aromatics extraction unit is usually blended into the gasoline pool, directed to a naphtha cracker, or used in aliphatic solvents.

While there are many different aromatics extraction processes in operation and offered by licensors, there are two main types: (i) processes that combine liquid–liquid extraction and extractive distillation (ED) and (ii) processes that solely use ED. All of these processes use a solvent that facilitates the separation of aromatics from nonaromatics. Processes that combine liquid–liquid extraction and ED will be referred herein as "conventional" aromatics extraction units while those that employ only extractive distillation will be referred to as "extractive distillation" units or "ED" units. These main types of aromatics extraction processes will be discussed in Sections 4.3.1 and 4.3.2, respectively.

In "conventional" aromatics extraction units, the solvent facilitates the separation of aromatics from nonaromatics in both the liquid–liquid extraction section and the ED section. In the liquid-liquid extraction section of the unit 4.3.1, aromatics are selectively dissolved in the solvent. The solvent is highly polar and aromatics are more polar than the other hydrocarbons in the feed. Highly polar compounds dissolve in highly polar solvents, or "like dissolves like." This is the fundamental principle of liquid–liquid extraction.

In the ED section of conventional aromatics extraction units (i.e. the stripper) and in solely extractive distillation units, the solvent increases the difference in relative volatility between the aromatic compounds and the nonaromatics. This facilitates the ability to separate by distillation compounds that "co-boil." As an example, it is impossible to separate benzene from normal hexane by conventional distillation because they boil so closely together. However, with the addition of a suitable solvent they can be separated rather easily by ED.

The two most important attributes of a solvent to be used for aromatics extraction are its capacity to absorb aromatics and its ability to differentiate between aromatics and nonaromatics. The solubility of aromatics in the solvent determines the capacity of the solvent to absorb aromatics. The ability to differentiate aromatics from nonaromatics is referred to as selectivity. The greater the capacity (solubility of aromatics) and selectivity of a solvent the better it performs in an aromatics extraction unit. The polar solvents that are used for aromatics extraction share the following characteristics and tendencies:

- The solubility of aromatics in the solvent is greater than that of paraffins, olefins, and naphthenes. In fact, due to their relative polarity, the order of solubility of the major hydrocarbon families is as follows: aromatics > naphthenes > olefins > paraffins. Therefore, from a solubility standpoint, the paraffins are the easiest to separate from the aromatics, followed by the olefins and finally the naphthenes.
- When hydrocarbons in the same family are compared, solubility decreases as molecular weight increases. Thus, for paraffins, pentanes are more soluble in the solvent than hexanes which are more soluble than heptanes, etc.
- The selectivity of a solvent decreases as the hydrocarbon content of the solvent phase (solvent loading) increases.
- In spite of these general similarities, various commercial solvents used for aromatics recovery have significant quantitative differences. Sulfolane demonstrates better aromatic solubilities at a given selectivity than any other commercial solvent.

- There are other important physical properties of the solvent, in addition to solubility and selectivity, that have a significant impact on plant investment and operating cost:
  - Solvent specific gravity – A high solvent specific gravity results in a large density difference between the hydrocarbon and solvent phases in the extractor. This minimizes the required diameter of the extractor. A high-density liquid phase in the stripper and ED column minimizes the size of that equipment as well.
  - Solvent specific heat – A low solvent specific heat reduces heat loads in the fractionators and minimizes the duty of solvent heat exchangers.
  - Solvent boiling point – A high solvent boiling point (significantly higher than that of the heaviest aromatic hydrocarbon to be recovered) facilitates the separation of solvent from the aromatic extract by conventional distillation in the recovery column.

4.3.1 Liquid–Liquid Extraction

Liquid–liquid extraction is a unit operation in which two (or more) liquid streams are intimately contacted in a countercurrent manner to facilitate mass transfer from one liquid to the other. In the case of aromatics extraction, one liquid stream is the aromatic-rich hydrocarbon feed stream, usually containing benzene, toluene, C_5–C_8 nonaromatics, and a small amount of C_8 aromatics. The other stream is the lean solvent. The lean solvent flows downward through the extractor vessel and becomes "rich" absorbing aromatics as it flows down the column contacting the feed stream. The nonaromatic components that are not soluble in the solvent continue to rise up the column.

The basic process flow through the extractor is illustrated in Figure 4.1. Lean solvent is introduced at the top of the extractor and flows downward. The lean solvent and hydrocarbon feed are two different liquid phases.

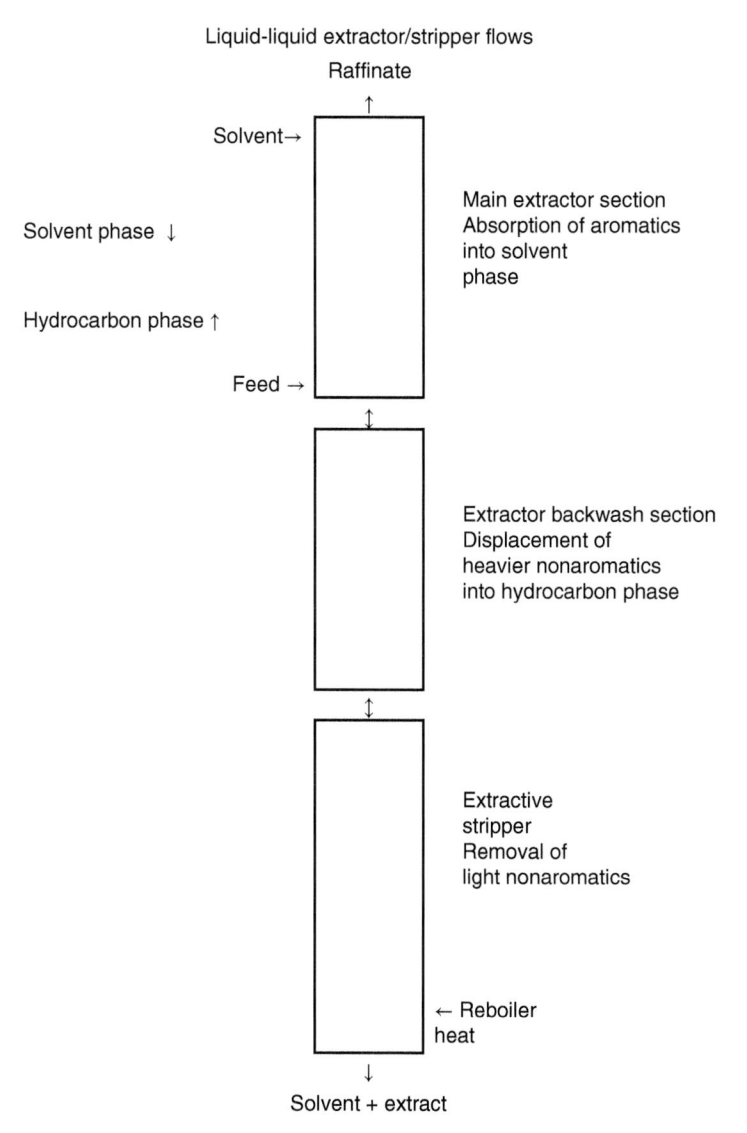

Figure 4.1 Liquid–liquid aromatics extractor and stripper sections and functions.

The hydrocarbon feed is introduced at the bottom and flows upward, countercurrent to the solvent phase. As the solvent phase flows downward, it is broken up into small droplets and re-dispersed into the hydrocarbon phase by each successive extractor stage. The solvent selectively absorbs the aromatic components from the feed. However, some of the nonaromatic hydrocarbon components are also absorbed. The bulk of the nonaromatic hydrocarbons remains in the hydrocarbon phase and leave the top of the extractor as the raffinate product.

The solvent phase, rich in aromatics, flows downward through the extractor. The stages in the extractor that are below the hydrocarbon feed point are referred to as backwash stages. In this section of the extractor, the solvent phase is contacted with a stream of light nonaromatic hydrocarbons from the top of the extractive stripper. The light nonaromatics which are more soluble in the solvent displace the heavier nonaromatic impurities from the solvent phase. The heavier nonaromatics then reenter the hydrocarbon phase and leave the extractor with the raffinate.

The rich solvent from the bottom of the extractor, containing only light nonaromatic impurities, is then sent to the extractive stripper for final purification of the aromatic product. The light nonaromatic impurities are stripped and removed overhead in the extractive stripper and recycled to the backwash stages of the extractor. The purified aromatics, or extract, are withdrawn as part of the solvent phase from the bottom of the extractive stripper. The solvent phase is then sent on to the solvent recovery column, where the extract product is separated from the solvent by distillation.

The most common solvents used in liquid–liquid-type aromatic extraction units are Sulfolane and morpholines. Sulfolane is an organosulfur compound, formally a cyclic sulfone, with the formula $(CH_2)4SO_2$, also known as tetramethylene sulfone or 2,3,4,5-tetrahydrothiophene-1, 1-dioxide. The Sulfolane solvent system was developed by Shell in the early 1960s and is still the most efficient solvent available for the recovery of aromatics. Most extraction units can operate at high purity and recovery by circulating more and more solvent. Sulfolane solvent exhibits higher selectivity and capacity for aromatics than any other commercial extraction solvent.

N-formylmorpholine (NFM) is the other primary solvent used for aromatics extraction. NFM is a nitrogen-based organic chemical compound having the chemical formula $C_5H_9NO_2$. A drawback of using NFM as the aromatics extraction solvent in an aromatics complex is the fact that it contains nitrogen, which is a poison to many catalysts used in transalkylation units and isomerization units. Carryover of the solvent into streams that go to these catalytic units can temporarily or permanently deactivate the catalysts.

The following is a discussion of the process flow of a Sulfolane liquid–liquid aromatics extraction unit. Figure 4.2 is an overall flow diagram of the process. The flow scheme of an extraction unit using NFM solvent is similar, although not exactly the same. Fresh feed enters the extractor and flows upward, countercurrent to a stream of lean solvent. As the feed flows through the extractor, aromatics are selectively dissolved in the solvent. A raffinate stream, very low in aromatics content, is withdrawn from the top of the extractor.

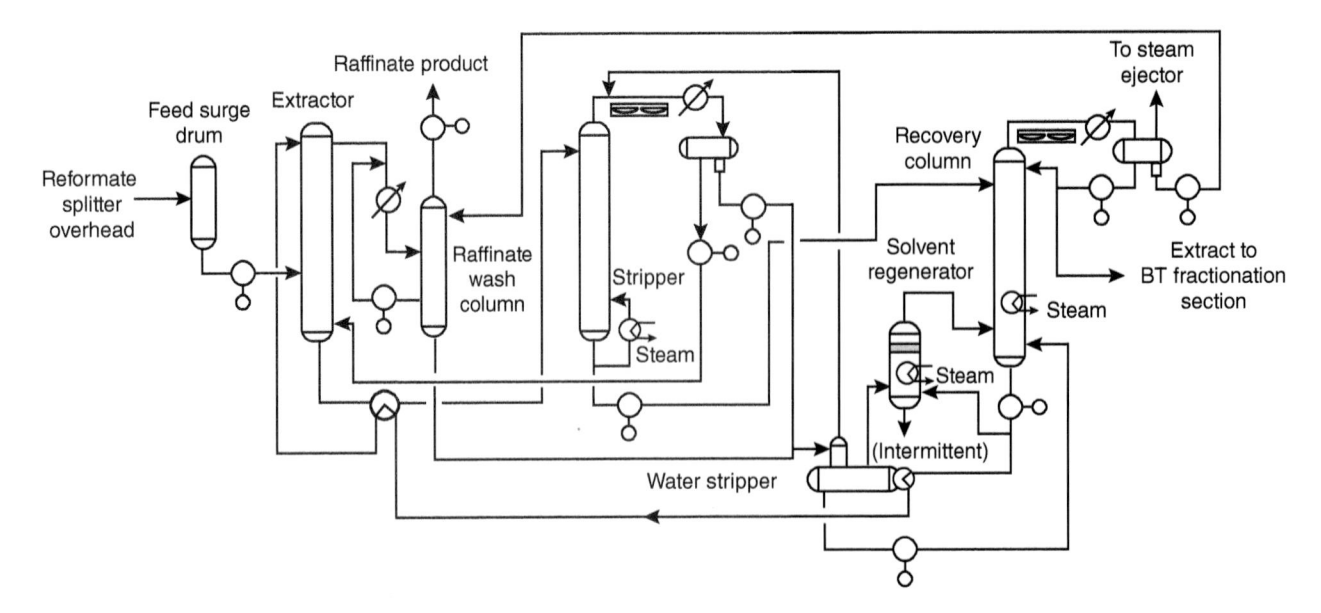

Figure 4.2 Sulfolane aromatics extraction unit flow diagram (liquid–liquid extraction type).

The rich solvent, loaded with aromatics, exits the bottom of the extractor and enters the stripper. The nonaromatic components having volatilities higher than that of benzene are completely separated from the solvent by ED and removed overhead along with some of the aromatics (>50 vol.%). This overhead stream is recycled to the extractor where the light nonaromatics displace the heavy nonaromatics from the solvent phase leaving the bottom of the extractor and the aromatics are re-extracted.

The bottoms stream from the stripper, substantially free of nonaromatic impurities, is sent to the recovery column, where the aromatic product is separated from the solvent. Because of the large difference in boiling point between the Sulfolane solvent and the heaviest aromatic component, this separation is accomplished easily, with relatively low energy input. Lean solvent from the bottom of the recovery column is returned to the extractor. The aromatic product, or extract, is recovered overhead and sent on to downstream distillation columns for separation and recovery of the individual aromatic products.

The raffinate stream exits the top of the extractor and is directed to the raffinate water wash column. In this column, the raffinate is contacted with water to remove dissolved solvent which is soluble in water. The solvent-rich wash water along with water from the stripper receiver is vaporized in the water stripper by exchange with hot circulating solvent and then used as stripping steam in the recovery column. Accumulated solvent from the bottom of the water stripper is pumped back to the recovery column.

The raffinate product exits the top of the raffinate water wash column. The amount of Sulfolane solvent retained in the raffinate is negligible. The raffinate product is commonly sent either to gasoline blending, a nearby naphtha cracking unit, or used for aliphatic solvent applications. Because it contains very little aromatics it has a relatively low research octane number (RONC), typically in the mid-60's (62–67).

Under normal operating conditions, Sulfolane solvent undergoes only minor oxidative degradation. A solvent regenerator is included in the design of the unit to remove degradation products due to air leakage into the unit. During normal operation, a small slip stream of circulating solvent is directed to the solvent regenerator for removal of oxidized solvent. At many facilities, the solvent regenerator is operated only intermittently, as required.

4.3.1.1 Operating Variables
4.3.1.1.1 Extractor Recycle Ratio
The extractor recycle ratio is the molar ratio of stripper overhead hydrocarbons to total aromatics in the feed to the extractor (ER/E). The extract rate can be closely approximated by assuming that all of the aromatics in the feed will be extracted. ER/E is typically in the range of 0.6–1.1 depending on the feed characteristics.

Solvent loading is the molar percent of hydrocarbon in the rich solvent. Solvent loading should not typically exceed 33 mol.%. If secondary solvent is used, the secondary solvent rate must be subtracted from the lean solvent rate.

$$\text{Solvent loading} = \frac{(E + ER)*100}{E + ER + LS} \quad (4.1)$$

where

E = total aromatics in the feed to the extractor (mol)
ER = extractor recycle (stripper net overhead hydrocarbons) (mol)
LS = lean solvent flow rate (mol)

4.3.1.1.2 Primary Solvent Temperature to Extractor
The temperature profile in the extractor affects the selectivity of the solvent. Higher temperature decreases the selectivity of the solvent. Lower temperature increases selectivity. The temperature of the primary solvent and fresh feed to the extractor and, to a lesser extent, the temperature of the extractor recycle and tertiary solvent (if any) affect the extractor temperature profile. Primary solvent is the solvent added at the top of the extractor. The temperature of the primary solvent changes the relative selectivity of the solvent for aromatics and nonaromatics and, to a minor extent, the capacity of the solvent. While a lower temperature increases selectivity, it decreases the capacity of the solvent for aromatics.

One of the primary means of energy recovery in the liquid–liquid extraction unit is via the lean-rich solvent exchanger. Hot lean solvent from the recovery column (via the water stripper) heats the rich solvent from the bottom of the extractor going to the stripper. The lean-rich solvent exchanger has a temperature-controlled bypass used to control the lean solvent temperature to the extractor. As the lean solvent temperature to the extractor is reduced (more heat transfer in the lean-rich solvent exchanger), the rich solvent temperature to the stripper increases, resulting in greater stripper feed flash.

Typically, the temperature of the lean solvent to the extractor (primary solvent) is in the range of 80–100 °C (176–212 °F). At constant extractor recycle ratio (ER/E), with increasing temperature, aromatics product purity decreases due to decreased solvent selectivity. At the same time, the recoveries of benzene and toluene also tend to decrease. Lower temperature lean solvent is better from the standpoint of product purity and recovery; however, at some point (less than about 80 °C) the lean-rich solvent exchanger requires more than four shells in series. Some designers feel that this is impractical, but it may be justified economically. Lighter feeds (less or no C_8's), requiring less stripping steam (and consequently having a higher lean-rich solvent exchanger inlet temperature) can operate with lower lean solvent temperatures.

The temperature profile in the extractor will affect the selectivity of the solvent. Higher temperatures increase the solvency and, consequently, decrease the selectivity of the solvent. Primary solvent temperature to the extractor, fresh feed temperature, and, to a lesser extent, extractor recycle and tertiary solvent temperatures affect the extractor temperature profile.

4.3.1.1.3 Number of Extractor Stages

The extractor typically contains six to nine theoretical stages, depending upon the required recoveries of the aromatic components and the required number of backwash stages. Theoretical stages above the feed point are responsible for aromatics component recovery. Typically, not more than six theoretical stages above the feed point are required. Tray efficiency is conservatively estimated to be about 12.5%.

Backwash stages are below the feed point and improve the purity of toluene and xylenes. For feeds containing only benzene, no backwash stages are required. The stripper is responsible for the purity of the benzene product. For feeds containing benzene, toluene, and xylenes, one to three backwash stages in the extractor may be required to achieve the desired C_7 and C_8 aromatic purities.

4.3.1.1.4 Number of Stripper Stages

The stripper generally has 16–18 theoretical stages as necessary to achieve benzene purity. Tray efficiency is conservatively estimated to be about 50%.

4.3.1.1.5 Stripper Temperature and Pressure

The stripper operating temperature is limited by the allowable temperature of the solvent to avoid thermal degradation. The bottoms temperature is typically held at about 174 °C (345 °F). This bottoms temperature and the bottoms composition then set the column pressure. In order to improve benzene purity, the flow of extractor recycle is increased with a corresponding increase in the reboiler duty.

4.3.1.1.6 Stripper Receiver Temperature and Pressure

The stripper receiver is generally operated at a temperature that can be achieved by air cooling only to avoid the use of a water-cooled trim condenser and its additional associated pressure drop. The vapor pressure of the extractor recycle must be less than the pressure in the receiver to avoid venting valuable benzene on a continuous basis. The operating pressure of the receiver is limited by the hydraulics of the steam coming from the water stripper column which is rather low. The receiver cannot be pressurized with a push–pull system because of this. Typically, the stripper receiver floats on the relief header to minimize its operating pressure. If the minimum temperature required to keep the vapor pressure below the operating pressure is not feasible with only air condensing, then a water-cooled condenser is used (typically without an air-cooled condenser) with an outlet temperature low enough such that the vapor pressure is satisfactory.

4.3.1.1.7 Secondary Solvent

Secondary solvent is lean solvent that is added to the stripper feed. Secondary solvent is used to increase the stripper temperature when the stripper is not pressurized, increasing the relative stripping factors for the nonaromatics and increasing the vapor/liquid ratio without decreasing solvent loading. It is generally not used when the stripper is pressurized, which is the case in most modern units. Secondary solvent may be sourced from a few different locations. Most commonly it is drawn from upstream or downstream of the lean/rich solvent exchanger. In some cases it may originate from upstream of the water stripper reboiler, although this is not preferred.

4.3.1.1.8 Tertiary Solvent

Tertiary solvent is lean solvent that is added to the fresh feed to the extractor. With rich feeds that contain more than 70 mol.% aromatics, tertiary solvent is used to limit primary solvent/raffinate molar ratio (PS/R) to about 10. When the PS/R > 10, the solvent carries excessive nonaromatics down the extractor, making it difficult to make aromatics product purity at reasonable ER/E. Phase separation at the top of the extractor also becomes more difficult, allowing more solvent to be carried over from the top of the extractor to the raffinate water wash column. When required, the tertiary solvent rate is gradually adjusted to keep the PS/R < 10. Because most feeds contain less than 70 mol.% aromatics, tertiary solvent is generally not required.

4.3.1.1.9 Extractor Recycle Drag

Extractor recycle drag is a small portion of the extractor recycle (usually <10%) that is directed to an upper feed point in the extractor in order to remove light olefins that may accumulate in the extractor recycle. Since the recycle contains benzene and toluene, feeding it too high in the extractor will result in a loss of these aromatics to the raffinate. The recycle drag allows for the reduction in the total recycle rate while keeping the aromatics purity constant.

4.3.2 Extractive Distillation

Aromatics extraction using ED eliminates the liquid–liquid extraction unit operation and uses just the extractive distillation unit operation. As discussed for liquid–liquid extraction, the solvent (most commonly Sulfolane or similar volatilities. As the hydrocarbon vapor flows up the ED column, countercurrent to the descending solvent, the aromatics are selectively absorbed. The basic process flow through the ED column as well as the function of each section is illustrated in Figure 4.3. Figure 4.4 is a flow diagram of an extractive distillation-type aromatics extraction unit.

The ED column takes the place of two columns in the conventional liquid–liquid extraction process, the extractor and the stripper. The ED column consists of three sections. The upper section consists of those stages above the solvent feed tray. The function of the upper section stages is to remove solvent from the overhead product. The overhead vapor is condensed and becomes the nonaromatic product which is referred to as the raffinate. A portion of the raffinate liquid is used as column reflux to rectify entrained solvent out of the overhead product. Overhead water is collected in the overhead receiver water boot and returned to the unit water circuit.

The middle section of the ED column consists of those stages between the solvent entry point and the feed entry point. The function of the middle section stages of the ED column is the absorption of the aromatics components by the solvent.

In the lower section of the ED column, the stages below the feed entry point, the nonaromatics are preferentially stripped out of the liquid and enter the middle portion of the column as a vapor phase due to the solvent selectivity, which has made the nonaromatic components relatively more volatile than the aromatic components. Again, because of finite selectivity, some aromatics, primarily benzene, are stripped into the middle section of the column where they must be reabsorbed. The lower section of the ED column serves the function of benzene purification.

The ED column is typically reboiled with steam, although it is possible to use a process stream or hot oil as the heating medium. The ED column bottoms contain solvent and highly purified aromatics. These materials are sent to the solvent recovery column (solvent stripper

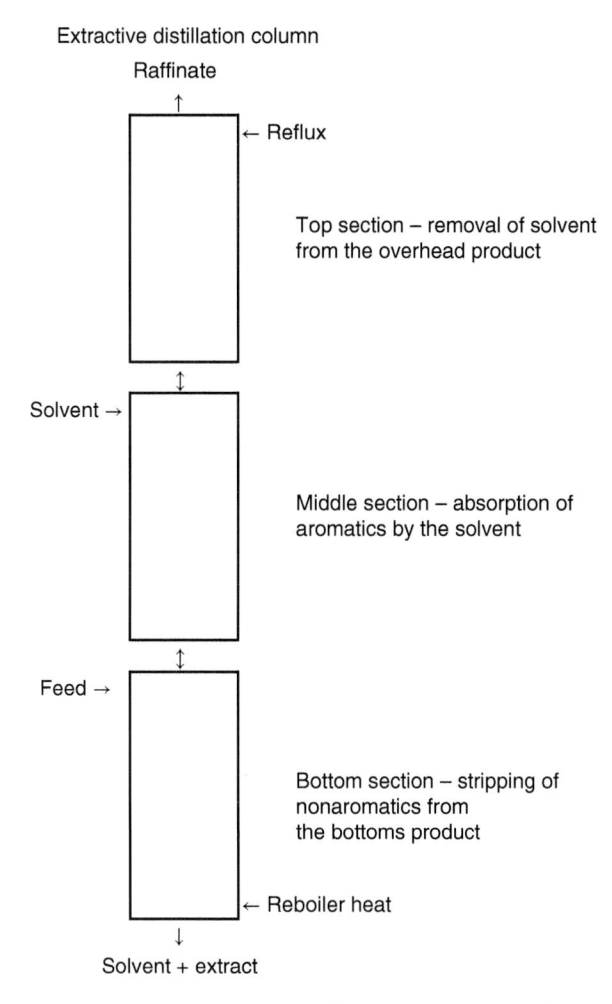

Figure 4.3 Extractive distillation column sections and their functions.

morphyline) alters the relative volatilities of the components to be separated due to the nonideal behavior of the mixture. The solvent renders the aromatic components to be less volatile than the nonaromatic components. Thus, in the presence of the solvent, the aromatics and nonaromatics can be separated by distillation, which would be impossible without the solvent due to their

Figure 4.4 Sulfolane aromatics extraction unit flow diagram (extractive distillation type).

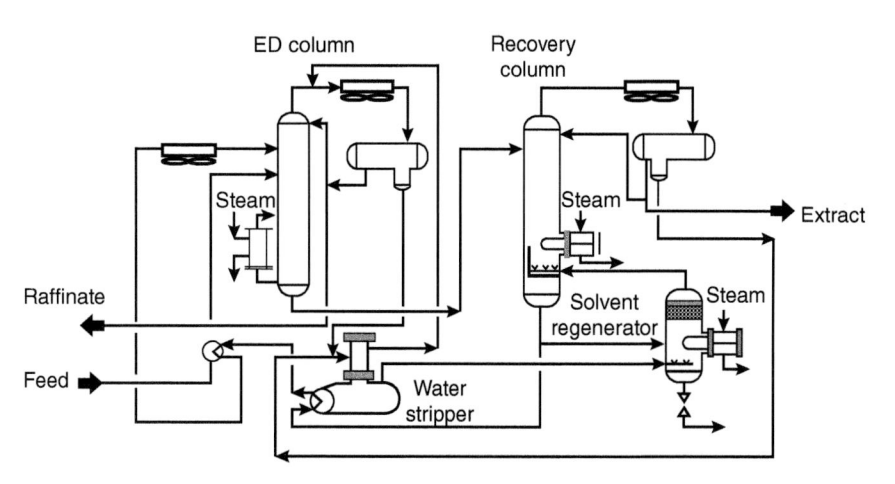

column). A reboiler preheater, using lean solvent as the heating medium, is a good choice for heat integration and improved thermal efficiency as it transfers some of the column heat input from near the top of the column to the bottom of the column where it is more efficiently used.

4.3.2.1 Operating Variables

4.3.2.1.1 Solvent: feed Ratio

Typically, with Sulfolane solvent, the solvent : feed ratio is about 2.5 : 1 on a mole basis or about 4 : 1 on a weight basis, although it can vary significantly. One might think that an aromatics extraction unit designed with a low solvent circulation rate (low solvent : feed ratio) would have lower capex and opex than one designed with a higher rate. However, this is not necessarily true. The only operating cost directly related to the solvent circulation rate is the electric power associated with solvent pumping, which is not a major operating cost for the unit. The solvent is not vaporized by the reboilers, so the reboiler duty does not change much with the solvent : feed ratio.

With respect to capital cost, a higher solvent rate will increase the diameters of the ED column and the recovery column but not the height, as the number of stages will be about the same in the ED column and the same in the recovery column regardless of the solvent rate. However, a higher solvent rate results in improved relative volatility between aromatics and nonaromatics (better selectivity) which leads to less severe column operation in the form of lower internal reflux required to accomplish the separation. Lower reflux results in lower reboiler duty and condenser duty, so it is likely that higher solvent : feed ratio will result in a smaller ED column reboiler and condenser. While a case-by-case economic evaluation is needed, it is often found that the capital costs are approximately equal and the operating costs lower when the unit is designed with a higher solvent rate and thus solvent : feed ratio.

4.3.2.1.2 Lean Solvent Feed Temperature

The lean solvent feed temperature is important for two main reasons:

1) It affects benzene recovery because lower temperature solvent absorbs nonaromatic components from a vapor phase better than a higher temperature solvent.
2) It affects the benzene purity because the quantity of heat that enters the column with the lean solvent impacts the amount of heat that can be added by the reboiler. Lower temperature lean solvent allows for greater reboiler duty which results in higher vapor rates and better stripping of nonaromatics on the trays below the fresh feed inlet. A lean solvent temperature of 77–95 °C (170–200 °F) is typical.

4.3.2.1.3 Fresh Feed Temperature

In the ED column, as in any distillation column, the temperature increases from stage to stage as one goes down the column. The ED column is provided with a reboiler at the bottom of the column, but no other external heat input above that, other than the heat that enters with the lean solvent and the fresh hydrocarbon feed. The heat needed to raise the temperature of the liquid flowing down the column comes from condensation of vapor within the column. The liquid flowing down the column is primarily solvent and requires a large amount of condensing hydrocarbon vapor to raise its temperature. Below the solvent inlet and above the fresh feed inlet the hydrocarbons in the column are mostly nonaromatics. If the quantity of nonaromatic condensation is excessive in order to satisfy the column temperature profile, the solvent may not be sufficient to dissolve all of the condensate. In this case, the liquid may split into two liquid phases, a solvent phase and a hydrocarbon phase. This is not desirable because it injects uncertainty in the simulation of the column, particularly the expected efficiency of the affected trays.

It is desirable to heat the fresh hydrocarbon feed via heat exchange with the lean solvent in order to introduce some heat lower in the column than the solvent feed location. The feed temperature should be such that some vaporization of the feed occurs. Depending on the composition of the feed, this temperature is in the range of 93–110 °C (200–230 °F).

It is not adequate to simply increase the temperature of the solvent feed, as this reduces the amount of heat that can be added by the reboiler which is needed to effect aromatic purification in the section of the column between the feed and the bottom.

4.3.2.1.4 ED Column Number of Stages

The ED column can be divided into three sections. The upper section above the lean solvent inlet stage; the middle section between the lean solvent inlet stage and the fresh feed stage; and the bottom section below the fresh feed stage. The number of stages above the lean solvent inlet stage affects the amount of reflux needed to reduce the solvent content of the overhead product (raffinate). The number of stages between the lean solvent inlet stage and the fresh feed stage affects the recovery of aromatics and the number of stages below the fresh feed stage affects the purity of the aromatics product.

The tray efficiency for the ED column is assumed to be about 50%. The total number of real trays in the ED column of a Sulfolane-based extractive distillation unit is typically in the range of 60–74 which is equivalent to 30–37 theoretical stages.

4.3.2.1.5 ED Column Reflux/Distillate Ratio

Some of the condensed overhead nonaromatic hydrocarbon is returned to the column as reflux to remove

Sulfolane solvent from the overhead raffinate product. The greater the reflux, the greater the required reboiler duty so the reflux should not be any greater than needed to achieve a reasonable level of Sulfolane in the raffinate and the desired aromatics purity. The reflux : distillate ratio depends on the feedstock composition, with higher reflux required for rich feedstocks (containing more aromatics and less nonaromatics) and lower reflux required for lean feedstocks. Reflux : distillate ratio can range anywhere from 0.2 to >1.0.

4.3.2.1.6 ED Column Bottoms Temperature

With Sulfolane solvent, the ED column bottoms temperature is typically about 174 °C (345 °F), with a range of 168–182 °C (335–360 °F). It is a function of the bottoms composition, including the water content of the solvent, and the column pressure. The ED column bottoms temperature is usually the same as the recovery column bottoms temperature.

4.3.2.1.7 ED Column Pressure

The ED column is operated above atmospheric pressure at as low a pressure as possible that allows the pressure control system to operate. The ED column operating pressure is usually determined by the bottoms temperature specified and the bottoms composition. Typically, the pressure is approximately 1.0–1.5 kg/cm^{2}g (14.2–21.3 psig).

4.3.2.1.8 Solvent Recovery Column

The solvent recovery column, also known simply as the recovery column, separates the aromatics extract product from the solvent. This column's design is essentially the same whether included in a conventional liquid–liquid-type aromatic extraction unit or an extractive distillation-type unit. The primary difference is the routing of water from the overhead of the recovery column. In a conventional liquid–liquid extraction unit, the water is used to wash the raffinate product to remove solvent before being sent to the stripping steam generator. Because there is no need to water wash the raffinate in an extractive distillation unit, the water recovered in the overhead receiver of the recovery column goes directly to the stripping steam generator.

4.3.2.1.9 Stripping Steam Ratio

Steam is introduced below the recovery column reboiler to help strip aromatics from the lean solvent. The stripping steam ratio is defined as the ratio of the moles of steam to the moles of lean solvent leaving the bottom of the column. The stripping steam ratio is typically between 0.09 and 0.12. The higher the carbon number of the aromatic components to be separated, the higher the stripping steam ratio needs to be, with benzene-only feeds requiring a stripping steam ratio of 0.09–0.10 and benzene–toluene feeds 0.11–0.12.

4.3.2.1.10 Recovery Column Number of Stages

For a Sulfolane-based unit, the tray efficiency for the recovery column is typically assumed to be 33%. When only a lower reboiler is used, the total number of real trays in the recovery column is usually 30, which is equivalent to 10 theoretical stages. As will be discussed in the next section, sometimes an upper reboiler is used in addition to the lower reboiler. With this configuration a few more trays are added to total 34 real trays or about 11 theoretical stages.

4.3.2.1.11 Recovery Column Reboilers

The recovery column always has a reboiler at the bottom, normally heated by steam, referred to as the lower reboiler. Sometimes the column is designed with a second reboiler, also heated by steam, located about halfway up the column, referred to as the upper reboiler. The purpose of the upper reboiler is to reduce the amount of stripping steam required. This is an economic decision, increased capex for second reboiler vs. less steam consumption, with evaluation to be carried out in the basic engineering design phase. Both reboilers are horizontal, stab-in-type exchangers.

4.3.2.1.12 Recovery Column Reflux/Distillate Ratio

The reflux : distillate ratio for the recovery column is determined by rigorous column simulation to achieve not more than 1 wt-ppm solvent in the overhead extract product. The resulting reflux : distillate ratio on a mole basis is typically in the range of 0.25–0.40.

4.3.2.1.13 Recovery Column Bottoms Temperature

The recovery column bottoms temperature is usually in the range of 174–182 °C (345–360 °F). The bottoms temperature is in the lower portion of the range when an ED column reboiler preheater is not used and in the higher portion of the range when an ED column reboiler preheater is used. The higher temperature of the recovery column bottoms (lean solvent) is to increase the driving force and thus heat exchange in the preheater exchanger.

4.3.2.1.14 Recovery Column Receiver Temperature

It is desirable to minimize the recovery column receiver temperature to minimize the loss of aromatics to vent (flare) as well as to minimize the load on the vacuum-producing equipment (ejector or vacuum pump). The design temperature is usually 38–40 °C (100–104 °F) and in most climates a water-cooled condenser or trim condenser is required to achieve this.

4.3.2.1.15 Recovery Column Pressures

The recovery column bottoms pressure is determined by the bottoms temperature and the composition of the bottoms material. Since there is only a very small amount of aromatics in the bottoms, it is composed of almost entirely solvent and water. The pressure is thus determined by the water content of the solvent at the bottom of the column and its temperature. The receiver pressure is then determined by subtracting the column tray pressure drop, overhead line loss, and overhead condenser pressure drop from the bottoms pressure.

4.4 Transalkylation

The transalkylation process is used to selectively convert toluene and C_9+ aromatics (A_9+) into benzene and xylenes. The term transalkylation or TA describes the conversion of a mixture of toluene and A_9+ into xylenes. The conversion of toluene alone into benzene and xylenes is called toluene disproportionation, or TDP. For simplicity, the unit that accomplishes both of these reactions is referred to as a transalkylation process unit. Incorporating a transalkylation process unit into an aromatics complex maximizes the yield of high-value benzene and *para*-xylene products and minimizes the production of lower-value toluene and heavy aromatic by-products.

Typically, the transalkylation process unit is integrated between the aromatics extraction and xylene recovery sections of the plant (Figure 4.5). Extracted or unextracted toluene is fed to the transalkylation unit rather than being blended into the gasoline pool or sold for solvent applications. The A_9+ material can also be fed to the transalkylation unit rather than blending it into the gasoline pool. Processing A_9+ in a transalkylation unit shifts the chemical equilibrium in the unit away from benzene production and towards the production of xylenes.

A transalkylation process unit normally consists of a reactor section and a fractionation section. The reactor section is similar to that found in most hydrotreating process units in that it consists of the same main components, although the particular design conditions and parameters differ. The reactor section typically includes six major pieces of equipment: a charge pump, a combined-feed effluent heat exchanger, a charge heater, a reactor, a product separator, and a recycle gas compressor. The function and pertinent design features of each of these equipment items will follow.

4.4.1 Process Flow Description

A transalkylation process unit uses a simple flow scheme consisting of a fixed-bed reactor and a product separation section (Figure 4.6). The fresh feed to the unit is first combined with hydrogen-rich recycle gas, preheated and vaporized by exchange with the hot reactor effluent, and then heated further in a fired heater where it is raised to reaction temperature. The hot feed vapor is then sent to the reactor, where it is sent downflow over a fixed bed of catalyst. The reactor effluent is then cooled by exchange with the combined feed, which is mixed with make-up gas to replace the hydrogen consumed by the reactions, condensed, and then sent to a product separator. Hydrogen-rich gas is taken off the top of the product separator, compressed in the recycle compressor, and recycled back to the feed upstream of the combined feed exchanger (CFE). A portion of the recycle gas is purged to remove accumulated light ends from the recycle gas loop, if required to maintain recycle gas hydrogen purity. Liquid from the bottom of the product

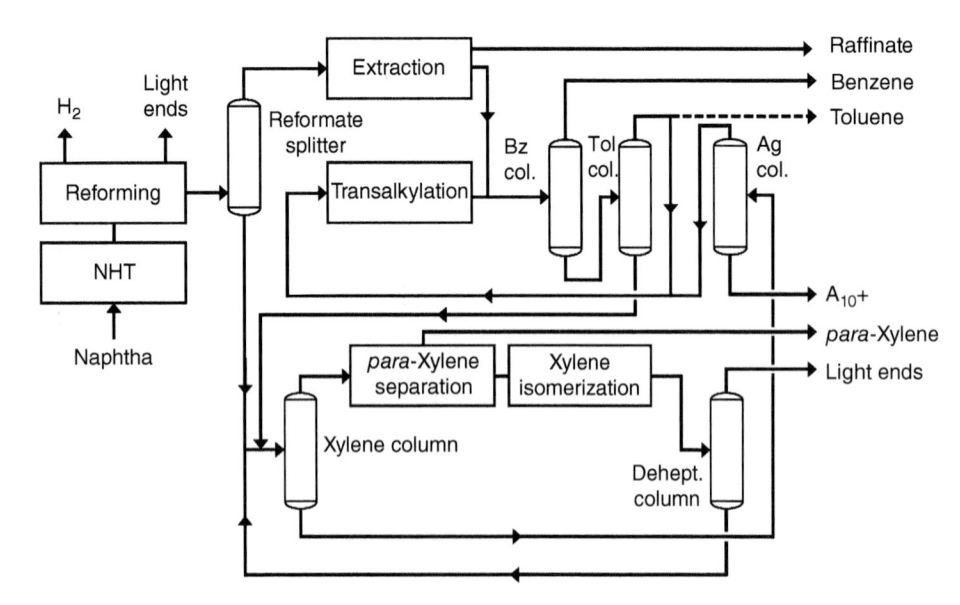

Figure 4.5 Location of transalkylation unit in an integrated aromatics complex.

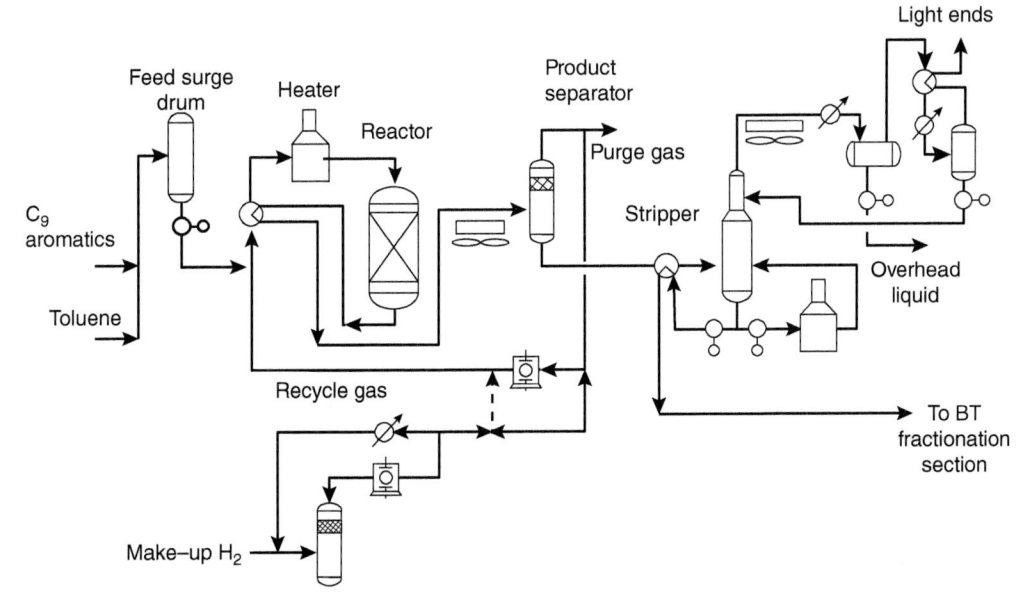

Figure 4.6 Transalkylation unit flow diagram.

separator is normally sent to a stripper column. The C_5- from the overhead of the stripper is cooled and separated into gas and liquid products. The stripper overhead gas is a light end stream consisting primarily of methane, ethane, and propane and is normally sent to the fuel gas system. The stripper net overhead liquid is composed of propane, butanes, pentanes, and C_6 compounds. It is usually combined with other similar light end containing liquids such as that from the isomerization unit or reforming unit and may be sent to the aromatics extraction unit to recover the benzene that it contains. The benzene and xylene products, together with the unreacted toluene and C_9+ aromatics, are taken from the bottom of the stripper and recycled back to the benzene column of the aromatics complex.

Depending on the relative rates of overhead liquid from the transalkylation unit and the xylene isomerization unit, the stabilization of these streams may be done in a stabilizer column in the transalkylation unit rather than the stripper column in the xylene isomerization unit. If a very high-purity benzene product is required to be produced by the aromatics complex, such as that needed to make cyclohexane, the benzene from the transalkylation unit will need to be sent to the aromatics extraction unit.

4.4.1.1 Combined Feed Exchanger

The CFE is also known as the combined feed-effluent exchanger. The CFE transfers heat from the hot reactor effluent to the cold combined feed (fresh liquid feed combined with recycle gas). It is by far the exchanger with the largest duty in the transalkylation unit and the exchanger that recovers the most heat from another stream in the unit. The CFE typically provides over 90% of the total feed heating duty (CFE duty plus charge heater duty). The combined feed is completely vaporized in the CFE. This is necessary because the material goes from the CFE to the charge heater to be heated to the desired reactor inlet temperature. The charge heater typically has multiple passes so the combined feed needs to be split evenly among the passes by the use of control valves. It is much easier to split flow with control valves if the material is single phase rather than multiphase.

The CFE also serves the dual purpose of cooling and partially condensing the reactor effluent. The reactor effluent comes to the CFE directly from the reactor outlet and is all vapor phase. There are three different types of heat exchangers used for CFE service in transalkylation units, as follows:

- Conventional multi-shell horizontal shell and tube-type heat exchangers
- Vertical shell and tube heat exchangers (also known as "Texas Towers")
- Welded-plate-type heat exchangers

The type of exchanger used in CFE service is based on economics and has changed over the years. Welded-plate exchangers allow for operation with a lower weighted MTD, and thus, greater heat recovery than does a vertical shell and tube, which can operate with a lower MTD and greater heat recovery than a multiple-shell horizontal shell and tube heat exchanger. Typically, the energy saving that is a result of the greater heat recovery easily justifies the higher capital cost of the welded-plate exchanger and thus this type of CFE has been used almost exclusively since the mid-1990s. Prior to the late 1980s, most

CFE's in transalkylation unit service were either conventional multiple-shell horizontal shell and tube heat exchangers or vertical shell and tube exchangers. Because transalkylation is a very clean service it is a very good application of welded-plate-type exchangers. Packinox, now owned by Alfa-Laval, is the predominant supplier of welded-plate exchangers in transalkylation service.

It is important to note that because the type of CFE affects its duty and temperature profile and therefore the temperature profile of the reactor section, the heat and weight balance of the transalkylation unit and the design of other equipment in the reactor section depend greatly on the choice of CFE type. Therefore, this must be decided early in the design basis phase of the project. Greater heat recovery in the CFE reduces the required duty of the charge heater and products condenser. Greater heat recovery in the CFE also increases the inlet temperature of the combined feed going to the charge heater and reduces the inlet temperature of the reactor effluent going to the products condenser.

Horizontal multi-shell shell and tube exchangers are generally not economical in transalkylation unit CFE service due to their high pressure drop that results in greater recycle gas compressor horsepower and reduced heat recovery capability. In addition, horizontal shell and tube exchangers require significantly more plot space than do either welded-plate exchangers or vertical shell and tube exchangers. However, this type of CFE may be found on some older transalkylation units. This is an obvious candidate for replacement with a welded-plate exchanger when revamping the unit for either increased capacity or improved energy efficiency as it may preclude the need to replace or augment the charge heater and/or the products condenser.

In a welded-plate exchanger or vertical shell and tube heat exchanger, the reactor effluent enters at the top of the exchanger and flows down exiting near the bottom of the exchanger. The combined feed enters a welded-plate exchanger or vertical shell and tube heat exchanger near the bottom and flows upward, countercurrent to the reactor effluent.

In very large transalkylation units multiple vertical shell and tube exchangers or welded-plate exchangers may be required. When this occurs, great care must be taken in the piping layout to assure good flow distribution to each of the exchangers. This is not a problem on the combined feed side as control valves can be used to divide the flows of liquid feed and recycle gas. It is more problematic on the reactor effluent side; however, due to the high temperature of the reactor effluent.

4.4.1.2 Charge Heater

The charge heater heats the vaporized combined feed from the CFE to the required reactor inlet temperature. The reactor inlet temperature increases from start-of-run (SOR) to end-of-run (EOR) as the catalyst in the reactor deactivates. The deactivation time, or catalyst life, depends on the feed composition, feed contaminants, and the catalyst itself. Typical SOR temperatures range from 340 to 370 °C (650–700 °F). The EOR reactor inlet temperature can be as high as 480 °C (900 °F). However, these temperatures may vary and are dictated by the catalyst supplier.

The charge heater is typically specified as an all-radiant, wicket-type fired heater. This is also known as an arbor-type heater. A wicket- or arbor-type heater has tubes that are in the shape of the wickets used in the game croquet. Each tube is in the shape of an arch, hoop, or an upside-down U. One end of each tube is connected to the inlet header (manifold) and the other end of each tube is connected to the outlet header (manifold). The inlet and outlet manifolds are at the bottom, parallel to each other, and perpendicular to the wickets (tubes). These heaters allow for many parallel passes and thus high capacity/high duty. The primary reason a wicket-type heater is specified is that it can operate with very low pressure drop of 0.2–0.3 kg/cm^2 (3–5 psi). This is in contrast to a vertical-cylindrical-type heater which operates with a higher pressure drop of at least 0.4 kg/cm^2 (6 psi). It is desirable to minimize the reactor circuit pressure drop, and thus the compressor head required, in order to minimize the size and horsepower (utility cost) of the recycle compressor.

Wicket-type heaters with bottom-mounted manifolds allow for the burners to be located in the floor as well as allowing for both gas and liquid fuel firing. A convection section is typically located above the radiant section for additional process heat recovery or steam generation.

Despite costing up to 25% more than a vertical-cylindrical heater, the lower pressure drop typically results in a payout of less than one year. Other advantages of the wicket-type heater are as follows:

- Tubes are spring supported so piping movements can be absorbed.
- Larger burner-to-tube clearance than vertical-cylindrical-type heater resulting in reduced flame impingement and therefore reduced maintenance and longer tube life.
- Easy field installation.

Due to the required operating and thus design temperatures, the tubes are typically specified to be 2-1/4Cr–1Mo alloy. One drawback of the wicket-type heater is that it requires somewhat more plot space than a vertical-cylindrical type.

The design duty of the charge heater needs to have enough margin for a couple of operating scenarios. First, even if the CFE can be designed to provide nearly all of the heat input to the combined feed, a portion of the total heat duty must be provided by the charge heater to allow for control of the reactor inlet temperature as that control cannot be provided by the CFE. This is particularly true if there is, or could be, a significant exothermic reaction in the reactor in which case the reactor effluent temperature could be high enough to "overheat" the combined feed above the desired reactor temperature. A CFE bypass is provided for this situation so that the inlet temperature to the charge heater is always maintained far enough below the desired reactor inlet temperature to allow the charge heater to be operated above the minimum firing level.

The second operating scenario that requires design margin in the charge heater is loss of performance of the CFE due to fouling. While the transalkylation process is relatively clean and significant fouling of the CFE is not expected, it is possible, especially in the presence of feed or recycle gas contaminants.

4.4.1.3 Reactor Design

The reactor in a transalkylation unit is typically a conventional downflow design with the inlet at the top and the outlet at the bottom. Because it is all vapor flow through the reactor, good flow distribution is relatively easy to achieve. Well-distributed flow is achieved by two primary means: an inlet flow distributor and sufficient reactor bed pressure drop. The inlet flow distributor is a baffled device that is mounted in the reactor inlet nozzle (see Figure 4.7). It is typically cylindrical with vanes and/or perforated plates that are designed to prevent a jet of flow from going straight down the center of the reactor. The flow is directed radially toward the walls of the reactor allowing it to distribute across the entire cross-section of the reactor. There must be adequate distance between the bottom of the inlet distributor and the top of the catalyst bed to allow the flow to distribute and prevent the top of the catalyst bed from being disturbed.

4.4.1.4 Catalyst Volume

The active catalyst volume required for the reactor is based on the specified weight hourly space velocity (WHSV) and is calculated using the following equation:

$$V = \frac{CF}{WHSV * \rho} \quad (4.2)$$

where

V = catalyst volume (ft^3)
CF = combined feed (lb/h)
ρ = catalyst density (lb/ft^3)
WHSV = weight hourly space velocity (per hour)

The WHSV is specified by the catalyst supplier. The combined feed flow rate is that of the total liquid feed to the unit, including recycle, but excluding recycle gas. The WHSV is essentially the inverse of residence time. As space velocity increases, catalyst volume decreases and residence time decreases.

4.4.1.5 Bed Pressure Drop

The pressure drop of the catalyst bed is calculated using the Ergun equation, as follows:

$$\Delta P = \frac{8.7 \times 10^{-7}(1-\varepsilon)^2 \mu \cdot G}{\varepsilon^2 D_p^2 \rho} + \frac{3.5 \times 10^{-10}(1-\varepsilon)G^2}{\varepsilon^3 D_p \rho} \quad (4.3)$$

where

ΔP = bed pressure drop (psi/ft)
ε = catalyst void fraction
D_p = effective catalyst diameter (inch)
G = fluid superficial mass velocity (lb/h-ft^2)
μ = fluid viscosity (cP)
ρ = fluid density (lb/ft^3)

It has been shown from computational fluid dynamic (CFD) flow modeling that the pressure drop of the catalyst bed and any catalyst support material, such as inert ceramic-alumina balls, causes the flow to distribute across the reactor cross-sectional area. Higher pressure drop results in better flow distribution. However, there are trade-offs. As discussed previously, high pressure drop results in greater power consumption by the recycle gas compressor and excessive velocity may result in disturbing the catalyst bed and causing the production of catalyst fines. Therefore, like most operating parameters, there is a happy medium. CFD flow modeling has shown that sufficient bed pressure drop results in good flow distribution (without excessive velocity and imposing too

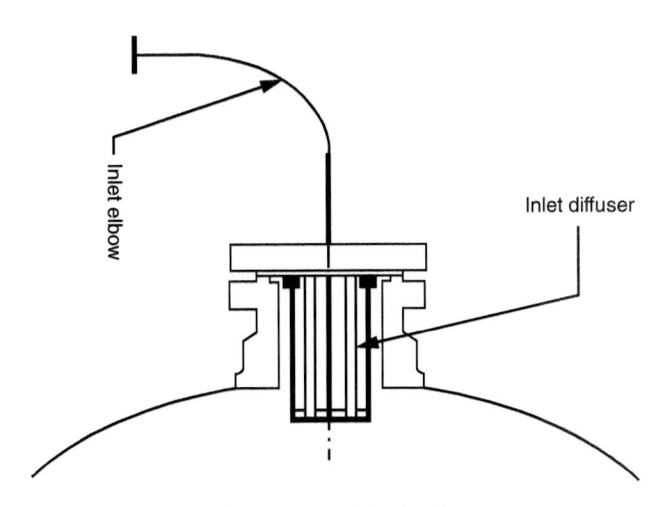

Figure 4.7 Transalkylation reactor inlet distributor.

much pressure drop). For a medium to large reactor, the overall bed pressure drop should be in the range of 0.28–0.7 kg/cm^2 (4–10 psi).

4.4.1.6 Reactor Bed Dimensions

The diameter of the reactor is based on achieving the desired bed pressure drop as discussed above. Once the diameter of the reactor has been calculated, the bed length may be determined to achieve the desired catalyst volume using the following equation:

$$L = \frac{V}{D^2 / 0.7854} \qquad (4.4)$$

where

L = bed length (height) (m)
V = catalyst volume (m^3)
D = reactor diameter (m)

To assure good flow distribution, the bed length to diameter ratio (L/D) should be no less than 0.8.

4.4.1.7 Products Condenser

The reactor effluent is cooled and partially condensed in the CFE from which it flows to a products condenser. In most transalkylation units the products condenser is an air-cooled heat exchanger. Due to the temperature of the reactor effluent to the products condenser and the desired temperature from the condenser, an air-cooled exchanger is usually the most practical and economic choice. The inclusion of a water-cooled trim condenser downstream of the air-cooled condenser depends on several factors, including the design ambient air temperature and the relative costs of electricity, cooling water, and fuel. Because it is desirable to minimize the loss of benzene to the vapor leaving the separator, the reactor effluent is typically cooled to the lowest temperature achievable by the air-cooled condenser, but not greater than 55 °C (130 °F). Unless located where the design air temperature is very high, a water-cooled trim condenser is typically not required.

4.4.1.8 Separator

A separator vessel is located downstream of the products condenser to separate the liquid and vapor phases of the cooled reactor effluent and to function as a surge vessel for the liquid feed to the stripper. In a transalkylation unit a simple vapor–liquid separation is required as there is no separate water phase present. The separator vessel is typically vertical with a horizontal inlet into the side of the vessel, a vapor outlet at the top, and a liquid outlet at the bottom. Besides separating vapor from liquid, the separator also usually serves as a knockout drum to protect the downstream recycle compressor. To eliminate liquid droplets, a mesh blanket or mist eliminator is employed near the top of the separator vessel.

For transalkylation unit operating conditions, phase separation of the cooled and condensed reactor effluent is governed by Stokes' Law:

$$\mu_t = \frac{g_c D_p^2 (\rho_p - \rho)}{18\mu} \qquad (4.5)$$

where

u_t = droplet terminal velocity (ft/s)
g_c = gravitational constant (32.2 ft/s^2)
D_p = droplet diameter (ft) (a diameter of 150 μm is typically used)
ρ_p = droplet density (lb/ft^3)
ρ = density of continuous fluid (lb/ft^3)
μ = viscosity of continuous fluid (lb/ft^2)

The diameter of the vessel must be such that the vapor velocity is less than or equal to the droplet terminal velocity (u_t) for the selected droplet diameter.

4.4.1.9 Recycle Gas Purity

Two parameters that have a major impact on catalyst performance and cycle length are recycle gas purity and hydrogen : hydrocarbon (H$_2$/HC) ratio. Together these parameters, along with reactor operating pressure, determine the hydrogen partial pressure in the reactor. For transalkyation units the recycle gas purity is typically required to be somewhere between 70 and 80 mol.% while the H$_2$/HC ranges anywhere from 3 : 1 to 6 : 1 depending on the catalyst. Recycle gas purity is controlled by venting gas from the separator. As the venting rate is increased, the amount of make-up hydrogen entering the reactor circuit increases to maintain the separator pressure and this then increases the recycle gas hydrogen purity.

4.4.1.10 Recycle Gas Compressor

A recycle gas compressor (aka recycle compressor) is employed to achieve the required H$_2$/HC ratio dictated by the reactor operating conditions. The recycle gas compressor pulls gas from the separator and sends it back to the CFE where it is mixed with fresh liquid feed and provides the driving force to circulate it through the reactor circuit on a continuous basis. In a commercial-scale transalkylation unit the recycle compressor is typically a centrifugal-type machine. Older, smaller transalkylation units have employed other types of compressors such as reciprocating machines. For reasons discussed elsewhere, centrifugal compressors are less maintenance intensive and therefore more reliable than reciprocating-type machines and are therefore preferred in modern transalkylation units.

4.5 Xylene Isomerization

The xylene isomerization process is used to maximize the recovery of a particular xylene isomer from a mixture of C_8 aromatic isomers (xylenes + ethylbenzene). The term "mixed xylenes" is used to describe a mixture of C_8 aromatics which contains a near-equilibrium mixture of *para*-xylene, *ortho*-xylene, *meta*-xylene, along with some ethylbenzene. The xylene isomerization process is most often applied to *para*-xylene recovery, but it can also be used to maximize the recovery of *ortho*-xylene or *meta*-xylene. In the case of *para*-xylene recovery, a mixed xylene feed is charged to a *para*-xylene separation unit where the *para*-xylene isomer is preferentially extracted. The raffinate from the *para*-xylene separation unit, almost entirely depleted of *para*-xylene, is then sent to the xylene isomerization unit. The xylene isomerization unit re-establishes an equilibrium distribution of xylene isomers, essentially creating additional *para*-xylene from the remaining *ortho*- and *meta*-isomers. The effluent from the xylene isomerization unit is then recycled back to the Parex unit for recovery of additional *para*-xylene. In this way, the *ortho*- and *meta*-isomers are recycled to extinction.

There are two different types of catalysts used in xylene isomerization processes. Both catalysts are used to re-establish an equilibrium mixture of xylene isomers, but they differ in the way ethylbenzene is processed. An EB isomerization catalyst uses an isomerization reaction mechanism to convert ethylbenzene into additional xylene isomers. The other type of catalyst uses a dealkylation mechanism to convert ethylbenzene into benzene. The choice of catalyst depends upon the product distribution desired from the aromatics complex, as well as possibly the availability of feedstock and other economic considerations such as the capital cost of the aromatics complex.

It is important to understand the differences and advantages of the two types of xylene isomerization catalysts. An "EB-dealkylation" catalyst converts ethylbenzene in the feed to a valuable benzene coproduct, while an "EB-isomerization" catalyst converts ethylbenzene into additional mixed xylenes. The proper selection of isomerization catalyst type depends on the configuration of the aromatics complex, the composition of the feedstocks, and the desired product slate. Using an EB-isomerization catalyst maximizes the yield of *para*-xylene from a given complex by converting EB to xylenes. An EB-isomerization catalyst is usually chosen when the primary goal of the aromatics complex is to maximize production of *para*-xylene from a fixed amount of feedstock. Alternatively, an EB-dealkylation catalyst debottlenecks the Parex unit by converting more EB per pass through the isomerization unit and eliminating the requirement for naphthenes' intermediate circulation around the Parex–Isomar recycle loop. Thus, using an EB-dealkylation catalyst minimizes the amount of capital required to produce a given amount of *para*-xylene by reducing the size of the xylene column, Parex, and Isomar units. However, this reduction in investment comes at the expense of lower *para*-xylene yields, since all the EB in the feed is being converted to benzene rather than additional *para*-xylene. Once again, the choice of isomerization technology and catalyst must be based on an analysis of the entire aromatics complex.

A xylene isomerization process unit normally consists of a reactor section and a fractionation section (see Figure 4.8). The reactor section is similar to that found in the transalkylation process unit and most hydrotreating process units in that it consists of the same main components, although the particular design conditions and parameters differ. The reactor section typically includes

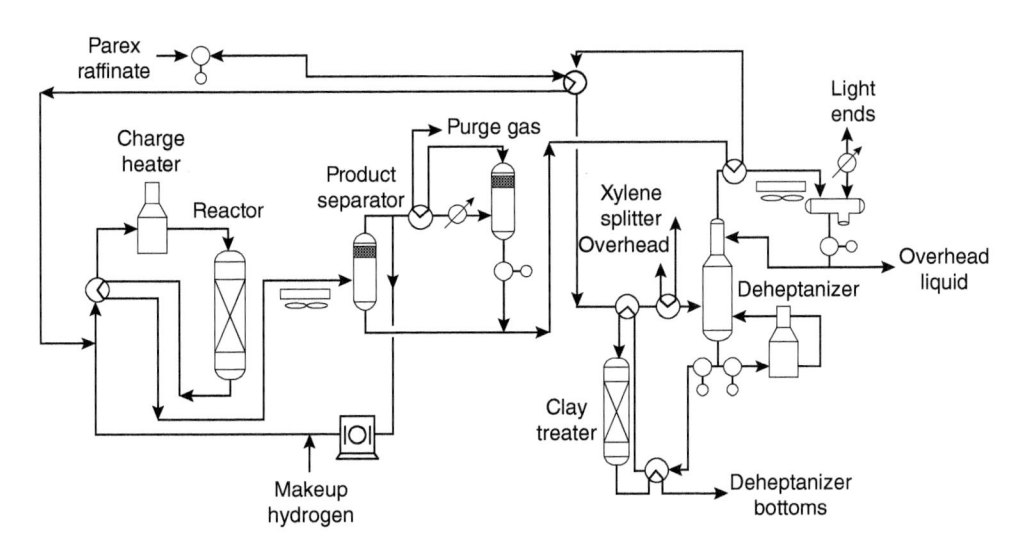

Figure 4.8 Xylene isomerization unit flow diagram.

six major pieces of equipment: a charge pump, a combined feed-effluent heat exchanger, a charge heater, a reactor, a product separator, and a recycle gas compressor. The function and pertinent design features of each of these equipment items will follow.

4.5.1 Combined Feed Exchanger

The following discussion is very similar to that included in Section 4.4. The CFE is also known as the combined feed-effluent exchanger. The CFE transfers heat from the hot reactor effluent to the cold combined feed (fresh liquid feed combined with recycle gas). It is by far the exchanger with the largest duty in the xylene isomerization unit and the exchanger that recovers the most heat from another stream in the unit. The CFE typically provides over 90% of the total feed heating duty (CFE duty plus charge heater duty). The combined feed is completely vaporized in the CFE. This is necessary because the material goes from the CFE to the charge heater to be heated to the desired reactor inlet temperature. The charge heater typically has multiple passes so the combined feed needs to be split evenly among the passes by the use of control valves. It is much easier to split flow with control valves if the material is single phase rather than multiphase.

The CFE also serves the dual purpose of cooling and partially condensing the reactor effluent. The reactor effluent comes to the CFE directly from the reactor outlet and is all vapor phase. There are three different types of heat exchangers used for CFE service in xylene isomerization units, as follows:

- Conventional multi-shell horizontal shell and tube-type heat exchangers
- Vertical shell and tube heat exchangers (also known as "Texas Towers")
- Welded-plate-type heat exchangers

The type of exchanger used in CFE service is based on economics and has changed over the years. Welded-plate exchangers allow for operation with lower approach temperatures, a lower "pinch" temperature, and a lower weighted MTD, and thus, greater heat recovery than does a vertical shell and tube, which can operate with a lower MTD and greater heat recovery than a multiple-shell horizontal shell and tube heat exchanger. Typically, the energy saving that is a result of the greater heat recovery easily justifies the higher capital cost of the welded-plate exchanger and thus this type of CFE has been used almost exclusively since the mid-1990s. Prior to the late 1980s, most CFE's in xylene isomerization unit service were either conventional multiple-shell horizontal shell and tube heat exchangers or vertical shell and tube exchangers. Because xylene isomerization is a very clean service, it is a very good application of welded-plate-type exchangers. Packinox, now owned by Alfa-Laval, is the predominant supplier of welded-plate exchangers in xylene isomerization service.

It is important to note that because the type of CFE affects its duty and temperature profile and therefore the temperature profile of the reactor section, the heat and weight balance of the xylene isomerization unit and the design of other equipment in the reactor section depend greatly on the choice of CFE type. Therefore, this must be decided early in the design basis phase of the project. Greater heat recovery in the CFE reduces the required duty of the charge heater and products condenser. Greater heat recovery in the CFE also increases the inlet temperature of the combined feed going to the charge heater and reduces the inlet temperature of the reactor effluent going to the products condenser.

Horizontal multi-shell shell and tube exchangers are generally not economical in xylene isomerization unit CFE service due to their high pressure drop that results in greater recycle gas compressor horsepower and reduced heat recovery capability. In addition, horizontal shell and tube exchangers require significantly more plot space than do either welded-plate exchangers or vertical shell and tube exchangers. However, this type of CFE may be found in some older xylene isomerization units. This is an obvious candidate for replacement with a welded-plate exchanger when revamping the unit for either increased capacity or improved energy efficiency as it may preclude the need to replace or augment the charge heater and/or the products condenser.

In a welded-plate exchanger or vertical shell and tube heat exchanger, the reactor effluent enters at the top of the exchanger and flows down, exiting near the bottom of the exchanger. The combined feed enters a welded-plate exchanger or vertical shell and tube heat exchanger near the bottom and flows upward, countercurrent to the reactor effluent.

In very large xylene isomerization units, multiple vertical shell and tube exchangers or welded-plate exchangers may be required. When this occurs, great care must be taken in the piping layout to assure good flow distribution to each of the exchangers. This is not a problem on the combined feed side as control valves can be used to divide the flows of liquid feed and recycle gas. It is more problematic on the reactor effluent side; however, due to the high temperature of the reactor effluent.

4.5.2 Charge Heater

The following discussion is very similar to that included in Section 4.4. The charge heater heats the vaporized combined feed from the CFE to the required reactor inlet temperature. The reactor inlet temperature increases from SOR to EOR as the catalyst in the reactor deactivates. The deactivation time, or catalyst life, depends on the feed composition, feed contaminants, and the catalyst itself. Typical SOR temperatures range from 340 to 370 °C (650–700 °F). The EOR reactor inlet temperature can be as high as 420–440 °C (788–824 °F). However, these temperatures may vary and are dictated by the catalyst supplier.

The charge heater is typically specified as an all-radiant, wicket-type fired heater. This is also known as an arbor-type heater. A wicket- or arbor-type heater has tubes that are in the shape of the wickets used in the game croquet. Each tube is in the shape of an arch, hoop, or an upside-down U. One end of each tube is connected to the inlet header (manifold) and the other end of each tube is connected to the outlet header (manifold). The inlet and outlet manifolds are at the bottom, parallel to each other, and perpendicular to the wickets (tubes). These heaters allow for many parallel passes and thus high capacity/high duty. The primary reason a wicket-type heater is specified is that it can operate with very low pressure drop of 0.2–0.3 kg/cm^2 (3–5 psi). This is in contrast to a vertical-cylindrical-type heater which operates with a higher pressure drop of at least 0.4 kg/cm^2 (6 psi). It is desirable to minimize the reactor circuit pressure drop, and thus the compressor head required, in order to minimize the size and horsepower (utility cost) of the recycle compressor.

Wicket-type heaters with bottom-mounted manifolds allow for the burners to be located in the floor as well as allowing for both gas and liquid fuel firing. A convection section is typically located above the radiant section for additional process heat recovery or steam generation.

Despite costing up to 25% more than a vertical-cylindrical heater, the lower pressure drop typically results in a payout of less than one year. Other advantages of the wicket-type heater are as follows:

- Tubes are spring supported so piping movements can be absorbed.
- Larger burner-to-tube clearance than vertical-cylindrical-type heater resulting in reduced flame impingement and therefore reduced maintenance and longer tube life.
- Easy field installation.

Due to the required operating and thus design temperatures, the tubes are typically specified to be 2-1/4Cr–1Mo alloy. One drawback of the wicket-type heater is that it requires somewhat more plot space than a vertical-cylindrical type.

The design duty of the charge heater needs to have enough margin for a couple of operating scenarios. First, even if the CFE can be designed to provide nearly all of the heat input to the combined feed, a portion of the total heat duty must be provided by the charge heater to allow for control of the reactor inlet temperature as that control cannot be provided by the CFE. This is particularly true if there is, or could be, a significant exothermic reaction in the reactor in which case the reactor effluent temperature could be high enough to "overheat" the combined feed above the desired reactor temperature. Xylene isomerization is only slightly exothermic, so this operating scenario is only a concern if the feed is contaminated with nonaromatics. Unlike transalkylation, a controlled bypass is not required because the exotherm is not significant.

The second operating scenario that requires design margin in the charge heater is loss of performance of the CFE due to fouling. While the xylene isomerization process is very clean and significant fouling of the CFE is not expected, it is possible, especially in the presence of feed or recycle gas contaminants.

4.5.3 Reactor Design

The reactor in a xylene isomerization unit is typically a radial flow design with the inlet at the top and the outlet at the bottom (see Figure 4.9). Flow from the inlet is forced toward the walls of the reactor, where it flows downward and then radially inward toward the center of the reactor. Typically, the flow is distributed by a set of "scallops" located around the periphery of the reactor.

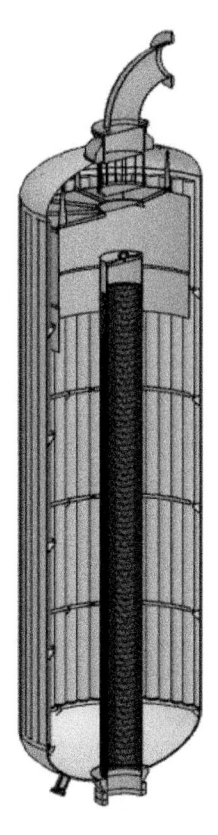

Figure 4.9 Radial flow reactor internals.

The purpose of the scallops is twofold, first to contain the catalyst from the outside, and second to help distribute the vapor flow of reactants evenly to the catalyst bed. A vertical centerpipe is located in the center of the reactor. The centerpipe also has two functions: first to contain the catalyst from the inside, and second to help distribute the flow. The catalyst is contained in the annular space between the scallops and the centerpipe. Because it is all vapor flow through the reactor, good flow distribution is relatively easy to achieve. Well-distributed flow is achieved by the pressure drop imposed by the scallops (minimal), the reactor catalyst bed, and the centerpipe. The largest of these three components of pressure drop is the catalyst bed. The scallops and centerpipe are usually constructed of a combination of stainless steel perforated plate and profile wire.

Radial flow reactor design is used because it results in lower pressure drop than a downflow reactor would in the same service. The space velocity for a modern xylene isomerization reactor is relatively high. The WHSV is typically in the range of 5–12 per hour. Due to the high space velocity, a downflow catalyst bed with proper geometry (L/D ratio) would have a very high pressure drop. To reduce the pressure drop to a reasonable and economical value, the L/D ratio would have to be too low to properly distribute the flow. The reactor bed would need to be quite shallow with a large diameter, looking something like a pancake. Flow modeling has shown that this bed geometry, with relatively low pressure drop, results in poor flow distribution. Therefore, a radial flow reactor design is employed.

4.5.4 Catalyst Volume

The active catalyst volume required for the reactor is based on the specified WHSV, and is calculated using Eq. (4.2).

4.5.5 Radial Flow Reactor Sizing

The sizing criteria and algorithms for radial flow reactor design are complex, depending on vapor flows, vapor properties, catalyst volume and properties, centerpipe sizing, accessibility, etc. The sizing of radial flow reactors for xylene isomerization units is beyond the scope of this book and will not be addressed.

4.5.6 Products Condenser

The reactor effluent is cooled and partially condensed in the CFE from which it flows to a products condenser. In most xylene isomerization units the products condenser is an air-cooled heat exchanger. Due to the temperature of the reactor effluent to the products condenser and the desired temperature from the condenser, an air-cooled exchanger is usually the most practical and economic choice. The inclusion of a water-cooled trim condenser downstream of the air-cooled condenser depends on several factors, including the design ambient air temperature and the relative costs of electricity, cooling water, and fuel. Because it is desirable to minimize the loss of benzene to the vapor leaving the separator, the reactor effluent is typically cooled to the lowest temperature achievable by the air-cooled condenser, but not greater than 55 °C (130 °F). Unless located where the design air temperature is very high, a water-cooled trim condenser is typically not required.

4.5.7 Separator

A separator vessel is located downstream of the products condenser to separate the liquid and vapor phases of the cooled reactor effluent and to function as a surge vessel for the liquid feed to the fractionator, typically a deheptanizer. In a xylene isomerization unit a simple vapor–liquid separation is required as there is no separate water phase present. The separator vessel is typically vertical with a horizontal inlet into the side of the vessel, a vapor outlet at the top, and a liquid outlet at the bottom. Besides separating vapor from liquid, the separator also usually serves as a knockout drum to protect the downstream recycle compressor. To eliminate liquid droplets, a mesh blanket or mist eliminator is employed near the top of the separator vessel.

For xylene isomerization unit operating conditions, phase separation of the cooled and condensed reactor effluent is governed by Stokes' Law as expressed in Eq. (4.5). The diameter of the vessel must be such that the vapor velocity is less than or equal to the droplet terminal velocity (u_t) for the selected droplet diameter.

4.5.8 Recycle Gas Purity

Two parameters that have a major impact on catalyst performance and cycle length are recycle gas purity and hydrogen : hydrocarbon (H_2/HC) ratio. Together these parameters, along with reactor operating pressure, determine the hydrogen partial pressure in the reactor. For xylene isomerization units the recycle gas purity is typically required to be at least 70 mol.% while the H_2/HC is 1 : 1 or greater, depending on the catalyst. Recycle gas purity is controlled by venting from the separator. As the venting rate is increased, the amount of make-up hydrogen entering the reactor circuit increases to maintain the separator pressure and this then increases the recycle gas hydrogen purity. It is not unusual that the production of light ends in the reactor is low enough that venting of gas from the separator is not required. That is, the recycle gas purity stays above the specified minimum value without any venting. Of course, some of the light ends are dissolved in the separator liquid and are therefore removed from the system with the liquid. Even with no venting, make-up gas flow is required to replenish the hydrogen that is consumed by the reactions.

4.5.9 Recycle Gas Compressor

A recycle gas compressor (aka recycle compressor) is employed to achieve the required H_2/HC ratio dictated by the reactor operating conditions. The recycle gas compressor pulls gas from the separator and sends it back to the CFE where it is mixed with fresh liquid feed and provides the driving force to circulate it through the reactor circuit on a continuous basis. In a commercial scale xylene isomerization unit, the recycle compressor is typically a centrifugal-type machine. Older, smaller xylene isomerization units have employed other types of compressors such as reciprocating machines. For reasons discussed elsewhere, centrifugal compressors are less maintenance intensive and therefore more reliable than reciprocating-type machines and are therefore preferred in modern xylene isomerization units.

4.6 para-Xylene Separation

The primary process used in the industry to separate and recover *para*-xylene from mixed xylenes is an adsorptive process using a selective, solid zeolitic adsorbent in a simulated moving bed system. The C_8 aromatic isomers (ethylbenzene, *para*-xylene, *meta*-xylene, and *ortho*-xylene) boil so closely together that separating them by conventional distillation is not practical. The adsorptive process provides an efficient means of recovering *para*-xylene using a solid zeolitic adsorbent that is selective for *para*-xylene. It is a continuous process that simulates the countercurrent flow of a liquid feed over a solid bed of adsorbent. Feed and products enter and leave the adsorbent bed continuously, at nearly constant compositions.

The UOP Parex process was the first to use this adsorptive process. Soon after it was introduced in 1971, the UOP Parex process quickly became the world's preferred technology for *para*-xylene recovery. Before UOP introduced the Parex process, *para*-xylene was produced exclusively by fractional crystallization. Today, Parex units are designed to recover over 97 wt.% of the *para*-xylene from the feed in a single pass while delivering *para*-xylene product purity of 99.9 wt.% or better.

Fractional crystallization is still used today, but the number of units and total capacity is far less than the adsorptive process. Fractional crystallization uses the large difference in freezing points of the C_8 aromatic isomers. *para*-Xylene has a relatively high freeze point of 13.2 °C while the freeze points of the other C_8 aromatic isomers are all well below zero Celsius (EB freeze point = −95 °C, *meta*-xylene freeze point = −48 °C, and *ortho*-xylene freeze point = −25 °C).

The quality of *para*-xylene demanded by the market has increased significantly since the 1970s. In 1970, the standard purity for *para*-xylene sold in the market was 99.2 wt.%. By 1992, the purity standard had become 99.7 wt.%, and the trend toward higher purity continues. Today, most *para*-xylene separation units are being designed to produce 99.8–99.9 wt.% pure *para*-xylene.

Figure 4.10 shows the flow diagram for a typical Parex unit. The separation occurs in the adsorbent chambers. Each adsorbent chamber is divided into a number of

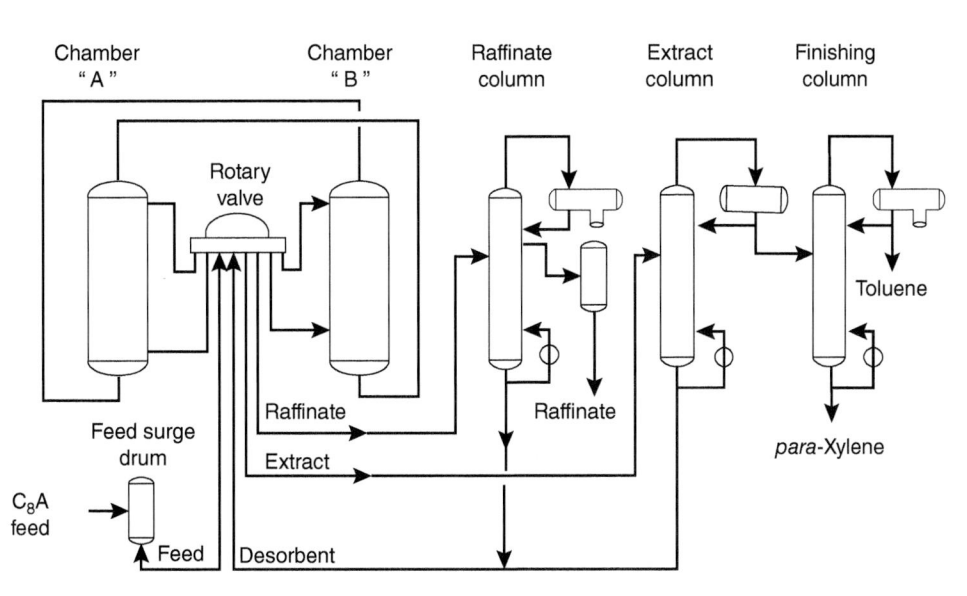

Figure 4.10 Parex process unit flow diagram.

separate adsorbent beds. A highly engineered bed support/distributor is located between each adsorbent bed, as well as above the top bed and below the bottom bed of each adsorbent chamber. The flow distributors between each adsorbent bed are used to inject or withdraw liquid from the chamber or to simply redistribute the liquid from above over the cross-sectional area of the adsorbent bed below. A device called a rotary valve is used to periodically switch the positions of the liquid feed and withdrawal streams as the composition profile moves down the chamber. Each flow distributor is connected to the rotary valve by a bedline.

A typical Parex unit has 24 adsorbent beds and 24 bedlines connecting each bed distributor with the rotary valve. Parex units consist of two adsorption chambers, connected in series with 12 adsorbent beds in each chamber.

At any given time, only a small number of bedlines are active, carrying the net streams into and out of the adsorbent chamber as well as flush streams that are designed to improve and assure product purity and recovery. A chamber circulation pump (aka push-around pump or pump-around pump) provides the liquid circulation from the bottom of one adsorbent chamber to the top of the other. In Parex, there are four major "net" streams that enter or leave the adsorbent chambers. These net streams are as follows:

- **Feed In** – Mixed xylenes feed to the unit;
- **Dilute Extract Out** – *para*-Xylene product diluted with desorbent;
- **Dilute Raffinate Out** – EB, *meta*-, and *ortho*-xylene diluted with desorbent;
- **Desorbent In** – Recycle desorbent from the fractionation section.

The principal advantage of the continuous adsorptive process (Parex) over crystallization technology is the ability to recover more than 97% of the *para*-xylene in the feed per pass. Crystallizers must contend with a eutectic composition limit that restricts *para*-xylene recovery to approximately 65% per pass. Figure 4.11 below clearly illustrates the implication of this difference, where an adsorptive complex producing 250 000 metric tons per annum (MTA) of *para*-xylene is compared with a crystallizer complex producing 168 000 MTA. The upper numbers in the figure indicate the flow rates through the adsorptive complex, while the lower numbers indicate the flow rates through a comparable crystallizer complex.

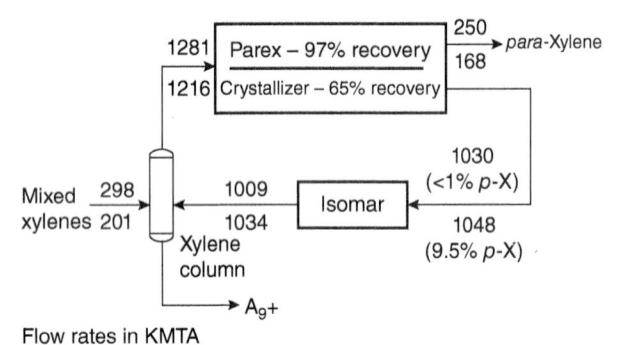

The adsorptive complex produces approximately 50% more *para*-xylene from a given capacity xylene column and xylene isomerization unit than a complex using crystallization. In addition, the yield of *para*-xylene per unit of fresh feed also improves, because a relatively smaller recycle flow means lower losses in the isomerization unit. The difference in production between 168 000 and 250 000 MTA of *para*-xylene is worth over $40 million per year (at a *para*-xylene price of $500/MT). Another method of comparing the technologies is keeping the *para*-xylene product rate constant. In this case, much larger xylene columns and isomerization units would be required to produce the same amount of *para*-xylene, increasing both the investment cost and utility consumption. This example clearly illustrates why it is important to compare different technologies only in the context of the entire aromatics complex.

The dilute extract from the rotary valve is sent to the extract column for separation of the extract from the desorbent. The overhead from the extract column is sent to a finishing column, where the highly pure *para*-xylene product is separated from any toluene which may have been present in the feed.

4.7 Process Design Considerations: Design Margin Philosophy

Design engineers must consider many factors when designing a process unit and its components. The amount of design margin applied to the entire unit and to individual components of it requires significant forethought.

The definition of design margin is as follows: the additional performance capability above required system parameters that may be specified by a system designer to compensate for uncertainties. In this section we will only address process design margin, not mechanical design margin.

Design margin, also referred to as design contingency, is included in the designs of almost all processes. There are many reasons to include design margin, including uncertainty in feedstock composition, uncertainty in thermodynamics, uncertainty of operating conditions, aging of equipment, and associated loss of performance and/or efficiency, as well as other factors and unknowns. The overriding purpose of including design margin is to

Figure 4.11 Comparison of Parex technology with crystallization for *para*-xylene production.

assure that the process meets its expected capacity and performance. Normally, a shortfall in capacity or performance cannot be tolerated.

How does one decide how much design margin to include? Should design margin be applied to all equipment components equally? The amount of design margin included depends on the maturity of the process, among other things. If the process unit being designed has a long history of successful implementation and operation, the amount of design margin applied can be minimal because the uncertainties are low. However, if the process unit is new and unproven, it would be prudent to include a relatively large design margin to account for uncertainty. Other factors to consider when deciding how much design margin to include are the expected increase in the cost of the unit or equipment item and the cost of not meeting the required capacity and/or performance of the unit or equipment.

If one assumes the typical cost associated with an increase in size/capacity of process equipment is related to the ratio of the capacities to the 0.6 power.

$$\frac{Cost_2}{Cost_1} = \left(\frac{Capacity_2}{Capacity_1}\right)^{0.6} \quad (4.6)$$

So, a design margin of 10% would have an expected cost impact of $(1.1)^{0.6} = 1.06$ or an increase of approximately 6%, whereas a design margin of 20% would have an expected cost impact of $(1.2)^{0.6} = 1.12$ or an increase of approximately 12%.

A common reason to include additional design margin in a process unit, or an entire complex, is if a future operating case is expected or planned. The future operating case may be at higher capacity, include an alternative feed, or have a different product slate. For any of these cases, heat and material balances would need to be performed and compared to the base design case to determine which equipment needs design margin, and how much, or if a complete alternative design case is required.

Another reason to include, or add, design margin to a process unit is to accommodate a possible catalyst change in the future. An alternate catalyst may require different operating conditions, such as a greater hydrogen/hydrocarbon ratio, which would result in a higher recycle gas flow, requiring a larger recycle compressor. An alternative catalyst may produce a different yield or product slate which may result in more light ends, which require larger vent gas system piping and valves and possibly a larger or more powerful recycle compressor.

Another factor to consider when deciding on how much design margin to include in a particular equipment type or item is previous operating experience with that equipment. For example, if it is clear that the process bottleneck in a particular unit or units has repeatedly been the same equipment item or equipment type, the design margin for that equipment should be greater than the design margin applied to other items in the unit.

It is the goal of the designer to design the unit so that all equipment reaches its maximum performance capability at the same unit capacity. That is, the charge pump and recycle compressor of a unit should both become bottlenecks at the same unit throughput. If the recycle compressor has 15% reserve capacity when the charge pump has reached its maximum capacity, it is very possible that money has been wasted on the recycle compressor because its maximum capacity will never be utilized. A smaller compressor may have been adequate. Therefore, different design margins may be appropriate for different equipment items.

4.7.1 Equipment Design Margins

The standard design margin included by the manufacturer in different types of process equipment varies. The supply of certain equipment types such as air-cooled heat exchangers is more competitive than other equipment types such as fired heaters or centrifugal compressors. Because of the competitive environment, the suppliers of air-cooled heat exchangers may include little or no design margin unless mandated to do so by the purchaser. On the other hand, suppliers of centrifugal pumps and compressors have to meet industry standards for operation above the design point. This lack of design margin included by air-cooled heat exchanger suppliers may also be due to difficulty in testing and certifying the equipment for the particular operating conditions for which it was designed. Centrifugal pumps and compressors must meet performance requirements depicted and stated by performance curves. No such performance curves are available and performance tests are not normally practical for air-cooled heat exchangers. Therefore, while it might be prudent to include 10% design margin in the design of the pumps and recycle compressor, a larger design margin of 20–25% may be prudent for the air-cooled heat exchangers.

4.7.1.1 Fired Heaters

The design margin for a fired heater in a reactor circuit varies with the type of process unit. The specific design margin depends upon factors such as the ratio of heat exchanger to heater duty, start-up requirements, magnitude of the heat of reaction, SOR versus EOR considerations, rich vs. lean feedstocks, clean vs. fouling feedstock, and other factors that may influence heater operation. This results in a design duty that is from 10 to 40% greater than the normal duty.

For aromatics units that contain a CFE and a charge heater, such as a transalkylation unit or a xylene isomerization unit, the charge heater should have enough design margin to make up for a possible shortfall in the duty of the CFE. Even though aromatics process units are relatively clean, the CFE may not meet original design duty due to fouling. Duty shortfall may also be a result of exchanger design issues, tube leakage, or physical damage. The minimum design duty of the charge heater may be considered to be 10% of the total of the CFE duty and the normal charge heater duty.

4.7.1.2 Process–Process Heat Exchangers and Water-Cooled Heat Exchangers

Generally, no design margin is applied to shell and tube heat exchangers in process vs. process services or water-cooled services. However, conservative fouling factors are generally applied which results in greater surface area being specified.

4.7.1.3 Air-Cooled Heat Exchangers

As discussed above, due to several factors such as the competitive nature of the air-cooled exchanger industry and difficulty in testing this type of equipment, air-cooled exchangers should be specified with between 10 and 25% design margin depending on the service. Air-cooled exchangers are typically designed based on the use of an ambient air temperature that is not exceeded more than 5% of the time in the warmest month of the year. Sizing of air coolers on this basis results in additional surface area, i.e. design margin, being available most of the time. Climate change trends should be considered when designing a plant, as the plant will have a life of 30 or more years.

4.7.1.4 Pumps

Centrifugal pumps are generally designed for 110% of normal flow, with some exceptions. Certain services must consider special cases such as start-up or regeneration which could require greater flow and/or head than normal operation. Reboiler pumps generally have no design margin because it is normal practice to keep the reboiler flow constant and change the vaporization rather than vary the flow. Proportioning pumps generally have a flow rate range of 10 to 1 with the normal capacity set in the middle of the range. This provides considerable design margin and no other margin is applied.

4.7.1.5 Compressors

The design margin for compressors depends on the type of compressor and the service. The design margin for a reciprocating compressor in a net gas or make-up gas service is normally 20%. This 20% design margin provides the amount of gas spilled back for process control. Reciprocating recycle gas compressors are generally sized for 110% of the normal flow requirements. Centrifugal recycle compressors are generally sized for 100% of the normal flow requirements. However, if there are yield or operating condition uncertainties, such as uncertainty in the optimum H_2 : HC ratio a centrifugal recycle compressor may be sized for 110% of the normal flow.

4.7.1.6 Fractionation Columns

For trayed columns, the design margin is expressed by the magnitude of several tray design criteria; namely percent of jet flood, percent of downcomer flood, and percent downcomer backup, among other criteria. For a new column design the percent of jet flood is generally limited to 75–82. The maximum operable jet flood is generally considered to be about 90%, so designing for 75% jet flood provides a design margin of about 20% and designing for 82% jet flood provides a design margin of about 10%. The downcomer flood is normally limited to no more than 75% and downcomer backup is limited to 50%. These criteria are consistent with a 10–20% design margin as well. Because the loads in a column vary from section to section, and tray to tray within a section, the tray flooding values can vary widely also. The tray design for a column section is usually based on the highest load within that section, so most of the trays actually have a greater design margin than the tray used for design. However, that really does not provide any additional capacity because even if flooding is limited to a single tray it will affect the operation and performance of the column.

4.7.1.7 Reactors

Reactor designs typically do not include any process design margin, although the catalyst bed volume may be increased by some amount to accommodate a greater loading for a future case or a future catalyst. In a downflow reactor this can be done by increasing the tangent length without changing the diameter. In this case, the design of other equipment in the reactor circuit must be considered for the future case as well, most notably the recycle compressor and charge pump, due to the increased pressure drop.

4.8 Process Design Considerations: Operational Flexibility

Flexibility is an important attribute of any chemical, petrochemical, or petroleum processing plant design. The flexibility of a particular piece of equipment, process unit, or entire plant (the system) is its ability to respond to internal or external changes affecting its operation, production, safety, and/or cost-effectiveness. The flexibility of the system can be defined as the ease with which the system can respond to uncertainties in key parameters, such as feedstock composition, feedstock conditions such as temperature and pressure, ambient conditions, catalyst performance, and equipment variabilities.

BTX aromatics process units and complexes need to have the ability to operate with varying feeds, varying product rates, with varying ambient conditions, and possibly varying utility conditions. Operational flexibility is a hallmark of good process design. On paper, a single point design without consideration for variability looks fine. As long as the design feedstock is available, the utilities supplied are as expected, and other conditions are as specified, the complex should make the specified products at the specified rates. However, as we all know, in the real world the unexpected happens. The feed supplier cannot supply the quantity or quality of feed promised or a lower cost supply is found. Other feeds, not meeting the original specifications, need to be substituted. The utilities supplied may not be at the conditions expected. If the steam supplied to reboilers and other exchanger services is at a lower pressure, the equipment will not provide the duty required. If the ambient air temperature is higher than expected (global warming?), the air-cooled exchangers will not be able to condense or cool products adequately. If the designer only designs for a single point, without consideration for variability in these operating parameters, the complex will not meet its nameplate production much of the time.

In many cases, the feedstock variability issue can be addressed by multiple feed cases. That is, the designer can include cases that have a range of feed characteristics to cover contingences. The primary design case is typically for the most likely feedstock the plant will obtain, or an average of the possible feedstocks the plant may purchase. This primary case may be a relatively "rich" or otherwise desirable feed with low amounts of contaminants and undesirable components. Other less optimum feedstocks that are less likely should be included as alternate feed cases. These feeds may have lower purity or lower concentrations of desirable components; such as a mixed xylenes feed with less mixed xylenes, greater quantities of undesirable components such as nonaromatics and A_9+ components.

In a typical aromatics complex this feed flexibility will affect some process units more than others. Variability of the nonaromatic content of the feed needs to be addressed in the size of the aromatics extraction unit design as it will see more or less feed with higher or lower feed aromatics content. Variability of the relative quantities of benzene, toluene, mixed xylenes, and A_9+ needs to be addressed by the fractionation column designs and individual unit designs as their feed rates and feed compositions can vary drastically. Because of the nature of the aromatics complex flow scheme with its conversion units and many recycle streams, an overall complex material balance is needed to determine the effect of feed changes on each particular unit and fractionation column. Discussion of design requirements for feed flexibility for each particular type of unit in an aromatics complex is included in the each unit's process design section.

4.9 Process Design Considerations: Fractionation Optimization

Process optimization is the subject of Chapter 16. It is not the intent to cover this subject thoroughly here, but rather to briefly discuss some of the important areas where fractionation split optimization should be employed and will have a great impact on the capital cost, energy consumption, and operating efficiency of an aromatics complex.

As we have discussed there are many fractionation columns in an aromatics complex. Several of these columns produce recycle streams that are continuously reprocessed. The splits that are chosen in these columns that produce recycle streams have an impact on the size and composition of the recycle stream(s) and thus size and capacity of downstream equipment. The following are some examples of these recycle stream optimizations:

Reformate splitter – In a typical reformate splitter where toluene and lighter material are taken overhead and C_8 aromatics and heavier are taken out from the bottom, the amount of C_8 aromatics allowed to go to the overhead affects the feed to the aromatics extraction unit and downstream fractionation, i.e. benzene column, toluene column, and xylene column.

Toluene column – The toluene column makes a split between toluene and C_8 aromatics. The toluene is generally sent to the transalkylation unit while the C_8 aromatics and heavier material goes on to the xylene column. The amount of C_8 aromatics allowed to go the overhead with the toluene affects the reactions and equilibrium in the transalkylation unit. The amount of toluene allowed to go to the bottom of the column and on to the xylene column affects the xylene column and the *para*-xylene separation unit (Parex or other).

Isomerization unit deheptanizer column – The deheptanizer column normally makes a split of the toluene coming to it, with a portion going with the bottoms back to the xylene column and a portion going overhead and on to a stripper column in the isomerization unit or elsewhere in the complex (such as the transalkylation unit or reforming unit). The amount that goes in each direction affects the equipment sizing and utilities of that portion of the complex. The optimum toluene split is not obvious and an economic analysis of the impact of different splits is justified.

The optimization of splits in columns that produce recycle streams is highly dependent on feedstock compositions and process yields. An optimization done for one case may not apply to another case unless the feeds and yields are very similar. Therefore, the split optimizations will likely have to be revisited for each design and possibly for each design case.

4.10 Safety Considerations

Like any hydrocarbon processing plant in the petroleum and petrochemical industries, safety is of utmost importance in the design and operation of an aromatics complex. When carrying out the design, the engineers must consider all potential hazards. This includes reducing exposure of operators and the public to hazardous materials, as well as eliminating, if possible, or minimizing to the greatest extent possible, the risk of any potential process hazards.

4.10.1 Reducing Exposure to Hazardous Materials

It is incumbent on the designers of aromatics complexes to include provisions to minimize the leakage and release of aromatic hydrocarbons to the atmosphere and to minimize the exposure of plant personnel to aromatics. Benzene is a known carcinogen. While there is insufficient information to determine whether or not toluene, xylene, and other C_7 through C_9 aromatic hydrocarbons are carcinogenic (i.e. they are not classifiable as carcinogens), it is believed these chemicals are possible human carcinogens or present other health hazards. Therefore, these chemicals must be carefully controlled. Features such as a closed aromatics drain header and closed drains for control valves, pumps, compressors, level gauges, and instruments must be included in the design of units and equipment handling these materials. Strict procedures for safely obtaining samples of materials containing benzene and other aromatics must be followed.

Typically, several formal reviews are undertaken to assure the safety of the complex and its individual process units. The first review is typically a process hazard analysis (PHA).

4.10.2 Process Hazard Analysis (PHA)

A PHA is a systematic assessment of the potential hazards that may be associated with the plant. The PHA is conducted with the purpose of analyzing the potential causes and consequences of occurrences such as over-pressuring, releases of flammable and/or toxic materials, spills, fires, explosions, etc. The PHA reviews equipment, instrumentation, operator actions, utility systems, and other external factors such as upstream or downstream valve failures, power failures, etc. The results of the PHA are used to modify the design, if necessary, and provide information to the owners and operators of the plant to reduce or eliminate potential hazards and improve safety.

The methodology used for a PHA of an aromatics complex depends on factors such as the complexity of the plant, whether a PHA or hazard and operability (HAZOP) study has been conducted on the types of units included in the complex, and the operating experience with the types of process units included. In some cases a failure mode and effects analysis (FMEA) is conducted and serves as the PHA or as part of the PHA. Alternately, a methodology using a "What If?" analysis or checklist is used. In some cases, if the design is similar to a previous design, the PHA for the previous design will be used as the starting point for the new PHA, with particular attention paid to the differences in the designs.

A PHA is usually only the first step in the overall safety review of the plant. It is usually conducted relatively early in the design phase, typically as part of the basic engineering design phase. At least one HAZOP, and frequently more than one, is conducted later in the detailed design phase, prior to procurement and construction of the plant. In some cases another HAZOP is conducted prior to the plant being commissioned and started.

4.10.3 Hazard and Operability (HAZOP) Study

A HAZOP is a more comprehensive review of a process unit or complex than a PHA. It is more structured in its methodology. The design of the plant (unit or complex) is broken into small and relatively simple sections called "nodes." Each node is reviewed individually. The nodes are defined with a given starting and ending point and are selected such that they are neither too small and simple nor too large and complex. The piping and instrumentation diagrams (P&IDs) are normally used to select and define the nodes.

The HAZOP is carried out by an experienced multidisciplinary team consisting of a facilitator or leader, a scribe or recorder, a process design engineer representing the design organization, a process engineer or operations engineer representing the owner or operator of the plant, and any specialists such as an instrument engineer or equipment specialists, as deemed necessary for the particular discussion topic or node evaluation. The facilitator or leader should be someone experienced in leading HAZOPs and should be familiar with the type of processes being reviewed. The facilitator is responsible for leading the team through the list of nodes, being the moderator of the discussions, and ensuring the clarity and accuracy of the notes and recommendations. The facilitator also identifies the individual(s) responsible for following up on action items.

The purpose of the HAZOP is to review the process and identify any design issues that affect safety or risks to personnel and/or equipment. Once the causes and effects of any design issues that present potential hazards have been established, the node or nodes being studied can be modified to eliminate or reduce the risk. A HAZOP of an aromatics complex is typically carried out unit-by-unit. Like a PHA, if a HAZOP has been previously done for a similar unit, the previous HAZOP can be used as a starting point with particular attention paid to the differences in the designs.

Further Reading

Ergun, S. (1952). Fluid flow through packed beds. *Chemical Engineering Progress* 48 (2): 89–94.

Meyers, R.A. (2016). *Handbook of Petroleum Refining Processes*, 4e. UOP Sulfolane Process, Chapter. New York: McGraw-Hill Professional.

5

Aromatics Process Revamp Design

5.1 Introduction

This chapter will delve into many aspects of process design for revamp projects. Almost every aromatics complex undergoes at least one major revamp project during its life, and some undergo more than one. The intent of the revamp is usually to increase capacity and thus revenue. Although, there are other reasons to revamp a plant, such as reducing utility consumption and thus operating costs, improving safety and/or reliability, or changing the complex's product slate.

The purpose of a revamp assessment is to identify the operating constraints of the plant and determine appropriate process and equipment modifications to overcome them. In addition, the assessment should aim to find ways to minimize capital costs via exploiting spare capacity available in design margin, modifying process conditions, if beneficial, and confirming and possibly increasing equipment limits. Effective methodology and guidelines introduced here will help engineers to efficiently find ways to remove constraints and possibly reduce the level of modifications to key equipment, which is a primary goal of a successful revamp project.

It is difficult to categorize revamps, as each revamp tends to be unique. In the most fundamental sense, every revamp is similar; there is needed modification to some parts of the plant, while usually retaining some of the existing equipment. The variations on this theme are more on the order of the extent of the revamp, not the concept.

A revamp evaluation and design is generally more challenging than grassroots design. Standard design templates and procedures are usually available for new grassroots designs. Designers make adjustments to the templates based on new feed rates, product specifications, and process conditions using a standard design procedure. In revamp assessment and design, one needs to be innovative to minimize modifications to existing equipment while maximizing improvements, such as capacity increase or operating cost reduction. At the same time, practical experience and engineering judgment play a critical role during revamp assessment.

Every revamp project should start with a study. The study can be stand-alone, a separate phase of the project, or incorporated into what is commonly called the "process" phase of the project.

If a study has been conducted prior to the start of design work, the project tends to be better defined and decision making is easier. If a study has been conducted, revamp decision making is usually confined to issues where the results of the study are not borne out by a more detailed look at the design (better hydraulic analysis, detailed exchanger evaluation, etc.) or the scope of the revamp has been changed to include equipment or plant sections that were not included in the original study.

The recommended approach for revamps is to separate the study and design phases. By separating these two phases of a revamp project, it allows a high degree of interaction during the study phase, when clarification is necessary and decisions must be made. There is much less interaction during the "design" phase, where the schedule for the project becomes more important. Also the design phase becomes better defined and straight forward since the designer and plant operator have an agreed-upon basis and both know the extent of the revamp being undertaken.

5.2 Stages of Revamp Assessment and Types of Revamp Studies

A revamp assessment typically goes through several stages or levels of study before the actual revamp project is finally implemented. These stages are typically classified as (i) scoping, (ii) feasibility, and (iii) process revamp, in order of increasing detail, accuracy, and engineering study cost. The revamp assessment studies range from a very preliminary type of study, often called a revamp scoping study to complete revamp basic engineering design, with a variety of intermediate engineering studies and services available.

The typical levels of revamp assessment are as follows:

- Revamp scoping study
- Revamp feasibility study

Efficient Petrochemical Processes: Technology, Design and Operation, First Edition.
Frank (Xin X.) Zhu, James A. Johnson, David W. Ablin, and Gregory A. Ernst.
© 2020 John Wiley & Sons, Inc. Published 2020 by John Wiley & Sons, Inc.

- Process revamp study (also known as a revamp process study or just a revamp study)
- Basic revamp engineering design package

The revamp scoping study is the most preliminary engineering evaluation of a potential revamp. It is generally used to make an initial determination of the extent of a revamp, and whether or not to proceed, or how to proceed with the next step. If the results of the revamp scoping study are favorable, the owner may choose to proceed with a more detailed engineering evaluation such as a revamp feasibility study or process revamp study, using the results of the scoping study as a guide.

A revamp scoping study is, by design, a very quick review of only the major equipment in a process unit, such as reactors, fractionators, fired heaters, combined feed-effluent exchangers, and compressors. Small vessels, most heat exchangers and pumps, instruments, piping, and relief valves are typically not considered in the review.

If the revamp looks attractive after the scoping study, the next step is identifying critical constraints and determining ways of overcoming them. This step is called a revamp feasibility study.

The purpose of the revamp feasibility study is to determine the practical, economic, and/or strategic viability of a revamp. A revamp feasibility study is a more detailed engineering evaluation than a scoping study. A revamp feasibility study can provide the type of information needed to determine whether or not to proceed with a particular revamp, or ascertain which of various alternatives under consideration are most attractive. During a feasibility study, improvement options will be evaluated in greater detail with more accurate estimates of costs and benefits based on more rigorous process simulations and more rigorous equipment evaluation.

In some cases a revamp improvement option may look very promising in the scoping stage but may fail to pass the feasibility study stage because unforeseen constraints or limitations could be uncovered in the feasibility study stage. These constraints or limitations, such as having to replace an expensive equipment item like a compressor, fired heater, or fractionator are sometimes called "show-stoppers."

While a revamp feasibility study is still preliminary in nature, and further engineering is required to confirm its findings, there can be a high level of confidence in the conclusions. If the results of the revamp feasibility study are favorable, the owner may choose to proceed with a more detailed engineering evaluation such as a process revamp study, or a basic engineering design package, using the results of the feasibility study as a guide.

The process revamp study is an even more comprehensive and detailed engineering evaluation of a potential revamp to identify the necessary operating conditions and required process and equipment modifications to meet a desired objective. All major equipment in the scope of the study are evaluated. This includes fired heaters, reactors, columns, vessels, heat exchangers, pumps, and compressors. The evaluation of instruments, piping, and relief valves may or may not be included in the scope of the process revamp study. These items will definitely be included in the basic engineering design package to determine their adequacy prior to detailed design and implementation of the project. In the process revamp study and basic engineering design stages, the process design and equipment evaluation are conducted based on rigorous equipment-rated process simulations.

Closer attention is paid to process optimization and equipment analysis in a process revamp study than is done in a revamp feasibility study. A process revamp study is a thorough investigation of a revamp scenario, as such it does not lend itself to the study of a variety of options. The revamp feasibility study may be used to evaluate processing alternatives, make an initial cut, and reduce the number of options. Results from the feasibility study are often used to set the basis for the more in-depth engineering analysis provided in the process revamp study.

The various levels of studies have varying levels of deliverables commensurate with the extent of detail of the study. Below is a typical list of study deliverables indicating what is typically included in each type of study:

- Executive summary of the results of the study (all types of studies).
- Statement of the basis, objectives, and scope of the study (all types of studies).
- Discussion of the results of the study, including a description of each case considered, a review of alternatives considered, recommended process flow changes, and necessary equipment modifications. The suitability of each major piece of equipment is summarized (all types of studies).
- Summary of revamp operating conditions, yields, and product specifications (all types of studies).
- A list of the major equipment evaluated in the study, identifying existing equipment which is suitable for the revamp, equipment which requires modification, and any new equipment required (all types of studies).
- Preliminary process flow sketch showing the basic process flow and the major equipment, highlighting areas of change (revamp scoping study).
- Process flow diagram (PFD) showing schematic flow and control, flagged with mass flows, temperatures, enthalpies, and selected pressures of major process streams. Also shown on the PFD is a material balance

with compositions and quantities of feed and product streams (revamp feasibility study and process revamp study).
- Budget equipment costs for new major equipment items included in the scope of the study. These costs generally have an accuracy of ±50% in revamp scoping studies, ±30% in revamp feasibility studies, and as low as ±10% for process revamp studies (all types of studies).
- Equipment data summary sheets showing operating conditions, design conditions, approximate sizes for new vessels, duties for heaters and exchangers, and capacities and estimated heads for pumps and compressors. For existing and modified equipment, these data sheets provide a comparison of the revamp operating conditions with the original design operating conditions (process revamp study).

The revamp basic engineering design package includes complete process design, process flow diagrams (PFD's), engineering design information, complete hydraulic tabulations, utility summaries, data sheets (aka specifications) for all new and existing equipment, piping and instrumentation diagrams (P&ID's), as well as specifications for piping, instrumentation, and relief valves and systems.

5.3 Revamp Project Approach

The difference between successful and unsuccessful revamp projects can usually be traced back to poor communication between the designer organization and the plant owner/operator. To help facilitate a successful revamp project, it is important to understand some of the more common revamp project types as described below. This will help in understanding and communicating the desired outcome of the revamp as well as to be able to keep the revamp project on track. The following examples give an insight into some revamp project variations and common drivers.

5.3.1 Specified Target Capacity

This type of revamp is not very common as the owner does not know what is required to meet the specified capacity. These types of revamps have an absolute production value that must be met, usually set by internal or market needs, by whatever means (including a new unit). This need can be driven by contractual commitments or expansion of a downstream unit requiring more feedstock. Also included here are revamps driven by government regulations or product purity requirements driven by the market. This type of revamp is relatively simple since once the requirement has been set, the equipment either fits (or can be modified) or you simply buy new equipment. This type of revamp tends to be the most costly; however, the market benefits usually support the cost. The design basis is fairly straightforward for these projects; the only consideration is the cost of the revamp compared to new unit costs and if the payout meets the plant operator's requirements. Typically, the throughput is guaranteed for this type of revamp and new unit criteria are used for evaluating equipment for replacement or modification. General characteristics of this type of revamp are:

- Firm production rate and/or product purity requirement.
- New-unit equipment design/evaluation criteria.
- Lowest engineering cost alternative since the equipment either fits, is replaced, or is modified – no process design iterations required.
- Throughput guarantees provided.
- Investment cost can approach new unit cost.

5.3.2 Target Production with Constraints

Many revamps fall into this category. The plant owner is either in a profitable market segment where the more they produce, the more they make, or they want to be positioned to take advantage of future opportunities. This type of revamp can be considered a pragmatic revamp in that the owner wants the maximum from a reasonable investment. General characteristics of this type of revamp are:

- Target production needs, frequently stated in terms of a range.
- Typically includes stipulation to retain certain major (expensive) equipment (recycle gas compressor, fired heaters, reactors, columns, etc.).
- Moderate interaction with the owner during design phase, discussing target changes based upon equipment replacement alternatives.
- Utilization of existing equipment design margins, based on vendor data and operating data, if available.
- Operating conditions may be adjusted to fit the existing equipment – i.e. some process design iterations may be required.
- Throughput targets provided, but may be modified during design based on major equipment limitations.

5.3.3 Maximize Throughput at Minimum Cost

In this type of revamp, the owner usually defines capital constraints, but wants to maximize profit (increased throughput, improved efficiency, etc.) with minimum investment. This can be looked upon as an opportunistic exercise in getting the most out of an existing facility. The owner is usually only interested in replacing minor equipment and possibly one or two major equipment items. These are challenging and interesting revamps. General characteristics of this type of revamp are:

- Very rough target production rates. Typically, the owner will want to maximize throughput without spending much capital.
- More extensive equipment replacement must be justified by very attractive economic payout.
- The interaction between the designer and the owner during the design phase is usually regarding equipment replacement alternatives and their impact on achievable production rates, as well as keeping the owner up to date on progress.
- Utilization of existing equipment design margins based on vendor data and operating data, if available. Willing to go up to the expected equipment maximum capacity without need for maintaining design margin.
- It may be necessary to adjust operating conditions to fit existing equipment. Often, several process design iterations are required.
- Throughput targets provided, but may be adjusted in the course of engineering design.

5.3.4 Identify Successive Bottlenecks

In this type of revamp the owner either is not sure of their capital constraints or wants a list of successively increasing cost/capacity options to present to their management before going forward. This can be looked upon as a strategic plan (list of future projects) for an existing plant. The owner may have an ultimate target in mind, but does not have enough capital to get there in a single step. These are usually the most challenging and interesting revamps. General characteristics of this type of revamp are:

- Capacity targets determined in the course of the work as successive bottlenecks are identified and eliminated by equipment replacement or addition.
- Although the project may proceed in a phased approach, if there is a fixed target capacity, any new equipment specified at any stage may be specified for the ultimate target. If this is the case, the capacity of any new equipment must not cause any limitations of other problems during its operation in the earlier phases of the revamp.
- Interaction between the designer and owner during the design phase is significant to steer the design work as each successive bottleneck is identified and cost estimates are developed.
- Utilization of existing equipment design margins based on vendor data and operating data, if available. Willing to go up to the expected equipment maximum capacity without need for maintaining design margin.
- It may be necessary to adjust operating conditions to fit existing equipment. Several process design iterations may be required.

5.4 Revamp Study Methodology and Strategies

If a revamp requires major changes to the plant such as replacing or adding principal equipment (e.g. a new reactor, fractionator, fired heater, or recycle compressor), the impact on the capital cost will be significant. In fact, in many cases the owner of the plant will not tolerate such changes and their associated costs, and the project may not move forward. It is possible that the replacement of major equipment may be avoided or the level of modifications reduced by employing a revamp assessment methodology and revamp strategies discussed herein. It is a goal of most revamp projects to achieve the goals of the revamp with minimum investment cost. A sound revamp methodology and using proven strategies will help to achieve that goal. A flow chart of the recommended methodology is shown in Figure 5.1 and discussed below.

The methodology consists of eight main steps, namely:

1) Setting of the revamp objectives, basis, and scope
2) Preliminary (pre-revamp or base case) process simulation
3) Revamp case process simulations
4) Equipment evaluation and rating
5) Heat integration analysis
6) Optimization of process conditions and flow scheme
7) Setting of proposed revamp process simulation and plant modifications
8) Review with owner and finalize report/deliverables

This methodology usually contains at least one recycle loop, between Step 6 and Step 3. However, the recycle loop may need to go all the way back to Step 1, although this would likely happen only once. After optimization of process conditions and possible flow scheme changes, it is necessary to redo the process simulation(s), equipment evaluation, and possibly heat integration analysis. This loop may be required to be repeated several times in order to zero in on the optimum solution or set of solutions.

This methodology has been applied to numerous revamp studies and projects with the following features and benefits:

- Clearly defined project objectives, scope, and basis.
- Reduced capital investment and operating costs.
- Increased possibility of project approval.

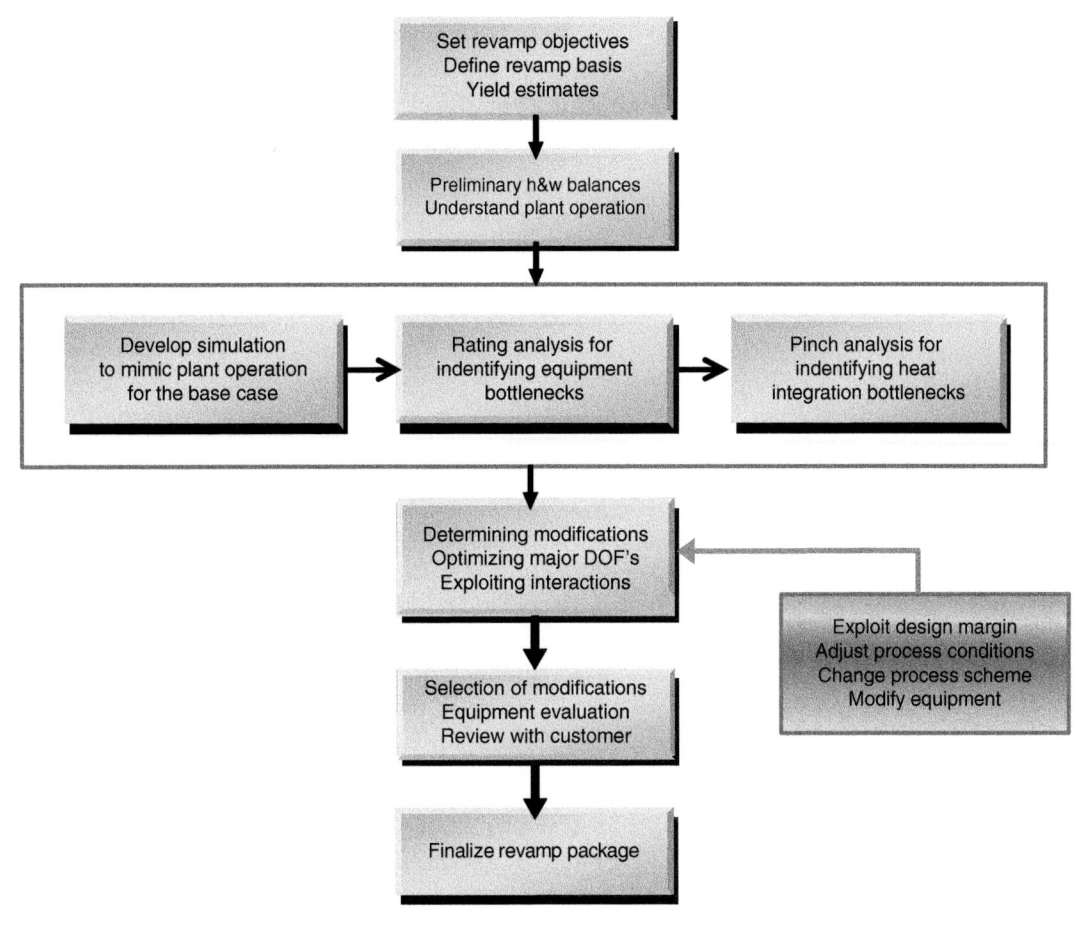

Figure 5.1 Recommended revamp methodology.

The purpose of the preliminary "pre-revamp" or "base case" process simulation is to gain good understanding of the current plant design and operation in terms of key operating parameters and equipment operation. A good process simulation of the current plant operation or test run that both the plant engineers and the design engineers agree upon can be very useful as a starting point from which to build the revamp simulations. The pre-revamp process simulation(s) can also provide the basis for rating of equipment such as heat exchangers, pumps, and compressors.

The key role of equipment rating analysis is to assess equipment performance and identify equipment limitations and spare capacity. Utilization of design margins and spare capacity can allow capacity expansion of 10–20% in general and accommodate revamp projects with low capital cost. In some cases, however, the plant has already utilized the spare capacity in their efforts to maximize throughput. It is expensive to replace major equipment items when they reach hard limitations. For example, a fractionation tower reaches the jet flood limit, a compressor is at the flow rate or head limit, and a furnace is at the heat flux limit. It is important to find ways, other than replacing these costly equipment items, to overcome these constraints.

Steps 5 and 6 apply heat integration methods (see Chapter 13) and process integration methods (see Chapter 12) to explore changes to process conditions, addition of equipment, changes to utilities, or flow scheme changes with the purpose of retaining existing equipment by shifting plant bottlenecks from more expensive to less expensive equipment. By capitalizing on heat integration and process integration, it is possible to utilize equipment spare capacity and push equipment to their true limits in order to avoid the need of replacing the existing equipment or installing parallel equipment. These are major features of this revamp methodology.

An example of application of heat integration and process integration is the addition of fractionation column feed preheat. The addition of heat to the column feed reduces the required reboiler duty. If a column reboiler reaches a duty limit, it is likely to be less expensive to add a feed preheat exchanger, possibly using available process heat, than replacing the reboiler. The column simulation must be redone to evaluate the effects on the separation, tray loadings, and condenser operation with increased feed preheat and reduced reboiler heat input as this change will affect the column temperature and composition profile. If the changes in column operation and separation are acceptable, this modification could avoid replacing the existing reboiler or installation of a new parallel reboiler.

Steps 5 and 6 employ integrated optimization and the driver is to take advantage of interactions between equipment, heat integration, process condition changes, and possibly flow scheme changes. Making changes to process conditions provides a major degree of freedom and will be discussed further in Section 5.6. One direct benefit of optimizing process conditions is that it can facilitate the utilization of the available capacity of the existing assets. Process redesign provides another major degree of freedom as it can increase heat recovery and relax equipment limitations. Integrated optimization can identify synergetic opportunities in achieving revamp objectives with greatly reduced revamp capital cost.

Process redesign includes process flow scheme redesign and equipment internals' redesign/replacement. Redesign of the process flow scheme may change routing of streams, provide improved heat integration, and change flow rates or conditions of streams to/from equipment. This may help shift capacity to other equipment to relax limitations with the purpose of avoiding replacement of major equipment. Examples of equipment redesign include changing column internals, which allows an existing column to operate at increased capacity with much lower capital cost than installing a new column; replacing heat exchanger tubes with enhanced surface area tubes allowing greater heat transfer, or replacing pump impellers or compressor wheels to increase capacity without replacing the entire equipment item.

Applying this revamp methodology provides the following benefits: alternative revamp options and/or improvement ideas will be identified and evaluated, and solutions to overcome or relax limitations will be obtained by taking advantage of process redesign and/or adjustments to operating conditions which optimize equipment performance and heat integration. This approach provides a pathway to help engineers find answers to the challenges of a revamp; making the revamp project technically and economically feasible.

5.5 Setting the Design Basis for Revamp Projects

Establishing a design basis for a revamp project is not a simple task. In most cases, revamp projects are conducted under a tight timeline. When the plant management approves a revamp project, they want to see the benefit to be captured as quickly as possible. However, sometimes a revamp study is hampered and may be unsuccessful due to a failure to define the objectives or the basis of the study properly and completely. If this occurs, the revamp study will likely embark on the wrong track and the result is that either the project deadline cannot be met due to rework, or worse, the results are not what the owner expected.

The best projects go through stages of studies before continuing on to the revamp design stage – some actually go through all study types: scoping, feasibility, and process revamp. The passage through these different studies can be viewed as stage-gate reviews, continuing to look at the viability of the project as it becomes better defined. Sometimes the project becomes more attractive; in some cases the project is dropped or redefined as the studies progress.

If the project is not well defined, a scoping study is a way to provide the owner with a rough idea of what is possible and what it will cost.

Regardless of how well defined a project is, there needs to be frequent contact between the owner and the engineers doing the revamp work, with discussions and updates regarding the progress of the project and sharing any "revelations" as soon as they are discovered. The scope and basis must be sufficiently detailed and documented to prevent "scope creep." If the basis and scope are not firm, the owner may try to change the objectives, feed basis, product slate, etc., as the study proceeds. This is just human nature.

The design firm's project manager or coordinator has an important role in the design basis phase of the project. He or she must ask the right questions and get the required information from the owner's engineers. He or she must have experience with the process unit(s) in the scope of the study. The plant owner should also assign a project manager or coordinator that is very familiar with the plant, able to provide the needed information to the design engineers, and understand and communicate the owner's goals and limitations for the revamp project. The owner's project manager must also work effectively with his team which usually consists of engineers of several different disciplines including, operation, process, maintenance, equipment specialists, control and instrumentation, and utility engineers.

The following are the areas that need to be discussed and defined to the extent possible before embarking on a revamp study, either scoping, feasibility, or process.

5.5.1 Agreement

No matter if the owner is paying for a study or receiving the services free of charge, a document defining the basis and scope of the study should always be prepared and agreed to by both the owner and the organization doing the revamp work before embarking on the study. This can be a formal agreement or documented in correspondence.

5.5.2 Processing Objectives

There are many reasons a refiner may want to revamp a process unit or complex. The following are some of the more common objectives:

- Capacity increase
- Feedstock change
- Product slate change or purity improvement
- Catalyst change
- Process type change
- Utilities reduction
- Updating for HS&E reasons

Many times the owner's objectives include more than one of the above. It is important to clearly understand, define, and document the processing objectives of a revamp in the agreement.

5.5.3 Define the Approach of the Study

There are several different approaches that can be taken when undertaking a revamp, as discussed in greater detail in Section 5.3. The approach that will be taken should be discussed and agreed to prior to the start of the study or in the design basis meeting.

- Specified Target Capacity
  In this approach either the feed rate or the product rate is specified and the study is to determine what is required to achieve the target, without constraints or limitations. This approach is the easiest, but this is not the most common type of study.
- Target Production with Certain Equipment and/or Cost Constraints
  With this approach, the objective is to determine if a target feed rate or production rate can be achieved and what is required to achieve it based on defined constraints such as utilizing an existing reactor, recycle gas compressor, fractionation column and/or other equipment/plant section, or limited capital expenditure.
- Maximize Capacity with Certain Equipment/Cost Limitations
  With this approach, the objective is to determine a maximum capacity and what is required to achieve it based on defined constraints such as utilizing an existing reactor, recycle gas compressor, fractionation column, and/or other equipment/plant section. This is the most common type of study approach.
- Identify Successive Bottlenecks
  In this approach, constraints may not be defined, but instead the owner wants to know what the maximum capacity is as each of several bottlenecks is reached and then removed. For example, the existing charge pump may limit the capacity to 120% of design. Assuming the charge pump is replaced, the next bottleneck is the stabilizer, which with the existing trays limits the capacity to 130% of design. After that bottleneck is removed, the recycle compressor limits capacity to 135% of design, and so on. This type of study is quite time consuming, especially if multiple feed cases or flow scheme cases are involved.

5.5.4 Feedstock and Make-Up Gas

The type of feedstock definition required for doing yield estimates and process design depends on the type of unit being revamped and in some cases the mode of operation of the unit. As an example, for reforming units the feed definition required depends on whether the unit is for motor fuel production or for aromatics production. For motor fuel production, the feed definition required includes the ASTM D-86 distillation, overall PNA, and API or specific gravity. However, for aromatics production the feed analysis should be either a PONA by carbon number or a componential analysis. For an aromatics complex including aromatics extraction, xylene isomerization, and transalkylation units, the feed definition should be a componential analysis.

5.5.5 Product Specifications

The owner's product specifications need to be identified and documented in the design basis and agreement. This includes purity requirements of all products, allowable specific contaminant levels, and whether the specs are standard ASTM product specifications or other specs. Specifications of by-products such as liquid petroleum gas (LPG), hydrogen, and fractionated by-products should also be identified.

5.5.6 Getting the Right Equipment Information

The request for equipment information should be definitive, listing the equipment and the types of drawings and data sheets needed to do the study. This is sometimes more difficult than one would expect, especially for older units where the information may be either not available or illegible.

The plant engineer that is assigned to collect the required information should be familiar with the plant, the location of the plant data, and should be dedicated to the task. When the owner's drawings and data sheets arrive, they should be checked for completeness of the information and the owner notified if any information is missing.

The following table lists the specific drawings and types of data sheets needed.

Fired heaters	API data sheets and coil arrangement drawings for both radiant and convection sections
Vessels	General arrangement (GA)-type drawings with major dimensions, metallurgy, and design conditions
Columns	GA drawings and layout drawings of trays, including no. of holes or valves, downcomer chord heights, weir heights, downcomer clearance, etc.
Reactors	GA drawings and drawings of internals
S&T heat exchangers	TEMA (or equal) data sheets including surface area, U value, pressure drop, and design conditions (if HTRI is to be used to evaluate certain heat exchangers, drawings of those exchangers are needed that detail tube and baffle arrangements)
Air-cooled exchangers	Vendor data sheets including surface area, U value, and air-side information
Pumps	API (or equal) data sheets including process conditions, efficiency, impeller size, driver data, and performance curves
Compressors	API (or equal) data sheets including process conditions, efficiency, driver data, and performance curves

Other design information needed includes the following:

- As-built (or as-exists) process flow diagrams
- As-built (or as-exists) piping & instrument diagrams and/or mechanical flow diagrams
- Utility data (steam, fuel gas, cooling water, nitrogen, etc.)
- Ambient air design temperature (may be different than the original design, i.e. less conservative)
- Maximum capacity achieved and current bottlenecks

5.5.7 Operating Data or Test Run Data

A recent capacity test run is recommended for assessing the "real" bottlenecks in the plant. The test run should be done at or near the maximum capacity of the unit or complex. This not only establishes a baseline for a capacity increase type revamp but also helps avoid a situation in which the maximum capacity of existing equipment is significantly underrated or overrated. Usually operating data or test run data are used only as a guide with vendor data sheets the primary source for rating equipment. A single-gauge pressure survey is very useful for evaluating reactor circuits that include compressors or multistage pumps. As a minimum, the compressor suction and discharge pressures should be obtained at a known throughput to help evaluate the reactor circuit hydraulics.

The engineers conducting the study will need to work closely with the plant engineers to get the appropriate operating data and/or carry out the test run to get the most useful information from it.

5.5.8 Constraints

As part of the design basis it needs to be determined and confirmed if there are any major constraints or limitations imposed on the project by the owner.

The following questions should be asked of the owner prior to the start of the study or in the design basis meeting:

- Is any equipment off-limits to replacement?
- Is any equipment off-limits to modification?
- Are there any plot area limitations? This may be handled be either a plant visit or by having the owner mark up a plot plan drawing showing available plot area. Open area on the plot plan may not necessarily be available for equipment as it may be needed for maintenance access or there may be underground piping/electrical equipment.
- Is welding allowed on existing vessels (e.g. new nozzles)? This may affect the ability to relocate the feed to columns and whether or not a side cut can be added, etc.
- Can equipment be re-rated for higher operating temperature or pressure? This may require a new code stamp which may require hydro testing. Some refiners may not want to or be equipped to do this.

5.5.9 Utilities

The availability of utilities is a critical aspect of most revamp projects. The following questions should be asked:

- What are the available quantities or limitations on the use of cooling water, steam, fuel gas, fuel oil, and hydrogen? (as required for the particular process units involved)
- What is the refiner's preference for cooling media, air or water?
- Is space available in the pipe racks for additional air coolers?
- What ambient temperature should be used for evaluating existing air coolers?
- What air temperature should be used for the design of new air coolers?

5.5.10 Replacement Equipment Options

During the design basis setting phase of the project, the specialized equipment that may be used to revamp the unit(s) should be discussed with the owner. This includes the use of specialized heat exchangers such as welded plate exchangers, high flux tube bundles to save existing reboiler and steam generator shells, and high-capacity trays to debottleneck columns.

The type of drivers for replacement rotating equipment should be discussed as well, i.e. electric motors or steam turbines.

5.5.11 Guarantees

The types of guarantees that are required by the owner or are to be offered by the organization doing the revamp work may have an impact on the outcome of equipment evaluation and revamp cost. If the revamp capacity is not to be guaranteed, the revamp engineers will be more willing to take existing equipment to its maximum capacity. On the other hand, if a capacity guarantee is to be made, some design margin will need to be considered and maintained when evaluating equipment. This may result in recommending replacement of equipment that would otherwise be retained, resulting in greater revamp cost to the refiner. This subject should be discussed with the owner prior to the start of any revamp study.

Yields and catalyst life relate to reactor space velocity, operating temperature window achievable, H_2/HC ratio, and potentially other operating parameters. If any of these parameters need to be compromised in the revamp operation, the licensor of the technology should be made aware of it to assess the impact on yields and guarantee position.

5.5.12 Economic Evaluation Criteria

Most studies require some type of economic evaluation to be performed, either as justification of the revamp or for selecting between alternatives. To perform these evaluations, the following owner specific information is needed:

- Feed and product values, including by-product values.
- Utility costs, including steam, fuel gas and fuel oil, electricity, and cooling water.
- The type of analysis the owner would like used – i.e. simple payback, NPV, or IRR.
- Interest rate, tax rate, and payback time period.
- Agreement of the basis to be used for equipment cost estimates, i.e. location, timing, installed cost, or equipment-only cost.

5.6 Process Design for Revamp Projects

The process design for a revamp project is almost never a once-through process like it is in a new unit (grassroots) design. As has been discussed, the process simulations are typically done for the pre-revamp case, sometimes referred to as the base case, to establish a starting point for the revamp, as well as to calibrate or rate the existing unit and its equipment. Depending on the type of revamp project (specified target capacity, target production with constraints, maximize throughput at minimum cost, or identifying successive bottlenecks) the next set of simulations will be based on some specified or estimated capacity point with operating conditions specified by the first pass yield estimate(s). Once this first "revamp" simulation is completed the first pass at major equipment evaluation can begin. Depending on the outcome of this first pass equipment evaluation and the type of revamp (see above), the process simulation(s) will be revised. As discussed in Section 5.6.1, the yield estimate(s) may need to be revised as well. In either case, a process simulation recycle loop has begun, which may require several iterations before the desired outcome and "final" simulation is performed. For a process engineer this is the fun part of the study, as it can be a challenge and somewhat like a puzzle, adjusting operating conditions and capacity, checking equipment, and trying to make it all fit. At the same time, incorporating ideas for utilities reduction, capital cost minimization, plant downtime minimization, and possibly safety and environmental improvements.

5.6.1 Adjusting Operating Conditions

The yield estimating model and process simulation model play a critical role in a revamp assessment. The effects of changing process conditions on yields are assessed in the yield estimating model. The effects of changing process conditions on equipment and utilities are evaluated in the process simulation model. Using both yield estimating and process simulation models allows engineers to conduct what-if analyses in order to effectively determine optimal revamp solutions.

It is rare that the yield estimating model(s) are available to operating plant personnel. That is one of the reasons why it is recommended that a revamp project be conducted by the technology licensor as they have the yield estimating capability, as well as design experience. The main criteria for selecting the organization to conduct a revamp study is process and operating knowledge of the unit(s) to be studied, yield estimating capability, and design experience and track record in conducting revamp projects.

In many cases it is advantageous to adjust unit operating conditions such as temperatures, pressures, reactor space velocity, hydrogen to hydrocarbon ratio ($H_2 : HC$), recycle gas composition, etc., to facilitate the most economic unit revamp. In order to do this, the designer must either understand the interrelationships between unit operating conditions and their effect on product quality, catalyst life, and other unit operations, or have access to someone that does. Below are some examples, not all inclusive, of how some operating conditions may be altered to facilitate a revamp project:

- Increase the reactor section separator pressure to save the recycle gas compressor.
- Reduce the recycle gas flow rate to save the recycle gas compressor.
- Accept a higher space velocity (reduced catalyst volume) to save a reactor.
- Operate at a lower reactor pressure to save equipment in the reactor circuit.
- Provide more feed preheat to a column to save a reboiler.
- Adjust the split on one column to save the downstream column(s). This may include adding a simple stripper to the overhead of a column or providing a pre-flash column ahead of another column.
- Provide advanced process control (APC) to allow reduced column reflux and reboiler duty.
- Use an alternate source of hydrogen, possibly with a different H_2 concentration or containing more or less or different contaminants

In many cases, especially when reactor operating conditions are considered to be modified, the licensor's process specialist(s) must be consulted to determine the acceptability of a change. In most cases the licensor is the primary party doing the revamp studies and basic revamp engineering design. In this case, the designer can work with the process specialists in the organization to change operating parameters and determine the effect on plant yields and operations. This is very beneficial as it allows quick mini-studies and "what-ifs" to be done to investigate alternatives.

If the revamp study or basic revamp engineering design is not done by the licensor, the designer needs to establish a working arrangement with the licensor or have guidelines and support from someone that has a strong understanding of the process and catalyst. If the designer has guidelines for acceptable operating parameter ranges and the effect of changes, it will allow for the quick mini-studies and "what-ifs" to be done.

5.6.2 Design Margin

Design margin implies great potential to people in the process industries because it is conventional wisdom that 10–20% additional capacity can be squeezed out of a plant with aggressive operation and little capital cost due to the original design margin included. This raises high expectations from plant management who wish to achieve the most out of the existing assets with the least investment. Many times, however, the plant has used up most of the design margin in efforts to push the capacity prior to embarking on a revamp project.

To understand the available equipment design margins and constraints, it is recommended that a rating survey of key equipment be conducted and the results summarized, indicating which equipment is limiting and which has spare capacity, and how much.

For example, a product fractionator is usually designed at 80–85% jet flood which provides about 10–15% spare capacity. For processes with reactor sections, the fired charge heater is designed with relatively large design margin, sometimes exceeding 20% for start-up, process control, and end of run purposes. This spare capacity may be used as normal operating capacity in a revamp situation. Almost all process pumps and compressors are designed based on rated flow rates, typically 110% of normal. Therefore, 10% spare capacity may be available for the revamp operation. Design margins are applied to off-site and utility equipment such as boilers, boiler feed water (BFW) pumps, wastewater pumps, storage tank pumps, air compressors, etc. There are many such examples of design margin built into grassroots plants which may be available in a revamp scenario.

Some design margins are not obvious, as they are not spelled out in the original design data sheet, but may still exist in the design. For example, many process fired heaters are designed based on a radiant flux limit or tube wall temperature (TWT) limit which has margin built into it. If the heater of interest operates below the allowable flux or TWT limits, the gap may provide potential additional capacity for the heater.

It may be possible to avoid the replacement or addition of a new reactor if reduced performance is deemed acceptable. Examples of reduced performance include reduced conversion, reduced catalyst cycle length, or a reduction in severity at the desired cycle length.

One must remember that when design margins are exploited, the equipment will be operated closer to its limits. For example, if a fractionator is operated near the jet flood condition, product quality may suffer due to poor fractionation efficiency or upsets may result in lost production. The risk of poorer fractionation may be acceptable if the fractionator provides a rough separation and the net streams are recycled within the complex. A furnace operating close to the flux limit or TWT limit leaves less room for safe operation. This risk may be managed by adopting better monitoring and control or using APC.

For the revamp design, the amount of acceptable design margin remaining in existing equipment or added to new equipment is dictated by the nature of the revamp, the owner's tolerance for risk, and whether or not any guarantees will be offered/required for the revamp from the licensor/designer. Typically, design margins that are less than those provided with a new unit design are acceptable for a revamp because equipment changes and costs are to be minimized. If a piece of equipment is to be replaced with new equipment, that equipment may have a larger design margin than other existing equipment.

5.7 Revamp Impact on Utilities

A revamp may have major impacts on utility consumption and/or production. The primary utilities used in an aromatics complex are fuel (usually fuel gas, sometimes fuel oil), steam (low, medium, and high pressure), cooling water, hydrogen, and nitrogen (for start-up, shutdown, and special operations).

A capacity increase revamp, without changes to product slate or flow scheme will typically have a direct effect on the utilities, i.e. a 30% capacity increase will require about 30% more utilities. However, in most cases the owner is looking for operating cost improvements with the capacity increase. Improvements such as better heat integration, greater recovery of valuable by-products such as benzene, LPG, and other light ends will change the utility balances significantly.

A process plant is said to be in a "fuel gas long" situation when fuel gas production on site is more than consumption. When surplus fuel gas cannot be exported, it may have to be flared, which will undo energy improvements and also increase CO_2 emissions. "Fuel gas long" could happen when energy efficiency improves.

In one example, the implementation of several large energy saving projects resulted in a "fuel gas long" situation. The fuel gas long scenario was not predicted in advance as the plant had a very poor fuel gas balance. The plant fuel gas balance underestimated the production of refinery fuel gas but overestimated fuel consumption. As a result, the plant had to cut down feed rates for major process units which hurt their economics. This was not enough to curb fuel gas long and thus the plant had to send large amounts of fuel gas to flare.

The lesson learned from this painful event is that the plant fuel gas balance needs to accurately predict the amount of fuel gas produced in individual process units as well as the amount consumed in fuel gas users, i.e. fired heaters and boilers. The balance should account for variations in feed rates and process conditions as well as weather changes. The variation could be very significant and thus modeling of variation in fuel gas production is very important in building the plant fuel balances.

The main feature of a good fuel gas balance is to be able to balance fuel gas to 97% closure and less than 1% variability across changes in feed and product slates, processing conditions, seasons, and off-site management as the plant feeds and yields vary over time. Processing different feeds could have significant impact on fuel gas production and hence the fuel gas balance. A similar effect occurs for dramatic seasonal weather changes.

When modeling for the fuel gas balance, it is important to understand both the rate and composition of the fuel gas. As changes in feed and product slates are implemented, the relative proportions of fuel gas from the different process units can change, thus causing the overall fuel gas composition to vary. Differences in fuel gas composition are often the result of variations in the fuel gas source.

Therefore, it is important for a plant to have a good fuel gas balance available and review the impact of operational changes and energy-saving projects on the fuel gas balance up front as part of the project evaluation.

If a fuel gas balance model predicts that the plant will be in a fuel gas long situation, a solution roadmap detailing what the plant can do in the short and long term should be developed. Efforts to reduce fuel gas production must attempt to understand the root causes at the source and determine the best choice for reducing it via operation changes and capital projects.

One option to consider is reduction in fuel gas make-up. This can be accomplished by operational changes including catalyst changes, increasing light ends recovery, reducing reaction severity, and finally reducing feed rate as the last resort. Another option is to use more fuel gas or steam in an effort to enhance production of desirable products and improve product quality.

The plan for capital projects should be developed as well if operational changes cannot bring the fuel gas long back into balance. Example includes removing column overhead cooling limits, which could minimize LPG lost to fuel gas. This could be achieved by using advanced heat transfer equipment to achieve greater cooling/condensation.

A fuel gas long situation also provides the opportunity to recover valuable products from the fuel gas. For example, a refinery plant might consider recovery of hydrogen from a hydro-processing gas source. Options for H_2 recovery include PSA and membrane systems with the selection usually based on the H_2 purity requirement. Alternatively, ethane and ethylene could be recovered from certain fuel production sources such as FCC offgas. On the other hand, a fuel gas long scenario could provide an opportunity to install a cogeneration facility as a long-term investment project to generate steam and power that might either reduce the plant's dependence on outside electricity or generate excess power that could be sold to a third party. Extra steam could be generated from boilers for steam turbines to drive rotating equipment that is currently using electric motors.

Overall, prevention of a fuel gas long situation is the best strategy. In the scenario of fuel gas long, a mitigation road map must be developed well in advance so that the plant has options in short, middle, and long terms and can be succeeded in tuning an unfortunate fuel gas long situation into a profitable opportunity.

The steam balance is also very important when implementing a revamp. There may be opportunities to reduce steam consumption, or change steam levels to some users, based on heat integration changes. Opportunities to reduce steam consumption, especially in older plants where the original design was based on much cheaper energy costs, may be significant. The design engineer must make sure any heat integration changes to reduce steam consumption are practical to implement and cost effective.

Increases in the consumption of cooling water can cause off-site capital cost expense in a revamp project if the resulting cooling water use exceeds the capacity of the cooling water system. There are ways to reduce cooling water consumption to avoid this scenario. Low temperature heat recovery may be possible if there is an appropriate process stream available. Shifting cooling capacity to air-cooled exchangers is a potential solution. Another option is to increase the rundown temperature of products to storage. This must be reviewed carefully with the owner and may only be possible for low vapor pressure materials. These strategies should be considered to avoid the addition of a new cooling tower and/or cooling water pump.

5.8 Equipment Evaluation for Revamps

The key task in any revamp study or assessment is the evaluation of existing equipment in the unit being studied. In the evaluation the equipment's capability to operate at the revamped conditions is assessed. This includes its appropriateness for the new service, performance capability, spare capacity, and other possible limitations. When equipment reaches hard limitations, such as a fractionator reaching its jet flood limit, a compressor at its flow rate or head limit, or a fired heater at its flux limit, the challenge of a revamp is to find ways to overcome the constraints so that expensive equipment can be saved and utilized in the revamp, if at all possible. Finding ways to avoid or relax plant limitations is one of the revamp design engineer's most important tasks.

Three general design constraints for any piece of equipment are its design pressure, design temperature, and metallurgy. If it is determined that equipment will operate at higher pressure than its design limit, the operating pressure needs to be reduced if possible. If it is determined that equipment will operate at higher temperature than its design limit, the operating temperature needs to be reduced if possible. In some cases, it may be possible to re-rate a piece of equipment for higher temperature or pressure, which will be discussed further in Section 5.8.2.2.4. These constraints may be resolved if the process conditions can be modified.

If the metallurgy of a major equipment item, such as a reactor or fired heater, is found to be inadequate for the revamp operation, it needs to be addressed as soon as possible as the cost of replacement may be prohibitive. Equipment replacement comes with a very high cost. Equipment modification is generally less costly than replacement and, therefore, should be investigated. Operating condition changes are generally the least costly alternative.

5.8.1 Fired Heater Evaluation

In most revamp cases the owner is not willing to replace an existing fired heater. Fired heaters are expensive and require significant down time to revamp. In addition, because a fired heater is a source of gaseous emissions, the refiner may have to deal with permitting issues if a fired heater is modified or replaced. In some cases the owner may be willing to make modifications such as replace burners, change the number of passes, or even retube the heater to make it suitable for the new process conditions.

5.8.1.1 Data Required
5.8.1.1.1 Data Sheets and Drawings
The vendor's data sheets and as-built heater drawings are needed to adequately evaluate an existing heater. The drawings should show the GA and layout of the coils in both the radiant and convection sections.

If detailed heater information is not available, any comments regarding the heater's suitability must be qualified by indicating that the evaluation is based on limited information and requires confirmation.

5.8.1.1.2 Operating Data
For critical service heaters such as reactor charge and interheaters, it is highly recommended that operating data for the heater(s) is collected and used to evaluate the actual operation of the heater rather than relying solely on design information.

5.8.1.2 Fired Heater Evaluation
Determine the duty, outlet temperature, and pressure requirements based on the governing heat and weight balance.

5.8.1.2.1 Preliminary Analysis
A preliminary analysis is all that is required when doing a revamp scoping study. Depending on how much a heater is being pushed beyond its original design, a

preliminary analysis may be all that is needed for a revamp feasibility study as well.

Most heaters are designed to either an average radiant flux rate or a TWT limit. A preliminary heater evaluation can often be completed quickly if the heater was designed to a flux limit. The original design average radiant flux rate for a heater can be found on the API or vendor's data sheets. To determine the radiant section flux rate in a radiant-convective heater, the original duty ratio between the radiant and convective sections can be used, or if no breakdown is available, a 75/25 ratio may be assumed. Flux limits for revamped fired heaters are discussed in Section 5.8.1.4. For a preliminary analysis, either flux rate or TWT, metallurgy (Section 5.8.1.5) and tube thickness/operating pressure (Section 5.8.1.7) may be the only criteria reviewed.

In processes such as transalkylation and xylene isomerization, the charge heater is designed with a significant amount of design margin due to limitations on the ratio of the combined feed exchanger (CFE) to the charge heater duty for start-up and control purposes. In a revamp all of the design margin should be considered available for normal operation.

5.8.1.2.2 Detailed Analysis

A detailed heater analysis should be performed when doing a process revamp study or basic revamp engineering design. Depending on the service and the extent of the changes in the operating conditions, a detailed analysis may also be required for a revamp feasibility study. A detailed heater analysis will include much more than a preliminary analysis and should be performed by a heater specialist. The following criteria are reviewed in a detailed analysis:

Radiant flux
TWT
Bridge-wall temperature (BWT)
Metallurgy
Coil thickness
Coil pressure drop (ΔP)
Burners
Stack
Air preheat (as required)
Fans (as required)

5.8.1.3 Heater Design Limitations

When a design engineer is reviewing a heater for a revamp, it is important to know the type of limitation for the original design, and for the revamp conditions. For capacity increase revamps, the type of limitation for the revamp is usually the same as for the original design. In conversion revamps where one type of process technology is converted to another (e.g. reforming to transalkylation), the type of heater limitation may be different for the new service.

The two types of heater design limitations are:

Flux
TWT

Heaters limited by flux are characterized by high process $\Delta P > 1.4 \text{ kg/cm}^2$ (20 psi). Most general service heaters such as reboilers fall into the flux-limited category.

TWT-limited designs are characterized by low process ΔP of about 0.14–0.42 kg/cm^2 (2–6 psi). The low ΔP heaters have low tube mass velocities, which result in low heat transfer coefficients, and thus high TWTs. Examples of TWT-limited heaters include reforming, Oleflex, and Cyclar reactor heaters.

5.8.1.4 Radiant Flux Limits
5.8.1.4.1 Single-Fired Heaters

UOP designs for new single-fired heaters use a maximum of 27,128 kcal/m^2-h (10,000 Btu/ft^2-h) radiant flux. Small heaters of <2.5 MM kcal/h (10 MM Btu/h) process duty and tube lengths <5.5 m (18 ft) are designed for lower radiant flux rates.

Table 5.1 shows the heater radiant flux limits for new designs and for revamps. For revamps, UOP limits single-fired heaters with process duties of 2.5 MM kcal/h. (10 MM Btu/h) or higher to a radiant flux of 32 553 kcal/m^2-h (12 000 Btu/ft^2-h). For small heaters, the revamp limit is 20% greater than the design radiant flux. These limits are based on field experience of what is practical on a day-to-day basis.

Estimate the revamp radiant flux for the revamp process duty:

$$\text{Revamp radiant flux} = \text{Original design flux} \times \frac{\text{Revamp process duty}}{\text{Original design process duty}}$$

Table 5.1 Heater radiant flux limits for single-fired heaters.

Process duty (MM Btu/h)	New design radiant flux (Btu/ft^2-h)	Revamp radiant flux (Btu/ft^2-h)
10 or higher	10,000	12,000
9	9000	10,800
8	8000	9600
7	7000	8400
6	6000	7200
5	5000	6000
4	4000	4800

5.8.1.4.2 Double-Fired Heaters

Double-fired heaters (burners located on both sides of the tubes) are usually TWT limited. Nevertheless, there are flux limits as well. The radiant flux limit for the revamp of a double-fired reforming reactor heater is 59,680 kcal/m^2-h (22,000 Btu/ft^2-h).

Consult with a heater specialist for other double-fired heater radiant flux limits.

5.8.1.5 TWT Limits

TWT-limited heaters usually occur in high temperature processes such as Platforming, Oleflex, Pacol, Unicracking, and Cyclar. The heater specialist calculates the TWT. Table 5.2 shows UOP's TWT limits for new designs, based on tube metallurgy.

The carbon steel (CS) tube limit is based on preventing graphitization in the tubes. The chrome (Cr) alloy tube limits are based on inhibiting tube oxidation. These values may be exceeded on revamps; however, doing so shortens the coil life. Consult a heater specialist for absolute maximums.

5.8.1.6 Metallurgy

5.8.1.6.1 Changes in Design Practice

Design practices have changed over the years. In some technologies the design metallurgy has changed from 1-1/4 Cr or 2-1/4 Cr to 9 Cr.

When reviewing an older heater for suitability in a revamp be aware that the metallurgy may be suitable even though it is not the current practice. In some cases it may be necessary to upgrade the coil metallurgy to meet current design practices. If there is any doubt about the suitability of the existing metallurgy, consult a heater specialist.

5.8.1.6.2 Tube Corrosion Rates

In general, aromatics heater services have low tube corrosion rates. Corrosion rates greater than 15 mil per year can cause plugging and pressure drop problems due to scale deposition in downstream equipment such as reactor beds. If the corrosion rate exceeds 15 mil per year, the coil metallurgy should be upgraded to austenitic stainless steel.

5.8.1.7 Tube Thickness

The required tube thickness is a function of the pressure in the tube, the TWT, and the corrosion allowance. This section gives two methods for finding the required tube thickness for the revamp conditions.

The first method involves the use of the API 530 procedures. This method is more rigorous and should be used for a detailed heater analysis.

The second method involves using tables that list pressure limits for various tube sizes and metallurgy to estimate the required tube thickness. This method may be used for preliminary heater analysis, such as for a scoping study.

For either method, compare the required tube thickness to the actual tube thickness shown in the heater vendor design data sheets or from the owner's inspection records.

5.8.1.7.1 Tube Thickness Using Look-up Tables

Tables 5.3–5.5 may be used to quickly check the required tube thickness for the revamp conditions. Tables 5.3 and 5.4 assume a 1/8 inch corrosion allowance, and Table 5.5 assumes a 0 to 1/16″ corrosion allowance.

In these tables, AW is average wall tube, with mill tolerance of −12.5 to +12.5%. MW is minimum wall tube, with mill tolerance of −0 to +28%. Due to the −12.5% tolerance, AW tubes have an actual wall thickness that is

Table 5.3 Estimated pressure limits for carbon steel heater tubes.

Thickness	Pressure limit, psig
CS Sch 40 AW	500
CS Sch 80 AW	1000

Note: The limits are at metallurgy design temperature of 800 °F for carbon steel.

Table 5.4 Estimated pressure limits for chrome heater tubes.

Thickness	Pressure limit (psig)
2-1/4 Cr Sch 40 AW	250
2-1/4 Cr Sch 80 AW	410
9 Cr Sch 40 AW	150
9 Cr Sch 80 AW	250

Note: The limits are at metallurgy design temperature of 1125 °F for 2-1/4 Cr and 1175 °F for 9 Cr.

Table 5.2 TWT limits for new unit designs.

Metallurgy	TWT limit (°F)
Carbon steel	800
1–1/4 Cr	1025
2–1/4 Cr	1100
5 Cr	1150
9 Cr	1175
Stainless steel	[a]

[a] For stainless steel (SS), process temperature limits usually occur before reaching the TWT limit.

Table 5.5 Estimated pressure limits for 347SS heater tubes.

Thickness	Pressure limit (psig)
Sch 40 AW	880
Sch 40 MW	1050
Sch 80 AW	1550
Sch 80 MW	1850
Sch 120 AW	2150
Sch 120 MW	2550
Sch 160 AW	2900
Sch 160 MW	3400

Note: The limits are for 6″ tubes at 1120–1125 °F TWT.

12.5% less than that of an MW tube of the same schedule number. Thus, for the same schedule number, the MW tubes can withstand a higher pressure.

Consult a heater specialist for tubes that are not represented in these tables.

5.8.1.8 Coil Pressure Drop
5.8.1.8.1 Reactor Circuit Heaters
Pressure drop (ΔP) is especially important for heaters in a reactor circuit.

1) Single-Phase Heater Pressure Drop
   Single-phase pressure drop can be estimated by use of the following equation:

$$\Delta P_2 = \Delta P_1 \left(\frac{G_2}{G_1}\right)^2 \left(\frac{\rho_1}{\rho_2}\right)\left(\frac{\mu_2}{\mu_1}\right)^{0.20} \quad (5.1)$$

   where

   G = mass flow
   ρ = average fluid density
   μ = absolute viscosity

   Ignore the viscosity term if the fluid viscosity for the revamp operation is not significantly different from the original design fluid viscosity.

2) Two-Phase Heater Pressure Drop
   For two-phase heaters, use Eq. (5.1) above using average vapor–liquid densities:

$$\rho_{ave} = \left(\frac{\text{mass of liquid} + \text{mass of vapor}}{\text{vol. of liquid} + \text{vol. of vapor}}\right) \quad (5.2)$$

3) New Number of Passes (Single- or Two-Phase)
   For heaters modified for a new number of passes, use the following equation to calculate the revamped heater pressure drop:

$$\Delta P_2 = \Delta P_1 \left(\frac{G_2}{G_1}\right)^2 \left(\frac{\rho_1}{\rho_2}\right)\left(\frac{\mu_2}{\mu_1}\right)^{0.20}\left(\frac{n_1}{n_2}\right)^3 \quad (5.3)$$

where

n_1 is the original number of passes
n_2 is the new number of passes.

5.8.1.8.2 Reboiler Pressure Drop
Most reboilers designed for 50 wt.% vaporization do not need additional flow to achieve an increased duty requirement. The percent vaporization may be increased up to 85% by volume to achieve the increased duty, keeping the flow rate to the heater the same as the original design. This has the added benefit of allowing the existing reboiler pumps to be used in the revamp operation. Estimate the new ΔP by multiplying the old ΔP by the ratio of the outlet vaporization rates (new vapor rate/existing vapor rate).

5.8.1.9 Burners
Burners are inexpensive to replace, typically representing only about 10% of the total heater cost. On revamps, new burners are often required to meet new emission requirements or increased duty.

5.8.1.9.1 Flame Length
Review the flame length to confirm the flame will fit within the existing firebox. This is extremely important on opposed fired U-tube heaters. Table 5.6 gives estimated flame lengths per heat released by one burner.

On opposed fired U-tube heaters, the flames from opposing burners are not expected to run into each other in the middle of the heater. On other heaters, the flame length is not expected to exceed 2/3 of the tube length.

Calculate the heat release of an individual burner for the revamp conditions as per the following equation:

$$\text{Burner heat release} = \frac{\text{Process duty}}{\text{Heater efficiency} \times \text{Number of burners}} \quad (5.4)$$

Table 5.6 Estimated flame lengths.

Burner type	Flame length (ft)/heat release (MMBtu/h)
Gas	1.5
Oil	2
Low NO$_x$ gas	2
Low NO$_x$ oil	2.5 to 3
Ultra-low NO$_x$ gas	2.5 to 3

5.8.1.9.2 Heat Release and Emission Requirements
In a detailed heater analysis, a heater specialist should check the suitability of the burners for heat release and emission requirements.

5.8.1.10 Stack
5.8.1.10.1 Revamp Studies
For revamp studies (scoping, feasibility, and process revamp studies) it is not usually necessary to check heater stack hydraulics.

5.8.1.10.2 Revamp Basic Engineering Design
For revamp basic engineering design, the heater specialist should check the stack hydraulics.

5.8.1.10.3 Revamp Recommendations for Stacks
Stack draft is inversely proportional to ambient temperature, i.e. lower ambient temperature results in greater draft. Therefore, stack height for new designs is set by the high ambient temperature. Most plants do not operate at or above the design high ambient temperature for more than about 5% of the year. Thus, additional draft is available most of the time.

It is extremely expensive to modify an existing stack. Thus, for revamps, it may be acceptable to use the existing stack even if it is short on draft (i.e. not tall enough) during periods of maximum ambient temperature. In this case, the following comment should be included in the revamp report and/or documentation: "Stack height may not provide sufficient draft at all operating conditions and ambient temperatures."

5.8.2 Vessels: Separators, Receivers, and Drums

Separators, receivers, and drums are in general relatively simple vessels that provide vapor liquid separation or liquid residence time. As such, they are in general not costly to replace unless they are unusually large (>3 m) or of exotic metallurgy. In many instances, these types of vessels can be reused by operating with less residence time or by augmenting the existing vessel separation with additional equipment.

5.8.2.1 Data Required
As-built vessel drawings are needed to adequately evaluate an existing separator, receiver, or drum. The drawings should show design temperature and pressure, GA of the vessel, as well as the dimensions and layout of internals such as mesh blankets or baffles. A listing of important vessel information is given below.

Design temperature and pressure
Diameter and tangent length
Materials of construction
Internals' dimensions and location
Nozzle sizes and flange ratings
Elevation

The vessel information required is a function of the type of study. The last two items are required only for a process revamp study and basic engineering design.

If complete vessel information is not readily available, but the diameter and tangent length and design temperature and pressure are known, the comments regarding vessel suitability should be qualified by indicating that the evaluation is based on limited information and requires confirmation.

5.8.2.2 Separator, Receiver, and Drum Evaluation
The vapor and liquid loads for the vessel are provided by the heat and weight balance for each revamp case. Comparison of the cases should allow elimination of the noncontrolling cases. The vessel should be evaluated based on the limiting requirements of downstream equipment or processes. These limiting requirements should take into consideration current operations as represented by test run data. If a revamp is to be guaranteed, the design margin used to evaluate each vessel should be established.

Review the following items to determine whether an existing vessel is suitable for a particular revamp service.

5.8.2.2.1 Vapor Liquid Separation
In the absence of any specific criteria agreed to with the owner, new unit separation criteria and diameter calculation for the appropriate separation should be used. In revamp service the calculated diameter based on the use of a mesh blanket can be reduced by approximately 25%. The equations for mesh blankets are based on using a vendor design factor of 1.0 (100%) to calculate limiting velocity. The allowable range of the vendor design factor is 0.3–1.1. For revamps, the design margin is minimized and so the design factor can be set as high as 1.1. This allows for a 10% increase in the limiting velocity over normal new unit design criteria.

When evaluating a separator, receiver, or drum, if the vessel does not appear to satisfy the vapor and liquid disengagement criteria, try to determine from the owner if they are experiencing any difficulties with the vessel separation or if it is adversely affecting downstream equipment.

If the vessel is close to satisfying the separation criteria, it may be possible to adjust the estimated bubble or droplet size. It is hard to justify replacing a separator, receiver, or drum if there are no significant consequences to a slightly poorer separation. If the vessel does not already have a mesh blanket, consider adding one to allow vessel reuse. In any event, the owner should be advised of the calculation results and, if necessary, the reason for replacement is recommended.

A coalescer can be added downstream of a separator to remove water entrained in the liquid hydrocarbon.

An additional separator or vent condenser can be added to collect valuable material from a separator vapor stream.

A change in the operating conditions of the vessel may also be a possible way to meet the separation requirements. Increasing the operating pressure by adding a pilot-operated relief valve or reducing the temperature by adding cooling are possible options.

5.8.2.2.2 Residence Time

Residence time in vessels and boots allows for continued operation during normal process fluctuations. For revamps, the residence times will generally decrease; however, unless the times are exceedingly short, they would not normally be expected to cause the replacement of the vessel. A decision to replace a vessel based on residence time should be reviewed with the owner.

5.8.2.2.3 Materials of Construction

The existing vessel materials of construction must suit the revamp service. Thus, changes in the process stream composition and/or temperature may require a different metallurgy than that of the existing equipment. To determine if the metallurgy of the existing vessel is acceptable, review any significant changes in operating temperature, sulfur content, H_2S content, hydrogen partial pressure, or stream contaminants.

Because of recent changes in API-recommended practice, vessels in high-pressure hydrogen service should be reviewed for potential high-pressure hydrogen attack. In particular, the operating conditions of vessels constructed of C-1/2Mo should be reviewed against the Nelson curve to confirm suitability for reuse. The owner should be notified of any special considerations regarding continued use of the equipment.

5.8.2.2.4 Design Pressures and Temperatures

The operating pressures and temperatures for the revamp service must not exceed the design pressure and temperature of the existing vessel. Sometimes, depending on metallurgy, flange ratings, and design pressure, it is possible to re-rate an existing vessel for a higher design temperature. This may require a corresponding decrease in design pressure. The original manufacturer or a qualified representative of the owner would be expected to re-rate existing vessels. Typically, re-rating will necessitate a new American Society of Mechanical Engineers (ASME) Pressure Vessel Code stamp or an equivalent stamp of the local or regional governing body.

5.8.2.2.5 Pressure Drop

In general, the pressure drop of a separator, receiver, or drum is not expected to be significant unless it contains special internals, such as adsorbent or alumina beds.

5.8.2.3 Process and Other Modifications

To avoid replacement of an existing vessel, the following modifications and/or process changes should be considered to make it suitable for the revamp.

5.8.2.3.1 Relax Separation Constraints

If a vessel is only required to do a rough separation, as is usually the case, and/or netting streams that are only recycled within a process unit, it may be possible to relax the separation requirements. Evaluate the extent of the change in the process operation to allow retaining the vessel. Reviewing the droplet or bubble size to be considered in the calculation may result in allowing reuse of the vessel. The owner will likely accept somewhat poorer separator performance if it means not having to replace the vessel. Sacrificing performance for a short period of time during the year due to climate or process cycles may be justifiable. In some cases, this may also require that the owner be consulted if relaxation of constraints leads to changes in specifications. Impact on the owner in areas such as increased maintenance, turnaround time, utilities, and increased chemicals or additives consumption such as desorbent or solvent used within the process should be considered.

5.8.2.3.2 Increase Operating Pressure

Increasing the operating pressure of a vessel by the addition of a relatively inexpensive pilot-operated relief valve may be a way to reuse an existing vessel. Line and equipment pressure drops generally increase significantly for revamps causing existing equipment to see higher operating pressures. Increasing the operating pressure on a separator can increase the vapor density and reduce the resulting pressure drop through a vapor or two-phase reactor circuit. Reducing the pressure drop can allow for additional operating capacity. In general, a pilot-operated relief valve can allow operation within 5% of design pressure or approximately 10 psi, as compared with 10% of design or 25 psi for a conventional relief valve. In cases where there may be a need to operate a separator, receiver, or drum at higher pressure, a pilot-operated relief valve may be a way to reuse the existing vessel.

5.8.2.3.3 Change Residence Time Constraint

Residence times for revamps will generally be less than for new units. The revamp typically just takes up original design margin and results in less flexibility in operations. In most cases this is not sufficient justification for replacing an existing vessel. For those instances where the residence times are extremely tight, it may be appropriate to recommend changing the normal operating level, changing the control range, or perhaps putting a positioner on the level control valve. These moves should improve process control to allow the "tighter" operation required to accommodate the revamp conditions. The reasoning behind making these changes must be made clear to the owner to be sure to get their acceptance of the "tighter" operation.

5.8.2.3.4 Provide Additional Equipment

If the vapor liquid separation or residence time is inadequate for the revamp service, consider adding an additional separator on the vapor or a coalescer on the liquid. If a vessel is being pushed to its limits, the carryover of liquids in the vapor or the heavier phase (e.g. water) in the hydrocarbon will increase. Installing additional equipment downstream may be the most cost-effective option provided there is plot space available. In the case of vapor losses of valuable material, a vent condenser should be considered.

If all of the above modifications are applied where possible and the vessel still exceeds the separation requirements for the revamp, it may be appropriate to consider the use of special vendor-supplied vapor liquid separation equipment.

5.8.2.4 Test Run Data

If operating or test run data for the unit are available, review them with respect to the performance of the separator, receiver, or drum. The operating data may indicate that the residence time is lower than what is normally used for a new unit design or that vapor liquid separation is suitable at the higher rates. These observations should be reviewed and discussed with the owner to avoid recommending a change that flies in the face of operating data.

5.8.2.5 Possible Recommendations

An evaluation of an existing separator, receiver, or drum may result in making one of the following recommendations. Whatever the recommendation to the owner, it must be cost effective and justifiable.

5.8.2.5.1 Use Existing Vessel

If the vapor liquid separation, residence time, design pressure, design temperature, and metallurgy satisfy the new process conditions, use the existing vessel without modification.

5.8.2.5.2 Modify or Add a Mesh Blanket to Existing Vessel

The existing vessel may need a mesh blanket or need to have the existing partial diameter mesh blanket increased in diameter to satisfy the new process conditions. The thickness or design of the mesh blanket may need to be modified. Alternatively, the material of construction of the mesh blanket may need to be upgraded due to degradation. This may be due to excessive carryover from the vessel to a compressor.

5.8.2.5.3 Modification or Addition of Ancillary Equipment

The existing vessel may satisfy the new process conditions if a new coalescer, vent gas condenser, or supplemental separator is installed downstream. Additional feed cooling or an increase in operating pressure through the addition of a pilot-operated relief valve may be sufficient to allow reuse of the vessel. If residence times are extremely short but do not cause any downstream concerns, the addition of a positioner on a control valve can reduce response time to allow suitable control. Alternatively, a change in the vessel normal liquid level through re-ranging or addition of a larger float may be adequate to meet process requirements.

5.8.2.5.4 New Vessel

If nothing can be done to save the existing vessel, consideration should be given to retiring the vessel and replacing it with a new, larger one or adding a new vessel in parallel with the existing vessel. There may be idle vessels available which can be evaluated for the new service.

If the vessel has a boot and it is too small, it may be possible to remove the boot and add a new larger one or connect it to another vessel with additional volume. Of course, this is a more extensive and expensive modification that may require that the foundation be checked for the increased weight as well as requiring more downtime or plot space.

5.8.3 Reactors

The engineering evaluation of existing reactors for a revamp is closely tied to the reactor section process design. The required catalyst volume and the number of reactors for many processes can be traded-off versus changes in reactor temperature, reactor pressure, and/or changes in hydrogen flow rate to the reactor. The process engineer is expected to work with the process specialist or yield estimator to optimize the reactor section process conditions, which are expected to be consistent with the capabilities of the existing equipment for the revamp.

For all types of studies, the catalyst type, the catalyst volume, the number of reactors, and the catalyst distribution should be established. For revamp feasibility study level and above, a reactor specialist should complete a hydraulic evaluation of the reactor(s) for the estimated process conditions established by the process engineer.

5.8.3.1 Data Required

As-built reactor drawings are needed to adequately evaluate an existing downflow or radial flow reactor. These drawings should include the following information:

Design temperature and pressure
Maximum allowable working pressure
GA, diameter, and tangent length
Nozzle sizes and flange ratings
Internals' details, dimensions, and locations
Materials of construction for the shell and internals
Current catalyst loading diagrams and catalyst volumes
Current inspection reports

The catalyst physical property information is also needed for the reactor evaluation. A listing of important catalyst information follows:

Nominal catalyst size
Shape (sphere, cylinder, trilobe, etc.)
Effective particle diameter
Average bulk density
Piece density
Void fraction
Friction factors (catalyst/catalyst and catalyst/reactor internals)

If original design drawings and information are used, the owner should verify that no changes have been made to the reactor or the internals since the original design.

If complete reactor information is not readily available but the diameter, tangent length, and design temperature and pressure are known, qualify any comments regarding reactor suitability by indicating that the evaluation is based on limited information and requires confirmation.

5.8.3.2 Reactor Process Evaluation

To begin the process work, yield estimates have to be developed based on the catalyst volume contained in the existing reactor(s). If the existing catalyst volume is not sufficient to meet the revamp objectives, then the yield estimates need to be rerun by increasing the catalyst volume, which normally requires the addition of another reactor for the expected revamp process conditions. As the process design progresses, an iterative process begins to then optimize the process conditions with the revamp objectives. It is expected that more than one yield estimate will be needed for each process case before the revamp process design is finished.

Heat and weight balances need to be prepared to establish the vapor and liquid loads for the reactor for each revamp case. If a revamp is to be guaranteed, the design margins used to evaluate each reactor should be established. If no guarantees are to be made, the reactor should be evaluated based on attainable process criteria that does not necessarily contain any design margin.

Review the following items to determine whether an existing reactor is suitable for a particular revamp service.

5.8.3.2.1 Flow Distribution

Good flow distribution within the catalyst bed is essential to obtaining optimal catalyst performance. For downflow reactors, the flow distribution will be a function of the inlet velocity head, the inlet distributor design, the mass flux, and the bed pressure drop. The inlet velocity head should be a small percentage of the bed pressure drop, and if found to be too high, can create bed maldistribution. The mass flux also must be above a minimum value to achieve good catalyst performance. However, if the mass flux/bed pressure drop is too high, the unit onstream factor may suffer due to shutdowns required to correct high bed pressure drop due to plugging of the top of the bed.

For complex downflow reactors with all-liquid or two-phase feeds, the evaluation of the existing inlet diffuser, the top liquid distribution tray, the quench zone internals, and the internal liquid distribution tray internals must be done to determine if any new or modified internals will be required.

For radial flow reactors, a complete flow distribution analysis is essential to evaluate the suitability of the reactor internals for the revamp conditions. This hydraulic analysis will include a calculation of the reactor inlet and centerpipe outlet velocity heads and the pressure drops of the scallop inlets, the catalyst bed, and the centerpipe holes. This analysis will then include an evaluation of the following:

- Seal fluidization – increases in volumetric flow rate and/or changes in catalyst type that create more bed pressure drop can increase a reactor's chance to fluidize the catalyst at the top of the bed. Modifications to the scallops, the seal area, and the cover deck can correct this problem.
- Vapor distribution – all radial flow reactors have uneven vapor flow radially across the bed from the top to the bottom of the reactor, and a difference of 5–8% for downflow centerpipe designs is acceptable. Greater maldistribution than this can hurt catalyst performance, and corrective action would be recommended, such as redesigning the reactor internals. The scallop riser pressure drop is established to get good circumferential vapor distribution versus the reactor inlet velocity head and to create a driving force for the down flow of vapor from the ventilated cover deck through the catalyst seal area. The centerpipe pressure drop is established to achieve good vapor distribution from the top to the bottom of the catalyst bed, taking into account the scallop and centerpipe outlet velocity heads and the bed pressure drop.

Pinning – for CCR reactors, the catalyst flows by gravity through the reactor. The horizontal force of the vapor flow on the catalyst can cause the catalyst to be "pinned" against the centerpipe if this force is greater than the gravitational force. Each reactor is evaluated for pinning, and if the pinning margin is too low, corrective action would be recommended. Normally, process changes would be needed, such as reduced recycle gas flow, increased operating pressure, or the introduction of a reactor bypass to bypass a portion of the reactor feed around the reactor, to increase the pinning margin by a significant amount for fixed revamp process conditions.

Void blowing – for non-CCR radial flow reactors, large increases in the volumetric flow rate together with a catalyst change to a higher pressure drop catalyst can result in portions of the catalyst bed being pushed away from the scallops. This higher flow results in the formation of voids between the scallops and the catalyst bed, which allows eddy currents to form. Within these eddies catalyst fines are generated, which then can locally plug the catalyst bed and cause vapor maldistribution. Corrective action needed for void blowing is similar to that for pinning.

5.8.3.2.2 Materials of Construction

The existing reactor materials of construction must suit the revamp service. Changes in the process stream composition and/or temperature may require a different metallurgy than that of the existing equipment. To determine if the metallurgy of the existing vessel is acceptable, review any significant changes in operating temperature, operating pressure, sulfur content, H_2S content, hydrogen partial pressure, or stream contaminants.

Because of recent changes in API-recommended practice, reactors in high-pressure hydrogen service should be reviewed for potential high-pressure hydrogen attack, and the operating conditions of reactors should be reviewed against the Nelson curve to confirm suitability for reuse. The owner should be notified of any special considerations regarding continued use of the reactor and associated equipment.

Any situation requiring the replacement of a reactor based on metallurgy should be reviewed with a metallurgist and process specialist.

5.8.3.2.3 Design Pressures and Temperatures

The operating pressures and temperatures for the revamp service must not exceed the design pressure and temperature of the existing reactor(s). Sometimes, depending on metallurgy, flange ratings, and design pressure, it is possible to re-rate an existing vessel for a higher design temperature. This re-rating may require a corresponding decrease in design pressure. The original manufacturer or a qualified representative of the owner would be expected to re-rate existing vessels. Typically, re-rating will necessitate a new ASME Pressure Vessel Code stamp or an equivalent stamp of the local or regional governing body.

5.8.3.2.4 Pressure Drop

For revamp feasibility studies, process revamp studies, and revamp basic engineering design, the final pressure drop of each reactor should be recalculated based on the final revamp process conditions.

5.8.3.3 Process and Other Modifications

Before recommending the replacement of an existing reactor, investigate the possibility for changes to the project objectives, an increase in the operating pressure, or modification of the reactor's internals to make the reactor(s) suitable for the revamp.

5.8.3.3.1 Relax Project Objective Constraints

If the owner has some flexibility in the project objectives and the capital budget is constrained, consideration should be given to relaxing the project objectives to fit the predicted performance of the existing reactor(s) if no additions or modifications are made. To evaluate this possibility, a complete understanding must be developed about how the revamped unit's performance affects the overall refinery operation. Some examples of potential reduced performance are as follows:

A reduction in throughput that reduces the project return on investment but may still allow for an attractive project
A reduction in conversion
A reduced catalyst cycle length
A reduction in severity

5.8.3.3.2 Increase Operating Pressure

Increasing the operating pressure of the reactor(s) by the addition of a relatively inexpensive pilot-operated relief valve into the reactor section may be a way to reuse the existing reactor(s). Line and equipment pressure drops generally increase significantly for revamps, causing existing equipment to see higher operating pressures. Increasing the operating pressure on a reactor can increase the vapor density and reduce the resulting pressure drop through a vapor or two-phase reactor circuit. Reducing the pressure drop can allow for additional operating capacity. In general, a pilot-operated relief valve can allow operation within 5% of design pressure, as compared with 10% of the design pressure for a conventional relief valve. In cases where there may be a need to operate a reactor at higher pressure, a pilot-operated relief valve may be a way to reuse the existing reactor.

5.8.3.3.3 Changes to the Reactor Internals

Changes to the reactor internals can be made to improve vapor distribution and to reduce pressure drop. Changes may also be possible to increase the catalyst volume, but this increase is usually not significant.

5.8.3.3.4 Provide an Additional Reactor

To increase the catalyst volume significantly, one or more reactors may have to be added. If pressure drop is available, adding a reactor in series with the existing reactors is the easiest alternative from a piping and flow distribution standpoint. If pressure drop is a limitation for all-vapor flow reactors, a reactor may be added in parallel but even flow distribution is a concern and must be addressed either by symmetrical piping or addition of valves.

5.8.3.4 Test Run Data

If operating or test run data for the unit are available, they should be reviewed with respect to the performance of the reactor and catalyst. The operating data may indicate that the catalyst performance is lower than what is normally expected or the pressure drop is above the calculated value. These observations should be reviewed and discussed with the owner to avoid recommending a change that is inconsistent with the operating data. Also, this operating data should be used by the yield estimator to establish a base line for the revamp yield estimate, especially when the revamp feed is similar to the test run feed.

5.8.3.5 Possible Recommendations

The revamp recommendations to the owner must be cost effective and justifiable. An evaluation of an existing reactor or reactor system may result in making one of the following recommendations:

5.8.3.5.1 Use Existing Reactor(s)

If the catalyst volume, flow distribution, pressure drop, design pressure, design temperature, and metallurgy satisfy the new process conditions, use the existing reactor(s) without modification.

5.8.3.5.2 Modify the Reactor Internals

The existing reactor(s) may require replacement, additions, or modifications to the existing internals to satisfy the new process conditions. These modifications are often required to improve flow distribution, to allow the addition of more catalyst, or to reduce pressure drop. There is great value in a revamp project to have the original licensor and/or process unit designer do the reactor revamp work (as well as entire unit revamp design) because the process parameters and performance can be adjusted via yield estimating capability. In addition, detailed hydraulic analysis of the reactors can be done confidently.

5.8.3.5.3 Modification of the Process Flow Scheme

To eliminate catalyst pinning and reduce pressure drop, the process flow scheme can be modified by using ideas such as the following:

Place two existing reactors in parallel.
Bypass a portion of the reactor combined feed around one reactor to the next reactor downstream when multiple series reactors exist.
Bypass a portion of the recycle gas around one reactor to the next reactor downstream when multiple series reactors exist

5.8.3.5.4 New Reactor

If nothing can be done to save the existing reactor, consideration should be given to replacing it with a new reactor. The owner may have an idle reactor available which may be suitable for the service.

5.8.4 Fractionator Evaluation

For most revamps, the owner is unwilling to replace an existing fractionation column, but may be willing to replace trays or modify ancillary equipment to make the column suitable for the new process conditions.

5.8.4.1 Data Required

As-built fractionator tray and vessel drawings are needed to adequately evaluate an existing column. The drawings should show the GA and layout of the trays including the number of holes for sieve trays or the number of valves, tray spacing, downcomer chord heights, weir heights, downcomer clearance, etc. If the original design drawings are provided, the owner should verify that no changes have been made to the vessel or the trays since the original design.

If tray and vessel information is not readily available from the owner, but the number of trays and column diameter is known, any comments regarding the column's suitability should be qualified with the statement that the evaluation is based on limited information and requires confirmation.

5.8.4.2 Fractionator Evaluation

A heat and weight balance must be developed for the column and the tray loadings determined. The designer must decide the tray efficiencies to be used for evaluation of a column. Tray efficiencies for new unit design are usually quite conservative, for good reason, to assure the column meets design throughput and separation requirements. For a revamp, the designer may choose to use efficiencies that have been determined from test run information for the column, known tray efficiencies for the column service, or new unit design tray efficiencies. This decision may depend on whether or not guarantees will be provided for the revamp operation.

Review the following items to determine whether an existing fractionator is suitable for a particular revamp service.

5.8.4.2.1 Jet Flood and Downcomer Flood
When evaluating a column, be sure to use the appropriate tray design type, i.e. sieve tray, valve tray, etc. Specific jet flood and downcomer flood limit criteria may be agreed to with the owner in the design basis phase of the project. If not, jet floods and downcomer floods of up to 90% are generally considered acceptable for a revamp. A program such as KG-TOWER by Koch-Glitsch may be used to evaluate the column at the revamp operating conditions.

If an existing column is being used in a new service and is considerably oversized, it may be necessary to blank off rows of holes or valves to increase vapor velocity and achieve expected tray efficiency.

5.8.4.2.2 Downcomer Backup
Generally, downcomer backup should not exceed 55%; however, for a revamp where other criteria such as downcomer flooding are acceptable, the downcomer backup criteria of 55% maximum can be exceeded, but certainly not beyond 80%.

5.8.4.2.3 Materials of Construction
The existing column's materials of construction must suit the revamp service. Thus, changes in the process stream composition and/or temperature may require a different metallurgy than that of the existing equipment. To determine if the metallurgy of the existing fractionator is acceptable, review any significant changes in operating temperature, sulfur content, H_2S content, hydrogen partial pressure, etc.

5.8.4.2.4 Pressure Drop
Estimate the pressure drop for the revamp service and evaluate how it affects the hydraulics of the column. Pressure drop across the column will impact the bottoms temperature which is important in determining the suitability of the reboiler.

5.8.4.2.5 Design Pressures and Temperatures
The operating pressures and temperatures for the revamp service must not exceed the design pressure and temperature of the existing vessel. Sometimes, depending on the metallurgy, flange ratings, and design pressure, it is possible to re-rate an existing vessel for a higher design temperature. This may require a corresponding decrease in design pressure. The original manufacturer would normally be expected to re-rate existing fractionators, which typically necessitates a new ASME Pressure Vessel Code stamp.

5.8.4.3 Retraying and Other Modifications
Before replacement of the existing trays in a column is recommended, one or more of the following modifications to the column's ancillary equipment should be investigated to make it suitable for the revamp.

5.8.4.3.1 Relax the Product Specifications
If a column is operating to make a split such that the product streams are recycled within the process unit or complex, relaxation of the column product specifications (column splits) should be considered. In some cases, this may require that the owner be consulted if relaxation of specifications results in the increased consumption of desorbent or solvent consumed.

5.8.4.3.2 Increase Operating Pressure
While increasing the operating pressure of a column will generally make the separation more difficult and result in increased reflux, vapor density will be higher at the higher pressure which could result in lower overall jet flood. In addition, the latent heat of vaporization of most hydrocarbons will decrease as the pressure is increased, thus reducing the overall reboiler and condenser duties.

If the column is normally operating at atmospheric pressure, it is possible to add a push–pull pressure control system to the overhead receiver. Increasing operating pressure could have the added benefit of reducing the surface area required for the overhead condenser because of an increase in the bubble point temperature in the receiver.

Operating pressure should only be increased to within about 25 psi of the design pressure of the column so as to avoid spurious lifting of the relief valves, unless pilot-operated relief valves are installed. Also, consideration must be given to the effect the higher operating pressure and, hence, increased bottoms temperature will have on the reboiler. If the reboiler is a fired heater, the impact of a higher bottoms temperature on the heater will be minimal. However, if steam or hot oil is used as the heating media, increasing the bottoms temperature could necessitate either replacement of the entire reboiler exchanger, replacing the tubes with enhanced surface such as High Flux tubes, or changing the heating medium to higher pressure steam.

5.8.4.3.3 Optimize the Column Design
Even though the total number of trays and feed tray(s) may be fixed, it is possible to optimize the design of the column by evaluating feed preheat and the location of

the feed tray(s), especially if a column is to be used in a new service.

Too much feed preheat can result in over vaporization of the feed at the feed tray and an increase in the amount of reflux required. If there is too much heat in the feed, look for ways to exchange the feed against some other process streams. A rule of thumb has been that the molar flow rate of vapor in the column feed should not exceed the distillate product molar flow rate.

If there is too little heat in the feed, an additional load may be placed on the reboiler which results in increased vapor traffic below the column feed point. Consider adding more shells of feed-bottoms exchange or heating the feed to the column with other process streams or by adding a steam or hot oil-heated preheat exchanger.

Optimize the feed tray location using a procedure in which the total number of trays and the number of trays in each section of the column are varied, optimizing the number and distribution based on the required reboiler duty. If the optimized feed tray location is considerably different from any available feed trays, consider one of the following options:

1) Remove one or two existing trays immediately above the optimized feed tray location to facilitate the addition of a new nozzle and feed distributor. Usually, it will be necessary to remove two trays and extend the downcomers from above to the new feed tray. Be sure to check what effect the removal of trays may have on the overall performance of the column.
2) Provide internal piping within the column from the existing feed tray to the optimized feed tray location.

5.8.4.3.4 Stub Column

If additional trays would help to reduce the reflux required and, therefore, reduce the reboiler duty and tray loadings, consideration could be given to the installation of a "stub" column in series with the existing column. Typically, a stub column is configured to add trays below the feed in the existing column so as to minimize the piping changes that may be required. Overhead vapor from the stub column is sent to the bottom of the existing column while liquid from the existing column is fed to the top tray of the stub column. If a stub column is proposed, consideration must be given to the location of the column and whether or not the existing reboiler can be used.

If all of the above modifications are applied where possible and the column still exceeds the jet flood and/or downcomer flood criteria, consideration should be given to retraying the column with higher capacity trays.

5.8.4.4 High-Capacity Trays

The MD tray by UOP is a high-capacity tray using multiple, terminated downcomers to increase liquid capacity and eliminating seal pans to increase vapor capacity. It is often applied in medium to high pressure/high liquid load applications due to the much greater length of the outlet weir, which allows for much higher liquid handling. The long total weir length also provides lower crest heights and permits closer tray spacing. These features allow for retrofitting a tower for increased capacity or reducing both height and diameter of a new column compared with conventional multi-pass trays.

For revamps, a typical capacity increase above conventional trays of 25–35% is typically achievable with MD trays. Enhanced capacity MD (UOP ECMD) trays may provide an additional 15–20% capacity above that of MD trays. Even greater capacity increases are possible based upon the existing column diameter (larger the better), existing tray spacing (greater the better), and revamp conditions (higher pressure and highly liquid loaded the better). The revamp strategy (1-for-1, 4-for-3, 3-for-2, and 2-for-1) is often selected based upon the relative efficiencies of MD trays versus conventional trays in that service. For most fractionation services, MD/ECMD trays will have lower efficiencies than conventional trays (in the 65–70% range versus conventional trays at 80%). This results in a 4-for-3 MD tray revamp, at 75% of the original tray spacing, such that the number of theoretical trays required can be met or exceeded. A multi-tray revamp strategy (4-for-3, 3-for-2, and 2-for-1) can also be selected if additional stages are desired.

Other high-capacity trays on the market include SUPERFRAC and ULTRA-FRAC high-performance trays from Koch-Glitsch.

For even greater capacity increases, the use of UOP SimulFlow tray devices may be considered. SimulFlow trays can more than double the capacity of an existing column shell.

5.8.4.5 Test Run Data

If operating or test run data for the unit are available, they should be reviewed with respect to the performance of the fractionator. The operating data may indicate that the tray efficiency is higher than what is normally used for a new unit design. A higher efficiency, approaching that shown by the operating data, could be used for the evaluation of the column under the new process conditions.

5.8.4.6 Possible Recommendations

An evaluation of an existing fractionation column may result in making one of the following recommendations:

5.8.4.6.1 Use Existing Column

If the jet flood, downcomer flood, design pressure and temperature, and metallurgy satisfy the new process conditions, use the existing column without modification.

5.8.4.6.2 Retray Existing Column

The existing column may need to be retrayed to satisfy the new process conditions. This may be due to excessive jet flood and/or downcomer flood.

5.8.4.6.3 Modification of Ancillary Equipment

The existing column may satisfy the new process conditions if a new feed tray is added, a feed preheater or feed exchanger is installed, or product specifications are relaxed.

5.8.4.6.4 New Column

If nothing can be done to save the existing column, consideration should be given to replacing it with a new larger column, adding a new smaller column in parallel with the existing column, or adding a stub column. If available, idle equipment should be evaluated for possible use.

If the top section of the column is swaged and it is the bottleneck, it may be possible to remove this section and replace it with a section of the same diameter as that of the bottom section. Of course, this is a more extensive and expensive modification that will require that the foundation be checked for the increased weight as well as requiring more downtime.

5.8.5 Heat Exchangers

The extent to which a design engineer evaluates an exchanger for a revamp depends upon the engineering product for which the evaluation is being conducted. A process revamp study and basic revamp engineering designs, for instance, require more detailed analyses than for revamp scoping studies or revamp feasibility studies.

5.8.5.1 Data Required

The heat exchanger vendor's Tubular Exchanger Manufacturers Association (TEMA) data sheet, or equivalent, is the most complete data sheet and should be used to evaluate a shell and tube exchanger. The surface area and calculated U value, among other data are required from the TEMA data sheet.

If the vendor's TEMA data sheets are not available, any comments regarding the exchanger's suitability should be qualified by indicating that the evaluation is based on limited information and requires confirmation.

5.8.5.2 Overall Exchanger Evaluation

Review the following items to determine whether an existing exchanger suits a particular revamp service.

5.8.5.2.1 Design Pressures and Temperatures

The operating pressures and temperatures for the revamp service must not exceed the respective design pressures and temperatures of the existing exchanger (both shell and tube sides). Sometimes, depending on the metallurgy, flange ratings, and design pressure, it is possible to re-rate an existing exchanger for a higher design temperature. This may require a corresponding decrease in design pressure. The original manufacturer is expected to re-rate existing exchangers, which typically necessitates a new ASME Pressure Vessel Code stamp.

5.8.5.2.2 Materials of Construction

The existing exchanger's materials of construction must suit the revamp service. Thus, changes in the process stream composition and/or temperature may require a different metallurgy than that of the existing equipment. To determine if the metallurgy of the existing exchanger is acceptable, review any significant changes in operating temperature, sulfur content, H_2S content, hydrogen partial pressure, etc.

5.8.5.2.3 Pressure Drop

Estimate the pressure drop for the revamp service and evaluate how it affects the process unit hydraulics. Replacing or augmenting an exchanger may be less expensive than replacing a compressor or multistage pump. Also, evaluate the pressure drop on the water side of a water-cooled exchanger and compare it to the available pressure drop of the existing cooling water system. Conduct this same type of analysis on the hot oil side of hot oil exchangers.

5.8.5.2.4 Surface Area

The revamp service needs sufficient heat transfer surface area. Although this may be obvious by inspection, it may require a detailed analysis. If analysis is required, there are several ways to evaluate the adequacy of the existing surface area. The method employed simply depends on the type of exchanger in question, the magnitude of change in the operating conditions, and the level of confidence required by the type of revamp engineering undertaken.

5.8.5.3 Thermal Rating Methods

5.8.5.3.1 Constant UA Method

Starting with the Fourier equation for heat transfer, $Q = UA$ (CMTD), the thermal rating may be simplified greatly if the U and A terms are assumed to be constant. If this assumption may be made, then the heat transferred, Q, is directly related to the corrected log mean temperature difference, CMTD. Obviously, if an existing exchanger will be used, the area, A, remains constant.

If the mass flows equal or exceed the design mass flows, and the fluid properties are nearly the same, the overall U for the revamp operation will equal or exceed that of the original design. In such cases, make a preliminary evaluation of the existing exchanger by assuming a constant UA. Do not assume a constant overall U value if the mass flows are less than the design mass flows, or if the fluid properties are significantly different from those of the original design. Instead, perform a more rigorous analysis to estimate the U expected for the revamp operation.

In many cases, the constant UA method sufficiently determines whether an existing exchanger has adequate surface area. Use this method of evaluation, when applicable, for revamp feasibility studies, process revamp studies, or revamp basic engineering design.

5.8.5.3.2 Using Key Variable Relationships

This method is based on determining the change in one or more of the five resistances that comprise the overall resistance to the heat transfer, $\mathbf{R}a$.

$$\mathbf{R} = r^{\text{film(shell)}} + r^{\text{fouling(shell)}} + r^{\text{tube wall}} + r^{\text{fouling(tube)}}\left(\frac{A_o}{A_i}\right) + r^{\text{film(tube)}}\left(\frac{A_o}{A_i}\right) \quad (5.5)$$

The reciprocal of $\mathbf{R}$ is the overall heat transfer coefficient, U.

Although this method technically applies only to sensible heat transfer, i.e. when there is no phase change, it may be used if the phase change of the revamp nearly equals that of the original design. If an exchanger will be used for the same service, assume that the fouling resistances are the same as those of the original design (unless operating data indicate otherwise). If the exchanger will be used in a different service, review and change the fouling resistances as necessary. Include the tube wall resistance in the calculation. Resistance values for typical tube sizes and materials are available on various websites.

Next, estimate the changes in the film resistances outside and inside the tubes. The film coefficients, h_o and h_i, are the inverse of the film resistances, r_o and r_i. Depending on the particular service of the exchanger in question, there are several ways to estimate the film coefficients. Section 5.8.5.4 explains how to rate various types of exchangers.

When one of the film resistances used for the original design is determined (or estimated), it is then possible to approximate the film resistances for the revamp operation by using the following relationships of key heat transfer variables:

Tube side:

$$\frac{h_2}{h_1} = \frac{r_1}{r_2} = \left(\frac{G_2}{G_1}\right)^{0.8}\left(\frac{k_2}{k_1}\right)^{0.67}\left(\frac{Cp_2}{Cp_1}\right)^{0.33}\left(\frac{\mu_1}{\mu_2}\right)^{0.47} \quad (5.6)$$

Shell side:

$$\frac{h_2}{h_1} = \frac{r_1}{r_2} = \left(\frac{G_2}{G_1}\right)^{0.6}\left(\frac{k_2}{k_y}\right)^{0.67}\left(\frac{Cp_y}{Cp_y}\right)^{0.33}\left(\frac{\mu_1}{\mu_2}\right)^{0.27} \quad (5.7)$$

where

G = mass velocity
k = thermal conductivity
Cp = heat capacity (specific heat)
μ = absolute viscosity
subscript 1 = original design
subscript 2 = revamp operation

If the exchanger's geometry stays the same, substitute mass flows for mass velocities. Use this method, when applicable, for revamp feasibility studies, process revamp studies, or revamp basic engineering design.

5.8.5.3.3 HTRI or HTFS Computer Programs

The Heat Transfer Research, Inc. (HTRI) and Heat Transfer and Fluid Flow System (HTFS) computer programs can be used to conduct more rigorous analyses of heat exchangers if the exchanger, using less rigorous methods, is found to be marginal, or the design engineer wants to confirm preliminary findings for critical services (e.g. combined feed-effluent exchangers) and the vendor data sheets provide the tube and shell construction information.

5.8.5.4 Rating Procedures

5.8.5.4.1 Water-Cooled Exchangers

Use the following formula to approximate the tube side film coefficient of a water-cooled exchanger (with water on the tube side):

$$h_i = 306(V_w)^{0.8} \quad (5.8)$$

where

V_w = water velocity, fps

The design shell side film resistance may be back-calculated, provided that the design tube side film coefficient, the design fouling factors, the tube wall resistance, and the design U are known. Use the shell side relationships shown in Section 5.8.5.3.2 to calculate a new shell side film coefficient and a new tube side (water) film coefficient. This makes it possible to determine the overall resistance, R, and the overall U value for the revamp.

The existing exchanger will suit the revamp if the calculated U value equals or exceeds the required U value.

For water-cooled exchangers, the shell side film coefficient controls the overall U value because it contributes more to the overall resistance, R. The h_i for water ranges from approximately 750–1250 Btu/h-ft^2-°F, which is three to four times greater than the shell side film coefficient for a hydrocarbon stream.

5.8.5.4.2 Process–Process Exchangers

The following equation applies to turbulent flow in tubes. Use it to estimate the tube side film coefficient for a process–process exchanger.

$$h_i = \frac{0.023 G^{0.8} k^{0.67} Cp^{0.33}}{\mu^{0.47} D^{0.2}} \quad (5.9)$$

where

G = mass velocity (lb/h-ft^2)
k = thermal conductivity (Btu/h-ft-°F)
Cp = heat capacity (Btu/lb-°F)
μ = viscosity (lb/ft-h) = cP × 2.42
D = tube ID, ft.

For exchangers with similar fluids on both sides, e.g. combined feed-effluent exchangers or fractionator feed-bottoms exchangers, estimate the change in the magnitude of the film coefficients for the revamp operation by making a simplification. If the tube side and shell side film coefficients are assumed equal, determine a value for the original design given the original U value, fouling factors, and tube wall resistance. Then, estimate the revamp film coefficients using the property relationships given in Section 5.8.5.3.2.

If the fluid properties and flow rates on both sides of the exchanger are similar, relate the revamp U value to that of the original design by using the following relationship of mass flows:

$$U_2 = U_1 \left(\frac{W_2}{W_1} \right)^{0.45} \quad (5.10)$$

where

W = mass flow rate

The flow rates for G_2 and G_1 may be tube side or shell side flows, as long as they are consistent.

5.8.5.4.3 Horizontal Thermosiphon and Kettle-Type Reboilers and Vaporizers

The overall heat transfer coefficient for steam-heated (condensing steam) horizontal thermosiphon reboilers and kettle-type reboilers and vaporizers does not change significantly with changes in the process side or steam side flow rates. Thus, the extent of possible capacity/duty increase is, for the most part, a function of the change in the LMTD of the original design.

Changes in the flow rate of hot oil will affect the overall heat transfer coefficient for hot oil-heated reboilers. Use the methods described in Section 5.8.5.3.2 to estimate the change in the tube side (hot oil) film coefficient.

The design engineer must check the heat flux rate and the approach temperature for the revamp conditions to ensure that they are within acceptable limits. The heat flux rate for steam shall be no less than 3 000 Btu/h-ft^2 and no greater than 15 000 Btu/h-ft^2 (2 000–12 000 Btu/h-ft^2 for hot oil). The approach temperature, when using conventional bare tubes, must be at least 25 °F. Consider using High Flux tubes for lower approach temperatures or higher flux rates (see Section 5.8.5.7.4).

Normally, 33% of the liquid entering horizontal or vertical thermosiphon reboilers vaporizes, which prevents an excessive pressure drop and unstable operation. For a revamp, up to 50% vaporization is permissible, provided that the consequent higher pressure drop across the exchanger can be tolerated.

For process revamp studies and revamp basic engineering design, evaluate the hydraulics of existing horizontal thermosiphon reboilers. Hydraulics usually are not evaluated for revamp scoping studies or revamp feasibility studies.

5.8.5.4.4 Air-Cooled Exchangers

Air-cooled exchangers are typically limited by the amount of air that the fans can deliver. Thus, to evaluate an existing air-cooled exchanger, first determine the air side flow and temperatures. This involves obtaining the design mass air flow rate from the vendor data sheet and calculating the outlet air temperature, given the inlet air temperature and desired revamp duty, using the basic equation for sensible heat transfer:

$$Q = W Cp \Delta T \quad (5.11)$$

where

W = air mass flow rate (lb/h)
Cp = heat capacity (Btu/lb-°F)
= 0.24 Btu/lb-°F for air below 300 °F
ΔT = air side temperature rise (°F)

Calculate the air side ΔT by rearranging the equation:

$$\Delta T = \frac{Q}{0.24 W} \quad (5.12)$$

Use the calculated outlet air temperature to calculate the linear mean temperature difference (MTD) for the revamp operation. If a phase change occurs, e.g. condensing service, determine the weighted MTD. Determine the MTD correction factor, f, by calculating the P and R values, knowing the configuration of the air cooler (number and type of passes) and using the appropriate figures from the air-cooled heat exchanger section of the GPSA Engineering Data Book, 14th edition, 2017. Over and under passes are the most common. Assume a correction factor of 1.0 for units with three or more passes.

After determining the CMTD, use the duty required and the available surface area to calculate the required U

value. Compare the U value required for the revamp with the U value used for the original design, which appears on the vendor data sheet. The design engineer must consistently use either the bare or "finned" surface area and the corresponding U value. The existing air cooler will suit the revamp if the required U value is less than or equal to the U value of the original design.

Because the air side film resistance is controlling, changes in the process side flow and fluid properties have less impact than they do in a typical shell and tube heat exchanger. Determine the effect of these changes using the same type of analysis that is used for S&T exchangers. First, estimate the air side film coefficient by calculating the air-side "face mass velocity" and using the appropriate figures from the GPSA Engineering Data Book. Then, determine the face mass velocity by dividing the pounds per hour of air by the air cooler face area in square feet, both of which are usually included on the vendor data sheet.

5.8.5.5 Pressure Drop Estimation

When conducting process revamp studies and revamp basic engineering design, estimate the pressure drop of an existing exchanger at the revamp operating conditions in terms of how it will affect the hydraulics of the circuit in which it is contained.

When conducting revamp feasibility studies, perform a preliminary hydraulic analysis of the circuits which contain major equipment (e.g. compressors or multistage pumps). This is often accomplished by estimating the pressure drop of the whole circuit rather than that of each piece of equipment.

If the overall circuit pressure drop is excessive, consider modifying or replacing an exchanger that accounts for much of the pressure drop. For single-phase flow, estimate the pressure drop for shell and tube exchangers based on the original calculated pressure drop from the vendor data sheet and by using the following relationship:

Tube side:

$$\Delta P_2 = P_1 \left(\frac{G_2}{G_1}\right)^{1.8} \left(\frac{\rho_1}{\rho_2}\right) \left(\frac{\mu_2}{\mu_1}\right)^{0.20} \quad (5.13)$$

Shell side:

$$\Delta P_2 = P_2 \left(\frac{G_2}{G_1}\right)^{1.85} \left(\frac{\rho_1}{\rho_2}\right) \left(\frac{\mu_2}{\mu_1}\right)^{0.15} \quad (5.14)$$

where

ρ = average fluid density

Ignore the viscosity term if the fluid viscosities for the revamp operation barely differ from the original design.

The tube side relationship also applies to the process side of air coolers. Use the above relationships for two-phase flow when the vapor/liquid ratio for the new conditions approximately equals that of the original design.

By using the following relationship for two-phase flow, the ratios of the revamp to the original exchanger inlet and outlet line ΔP's may be used to determine the exchanger ΔP for the revamp:

$$\Delta P_2 = \frac{\Delta P_1}{2}\left[\left(\frac{\Delta P / 100 \text{ ft}_2}{\Delta P / 10 \text{ ft}_1}\right)_{\text{inlet}} + \left(\frac{\Delta P / 100 \text{ ft}_2}{\Delta P / 100 \text{ ft}_1}\right)_{\text{outlet}}\right]$$

(5.15)

Avoid excessive pressure drops caused by high fluid velocities, even if they are tolerated by the circuit hydraulics. High fluid velocities may cause vibration and/or erosion problems. For water and similar liquids, velocities of 10 ft/s and above are considered excessive.

5.8.5.6 Use of Operating Data

If operating or test run data for the unit is available, review it with respect to the performance and pressure drop of major heat exchangers. The operating data may indicate that the U value or pressure drop for a particular exchanger is different than what is shown on the vendor data sheet. This may be due to excessive fouling, maldistribution, conservative design, etc. If the operating data seems reliable, factor it into the evaluation. However, if the operating data indicate that an exchanger has excessive fouling, discuss and verify this with the owner's operating personnel, if possible.

Based on discussions with the owner, the design engineer may choose to rate the exchanger based on the design condition shown on the TEMA data sheets, or the fouled condition indicated by the operating data. The study report should state the assumptions used in the evaluation and the rating basis.

5.8.5.7 Possible Recommendations

An evaluation of an existing exchanger may result in UOP making one of the following recommendations.

5.8.5.7.1 Use Existing Exchanger

If the design pressure and temperature, metallurgy, heat transfer surface area, and pressure drop satisfy the new process conditions, the existing exchanger may be used without modification.

5.8.5.7.2 Replace Existing Exchanger

The existing exchanger may need to be replaced to satisfy the new process conditions. This may be due to inadequate design conditions, unacceptable metallurgy, excessive pressure drop, lack of surface area, or some combination of the above.

5.8.5.7.3 Add Additional Exchanger Shell(s)

If the design conditions and metallurgy are acceptable, but the existing exchanger has insufficient surface area, recommend an additional shell, or in the case of an air-cooled exchanger, an additional section or air bay. The additional shell may be added in series or in parallel with the existing exchanger. If the pressure drop permits, it is better to place the new shell in series with the existing exchanger shells. This eliminates flow distribution concerns and it may increase the MTD correction factor. However, in many cases pressure drop considerations require that the additional shell be added in parallel. In this event, use a duplicate of the existing exchanger and symmetrical piping to provide for even flow distribution. This is particularly important for two-phase flow to the exchanger. For air-cooled exchangers, air bays are almost always added in parallel using symmetrical piping to provide good flow distribution.

5.8.5.7.4 High Flux Tubing

For boiling services where the design conditions and shell metallurgy are acceptable, but the exchanger has insufficient surface area for the revamp, it may be economical to replace the existing tube bundle with UOP High Flux tubes. This may result in an overall U value that is two to four times greater than that obtained with conventional bare tubes. Installing High Flux tubes in an existing shell does not require additional plot space and it minimizes downtime.

Determine the pressure drop and outlet nozzle size of reboilers revamped with High Flux tubes. It may be necessary to replace nozzles if the velocity is too high with the existing nozzles.

Consult with the High Flux tube supplier regarding the applicability of High Flux tubes for specific revamp services.

5.8.5.7.5 Rearrangement and Other Modifications

1) Consider using existing exchanger shells (or air bays) in a different service if they do not suit the revamp of their previous service.
2) Consider rearranging the existing shells from series flow to parallel operation when the pressure drop is excessive and the required duty and operating temperatures permit such a change.
3) If sufficient water is available, reduce the water side pressure drop for water-cooled exchangers with multiple shells in series by changing the water side from series flow to parallel flow.
4) If the surface area is adequate, but the pressure drop on the tube side of a shell and tube exchanger is excessive, try to reduce the number of tube passes by removing one or more pass partitions from the channel. This may be suggested in the study phase of a project but the contractor or vendor must confirm its practicality.
5) Replacing existing fans with higher efficiency fans may increase the capacity of older air-cooled exchangers by 10–15%. Other possible ways of increasing air cooler capacity include changing fan blade pitch or increasing fan speed by replacing the drivers. Again, this may be suggested in the study phase of a project but the contractor or vendor must confirm its practicality.

5.8.5.8 Special Exchanger Services

Certain heat exchangers require special consideration when they are evaluated for a revamp service. Some examples are listed below.

5.8.5.8.1 Vertical Combined Feed Exchangers

These exchangers have a single tube-side pass, which means they approach "true countercurrent" flow. However, for no-tube-in-the-window (NTIW)-type designs, the long baffle spacing makes it necessary to apply a correction factor of approximately 0.9 to the weighted MTD. This will account for the cross-flow nature of the shell-side fluid. Also, if the combined feed liquid/vapor mixture is distributed unevenly to the tubes, these exchangers may suffer from poor performance with operating heat transfer coefficients that are less than that of the design.

5.8.5.8.2 Welded-Plate-Type Exchangers

Consider using welded-plate-type combined feed-effluent exchangers in revamps of reforming, transalkylation, and xylene isomerization process units. A single welded-plate exchanger can replace existing shell and tube exchangers while providing greater heat transfer efficiency and reducing reactor circuit pressure drop. In revamp scenarios this may make it possible to reuse the existing charge heater and recycle gas compressor. Alfa Laval Packinox is the preferred supplier of welded-plate heat exchangers.

5.8.5.8.3 Compabloc and Ziepack Exchangers

Compabloc and Ziepack heat exchangers have benefits that make them useful in revamp services. These exchangers are compact, have reduced plot space requirements, and reduced pressure drop compared to conventional heat exchangers. Compabloc are welded-plate heat exchangers that can replace several conventional shells in series with a single unit.

Ziepack heat exchangers are compact in-column condensers, stab-in reboilers, and gas–gas heat exchangers with very low pressure drop. Both Compabloc and Ziepack heat exchangers are made by Alfa Laval.

5.8.5.8.4 Strength-Welded Tube to Tubesheet Joints

Certain exchanger services require strength-welded tube to tubesheet joints (SWTTJ) to prevent or minimize the possibility of cross leakage. If an existing exchanger without SWTTJ will be used in a service that requires SWTTJ, replace the tube bundle with a strength-welded U-tube bundle.

5.8.5.9 Overpressure Protection

Consider whether exchangers need overpressure protection when the design pressure of the low-pressure side is less than that of the high-pressure side.

5.8.6 Pumps

The extent to which the design engineer evaluates the pumps in a unit depends upon the level of revamp study or engineering for which the evaluation is being conducted. Process revamp studies and revamp basic engineering design, for instance, require more detailed analyses than revamp scoping studies or revamp feasibility studies. For these lower level studies, hydraulics are generally not evaluated, so the extent of pump evaluation is normally less rigorous. Typically, pumps are not included in the scope of revamp scoping studies.

5.8.6.1 Data Required

The vendor's API data sheets (or equivalent) and the pump's characteristic performance curve (for centrifugal pumps) are needed to evaluate a pump. The installed impeller size, driver size, efficiency, minimum flow, and Net Positive Suction Head Required (NPSHR) will be needed from the vendor's data sheet.

If vendor data sheets and performance curves are not available, the licensor's project specification (data sheet) and the estimated pump capacity and head requirements can be used to make rough evaluations. Qualify any comments regarding the pump's suitability by indicating that the evaluation is based on limited information and requires confirmation.

5.8.6.2 Centrifugal Pump Evaluation

Review the following items to determine whether an existing pump suits a particular revamp service.

5.8.6.2.1 Flow and Head

The primary criteria that determines whether or not an existing pump can be used in a revamp service is a combination of the flow rate the pump can deliver and the head (pressure) at which it can be delivered. The relationship between flow and head for a centrifugal pump is defined by the pump's characteristic performance curve. Therefore, having the performance curve is necessary to do a proper evaluation of an existing centrifugal pump.

The typical procedure for determining a centrifugal pump's suitability from a flow vs. head standpoint is outlined below:

1) Determine required flow rate for revamp service.
2) Determine head from pump curve at required flow rate.
3) Determine pump suction pressure for revamp operation.
4) Calculate available pump discharge pressure based on head from curve, fluid specific gravity, and pump suction pressure [see Eq. (5.16) below to calculate differential pressure from head].
5) Compare calculated pump discharge pressure to required discharge pressure. This requires that some preliminary hydraulics be done to estimate the required discharge pressure.

$$\text{Differential pressure (psi)} = \text{Head (ft)} \times \text{sp.gr.}/2.31 \quad (5.16)$$

While ideally the operating point should be kept close to the best efficiency point on the curve, substantial variation in flow to either side of the best efficiency point is usually acceptable, especially in a revamp when trying to maximize the use of existing equipment. Be sure that the new operating point is above the minimum flow of the pump, which is usually identified on the performance curve. If it is not, the minimum flow can be estimated where the NPSHR and/or efficiency curves end.

Most pumped circuits contain a control valve in the pump discharge line. The pressure drop across this control valve can often be reduced from the design value to accommodate an existing pump. An instrument/controls engineer should be consulted to determine the minimum required control valve pressure drop for a given service.

5.8.6.2.2 Driver Power

The driver type and size is typically given on the pump's API data sheet. Sometimes the maximum driver power is shown on the characteristic performance curve. In addition to the head vs. flow curve, efficiency curves are also usually provided on the performance curves.

1) For centrifugal pumps, calculate the driver power required for the revamp operation from the equation below using the head and efficiency from the performance curve:

$$\text{Brake HP} = \frac{\text{gpm} \times \text{Head (ft)} \times \text{sp.gr.}}{3960 \times \text{efficiency}} \quad (5.17)$$

2) Compare the calculated required power to the maximum driver power.

5.8.6.2.3 NPSH

The Net Positive Suction Head Available (NPSHA) must be greater than or equal to the NPSHR, which is shown on the API data sheet for the rated capacity of the pump, or plotted on the characteristic pump performance curve. Note that the NPSHR increases with increasing capacity (flow).

5.8.6.2.4 Materials of Construction

The materials of construction of the existing pump must suit the revamp service. Changes in the process stream composition and/or temperature may require a different metallurgy than that of the existing equipment. To determine if the metallurgy of the existing pump is acceptable, review any significant changes in operating temperature, sulfur content, H_2S content, chloride content, etc.

5.8.6.2.5 Seal Type

The type of seals employed for centrifugal pumps must be consistent with current industry practices. If information on the existing seals is not available, it should be stated in the report "Information on the existing pump seal(s) was not provided. Further evaluation of the pump(s) to confirm that the appropriate type of seal is installed will be required."

5.8.6.3 Proportioning Pumps

Proportioning pumps are positive-displacement-type pumps. These pumps typically either have a variable speed drive or stroke adjusting mechanism to vary the flow. Turndown is typically 10 : 1 and the maximum and minimum flows are given on the data sheet. A positive displacement pump will produce whatever head is required by the pressure drop imposed on the discharge side, up to the relief valve set pressure or the maximum driver power available. The relief valve setting may be increased to accommodate an increased discharge pressure requirement but is limited by the design pressure of downstream piping and equipment.

The existing pump's materials of construction must suit the revamp service. Changes in the process stream composition and/or temperature may require a different metallurgy than that of the existing equipment. To determine if the metallurgy of the existing pump is acceptable, review any significant changes in operating temperature, sulfur content, H_2S content, chloride content, etc.

5.8.6.4 Use of Operating Data

If operating or test run data for the unit are available, review them with respect to the hydraulics of circuits that contain major pumps. If the operating data seems reliable, factor them into the evaluation.

5.8.6.5 Possible Recommendations

An evaluation of an existing pump may result in one of the following recommendations.

5.8.6.5.1 Use Existing Pump

If the flow, head, driver, seals, and metallurgy satisfy the new process conditions, use the existing pump without modification.

5.8.6.5.2 Replace Impeller

If the flow and/or head required are greater than that can be delivered by the existing pump (with the installed impeller), it may be possible to achieve sufficient increases in flow and/or head by replacing the impeller. API 610 requires that the casing of a new pump be large enough to allow for a larger impeller to provide at least a 5% head increase over that specified. Therefore, installation of a larger impeller is nearly always an option. The API data sheet usually shows the installed impeller size and the maximum impeller size that will fit in the casing. Pump curves for larger impellers are sometimes included on the performance curve provided. If not provided, the relationships for capacity, head, and horsepower with changes in impeller diameter are approximately as follows:

$$Q_2 = Q_1 \left(\frac{D_2}{D_1}\right) \tag{5.18}$$

$$H_2 = H_1 \left(\frac{D_2}{D_1}\right)^2 \tag{5.19}$$

$$BHP_2 = BHP_1 \left(\frac{D_2}{D_1}\right)^3 \tag{5.20}$$

where

D = impeller diameter
Q = flow rate
H = head
BHP = brake horsepower

5.8.6.5.3 Replace Driver

If the pump itself is acceptable, but the driver has insufficient power, recommend that the driver be replaced. This may entail more changes than just the driver, however. In some cases the motor control center (MCC) and/or switchgear may need modification or replacement. While this electrical equipment is typically outside the scope for the revamp study, the design engineer should be aware that a driver change may entail more than just replacing the motor.

5.8.6.5.4 Operate Two (or More) Pumps in Parallel

A third identical pump can be added to allow two existing pumps to be operated in parallel, resulting in a three-pump hook-up, with two operating and one spare.

5.8.6.5.5 Replace Pump and Driver

If a larger impeller cannot achieve the required flow and/or head, a new pump and driver will be required. Keep in mind that the replaced pump will become available and may be usable in another service.

5.8.6.5.6 Use an Existing Pump in a Different Service
The revamp design engineer should always look for opportunities to use pumps that are no longer suitable for their original service in a different service, as required by the revamp.

5.8.6.6 Tools
5.8.6.6.1 Pump Evaluation Spreadsheet
A simple spreadsheet tool can be prepared for evaluating centrifugal pumps in a revamp. This tool may be set up such that the revamp flow rate and head from the performance curve are input along with the suction pressure, specific gravity of the fluid, and efficiency of the pump. The discharge pressure and brake horsepower (BHP) are calculated to allow the user to determine if the pump can meet the process requirements.

5.8.6.7 Special Pump Services
Certain pumps require special consideration when they are evaluated for a revamp service. Some examples include high head multistage pumps. The design engineer should consult a pump specialist if they are unsure of a pump's suitability or if there are questions regarding any special features required by a given pump service (Figure 5.2).

5.8.7 Compressors
Since compressors are some of the most expensive pieces of equipment in a process unit, they are normally evaluated in every type of revamp study, including scoping studies. The level of detail to which the revamp design engineer evaluates the compressors depends upon the study level for which the evaluation is being conducted. For revamp scoping and revamp feasibility studies, detailed hydraulics are generally not done, but the engineer must perform preliminary, less rigorous hydraulics on the circuits which contain compressors so that a

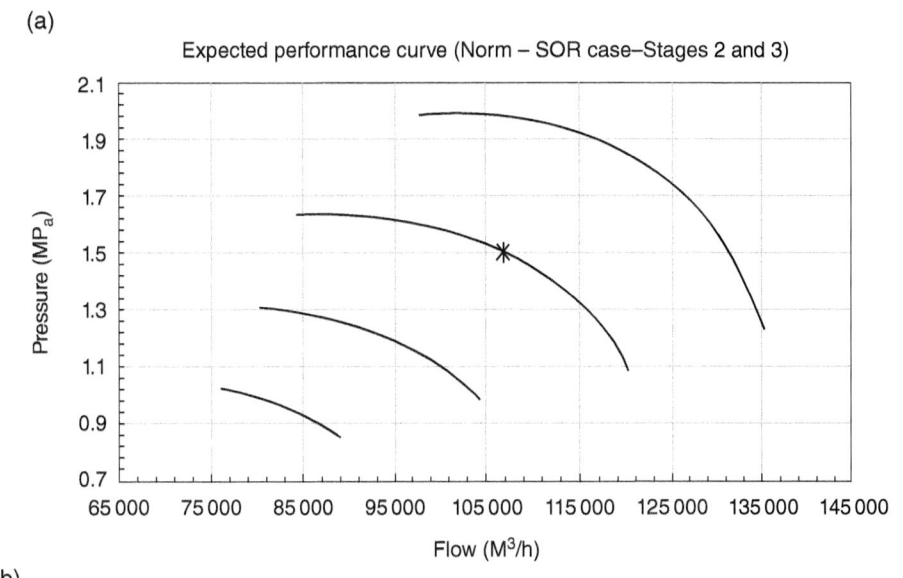

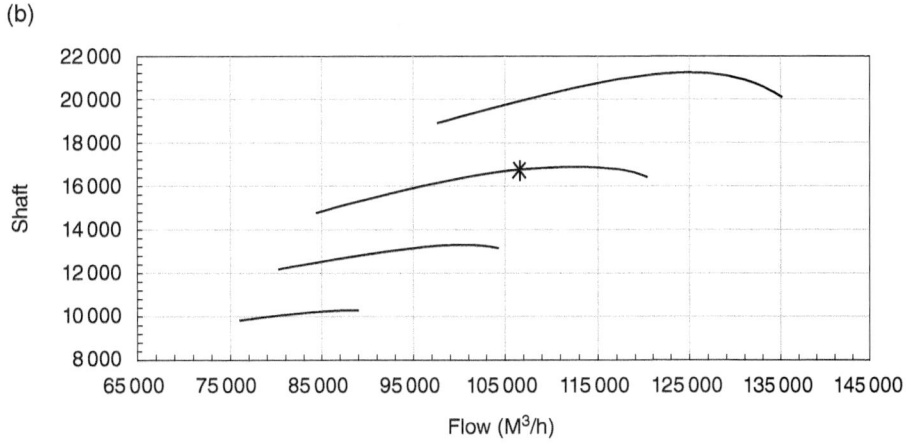

Figure 5.2 Typical pump performance curves.

determination can be made as to the acceptability of the compressor in the revamp service. Process revamp studies and revamp basic engineering design require more detailed hydraulics and analysis.

5.8.7.1 Data Required

The vendor's data sheets and the compressor characteristic performance curve (for centrifugal compressors) are needed to evaluate a compressor. For reciprocating-type compressors, a performance curve is not needed since it is a constant volume machine.

If vendor data sheets and performance curves are not available, the licensor's project specification (data sheet) and the estimated compressor capacity and head requirements can be used to make rough evaluations. Qualify any comments regarding the compressor's suitability by indicating that the evaluation is based on limited information and requires confirmation.

5.8.7.2 Centrifugal Compressor Evaluation

Centrifugal compressors are typically used in recycle gas services. The primary criteria that determines whether or not an existing centrifugal compressor can be used in a revamp service is a combination of the flow rate the compressor can deliver and the head at which it can be delivered. The relationship between flow and head for a centrifugal compressor is defined by the compressor's performance curve. Therefore, having the performance curve is mandatory in order to do a proper evaluation of an existing centrifugal compressor. If the compressor is driven by a variable speed drive such as a steam turbine, there will typically be a family of curves at speeds ranging from 70 to 100 or 105% of the rated speed. (See Figure 5.3 for an example compressor performance curve.) A fixed speed centrifugal compressor will have a single performance curve. Fixed speed motor-driven compressors will normally have a suction throttling valve for flow control.

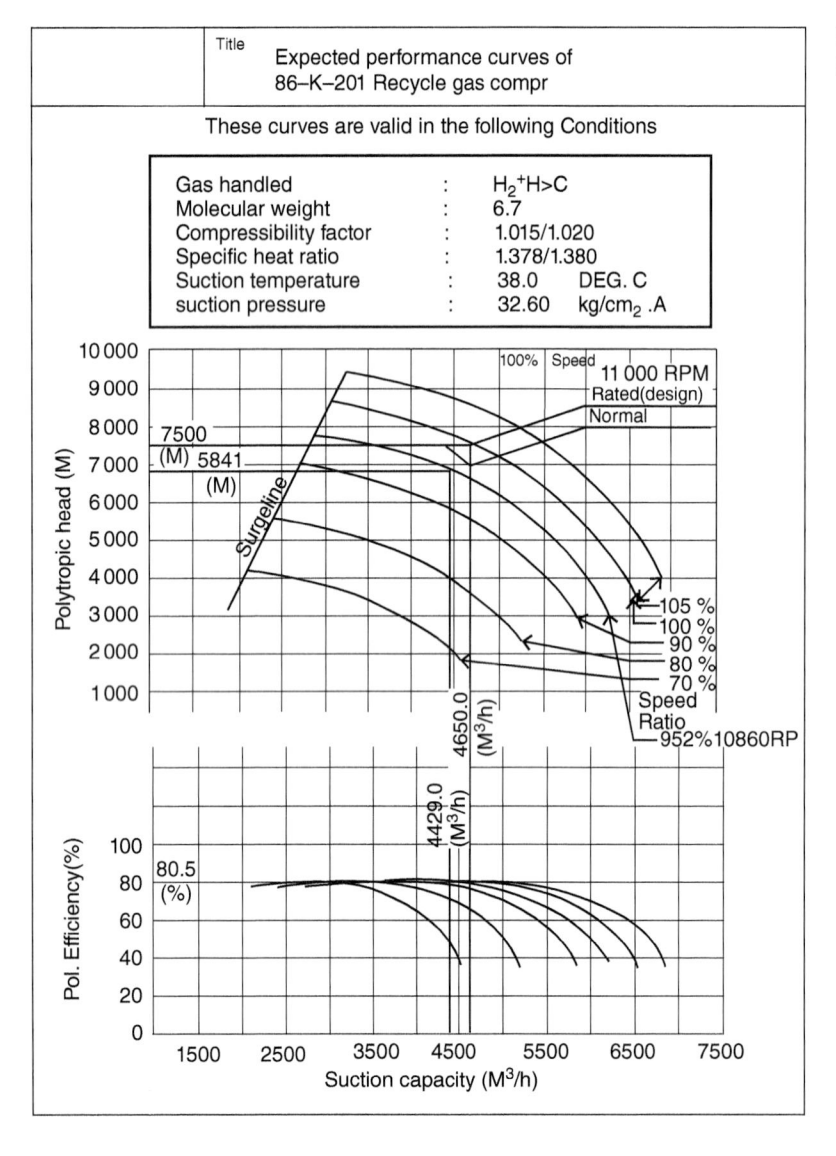

Figure 5.3 Typical centrifugal compressor performance curves.

Title	Expected performance curves of 86–K–201 Recycle gas compr

These curves are valid in the following Conditions

Gas handled	: H_2+H>C
Molecular weight	: 6.7
Compressibility factor	: 1.015/1.020
Specific heat ratio	: 1.378/1.380
Suction temperature	: 38.0 DEG. C
suction pressure	: 32.60 kg/cm$_2$ .A

5.8.7.2.1 Flow and Head

The typical procedure for determining a centrifugal compressor's suitability from a flow versus head standpoint is outlined below:

1) If the compressor performance curve is not in terms of suction flow rate versus polytropic head, convert it to these terms. Compressor curves that show suction flow versus discharge pressure or suction flow versus compression ratio cannot be used unless the suction pressure and recycle gas density (suction) are the same for the rating case as the original design. If more than one performance curve is provided, use the one that has the highest flow versus head.
2) Determine the required flow rate and compressor suction pressure for the revamp operation.
3) Estimate the required compressor discharge pressure based on evaluation of the circuit hydraulics for the revamp operation.
4) Calculate the polytropic head using Eq. (5.21) given below. The polytropic efficiency that is needed to calculate the head is usually provided on the performance curves.

$$\text{Head}_{\text{Poly}} = \left(\frac{n}{n-1}\right)\left(\frac{Z_a R T_1}{MW}\right)\left[\left(\frac{P_2}{P_1}\right)^{\frac{n-1}{n}} - 1\right] \quad (5.21)$$

where

n = polytropic efficiency
Z_a = average gas compressibility factor
R = gas constant
T_1 = suction temperature (°R)
MW = gas molecular weight
P_2 = discharge pressure, kPa (psia)
P_1 = suction pressure, kPa (psia)

5) Plot the required suction flow versus head on the compressor performance curve and determine if it is at an operable point. The suction flow rate should be at least 10% greater than the flow rate at the surge line (left end of the family of curves) at the operating speed of the compressor. The operating point should be on or below the maximum continuous speed curve. The maximum continuous speed is typically 105% of the design speed and the minimum speed is 70% of the design speed. Operation at points all the way to the choke line (right end of the family of performance curves) is acceptable, but not beyond the choke line.

5.8.7.3 Reciprocating Compressor Evaluation

Reciprocating compressors are positive-displacement-type machines. They are typically used in low flow, high head applications such as make-up hydrogen services or net gas compression services. Reciprocating compressors are essentially constant volume machines and are limited by suction flow rate and discharge temperature. Calculate the revamp operation suction flow rate and discharge temperature. The discharge temperature can be calculated using the following formula:

$$T_d = T_s \left(\frac{P_d}{P_s}\right)^{(k-1)/k} \quad (5.22)$$

where

T_d = discharge temperature (°R)
T_s = suction temperature (°R)
P_d = discharge pressure (psia)
P_s = suction pressure (psia)
k = specific heat ratio (C_p/C_v)

It is recommended to limit discharge temperature for new unit designs to 120 °C (250 °F) to reduce maintenance and failures. API 618 limits discharge temperature to 135 °C (275 °F). The discharge temperature for a revamp service can exceed 120 °C (250 °F), but should not exceed 135 °C (275 °F).

5.8.7.4 Driver Power

Calculate the driver power required for the revamp operation using the appropriate equation depending on the compressor type. The driver type and size are typically given on the vendor's data sheet. Sometimes the maximum driver power is shown on the performance curve. Compare the calculated required power to the maximum driver power.

5.8.7.5 Materials of Construction

The materials of construction of the existing compressor must suit the revamp service. Changes in the process stream composition and/or temperature may require a different metallurgy than that of the existing equipment. To determine if the metallurgy of the existing compressor is acceptable, review any significant changes in operating temperature, sulfur content, H_2S content, chloride content, etc., and have a metallurgy specialist or the compressor vendor review the service and the compressor metallurgy.

5.8.7.6 Use of Operating Data

For circuits that include compressors, such as reactor circuits, it is very useful to get a single-gauge pressure survey at a capacity at or near the historical maximum to use as a benchmark for determining circuit pressure drop. As a minimum, the compressor suction and discharge pressures should be obtained at a known set of operating conditions.

5.8.7.7 Potential Remedies

An evaluation of an existing compressor may result in one of the following recommendations.

5.8.7.7.1 Use Existing Compressor

If the flow, head, driver, and metallurgy satisfy the new process conditions, use the existing compressor without modification.

5.8.7.7.2 Change Operating Conditions

For a centrifugal compressor, if the operating point for the revamp operation is above the maximum continuous speed curve or beyond the choke line, check to see if the operating pressure can be increased, or the H_2/HC ratio decreased, or both. If neither of these is possible, or if the new operating point is still outside the operable range of the compressor, modifications to the compressor will be required.

For a centrifugal compressor, if the revamp operating point is below the minimum speed line or to the left of the surge line, consider increasing the flow through the compressor. Increased flow may be achieved either by increasing H_2/HC ratio or by adding a spillback.

Reciprocating compressors and some centrifugal compressors are equipped with a spillback. An easy way to increase capacity is to reduce the spillback flow.

For a reciprocating compressor in which a higher discharge pressure is required, investigate ways of increasing the suction pressure.

5.8.7.7.3 Replace or Add Wheel(s): Centrifugal Compressors

If the flow and/or head required are greater than that can be delivered by the existing compressor (with the installed wheels), it may be possible to achieve sufficient increases in flow and/or head by replacing the wheels or adding wheel(s). The vendor's data sheet may or may not show the installed wheel size, the number of wheels, and the maximum number of wheels that will fit in the casing. If not, the design engineer must consult with a compressor specialist and/or vendor to determine if the compressor can be modified to meet the requirements of the revamp operation.

5.8.7.7.4 Increase Cylinder Liner Bore: Reciprocating Compressors

It may be possible to increase the capacity of a reciprocating compressor marginally by increasing the diameter of the cylinder(s) by machining the casing liner. A compressor specialist and/or the compressor vendor should be contacted to determine what is possible.

5.8.7.7.5 Replace Driver

If the compressor itself is acceptable, but the driver has insufficient power, recommend that the driver be replaced. This may entail more changes than just the driver, however. In some cases the MCC and/or switchgear may need modification or replacement. While this electrical equipment is typically outside of the scope for the revamp study, the design engineer should be aware that a driver change may entail more than just replacing the motor.

5.8.7.7.6 Operate Two (or More) Compressors in Parallel

Especially, if the existing machines are reciprocating type, a third (or forth) identical compressor can be added to allow two (or three) existing compressors to be operated in parallel.

5.8.7.7.7 Replace Compressor and Driver

If none of the above remedies can achieve the required flow and/or head, a new compressor and driver will be required. Keep in mind that the replaced compressor will become available and may be usable in another service.

5.8.7.7.8 Use an Existing Compressor in a Different Service

The design engineer should always look for opportunities to use equipment that is no longer suitable for its original service in a different service, as required by the revamp.

5.8.8 Hydraulics/Piping

Critical piping circuits and hydraulics need to be evaluated when included in the scope of a revamp study or revamp basic engineering design. The following are guidelines for piping evaluation and when piping needs to be replaced. Many of the evaluation criteria and guidelines are process specific. Process specialists should be consulted on process-specific requirements/guidelines.

5.8.8.1 New Unit Line Sizing Criteria Are Generally *Not* Applicable

First and foremost, it should be understood that new unit criteria generally do not apply to revamps. New unit line sizing criteria are primarily based on economics (capex vs. opex) and historic pipe sizing criteria. In most cases, piping changes are costly and undesirable in a revamp.

5.8.8.2 Pressure Drop Requires Replacement of Other Equipment

If excessive piping pressure drop causes the need to replace a pump or compressor, it may be more cost effective to replace the piping. Because piping replacement cost is very site specific, this should be discussed with the owner. Some level of economic evaluation may need to be done by the design engineer or contractor.

5.8.8.3 Approaching Sonic Velocity

For vapor-phase service, the piping should be replaced if the velocity is approaching sonic velocity. For a revamp, the velocity should be kept below about 90% of sonic.

5.8.8.4 Erosion Concerns

For two-phase flow and fluids that contain solids, excessive velocity can cause excessive erosion of the piping, valves, and fittings. The characteristics of the solids are process specific, so velocity limits are process specific for these services. Guidelines for two-phase flow and flashing of fluids across a control valve to a significantly lower pressure is also process specific and should be discussed with a process specialist.

5.8.8.5 Pressure Drop Affects Yields

In some processes a relatively low reactor pressure is critical to achieving good yields, e.g. reforming. In cases where excessive piping pressure drop causes high reactor pressure, some or all of the reactor circuit piping may need to be replaced.

5.8.8.6 Pressure Drop Affects Fractionator Operation or Utilities

In cases where excessive piping pressure drop causes high fractionator pressure, some or all of the column overhead and/or net overhead vapor piping may need to be replaced. This is the case when the column pressure is governed by the destination of the overhead vapor, such as if it is floating on the flare header, vent header, or venting to the fuel gas system, and high column pressure is detrimental to achieving the desired separation or the reboiler is adversely affected by higher pressure operation.

5.9 Economic Evaluation

Although not always required as part of a revamp study report, some form of economics is useful as part of a revamp project. Performing this evaluation helps the revamp design engineer and the owner have an understanding of the benefits of implementing the project. If the evaluation is only done internally for making process decisions, it may not have to be fully documented, but understanding the basic project economics still provides the basis for process decisions. Economics also provides a basis to discuss project implementation.

Generally, economic analyses done for revamps are "incremental analyses." That means that the focus is on just the changes made to the existing plant to implement the revamp design. Generally, this is different than in a new unit analysis, where a large investment for a process unit is justified based on an overall price structure and overhead. In an incremental analysis, we need a base case (usually current operations) and account for the things that change.

This section discusses the basic components of a cost/benefit analysis, some of the data needed to perform the analysis, and some of the tools which can be used. Generally, it is a good idea to make a list of all the changes caused by implementing the revamp, sort them into costs or benefits, and then try to assign a value to the change. (A partial list is included as Table 5.1. This list can be used as a starting point to create a specific list of costs/benefits for a project.)

5.9.1 Costs

There are several types of project costs which affect an owner's implementation decision. Equipment information (new or modified equipment) must be converted into capital cost. Cost estimating engineering must be a part of the team doing the revamp study.

5.9.1.1 Capital Costs

Capital costs for a typical revamp project include costs for new or modified vessels, including columns and reactors, heat exchangers, pumps, compressors, and fired heaters. Improper estimates of capital cost can lead to faulty conclusions regarding the economics of the revamp project.

Generally, there are three main types of capital cost estimates:

Equipment costs (±30%)
Curve costs (±30–40%)
Factored equipment costs (±50%)

It is difficult to estimate the cost of a revamp because there are a lot of factors that affect the cost that are not well known at the process engineering stage. These factors include: mechanical integrity, underground electrical, plot area restrictions, instrumentation upgrades, etc. However, cost estimating engineers can generally make a ±30% accuracy estimate of the cost of a new piece of equipment with their tools, without going to vendors for quotes. To estimate the installed cost, they generally use a correlation based on the number of pieces of equipment and their average cost. The installation factor is applied to the total cost of the individual equipment items. This factor is based on a set of assumptions, such as a clear plot area, use of an outside contractor, etc., which may or may not be accurate, but can be a reasonable starting point. This estimate is the "Factored Equipment Cost" mentioned above. These factored estimates assume "full home office expense" for contractors. In most cases this is a good assumption, but may not be correct for small projects in which the owner may have their own engineering and maintenance departments do the work, in which case "full home office expense" may not apply.

In general, the values for the revamp installation factor may be several times higher than that used in grassroots design. This is because in revamp scenarios, demolishing, removal, piping, and other infrastructure costs are a major cost component. In addition, the penalty for plant shutdown and reduced production must be considered. The installation factors can range from as low as two to as high as eight.

To estimate the cost of any new sections of the plant (such as new fractionators or reactor sections), it is generally easier to get a curve cost rather than cost up each equipment piece and factor the whole estimate.

1) Shortcuts for Obtaining Capital Costs
   Time is often a problem in developing the cost estimate, since the data required to develop the cost estimate are often only known at the end of the study. Here are some shortcuts to help get a handle on the capital costs, when time is short. However, use of these methods will decrease the accuracy of the estimate.

   - Previous cost estimates

   Previous costs can be scaled based on scope or size to get a rough estimate. An example might be to scale an exchanger cost based on $/square foot of surface area.

   - Fraction of new unit cost

   New unit curve costs with the same flow scheme can be factored based on the feed flow rate and extent of changes/new equipment.

2) Revamped Equipment
   Occasionally, only a portion of an equipment item is needed to be replaced, such as a compressor rotor or trays for a tower. The best way to get a cost is to consult the manufacturer. However, if the manufacturer is no longer in business or cannot supply the information, another vendor may be able to. Other options to estimate these kinds of costs are to use a percent of the new equipment cost or consult with equipment specialists.

In general, the capital cost estimates for revamp projects are lower than they should be. One of the primary reasons for this is that revamp capital cost estimates many times do not include accurate erection costs and OSBL costs, which could be as much as the ISBL costs. Therefore, it is prudent to add a reasonable "contingency" to the capital cost estimate, especially if shortcut methods are used.

5.9.1.2 Operating Costs

Other than the capital costs, we often need to consider the operating costs. Usually, a revamp has some effect on the operating costs, either positive or negative. Some examples might be: decreased utility consumption, lower feed consumption, or lower by-product production. Customers will sometimes accept a higher operating cost if it means they are able to increase capacity with minor equipment changes.

5.9.1.3 Downtime

Downtime is very expensive and shortening it can sometimes justify a complete equipment replacement, instead of modifying equipment. It can also affect the equipment layout and implementation strategy. Sometimes the high cost of downtime can justify replacing a reactor, rather than repairing it. This allows it to be redesigned and specify a larger catalyst volume. A common strategy to minimize downtime is to take a quick shutdown to install just the tie-ins and install the rest of the equipment later.

5.9.1.4 ISBL Vs. OSBL

The cost of adding a new cooling tower or boiler (OSBL costs) can seriously affect the project economics and should be considered in the overall capital cost estimate. This is usually done during the mechanical design phase of the project, when the process requirements are already established. If the owner is aware of utility limitations, these should be communicated during the design basis meeting so an effort can be made to live within the constraints, if possible. It may be possible to trade-off heating and cooling sources if the limits are known.

OSBL costs can drive a revamp decision. When considering a driver replacement for a recycle compressor, an electric motor appeared to be the correct choice because electric power was priced significantly lower than the equivalent quantity of high-pressure steam. Based on the cost estimate for installation of the new motor (ISBL), it appeared to be a good payback of less than three years. However, during a field visit to the process unit, the question was raised: From where will the 5 MW of electric power for the motor driver be supplied? The answer was that there is a substation nearby. Okay, but does the substation have sufficient spare capacity to supply the extra 5 MW? Once the information about the substation was available, it was determined that it did not have sufficient capacity. The OSBL cost of a new substation and supporting infrastructure made the cost of switching to a motor driver very uneconomical. The moral of this story is that OSBL costs can be quite large and should not be overlooked when evaluating a potential revamp option or project.

5.9.1.5 Other Costs

In addition to the items in Table 5.1, there is sometimes a concern about how royalties for the revamp, if any, would be accounted. In general, these are defined in the original license agreement; however, they may be negotiated as part of the revamp project basic engineering design phase.

Another common issue is how to account for precious metals. Some owners lease their platinum (an operating expense) and some purchase it (working capital). Also, some owners have certain precious metals in inventory at a historical price. Another complication is that there is sometimes a precious metal investment in the existing catalyst load that can be "monetized" when the catalyst is replaced, to help finance the new catalyst load. Due to the variation in approaches, the design engineer must make a judgment whether it is worthwhile to consider it in the economic evaluation or just report the amount required and let the owner factor it into their calculation later.

Typically, the cost of demolition of existing equipment and its removal is ignored in the early stages of a project (revamp study or assessment phase) and may be accounted for by the contractor and/or owner in the detailed design phase.

5.9.2 Benefits

A general list of project benefits can be seen in Table 5.1. Estimating the potential benefits from the project can help sell the implementation. The benefit stream is the driving force for the project and if the benefits are large enough, some increases in the project cost can be tolerated and still have an attractive project.

5.9.2.1 Increased Product
This is the most common type of benefit and can be obtained by improved catalyst yields, improved recovery, or increased throughput. In some cases the benefits of increased product are due to improved catalyst as well as revamping for increased throughput. A small amount of yield improvement can create a large benefit, so that a new catalyst load can often be justified based on yield improvements.

5.9.2.2 Lower Cost Feed
Another way plant operators improve their economics is by purchasing lower cost feeds. These feeds may be lower quality, higher in contaminants, leaner in desirable components, or in some way less ideal. The owner may be hesitant to reveal their feed cost, so the cost savings may be difficult to estimate. Feed costs are typically 80% of the owner's operating cost, so a reduction can be significant in the overall economics.

5.9.2.3 Higher Value Product
Higher value products may be a result of a product slate change and include things like producing higher purity benzene, converting more benzene or toluene to xylenes, adding a product such as *ortho*-xylene or *meta*-xylene to the slate, etc.

5.9.2.4 Lower Operating Cost
Operating costs are often focused on by plant management as something that they can control. A reduction of utilities is the primary way to lower operating costs, but other cost reductions such as manpower are included in this category.

5.9.3 Data Requirements

One of the challenges of the revamp design engineer is to get the information needed for economic evaluations in a timely fashion.

5.9.3.1 Feed/Product Pricing
Although one must ask for this information, it is usually not provided by the owner or plant personnel. This could be because it is difficult to obtain or because the owner considers it confidential. In any case, even if the owner is not able to supply it, a price structure based on typical international values should be assumed and the assumptions reported. This will at least provide a starting point for discussions. Typical pricing can usually be obtained from the literature, previous projects, industry sources, etc.

Since prices fluctuate with the market, it is often difficult to get a complete set. It is not important that they are absolutely correct. Since revamp analyses are usually delta analyses, the key thing is the deltas between prices in the set and the consistent use of the prices across the cases.

5.9.3.2 Utility Pricing
It must be emphasized that the method used for evaluating utility costs has a dramatic effect on project economics, and therefore the investment decision. Improper utility pricing can lead to bad decisions because good projects may be discarded, and bad projects may be implemented. Regrettably, this is relatively common. To avoid such mistakes, it is imperative that plant engineers and managers use appropriate methods for steam pricing, taking into account all the parameters that impact energy costs – fuel, condensate, power generation, and cooling water – when evaluating proposed projects.

For many companies, the reported cost of boiler steam is the average cost of generation at a particular production rate. The total operating costs, including fuel, power, water, chemical additives, labor, maintenance, depreciation, interest, and administrative overheads, are divided by the total amount of steam produced. Then, the costs of medium and low pressure steam are usually determined based on their enthalpy values in relation to the boiler steam. However, these prices do not reflect the true costs of steam production. Use of these steam prices as the basis can lead to a faulty economic evaluation. As a consequence, a good energy project may be rejected, and a bad project may be approved for implementation. So how to determine the correct steam prices?

The method based on marginal analysis (see Chapter 14) can indicate true steam costs at point of use. The reason why the marginal price method can provide the true steam cost is that it is based on the last incremental amount of steam saved or generated. Determination of marginal steam prices relies on an overall steam and power balance which takes into account the steam balance, steam pressure conditions, boiler and turbine efficiency, condensate return and BFW make-up, and availability of multiple fuels.

If pricing is provided by the owner for only fuel and not for either steam or electricity, it is possible to convert between fuel, steam, and electricity using the following assumptions:

A fuel oil net heating value of 9 444 kcal/kg (17 000 Btu/lb). Steam values can be based on an assumed boiler efficiency of 90% and let-down turbine efficiency of 75%. An electricity value can be based on 10 000 Btu/kW-h of power produced.

5.9.3.3 Catalyst/Adsorbent
Catalyst and adsorbent quantities can be obtained or calculated from the yield estimate. An estimate of the cost may be obtained from the licensor.

5.9.3.4 Other Info
Economic data for other costs/benefits may be hard to quantify; however, a list of unquantified factors might be included in the study report. Factors such as the value of an additional day of downtime might be pursed with the owner or plant management personnel.

5.9.4 Types of Economic Analyses

The type of economic analysis depends on the time available, data available, and owner expectations. There is no sense in doing a detailed economic analysis, if there are so many assumptions that the conclusions are meaningless. Economic analyses at early stages in a project are primarily useful as a screening tool. More detailed analyses can be completed when the project is better defined. There are four basic types of economic analysis used for evaluating a revamp project: basic comparison of alternatives, simple payback, net present value (NPV), and internal rate of return (IRR).

5.9.4.1 Basic Comparison of Alternatives
This basic analysis can be done even without economic data. A compilation of feed and product rates and the status of major equipment items can give a rough idea of the costs and benefits of alternatives. This basic analysis is useful when a scoping or feasibility study is performed and preliminary economic factors such as feed, product, and utility prices are not known and must be assumed. As long as the same set of assumptions is used for each alternative case, the analysis should be useful in selecting between alternatives. However, if economic data are available, the economic benefit can be quantified more accurately by using any of the more detailed methods described below.

5.9.4.2 Simple Payback
After a basic analysis of alternatives, the simple payback technique is the simplest and most common form of analysis. The basic form is:

$$P_B = \frac{I}{B} \tag{5.23}$$

where

P_B = Payback time (years)
I = Investment costs ($)
B = Net benefits ($/year)

Investment costs are primarily considered to be installed equipment costs, i.e. capitalized project costs. Net benefits are the annual increase in profits from the project, based on change in benefits minus change in operating costs. Payback is usually expressed in years, with a value of two to three usually a reasonable investment threshold for a revamp project. However, some companies use as many as seven years as the payback criteria. This indicator does not take into account the time value of money. The NPV and IRR methods do account for this.

5.9.4.3 Net Present Value (NPV)
The NPV method involves assigning a rate of return(r) that is specific to the project and then computing the present value of the expected stream of revenues (R_i). It provides indication of what the present value of the investment is considering the investment cost while realizing the future revenue from the investment at the present time. The NPV is, simply, the present value of future cash flows minus the investment price, taking inflation and returns into account. Since the investment is initially expended, it is counted as negative revenue. A NPV formula is expressed as

$$NPV = -I + \frac{R_1}{(1+r)} + \frac{R_2}{(1+r)^2} + \frac{R_3}{(1+r)^3} + \cdots + \frac{R_N}{(1+r)^N} \tag{5.24}$$

where

R_1 is the first year revenues
R_2 is the second year revenues, and so on

N is the last year to account for revenue (R_N) from the investment (I)
I is the investment amount.

The analysis uses a minimum acceptable return on investment (r). This NPV indicator is defined as the difference between the present value of the expected returns (r) and the initial investment required to generate the returns. If the NPV is positive, the NPV will add value to the business. If it is negative, the investment will subtract value from the business.

Two tasks are required to conduct an NPV analysis. First, investment and revenues must be estimated. For a revamp project, the total capital cost required becomes the investment. The total cost should include all possible major capital cost-related items such as equipment, installation, infrastructure, downtime, etc. The revenues are the net revenues or net positive cash flows expected in future. Second, an appropriate rate of return must be identified. Most investments undertaken by companies are financed with retained earnings with profits from previous activities instead of borrowing. Thus, once a company approves and undertakes one investment, it cannot execute other investments at the expense of the approved investment, and the interest rate has to account for the internal corporate value of funds. As a result of these factors, interest rates of 10–20% are common for evaluating the NPV of projects.

If one assumes a constant revenue over time, Eq. (5.24) reduces to

$$NPV = -I + R \times \sum_{i=1}^{N} \frac{1}{(1+r)^i} = n \times R - I \quad (5.25)$$

where

n is the factor converting revenue across N number of years into present value.

$$n = \sum_{i=1}^{N} \frac{1}{(1+r)^i} = \frac{1}{r}\left(1 - \frac{1}{(1+r)^N}\right) \quad (5.26)$$

Typically, n takes a value within the range of 5–10. For example, if $N = 10$ years and $r = 10\%$, $n = 6.14$. If $N = 20$ years and $r = 10\%$, $n = 8.51$. If $N = 10$ years and $r = 15\%$, $n = 5.02$.

A revamp project reduces net operating expenditure (OpEx-Saving) by \$10 MM per year with capital expenditure (CapEx) of \$30 MM. Assuming capital is spent initially, the present value (PV) for the OpEx saving across 10 years at a discounted rate of 10% is

$$PV(OpexSaving) = 6.14 \times OpexSaving = \$61.4 \text{ MM} \quad (5.27)$$

$$NPV = PV(OpexSaving) - CapEx$$
$$= 61.4 - 30 = \$31.4 \text{ MM} \quad (5.28)$$

This project should go ahead based on the NPV approach as it has a high NPV. However, the simple payback is three years. The simple payback approach would make this a borderline project since the two to three years payback period is typically used as a guideline for capital investment projects.

For the above revamp project, if the CapEx is spent over the next three years in four installments, \$10 MM initially (year zero), eight in the first year, seven in the second year, and the last five in the third year,

$$PV(CapEx) = 10 + \frac{8}{(1+10\%)} + \frac{7}{(1+10\%)^2} + \frac{5}{(1+10\%)^3}$$
$$= \$26.8 \text{ MM} \quad (5.29)$$

$$NPV = PV(OpexSaving) - PV(CapEx)$$
$$= 61.4 - 26.8 = \$34.6 \text{ MM} \quad (5.30)$$

In some cases when a grassroots design is improved via design changes, these changes could lead to reduction in both Opex and CapEx. In these cases, CapEx (I) takes a negative value when applying Eq. (5.30). Thus, CapEx savings contribute to NPV as well.

For example, an improvement idea reduces both OpEx by \$3 MM per year and CapEx by \$5 MM. Assume this OpEx saving spreads across 10 years with discount rate of 10% and CapEx saving is obtained initially,

$$PV(OpExSaving) = 6.14 \times OpExSaving$$
$$= \$18.4 \text{ MM} \quad (5.31)$$

$$NPV = PV(OpExSaving) - CapExSaving$$
$$= 18.4 - (-5) = \$23.4 \text{ MM} \quad (5.32)$$

5.9.4.4 Internal Rate of Return (IRR)

The IRR method is similar to NPV. The difference is that rather than assume an acceptable return on investment, the IRR approach solves the equation NPV = 0 to determine an effective interest rate or IRR. In other words, IRR is the interest rate at which the NPV of all the cash flows (both positive and negative) from an investment equal zero,

$$0 = -I + \frac{R_1}{(1+IRR)} + \frac{R_2}{(1+IRR)^2} + \frac{R_3}{(1+IRR)^3} + \cdots + \frac{R_N}{(1+IRR)^N} \quad (5.33)$$

If the IRR determined from Eq. (5.33) exceeds a company's required rate of return, that project is desirable. Otherwise, if IRR falls below the required rate of return, the project should be rejected.

When the question is whether to undertake a project or not, NPV is a stronger metric than simple payback and IRR for making investment decisions. This is because NPV determines the present value of making the investment and not just the amount of time needed to realize the investment. For this reason, NPV has become the most common approach to investment decisions. The IRR approach provides another metric to rate the profitability of the project if it agrees with the NPV result.

The simple payback approach indicates how many years an investment project will be run before the investment is realized and profitability is reached. The problem with the payback period is how to decide between projects. The simple payback approach would select the project with the faster payback, but that may only make a small amount of money very quickly. This may not be the best choice. However, the choice is clear with NPV – select the project with the higher NPV because it has the highest value and produces the strongest benefit for the company.

5.9.4.5 Issues

Accuracy/Uncertainty – when performing a revamp study, there are a lot of things unknown to the engineers. This is especially true if the scope of the project and the effect on off-sites is unknown. If warranted, some of the issues can be dealt with by doing various cases (scenarios) and then assigning probabilities to them. This gives a probability weighted scenario economic analysis.

Scope – the scope of a revamp study is usually limited to the process area; however, if the revamp has an impact on the utility system or tankage, the OSBL investment may not be known until later when the off-sites and utility systems are studied. Therefore, the study may not show a complete picture of the economics of the revamp (Table 5.7).

5.10 Example Revamp Cases

The following are three typical revamp cases which highlight some of the strategies used in an aromatic complex or unit revamp, with a common theme, minimize changes and cost. In all cases the revamp methodology discussed in this chapter and some of the basic strategies, such as utilizing design margin, are employed, although not stated specifically below.

Table 5.7 Potential costs/benefits for a revamp project.

Costs
New equipment
Downtime
New catalyst/adsorbent
Additional platinum investment
Less byproduct yield
New instrumentation
Additional feed
Use of plot area
New column internals
Equipment modifications

Benefits
Improved product quality
Additional product produced
Higher onstream efficiency
Less manpower required
Less catalyst/chemical consumption
Longer catalyst cycle length
Less utility consumption
Higher feed contaminant tolerance
Use of lower cost feedstock
Lower platinum investment
Less/easier maintenance
Easy future expansion
Flexibility to process lower quality feeds
Easier operation

5.10.1 Aromatics Complex Revamp with Adsorbent Reload

A major petrochemical company sought to increase the capacity of an existing aromatics complex. The intent of the revamp was to maximize throughput, i.e. increase production of *para*-xylene and benzene. The original Parex adsorbent was near the end of its useful life and the next generation of adsorbent would be loaded for the revamp operation. As is the case with most revamp projects, the refiner wanted to maximize the use of existing equipment and minimize cost. In this type of revamp it is important to identify in the design basis phase, the equipment that the owner is not willing to replace. The owner stated that replacement of fired heaters, reactors, Parex adsorbent chambers, large fractionation columns, and large recycle compressors was off-limits. These equipment items could be modified but not replaced.

It was determined during the study that the capacity limit, based on the constraints imposed by the owner, was set by the volume of adsorbent in the existing adsorbent chambers. The new adsorbent would allow for a capacity increase of approximately 30%. With this limitation, the challenge was now to find ways of minimizing other costly changes.

The following is a summary of the other changes made to the equipment and operating conditions to allow achievement of the capacity increase with minimum major equipment changes and associated costs:

1) The hydrogen: hydrocarbon (H_2/HC) ratio in both the transalkylation and xylene isomerization units was reduced in order to save the recycle gas compressors. This change was facilitated by replacing the catalyst in both units with new, more stable catalyst that could operate with reduced H_2/HC. These catalysts were nearing the end of their life, so would need to be replaced in the near term in any case.
2) The vertical shell and tube CFE's in both the transalkylation and xylene isomerization units were replaced with welded-plate exchangers that allow for greater heat recovery. This helped to save the charge heaters and condensers in these unit's reactor sections.
3) Several large columns in the complex, including the reformate splitter, xylene column, toluene column, and benzene column were retrayed with high-capacity trays, allowing the existing column shells, foundations, piping, and structures to be saved.
4) A feed preheat exchanger was added to the reformate splitter. This reduced the required duty of the reboiler, allowing its reuse.
5) Several reboilers were retrofitted with High Flux tubing to allow reuse of the shells. Some of the shells needed modification to increase the size of nozzles due to the higher throughput. But this was found to be less costly than installing new heat exchangers.
6) New, larger impellers were installed in several pumps, along with new larger motors in some of the services. This saved not only the cost of new pumps but also provided saving in piping, foundations, and installation costs.
7) The rundown temperature to storage for low vapor pressure products (*para*-xylene and heavy aromatics) was increased from the original design 104 °F (40 °C) to 113 °F (45 °C). This avoided adding new cooling water exchangers for these services and reduced the consumption of cooling water.

5.10.2 Aromatics Complex Revamp with Xylene Isomerization Catalyst Change

A major international refinery and petrochemical company sought to increase the capacity of an existing aromatics complex. The refiner wanted to increase feed rate and yields to increase *para*-xylene production by a fixed amount of 30%. The refiner wanted to maximize the use of existing equipment and minimize cost.

Because there was a firm basis for the revamp, a process revamp study was performed on eight process units including the naphtha hydrotreating unit, reforming unit, xylene fractionation, aromatics extraction unit, benzene–toluene fractionation unit, transalkylation unit, xylene isomerization unit, and *para*-xylene separation (Parex) unit. The study identified the appropriate operating conditions, the estimated yields, the required equipment modifications and additions, and provided budget cost estimates for the new equipment. One of the major changes to the complex included in the revamp was changing the xylene isomerization catalyst from an ethylbenzene (EB) isomerization type to an EB dealkylation type. The dealkyation of EB, which converts EB to benzene, reduced the flows through the xylene fractionation, Parex, and xylene isomerization units by more than 10%. This helped to debottleneck these units and avoided significant equipment replacement. Other changes similar to those described in the example in 5.10.1 above were also required, but the catalyst change was the primary facilitator of the revamp.

The refiner used the results of the study to develop a preliminary cost estimate and justification to request funding for the project. After the project was approved, basic revamp design engineering was prepared and the project was implemented, achieving the refiner's original goals for the unit.

In this case it was possible to go directly from the process revamp study to the basic revamp engineering design because the basis and scope of the project were well defined and the economics of the revamp were attractive.

5.10.3 Transalkylation Unit Revamp

An aromatics producer wanted to increase the capacity of an existing transalkylation unit by 25% without replacing major equipment such as the charge heater, recycle compressor, etc. The initial heat and weight balance indicated that neither the charge heater, CFE, nor products condenser would be able to meet the revamp duty. The transalkylation unit was an older unit equipped with a vertical shell and tube CFE. Reviewing the current operating conditions showed that the existing CFE had a weighted MTD of about 75 °F (48 °C). Replacing the CFE with a new welded-plate-type exchanger with an MTD of 50 °F (28 °C) could recover significantly more heat from the hot reactor effluent and allow reuse of the existing charge heater and products condenser. This not only saved several million dollars of capital but also reduced the fuel consumption of the charge heater by more than 20%. These two factors made the revamp economically attractive and the owner was able to justify and implement the project.

Further Reading

Alanis, F.J. and Sinclair, I.J.C. (2002). Understanding process and design interactions: the key to efficiency improvements and low cost revamps in ethylene plants. *Fourth European Petrochemicals Technology Conference*, Budapest, Hungary (26–27 June).

Kumana, J.D. (2003). How to Calculate the True Cost of Steam. A DOE Report, September.

Lee, J.J., Ye, B.H., Jeong, H.Y. et al. (2007). Reduce revamp costs by optimizing design and operations. *Hydrocarbon Processing* 86 (4): 77–81.

Zhu, X.X. and Martindale, D. (2007). *Energy Optimization to Enhance Both New Projects and Return On Existing Assets*. San Antonio, TX: NPRA.

Zhu, X., Maher, G., and Werba, G. (2011). Spend money to make money via revamp. *Hydrocarbon Engineering*: 33–38.

Part III

Process Equipment Assessment

6

Distillation Column Assessment

6.1 Introduction

Distillation is the core of a process unit for converting multicomponent streams into desirable products and accounts for the majority of energy consumptions. Improving energy utilization, reducing capital costs, and enhancing operational flexibility are spurring increasing attention to distillation column optimization during design and operation. A good understanding of distillation fundamentals, feasible operation, and equipment constraints will enable process engineers gain insights for the distillation performance.

Obtaining good evaluation of separation systems can *provide* insights into the complex interactions in the system and understand how well the system operates. Although it could be a challenging task, doing it properly could yield huge benefits in enhancing operating margin as well as generating a wealth of process knowledge that can be invaluable for the operation of the system.

6.2 Define a Base Case

The first step for tower performance evaluation is to simulate the original tower design because it is uncommon that the original tower data sheets are unavailable or inaccurate. To do this, selection of a proper VLE calculation package is critical. For hydrocarbon separation, the Peng–Robinson equation of state model is a common choice. By providing process data including feed and product flows and compositions, together with tower data including temperature and pressure, feed tray, the number of theoretical stages, and reflux rate, the simulation will generate mass and composition balances as well as heat balances indicating reboiling and condensing duties.

Once the process simulation is developed, it is desirable to verify the simulation fidelity using different process conditions. The predicted product rates and purity and compositions as well as key operating parameters such as reflux rate and reboiling/condensing duties can then be compared with measurement. In some cases, performance tests are required to gather key data to compare with simulation for the accuracy and reliability of the simulation. To do this, the performance tests must be conducted under steady and smooth conditions to mimic steady-state operations.

If the simulation fidelity is proven to be sufficient enough, it is ready to move to the next task which is evaluation of the tower performance because the purpose of reproducing the original design data is to understand the tower hydraulic and thermal performances of the base case.

An important aspect of defining the base case is gathering all the important data for the material and heat balances in one single sheet for a tower of interest. It would be very informative to have important mass flows, temperature, pressure, and composition data in one table so that a snap shot of the tower performance can be seen at a glance. Such an example is a heat-pumped C_3 splitter shown in Figure 6.1 and Table 6.1. In building such a table, it is a good practice to include tag number of the instrument for each parameter so that the data can be retrieved readily from historian to produce the table with snap shots of different times for evaluation of tower performance in future. The "accuracy" column shows the high and low values of the corresponding parameters. The accuracy is determined by recording (or observing) operation during the steady-state period and noting the average high and low values of the various instruments during this period. This information could be very helpful when establishing heat and mass balances with indication of closure percentage. Typically, smaller flows than feed streams can have a higher inaccuracy than larger streams and not severely affect the material balance. Therefore, it is good to know which streams have the highest reliability when determining the material or heat balance. It is also important to record the date and the time period that the data were taken for future reference. A ready reference of what data are needed in a typical tower evaluation can be seen here.

Efficient Petrochemical Processes: Technology, Design and Operation, First Edition.
Frank (Xin X.) Zhu, James A. Johnson, David W. Ablin, and Gregory A. Ernst.
© 2020 John Wiley & Sons, Inc. Published 2020 by John Wiley & Sons, Inc.

Figure 6.1 Heat-pumped C_3 splitter.

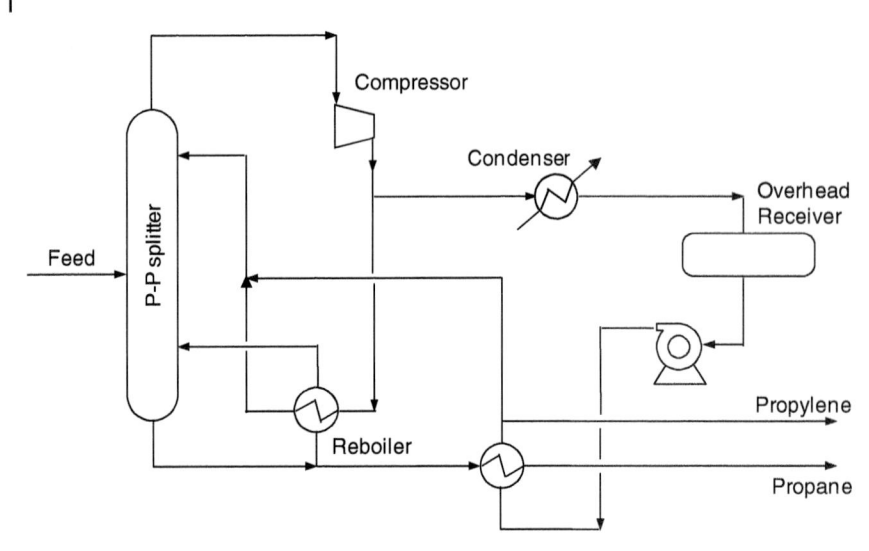

Table 6.1 Major data set for a heat-pumped C_3 splitter.

Data	Units	Tag no.	Value	Accuracy
Feed rate	Barrels per day	FE-8854	4975	±50
Feed temp.	°F	Pyrometer	87	—
Top pressure	psig	PI-8831	100	—
Delta-P	psi	PDI-8827	9.2	—
Top temp.	°F	TI-8774	53	—
Bottom temp.	°F	Pyrometer	73	—
Comp. suction temp.	°F	Pyrometer	55	—
Comp. discharge press	psig	PC-8832	230	—
Comp. discharge temp.	°F	TI-8776	119	—
Comp. discharge temp.	°F	Pyrometer	135	—
Main reflux rate	MSCFD	FT-8858	34 550	Too low
Main reflux temp.	°F	Pyrometer	74	—
Trim reflux rate	Barrels per day	FT-8857	600	—
Trim reflux temp.	°F	Pyrometer	99.5	—
Bottoms flow	Barrels per day	FT-8864	1 060	±50
Propylene product temp.	°F	Pyrometer	110	—
Propylene flow rate	Barrels per day	FT-8860	3 840	±100
Overhead composition	vol.% $C_3=$	AR 869-3	92.1	±0.5
Bottoms composition	vol.% C_3-	AR 869-2	97.1	±0.1

Source: From Summers (2009), reprinted with permission by AIChE.

Defining a base case is to determine the base case operation of the tower of interest. This requires extracting two kinds of data. One kind is process data in terms of feed and product conditions such as flows and compositions, while the other is tower operating data including temperature, pressure, and reflux rate. The former defines the mass and composition balances and the latter sets the heat balance around the tower with Table 6.2 giving such an example of C_2 splitter column.

Due to the importance of developing a reliable base case as the basis for evaluation, Summers (2009) gives excellent discussions for this topic. For understanding the difference between simulation and measurement, readers can refer to Kister (2006).

Table 6.2 Heat and mass balances for a C$_2$ splitter.

Composition wt.%	Feed	Vent	Ethylene product	Dilute ethylene product	Ethane bottoms
Hydrogen	0.0016%	0.26%	0.21 ppm	0	0
CO$_2$	0.0001%	0.0006%	0.0002%	0.61 ppm	0
Methane	0.091%	14.45%	0.007%	0.007%	0
Ethylene	77.77%	85.28%	99.98%	80.44%	1.55%
Ethane	21.66%	0.0002%	0.0109%	19.56%	96.17%
Propylene	0.291%	0	0	0.002%	1.37%
Propane	0.0071%	0	0	0	0.033%
Isobutane and heavier	0.187%	0	0	0	0.88%
Total	100.000[a]	588	71.853	6.292	21.265
Phase	Vapor	Vapor	Liquid	Liquid	Liquid
Temperature (°C)	−13.0	−43.4	−29.8	−26.1	−70
Pressure (psig)	340	250	270.5	276.2	279.7

DA-2410 condenser pressure	250	psig
DA-2410 top pressure	251	psig
DA-2404 condenser pressure	269.9	psig
DA-2404 top pressure	269.9	psig
Vent condenser duty[b]	0.73	MMBTU/h
Condenser duty[b]	49.87	MMBTU/h
Reboiler duty[b]	23.18	MMBTU/h
Side reboiler duty[b]	13.17	MMBTU/h
Reflux rate to DA-2410[a]	4.730	lb/h
DA-2410 reflux temperature	−43.4	°C
DA-2410 top temperature	−36.1	°C
Vapor rate to DA-2410[a]	5 318	lb/h
DA-2404 reflux rate[a]	349 370	lb/h
DA-2404 reflux temperature	−33.7	°C
DA-2404 top temperature	−30.4	°C

Source: From Summers (2009), reprinted with permission by AIChE.
[a] All flows adjusted to a 100 klb feed basis to mask the true capacity of the unit.
[b] All duties adjusted to a 100 klb feed basis.

6.3 Calculations for Missing and Incomplete Data

Plant historian data is the best source, but they are usually incomplete. This is particularly true for old process units. In order to avoid wasted time and rework, you need to make sure critical meters are working properly, which the instrument engineers can help verify. In most cases, design and operating data is of interest and the key is to understand the difference and reasons. For example, the knowledge about heat exchanger fouling can help evaluation of current operation performance considerably. Major consumption and any critical inputs must be verified carefully. The first stage of verification is to compare design data with operating data and perform some adjustments. This first pass verification can separate the important from the trivial data so that the effort for chasing high precision and gathering miniature data and nit-gritty details can be avoided.

Since most correlations for heat exchangers are empirical based, the heat transfer calculations for exchangers are only accurate to about 85–90% when all the necessary

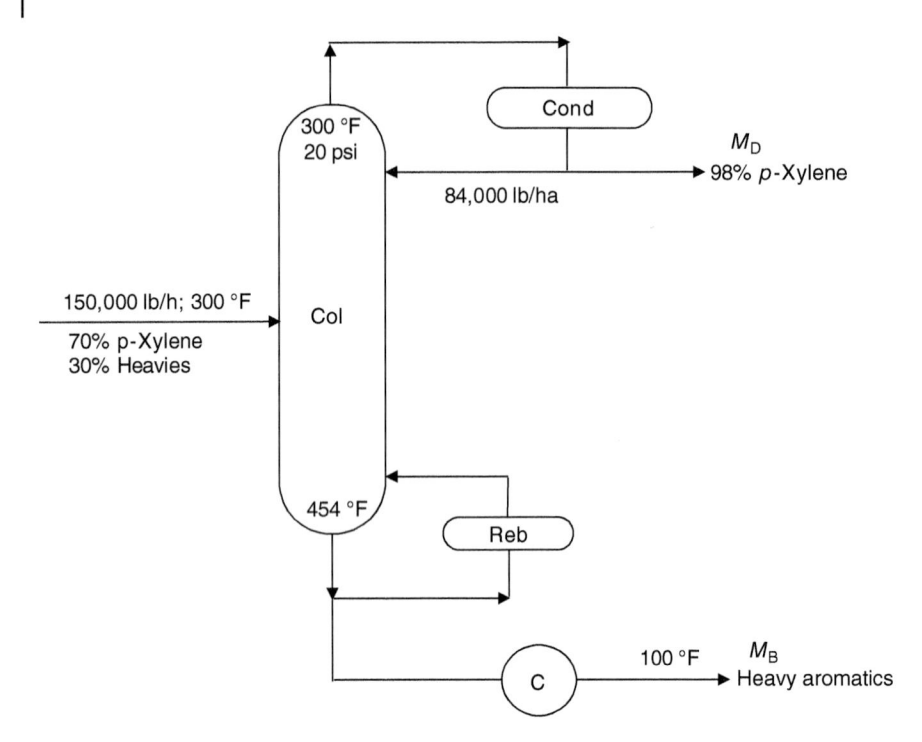

Figure 6.2 Use of heat/mass balances to obtain missing data.

data are known. When some data have to be estimated, the accuracy gets worse. However, this accuracy is sufficient to tell if a heat exchanger is functioning as expected or not.

In many cases, shortcut calculations can fill in the gaps. An example used in Kenney's book (1984) gives good illustration for how to do it. This example is revised to reflect the reality. Consider the tower in Figure 6.2. As for many plants, cooling water rates are not measured and overhead product comes off on level control. However, since feed rate and composition and overhead product composition are known, much of the missing data can be inferred by energy and mass balances and the primary heat transfer equation.

In this problem, p-xylene is to be recovered from a stream containing heavier aromatics. Neither product rate is measured, but feed rate and reflux rates and the p-xylene content of the overhead are. No heat exchanger duties are measured. With some data from a readily available source, the energy use for the tower can be estimated.

Example 6.1 Obtaining Missing Data Above "Given"

Given
For p-xylene: Normal boiling point = 138.5 °C; latent heat of vaporization = 146.2 Btu/lb; specific heat = 0.38 Btu/°C lb at 0 °C and = 0.43 at 41 °C and = 0.55 extrapolating to 140 °C. For heavier aromatics: specific heat = 0.4 Btu/°C lb. for naphthalene at 87 °C and = 0.5 for pentadecane at 50 °C = 0.8 extrapolating to 230 °C.

Calculate Missing Data
p-xylene product rate; heat duty for the overhead condenser, bottom cooler, and reboiler.

Solution

i) Calculate p-xylene product rate:
Applying component balance on p-xylene and mass balance on tower gives:

$$70\% \times F = 98\% \times m_D; \quad m_D = \frac{0.70 \times 150\,000}{0.98}$$
$$= 107\,143 \text{ lb/h}$$

$$m_B = F - m_D = 150\,000 - 107\,143 = 42\,857 \text{ lb/h}$$

where
F is feed rate.

ii) Calculate the bottom cooler duty:
The heat rejected in the bottom cooler is

$$Q_{\text{bottom cooler}} = m_B \times Cp \times \Delta T$$
$$= 42\,857 \times 0.8 \times (454 - 150)/10^6$$
$$= 10.4 \text{ MMBtu/h}$$

If the heat capacity data is in error, the calculated duty would vary ±0.1 Btu/(lb °F). This is within the precision of other data.

iii) Calculate the overhead cooler duty:

$$m_{\text{overhead}} = 107\,143 + 84\,000 = 191\,143 \text{ lb/h}$$

Case 1: no subcooling, the condenser duty is

$$Q_{condenser} = m_{overhead} \times q_{latent} = 191143 \times 146.2 / 10^6$$
$$= 27.9 \text{ MMBtu/h}$$

Case 2: Assuming 30 °F subcooling, the condenser duty is the summation of latent heat duty and subcooling duty, which can be calculated as

$$Q_{condenser} = m_{overhead} \times q_{latent} + m_{overhead} \times Cp \times \Delta T$$
$$= 191143 \times (146.2 + 0.55 \times 30) / 10^6$$
$$= 31.1 \text{ MMBtu/h}$$

iv) Calculate the reboiler duty:
Applying the energy balance around the tower indicates that the reboiler duty is the summation of the condenser duty and the heat required to raise the bottom from 300 to 454 °C.
Case 1: no subcooling in the overhead, the reboiler duty is

$$Q_{reboiler} = Q_{condenser} + m_B \times Cp \times \Delta T$$
$$= 27.9 + 42\,857 \times 0.8 \times (454 - 300) / 10^6$$
$$= 33.2 \text{ MMBtu/h}$$

Case 2: 30 °F subcooling in the overhead, the reboiler duty is

$$Q_{reboiler} = Q_{condenser} + m_B \times Cp \times \Delta T$$
$$= 31.1 + 42\,857 \times 0.8 \times (454 - 300) / 10^6$$
$$= 36.4 \text{ MMBtu/h}$$

With these approximations, the heat duties on condenser, reboiling, and cooler are established which provides the basis for the process simulation.

6.4 Building Process Simulation

A separation column simulation is conducted in a process simulation tool based on tray-by-tray equilibrium calculations for mass and heat balances. Given the data for feed and products in terms of flow rates and compositions, as well as column operating conditions in terms of pressure and temperature, the column simulation can mimic the mass and heat balances for the current operation. Table 6.2 shows an example of the data required for conducting simulation of a C_2 splitter column.

For some processes which involve a process stream with many components, it could be too difficult to gather all the components for simulation. In this case, the concept of pseudo-components is applied so that a group of components is lumped together into a pseudo-component with similar physical properties. For oil refining processes, crude oils and refining products are a mixture of many different chemical compounds. They cannot be evaluated based on chemical analysis alone. In order to characterize any crude oil and refining products, the petroleum industry applies a shorthand method of describing hydrocarbon compounds by the number of carbon atoms and unsaturated bonds in the molecule and uses distillation temperatures and properties to define crude and products. For example, commercial jet fuel can be represented by an ASTM D-86 distillation temperature plot with the kerosene boiling range of 401 °F at 10% and 572 °F endpoint while naphtha jet fuel, also called aviation gasoline, can be represented by a shorter distillation range of 122 °F at 10% and 338 °F endpoint.

The first step is feed simulation. If detailed feed analysis is available which includes composition and conditions, a feed can be readily defined in simulation. Otherwise, the feed can be simulated from back-calculation as the summation of all products for which compositions and conditions are provided as part the of mass and heat balance data.

The second step is to determine feed tray position. Theoretical stages should be used in simulating a column. If tray efficiency is known, the feed tray in terms of theoretical stage can be determined from the actual feed tray and tray efficiency. However, tray efficiency is usually unknown. In this case, a sample lab test may be warranted. It is recommended to take a side sample one tray away from the feed tray. The feed point is one stage away from the theoretical stage which matches the sample composition the best. Taking the sample from the feed tray would give compositions which are highly influenced by the feed and hence cannot truly represent the internal compositions inside the column.

The third step is to determine the number of theoretical stages required. With the feeds defined and product conditions given in the tabulated data, a column simulation can be established. For a simple column with two products, one from the overhead and from the bottom, the number of theoretical stages can be determined from the measured reflux rate. For a given reflux rate, the required number of theoretical stages is the one which can match the product specifications. As the reflux rate defines the reboiler duty and hence the column heat balance for a simple column, the heat balance determines the product specifications for given column conditions. For a complex column involving side draws and pump arounds, the column should be simulated section-by-section because the column heat balance is defined by reflux rate together with the column pump arounds. It is recommended to simulate a complex column from top to bottom. The top section has an overhead product, a side draw, and a pump around next to the side draw. For a given reflux rate and pump around duty, the number of theoretical stages in the top section is determined by matching the given product specifications. The section next to the top section is then simulated similarly.

The simulation can provide the sound basis for conducting other assessment tasks, which are discussed below.

6.5 Heat and Material Balance Assessment

One of the early steps of assessing the fractionation system is to obtain good material and energy balances. Otherwise, it could be possible that assessment yields misleading conclusions.

The material and energy balances can be built based on the input of feeds and energy as well as outputs of products and energy in operation. The purpose of conducting a column material balance is to make sure feeds and products are measured accurately and desirable products are obtained. The energy balance is to verify if all major sources of energy input are accounted and if efficient use of energy is achieved.

Heat input is the driving force for fractionation. For a simple fractionation column, heat input comes from feed and bottom reboiling while heat is removed from products and overhead condenser. For a complex fractionation column, multiple products are produced while pump arounds are located to remove excess heat in the column and recover this heat for process usage.

Both material and energy balances should be conducted on the basis of steady-state operation as this is stable operation away from any transient excessive flooding or weeping operation. The steady-state operation can be viewed from historian when process data remain virtually the same within a very narrow band. On the other hand, steady-state operation can be obtained in operation after a minimum time from making operating adjustment to a tower. This minimum time can be expressed as:

$$t_{\min} = \frac{M_{\text{hold}} R_f}{F} \quad (6.1)$$

where

M_{hold} is the summation of material holdup in sump and receiver drum
R_f is the reflux ratio
F is the tower feed rate

For any fractionation column, there are overall mass balance and component mass balances which follow mass conservation law.

For overall mass balance:

Total mass input = Total mass output   (6.2)

For component balance:

Total input of component j = Total output of component j

(6.3)

Similarly, heat balance follows energy conservation law:

Total heat input = Total heat output   (6.4)

6.5.1 Material Balance Assessment

Good understanding of a material balance and key component balance can give insights for maximizing desirable product yields while minimizing undesirable products. Material flows are measured for feed and products, which are readily available online. Component measurement is usually obtained from lab tests for key components. Samples are taken daily on most towers and analyzed in the plant's local laboratory. However, these laboratories are typically setup to measure for certain key compounds that can contaminate the final product and do not have the capability of measuring the full spectrum of multicomponents involved in feed. Therefore, unless the tower of interest has only a few components in the feed, a complete component balance will typically need special laboratory assistance which more than likely will come from outside the local plant. In this case, be very careful with compositions and understand the units of measurements that are provided by the laboratory. However, the component balance could be difficult to obtain for hydrocarbon separation towers due to the fact that individual components cannot be fully characterized. But for most chemical and natural gas separation towers, it is possible to establish individual component balances.

When the material balances achieve at least ±10% offset ([total mass input − total mass output]/total mass input), it is acceptable for tower evaluation (Summers 2009). For hydrocarbon separation processes, the closure could be as high as ±5%. It is common that a poor mass balance is caused by transmission error between pressure drop and flow rate for some of the material streams. Most flow meters are pressure drop-based devises and they could give wrong readings if physical properties are not used properly for converting pressure drop to flow rate. In some cases, wrong readings can be corrected by meter calibration including proper zeroing and spanning. Consult with instrument engineers and they can help resolve the meter-related issues.

Example 6.2 *Overall Mass Balance Assessment*
This example comes from the main fractionation tower in a hydrocracking process which is operated to make naphtha, kerosene, and diesel products. The bottom product is called unconverted oil, part of which is recycled back to reaction for further conversion and the rest sold as fuel oil to the market. Table 6.3 shows the material balance for the fractionator indicated by expected yields versus the actual yields as well as the flow rates for feed and products measured online.

Let us look at the overall material balance for the tower. The measured product rates in barrel per day are 18 130 of naphtha, 12 200 of kerosene, 9 968 of diesel, and 9 968 of unconverted oil, which gives a total of 50 266 barrel per day. The difference between the measured total of 50 266 and feed rate of 51 548 is 3 vol.%. As the light end products are not measured, it usually accounts for around 3 vol.% of feed. Thus, the material balance for the tower is in a good closure at less than 1% of uncertainty.

What could we learn from this material balance? The first observation is that 1.6% extra kerosene is produced than expected. This has a simple explanation because the tower was operated to maximize kerosene production as kerosene was more valuable in the local market at the certain time. In contrast, 4.3% of less diesel was made, which was surprising. Two plausible causes were thought of by the process engineer responsible for the tower operation. One reason was that kerosene cuts 1.6% deep into diesel while the other was that 2.7% (4.3 − 1.6%) diesel slumps into the bottom unconverted oil. If the latter was true, it could be a significant yield loss and should be resolved.

This was only a hindsight which could be wrong. The expected yield comes from yield estimates based on empirical correlations which could give inaccurate estimates sometime. Distillation temperature for the bottom product could provide the answer to this question. Thus, a lab test was conducted for the bottom unconverted oil, which shows that the 5% distillation temperature is 720 °F. This temperature cut should belong to the diesel range. It was clear that the unconverted oil contains a good portion of diesel which could have been sold to the market at a premium price. The diesel price was at 2.45/gal, and thus $143 000/day was lost under this tower operation, or $50 MM/year could be lost if this problem was not resolved. This price tag rang an alarm large enough to secure swift actions for troubleshooting.

In summary, the investigation revealed two root causes. The first one was to do operation that stripping steam at the bottom of the tower was insufficient as it was put on constant control based on the original design point. Although the throughout was increased by 10% over the years and feed became heavier, the stripping steam did not change. After adjusting it accordingly, the diesel recovery was improved. The reason is more stripping steam into the tower reduces hydrocarbon partial pressure which helps to vaporize or lift more diesel components from the heavies in the bottom. The second cause was lack of trays in the bottom section as the tower was designed for dealing with lighter feed than what it is handling now. A revamp project to add a few trays in the bottom section was scheduled for the turnaround.

Comment
A simple mass balance identifies a major yield loss.

Example 6.3 *Component Mass Balance Assessment*
A stripper column in a naphtha hydrotreating process unit needs to remove H_2S which is corrosive and could poison the catalyst in a downstream naphtha reforming unit. Another objective is to remove as much C_5 as possible from the stripper bottom which is the feed to the naphtha reforming unit.

The stripper was operated with these two objectives in mind. However, the lab test showed the C_5 component distribution. The stripper bottom contained 2 mol.% C_5 which exceeds the targeted C_5 removal from the bottom, which is undesirable. Two negative results were observed. As C_5 does not involve in the catalytic reforming reaction,

Table 6.3 Mass balance around a fractionation tower.

	Vol.% of feed (yield expected)	Vol.% of feed (yield produced)		Barrel per day (produced)		Barrel per day
	Feed	Distillate	Bottom	Distillate	Bottom	Feed
Naphtha	35.2	35.2	—	18 130	—	—
Kerosene	22.1	23.7	—	12 200	—	—
Diesel	23.6	19.3	—	9 968	—	—
Unconverted oil	19.1	—	21.8	—	9 968	—
Total	100	78.2	21.8	40 298	9 968	51 548

C_5 material not only occupied the space in reactor and hence reduced reaction throughput but also consumed extra heat in the feed heater for the reforming reactors.

Once the problem was identified, the process engineer discussed it with the control engineer who quickly changed the set point for the reboiling duty and reflux rate. With increased reboiling duty and consequently increased reflux rate, better fractionation in the top section and hence more C_5 was stripped out of the bottom. However, the reflux rate is controlled on minimum overflash so that no extra energy than necessary was consumed.

Comment

Therefore, the component mass balance could help determine the desirable locations for key components from separation. Improper separation could cost not only the energy but also have negative effect on yields.

6.5.2 Heat Balance Assessment

A major part of tower heat balance is checking reflux rate and temperature, which determines both the condenser and reboiler duties. It is important to measure reflux temperature as it affects the heat balance significantly when the reflux is subcooled. Example 6.2 calculations above demonstrated the effect of reflux subcooling.

The common problem with measuring reflux rate is that reflux meters are typically set at startup and then never adjusted again. Therefore, the reflux flow rate is typically not reliable. The reflux ratio is checked and monitored as an important operating parameter but the absolute value of the reflux rate is rarely monitored.

However, to have a correct heat balance, the reflux flow meter must be checked and calibrated in order to achieve at least ±5% closure of heat balance ([total heat input − total heat output]/total heat input). Only with this accuracy of heat balance, tray efficiency or packing HETP can be accurately determined (Summers 2009).

6.6 Tower Efficiency Assessment

A benchmarking efficiency for a tower should be established. By comparing the actual efficiency with the benchmark efficiency, it is important to obtain a trend of efficiency over time and see the sign of poor separation. A good efficiency indicates a healthy operation of the tower in general while a poor efficiency identifies signs of unstable operation, which warrants a tower rating assessment to reveal root causes of abnormality and thus determine actions for corrections.

Calculation of distillation efficiency requires process simulation. From the number of theoretical stages simulated for each section, column section and overall efficiency can be determined, respectively. For a simple tower with two products, one from top and one from bottom together with one condenser and one reboiler, the separation efficiency can be calculated via

$$\eta_o = \frac{N_{eq}}{N_{act}} \quad (6.5)$$

As an example for illustration, McCabe–Thiele diagram (McCabe and Thiele 1925) in Figure 6.3 indicates 12 actual stages required in comparison with eight theoretical

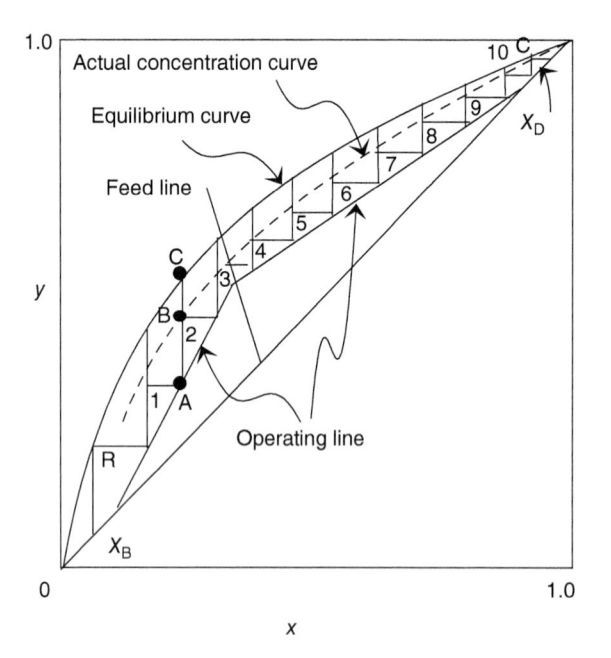

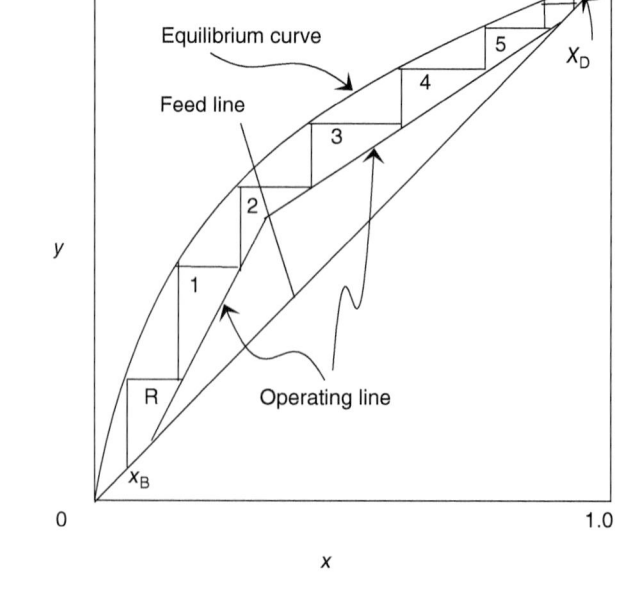

Figure 6.3 McCabe–Thiele diagram.

stages in the tower. Partial condenser and partial reboiler are counted in both the theoretical stages and actual stages. Thus, the overall tower efficiency is 72%.

For a complex column, Eq. (6.5) cannot give the right answer as it is only applied to each section. Instead, fractionation correlation plots by O'Connell (1946) is widely used for overall tower efficiency, which is the standard of the industry for industrial tower efficiency. Lockett converted the O'Connell's plots into an equation form as

$$\eta_o = 0.492(\mu\alpha)^{-0.245} \qquad (6.6)$$

where μ is the viscosity of liquid and α is the relative volatility which are calculated based on average temperature and pressure between column top and bottom.

Thus, O'Connell correlation states that higher viscosity leads to lower efficiency due to greater liquid-phase resistance while higher relative volatility also reduces efficiency as it increases the significance of the liquid-phase resistance.

However, O'Connell's correlation plots and Lockett's Eq. (6.6) are developed based on efficiency data points for industrial towers and do not reveal fundamental reasons for what to do, why, and how in order to improve efficiency.

Thus, the natural question is: What things affect tower separation efficiency? Mainly, there are three kinds of parameters. The first is flow properties such as relative volatility and viscosity, which are intrinsic. The second one is tray layout such as tray deck type (sieve or valve), flow path length, tower diameter, tray spacing, and weir length, which affect the liquid and vapor distribution and flow regime and are determined by design. The third one is process conditions such as tower feed rate and reboiling duty. The common effect of these parameters is in impacting the balance between vapor and liquid loadings.

Efficiency varies very little in the region of stable operation while efficiency falls off the cliff outside the feasible region. Figure 6.4 shows a typical trend of tower efficiency dependent on the balance of vapor and liquid rates. In the middle of the efficiency curve corresponding to stable operation, there is a relative flat region although with marginal variation. Trays with good turndown features such as valve tray comparing with sieve tray have wider flat or stable operating region. On the either side of the curve, efficiency drops off dramatically. Efficiency declines under low feed rate corresponding to turn down operation and falls off the cliff when dumping occurs. On the hand, efficiency reduces at excessive entrainment and thus plummets when spray or flooding happens. Optimization in design and operation tends to push the tower toward the boundary of stable operation. Understanding of these controlling mechanisms can

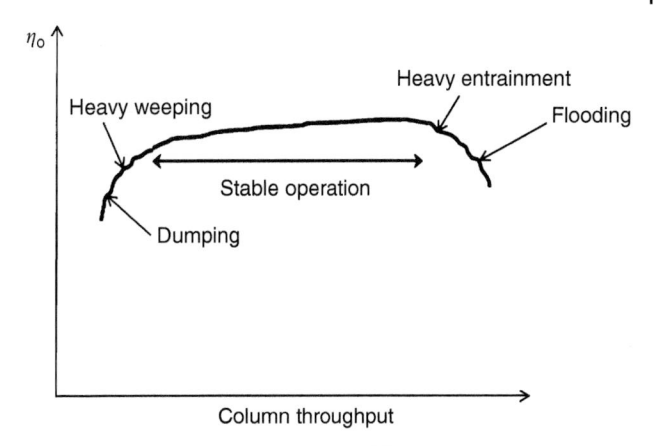

Figure 6.4 A typical trend of tower efficiency.

shed insight into how to optimize tower design and operation while achieving stable operation.

Efficiency assessment can detect the section(s) with poor efficiency from which root causes can be found. A section or whole column could be flooded due to too high vapor or liquid loading. This could be caused by changes in the conditions of feed and products in terms of rates, compositions, and product specifications. It could be also caused by too high reboiler duty, high feed temperature, and low column pressure, or combination of these. For the case of changes in feed compositions, it could be traced back to processing issues in upstream.

From retrofit point of view when dealing with too high liquid loading, enhanced capacity trays could be used such as UOP MD/ECMD or Shell HiFi or Sulzer high capacity or Koch-Glitsch high-performance trays. For the case of too low liquid loading, packing could be the cure. For too high vapor loading, valve trays could be considered. Tray damage could also cause malfunction of a column operation. In whichever cases, identification of low fractionation efficiency triggers the search for the root causes and solutions.

Example 6.4 Overall Efficiency Estimate Using O'Connell's Correlation

This example (Wankat 1988) was revised. A sieve tray distillation column is separating a feed that is 50 mol.% n-hexane and 50 mol.% n-heptane. The feed is a saturated liquid. Tray spacing is 24 inch. The average column pressure is 114.7 psia. Distillate composition is 99.9 mol.% of n-hexane and 0.1 mol.% of n-heptane. Feed rate is 1000 lb mol/h. Internal reflux ratio L/V is 0.8. The column has a total reboiler and total condenser. Estimate the overall efficiency.

Solution

To apply Eq. (6.6), we need to estimate α and μ at the average temperature and pressure of the column. The column temperature can be obtained from the modified DePriester chart (Dadyburjor 1978) as shown below.

x_{C6}	0.000	0.3	0.398	0.500	1.000
y_{C6}	0.000	0.545	0.609	0.700	0.700
T °C	98.4	85.0	83.7	80.0	80.0

Relative volatility is $\alpha = (y/x)/[(1-y)/(1-x)]$. The average temperature can be estimated in several ways:

Average temperature $T = (98.4 + 69.0)/2 = 83.7$; x and y at $T = 83.7\,°C$ can be interpolated based on the above table. Thus, $\alpha = 2.36$ at $T = 83.7\,°C$. If average at $x = 0.5$, $T = 80$, $\alpha = 2.33$ at $T = 80\,°C$. Not much difference. Use $\alpha = 2.35$ corresponding to $T = 82.5\,°C$.

The liquid viscosity of the feed can be estimated (Reid et al. 1977) from

$$\ln \mu_{mix} = x_1 \ln \mu_1 + x_2 \ln \mu_2 \qquad (6.7)$$

The pure component viscosities can be estimated from

$$\log_{10} \mu = A\left(\frac{1}{T} - \frac{1}{B}\right) \qquad (6.8)$$

where

μ is in cP
T in K (Reid et al. 1977).
nC_6: $A = 362.79$; $B = 207.08$; nC_7: $A = 436.73$; $B = 232.53$.

The above two equations for μ_{mix} and μ give $\mu_{C6} = 0.186$, $\mu_{C7} = 0.224$, and $\mu_{mix} = 0.204$. Thus, $\alpha\mu_{mix} = 0.479$. Applying Eq. (6.6) gives $\eta_o = 58.9\%$ which agrees well with $\eta_o = 59.0\%$ obtained from O'Connell correlation plots. The lower value should be used for conservative purpose.

6.7 Operating Profile Assessment

Another simple assessment method is based on tower profiles generated from simulation, which include flow, temperature, pressure, and composition profiles. What can we learn from these profiles? In a nut shell, tower profiles can allow us to observe what is going on inside the tower, like X-ray photos by vision.

The flow profile shows internal liquid and vapor flows across the column, which can vary from tray to tray with sudden change at feed stage and withdraw stages. In general, in the rectifying section above the feed stage up to condenser, the vapor flow is higher than the liquid flow while it is opposite in the stripping section below the feed stage. As part of the flow estimates, the feed is flashed at the feed tray conditions. The importance of flow estimates is not so much the absolute values but the ratio of L/V which determines the internal reflux and the slope of the operating line. This behavior can be observed in Figure 6.5.

The temperature and pressure profiles show a general trend of monotonic reduction in both temperature and pressure from the reboiler to the condenser. Figure 6.6 shows an example temperature profile where the steep parts of the curve would be where light and heavy keys are significantly separating. In some cases, temperature profiles feature plateaus in certain trays where little temperature change occur. In these flat regions, it indicates virtually no separation taking place although non-key components are being distributed. When there are a large number of stages, these plateaus can be more self-evident. These stages represent the pinch region where the operating line is very close to the equilibrium curve.

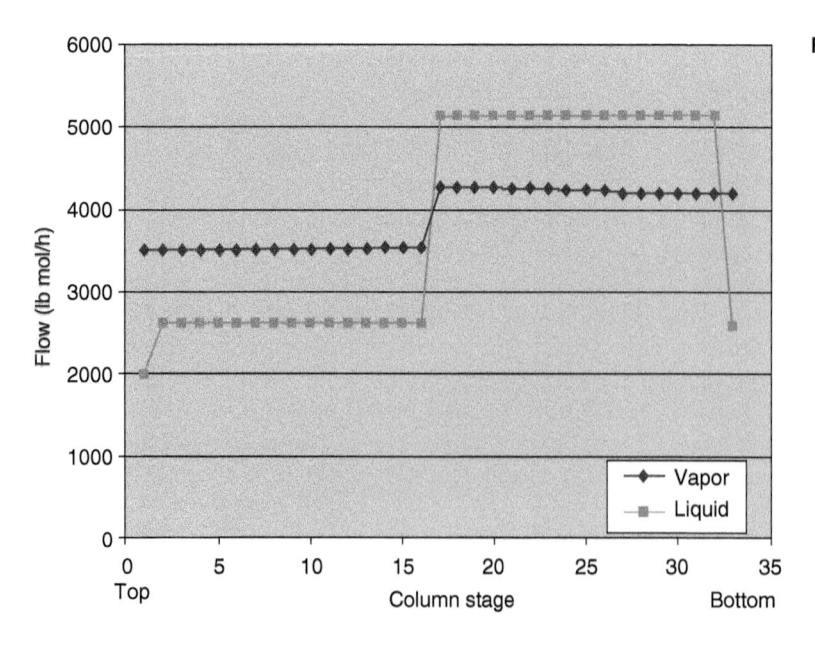

Figure 6.5 Example column flow profile.

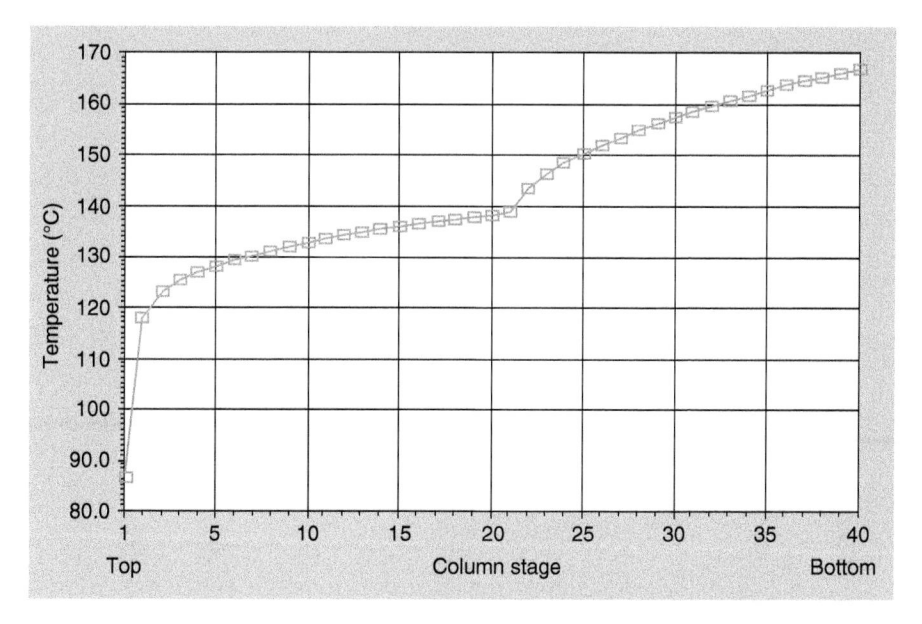

Figure 6.6 Example column temperature profile for benzene–toluene separation.

In this pinch region, the ratio of relative volatility between key components is very small corresponding to a difficult fractionation.

Obviously, when a column or a section is flooded or dumping, a flat temperature profile can be obtained since there is no fractionation taking place.

A tower *pressure drop profile* can also indicate abnormal operation. A too low pressure drop across a tower or a section indicates potential dumping or flow channeling while a too high pressure drop manifests flooding operation.

Lieberman (1991) recommends a simpler method for assessing flooding condition based on his operation experience. Lieberman's method indicates occurrence of flooding when

$$\frac{\Delta P}{Sp_L N_T H_S} \geq 22-25\% \qquad (6.9)$$

where

ΔP, inches of water, is overall column pressure drop between column overhead and reboiler outlet or section pressure drop
Sp_L is average specific gravity of liquid on tray
N_T is the number of trays
H_S is tray spacing (inch).

To troubleshooting column pressure problems, Kister and Hanson (2015) provided a simple retrofit method for column pressure control. A survey by Kister (2006) identified the poorly designed hot-vapor bypass control as the most troublesome pressure and condenser control method, which causes unstable pressure within column. Unstable pressure results in an unsteady column as pressure affects column vaporization, condensation, temperature, volatility, etc.

The simple method which Kister and Hanson proposed is to add a throttle valve in the condensate outline. Ideally, the valve should have a pressure drop larger than 3–4 psi and should be installed more than 10 inch in diameters away from the reflux drum liquid inlet to minimize turbulence at the drum inlet. Another practice is to install a horizontal baffle in front of the vapor nozzle to disperse the vapor flow and prevent it from impinging on the liquid surface upon intensification.

The *composition profile* can reveal the details of separation taking place inside the tower. For component balances, it is highly important to know the composition of the feed and product streams around the tower. Samples are taken daily on most towers and analyzed in the plant's local laboratory. However, these laboratories are typically setup to measure for certain key compounds that can contaminate the final product and do not have the capability of measuring the full spectrum of a multicomponent columns feed. Therefore, unless the tower of interest has only a few components in the feed, a full component balance will typically need special laboratory assistance which more than likely will come from outside the local plant. One needs to understand the units of measurements for compositions by the laboratories. Frequently they will not provide the units such as molar, volume, or weight percentages.

This example is about separation of toluene (light key) from ethylbenzene (heavy key). Figure 6.7 shows the liquid mole fraction for these four components. Stage 21 is the feed stage.

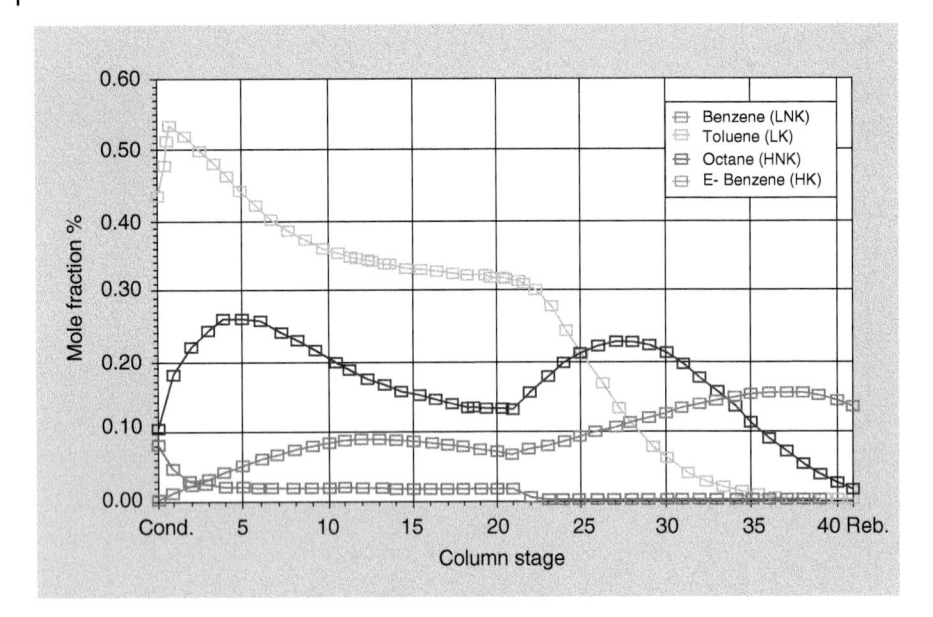

Figure 6.7 Example composition profile for toluene–ethylbenzene separation.

Let us follow the toluene mole fraction curve, which is more obvious. Concentration of LK toluene increases through the tower in a monotonous manner until it peaks at the top of tower but dips at the receiver. This is because benzene, the non-key light component, is the most volatile component and peaks at the receiver; but should not be in the bottom.

On the other hand, the concentration of the heavy key ethylbenzene enriches toward the bottom of the tower monotonously through the tower and peaks a few stages above the reboiler because it is the less volatile component.

The concentration profile for the non-heavy key octane is the most confusing one as it goes up and down throughout the tower with two maxima because of the competition with other compounds on those trays and how they concentrate. From the feed Stages 21–28, the separation takes place between the LK toluene versus HK ethylbenzene and HNK octane. As the result, the concentrations of both ethylbenzene and octane increase while toluene concentration reduces. From Stages 29–40 where LK toluene concentration decreases to very low, the separation takes place mainly between HK ethylbenzene and HNK octane. In this section, octane concentration reduces steeply and ethylbenzene concentration increases. In summary, octane concentration increases from the feed stage until it peaks at Stage 28 and then starts to decrease toward the bottom. This creates the first maxima in octane concentration.

From Stages 15–5, LK toluene is separated from HK ethylbenzene. At the same time, the separation between HK ethylbenzene and HNK octane occurs where ethylbenzene concentration goes down as octane concentration steps up until octane peaks at Stage 5. Above Stage 5 toward the top of the tower, the separation takes place between LK toluene against both HK ethylbenzene and HNK octane where toluene concentration climbs and peaks at the top. In contrast, HNK octane concentration plummets and HK ethylbenzene concentration reduces to distinction. In summary, octane concentration increases from Stage 15 until it peaks at Stage 5 and then it reduces sharply. This creates the second maxima in octane concentration.

6.8 Tower Rating Assessment

Tower simulation is only mimicking the current operation and can provide the vapor and liquid loadings as the basis for rating assessment while rating assessment will tell how the tower is operating under current conditions in relation to the feasible operating window.

Tower rating is applied to assess the effects for changing process conditions, in particular feed rate change, on tower performance. It can also be applied when an existing tower is considered to be used for a new service. Briefly, a tower rating can answer three questions:

1) Can the tower operate with increased throughput or with changing process conditions within the feasible operating window? The calculations for the operating window were shown earlier in this chapter. With the internal L/V from simulation, we can determine the current operating point in relation to the operating window.
2) What is the hydraulic performance of the tower under new conditions? The operating pressure drops can be

calculated based on hydraulic calculations discussed in Chapter 12 (Zhu 2014).
3) What are the limiting factors of the tower under new conditions? The limitations could come from the size of the tower, tray spacing, downcomer geometry, etc.

The criteria can be established as necessary and sufficient conditions for the suitability of an existing tower for changing conditions or new services:

1) The operating point must fall within the operating window. For example, the actual vapor and liquid rates should be less than the maximum limits.
2) Operating pressure drop must be less than allowable pressure drops.

When these two conditions are fulfilled, an existing tower is suitable for different conditions for which it is rated. When the process conditions undergo significant changes, the rating assessment should be performed to make sure the tower can perform the task satisfactorily under new conditions. Otherwise, either operating conditions should be altered or modifications to the existing tower need to be implemented.

Why would we want to use a tower for a service that it was not designed for? The main reason is that it is less expensive and quick to modify an existing tower than to purchase a new one. It is rare that the existing tower provides a perfect fit to a new service. But, engineers are keen to take the challenge of modifying existing equipment as it is their second nature of seeking the most economical solution with quick turnaround.

Tower rating assessment can be conducted using tower evaluation software by vendors (e.g. Sulzer's Sulcol tool). A tower simulation provides basic data required for rating. In generating data from simulation to rating assessment, a tower is divided into sections and the stage with highest vapor loading in each section is selected to represent this section as this tray is the most constrained tray for the whole section. Thus, the data for this stage is entered into the tower rating software. The input data include (i) vapor and liquid loadings and physical properties for both vapor and liquid, which are obtained from simulation; (ii) tower geometry layout (e.g. tower diameter). Execution of the rating software will give percentages of tray flooding, downcomer backup and dumping, vapor maximum capacity, liquid maximum capacity, froth/spray transition, pressure drop, dry tray pressure drop, downcomer velocity, and weir loading.

The rating assessment software will indicate the current operating point in relation to operating window and thus reveal what operating limits the tower may have gone beyond, which are the root causes for sudden decrease of tower efficiency. Some commercial rating tools can generate operating window or performance diagrams (Summers 2004). Performance diagrams, if plotted with vapor and liquid volume loadings, can represent tray performance independent of operating pressure and composition.

6.9 Guidelines for Existing Columns

The guidelines discussed below are recommended by Wankat (1988) and the following things in order of increasing costs can be explored when the existing column cannot produce desired product purities:

- Find out whether the product specifications can be relaxed. A purity of 99.5% is much easier to obtain than 99.99%.
- Increase reflux rate and see if it can meet product specifications. Remember to check if column vapor capacity is sufficient as flooding could be an issue with an increased reflux rate. Also check if existing reboiler and condenser are large enough. If the tower can make purer products, usually reducing reflux rate can make product back to specification, which also reduces operating cost.
- Change the feed temperature. This change may require altering of feed stage and could result in an optimal feed location.
- Will a new feed stage at the optimal stage allow meeting product specification?
- Consider replacing the existing column internals with more efficient or tighter spaced trays or new packing. This is relatively expensive but is cheaper than a new column.
- Add a stub column to increase the total number of trays.

If the column vapor loading is more than the limit implying the existing column diameter is not large enough, engineers can consider:

- Operating at a reduced reflux ratio, which reduces vapor loading; but this could make it difficult to meet product specifications.
- Operating at a higher pressure, which increases vapor density. Need to check if the column can operate at the increased pressure.
- Using two columns in parallel.
- Replacing the existing downcomers with large ones.
- Replacing the trays or packing with higher capacity ones.

On the other hand, if the column diameter is too large, vapor velocities will be too low. Trays will operate too lower efficiency and in severe cases they may not operate since liquid may dump through the holes. Engineers can consider:

- Decrease column pressure to decrease vapor density and hence vapor velocity.
- Increase reflux ratio.
- Recycle some distillate and bottom products.

Using existing columns for new services often requires creative solutions. Thus, it can be both challenging and fun; they are also often assigned to engineers just out of school but under supervision of experienced engineers.

Nomenclature

C_p	specific heat
F	feed rate
H_S	tray spacing
M	mass flow
N_T	total number of trays
N_{eq}	number of theoretical trays
N_{act}	number of actual trays
ΔP	pressure drop
q	latent heat
Q	heat content
R_f	tower reflux ratio
Sp_L	average specific gravity of liquid on tray
t	time
T	temperature
ΔT	temperature difference

Greek Letters

α	relative volatility
μ	liquid viscosity
η_o	overall tower efficiency

References

Dadyburjor, D.B. (1978). SI units for distribution coefficients. *Chemical Engineering Progress* 74: 85. AIChE.

Kenny, W.F. (1984). *Energy Conservation in the Process Industries*. Academic Press Inc.

Kister, H.Z. (2006). *Distillation Troubleshooting*. AIChE-Wiley.

Kister, H.Z. and Hanson, D.W. (2015). Control column pressure via hot-vapor bypass. *Chemical Engineering Progress* 111: 35–45.

Lieberman, N. (1991). *Troubleshooting Process Operation*, 3e. PennWell.

McCabe, W.L. and Thiele, E.W. (1925). Graphic design of fractionation columns. *Industrial and Engineering Chemistry* 17: 605–611.

O'Connell, H.E. (1946). Plate efficiency of fractionating columns and absorbers. *Transactions of the American Institute of Chemical Engineers* 42: 741.

Reid, R.C., Prausnitz, J.M., and Sherwood, T.K. (1977). *The Properties f Gases and Liquids*, 3e. New York: McGraw-Hill.

Summers, D.R. (2004). Performance diagrams: all your tray hydraulics in one place. *AIChE Annual Meeting: Distillation Symposium*, Austin (7–12 November 2004), paper 228f.

Summers, D.R. (2009). How to properly evaluate and document tower performance. *AIChE Spring Meeting*, Florida (27 April 2009).

Wankat, P.C. (1988). *Equilibrium Staged Separations*. Upper Saddle River: PTR Prentice Hall.

Zhu, X.X. (2014). *Energy Optimization for the Process Industries*. Wiley/AIChE.

7

Heat Exchanger Assessment

7.1 Introduction

Often exchangers do not perform as they should and their performance deviates from optimum. Sometimes, they do not accomplish what they are capable of and other times they are asked to perform what they are not capable of. The primary purpose of heat exchanger assessment is to identify the root causes if it is due to poor design, or excessive fouling, or mechanical failure, and determine the required actions to improve the performance.

This chapter will provide the basic understanding of heat exchange assessment supported with examples considering whether the exchanger is designed correctly, evaluation of operating performance, evaluation of fouling, and its effect on heat transfer and pressure drop. On this basis, methods for improving exchanger performance are provided using examples associated with different application scenarios. The methods discussed in this chapter focus on shell-and-tube exchangers as they are the most commonly used in the process industry although the assessment methodology can be applied to other types of heat exchangers. Detailed assessment of heat exchange performance may be conducted using commercial software.

7.2 Basic Calculations

As it is well known, the primary equation for heat exchange between two fluids is the Fourier equation expressed as:

$$Q = UA\Delta T_M \quad (7.1)$$

where

Q = heat duty (MMBtu/h)
A = heat transfer surface area (ft^2)
U = overall heat transfer coefficient [Btu/(ft^2 °F h)]
ΔT_M = effective mean temperature difference (EMTD) (°F)

Let us define U-value first based on Figure 7.1 where h_i and h_o are film coefficients for fluids inside and outside of the tube and they can be calculated from the physical form of heat exchanger, physical properties of streams, and process conditions of streams. Thus, clean overall heat transfer coefficient (U_C) can be determined based on

$$\frac{1}{U_C} = \frac{1}{h_o} + \frac{1}{h_i}\left(\frac{A_o}{A_i}\right) + r_w \quad (7.2)$$

where

r_w is the conductive resistance of the tube wall
A_o and A_i are outside and inside tube surface area with subscripts i and o denoting inside and outside of tube

In reality, heat exchangers operate under fouled conditions with dirt, scale, and particulates deposit on the inside and outside of tubes. Allowance for the fouling must be given in calculating overall heat transfer coefficient. The graphical description of fouling resistances (R_o, R_i) and film coefficients (h_o, h_i) inside and outside of tube is provided in Figure 7.1. Conceptually, R_i and h_i are equivalent to R_t and h_t (t for tube side) while R_o and h_o are for R_s and h_s (s for shell side).

The overall fouling resistance is then defined as

$$R_f = R_o + R_i\left(\frac{A_o}{A_i}\right) \quad (7.3)$$

By adding the overall fouling resistance to U_C, actual U_A is defined as

$$\frac{1}{U_A} = \frac{1}{U_C} + R_f \quad (7.4)$$

More detailed discussions for U-values are provided later in this chapter. Now, let us turn our attention on ΔT_M or EMTD. Several temperature differences can be used to calculate ΔT_M including inlet temperature

---

Efficient Petrochemical Processes: Technology, Design and Operation, First Edition.
Frank (Xin X.) Zhu, James A. Johnson, David W. Ablin, and Gregory A. Ernst.
© 2020 John Wiley & Sons, Inc. Published 2020 by John Wiley & Sons, Inc.

7 Heat Exchanger Assessment

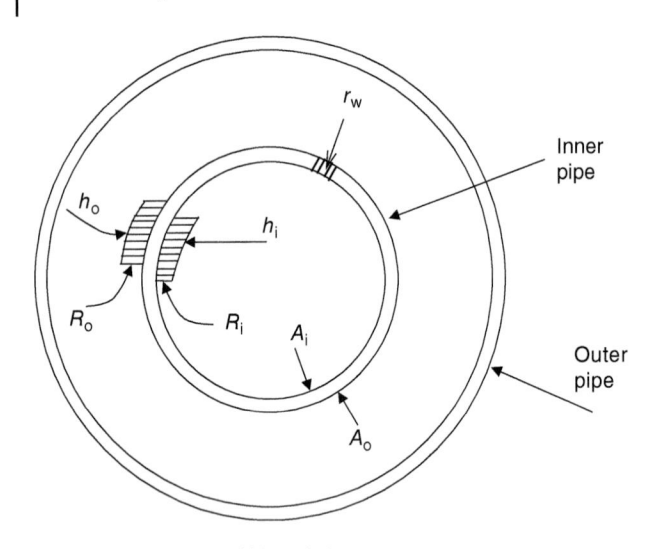

Figure 7.1 Location of h's and R's.

difference, arithmetic temperature difference, and logarithmic mean temperature difference (LMTD). Figure 7.2 is used for illustration.

Inlet temperature difference can be expressed as

$$\Delta T_1 = T_1 - t_2 \quad \text{(for countercurrent)} \quad (7.5a)$$

$$\Delta T_1 = T_1 - t_1 \quad \text{(for cocurrent)} \quad (7.5b)$$

This temperature difference could lead to a gross error in estimating true temperature difference over the entire pipe length.

Arithmetic mean temperature difference is defined as

$$\Delta T_A = \frac{\Delta T_1 + \Delta T_2}{2}$$
$$= \frac{(T_1 - t_2) + (T_2 - t_1)}{2} \quad \text{(for countercurrent)} \quad (7.6a)$$

$$\Delta T_A = \frac{\Delta T_1 + \Delta T_2}{2}$$
$$= \frac{(T_1 - t_1) + (T_2 - t_2)}{2} \quad \text{(for cocurrent)} \quad (7.6b)$$

This temperature difference could give erroneous estimate of true temperature difference when ΔT_1 (hot end approach) and ΔT_2 (cold end approach) differ significantly.

The LMTD is defined as

$$\Delta T_{LM} = \frac{\Delta T_1 - \Delta T_2}{\ln(\Delta T_1 / \Delta T_2)} \quad (7.7)$$

ΔT_{LM} represents a true temperature difference for a perfect countercurrent as well as cocurrent heat exchange (Figure 7.2).

At this point, the first question is: Why the countercurrent pattern is widely adopted in shell-and-tube exchangers? The answer is that the LMTD for countercurrent is always greater than the cocurrent LMTD. An example corresponding to Figure 7.2 is shown below.

Countercurrent			**Cocurrent**		
Hot fluid	Cold fluid		Hot fluid	Cold fluid	
$T_1 = 350°$	$t_2 = 230°$	$\Delta T_1 = 120°$	$T_1 = 350°$	$t_1 = 150°$	$\Delta T = 200°$
$T_2 = 250°$	$t_1 = 150°$	$\Delta T_2 = 100°$	$T_2 = 250°$	$t_2 = 230°$	$\Delta T_2 = 20°$
		LMTD = 109.7°			LMTD = 78.2°

The second question is: What should be done if a heat exchange is not a perfect countercurrent? In fact, the flow pattern in most shell and tube exchangers is a mixture of cocurrent, countercurrent, and crossflow. In these cases, EMTD ≤ LMTD. Thus, a LMTD correction factor F_t must be introduced,

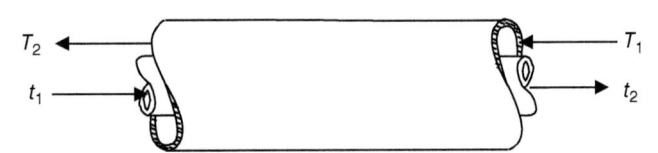

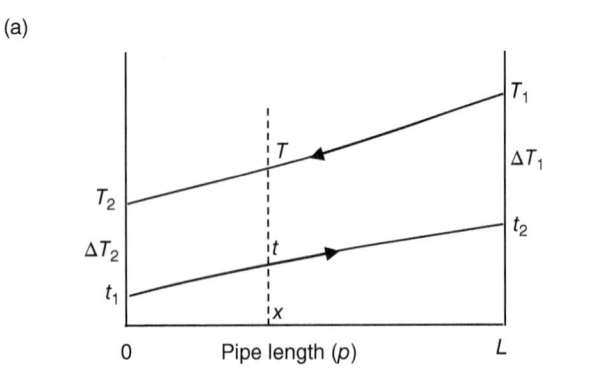

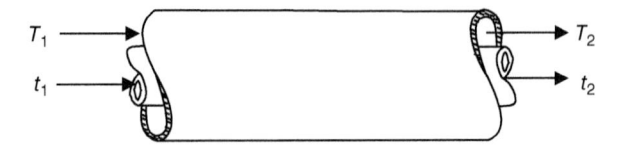

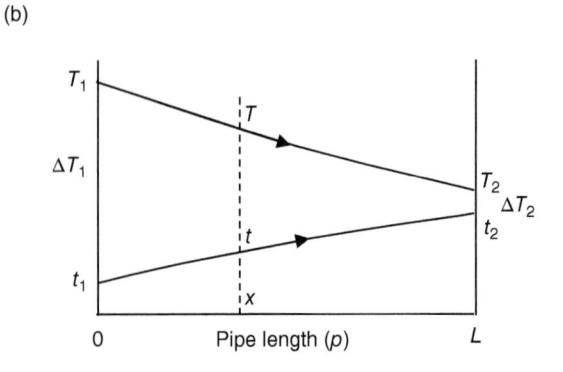

Figure 7.2 (a) Countercurrent flow. (b) Cocurrent flow.

$$\text{EMTD} = F_t \times \text{LMTD} \tag{7.8}$$

$F_t = 1$ for a true countercurrent heat exchange; otherwise, $F_t < 1$.

As a short summary, EMTD is obtained by calculating LMTD based on Eq. (7.7) first and then applying F_t to account for non-perfect countercurrent flow.

F_t can be obtained via equations or charts (Shah and Sekulić 2003). For example, F_t for 1–2 heat exchangers can be numerically calculated by

$$F_t = \frac{\sqrt{R^2+1}\ln\left[(1-P)/(1-R\times P)\right]}{(R-1)\ln\dfrac{2-P\left(R+1-\sqrt{R^2+1}\right)}{2-P\left(R+1+\sqrt{R^2+1}\right)}} \tag{7.9}$$

where

P is temperature efficiency
R is the ratio of heat flow, which are defined as

$$P = \frac{t_2 - t_1}{T_1 - t_1} \tag{7.10}$$

$$R = \frac{T_1 - T_2}{t_2 - t_1} = \frac{m \times \text{cp}}{M \times \text{Cp}} \tag{7.11}$$

F_t for 1–2 type (one shell pass and two tube passes) exchangers can also be found in the chart as shown in Figure 7.3. It can be observed from the figure that F_t values drop off rapidly below 0.8. Consequently, if a design indicates a F_t less than 0.8, it probably needs to redesign to get a better approximation of countercurrent flow and thus higher F_t value. Different F_t charts are available for each exchanger layout (TEMA 1–2, 1–4, etc.).

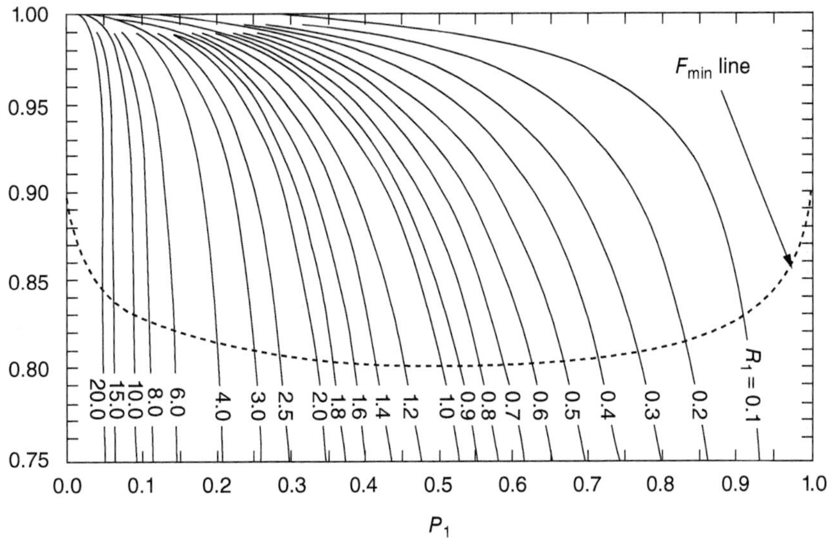

7.3 Understand Performance Criterion: U-Values

The critical question for operation assessment is: What is the "performance" indicator for heat exchanger? As a heat exchanger is used to transfer heat, someone may naturally consider heat exchanger duty as the performance indicator. For verification, let us look at an example of a reaction effluent cooler. The operating data were obtained which are shown against the design data (Table 7.1). Operating temperatures can be measured from instrumentation. Then, LMTD is calculated by Eq. (7.7) and F_t factor is obtained from F_t charts. Then, UA is calculated via Eq. (7.1) and U is derived for a given surface area A of the cooler. The calculation results for operating data are shown in Table 7.2. The design data were obtained from the exchanger data sheet.

As can be observed in Table 7.2, the heat exchanger duty in operation is a little higher and the temperature changes are similar for both operation and design. What

Table 7.1 Gathered data for a reaction air cooler.

	Design	Operation
Q (MMBtu/h)	16.3	18.0
$M_{\text{(effluent)}}$ (lb/h)	154 447	162 919
T_1 (°F)	296.6	341.6
T_2 (°F)	105.8	149.0
t_1 (°F)	89.6	81.0
t_2 (°F)	104.0	95.0
$\Delta T(T_1 - T_2)$ (°F)	190.8	192.6
$\Delta t(t_2 - t_1)$ (°F)	14.4	14.0

Figure 7.3 F_t factor for 1–2 TEMA E shell-and-tube exchangers. *Source:* Shah and Sekulić (2003), reprint with permission by John Wiley & Sons.

Table 7.2 Calculation results for a reaction air cooler.

	Design	Operation
Q (MMBtu/h)	16.3	18.0
$M_{(effluent)}$ (lb/h)	154 447	162 919
T_1 (°F)	296.6	341.6
T_2 (°F)	105.8	149.0
t_1 (°F)	89.6	81.0
t_2 (°F)	104.0	95.0
ΔT_1 (°F)	192.6	246.6
ΔT_2 (°F)	16.2	68.0
ΔT_{1m} (°F)	71.3	138.7
F_t	0.85	0.97
UA (Btu/°F-h)	269 896	133 759
A (ft²)	5167	5167
U (Btu/ft²-°F-h)	52	26
$U_{operation}/U_{design}$		0.50

do you think of the performance of this exchanger in operation? If based on the exchanger duty, we could conclude it is performing fine or at least not worse than design performance. However, if comparing overall heat transfer coefficient, surprisingly, the operation U-value is only half of the design U-value, although the heat duty in operation is 10% higher. If the operation U-value could maintain similar to design U-value, the heat duty could be increased much higher than 10%!

This example concludes that U-value is a true performance indicator for heat exchanger under any process conditions. *The higher the U-value, the better performance that a heat exchanger achieves.*

Clearly, good understanding of U-value is of paramount importance for appropriate assessment of heat exchanger performance as it is the most important characteristic of heat exchanger representing its heat transfer capability. In view of the fact that many engineers are confused about the terminologies related to U, it is essential to get the basic understanding right before diverging into details of assessment methods.

7.3.1 Required U-Value (U_R)

The need of heat exchanger is to satisfy process requirement in terms of heat duty (Q) and temperatures (LMTD or ΔT_{LM}). Thus, heat exchanger is designed to have a certain surface area (A) in order to fulfill the process requirement. Based on process temperature requirement for reaction and fractionation, a certain amount of heat duty must be transferred. Under the basis of heat transfer duty and process temperatures, required U-value can be calculated from the Fourier Eq. (7.1):

$$U_R = \frac{Q}{AF_t\Delta T_{LM}} \tag{7.1}$$

7.3.2 Clean U-Value (U_C)

Independent of the required U based on thermodynamics stated as in Eq. (7.1), U-value can be calculated based on transport considerations without taking into account of fouling resistances. In other words, transport-based U is a function of film coefficients (h_t for tube side and h_s for shell side in Btu/h-ft²-°F) as expressed as in Eq. (7.2):

$$\frac{1}{U_C} = \frac{1}{h_o} + \frac{1}{h_i}\left(\frac{A_o}{A_i}\right) + r_w \tag{7.2}$$

This U-value is called as clean U-value as fouling resistances (R_i, R_o) are not taken into account in Eq. (7.2). The film coefficients (h_t for tube side and h_s for shell side) can be calculated based on fluids' physical properties and geometry of heat exchanger. For example, for U-tube exchangers with streams all liquid or all vapor (no boiling and condensing), the correlation (Dittus and Boelter 1930) is used to estimate the tube-side Nusselt number, Nu_t, and then tube-side film coefficient, h_t:

$$Nu_t = 0.027\left(\frac{C_p\mu}{k}\right)^{\frac{1}{3}}\left(\frac{\rho u d_i}{\mu}\right)^{0.8} \tag{7.12}$$

$$h_\mathrm{t} = \frac{k}{d_i}\mathrm{Nu}_\mathrm{t} = 0.027 \frac{\left(C_\mathrm{p} k^2\right)^{\frac{1}{3}} (\rho u)^{0.8}}{d_i^{0.2} \mu^{\frac{7}{15}}} \qquad (7.13)$$

where

C_p = fluid heat capacity [Btu/(lb °F)]
d_i = inner diameter of the tube (ft)
k = fluid thermal conductivity [Btu/(h ft °F)]
u = fluid velocity (ft/h)
ρ = fluid density (lb/ft³)
μ = fluid viscosity [lb/(ft h)]

Equation (7.13) states that physical properties of tube-side stream (namely conductivity k, specific heat capacity Cp) and mass velocity u have positive effect on tube-side film coefficient h_t. In contrast, viscosity μ and tube inside diameter d_i have negative effect.

The Kern's correlation (1950) is used to estimate the shell-side Nusselt umber, Nu_S, and then shell-side film coefficient, h_s:

$$\mathrm{Nu}_\mathrm{S} = 0.36 \left(\frac{C_\mathrm{p}\mu}{k}\right)^{\frac{1}{3}} \left(\frac{D_\mathrm{e}\rho u}{\mu}\right)^{0.55} \left(\frac{\mu}{\mu_\mathrm{w}}\right)^{0.14} \qquad (7.14)$$

$$h_\mathrm{S} = \frac{k}{D_\mathrm{e}} \mathrm{Nu}_\mathrm{S} = 0.36 \frac{\left(C_\mathrm{p} k^2\right)^{\frac{1}{3}} (\rho u)^{0.55}}{D_\mathrm{e}^{0.45} \mu^{0.08} \mu_\mathrm{w}^{0.14}} \qquad (7.15)$$

where

$$D_\mathrm{e} = \frac{4\left(p^2 - \frac{\pi d_\mathrm{o}^2}{4}\right)}{\pi d_\mathrm{o}} \quad \text{(for spare tube pitch)} \qquad (7.16)$$

where

μ_w = water viscosity [lb/(ft h)]
D_e = shell-side equivalent diameter (ft)
d_o = outer diameter of the tube (ft)
p = tube pitch (ft)

Equation (7.15) states that physical properties of shell-side stream (namely conductivity k, specific heat capacity Cp), velocity u, and tube outside diameter d_o have positive effect on shell-side film coefficient h_t. In contrast, viscosity μ and tube pitch p have negative effect.

The above heat transfer equations provided the well-known observations: heat transfer coefficient on tube side is proportional to the 0.8 power of velocity, the 0.67 power of thermo conductivity, and the −0.47 power of viscosity,

$$h_\mathrm{t} \propto u^{0.8} \qquad (7.17a)$$

$$h_\mathrm{t} \propto k^{0.67} \qquad (7.17b)$$

$$h_\mathrm{t} \propto \mu^{-0.47} \qquad (7.17c)$$

That is the reason why cooling water has a very high heat transfer coefficient, followed by hydrocarbon and then hydrocarbon gases because of the values of thermo conductivities for these fluids. Hydrogen is an unusual gas due to its extremely high thermo conductivity (greater than that of hydrocarbon liquids). Thus, its heat transfer coefficient is toward the upper limit of the range for the hydrocarbon liquid. The heat transfer coefficients for hydrocarbon liquids vary in a large range due to the large variations in viscosity, from less than 1 cP for ethylene to more than 1000 cP for bitumen. Heat transfer coefficients for hydrocarbon gases are proportional to pressure because higher pressure generates higher gas density resulting in higher gas velocity.

7.3.3 Actual U-Value (U_A)

In reality, heat exchangers operate under fouled conditions with dirt, scale, and particulates deposit on the inside and outside of tubes. The overall fouling resistance is defined in Eq. (7.3) as

$$R_\mathrm{f} = R_\mathrm{o} + R_i \left(\frac{A_\mathrm{o}}{A_\mathrm{i}}\right) \qquad (7.3)$$

By adding the overall fouling resistance to U_C, actual U_A is defined in Eq. (7.4) as

$$\frac{1}{U_\mathrm{A}} = \frac{1}{U_\mathrm{C}} + R_\mathrm{f} \qquad (7.4)$$

Clearly, U_C is the heat transfer capability that the exchanger can deliver when no fouling is included while U_A takes into account of fouling resistances. U_A can be thought of predicted or expected overall coefficient for actual heat transfer including the design fouling resistances.

Fouling resistances for streams are based on the physical properties of the streams and the average fouling factors are documented in TEMA (2007). For illustration purpose, Table 7.3 shows typical overall fouling resistances

Table 7.3 Liquid fouling factors.

API gravity	R_f (ft²-h-°F/Btu)
>40	0.002
<40	0.003
<15	0.004
<5	0.005

for hydrocarbon liquids based on the API gravity of the streams.

The U-values should follow the order: $U_C \geq U_A \geq U_R$. The main reasons for the inequality are the practical considerations of fouling, process variations, as well as inaccuracy in physical properties estimates and heat transfer calculations.

7.3.4 Overdesign (OD$_A$)

For a heat exchanger to satisfy process requirement under changing process conditions, U_A must be greater than or equal to U_R. Actual overdesign or design margin can be defined as

$$\%OD_A = \left(\frac{U_A}{U_R} - 1\right) \times 100 \tag{7.18}$$

Overdesign is provided in the design stage beyond fouling factors in order to account for operation variations in fluid rates and properties as well as calculation inaccuracy for heat transfer and pressure drops. Some designers may use 5–10% overdesign for new heat exchangers if the designers have confidence in fluid properties and heat transfer calculation accuracy. Otherwise, 10–20% or higher overdesign might be used. In contrast, near-zero overdesign could be used for services with well-known fluid properties and accurate heat transfer calculations.

Statistically, heat exchangers are often designed with large overdesign intentionally because the designer wants to make sure it will satisfy process demand no matter whatever occurs in operation. There are several uncertain factors that the designer has to consider in design stage (Bennett et al. 2007).

Firstly, the uncertainty is the accuracy in estimating fouling resistances to reflect the actual fouling. Furthermore, fouling resistances are static values, which are used in computation. In reality, fouling is a dynamic mechanism. The designer uses overdesign to account for this fouling dynamics based on his/her experience or company's best practices so that the exchanger can still satisfy the process demand under more severe fouling scenarios than estimated fouling resistances. The second factor is variations in process conditions. In particular, increasing feed rate is common as companies want to generate additional revenue using existing equipment. The designer provides overdesign to accommodate operating scenarios with increased feed rate. Thirdly, the designer uses overdesign to account for the effects of inaccuracy in fluid properties and heat transfer calculations. These uncertainties become the basis for the designer to provide overdesign.

However, excessive overdesign can cause fouling and other problems with the exchanger. When too much overdesign in surface area is added, velocity reduces which makes it easier for fouling deposits to accumulate. In some cases, a temperature controlled bypass line may be required for critical services to avoid too much heat transfer than process requirements in the start of run. Bypass operation could enhance fouling as fluid velocity reduces.

7.3.5 Controlling Resistance

If the actual film coefficient of one side is much larger than the other, this side is referred to controlling side of resistance. In design and operation, special attention is devoted to this controlling resistance as any incremental decrease to this controlling resistance will greatly increase the overall U-value. On the other hand, incremental change to the non-controlling side film coefficient has very little effect on the overall U-value.

One way to minimize the adverse effect of controlling resistance is to use extended surface area to offset the effect. Another way is to increase the velocity on the controlling side. Furthermore, the most heavily fouling stream should be placed on the tube side for ease of cleaning. Use of fouling mitigation methods, such as fluid treatment, antifouling additive, and regular cleaning, to prolong the "clean" operation can help maintain high U-value.

7.4 Understand Fouling

Fouling is accumulation of undesirable materials as deposits on heat exchanger surfaces. Fouling deposits come in many different causes and forms. No matter what material is contained in heat exchanger fouling deposits, it leads to similar consequences, which are reduction in thermal performance and an increase in pressure drop. There are complex factors causing heat exchanger fouling such as physical and chemical properties of process streams, operating conditions, heat exchanger design, and operation. Since complex factors affect the choice of methods to reduce and prevent fouling, identification of root causes could derive more effective solutions.

7.4.1 Root Causes of Fouling

Since there are a great variety of fouling phenomena, it is useful to group them into six types of fouling mechanisms for better understanding (Melo et al. 1988):

1) Crystallization fouling: precipitation and deposition of dissolved salts, which are supersaturated at the heat transfer surface. Supersaturation may be caused by
   - Evaporation of solvent.
   - Cooling below the solubility limit for normal solubility (increasing solubility with decreasing temperature,

such as wax deposits, gas hydrates, and freezing of water/water vapor). The precipitation fouling occurs on the cold surface (i.e. by cooling the solution).
- Heating above the solubility limit for inverse solubility (increasing solubility with increasing temperature, such as calcium and magnesium salts). The precipitation of salt occurs with heating the solution.
- Mixing of streams with incompatible compositions.
- Variation of pH which affects the solubility of CO_2 in water.

2) Particulate fouling: accumulation of particles from heat exchanger working fluids (liquids and/or gaseous suspensions) on the heat transfer surface. Most often, this type of fouling involves deposition of corrosion products dispersed in fluids, clay and mineral particles in river water, suspended solids in cooling water, soot particles of incomplete combustion, magnetic particles in economizers, deposition of salts in desalination systems, deposition of dust particles in air coolers, particulates partially present in fire-side (gas-side) fouling of boiler, and so on. If particular fouling is of gravitational settling of relative large particles onto horizontal surfaces, this phenomenon is also called sedimentation fouling.

3) Chemical reaction fouling: deposit formation (fouling precursors) at the heat transfer surface by unwanted chemical reaction (such as polymerization, coking) within the process fluid, but heat transfer surface material itself does not involve in the chemical reaction. Thermal instability of chemical components such as asphaltenes and proteins can become fouling precursors. Usually, this type of fouling starts to form at local hot spots in a heat exchanger. It can occur over a wide temperature range from ambient to over 1000 °C (1832 °F) but is more pronounced at higher temperatures.

4) Corrosion fouling: the heat transfer surface itself reacts with chemical species present in the process fluid. Its trace materials are carried by the fluid in the exchanger, and it produces corrosion products that deposit on the surface. The thermal resistance of corrosion layers is low due to high thermal conductivity of oxides.

5) Biological fouling: deposition and growth of macro and micro organisms on the heat transfer surface. It usually happens in water streams.

6) Freezing fouling: it is also called solidification fouling which occurs due to freezing of a liquid or some of its constituents to form deposition of solids on a subcooled heat transfer surface. For example, formation of ice on a heat transfer surface during chilled water production or cooling of moist air, deposits formed in phenol coolers, and deposits formed during cooling of mixtures of substances such as paraffin are some examples of solidification fouling. This fouling mechanism occurs at low temperatures, usually ambient and below.

There is no single unified theory to model the fouling process because combined fouling occurs in many applications and no single solution exists for fouling control. Appropriate theories and methods must be selected in order to tackle fouling issues for each application.

7.4.2 Estimate Fouling Factor R_f

The fouling factor has to be determined from actual heat exchanger performance based on online measurement taken from a process unit test run. Heat exchanger clean performance is obtained from process flowsheet simulation software (e.g. HYSYS by Aspen Tech or UniSim by Honeywell) while dirt performance from exchanger rating software (e.g. HTRI by Heat Transfer Research Institute).

First, heat exchanger heat balance calculations are conducted in a flowsheet simulation software which has adequate thermal data and can describe process streams according to their physical properties and operating conditions. By providing measured temperatures, the simulation can determine heat transfer duty from $Q = m \cdot C_p \cdot \Delta T$ under design mode. At the same time, the simulation calculates heat transfer capability UA as the product of overall heat transfer coefficient and surface area as $UA = Q/\Delta T_{LM}$. UA is also called effective surface area.

Secondly, heat exchanger performance calculations are performed in exchanger rating software. The thermal and physical property data for process streams are transferred from the flowsheet simulation and the dimensions and geometry of the heat exchanger are entered into the rating software based on the manufacturing data sheet.

The rating software calculates two U-values, namely required and actual U. For given surface area (A), process heat duty (Q), and temperatures, the required U-value is obtained according to Eq. (7.1): $U = Q/(A \cdot \Delta T_{LM})$. At the same time, the software calculates actual U-value. ($U_A - U_R$), where A indicates the loss of effective area due to fouling. The fouling factor can then be calculated by Eq. (7.4).

7.4.3 Determine Additional Pressure Drop Due to Fouling

As the tube wall thickness increases with fouling deposits, pressure drop measurement must be conducted and used as the basis for pressure drop rating calculations. In doing so, the tube wall thickness including fouling deposits are assumed and iterated until the calculated pressure drops from the rating software converges with measured ones.

Typical fouled exchanger pressure drops are 1.3–2 times that of clean exchangers (Barletta 1998). For extreme cases, fouled exchanger pressure drops are much higher than that of clean exchangers.

It is recommended that hydraulic calculations should be conducted in exchanger rating software (e.g. HTRI) as the rating software is more rigorous in pressure drop calculations than flowsheet simulation software.

7.5 Understand Pressure Drop

In technical discussions on heat exchangers, pressure drop will naturally become an important topic. Process engineers usually prefer to keep pressure drop as low as possible in order to maintain sufficient suction pressure downstream of the heat exchanger and reduce pump power consumption and avoid process issues. For example, high pressure drops could cause feed flashing before fired heaters downstream. In contrast, reliability and design engineers would like to maintain pressure drop as high as possible in order to reduce fouling and improve film coefficients. This helps to avoid operation issues and minimize overdesign.

Basically, heat exchanger pressure drop is a function of velocity, i.e. tube velocity for tube-side pressure drop and bundle velocity for shell-side pressure drop.

7.5.1 Tube-Side Pressure Drop

Pressure drop for tube side can be expressed as

$$\Delta P_t = \frac{1}{2}\rho(u_t)^2 \frac{4L}{d_i} f_t \quad \{f_t = f(\text{Re})\} \quad (7.19)$$

where

u_t = tube velocity (ft/h)
f_t = tube-side friction factor [(ft^2 °F h)/Btu]

From Eq. (7.19), we can observe that the major parameters affecting the tube-side pressure drop include tube diameter and length, fluid density, viscosity, and velocity,

$$\Delta P_t \propto u^2 \quad (7.20a)$$
$$\Delta P_t \propto \rho \quad (7.20b)$$
$$\Delta P_t \propto L \quad (7.20c)$$
$$\Delta P_t \propto f_t \quad (7.20d)$$
$$\Delta P_t \propto d_i^{-1} \quad (7.20e)$$

7.5.2 Shell-Side Pressure Drop

The shell-side flow path is more complex than that for tube; hence, the calculation of shell-side pressure drop is more difficult. More accurate calculation of shell-side pressure drop could be obtained by the Bell–Delaware method (1973). For the purpose of providing explanation of shell pressure drop conceptually, Kern's correlation (1950) is used here. Based on bundle velocity, Kern's correlation for shell-side pressure drop Eq. (7.21) mirrors Eq. (7.19) for tube-side pressure drop:

$$\Delta P_s = \frac{1}{2}u^2 \frac{4D_s(N_B+1)}{\rho D_e} f_s \quad \{f_s = f(\text{Re})\} \quad (7.21)$$

where

u = shell-side crossflow velocity (ft/h)
D_s = shell diameter (ft)
D_e = equivalent shell diameter (ft)
N_B = number of baffles
f_s = shell-side friction factor [(ft^2 °F h)/Btu]. f_s is a function of Reynolds number and f_s charts are available in Hewitt et al. (1994).

To transform the friction factor to a shell-side pressure drop, the number of the fluid crossing the tube bundle should be given. As the fluid crosses between baffles, the number of "crosses" will be one more than the number of baffles, N_B. If the number of baffles is unknown, it can be determined using the baffle spacing P_B and tube length L:

$$N_B + 1 = \frac{L}{P_B} \quad (7.22)$$

Equation (7.21) is then reduced to

$$\Delta P_s = \frac{1}{2}u^2 \frac{4D_s L}{\rho D_e P_B} f_s \quad (7.23)$$

Clearly, Eq. (7.23) indicates major parameters affecting shell-side pressure drop, which include baffle spacing, tube length, and fluid density, velocity, and viscosity. Some of the important observations are

$$\Delta P_s \propto u^2 \quad (7.24a)$$
$$\Delta P_s \propto L \quad (7.24b)$$
$$\Delta P_s \propto N_B \quad (7.24c)$$
$$\Delta P_s \propto D_s^{-1} \quad (7.24d)$$

7.6 Effects of Velocity on Heat Transfer, Pressure Drop, and Fouling

Examination of Eqs. (7.13) and (7.15) for heat transfer, and Eqs. (7.19) and (7.23) for pressure drop indicates that for given heat exchanger and fluids, the fluid velocity is the most important parameter effecting pressure drop on both tube and shell sides. Thus, with increasing velocity, both pressure drop and heat transfer coefficient increase. The rate of pressure drop increase is faster than that of heat transfer coefficient. Since pressure drop is supplied by pumping (for liquid) or compression (for gas), higher pressure drop is at the expense of extra power cost while increased heat coefficient results in smaller surface area.

Learning from the above equations can lead to the conclusion that a short and wide heat exchanger could have a low pressure drop but a low heat transfer coefficient for both tube and shell sides. Clearly, higher pressure drop (ΔP value) forces the fluids through the heat exchanger at higher velocity leading to higher overall heat transfer coefficient (U-value). But this is at the cost of high pump power. On the other hand, for a large surface area, the U and ΔP do not need to be so high, but this is at expense of a larger heat exchanger. Therefore, there is an optimal velocity for each side in a heat exchanger which can be obtained from the trade-off between the capital cost of a heat exchanger in terms of size and the operating cost in terms of power.

One common case is that actual pressure drop could be less than allowable pressure drop on either tube or shell side. This opportunity may be used to enhance the U-value via increasing the fluid velocity. Velocity increase can be achieved by increasing flow passes on either tube side or shell side depending on which side is controlling side on U-value. Due to the fact that tube-side pressure drop rises steeply with increase in tube passes, it often happens that pressure drop is much lower than allowable value for a given number of tubes and two tube passes, but it exceeds the allowable value with four passes. In this case, the tube diameter and length could be varied to increase pressure drop with the result of a higher tube-side velocity obtained.

Another common scenario is that hydraulics can impose constraints when a heat recovery opportunity is implemented. In this case, the fluid velocity could be reduced via parallel arrangement of new and existing heat exchangers by splitting a total flow into two flows. Assuming that the flow split is equal, the fluid velocity for each branch flow is reduced by half while pressure drops on both sides are reduced by four times.

Fouling has to be addressed in heat exchanger design and operation. When heat exchanger is fouling, the fouling deposits build up additional resistance to heat transfer. At the same time, fouling deposits reduce cross-sectional flow area and increase pressure drop. Plugging could also reduce cross-sectional flow area and it could be treated the same as fouling in its effect on pressure drop. Fouling in liquids reduces heat transfer coefficient more rapidly than increase in pumping power. In contrast, fouling in gases reduces heat transfer in the range of 5–10%, but it increases pressure drop and fluid pumping power more steeply.

Increasing fluid velocity also reduces fouling tendency. Bennett et al. (2007) provided design guidelines for heavy fouling services with fluid velocity for shell-and-tube exchangers: tube-side velocity ≥2 m/s (6.5 ft/s) and shell-side B-stream (the main crossflow stream through the bundle) ≥0.6 m/s (2 ft/s).

7.6.1 Heat Exchanger Rating Assessment

When an evaluation is performed to assess the suitability of an existing heat exchanger for given process conditions or for new conditions, this exercise is called heat exchanger rating. Applications of rating can be for operational performance, for changes in process conditions or in process design. There are three fundamental points in determining if a heat exchanger performs well for given operating conditions or for a new service:

1) What actual coefficient U_A value can be "performed" by the two fluids as the result of their flow rates, individual film coefficients h_t and h_s, and fouling resistance?
2) From the heat balance: $Q = M \cdot C_p \cdot (T_1 - T_2) = m \cdot c_p \cdot (t_2 - t_1)$, known area A, and actual temperatures, required U-value (U_R) can be calculated based on Fourier's Eq. (7.1).
3) The operating pressure drops for the two streams passing through the existing heat exchanger.

The criteria can be established for the suitability of an existing exchanger for given or new services as two necessary and sufficient conditions:

a) U_A must exceed U_R to give desired overdesign (%OD) so that the heat exchanger can meet changing process conditions for a reasonable period of service continuously.
b) Operating pressure drops on both sides must be less than allowable pressure drops.

When these two conditions are fulfilled, an existing exchanger is suitable for the process conditions for which it was rated. When the process conditions undergo significant changes, a rating should be performed to make sure the exchanger can perform the task satisfactorily under the new conditions.

7.6.2 Assess the Suitability of an Existing Exchanger for Changing Conditions

When it is considered to use an existing exchanger for changing conditions or new services, rating assessment must be conducted well in advance for the suitability of existing exchangers for such services.

Example 7.1 Rating of an existing naphtha–heavier naphtha exchanger to operate under small changes in flow rates. 124 600 lb/h (versus 122 500 in design) of a 56.3°API heavy naphtha leaves the naphtha splitter tower at 276 °F and is cooled to 174 °F by 193 000 lb/h (versus 188 000 in design) of 69°API naphtha feed at 116 °F and heated to 170 °F. There is 6.3% vapor in the naphtha at 170 °F. 10 and 5 psi pressure drops are permissible on tube and shell sides, respectively. Can this exchanger operate satisfactorily under new conditions?

The exchanger is TEMA type AES (see Appendix 7.A) with 21-inch shell ID having 268 tubes with 3/4-inch tube OD, 14 BWG thickness, and 20 ft long, which are laid out on 1-inch triangle pitch. There are four tube-passes and one shell-pass with baffles spaced 11¼ inch apart and baffle cut 32% of shell diameter. The hot heavy naphtha is on the tube side.

Solution
1) Heat balance
   For naphtha feed:

$$Q = M \times C \times \Delta T + q \times M \times \text{vapor\%}$$
$$= 193\,000 \times \left[0.53 \times (170 - 116) + 141.6 \times 6.3\%\right]$$
$$= 7.23 \text{ MMBtu/h}$$

   For heavy naphtha:

$$Q = m \times c \times \Delta t$$
$$= 124\,600 \times 0.57 \times (276 - 174)/10^6$$
$$= 7.23 \text{ MMBtu/h}$$

2) ΔT_{LM} and F_t

Tube side		Shell side		
Hot stream		Cold stream	Difference	
276	Higher temperature	170	106	ΔT_1
174	Lower temperature	116	58	ΔT_2
102	Differences	54	48	$\Delta T_1 - \Delta T_2$
ΔT		Δt		

$$\Delta T_{LM} = \frac{\Delta T_1 - \Delta T_2}{\ln(\Delta T_1/\Delta T_2)} = \frac{48}{\ln(106/58)} = 79.6$$

$$R = \frac{\Delta T}{\Delta t} = \frac{102}{54} = 1.89$$

$$P = \frac{\Delta t}{T_1 - t_1} = \frac{54}{276 - 116} = 0.34$$

$$F_t = 0.83$$
$$F_t \Delta T_{LM} = 66.1 \text{ °F}$$

3) Rating summary

U_C	166.1
U_A	124.7
U_R	115.1
Overdesign	8%
R_f calculated	0.003
R_f required	0.002
ΔP_s calculated	3.8
ΔP_s allowable	5
ΔP_t calculated	9.9
ΔP_t allowable	10

The allowable fouling factor of 0.002 is assumed based on Table 7.3. UA is calculated by taking fouling into account. The heat exchanger has 8% of overdesign over normal fouling conditions. Pressure drops on both sides of the exchanger are less than allowable pressure drops. Thus, this exchanger meets the two criteria. Therefore, it can operate satisfactorily to fulfill the new flow conditions.

From time to time, a process plant wishes to increase feed rate and/or make different product yields due to economic drivers. In feasibility evaluation, it is essential to assess the suitability of existing heat exchangers for new process conditions and find the most economic ways to handle significant changes.

Example 7.2 Rating of an existing naphtha–diesel exchanger to handle large increase in flow rate. A refinery plant plans to increase diesel production by 20% via revamping the hydrocracking unit. This is because diesel is highly desirable commodity in today's energy market. Currently, the naphtha–diesel exchanger is located downstream of naphtha–heavy naphtha exchanger which is discussed in Example 7.1. Via the naphtha–diesel exchanger under the scenario of increased diesel production, 193 000 lb/h naphtha feed to the tower will increase vaporization up to 29.8% from 6.3% by 121 500 lb/h diesel product at 351 °F cooled to 260 °F. A 5 psi pressure drop is permissible on both sides. Can the current exchanger operate satisfactorily under new conditions?

The exchanger is TEMA type AES with 16-inch shell ID having 130 tubes with 3/4-inch tube OD, 14 BWG thickness, and 16 ft long, which are laid out on 1-inch triangle pitch. There are two tube-passes and one shell-pass with single segmental baffles spaced 14 inch apart and baffle cut 40% of shell diameter. The hot diesel is on the tube side.

Solution

1) Heat duty
   Naphtha feed:
   $$Q = W \times q \times \Delta \text{vapor\%}$$
   $$= 193\,000 \times 141.7 \times (29.8 - 6.3\%)$$
   $$= 6.4 \text{ MMBtu/h}$$

   Diesel product:
   $$Q = w \times c \times \Delta t$$
   $$= 121\,500 \times 0.587 \times (351 - 260)/10^6$$
   $$= 6.4 \text{ MMBtu/h}$$

2) ΔT_{LM} and F_t

Tube side	Shell side			
Hot stream		Cold stream	Difference	
351	Higher temperature	171	180	ΔT_1
260	Lower temperature	170	90	ΔT_2
91	Differences	1	90	$\Delta T_1 - \Delta T_2$
ΔT		Δt		

$$\Delta T_{LM} = \frac{\Delta T_1 - \Delta T_2}{\ln(\Delta T_1 / \Delta T_2)} = 129.8$$

$$R = \frac{\Delta T}{\Delta t} = 91$$

$$P = \frac{\Delta t}{T_1 - t_1} = \frac{1}{351 - 170} = 0.01$$

$$F_t = 0.99$$

$$F_t \Delta T_{LM} = 128.5 \text{ °F}$$

3) Rating summary

U_C	135.4
U_A	106.6
U_R	123.5
Overdesign	−14%
ΔP_s calculated	10.7
ΔP_s allowable	5
ΔP_t calculated	5.6
ΔP_t allowable	5

The above rating calculations show that the existing exchanger alone cannot handle 20% increase in diesel flow rate. This is because surface area is not sufficient as well as the shell-side pressure drop in particular is too large to be allowed. Thus, it violates the criteria for the suitability of an existing exchanger to fulfill changing process conditions. The following discussions will show how to assess practical solutions by use of spare heat exchangers.

7.6.3 Determine Arrangement of Heat Exchangers in Series or Parallel

In some plants where a large number of exchangers are used, certain size standards are usually established in-house for 1–2 type of exchangers so that future services can be satisfied by making arrangement of standard exchangers in series or in parallel. Use of standard exchangers could come at a price because of impossibility of utilizing the standard equipment in the most efficient manner. However, it does offer a great advantage of reducing spare parts, tubes, and tools for replacement. When tube bundles are retubed, the standard exchangers can provide services as new ones to meet process conditions.

There are two basic arrangements of exchangers, namely series and parallel arrangements. When use of a single 1–2 exchanger could not satisfy new process conditions or lead to a severe temperature cross-signaled by a low F_t factor, it may be necessary to use two 1–2 exchangers in series. On the other hand, when hydraulic limitation could be an issue for a 1–2 exchanger, placing multiple 1–2 exchangers in parallel could resolve the issue.

Example 7.3 (Continuing from Example 7.2) The above rating assessment for the existing naphtha–diesel exchanger showed that the single 1–2 exchanger is not sufficient to meet 20% increase in diesel flow rate. A spare 1–2 exchanger was considered to add to the existing naphtha–diesel exchanger. What is the proper arrangement of this spare 1–2 exchanger in relation to the existing naphtha–diesel exchanger to handle the large increase in diesel flow rate?

Under increased diesel production, 193 000 lb/h naphtha feed (placed on the shell side of the exchanger) to the tower will increase vaporization up to 29.8 from 6.3% by 121 500 lb/h diesel product at 351 °F cooled to 260 °F. 5 psi pressure drop is permissible on both sides.

Based on the rating assessment in Example 7.2, it is observed that pressure drop on the shell side is too large to be allowed. Thus, a parallel arrangement is considered in this assessment as shown in Figure 7.4. The two exchangers are accounted as one exchanger unit for the rating calculations below.

Solution

1) Heat duty
   Naphtha feed:
   $$Q = W \times q \times \Delta \text{vapor\%}$$
   $$= 193\,000 \times 141.7 \times (29.8 - 6.3\%)$$
   $$= 6.4 \text{ MMBtu/h}$$

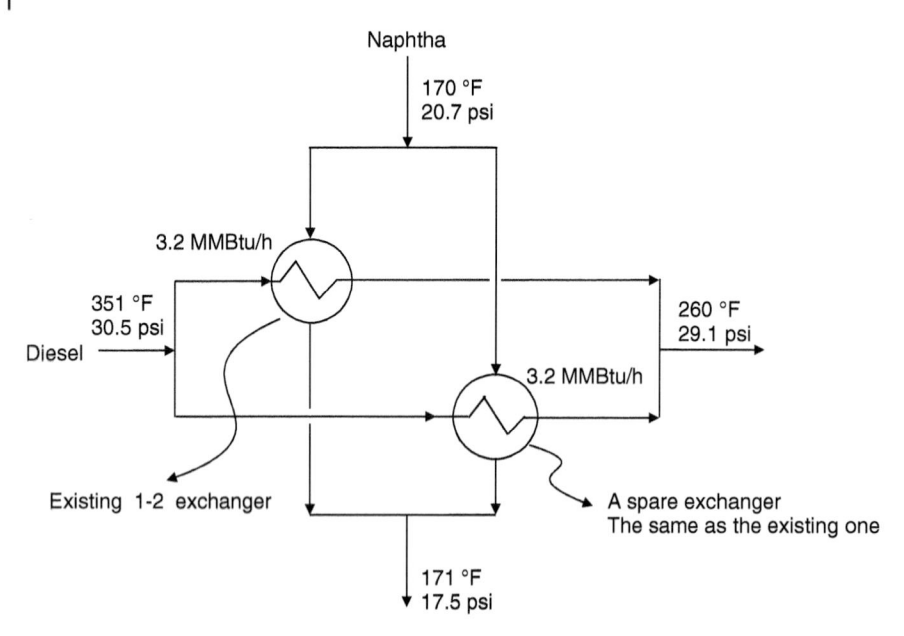

Figure 7.4 A parallel arrangement of two 1–2 exchangers.

Diesel:

$$Q = w \times c \times \Delta t$$
$$= 121\,500 \times 0.587 \times (351 - 260) / 10^6$$
$$= 6.4 \text{ MMBtu/h}$$

2) ΔT_{LM} and F_t

Tube side	Shell side		
Hot stream		Cold stream	Difference
351	Higher temperature	171	180 ΔT_1
260	Lower temperature	170	90 ΔT_2
91	Differences	1	90 $\Delta T_1 - \Delta T_2$
ΔT		Δt	

$$\Delta T_{LM} = \frac{\Delta T_1 - \Delta T_2}{\ln(\Delta T_1 / \Delta T_2)} = 129.8$$

$$R = \frac{\Delta T}{\Delta t} = 91$$

$$P = \frac{\Delta t}{T_1 - t_1} = 0.01$$

$$F_t = 0.99$$

$$F_t \Delta T_{LM} = 128.5 \text{ °F}$$

3) Rating summary

U_C	78.7
U_A	68.0
U_R	61.8
Overdesign	10%
R_f calculated	0.003
R_f required	0.002
ΔP_s calculated	3.2
ΔP_s allowable	5
ΔP_t calculated	1.4
ΔP_t allowable	5

Two 1–2 heat exchangers in parallel are adequate to satisfy process heat transfer requirement with 10% overdesign. Pressure drops on both sides of the exchanger are less than allowable pressure drops.

Example 7.4 Use of spare exchangers in series to an existing acetone–acetic acid exchanger. Acetone at 250 °F is to be sent to storage at 100 °F and at a rate of 60 000 lb/h. The heat will be received by 185 000 lb/h of 100% acetic acid coming from storage at 90 °F and heated to 150 °F. Pressure drops of 10.0 psi are available for both fluids, and an overall fouling factor of 0.004 should be provided.

Available for the service are several 1–2 exchangers having 21¼ inch shell ID, having 270 tubes with ¾ inch tube OD, 14 BWG, 16′0″ long and laid out on 1-inch square pitch. The bundles are arranged for two tube passes with segmental baffle spaced 5 inch apart.

Determine the suitability of these 1–2 exchangers for the specific service.

Solution

1) Exchanger data

Shell side	Tube side
ID = 21¼ inch	Number and length = 270, 16′ 0″
Baffle spacing = 5 inch	OD/BW G/pitch = 3/4 inch/14 BWG/1 inch square
Shell passes = 1	Tube passes = 2

2) Heat balances
   Acetone $Q = 60\,000 \times 0.57(250-100) = 5\,130\,000$ Btu/h.
   Acetic acid $Q = 168\,000 \times 0.51(15-90) = 5\,130\,000$ Btu/h.

3) F_t factor

Shell side	Tube side			
Hot stream		Cold stream	Difference	
250	Higher temperature	150	100	ΔT_1
100	Lower temperature	90	10	ΔT_2
150	Differences	60	90	$\Delta T_1 - \Delta T_2$
ΔT		Δt		

$\Delta T_{LM} = 39.1\,°F$

$R = \dfrac{150}{60} = 2.5$

$P = \dfrac{60}{250-90} = 0.375$

Thus,
one 1–2 exchanger, F_t is not on F_t charts
two 1–2 exchangers, F_t = 0.57 (too small)
three 1–2 exchangers, F_t = 0.86 (OK)
$F_t \times \Delta T_{LM} = 33.6\,°F$

To permit the heat transfer with the temperatures given by the process, a minimum of three 1–2 exchangers is required. If the sum of the surface area in three exchangers is insufficient, a greater number of 1–2 exchangers are required.

4) U_C: Calculated from heat exchanger rating software
   $h_t = 194$ Btu/h ft² °F and $h_s = 242$ Btu/h ft² °F
   $U_C = h_t \times h_s/(h_t + h_s) = 194 \times 242/(194+242) = 107.7$ Btu/h ft² °F
   U_R: $U_R = Q/(AF_t\,LMTD) = 5\,130\,000/(2540 \times 34.4) = 58.8$ Btu/h ft² °F

5) R_f: $R_f = (U_C - U_R)/U_C U_R = (107.5 - 58.8)/(107.5 \times 58.8) = 0.0077$ h ft² °F/Btu

6) ΔP_s and ΔP_t: Calculated from heat exchanger rating software

7) ΔP_s = 10.4 psi (allowable ΔP_s = 10.0 psi) and ΔP_t = 5.2 psi (allowable ΔP_t = 10.0 psi)

8) Rating summary

U_C	107.5
U_A	75.2
U_R	58.8
OD%	28%
R_f calculated	0.0077
R_f required	0.004
ΔP_s calculated	10.4
ΔP_s allowable	10
ΔP_t calculated	5.2
ΔP_t allowable	10

Conclusion
Three 1–2 exchangers are more than adequate for heat transfer even though the pressure drop on the shell side is slightly higher than allowable. Fewer exchangers cannot fulfill the process requirement.

7.6.4 Assess Heat Exchanger Fouling

Heat exchanger fouling occurs during operation and cause exchanger performance to deteriorate over time. In some cases, it requires cleaning several times before the entire process is shut down for turnaround maintenance. In extreme cases, it may not be possible to remove tube bundles which must be replaced. It is an economic decision for selecting fouling mitigation methods and when to apply. Heat exchanger rating can determine the level of fouling and if the heat exchanger in question requires attention.

Example 7.5 Calculation of heat transfer performance for an existing exchanger. The cold stream of 710 000 lb/h going through the tube side of the exchanger is heated from 359 to 375 °F by 213 500 lb/h of hot stream entering at 503 °F and cooled to 449 °F. How is this heat exchanger performing?

The exchanger is TEMA type AES with 48-inch ID shell having 964 tubes with 1-inch OD, 12 BWG thickness, and 24 ft long, which are laid out on 1¼-inch square pitch. There are four tube-passes and one shell-pass with baffles spaced 9.5 inch apart and baffle cut 15%. Fouling factors of 0.003 and 0.01 are provided for crude and vacuum residue, respectively.

Solution

1) Heat duty
   Crude oil: $Q = 730\,000 \times 0.60 \times (375 - 359) = 7.0\,\text{MMBtu/h}$.
   Vacuum residue: $Q = 213\,500 \times 0.61 \times (503 - 449) = 7.0\,\text{MMBtu/h}$.

2) ΔT_{LM} and F_t

Tube side	Shell side			
Hot stream		Cold stream	Difference	
503	Higher temperature	375	128	ΔT_1
449	Lower temperature	359	90	ΔT_2
54	Differences	16	38	$\Delta T_1 - \Delta T_2$
ΔT		Δt		

$$\Delta T_{LM} = \frac{\Delta T_1 - \Delta T_2}{\ln(\Delta T_1 / \Delta T_2)} = 107.9$$

$$R = \frac{\Delta T}{\Delta t} = 3.38$$

$$S = \frac{\Delta t}{T_1 - t_1} = 0.11$$

$F_t = 0.98$

$F_t \Delta T_{LM} = 105.8\ °F$

3) Initial assessment

U_C	49.6
U_A	30.2
U_R	11.1
OD%	171%
R_f calculated	0.070
R_f required	0.013

The required U-value to achieve 7.0 MMBtu of heat transfer is only 11.1 in comparison with the actual U-value of 30.2 based on fouling factors of 0.01 for the hot stream and 0.003 for the cold stream. In other words, the heat exchanger only accomplishes one-third of the heat transfer capability offered by the heat exchanger, which warrants a more detailed investigation.

4) More detailed assessment
   As a follow-up, engineers conducted the performance comparison between operation and design and the results are given in the table below. It can be observed that the flow rates in operation are higher than those in design. The higher flow rates should have corresponded to a higher U-value in operation. However, in this case, the U-value in design is 41% higher than in operation.

	Design	Operation
Q (MMBtu/h)	8.9	7.0
W(resid) (lb/h)	167 250	213 500
T_1 (°F)	505	503
T_2 (°F)	416	449
w(crude) (lb/h)	665 000	710 000
t_1 (°F)	360	359
t_2 (°F)	382	375
ΔT_1 (°F)	123	128
ΔT_2 (°F)	56	90
ΔT_{LM} (°C)	85	108
F_t	0.93	0.98
UA	112 658	66 074
A	5936	5936
U	18.98	11.13
$U_{operation}/U_{design}$		0.59

Field inspection was performed and pressure drop was measured. It was found the pressure drop on the crude (tube side) was around 60 psi versus 6.8 psi under normal fouling conditions. It was concluded that the heat exchanger suffers severe fouling with loss of more than half the heat transfer capability. In addition, the much higher pressure drop caused crude feed flashing before the charge heater, which could be the potential safety issue for the heater. Thus, it was decided to clean the exchanger immediately online by means of by-pass arrangement. After cleaning, dedicated investigation was conducted to identify the root causes of this fouling.

It was found from rating assessment that the tube-side crude velocity is a bit too low, which is 5.3 ft/s. It should be 7 ft/s for this hot crude heating service because precipitation fouling becomes more active under high temperature. The change was made to the number of tube passes from four to six. As the result, the tube velocity was increased to 7.9 ft/s, but at the expense of higher pressure drop on the tube side. The tube-side pressure drop was increased to 21 psi versus 6.8 psi with four tube passes. This change helped reducing tube fouling and prolonged the operation of the exchanger between cleanings.

The lessons learnt from this investigation indicate that fluid flow rate affects fouling behavior significantly. Flow rates much lower than design result in lower velocity, which can promote accumulation of fouling deposits. High temperature is another major cause for promoting fouling. The heat exchangers in the high-temperature region are more prone to be fouled due to inherent thermal coking tendency.

Threshold conditions in terms of velocity and temperature should be identified beyond which fouling occurs in a faster pace.

7.7 Improving Heat Exchanger Performance

The objective of heat exchanger operation management is to maintain good performance to fulfill process requirements for desirable periods of time. Basically, there are three major reasons why exchanger operation could deviate from design: poor design, excessive fouling, and mechanical failure. In any event, heat exchangers can deliver trouble-free services while meeting process requirements if the heat exchanger is designed well thermally and mechanically, stored carefully before use, installed correctly, operating within its design limits, and cleaned periodically depending on fouling formation. In contrast, it can be stressful if heat exchangers do not perform as expected in meeting process requirements. In the worst case, mechanical and performance failure of heat exchangers could cause undesirable unit shutdowns.

The methods for monitoring and troubleshooting are provided here with the focus on thermal and hydraulic performance. Fijas (1989) provides good discussions on mechanical problems often encountered with heat exchangers. With exchanger performance, the priority issues are good knowledge of fouling resistances to avoid poor design, continuous monitoring of U-value to maintain good performance, pressure drop survey for troubleshooting, managing of two-phase flow, fouling mitigation, and heat transfer enhancements. The general methodology for improving heat exchanger performance is: monitor performance trends, identify opportunity, and develop and implement solutions.

7.7.1 How to Identify Deteriorating Performance

7.7.1.1 Fouling Resistances

Inappropriate estimate of fouling resistances results in either too much or too little overdesign. Although the TEMA fouling resistances were originally only considered to be rough guidelines for heat exchanger design, they are often treated as accurate values. This may cause considerable errors because the transient character of the fouling process is neglected. Conditions in initially over-designed heat exchangers often promote fouling deposition, thus making fouling a self-fulfilling prophecy. Thus, one needs to be critical of the fouling resistances listed in the public domain and make proper adjustment based on historical fouling data. For existing services, obtain historical fouling trends and assess the characteristics of the system to determine the root causes for fouling and design accordingly.

7.7.1.2 U-Value Monitoring

Due to the fact that heat exchanger performance varies with flow rates, compositions, and fouling conditions, heat exchanger assessment must be conducted on a regular basis so that a performance trend over time can be measured and problems can be detected at an early stage. A single rating of an exchanger is good for getting a baseline data on its performance but it must be done on a regular basis to define trends. From a single point of rating, you can calculate a single U-value, pressure drops of shell side and tube side, and calculate a single value of heat duty. However, single-point assessment cannot provide insights into fouling evolution over time and sudden changes in U-value due to process variations. But a U-value trend can help you with these operating issues.

The most important thing of a U-value trend is its capability of showing the fouling behavior. The purpose of U-trend monitoring is to identify any abnormal fouling behavior. In general, fouling accumulation in heat exchanger depends on the type of fouling, the service (fluid compositions, temperature, and pressure), the exchanger design, and so on. Under normal operation, a U-trend should display gradual changes in U-value. However, operation changes could affect fouling, which include feed rate variations, fluid composition change, bypass operation, hydraulic head change, etc. If an operation change suddenly distorts the normal fouling behavior (e.g. U-value reduces sharply), this change must be investigated and appropriate actions must be taken.

When an exchanger is new, a detailed performance evaluation is warranted and it should be repeated after six months or so. One should trend data in-between and afterward. Process temperatures, pressures, and flows around the heat exchanger are measured in daily averages. It is recommended to use distributed control trend logs for data collection. It is important to keep the data and the calculations for reference in future.

7.7.1.3 Pressure Drop Monitoring

The importance of pressure drop calculations cannot be overemphasized as it can help with analyzing performance problems and troubleshooting of heat exchanger malfunction. Calculated pressure drop for single-phase flow can be reasonably close to measured pressure drop if there is no fouling. For two-phase flow, calculated pressure drop can also be reasonably close to measured pressure drop if pressure drop zones are used and flow patterns are considered. With these two assumptions as the basis, pressure drop calculations can be used as a tool for identifying problems.

If measured pressure drops are significantly lower than calculated drops, this might indicate fluid bypassing, which could occur either on tube side or shell side. On the other hand, if measured pressure drop is too high, this is often caused by severe plugging or fouling, or freezing or slug flow for two-phase flows.

7.7.1.4 Avoid Poor Design

Chemical and petroleum industries have been plagued for decades with poorly operating and occasionally inoperable heat exchangers. One of the common causes is usually traced to poor design, which should be avoided in the design stage by all means. Careful considerations of major design choices must be made in order to obtain an "optimal" heat exchanger design. The design issues include fouling considerations, tube-side design (tube counts, tube passes, tube length, tube pitch, and tube layout), shell-side design (shell diameter), shell types (including TEMA types E/F/G/H/J/K/X, shell flow distribution), and baffle design (baffle types, segmental baffle including single/double/triple, baffle spacing, and baffle cut).

The essential design task is to optimize velocity in both tube and shell sides by the best use of allowable pressure drop available. For example, when the number of tube-passes is increased from one to two passes, the velocity could be twice that of one-pass velocity as the travel distance is doubled. Then the heat tube-side transfer coefficient will increase according to the 0.8 power of velocity. At the same time, the tube-side pressure drop will increase according to the square of velocity and to the travel distance. Therefore, pressure drop will rise to the cubic of the increase in tube passes for a given tube counts and tube-side flow rate. When the pressure drop is higher than allowable one, reduction of tube length could reduce pressure drop. The tube outside diameter, tube pitch, tube counts, and layout are the important design choices beside the number of tube passes and tube length.

Another example is design choices available to reduce shell-side pressure drop. The number of baffles (N_B) is proportional to the baffle spacing. Baffle spacing and baffle cut have profound effect on shell-side pressure drop. In many cases, the shell-side pressure drop is still too high with single segmental baffles in a single-pass shell even after increasing the baffle spacing and baffle cut to the highest values recommended. These cases may accompany with very high shell-side flow rate. The next design choice is to consider double segmental baffles. When double segmental baffles at relatively high baffle spacing cannot satisfy shell-side allowable pressure drop, a divided-flow shell (TEMA J) with single segmental baffles could be considered. Since pressure drop is proportional to the square of velocity (u^2) and to the length of travel (L), a divided-flow shell could have one-eighth the pressure drop in an identical single-pass exchanger. This discussion can go on as there are other design choices which are available to deal with high shell-side pressure drops, for which Mukherjee (1998) provides detailed explanations.

7.A TEMA Types of Heat Exchangers

Nomenclature

A	surface area (ft^2)
C, c	fluid heat capacity of hot and cold streams [Btu/(lb °F)]
d_i	inner diameter of the tube (ft)
d_o	outer diameter of the tube (ft)
D_e	equivalent shell diameter (ft)
D_s	shell diameter (ft)
f_s	shell-side friction factor [(ft^2 °F h)/Btu]
f_t	tube-side friction factor [(ft^2 °F h)/Btu]
F_t	LMTD correction factor (fraction)
h_t	tube-side file coefficient [Btu/(ft^2 °F h)]
h_s	shell-side file coefficient [Btu/(ft^2 °F h)]
k	fluid thermal conductivity [Btu/(h ft °F)]
L	tube length (ft)
LMTD	ΔT_{LM} (°F)
M, m	mass flow rate for hot (cold) streams (lb/h)
N_B	number of baffles
p	tube pitch (ft)
ΔP	pressure drop (psia)
Q	heat duty (MMBtu/h)
R	temperature ratio: $(T_1 - T_2)/(t_2 - t_1)$, dimensionless
R_e	Reynolds number, dimensionless
R_f	overall fouling resistance = $R_t + R_s$ [(ft^2 °F h)/Btu]
R_t, R_s	tube- (shell-) side fouling resistance [(ft^2 °F h)/Btu]
P	temperature ratio: $(t_2 - t_1)/(T_1 - t_1)$, dimensionless
r_w	resistance of the inner tube referred to the tube outside diameter [(ft^2 °F h)/Btu]
T_1, t_1	supply temperature of hot (cold) stream (°F)
T_2, t_1	target temperature of hot (cold) stream (°F)
ΔT_1	hot end temperature approach (°F)
ΔT_2	cold end temperature approach (°F)
ΔT_{LM}	logarithmic mean temperature difference (LMTD) (°F)
u	shell-side crossflow velocity or tube velocity (ft/h)
U	overall heat transfer coefficient [Btu/(ft^2 °F h)]
V_S	superficial gas velocity (ft/s)

Greek Letter

ρ	fluid density (lb/ft^3)
μ	fluid viscosity [lb/(ft h)]
μ_w	water viscosity [lb/(ft h)]
τ_0	sheer stress [lb/ft^2]

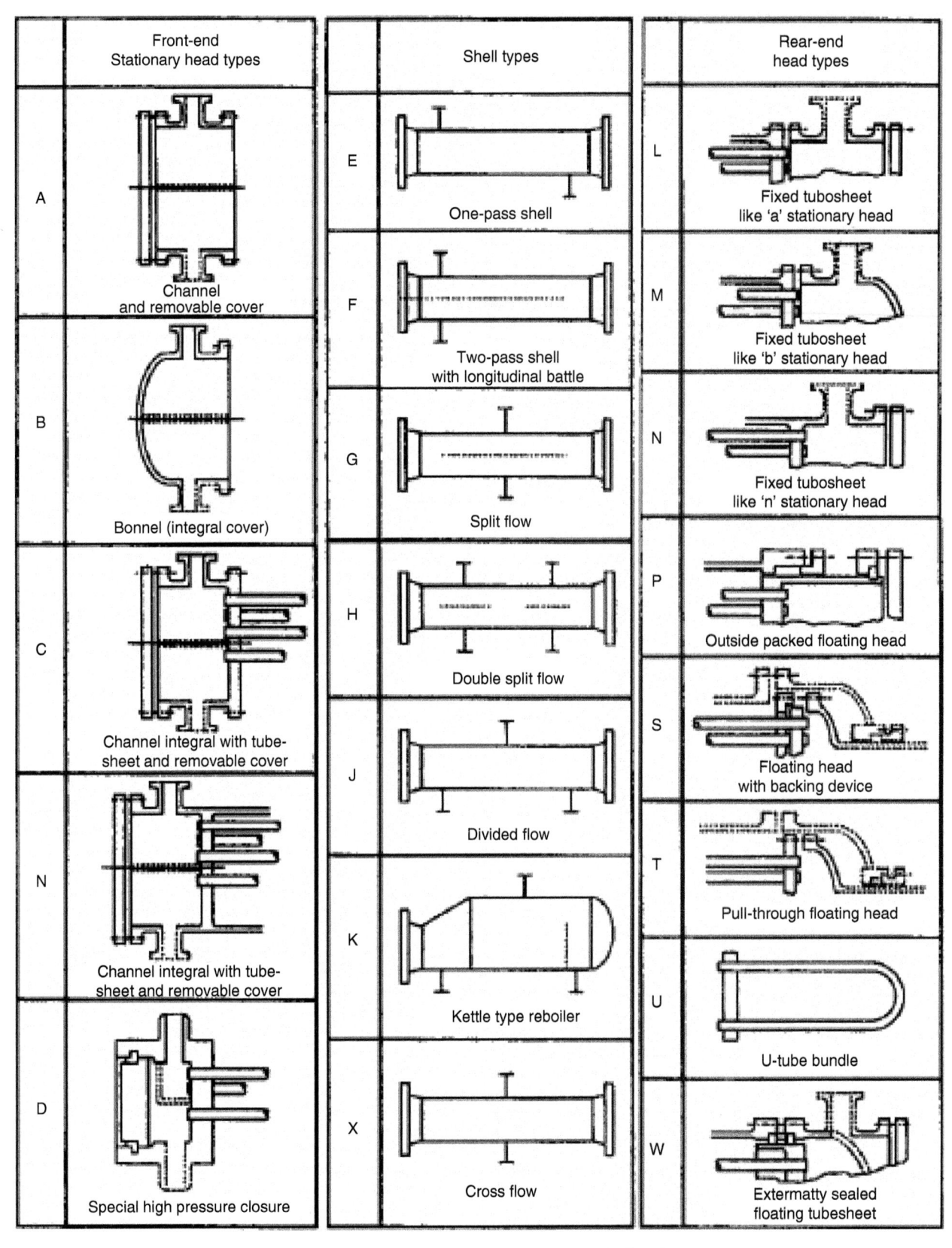

	Front-end Stationary head types		Shell types		Rear-end head types
A	Channel and removable cover	E	One-pass shell	L	Fixed tubesheet like 'a' stationary head
B	Bonnel (integral cover)	F	Two-pass shell with longitudinal baffle	M	Fixed tubosheet like 'b' stationary head
C	Channel integral with tube-sheet and removable cover	G	Split flow	N	Fixed tubesheet like 'n' stationary head
		H	Double split flow	P	Outside packed floating head
N	Channel integral with tube-sheet and removable cover	J	Divided flow	S	Floating head with backing device
		K	Kettle type reboiler	T	Pull-through floating head
				U	U-tube bundle
D	Special high pressure closure	X	Cross flow	W	Extermatty sealed floating tubesheet

Figure 7.A.1 TEMA standard shell types and front and rear-end head types. *Source:* TEMA (1999).

Subscript and Superscript
A actual
C clean
D design
e equivalent
f friction
G gas

i   inside of tube or shell
L   liquid
lm logarithmic mean
o   outside of tube or shell
s   shell side
t   tube side

References

Barletta, A.F. (1998). Revamping crude units. *Hydrocarbon Procession* 77: 51–57.

Bell, K.L. (1973). Thermal design of heat transfer equipment. In: *Chemical Engineers Handbook*, 5e (eds. R.H. Perry and C.E. Chilton), 10. McGraw-Hill.

Bennett, C.A., Kistler, R.S., Lestina, T.G., and King, D.C. (2007). Improving heat exchanger designs. *Chemical Engineering Progress* 103: 40–46.

Dittus, F.W. and Boelter, L.M.K. (1930). *Publications on Engineering*, vol. 2, 443. Berkley: University of California.

Fijas, D.F. (1989). Getting top performance from heat exchangers. *Chemical Engineering* 96: 141–145.

Hewitt, G.F., Shires, G.L., and Bott, T.R. (1994). *Process Heat Transfer*, 275–285. CRC Press.

Kern, D.Q. (1950). *Process Heat Transfer*, 148. McGraw-Hill.

Melo, L.F., Bott, T.R., and Bernardo, C.A. (eds.) (1988). *Advances in Fouling Science and Technology*. Kluwer Academic.

Mukherjee, R. (1998). Effectively design shell-and-tube heat exchangers. *Chemical Engineering Progress* 94: 21–37. AIChE.

Shah, R.K. and Sekulić, P. (2003). *Fundamentals of Heat Exchanger Design*. Wiley.

TEMA (1999). *Standards of TEMA*, 8e. New York: Tubular Exchanger Manufacturers Association.

TEMA (2007). *Standards of TEMA*, 9e. New York: Tubular Exchangers Manufacturer Association.

8

Fired Heater Assessment

8.1 Introduction

Fired heaters are used to provide high-temperature heating when high-pressure steam is unable to satisfy process heating demand in terms of temperature. The primary role of industrial fired heaters is to provide heat required for reaction and separation processes. In a fired heater as shown in Figure 8.1, the process fluid enters the tubes at the top of the convection section and flows down countercurrent to the flue gas flow. The fuel mixes with the combustion air in the burner and provides the heat to heat up the process steam. The hot combustion gases need residence time to transfer the heat to the tubes. The shock tubes are often the hottest tube in the fired heater. The shock tubes receive the full radiant heat transfer of around 10 000 Btu/h-ft^2, plus the hot gases flowing over the tubes results in an additional convective heat transfer rate of around 5 000 Btu/h-ft^2. Since the firebox operates in very high temperature, refractory lining is required to prevent heat loss to the atmosphere.

Due to the fact that fired heaters operate under severe conditions, they are designed with careful considerations of high-temperature characteristics of the alloy. With proper maintenance and operation, a fired heater can have a long operating life. However, the life of a fired heater can be greatly shortened due to creep, fatigue, corrosions, and erosion by lack of maintenance and reliability considerations. Fired heater failure could not only result in significant production loss; in the worst case, it could cause damage to human life.

Therefore, maintaining fired heater in reliable operation is the highest priority. With this priority in place, process plants strive to maximize fired heater efficiency and hence reduce its operating cost. This is because of a simple fact: fired heaters are the largest energy consumers in process plants and account for majority of total energy use.

8.2 Fired Heater Design for High Reliability

The eventual measure of fired heater reliability is availability and the goal is that a fired heater needs to be online almost 100% of time. What does it take for a fired heater to achieve this high reliability from design point of view?

In this section, we will discuss the critical issues that a highly reliable fired heater must acquire and shed light on fundamentals for these features so that they can be used as the benchmarking for assessing a fired heater. The critical reliability issues are discussed below.

- Flux rate
- Burner to tube clearance
- Burner selection
- Fuel conditioning system

8.2.1 Heat Flux Rate

Radiant heat flux is defined as heat intensity on a specific tube surface. Thus, heat flux represents the combustion intensity and is analogy to "how hard a fired heater is run." More specifically, to keep firing rate within safe limit is equivalent to maintain the peak heat flux being less than the design limit because high firebox temperatures could cause tubes, tube-sheet support, and refractory failures. What is the peak flux and why it is so important to keep it within the limit? These questions will be answered below.

Flux rate is influenced by combustion characteristics and heat distribution. While the combustion characteristics can be described by combustion intensity, the heat distribution is explained by heat flux. The heat flux is defined as the heat transferred to the process feed while combustion intensity is the heat released from flame divided by flame's external surface area. Clearly, combustion intensity is related to combustion flame while

Efficient Petrochemical Processes: Technology, Design and Operation, First Edition.
Frank (Xin X.) Zhu, James A. Johnson, David W. Ablin, and Gregory A. Ernst.
© 2020 John Wiley & Sons, Inc. Published 2020 by John Wiley & Sons, Inc.

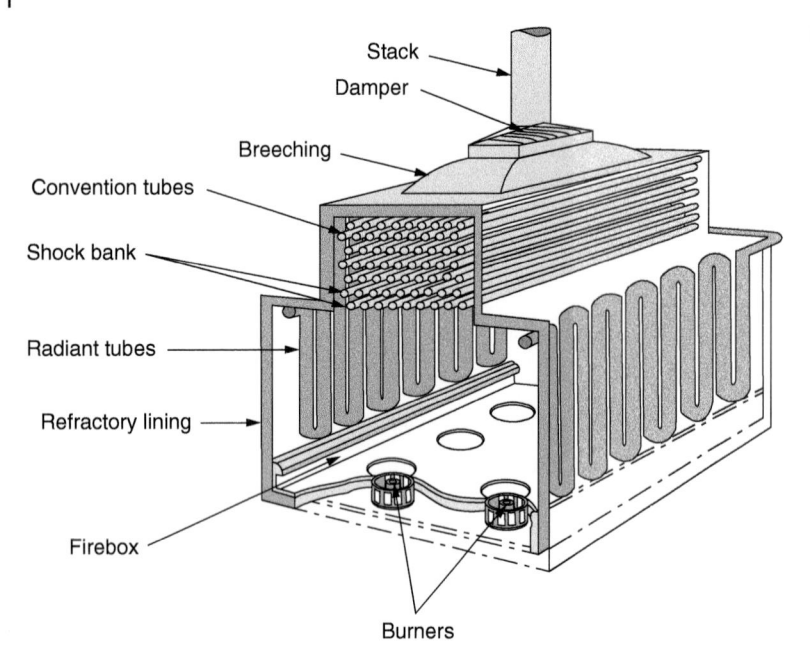

Figure 8.1 Schematic view of a typical process fired heater.

heat flux to process. Another difference is that combustion intensity is inevitably an average value while heat flux is either average or local values. Local heat flux requires more attention in design and operation.

Figure 8.2 shows a typical pattern of a heat flux profile. It can be observed that the flux distribution is not uniform. The heat intensity nearest to flames is the highest (peak flux) and declines away from flames. The peak flux can exceed twice the average value and should be maintained below the flux limit all the time for safe operation. The nonuniformity is described by the ratio of peak flux to average flux. A good heater design and operation should have a heat flux profile featuring a high average and low peak flux values.

Jenkins and Boothman (1996) reported an operating case with average flux at only 700 Btu/h-ft^2 but the peak

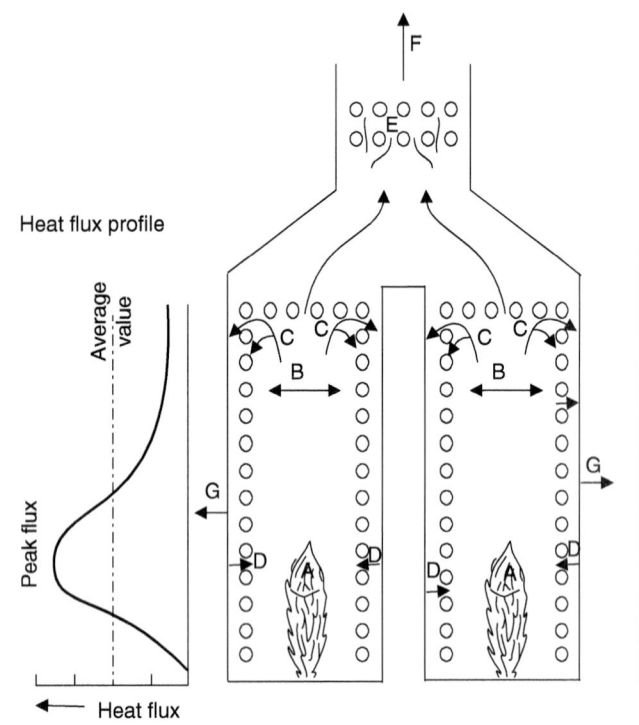

Typical heat distribution In a heater

A	Direct heat transfer through flame radiation	14%
B	Direct heat transfer through gas radiation	28%
C	Direct heat transfer through gas convection	6%
D	Indirect heat transfer through refractory reflection	12%
	Total heat transfer in radiant section	**60%**
E	Direct heat transfer through convection in convection bank	25%
	Total useful heat	**85%**
F	Flue gas loss	13%
G	Radiation loss	2%
	Total	100%

Figure 8.2 Flux profile and heat distribution in a heater.

flux was over 20 000 Btu/h-ft² nearly three times of the average. The peak flux exceeded the safe limit of the process, and residual oil in the tubes in the peak flux area was cracked and coke was deposited in the inner tube surface. As the consequence, the coke acted as an insulating layer and caused the tubes to overheat, which was measured by tube wall temperature (TWT). This was dangerous as the tubes in the peak flux area could rapture. To mitigate the safety risk, the plant operators had to reduce the process feed rate and hence cut down the firing rate in order to keep the peak flux below the safe limit. Furthermore, the heater had to be shut down at regular intervals so that tubes could be cleaned to remove the carbon deposits. These mitigation actions resulted in significant cost to the plant. The problem was solved fundamentally by improving air distribution between the burners together with changes to the burner gas nozzle and flame stabilizer. These changes result in a lower air pressure loss and improved fuel and air mixing. After these changes, the burners produced a more even flux profile: the average flux was increased from 700 to 10 000 Btu/h-ft² while the peak flux reduced to 18 000 Btu/h-ft². Coking inside tubes was eliminated, which allowed the feed rate to increase by 4% and the heater could run continuously between scheduled outages.

The flux distribution around the tube is not uniform as well. As indicated in Figure 8.3, the radiating plane is the flame. The diagram on the left shows the flux profile for a single fired heater. The front of the tube facing the fire picks up most of the heat. The diagram on the right shows the profile for a double fired heater with flames on both sides of the tubes. The flux pattern is close to uniform.

The nonuniformity is described by the circumferential flux factor, which is the ratio of peak flux to average flux. Peak flux determines the maximum TWT. The peak flux is typically 1.5–1.8 times of the average for a single fired heater while it is 1.2 times of the average for a double fired heater. That explains why the double fired heater has longer run length as it has lower flux rate and hence lower TWT than the single fired heater.

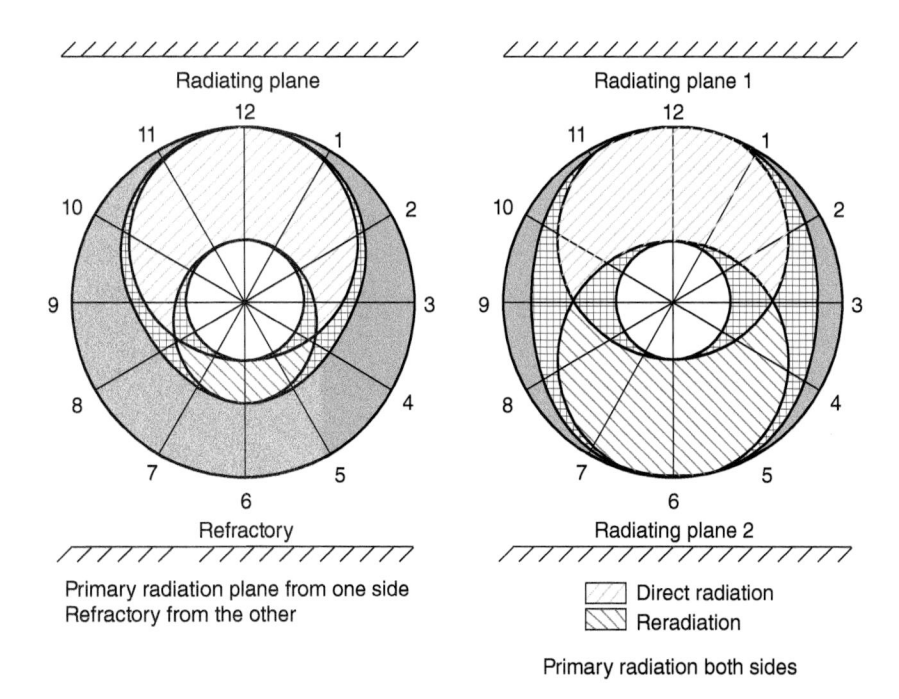

The tube thinning follows the same pattern of flux distribution. Figure 8.4 shows a fired heater tube with severe thinning creep caused by internal coking especially on the fireside of the tube. The internal coking follows the same pattern with much greater coke thickness at the front face facing the flame. This is why inspectors concentrate their tube inspections on the fireside of the tube. As a reference, Table 8.1 gives the typical maximum heat flux.

Heat distribution throughout the fired heater is not even. Radiation section makes up 70–75% of total process heat transfer while convection section accounts for 25–30%, which can be observed in Figure 8.2. Different fuels have different heat distribution. For a gas fired heater, 1/3 of the heat transfer in the radiant section is flame radiation and 2/3 is hot gas radiation. If the flame height is too high, there is not enough residence time for the hot gas cloud represented as "B" in Figure 8.2 to transfer heat to the tubes. This situation occurs when a long flame burner is placed in a short firebox. Oil firing is different. The oil flame has very high flame radiation, so approximately 2/3 of the heat transfer in the radiant section is flame radiation and 1/3 is hot gas radiation.

Figure 8.3 Flux distribution around fired heater tube.

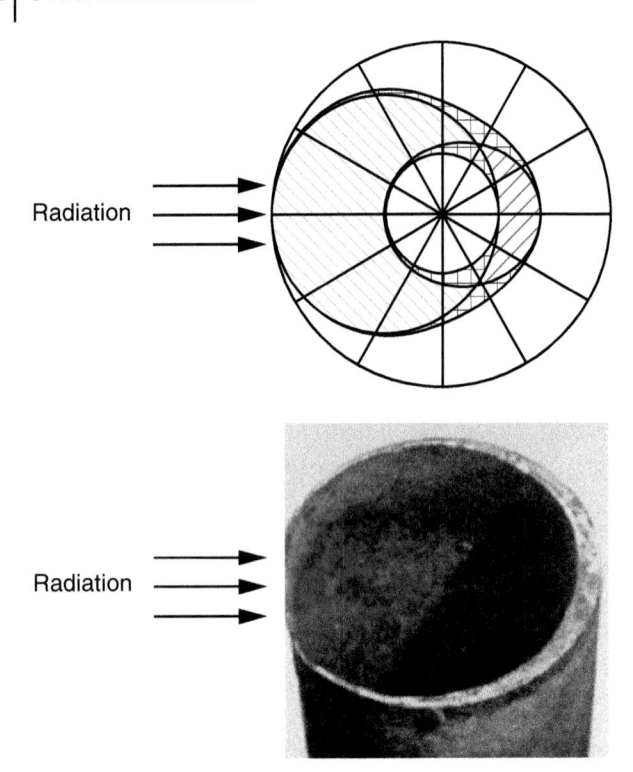

Figure 8.4 Tube thinning follows the flux distribution.

Table 8.1 Maximum flux rate used in an operating company, Btu/h-ft^2.

Vertical cylindrical with tube length 20–30 ft	12 000
Vertical cylindrical with tube length > 30 ft	13 000
Cabin	14 000
Double fired U-tube	22 000

Oil and gas firing have different combustion characteristics. Oil firing is governed by flame radiation with the presence of visible flame light waves. In contrast, hot gas radiation produced by combustion is governed by gas firing. Oil has high emissivity close to one and thus be able to drive the heat through the ash resistance.

8.2.2 Burner to Tube Clearance

Burner to tube clearance is very important in heater design because flame radiation is directly proportional to the square of the distance to the tube. Small burner to tube clearance can result in flame impingement, hot spots, and tube failure. That is why most heater failures can be traced to flame impingement due to burners placed too close to the tubes. For example, consider a 5′-0″ burner to tube clearance versus 3′-0″ spacing, the smaller spacing case results in 2.8 [=(5/3)^2] times of the flame radiation as the larger spacing.

8.2.3 Burner Selection

There are four types of burners, namely standard, premixed, staged air/fuel (low NO$_x$), and next generation (ultralow NO$_x$). Standard gas and premixed burners have luminous flames. The combustion reaction occurs within the visible flame boundaries. Ultralow NO$_x$ and next-generation burners have nonluminous flames and much of the combustion reaction is not visible.

8.2.3.1 NO$_x$ Emission

NO$_x$ emission is an important environmental issue for the process industry today. NO$_x$ is formed by nitrogen and oxygen reacting at the peak temperatures of the flames. A standard gas burner produces 100 ppm NO$_x$; staged air gas burner 80 ppm; staged gas burners 40 ppm; ultralow NO$_x$ gas burners 30 ppm; and the latest generation ultralow NO$_x$ gas burners produce 8–15 ppm NO$_x$. SO$_x$ is controlled by the sulfur in the fuel. Many plants have sulfur limits that require burning low sulfur fuel oil. CO should be less than 20 ppm.

8.2.3.2 Objective of Burner Selection

The objective of burner selection is to determine burner type and configuration in order to obtain the desired heat flux profile to meet process heating demand. The combustion space and shape may be determined by physical, mechanical, or structural factors, but that space must be able to accommodate efficient aerodynamic mixing and combustion of the fuel, and generate the desired heat flux profile for the product. The heat release and hence heat flux generated from burner flames is not even. It is generally high in the region near to the burner port, where fuel and air are plentiful, and reduces as the flame develops, owing to the depleting fuel content, and by losing heat to its surroundings. The burner designer can adjust this profile from burner type and configuration and flame envelope although it never achieves uniform flux distribution.

8.2.3.3 Flame Envelope

The flame envelope is defined as the visible combustion length and diameter. The flame length should be 1/3 to 1/2 of the firebox height. The hot combustion gases need residence time to transfer the heat to the tubes. Many burners have flame diameters that are between 1 and 1.5 times the diameter of the burner tile. Since the tile diameters are often larger for ultralow NO$_x$ and latest generation burners, the flame diameters at the base of the flame may be slightly larger. The flame diameter often expands, giving a wider flame at the top.

Ultralow NO$_x$ and latest generation burners have longer flame lengths than conventional burners. Longer flame lengths change the heat transfer profile in the firebox and can result in flame impingement on the tubes.

8.2.3.4 Physical Dimension of Firebox

Optimized designs have burner spacing that is designed to have gaps between the flame envelopes. Since the tile diameters are often larger for ultralow NO_x and latest generation burners, retrofits can result in closer turner-to-burner spacing and flame interaction. Flame interaction can produce longer flames and higher NO_x. Flame interaction can interrupt the flue gas convection currents in the firebox, reducing the amount of entrained flue gas in the flame envelope. This condition increases the NO_x levels. Ultralow NO_x and latest generation burners should be spaced far enough apart to allow even flue gas recirculation currents to the burners.

The burner centerline to burner centerline dimension is one of the most important dimensions in the firebox tube. Many tube failures are caused by flame and hot gas impingement. When ultralow NO_x and latest generation burners are being retrofitted, the larger size of the flame envelope must be evaluated. Firebox convection currents can push the slow burning flames into the tubes.

Flame impingement on refractory often causes damage. When ultralow NO_x and latest generation burners are being retrofitted, the larger burner diameter may result in the burners being spaced closer to the refractory. Unshielded refractory may require hot face protection.

Many heaters are designed for flame lengths that are 1/3 to 1/2 the firebox height. Ultralow NO_x and latest generation burners typically have flame heights of 2–2.5 ft/million Btu (2–2.5 m/MW). Longer flame heights from ultralow NO_x and latest generation burners may change the heat transfer profile in the firebox. The longer flames may result in flame or hot gas impingement on the roof and shock tubes. In this case, the solution is to change burners. Some older heaters have very short firebox heights and may not be suitable for retrofits to ultralow NO_x and latest generation burners.

8.2.3.5 Process-Related Parameters

Ultralow NO_x and latest generation burners have longer flames that change the heat flux profile. This is especially important on thermal cracking heaters such as cokers and visbreakers in oil refineries. The longer flames may increase the bridge wall temperature (BWT) and change the duty split between the radiant section and convection section.

The location of the maximum tube metal temperature (TMT) changes as the heat flux profile changes. Retrofitting ultralow NO_x and latest generation burners in short fireboxes can result in high metal temperatures for roof and shock tubes.

Ultralow NO_x and latest generation burners may have less turndown capability than conventional burners. High CO levels can occur when firebox temperatures are below 1240 °F. Flame instability and flameout can occur when firebox temperatures are below 1200 °F. Since ultralow NO_x and latest generation burners are often designed at the limit of stability, a fuel composition change may cause a stability problem.

The proper design basis for the burner selection is extremely important. Sometimes the process requirements have changed significantly since the fired heater was designed. Important design basis items include: (i) emission requirements; (ii) process duty requirements; (iii) turndown requirements; (iv) fuel composition ranges; (v) fuel pressure; and (vi) start-up considerations.

The guideline for burner selection is to select the most appropriate burner technology while meeting the NO_x emission limit. Reliability should be placed as higher priority than cost in burner selection because industrial applications show that 90% of fired heater problems come from poorly maintained and operated burners. Although it could be more expensive with the best burner technology, the money spent is worthwhile as burners cost only 5–10% of fired heater overall cost but it could avoid 90% of fired heater problems.

8.2.4 Fuel Conditioning System

Poor fuel conditioning could cause problems in burners and combustion. While many conventional burners have orifices 1/8″ (3 mm) and larger, ultralow NO_x and latest generation burners often have tip drillings of 1/16″ (1.5 mm). These small orifices are extremely prone to plugging and require special protection. Most fuel systems are designed with carbon steel piping. Pipe scale forms from corrosion products and plugs the burner tips. Although tip plugging is unacceptable for any burner, it is even more important not to have plugged tips on ultralow NO_x and latest generation burners because plugged tips can result in stability problems and higher emissions.

Many companies have installed austenitic piping downstream of the fuel coalescer/filter to prevent scale plugging problems:

- Coalescers or fuel filters are required on all ultralow NO_x and latest generation burner installations to prevent tip plugging problems. The coalescers are often designed to remove liquid aerosol particles down to 0.3–0.6 μm. Some companies install pipe strainers upstream of the coalescer to prevent particulate fouling of the coalescing elements.
- Piping insulation and tracing are required on fuel piping downstream of the coalescer/fuel filter to prevent condensation in fuel piping. Some companies have used a fuel gas heater to superheat the fuel gas in place of pipe tracing. Unsaturated hydrocarbons can quickly plug the smaller burner tip holes on ultralow NO_x and latest generation burners.

8.3 Fired Heater Operation for High Reliability

Fired heater capacity for critical processes is usually pushed hard for more production and thus the fired heaters are operated near or at the operation limits. It is essential to make sure the fired heater is running in a safe and reliable manner with the following key operating reliability parameters within acceptable limits:

- Draft control: Avoid positive pressure to prevent safety hazards and provide sufficient primary air for burners.
- High BWT: BWT directly relates to flux rate and indicates how hard a heater is running.
- TWT or TMT: Identify root cause for high TWT operation.
- Flame impingement: The most common reliability hazard for fired heaters.
- Excess air or O_2 content: Optimal O_2% is the balance between reliability and efficiency.
- Flame pattern: Visualize the flame shape, height, and color to identify abnormal combustion problems.

8.3.1 Draft

There are two types of draft: one is natural draft and the other is forced draft. For natural draft, the draft depends on the density difference between hot flue gas and ambient air. Thus, stack height must be sufficient in order to provide adequate draft while stack damper opening must be adjusted properly in operation at the same time. For forced draft, stack height can be short as fan is used for providing air. Thus, stack height is only set based on dispersion requirements. Similar to natural draft, stack opening must be adjusted properly in operation. The key objective of draft control for both natural and forced draft is to avoid positive pressure inside the heater to prevent damage or safety hazards and provide sufficient combustion air (primary air) at the same time. A proper draft control is to maintain the draft at the range of 0.1–0.2″ WC (water column) vacuum measured underneath the convection tubes or at the bridge wall (line Y; Figure 8.5a). This can be achieved by adjustment of both stack damper and air register. With this draft, sufficient air can be drawn in through the burners as primary air to obtain flame stability while secondary air is provided by air register for O_2 control. However, too high or low draft must be avoided. A too high draft could occur when the damper is widely open and register fully closed. This could result in a too high vacuum in the stack and could increase cold air leakage into the heater (Figure 8.5b). Excessive draft could cause flame liftoff the burners touching the tubes and this could lead to serious damage to the heater. A too low draft corresponds to the case when the damper is almost closed and the air register widely open. In this case, positive pressure at bridge wall could be developed which forces hot flue gases flowing outward through leaks in the convection section (Figure 8.5c). This could lead to serious structural damage. The draft profiles for these three cases are provided in Figure 8.5d.

Weather change could cause draft fluctuation. For example, when strong winds occur and cause draft fluctuation, the damper opening should be increased gradually to maintain flame stability. On the other hand, in a windy weather, if the heater faces toward the wind with the highest static atmospheric pressure, this may result in a too high draft. In this case, the damper should be closed slightly.

8.3.2 Bridge Wall Temperature

A high BWT measurement indicates the heater operates at high radiant flux rates. BWT is the key reliability parameter for fired heater as high BWT or TWT can cause mechanical failures on tube sheet supports and refractory. Majority of heater failures are accompanied by high BWT. The general guideline for BWT is not to exceed the mechanical design limit that tube sheet supports and the BWT limit depends on design. High BWT could be caused by long flames, not enough flue gas residence time, and external fouling on the tubes.

8.3.3 Tube Wall Temperature

The skin temperature is the process temperature inside the tube, plus the temperature differences across the film and metal resistances. The film resistance is usually larger than the metal resistance. It is calculated by taking the peak flux and dividing by the heat transfer coefficient. The heat transfer coefficient is usually 200–500 Btu/(h·ft^2·°F), which provides a typical film resistance of 45–80 °F. The metal resistance is much smaller than the film resistance. It is calculated by taking the peak flux and dividing by the thermal conductivity of the metal. The thermal conductivity is usually 12–16 Btu/(h·ft·°F), which results in a typical metal resistance of 15—20 °F (8–11 °C). The exception is for thick-walled tubes which could have a metal resistance as high as 80 °F (44 °C).

Tube wall or skin temperature is an important reliability parameter and should be closely monitored and guidelines for tube life can be developed. Guidelines should be effectively communicated to operators so that appropriate tube temperature can be determined that could meet the production requirement while minimizing the risk of tube damage. It is important for operators to know that overfiring is the main cause of tube damage.

8.3 Fired Heater Operation for High Reliability | 195

(a)

Damper

Negative draft
0.1–0.2′ WC vacuum

Y

Positive draft

Air registers

(b)

Damper wide open

Cold air sucked in

Too high vacuum

Y

Positive pressure

Air registers wideclosed

(c)

Damper mostly closed

Hot flue gas blown out

Y

Air registers wide open

(b) Too high draft
(a) Proper draft
(c) Too low draft

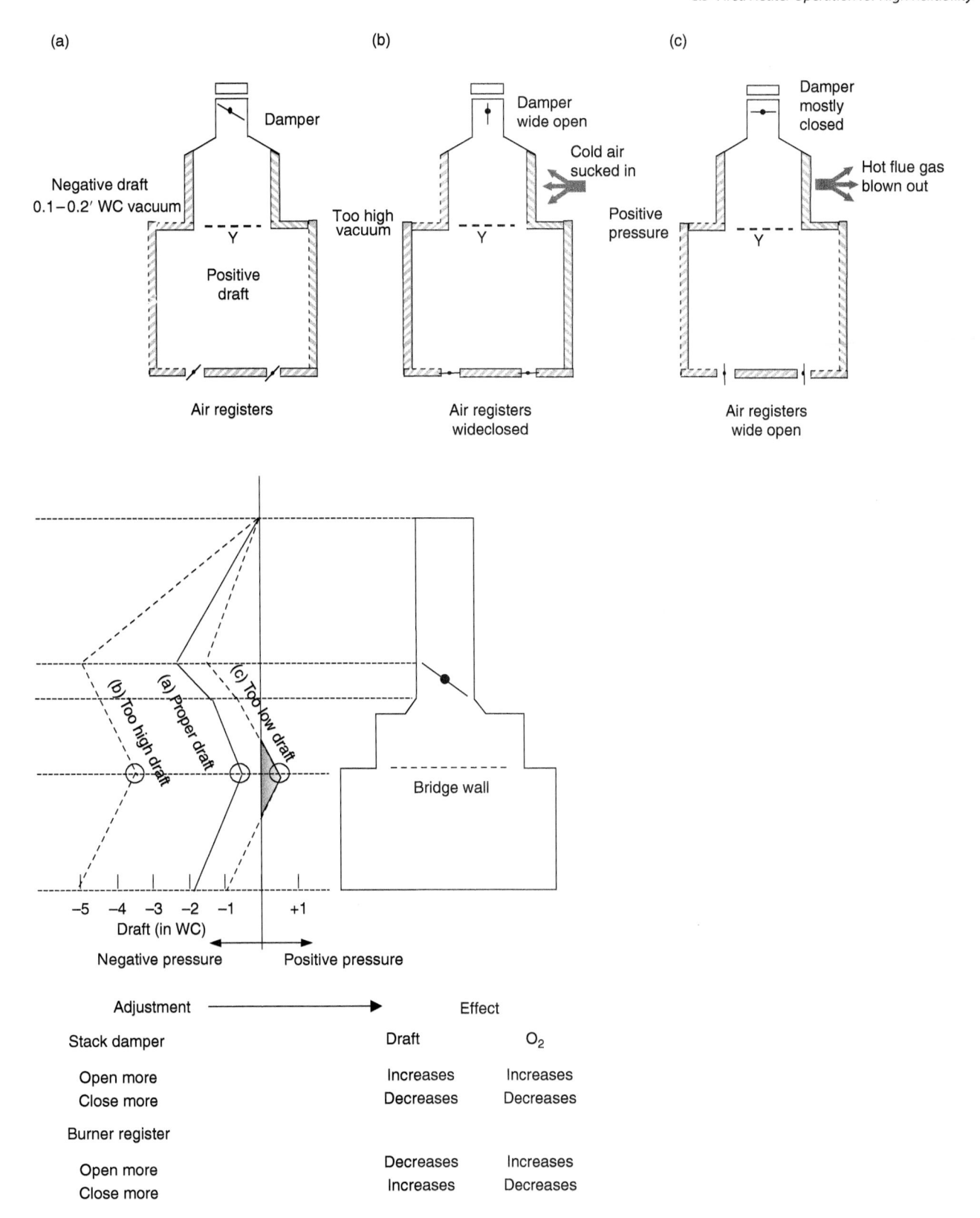

Bridge wall

−5 −4 −3 −2 −1 +1
Draft (in WC)

Negative pressure | Positive pressure

Adjustment	Effect	
Stack damper	Draft	O_2
Open more	Increases	Increases
Close more	Decreases	Decreases
Burner register		
Open more	Decreases	Increases
Close more	Increases	Decreases

Figure 8.5 (a) Proper draft control. (b) Too high draft. (c) Too low draft. (d) Correct and incorrect draft.

Process plants use skin thermocouples and infrared pyrometers to monitor TWTs.

It is very important to monitor the amount of scale on tubes in order to measure coking/fouling/corrosion rates. This can be achieved by thermocouple and infrared pyrometer monitoring program. The scales on tube increase TWT or skin temperature. 10 mils (0.01″) scale on tube could raise tube surface temperature by 100 °F. The common ways to get rid of scale is to sandblast the scale off the tubes while ceramic coating on tubes is a preventive measure; but it is expensive.

8.3.4 Flame Impingement

Flame impingement could be caused by low air as well as burner tip fouling, which could be avoided by adjusting excessive air and fuel pressure. Figure 8.6 shows a fired heater operating with severe flame impingement in which a long flame reaches tubes and the tube front receives almost six times as much heat as the back side of tube does. The best way to know if hard flame impingement is formed is to view the firebox using the glasses especially for that purpose. These glasses eliminate the glare and bright haze and make it possible to view real flame positions.

The following guidelines for better mixing could be used to determine the root causes for flame impingement:

- Primary air is used for achieving flame stability while secondary air for O_2/NO_x control. Thus, primary air should be increased via damper opening to a limit beyond which the fame will liftoff the burners. Excess air is provided by adjusting secondary air via register. Too much and too little secondary air gives poor combustion. This is because a minimum excess air is required for flame stability and too much excess air reduces flame temperature and hence efficiency drops and NO_x increases as the result.
- Close ignition ports, peep doors, and other holes around burners. Combustion air only mixes well with fuel gas when it flows through the air registers.
- At turndown operation, some of the burners may be blanketed off and do not forget to close the air registers for the idle burners. Burners work more satisfactory close to design capacity.
- Plugged burners require more excess air for combustion but too much excess air could liftoff flame. Sulfur deposits is the common cause of burner plugging and a solution is to prevent oxygen from entering the fuel gas system as it could combine with hydrogen sulfide in the fuel gas to form NH_3Cl.

8.3.5 Tube Life

Realistic average tube life can be assessed based on creep measurement and metallurgic examination. The guidelines derived from assessment should be illustrated to operators for the serious damage that could occur by operating a fired heater over the TWT limit. In general, 18 °F increase over the TWT limit could half the life of a heater. 30 °F over the TWT limit could shorten a heater's life substantially and cause rapid failure when a heater is in the creep range. It is important to know that it is the peak TWT which should not exceed the limit instead of the average TWT.

A fired heater is not operated uniform over the entire run as it could run light in turndown operation and harder in full capacity and toward end of run for reaction heaters. To estimate the effects of changing TWT, corrosion rates, and pressure, metallurgic examination can be applied to estimate the remaining life of tubes. Knowing the tube life not only prevents premature tube failure but also identifies the need of metal upgrade if operating skin temperature increases over time.

8.3.6 Excess Air or O_2 Content

It must be stated that optimal O_2% is the balance between safety and efficiency. There are several signs visible when

Figure 8.6 An example of flame impingement.

a firebox is short of combustion air: a hazy flame; regular thumping sound; and long flame touching the tubes.

One of reasons causing insufficient air is aggressive $O_2\%$ management regardless of burner conditions. Another root cause of insufficient air is the O_2 measurement based on the flue gas sample taken from the stack. This measurement is not accurate representative of the oxygen available in the firebox. Leaks in the convection section allow air to bypass the firebox and exit in the stack and contribute to the $O_2\%$ measured in the stack. When air registers are adjusted based on the oxygen level measured from stack, the firebox could be in short of air. On the other hand, air leak is waste of hot flue gas for heating up cold air that is sucked into the convection section.

The cost-effective activities include seal welding of casing, mudding up header boxes, using high-temperature sealants. Leaks through roof penetration are also a major source of air leak, which should be inspected during turnaround. These activities are especially important for NO_x control.

8.3.7 Flame Pattern

Proper control of combustion air is the key to make complete combustion and stable flame and thus avoid flame impingement. Lower fuel pressure also helps to avoid flame impingement. When the amount of excess air is appropriate, flame is orange and flue gas from stack is light gray. With sufficient air, if flame is long with much smoke, burners may have problems.

Figure 8.7 shows a good combustion with orange color and a proper flame height of about 1/3 to 1/2 of the firebox height. In contrast, Figure 8.8 displays a poor combustion with plugged gas tips on the first burner. There is a strong haze from the flame of the first burner indicating incomplete combustion. The burner tip plugging could be reduced by using fuel gas coalescer and steam heater.

8.4 Efficient Fired Heater Operation

Operators understand the importance of maintaining fired heaters in safe and reliable operation. The response from operators to this priority could go to another extreme: run fired heaters with too much excess air. The result of much excess air is much reduced flame length and thus the risk of flame impingement is minimized. However, the price for too much excess air is the higher operating cost from burning extra fuel. Therefore, there is an optimization need for excess air.

Too much excess air is waste of fuel as cold air needs to be heated up from ambient to stack temperature. Figure 8.9 shows the fuel saved by dropping 1% of O_2 from reduced excess air. For example, for a heater with operating duty of 200 MMBtu/h with stack temperature at 500 °F, a reduction of 1% oxygen saves 1 MMBtu/h of fuel which is worth $72/day or $26 280/year for fuel price at $3/MMBtu. Reducing $O_2\%$ from 7 to 3%, the saving could worth around $100 000/year. If fuel is priced at $6/MMBtu, dropping 4% of O_2 could save $400 000/year.

Figure 8.7 Good flame color and height.

Figure 8.8 Poor flame pattern from the first burner.

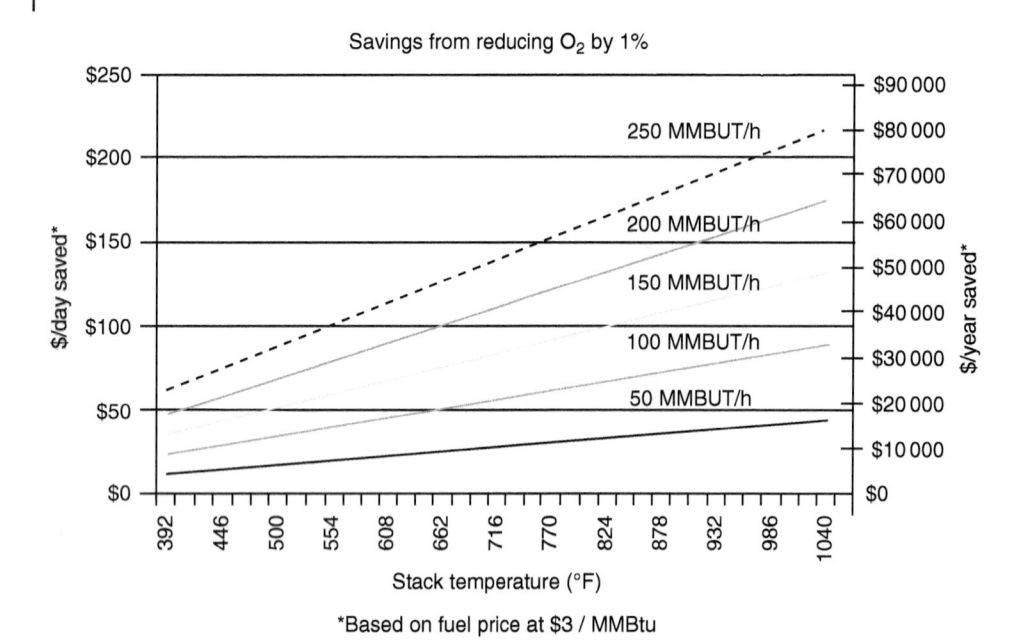

Figure 8.9 Dollar value for reducing O$_2$% by 1%.

Three percent of O$_2$ is used as the basis for benefit calculation here as 3% is a typical limit for industrial fired heaters. However, do not start O$_2$% reduction before burners are in good working conditions and O$_2$ analyzers are installed and calibrated with corrected readings.

Similarly, reducing stack temperature could improve heater efficiency more than O$_2$% optimization. Every 40 °F increase in stack temperature is equivalent to 1% fuel efficiency improvement. For example, a small heater with duty of 50 MMBtu/h does not have a convection section and stack temperature is at 1250 °F. If the flue gas is routed to the convection section of a large heater in a close location, the stack temperature could be reduced to 500 °F. Capture of this waste heat could worth $60/day and $220 000/year for fuel price at $6/MMBtu. In general, reducing stack temperature is more of a design issue, for example, installing steam generator and economizer in the convection section to recover waste heat. In contrast, O$_2$ level is an operation issue which can be controlled by adjusting secondary air via air register.

8.4.1 O$_2$ Analyzer

Fired heaters have either forced draft fans or induced draft fans to control air to the burners. This allows control of oxygen amount by direct measurement of air and fuel flow rates. Large and efficient process fired heaters with natural draft burners usually have induced draft fans. It is desirable to have control systems devised to maintain the desired amount of excess air. With O$_2$ analyzers, the control system adjusts damper openings automatically to control O$_2$ subject to a limitation on absolute draft level. Relative small fired heaters can also justify O$_2$ analyzers for energy saving.

To obtain more uniform O$_2$ reading, every 30 ft should have one sample point and sample points should be installed downstream from the convection section. The requirement is that there should be minimum air leakage into the convection section to avoid false O$_2$ readings. In general, sample points should not be located in the radiation section for the reason that flue gas from different burners are not well mixed. Otherwise, the O$_2$ reading would mainly reflect the operation of the burners close to the sample points. The exception to placing the oxygen analyzer downstream of the convection section is for fired heaters with high tube temperatures in the convection section. This is because it is desirable to monitor radiant section oxygen to avoid afterburning.

8.4.2 Why Need to Optimize Excess Air

In an ideal combustion of fuel purely based on stoichiometric conversion, fuel is burnt to CO$_2$ and H$_2$O 100% with 0% excess air so that there is no oxygen left in the combustion flue gas. However, in reality, industrial fired heaters require excess air. To achieve complete combustion, minimum 10–15% excess air (2–3% O$_2$ in flue gas) is required for fuel gas. Otherwise, carbon monoxide and unburned hydrocarbon could appear in flue gas leaving stack. Fuel oil usually requires 5–10% higher excess air than fuel gas. In other words, minimum 15–20% excess air (3–4% O$_2$) is required for fuel oil for complete combustion.

Figure 8.10 Optimizing excess air.

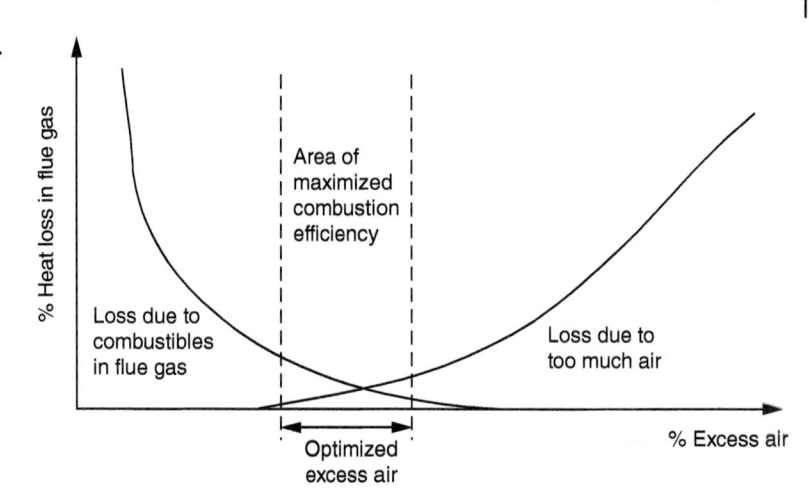

Older heater with poor burner conditions could have O_2% higher than 5%. This is because many older heaters are not designed for low O_2 operations. The burner flame will be very poor below the 0.15″ H_2O. High excess air is required for these operations. For fuel oil used as fuel, black smoke is visible from stack under incomplete combustion. For fuel gas and natural gas, smoke is not visible from stack, but incomplete combustion can be measured by CO concentration in the flue gas.

Typically, 1% CO measure in the stack flue gas implies that 3–4% of fuel is wasted. Because O_2% is measured online, thus, efficient and reliable operation of heaters should maintain O_2% as close (but not less than) to the limit as possible. It is important to make sure that O_2% is not a false indication as air coming from leaking could contribute to the O_2% measured. Too little excess air available for combustion could cause flame impingement to tubes and result in local hot spots, and coking on tube eventually causes severe tube damage. Another consequence of too little excess air is afterburning in the convection section which could result in elevated tube temperature, which is the root cause for premature tube failure and sagging of horizontal tubes. This is because the fired heater undergoes incomplete combustion and thus the combustibles or CO in the flue gas increases. The incomplete combustion makes lazy flames. These long flames can reach tubes in the radiation section and even convection section. In the worst case, flames could reach the exit of stack.

So what is optimal excess air or O_2%? The basis is to achieve complete combustion. For reliability considerations, optimal O_2% should be determined with a safety margin on top of minimum excess air when burners are under good conditions. The safety margin depends on specific technology, design, and conditions for each heater as well as measurement. Figure 8.10 is commonly used to explain qualitatively the existence of optimal excess air.

The more rigorous way than O_2 measurement is to measure CO in the flue gas. This can be accomplished by measuring combustibles in the flue gas. Combustibles here refer to the products of incomplete combustion including carbon monoxide (CO), hydrogen, and trace hydrocarbons while CO accounts for the majority of combustibles. For consistency with O_2 measurement, the combustibles measurement should be taken in the same location as the O_2 analyzer. With reliable combustibles measurement available for ppm concentration, it allows the O_2% level to be reduced safely (safety margin) until the combustibles start to increase (Figure 8.11). This is the optimal O_2% for the heater.

8.4.3 Draft Effects

Efficient heater operation requires that excess air entering the convection section be minimized, which is indicated by a very small negative pressure at the convection section inlet. To achieve this, it should have a well-balanced draft pressure profile between firebox and stack. The hot gas pushes so that the pressure is always greatest at the firewall while the stack draft pulls. When

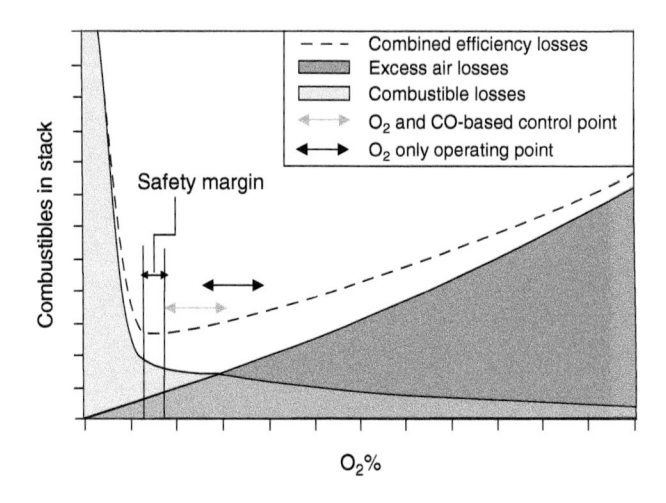

Figure 8.11 Determining optimal O_2% level.

this draft is correctly balanced, the pressure at the bridge wall should be around 0.1–0.2′ WC. Too much draft allows cold air leakage into fired box resulting in wasted fuel.

8.4.4 Air Preheat Effects

Air preheating is a classic example of upgrading low-valued heat. This is done by providing heat to raise the combustion air temperature from the ambient temperature using waste heat. Air preheat can be accomplished via low-pressure steam or flue gas. Typically, air preheat can increase fired heater efficiency up to 5%, which is more significant than reducing O_2%.

8.4.5 Too Little Excess Air and Reliability

Too little excess air could result in flame impingement and afterburn in the convection section, which impose reliability risks. With too little excess air, incomplete combustion occurs and reduces flame temperature, which might encourage operators to increase fuel flow in order to increase heater duty. Increased fuel with too little excess air enhances afterburn and could be dangerous.

8.4.6 Too Much Excess Air

This is inefficient operation and should be avoided. According to Kenney (1984), the common causes of too much excess air are:

- Improper draft control
- Air leakage into the convection section
- Improper calibration of O_2 analyzer
- Faulty burner operation: (i) dirt burners; (ii) poor maintenance on air doors; and (iii) dual fuel burners needed

8.4.7 Availability and Efficiency

Making fired heater in high availability is desirable for continuous production without interruption. As the consequence, a plant can achieve high profit and high energy efficiency at the same time. Experience from the industry indicates that high availability is the major contributor to improved energy efficiency.

8.4.8 Guidelines for Fired Heater Reliable and Efficient Operation

Draft and excess air control should be considered together in operation. This is because the draft provides primary air while air register delivers secondary air for burners. As discussed before, both air supply could have an effect on reliability and efficiency. The systematical method for optimizing draft and excess air together is proposed in Figure 8.12.

We feel the need to provide additional comments on excess air as many plants have an O_2 reduction program. O_2 reduction (or minimum excess air) must be built upon the basis of proper draft control. Minimum excess air for the fired heater can be obtained when it is reduced to the point where combustibles begin to appear in the stack. For modern fired heaters, this occurs at 8% excess air equivalent to 1.8% of oxygen level in the flue gas. However, practical constraints prevent achieving this minimum excess air in operation, and these constraints include variations in fuel quality, feed rates, and other process variables. Thus, operation without flame impingement sets the limit for practical minimum excess air. The optimal flue gas O_2 concentration depends on the heater duty, burner design, types of fuel, and burner performance.

To achieve the limit, the first step is monitoring O_2%. O_2 measurement must reflect the true amount of excess air and air leaks must be eliminated. The following guidelines can be used for operation reference.

- O_2 analyzers should be installed below the convection section instead of stack. If not, a correction factor must be developed for the readings with a portable analyzer. O_2 analyzers should be calibrated once per week.
- Efficiency based on stack temperature and corrected O_2% should be reported daily.
- Draft should be monitored and maintained as required for the specific fired heater design. Even fired heaters without draft control should be periodically checked.
- Convective section air leakage should be measured once per shift and determined as the difference in convective inlet and outlet O_2. The source of leakage should be identified via inspection and eliminated. Ideally, all oxygen should enter the fired heater through the burners.
- Coil flow paths should be balanced within ±5% accuracy once per shift in order to obtain equal outlet temperatures. On large fired heaters, this may be as often as every two hours, or continuously with control systems.
- In cases of turnarounds and large load changes, flue-gas parameters (draft, O_2%, etc.) should be checked and adjusted as necessary.
- Soot blowers on oil-fired fired heaters and boilers should be activated once a shift. The operator should observe which ones actually rotate and report (in writing) those soot blowers that have failed. Where operability of soot blowers is less than 70%, an alternative

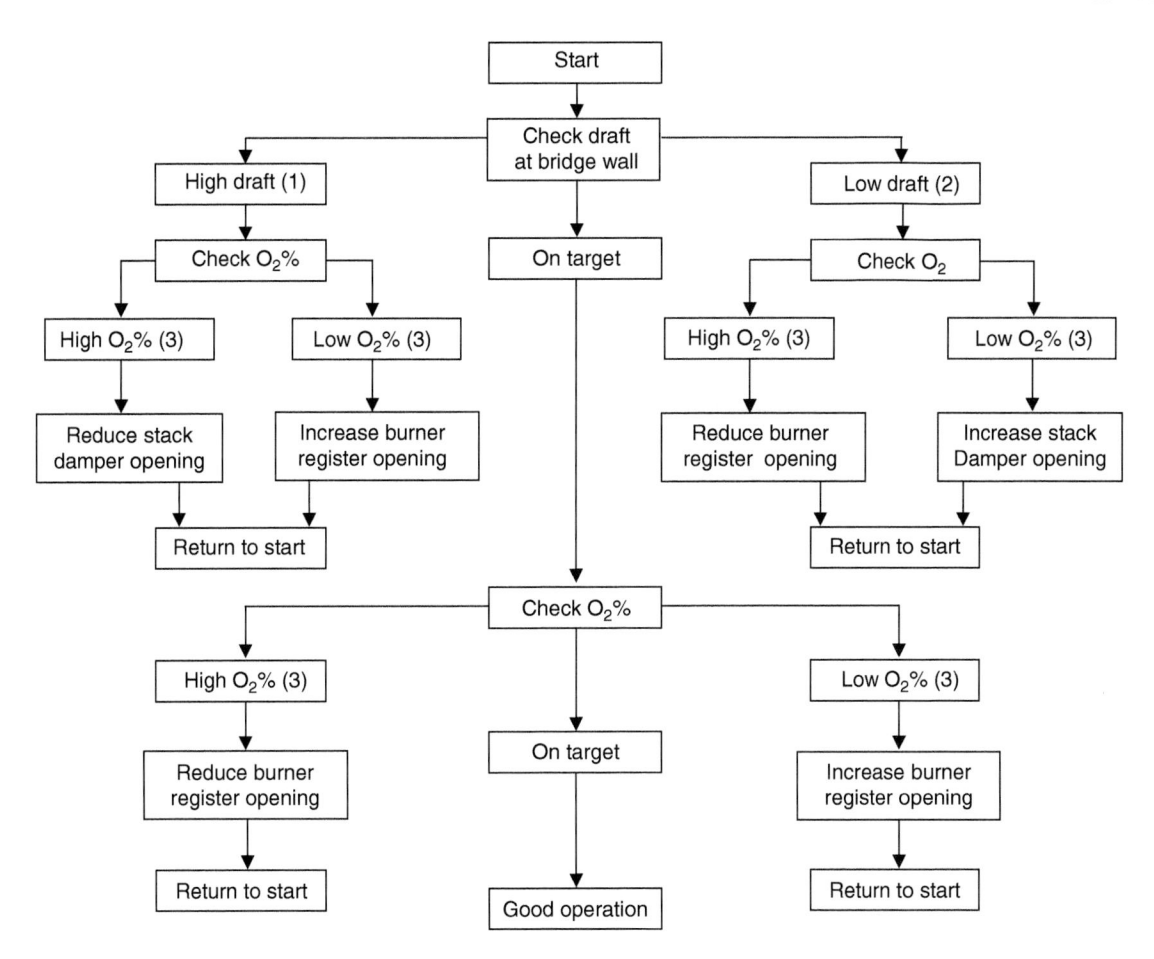

```
                          Start
                            │
                            ▼
                    Check draft
                    at bridge wall
        ┌───────────────┼───────────────┐
        ▼               ▼               ▼
   High draft (1)   On target      Low draft (2)
        │                               │
        ▼                               ▼
    Check O₂%                       Check O₂
     ┌──┴──┐                         ┌──┴──┐
     ▼     ▼                         ▼     ▼
  High O₂% Low O₂%              High O₂%  Low O₂%
   (3)     (3)                    (3)      (3)
    │       │                      │        │
    ▼       ▼                      ▼        ▼
 Reduce   Increase              Reduce    Increase
 stack    burner                burner     stack
 damper   register              register   Damper
 opening  opening               opening    opening
    │       │                      │        │
    ▼       ▼                      ▼        ▼
  Return to start              Return to start
                            │
                            ▼
                       Check O₂%
        ┌───────────────┼───────────────┐
        ▼               ▼               ▼
   High O₂% (3)     On target       Low O₂% (3)
        │               │               │
        ▼               ▼               ▼
    Reduce burner   Good operation  Increase burner
    register opening                register opening
        │                               │
        ▼                               ▼
   Return to start                 Return to start
```

(1) High draft – fire box pressure more negative.

(2) Low draft – fire box pressure more positive.

(3) Low or high O₂%– O₂% is above or below target.

Figure 8.12 Integrated draft and O$_2$ control.

plan for cleaning should be prepared and executed. This may include onstream water-washing.
- The need for onstream cleaning of outside tube surface should be evaluated. This may include water-washing of both convective and radiant sections.

8.5 Fired Heater Revamp

In general, fired heaters are revamped for capacity expansion, process conversion changes, energy efficiency, and NO$_x$ reduction. For capacity-expansion revamps, the type of limitation for the revamp is usually the same as for the original design. In conversion revamps, one type of process technology is converted to another. Thus, in conversion revamps, a heater designed for one service may be used in a new service. Therefore, the type of heater limitation may be different for the new service.

Heaters encounter with four major design limitations: heat flux, process pressure drop, TWT, and BWT. Heat flux-limited heaters are usually characterized with high pressure ΔP (>20 psi) and most general service heaters fall into this flux-limited category. Typically, the flux limit for single fired heaters is around 10 000 Btu/ft^2-h. Small heaters (<10 MM Btu/h process duty) have lower flux rates. For revamps, the heat flux limit can go up to 12 000 Btu/ft^2-h.

Double fired heaters are usually TWT-limited. However, flux limits are specified for revamps, which depend on specific services. The limits are provided by heater specialists.

TWT-limited heaters are characterized by low process ΔP (2–6 psi). Because of low ΔP, the heaters have low tube mass velocities, which result in low heat transfer coefficients, and thus high TWTs. TWT-limited heaters usually occur in high-temperature processes. For example, TWT of 800 °F is used for killed carbon steel heaters. The chrome (Cr) limits are based on inhibiting tube oxidation and use a limit of 1075–1100 °F per recent data. For stainless steel (SS), process temperature limits usually occur before reaching the TWT limit. There can be exception for high-pressure heaters.

On the other hand, BWT-limited heaters are usually encountered when a heater with conventional burners is replaced with low NO_x/ultralow NO_x/new generation burners or the revamp requires higher turndown ability. Flame instability and flameout can occur at low BWT. For fuel gas firing with ultralow NO_x and next-generation burners, BWT should be greater than 1200 °F. For low NO_x burners, the BWT should be greater than 1000 °F. For oil firing in combination burners, the BWT shall be greater than 1200 °F. When determining the BWT limit, the burner spacing and BWTs at normal and turndown operations must be investigated on the stability and operation of burners.

There is much more to discuss about heater revamp which is beyond the scope of this chapter. As a general recommendation, heater revamp projects should be conducted by heater specialists who not only have good knowledge of the heaters but also process that the heaters serve.

Nomenclature
BWT   bridge wall temperature
TMT   tube metal temperature
TWT   tube wall temperature

References

Jenkins, B.G. and Boothman, M. (1996). Combustion science: a contradiction in terms. *Petroleum Technology Quarterly* Autumn: 71–76.

Kenney, W.F. (1984). *Energy Conservation in the Process Industries*. Academic Press.

9

Compressor Assessment

9.1 Introduction

Compressors are important equipment in the process industries. Their primary purpose is to compress air or gas into a smaller volume and thus simultaneously raise the pressure and also the temperature. The basic principles of compression are summarized in Figure 9.1.

Compressors are very expensive and they account for major part of capital costs in the overall process. Selecting the right type of compressor for a specific application is important for both cost and reliability considerations. It has been realized that one of the most common reliability issues for compressor is caused by improper selection of a compressor.

There are numerous types of compressors. They can be categorized under two basic types: positive displacement and dynamic. Positive displacement compressors include piston or reciprocating, screw, vane, and lobe compressors. Axial and radial (or centrifugal) compressors belong to the dynamic type as the required pressure rise and flow are imparted to the fluid by transferring kinetic energy to the process gas.

Positive displacement compressors are of constant volume. The volumetric flow rate is not affected by changes in gas characterizations (pressure, temperature, or molecular weight [MW]). In contrast, with dynamic compressors, volumetric flow rate is affected by changes in gas characterizations.

Positive displacement compressors are generally suitable for gases with low flow, low MW gases, and requiring high compression ratios. Centrifugal compressors can handle higher flow rates. Centrifugal compressors are head-limited. Head is a function of compression ratio and MW. The gas MW will define the allowable compression ratio per centrifugal casing. Axial flow-type compressors are used for high flow and low head applications.

Multistage centrifugal and reciprocating compressors are commonly used in the refinery for recycle gas, net gas, and hydrogen make-up services. The other types of compressors previously mentioned are used for more specialty applications. Multistage centrifugal compressors are the most common type used since they have wide operating range, are reliable and efficient, and are less affected by performance degradation due to fouling compared to reciprocating type. Furthermore, reciprocating compressors have moving and wearing parts so they are typically spared which raises the cost compared with centrifugal compressors. Therefore, in this chapter, multistage centrifugal compressor is the focus of discussions. For readers who like to dig deeper into the subject, detailed discussions can be found in the work of Sorokes (2013).

9.2 Types of Compressors

Centrifugal compressors can be beam type or integrally geared and both types of compressors could be single stage or multistage.

9.2.1 Multistage Beam-Type Compressor

For volumetric flow rates between 1 000 and 100 000 ACFM and polytropic heads under 120 000 ft (recycle gas, wet gas, and net gas applications), a multistage beam-type centrifugal compressor is used and is typically unspared.

Beam-type compressor casings can be horizontally or vertically split. A horizontally split compressor has a casing which is divided into upper and lower halves along the horizontal centerline. With this arrangement, all that is necessary is to lift the upper casing and gain access to the internal components without disturbing the rotor to casing clearance or bearing alignment (Figure 9.2). API 617 "Axial and Centrifugal Compressors and Expander-compressors" states that when the partial pressure of hydrogen is over 200 psig, casings with vertical or radial splits shall be used. This is to prevent leakage of light gases along the casing split.

In the vertically split compressor, the casings are formed by a cylinder closed by two end covers: hence the name "barrel" compressor. With one end cover being

Efficient Petrochemical Processes: Technology, Design and Operation, First Edition.
Frank (Xin X.) Zhu, James A. Johnson, David W. Ablin, and Gregory A. Ernst.
© 2020 John Wiley & Sons, Inc. Published 2020 by John Wiley & Sons, Inc.

9 Compressor Assessment

Inlet (P_1, T_1, Q_1, m_1)

Discharge (P_2, T_2, Q_2, m_2)

Compressor

Discharge vs. Inlet
$P_2 > P_1$
$T_2 > T_1$
$Q_2 < Q_1$
$m_1 = m_2$

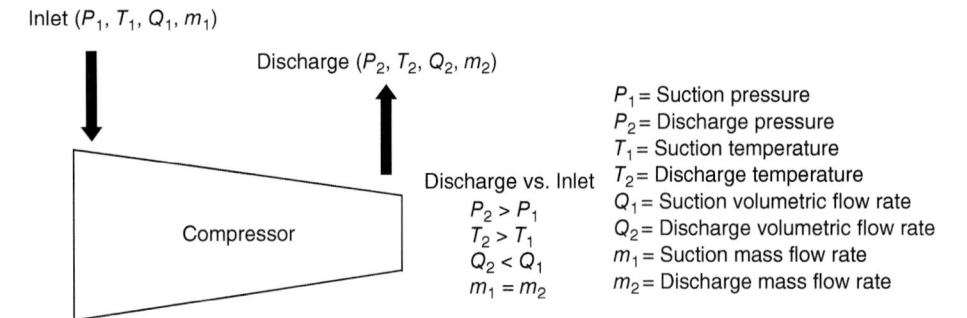

P_1 = Suction pressure
P_2 = Discharge pressure
T_1 = Suction temperature
T_2 = Discharge temperature
Q_1 = Suction volumetric flow rate
Q_2 = Discharge volumetric flow rate
m_1 = Suction mass flow rate
m_2 = Discharge mass flow rate

Figure 9.1 Basic principles of compressor.

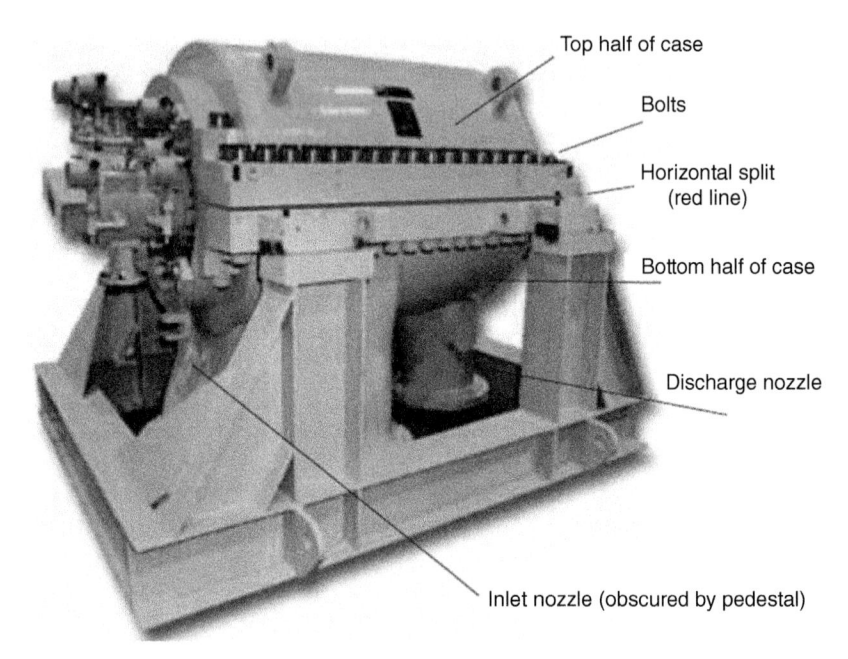

Figure 9.2 Centrifugal multistage horizontal split. *Source:* From Sorokes (2013), reprinted with permission by AIChE.

removable, it allows access to the inner casing with the internal components (Figure 9.3). Inside the casings, the rotor and diaphragms are essentially the same as that of horizontal split compressors.

Effective sealing is important to prevent leakages. In particular, it is the shaft end seals, which keep the process gases from leaking to atmosphere. Dry gas seals have gained wide acceptance and are the choice for most applications.

9.2.2 Multistage Integral Geared Compressors

This type of compressors has a low-speed bull gear that drives multiple high-speed gears (pinions) which connect impellers. Integrally geared compressors achieve the required head with smaller and fewer number of higher speed impellers than a beam-type compressor which can have up to 10 impellers. Impellers are mounted at one or both ends of each pinion (Figure 9.4). Each impeller has its own casing that is bolted to the gear casing. The gear casing is usually horizontally split to allow access to the gears. These are used for some lower flow applications and may be less expensive than a multistage beam-type compressor. They are used predominantly in gas processing plants.

In the integrally geared compressor, an impeller receives the gas from the first-stage inlet nozzle, compresses it, and discharges it to the diffuser where the velocity is converted to pressure. Then, the gas exits the first stage via discharge nozzle, and enters an intercooler if required, and is then piped to the second stage. The discharge from the second stage enters an intercooler if required, to keep gas temperatures within limits and it enters the third stage, and so on.

Integrally geared compressor can be single staged as well, which is mainly used for relative low-pressure ratio applications. Single-stage integral gear compressor is also called "Sundyne" compressor. For a Sundyne-type compressor, it is an in-line type similar to a pump, usually

Figure 9.3 Centrifugal multistage radially split compressor. *Source:* From Sorokes (2013), reprinted with permission by AICHE.

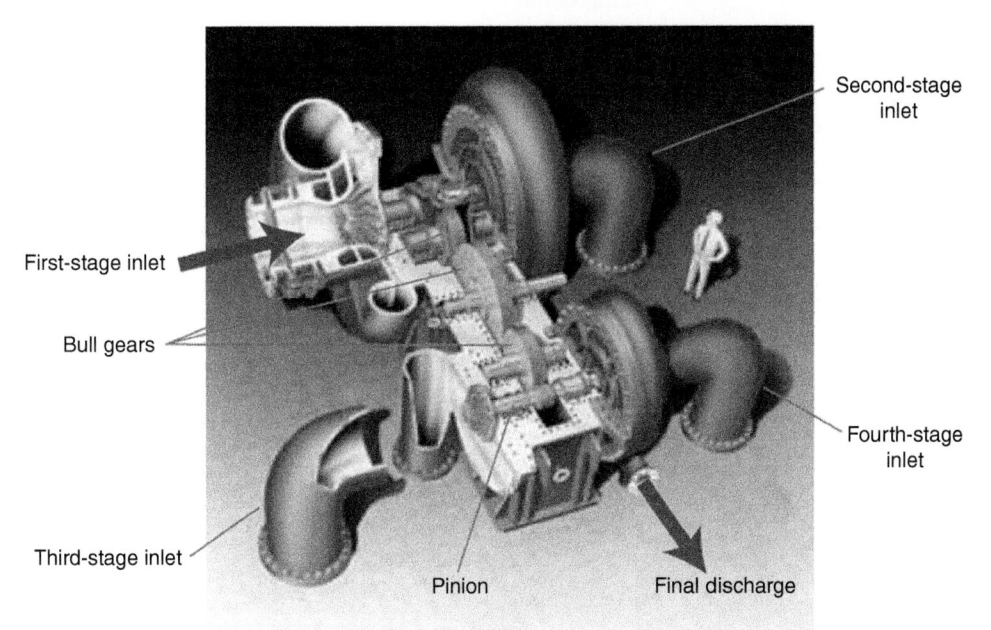

Figure 9.4 Integrally geared centrifugal compressor. *Source:* From Sorokes (2013), reprinted with permission by AICHE.

driven by motor through an integrally mounted gear box. The Sundyne compressors feature low flow and high head and are used in many applications that used to be served by positive displacement compressors.

9.3 Impeller Configurations

The most critical component in a centrifugal compressor is the rotor which is the shaft plus the impellers. Impellers provide 100% kinetic energy to the gas, which is responsible for around 70% of static pressure rise in a compression stage. Well-designed impellers are very energy efficient and only 4% energy expanded is lost. The losses in stationary parts in a compressor reduce overall energy efficiency. The type of impellers chosen depends on required pressure ratio, gas compositions, operating speed, equipment cost, etc.

Multistage centrifugal compressors have two types of impeller configurations: between-bearing (for beam type) and integrally geared.

9.3.1 Between-Bearing Configuration

Impellers in the between-bearing compressor are mounted on a single shaft. Between-bearing compressors are available with horizontally or vertically split casing. A driver rotates the shaft and impellers at a common speed. Between-bearing compressors come with two categories of configurations: straight-through and back-to-back. The straight-through arrangement is typically used for vertically split compressors in which the fluid flows at the one end and exits at the opposite end of the compressor (Figure 9.5). A balance drum or balance piston is required to absorb axial thrust.

In the back-to-back arrangement which is typically used for horizontally split compressors, impellers are arranged back-to-back with the exit flow at the center. In other words, the impellers are allocated in opposite directions (Figure 9.6). In this design, the main inlet is at the both ends of the rotor and the impellers guide the flow toward the center of the compressor. In this configuration, the axial thrust is self-correcting and balanced so that the force on the thrust bearings is reduced.

9.3.2 Integrally Geared Configuration

In an integrally geared configuration, multiple shafts may be used in which the impellers are mounted at the ends of multiple pinions that can rotate at different speeds depending on the gear ratio between the individual pinions and the bull gears. The number of impellers and the number of pinions vary depending on applications. Typical integrally geared compressors have two to four pinions with one or two impellers mounted at the ends of each pinion.

Instead of circumferential arrangement as in the between-bearing compressors, axial flow inlet arrangement is obtained in the integrally geared compressors. This is achieved by flow entering the first impeller via an axial or straight run of pipe and the flow at the volute (or collector) is piped to the axial inlet for the next impeller. This eliminates the flow inefficiency incurred from the inlet bend, return bend, and return channel used in the between-bearing design.

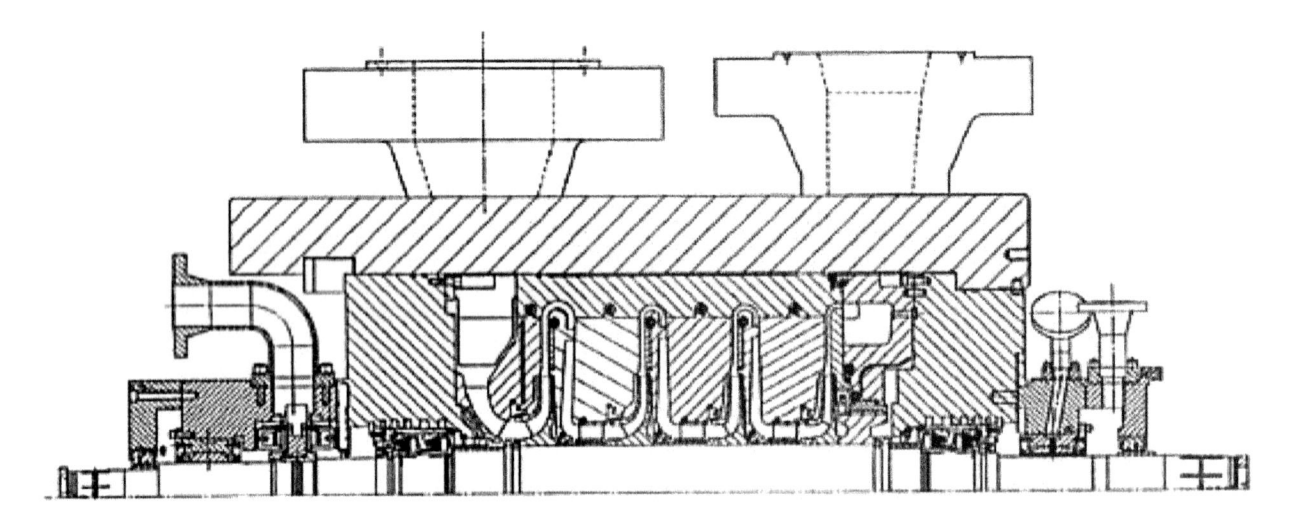

Figure 9.5 Straight flow-through compressor. *Source:* From Sorokes (2013), reprinted with permission by AICHE.

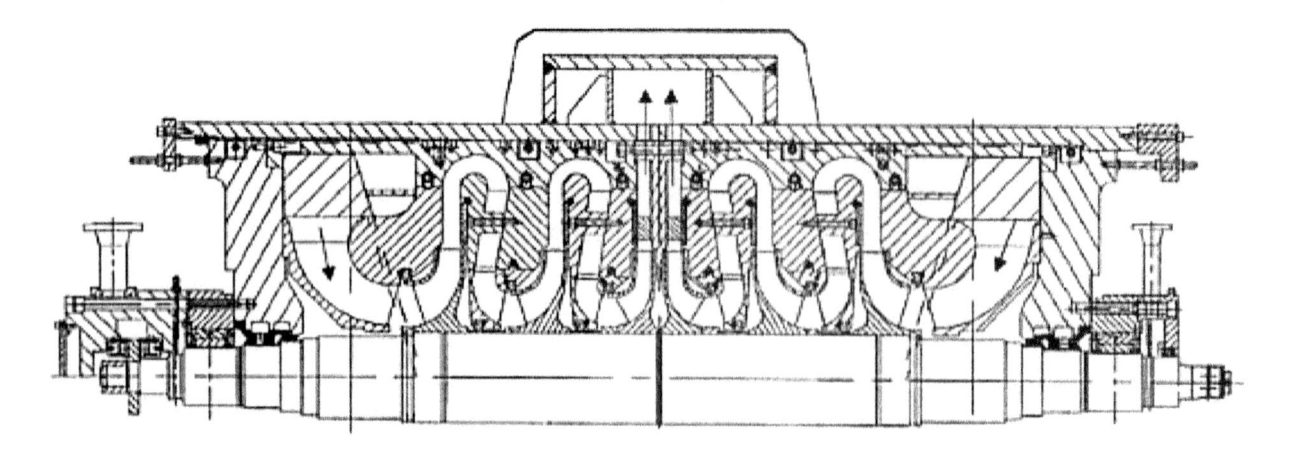

Figure 9.6 Back-to-back compressor with double flow inlet. *Source:* From Sorokes (2013), reprinted with permission by AICHE.

9.4 Type of Blades

The blades can be two-dimensional (2D) or three-dimensional (3D) (Figure 9.7). 3D impellers have better aerodynamic efficiency than 2D impellers but are more expensive. Selection of the type of impellers depends on flow coefficient (FC), operating speed, desired pressure ratio, efficiency, and equipment cost.

Figure 9.7 2D blades with circular arc shape (top) or 3D blades with complex shape (bottom). *Source:* From Sorokes (2013), reprinted with permission by AICHE.

9.5 How a Compressor Works

A centrifugal compressor (Figure 9.8a) consists of four basic parts: inlet nozzle for inlet flow guidance, impeller for increasing gas velocity, diffuser for converting velocity energy to pressure energy, and volute (or scroll) for existing flow guidance.

In the impeller, the gas velocity is increased due to the centrifugal action of the rotating blades. Then, this velocity is converted to pressure in the diffuser. In this mechanism, the principle of centrifugal compressor is very similar to that of a centrifugal pump.

An impeller consists of a hub and a number of rotating blades that impart mechanical energy to the gas via increasing the velocity of the gas (Figure 9.8b). The gas leaves the impeller with increased velocity and enters the diffuser. The diffuser gradually reduces the velocity of the gas in order to increase gas pressure. In this manner, the diffuser converts the velocity energy to a higher pressure. In a single-stage compressor, the gas leaves the diffuser and enters a volute before it exits the compressor through the discharge nozzle. The volute collects the exiting gas and reduces the gas velocity further through increased cross-sectional area. Thus, it gives additional pressure rise. In a multistage compressor, the gas leaves the diffuser and enters return vanes that direct the gas into the impeller of the next stage.

The velocity of the gas is the key to understand how a centrifugal compressor can compress the gas dynamically. When the gas enters the impeller, it flows into the narrowe passage between blades. The gas velocity in the passage relative to the blades is called relative velocity (V_R) and also called radial velocity as it occurs in a radial direction. The gas has a higher relative velocity in a long and narrow flow passage. The opposite is true that the gas's relative velocity is lower in a short and wide flow

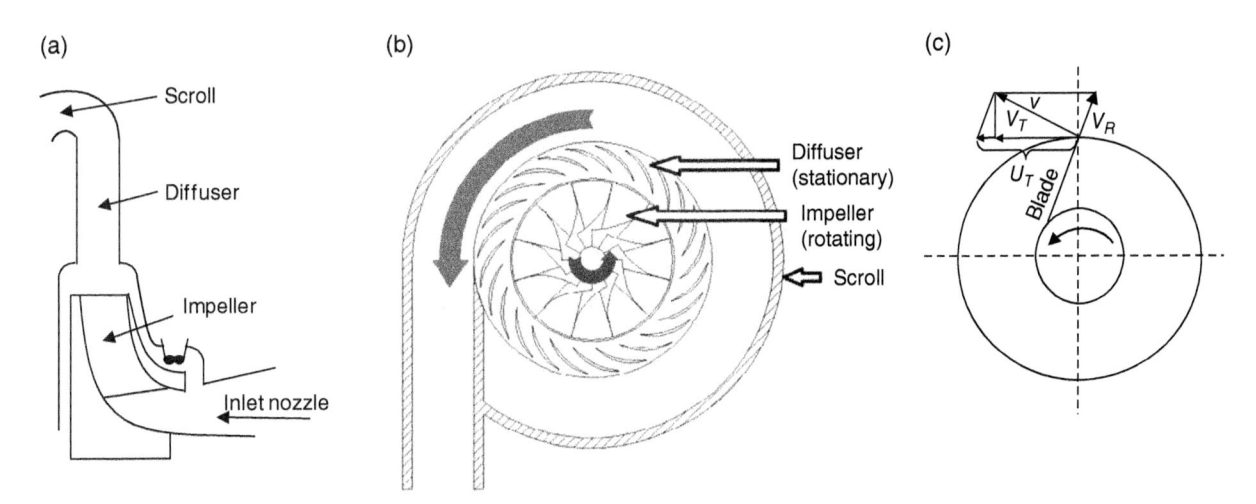

Figure 9.8 Key components of centrifugal compressor. (a) Plain view, (b) cross-section view, and (c) blade flow velocity view.

passage. As soon as the gas flows out of the flow channel between blades, the tangent velocity of the gas increases from V_T to the blade tip speed U_T as shown in Figure 9.8c. The blade tip speed is the product of blade tip diameter and RPM. As the result of increased tangent velocity, the gas energy increases in proportion to the net velocity (V) (also called exit velocity) that is the vector sum of the relative velocity (V_R) and the blade tip speed (U_T). Therefore, the blade tip speed and the relative velocity determine the pressure ratio, i.e. the performance of the compressor.

9.6 Fundamentals of Centrifugal Compressors

Flow coefficient (FC), φ_G, is the most important parameter for selection of centrifugal compressor types. φ_G relates flow rate with both impeller size and speed and is defined as

$$\varphi_G = \frac{G}{ND^3} \qquad (9.1)$$

where G is impeller's volumetric flow rate, N impeller's speed, and D impeller's exit diameter.

Centrifugal impellers can be classified into two categories based on FC: low FC and high FC impellers. The former features long and narrow passages with higher pressure ratio. In contrast, the latter has much wider flow passages to accommodate higher flows but with lower pressure ratio. Figure 9.9 shows impellers with different FC's. The impeller with the highest FC is located at the right end of the rotor with remaining impellers progressively narrower in flow passages are from the right end towards the center. In contrast, the impellers with the low FC are located at the left of the rotor with the lowest FC impeller located closest to the center from the left. As what can be observed, these low FC impellers have much narrow flow passages than the high FC impellers on the right. In this configuration, fluid pressure increases beginning with the high FC impellers on the right with relative lower pressure ratio and then to the medium FC impeller in the center on the right-hand side with medium pressure ratio. Lastly, the fluid pressure continues to increase with relative higher pressure ratio from the low FC impellers on the far left toward the highest pressure in the center on the left-hand side.

Low FC impellers have 2D blades (simpler blade design). In contrast, high FC impellers have 3D blades (complex blade design), which can be observed in Figure 9.7. Due to their narrow passages and simpler blades, low FC impellers have lower aerodynamic efficiency than high FC ones.

Both the straight-through and back-to-back designs can allow inter-cooling which keeps the temperature of the material below the strength limit as well as reduces the shaft power requirement. If intercooling is required, the gas is discharged from the compressor after travelling through half the impellers, cooled and injected back into the compressor before travelling through the remaining impellers.

The integrally geared design has several advantages over the between-bearing design. The most important is the aerodynamic advantage due to axial flow arrangement, which eliminates flow turning and thus reduces pressure losses. Furthermore, impellers can have different speeds or diameters. These two features make integrally geared design more energy efficient than the between-bearing design.

However, the disadvantage of the integrally geared design is the complex mechanical arrangement because

Figure 9.9 Different FC impellers: from low at the left to high at the right. *Source:* From Sorokes (2013), reprinted with permission by AICHE.

it contains a large number of bearings and seals. Vibration could be a problem in operation for integrally geared design due to the fact that impellers are located outside the bearing supports (overhung) and they can vibrate if not mounted properly.

To relate the pressure head with impeller geometry and speed, use head or pressure coefficient μ_p, as defined below:

$$\mu_p = K \frac{H_p}{N^2 D^2} \quad (9.2)$$

where K is constant and it can be calculated as

$$K = \frac{g}{\left(\frac{\pi}{720}\right)^2} \quad (9.3)$$

where g is gravitational constant.

High head impellers have a narrower flow range than those of low head impellers. High head impellers are also more prone to surge than those of low head impellers. Compressor surge is when the compressor cannot overcome the discharge pressure. This causes reverse flow through the compressor and damage to the seals, bearings, and rotor. To prevent surge, the user often specifies an anti-surge spillback.

9.7 Performance Curves

Figure 9.10 shows a typical compressor performance curve at a design speed. The curve shows that the centrifugal compressor has limited head capability, with variable volume characteristic.

Head is a process condition and is a function of compression ratio and MW. The curve sets the flow rate.

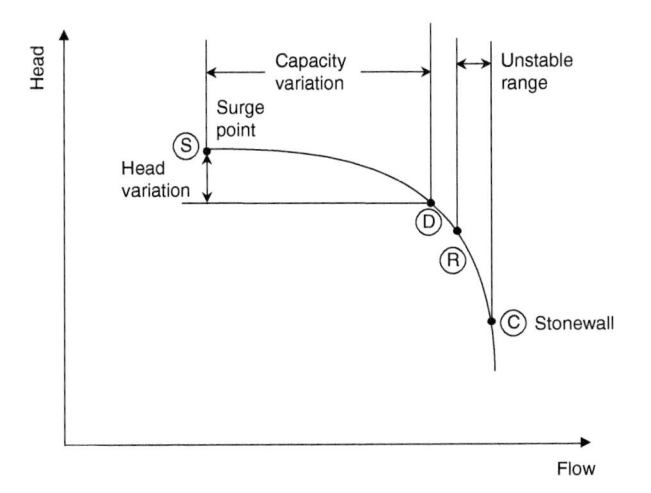

Figure 9.10 Performance curve for a centrifugal compressor.

If process conditions do not change, raising the compressor speed will raise the flow rate. The required flow rate is obtained by changing the speed of a variable speed compressor or throttling the suction valve for a constant speed compressor.

9.7.1 Design Point

"D" is the design point for the compressor at the given speed. The section between "S" and "D" is the normal operating range. The curve shows that the compressor discharge pressure is balanced with relatively large changes in volume flow. The compressor can operate at the rated condition represented by point "R"; but the region between points "R" and "C" is unstable. For optimum compressor size selection, the compressor should be selected to have the design point close to the right-hand side of the curve, such as point "D," but not too close to the stonewall or choke point "C." If the design point is located too far to the left from point "D," the compressor is sized too large, the compressor is too small if farther to the right of point "D."

9.7.2 Surge

Surge is characterized by intense and rapid flow and pressure fluctuations throughout the compressor and is associated with stall involving one or more compressor stages. It is accompanied by strong noise and violent vibration which can damage the compressor. Surging occurs at a minimum suction flow with point "S" in Figure 9.10 marked as the surge point. When the discharge pressure in the centrifugal compressor increases, the mass flow rate decreases. There is a minimum flow limit. Below this limit, the compressor operation becomes unstable. When the compressor cannot overcome the discharge pressure, the easiest path for the gas is back through the compressor. After the back flow slug has been discharged, the compressor still faces the problem of insufficient gas flow and the back flow reoccurs. This unstable operation manifests itself in the forms of pressure and flow oscillation.

Manufacturer provides operating limits to avoid surge for multistage compressors.

If the compressor is to be operated below the minimum flow, the compressor must be equipped with a low flow spillback. The details for partial control can be seen below.

9.7.3 Choking

Choking is the opposite of surge in the centrifugal compressor. It occurs at point "C" under which the gas flow is too large and that is more than what the impeller can handle. At the choking point "C," the fluid reaches sonic conditions and there will be abrupt decrease in the compressor performance. The occurrence of choking depends not only on the high flow condition but also on the fluid thermodynamic properties. For example, choking can occur to compressors operating with fluids of high MW.

Many industrial compressors normally operate at conditions far away from choking. For these compressors, the maximum flow limit is usually defined as the flow corresponding to a sharp reduction in efficiency based on company's best practice.

If variation of impeller speed is considered, there will be a family of compressor performance curves as shown in Figure 9.11. An increase in rotor speed or RPM increases the compressor flow rate. At a particular rotor speed, a decrease in flow rate can be obtained by increasing the compression ratio. Minimum and maximum permissible flow rates at constant RPM are termed surge and choke (stonewall) limits. A line tracing the stall points of all the constant RPM lines is called a surge line.

Flow Coefficients and pressure coefficients can be used to determine various design characteristics. Use of FC enables selection of impeller type while knowing the head pressure coefficient helps to determine the hydraulic performance. For example, Figure 9.12 shows simplified performance curves for low and high head coefficient impellers. As can be observed, the low head coefficient impeller has a steeper rise-to-surge slope than the high head coefficient impeller. Therefore, comparing with the high head coefficient impellers, the low head coefficient impellers have much higher pressure sensitivity from the flow change and thus is easier to measure the pressure condition to detect how close the compressor is to the surge limit.

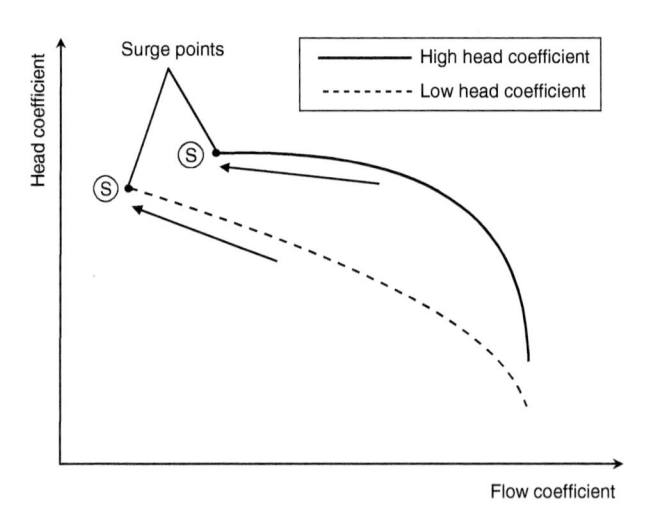

Figure 9.12 Impeller with higher head coefficient has a smaller rise-to-surge.

A performance curve is obtained under new and clean conditions. After operating for a period of time, however, a compressor will deteriorate in performance. Even after a full maintenance, a compressor will rarely retain its original performance.

9.8 Partial Load Control

As mentioned above, surge could lead to large and violent flow oscillation and thus cause damage to the compressor. Hence, it should be avoided by all means. The surge line is established during manufacturing shop performance testing. Instrumentation will open a recycle valve before the compressor goes into the surge region.

9.8.1 Recycle or Surge Control Valve

At constant speed the head–flow relationship will vary in accordance with the performance curve (see Figure 9.11). For a constant compressor speed, a recycle valve is needed for surge control if process conditions (rising compression ratio and dropping MW) push the operating point into surge.

9.8.2 Variable Speed Control

A performance curve is established for each speed, as shown in Figure 9.13. If the compressor-drive system (compressor, driver, and gear) is designed for 90–110% speed variation as shown in Figure 9.13, by varying the compressor speed, the centrifugal compressor can be operated at any partial load point on the right-hand side of the surge line. The speed control is actually

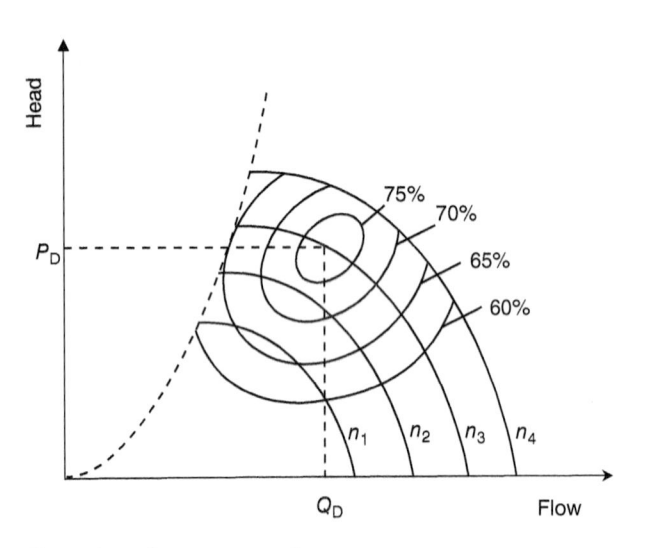

Figure 9.11 Compressor performance curves.

shifting the compressor curve until it reaches the new head and flow requirement of the compressor under the partial load conditions. The compressor can even handle the head higher than the design head by increasing the speed of the compressor. However, the compressor will surge if the flow is less than the surge point, unless it is equipped with a recycle control valve.

The partial load performance can be presented by a horizontal line D–F in Figure 9.13 if the compressor is operated under constant head mode. The point D is design point. The minimum partial load is about 73% at the surge point of "F." If the compressor is operating under decreasing head mode, then the performance line is D–E. The minimum partial load without surge is about 67% at the surge point of "E."

9.8.3 Inlet Guide Vane (Prerotation Vane)

Inlet guide vanes are primarily used for integrally geared compressors. Located at the compressor inlet, the guide vanes change the direction of the velocity entering the first-stage impeller. By changing the angle of flow, these vanes direct the flow into the impeller, and consequently the shape of the performance curve is changed. With velocity change to the inlet gas by the guide vanes, the performance curve steepens with very little efficiency loss.

Figure 9.14 shows the typical performance curve for the centrifugal compressor equipped with inlet guide vane control for constant speed driver. The performance line of D–A–G–H is the maximum head capability of the compressor. The inlet guide vane allows the centrifugal compressor to operate any point and any conditions

Figure 9.13 Typical variable speed control compressor performance curves.

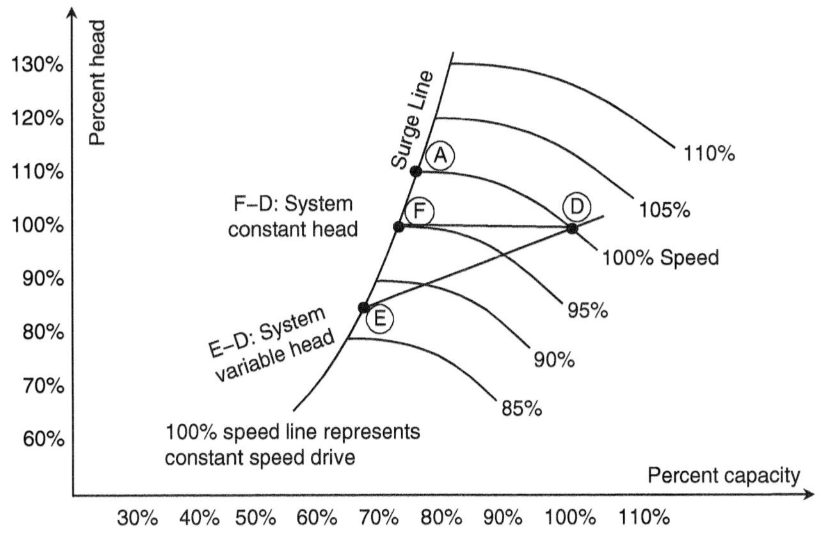

Figure 9.14 Performance curves for inlet guide vane control with constant speed driver.

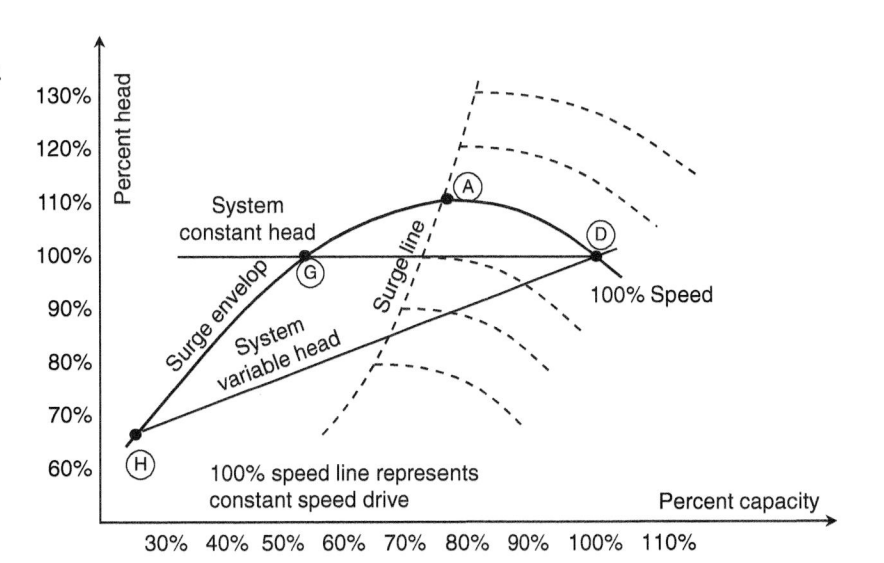

below the envelop of D–A–G–H without surge. As compared to variable speed control as shown in Figure 9.13, the surge limits are extended and the partial load capability of the compressor is greatly increased without change of compressor speed.

Figure 9.15 is used to explain the performance curves with inlet guide vane openings at constant speed. The compressor is operating with the vane fully open on the line of B–D–A. The inlet guide vane is adjusting and changing its position for the compressor to operate in other areas away from full load. The compressor will exceed its head capability if it is operating above the performance curve of D–A. Hot gas bypass is required if the operating conditions of the compressor are located in the area of the left-hand side of the surge limit of A–G–H.

9.9 Inlet Throttle Valve

Between-bearing compressors with constant speed electric motor drivers are controlled with suction throttle valves. Throttling at the suction instead of the discharge takes advantage of volume reduction so less pressure is throttled and less energy is used when operating at off-design conditions.

When throttling at the discharge, the suction pressure does not change so mass flow rate (G) is proportional to volumetric flow rate (ACFM). When throttling at the suction, the compressor suction pressure drops, the same G is obtained at a higher ACFM [see Eq. (9.4)]. Thus, less throttling is required and less energy is used by using inlet throttle valve.

$$\text{ACFM} \approx \frac{G}{P_1} \qquad (9.4)$$

9.10 Process Context for a Centrifugal Compressor

The objective of using a compressor is to compress a certain amount of gas to a desired pressure. A typical process involving a compressor is shown in Figure 9.16. The responsibility of an engineer is to provide process conditions as the basis for determining the type of compressor and the number of compression stages. The process data

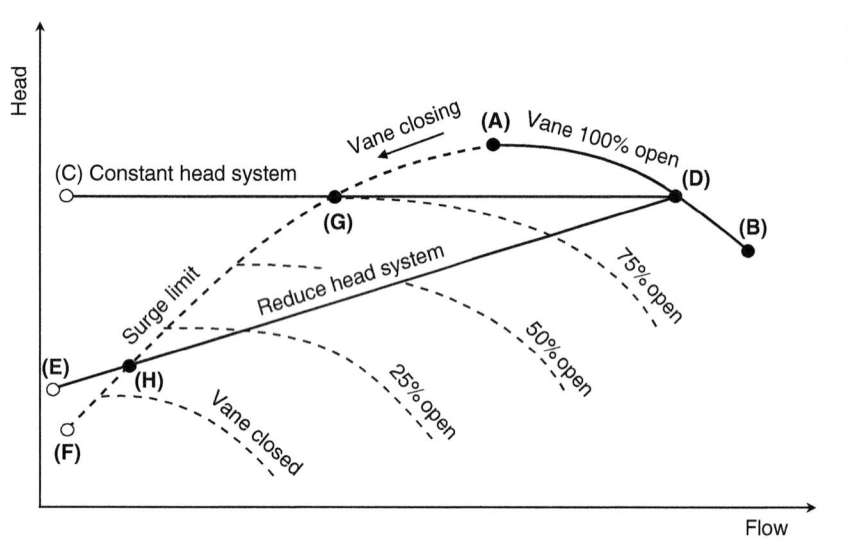

Figure 9.15 Inlet guide vane control–constant speed driver.

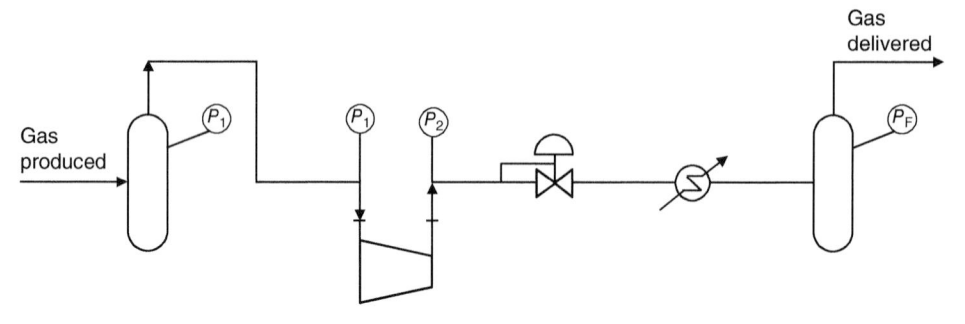

Figure 9.16 Typical process involving a compressor.

that the engineer will provide include mass flow, inlet and discharge pressure, temperature, and gas compositions. Then, the manufacturer will calculate actual flow, the type of compressor, horsepower, and efficiency.

Firstly, the gas flow range that the compressor must handle should be provided. The minimum flow corresponds to turndown operation while maximum flow is for rated operation. Secondly, the fluid conditions (temperature and pressure) and property (composition) should be specified for several process scenarios so that the selected compressor can handle all the variations. Next, inlet and discharge pressure must be specified.

We need to differentiate volumetric flow, mass flow, and standard flow. The compressor is sized on volumetric flow rate. For a compressor with compression ratio of two, one actual cubic feet of gas per minute will be compressed to a discharge volume of exactly one half of a cubic feet per minute assuming no increase in temperature in the compressor and that the gas is dry because the gas pressure is doubling.

Standard volume has only one volume referenced to the same pressure and temperature. At default, the standard conditions are defined at 14.7 psia (atmospheric pressure) and 60 °F. A measurement in standard cubic feet is the ratio of the actual pressure to the referenced standard pressure multiplied by the actual volume. Referring back to the previous example of a compressor with compression ratio of two and no change in temperature, the standard volume from this compressor would remain the same because the pressure remains constant at 14.7 psia. Even though the actual volume of the gas decreases by one half, the discharge standard volume is the ratio of the discharge pressure to the standard atmospheric pressure multiplied by the discharge volume. This will result in the same discharge standard volume as the inlet one.

Mass flow is the product of the actual volume flow and the density of the specific gas. Same as the standard volume, mass flow through the aforementioned compressor will remain the same provided the gas is dry.

Both standard volume and mass flow are used to describe process capacity and in horse power calculations. The gas price is based on the standard volume or mass flow.

9.11 Compressor Selection

Based on the actual flow rates provided, FC as defined in Eq. (9.1) can be calculated, which is used to select impeller blade type: low FC or high FC impellers. As aforementioned, the major difference between these types is that low FC impellers are characterized by long and narrow passages, while high FC impellers are featured with wide passages to accommodate high flow rate.

The pressure coefficient defined in Eq. (9.2) can be used to determine either low or high head impellers. To prevent surge, the user often specifies a minimum rise-to-surge limit for compressor selection.

Based on the discharge pressure specified, the pressure head (H_p) raised by the compressor can be determined via

$$H_p = \frac{\gamma}{\gamma-1} ZRT_1 \left[\left(\frac{P_2}{P_1}\right)^{\frac{\gamma-1}{\gamma}} - 1 \right] \quad (9.5)$$

where $\gamma = C_p/C_v$ in which C_p is the specific heat capacity at constant pressure and C_v is the specific heat capacity under constant volume; Z is the compressibility of the gas; R is the gas constant in ft-lb$_f$/(lb-mol)(°R); T_1 is the inlet temperature in °R; P_1 is the inlet pressure in psia and P_2 is the discharge pressure in psia.

Knowing the mass flow (G) and pressure increase that the compressor must deliver, polytropic efficiency can be determined via

$$\eta_P = \frac{\text{Work Out}}{\text{Work In}} = \frac{\gamma-1}{\gamma} \times \frac{\ln(P_2/P_1)}{\ln(T_2/T_1)} \quad (9.6)$$

Then, horsepower requirement can be calculated via

$$W_{HP} = \frac{G \times H_p}{\eta_P} \quad (9.7)$$

Obviously, the compressor with highest efficiency will require least power and hence lowest operating cost. Horsepower is linearly proportional to flow rate and head increase.

When selecting a compressor, the best practice for ensuring reliability is to preselect compressor casing type (horizontally split, vertically split, or integrally geared), impeller type (open or closed), the number of impellers allowed in each casing based on head per stage limits, and shaft stiffness. One should request vendors' experience references for installed compressor with similar design parameters.

Reference

Sorokes, J. (2013). Selecting a centrifugal compressor. *Chemical Engineering Progress (CEP)* 109: 44. (June), AIChE, New York.

10

Pump Assessment

10.1 Introduction

The pump's role is to move the fluid through the system at desired flow rate and pressure. Pumps include centrifugal type and positive displacement types. The latter usually includes reciprocating and rotary pumps. The following discussions focus on the centrifugal pump as it is the most common type used in the process industry.

There are two basic tasks for pump selection. The first is to determine the pump head required for given process requirement while the second is to select the pump that can deliver the desired flow rate through a pump system under required head. The required head depends solely on the process characteristics (suction pressure, discharge pressure, and liquid density) while the flow rate relies on the pump characteristics (impeller size and speed). As a process engineer, your role is to find the best match between these two characteristics and make sure the pump selected can satisfy process requirement in the most reliable and efficient manner.

In selecting and operating a pump, it is essential that you have good knowledge of the process requirement in terms of pump head and flow rate required, how the pump works, and guidelines for pump selection and operation. In particular, adequate knowledge of the process conditions and compositions is the most important aspect for optimizing pump selection. Process constraints must be considered such as operation flexibility, turndown, start-up, shutdown, etc. With good understanding of process conditions and constraints as well as pump characteristics, application of API Standard 610 for centrifugal pumps and Standard 682 for mechanical seals will result in improved reliability and extended on-stream operation.

10.2 Understanding Pump Head

Why is pump performance always measured in flow vs head rather than flow vs pressure? This is the myth we want to clarify through discussions here. When the pump operates in a process unit, the process requires the liquid at a desired flow rate to be delivered from the suction pressure (P_S) to discharge pressure (P_D). If the pump is capable, it will move fluid forward at the required discharge pressure for given suction pressure. Otherwise, if the pump cannot create enough head to move the fluid forward at the discharge pressure, it will operate at no flow.

The relationship between the pump head and differential pressure is described by the following formula:

$$H_T = \frac{2.31(P_D - P_S)}{SG} \quad (10.1)$$

where

H_T = pump head (ft);
P_D = discharge pressure (psig);
P_S = suction pressure (psig);
specific gravity $SG = \rho/\rho_{H_2O}$;
ρ is the density of the liquid;
standard water density ρ_{H_2O} = 62.4 lb/ft³ (1000 kg/m³) at temperature 4 °C (39.2 °F);
2.31 is the conversion factor.

Note that Eq. (10.1) will be derived later.

By measuring pump head as opposed to differential pressure, the density or specific gravity of the fluid is already accounted for. Consider Figure 10.1 for a visual interpretation where three identical pumps are pumping three fluids of different specific gravity or density under the same suction atmospheric pressure. The head or height of the fluids is the same for three cases even though the discharge pressures and power requirements are different. Thus, by using pump head, the pump performance depends on the pump mechanical design characteristics instead of types of liquids. As the result, the pump performance curve based on pump head versus flow rate will remain constant for any liquids. This explains the reason why the pump performance is described by pump head instead of pressure.

Otherwise, use of discharge pressure could be problematic for specifying a pump because the discharge

---

Efficient Petrochemical Processes: Technology, Design and Operation, First Edition.
Frank (Xin X.) Zhu, James A. Johnson, David W. Ablin, and Gregory A. Ernst.
© 2020 John Wiley & Sons, Inc. Published 2020 by John Wiley & Sons, Inc.

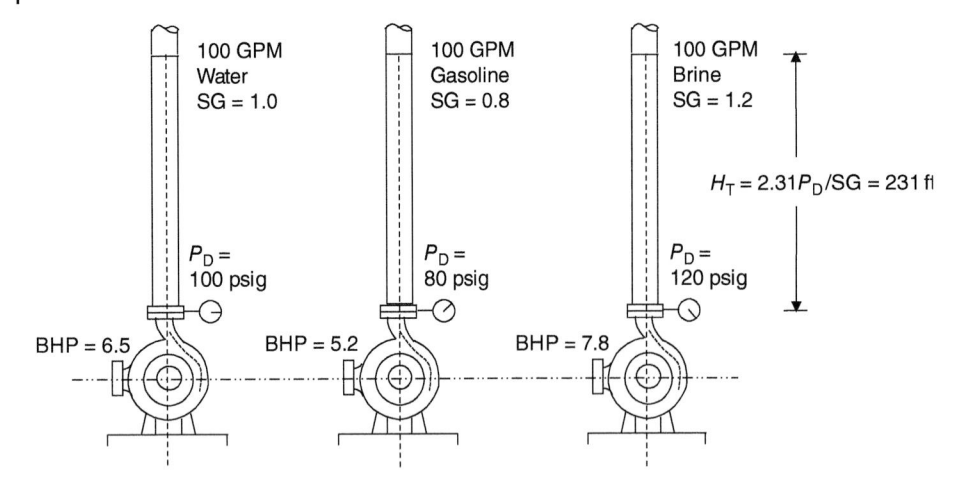

Figure 10.1 Pump head applies to any liquid (pump operating under no flow condition).

pressure depends upon the suction pressure and the specific gravity of the liquid being pumped. The specific gravity changes with temperature, type of fluid, and fluid composition. But the pump manufacturer does not know these process parameters in prior and use of pump head for pump selection can avoid these uncertainties.

10.3 Define Pump Head: Bernoulli Equation

After you have some understanding about the pump head, you may want to know how the head is defined and calculated. This is the question we will focus in this and the next sections.

Let us consider Figure 10.2. The principle of Energy Conservation states that energy is neither created nor destroyed but is simply converted from one form of energy to another. Bernoulli equation is the most well-known expression for this principle.

Thus, applying the Bernoulli equation to the liquid at Points A and B in Figure 10.2 gives

$$\frac{P_A}{\rho} + \frac{u_A^2}{2g} + Z_A = \frac{P_B}{\rho} + \frac{u_B^2}{2g} + Z_B \quad (10.2)$$

where three energy components are involved as:

- Potential energy from elevation: Z in (ft) using the pump suction centerline as the datum.
- Pressure energy: P/ρ in (ft) as P in (lb/ft^2) and density ρ in (lb/ft^3); P is the surface pressure acting upon the liquid surface.
- Kinetic energy from flow velocity: $u^2/2g$ in (ft) as velocity u in (ft/s) and acceleration g is 32.2 ft/s^2.

If a pump is placed between A and B as shown in Figure 10.3, the net liquid column in terms of pump head (H_T) should be added to the left-hand side (LHS) of Eq. (10.2) to account for the energy added by the pump, which yields

$$\frac{P_A}{\rho} + \frac{u_A^2}{2g} + Z_A + H_T = \frac{P_B}{\rho} + \frac{u_B^2}{2g} + Z_B \quad (10.3)$$

In reality, a process will have piping and fittings, heat exchangers, and control valves between the source and

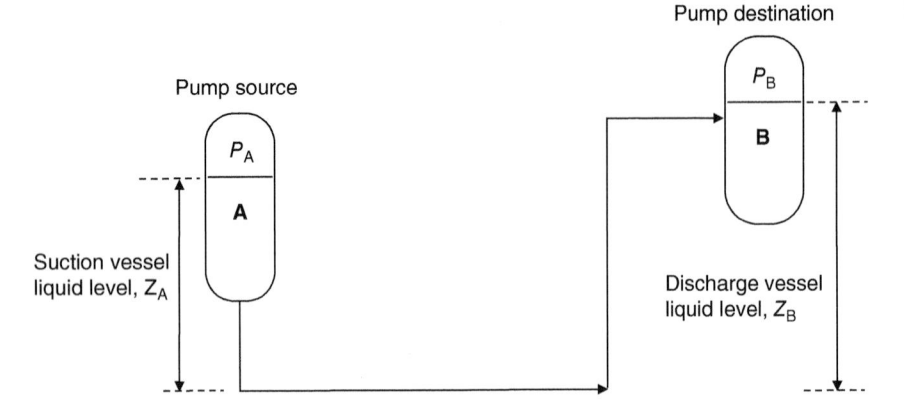

Figure 10.2 A simple process system.

Figure 10.3 A practical process system.

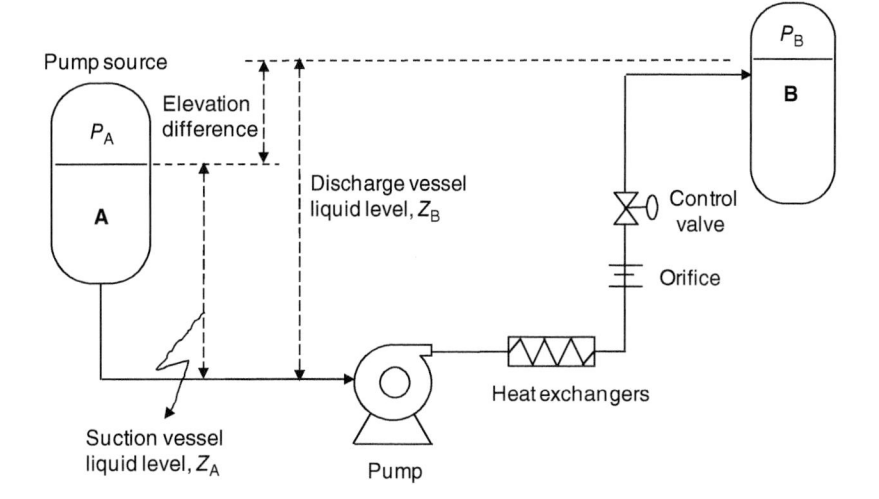

destination. Thus, friction losses occur in both sides of the pump: $H_{A,f}$ represents the friction loss in the suction side while $H_{B,f}$ is the friction loss in the discharge side. To count for these losses, $H_{A,f}$ must be deducted from the LHS of Eq. (10.3) as it makes negative contribution to the suction energy. In contrast, the friction loss $H_{B,f}$ must be added to the right hand side (RHS) of Eq. (10.3) as it increases the energy requirement in the discharge side. Thus, Eq. (10.3) becomes

$$\frac{P_A}{\rho} + \frac{u_A^2}{2g} + Z_A - H_{A,f} + H_T = \frac{P_B}{\rho} + \frac{u_B^2}{2g} + Z_B + H_{B,f} \tag{10.4}$$

The pump head is derived as

$$H_T = \frac{P_B - P_A}{\rho} + (H_{B,f} + H_{A,f}) + + \frac{u_B^2 - u_A^2}{2g} + (Z_B - Z_A) \tag{10.5}$$

If the surface pressure P_A, P_B are expressed in psi instead of lb/ft², the conversion of pressure to liquid head is

$$H_P = \frac{144(P_B - P_A)}{\rho} = \frac{2.31(P_B - P_A)}{SG} \tag{10.6}$$

Thus, the pump head in Eq. (10.6) can be expressed as

$$H_T = \frac{2.31(P_B - P_A)}{SG} + (H_{A,f} + H_{B,f}) + \frac{u_B^2 - u_A^2}{2g} + (Z_B - Z_A) \tag{10.7a}$$

or

$$H_T = \frac{2.31(P_B - P_A)}{SG} + H_f + \frac{u_B^2 - u_A^2}{2g} + (Z_B - Z_A) \tag{10.7b}$$

H_T is called total dynamic head required from a pump while the terms in the order of sequence in RHS in Eq. (10.7b) are static surface pressure head, friction head, velocity head, and static elevation head, respectively. Equation (10.7b) states that a pump must overcome the total head H_T to deliver a desired process fluid rate in order to satisfy the process requirement.

Let us consider the following example to illustrate the Bernoulli equation as it is always beneficial to walk through theory with practical examples.

Example 10.1
Pump Water to a Higher Location (Figure 10.4)
According to Eq. (10.7b), the total pump head is

$$H_T = \frac{2.31(P_2 - P_1)}{SG} + H_f + \frac{u_2^2 - u_1^2}{2g} + (Z_2 - Z_1)$$

where the differential static head $(Z_2 - Z_1)$ is 31 m. P_1 and $P_2 = 0$ because the water surfaces open to atmosphere. $u_1 = 0$ because the inlet reference is the reservoir water surface (the decline in the large reservoir surface is negligible). u_2 can be calculated as

$$u_2 = \frac{Q}{A} = \frac{4 \times (0.102)}{\pi (0.295)^2} = 1.49 \text{ m/s}$$

Thus, the velocity head $u_2^2/2g$ is 0.11 m.
The flow resistances occur in screen, three elbows, and piping, resulting in total friction loss as

$$H_f = H_{f,screen} + 3H_{f,elbow} + H_{f,pipie}$$

For the 12-inch pipe made of PVC, the friction coefficient can be found in engineering design manual to be 0.0141. The piping friction loss can be calculated as

$$H_{f,pipe} = f \frac{L}{D} \frac{u^2}{2g} = 0.0141 \frac{1530}{0.295} \times 0.11 = 8.04 \text{ m}$$

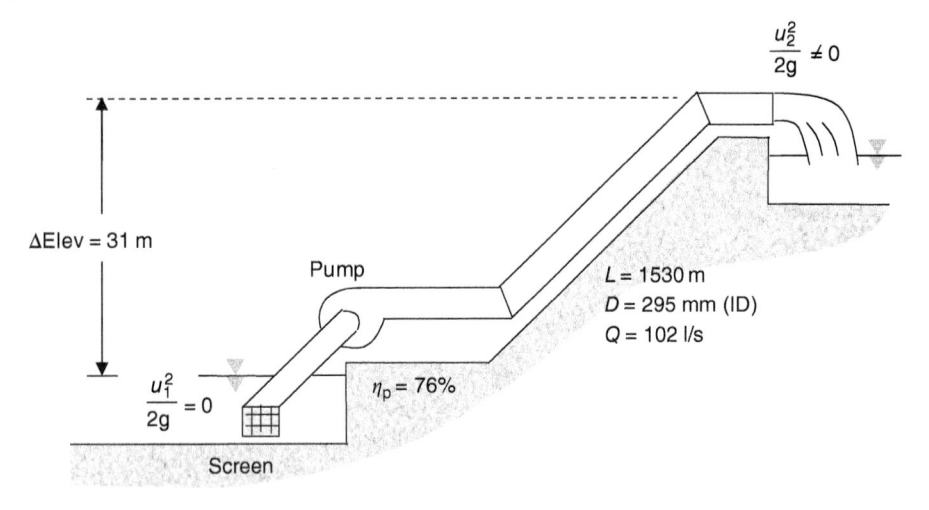

Figure 10.4 Illustration of Bernoulli equation (10.7).

For a 12-inch pipe and 45° flanged elbow, the friction coefficient is 0.15. Thus, the elbow friction loss can be calculated via

$$H_{f,elbow} = K\frac{u^2}{2g} = 0.15 \times 0.11 = 0.0165 \text{ m}$$

By assuming the screen friction loss as 0.2 m, the total friction loss becomes

$$H_f = H_{f,screen} + 3H_{f,elbow} + H_{f,pipe}$$
$$= 0.2 + 3 \times 0.0165 + 8.04 = 8.26 \text{ m}$$

Thus,

$$H_T = \frac{2.31(P_2 - P_1)}{SG} + H_f + \frac{u_2^2 - u_1^2}{2g} + (Z_2 - Z_1)$$
$$= 0 + 8.26 + 0.11 + 31 = 39.37 \text{ m}$$

The above calculations give one total pump head required for delivering the flow rate of 102 lb/s.

When the flow rate varies, the calculation of the pump head can be repeated as above. Thus, the relationship of pump head and flow rate can be developed to form the system curve as shown in Figure 10.5 and the characteristics of the system curve will be discussed later. In conclusion, total head is a function of the process as it is comprised of suction pressure, discharge pressure, liquid specific gravity, friction losses, and elevation. All of these are process conditions.

10.4 Calculate Pump Head

Let us take another look at Eq. (10.7a) and it can be rearranged and expressed as

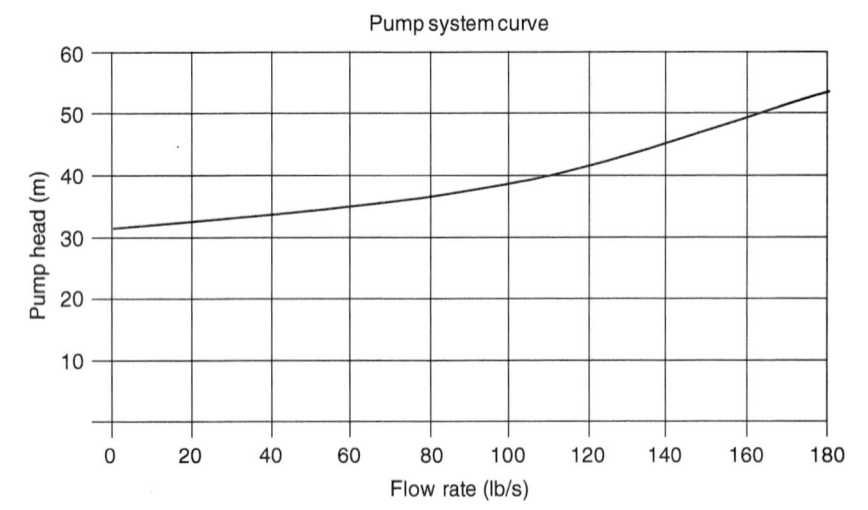

$$H_T = \left[\frac{2.31P_B}{SG} + H_{B,f} + \frac{u_B^2}{2g} + Z_B\right] - \left[\frac{2.31P_A}{SG} - H_{A,f} + \frac{u_A^2}{2g} + Z_A\right] \quad (10.8)$$

Figure 10.5 Example 10.1 pump system curve.

Or simply (if using D denoting discharge or the destination and S for suction or the source)

$$H_T = H_D - H_S \tag{10.9a}$$

where

$$H_D = \frac{2.31 P_{D,SP}}{SG} + H_{D,f} + \frac{u_D^2}{2g} + Z_D \tag{10.9b}$$

$$H_S = \frac{2.31 P_{S,SP}}{SG} - H_{S,f} + \frac{u_S^2}{2g} + Z_S \tag{10.9c}$$

where SP stands for surface pressure.

If both suction and discharge head are expressed in pressure (psig), Eq. (10.9a) becomes

$$H_T = \frac{2.31(P_D - P_S)}{SG} \tag{10.1}$$

where

$$P_D = \frac{SG \times H_D}{2.31}$$

and

$$P_S = \frac{SG \times H_S}{2.31}$$

H_D and H_S are expressed as Eqs. (10.9b) and (10.9c).

Following Eq. (10.9b), total discharge head (H_D) can be expressed in head as

$$H_D = H_{D,SP} + H_{D,f} + H_{D,v} + H_{D,E} \tag{10.10}$$

where

$$H_{D,SP} = 2.31 \frac{P_{D,SP}}{SG}$$

$$H_{D,v} = \frac{u_D^2}{2g}$$

$$H_{D,E} = Z_D$$

and subscript D stands for discharge, v for velocity head, and E for elevation head.

Similarly, total suction head (H_S) can be expressed in head based on Eq. (10.9c) as

$$H_S = H_{S,SP} - H_{S,f} + H_{S,v} + H_{S,E} \tag{10.11}$$

where

$$H_{S,SP} = 2.31 \frac{P_{S,SP}}{SG}$$

$$H_{S,v} = \frac{u_S^2}{2g}$$

$$H_{S,E} = Z_S$$

and subscript S stands for suction, v for velocity head, and E for elevation head.

In simplicity, you can think of total suction head as if you stand right at the pump suction flange and look back from the pump toward the suction tank. With this perspective, you can understand why the suction friction head takes a negative sign because the suction friction head negatively contributes to the total suction head. Similarly, you can think of total discharge head as if you stand right at the pump discharge flange and look forward from the pump toward the discharge terminal. Therefore, the pump must overcome the surface pressure, elevation, and friction losses in the discharge side.

By expanding Eq. (10.9a) based on Eqs. (10.10) and (10.11), we have

$$H_T = (H_{D,SP} - H_{S,SP}) + (H_{D,f} + H_{S,f}) + (H_{D,v} - H_{S,v}) + (H_{D,E} - H_{S,E}) \tag{10.12}$$

Equation (10.12) indicates that we calculate the total head based on the energy difference between the suction and discharge sides.

10.5 Total Head Calculation Examples

Several examples below show how the total head is determined based on Eq. (10.12). The essence of Eq. (10.12) is to divide the pump system into the suction side and discharge side. It can be demonstrated from the examples below that with this decomposition, the calculations of the total head can be simplified while better understanding of the key components contributing to the total head can be obtained.

Example 10.2
Calculate Total Head (Figure 10.6)
The process conditions and equipment elevations are shown in Figure 10.6. The maximum and minimum liquid levels in the column on the suction side are 32 and 22 ft, respectively. The specific gravity of the liquid is 0.488. The elevation of the pump is 3 ft while the liquid level in the column on the discharge side is 72 ft. The pressure drop in the suction piping is 1 psi. In the discharge side, the pressure drop through 6 inch pipe at 500 gallon per minute (GPM) is 10 psi while the pressure drops through orifice and control valve are 2 and 30 psi, respectively. The pressure drop through the exchangers is 15 psi. Calculate the pump head.

Solution
i) Total suction head:
We will begin with the total suction head calculation.

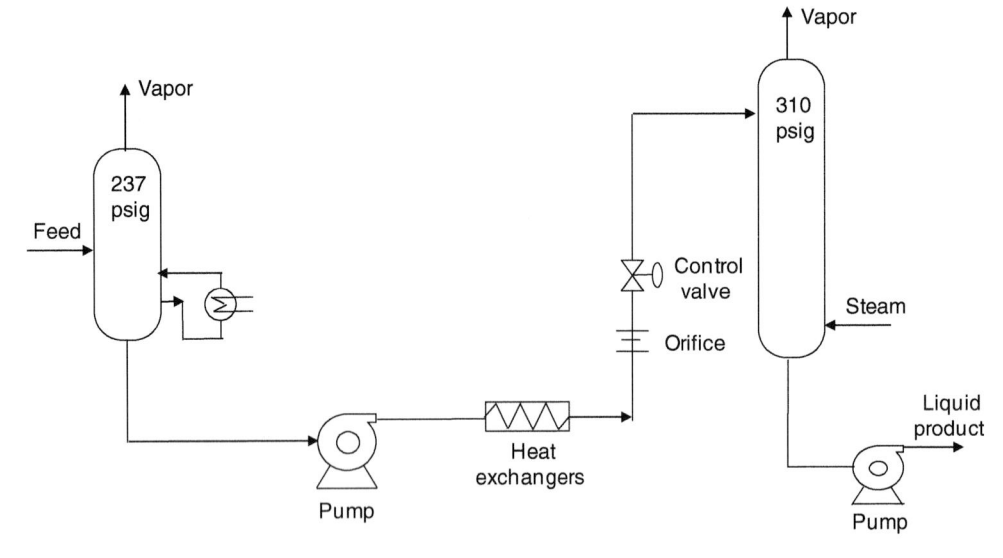

Figure 10.6 Process system.

- The suction surface pressure is 237 psig. Thus, $H_{S,SP} = 2.31 \times 237/0.488 = 1121.86$ ft
- The pressure drop in suction piping due to friction loss at 500 GPM is 1 psi. Thus, $H_{S,f} = 2.31 \times 1/0.488 = 4.73$ ft
- The velocity head $H_{S,v}$ is almost zero as the suction reference in this case is the surface of the suction tank.
- The suction elevation head $H_{S,E} = 22 - 3 = 19$ ft where the pump is 3 ft above suction centerline. For being conservative, the minimum suction vessel level of 22 ft is used.

Therefore, the total suction head is

$$H_S = H_{S,SP} - H_{S,f} + H_{S,v} + H_{S,E}$$
$$= 1121.8 - 4.7 + 0 + 19 = 1136.1 \text{ ft}$$

ii) Total discharge head:

- Discharge surface pressure = 310 psig. Thus, $H_{D,SP} = 2.31 \times 310/0.488 = 1467.4$ ft
- The total discharge friction head ($H_{D,f}$) is the sum of all the friction losses in the suction line:
- Pressure drop through 6 inch pipe at 500 GPM is 10 psi.
  - Pressure drops through orifice and control valve are 2 and 30 psi, respectively.
  - Pressure drop through exchangers is 15 psi.
  - Thus, the total discharge friction head is the sum of the above losses, $H_{D,f} = 2.31 \times (10+2+30+15)/0.488 = 269.8$ ft at 500 GPM.
- The velocity head $H_{D,v}$ is almost zero as the discharge reference in this case is the surface of the discharge tank.
- The discharge elevation head $H_{D,E} = 72 - 3 = 69$ feet.

Therefore, the total discharge head is

$$H_D = H_{D,SP} + H_{D,f} + H_{D,v} + H_{D,E}$$
$$= 1467.4 + 269.8 + 0 + 69 = 1806.2 \text{ ft gauge}$$

iii) Thus, total system head:

$$H_T = H_D - H_S = 1806.2 - 1136.1 = 670.1 \text{ ft for 500 gpm}$$

iv) Alternatively, since the pressure drops for the piping, fittings, and exchangers are given in this example, the pump head required can be calculated based on the total pressure drop between the discharge and suction flanges via

$$H_T = \frac{2.31(P_D - P_S)\big|_{\text{at flanges}}}{SG} \quad (10.1)$$

The pressure at the suction flange is

$$P_S = P_{S,SP} - \Delta P_{S,f} + P_{S,E} = 237 - 1 + \frac{(22-3) \times 0.488}{2.31} = 240 \text{ psi}$$

The pressure at the discharge flange is

$$P_D = P_{D,SP} + \Sigma \Delta P_{D,f} + P_{D,E} = 310 + (10+2+30+15)$$
$$+ \frac{(72-3) \times 0.488}{2.31} = 381.6 \text{ psi}$$

Thus, the total pressure drop between the discharge and suction flanges is

$$(P_D - P_S)\big|_{\text{at the flanges}} = 381.6 - 240 = 141.6 \text{ psi}$$

Total pump head required can then be calculated based on Eq. (10.1) as

$$H_T = \frac{2.31 \times 141.6}{0.488} = 670.1 \text{ ft for 500 gpm}$$

10.6 Pump System Characteristics: System Curve

Bernoulli Eq. (10.8) defines the total head requirement and describes the pump system characteristic. The equation can be represented graphically by the system curve as shown in Figure 10.7, which describes the relationship between system total head and flow rate for a given pump system. A few important facts for the pump system curve are summarized below:

- There is only one system curve for a given pump system design and control valve openings.
- The system curve changes if the pump system conditions vary (e.g. degree of opening varies in control valves).
- The shape of the system curve is parabolic because the terms for friction losses and velocity head take the exponent of 2.
- The system curve is independent of the pump mechanical characteristics.
- At zero flow, the curve will be vertically offset due to static head or elevation difference. At the zero flow, Eq. (10.12) becomes

$$H_T \big|_{Q=0} = \left(H_{D,E} - H_{S,E} \right) \quad (10.13)$$

10.6.1 Examples of System Curves

For illustration purposes, system curves for different pump system designs are provided below.
Case 1: All friction losses and no static lift (Figure 10.8).
Case 2: More static lift and less friction losses (Figure 10.9).
Case 3: Negative static lift (Figure 10.10).
Case 4: Two different static lifts in a branching pipe system (Figure 10.11).

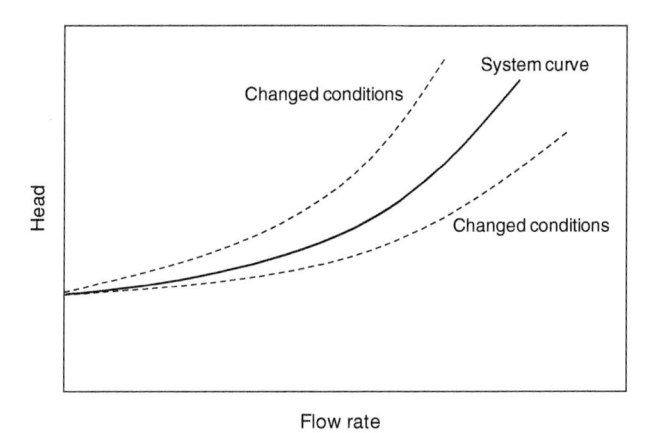

Figure 10.7 Pump system curve.

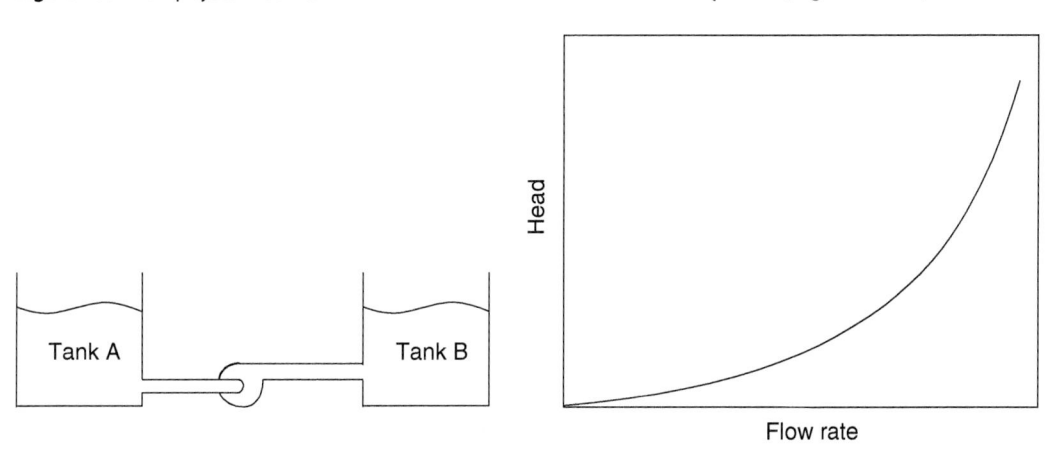

Figure 10.8 System curve for no static lift.

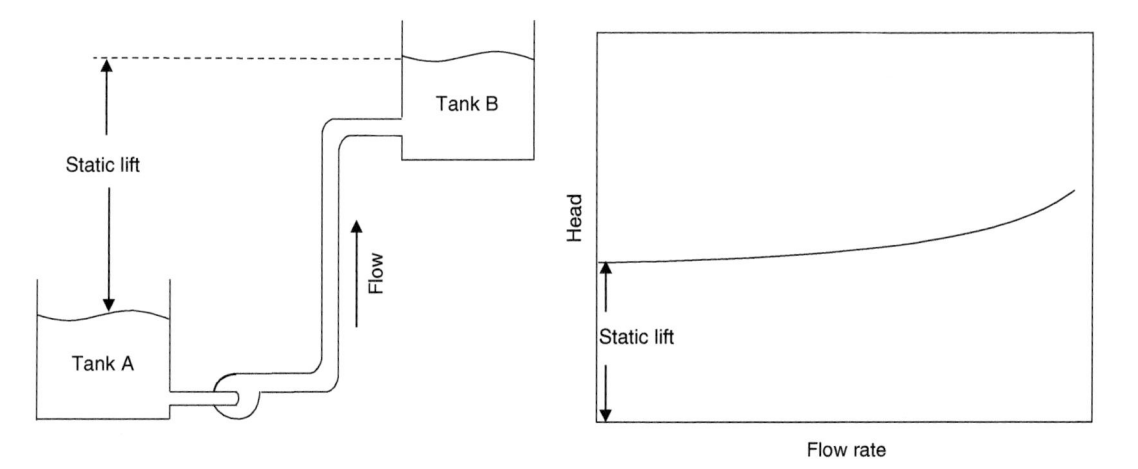

Figure 10.9 System curve for small friction losses.

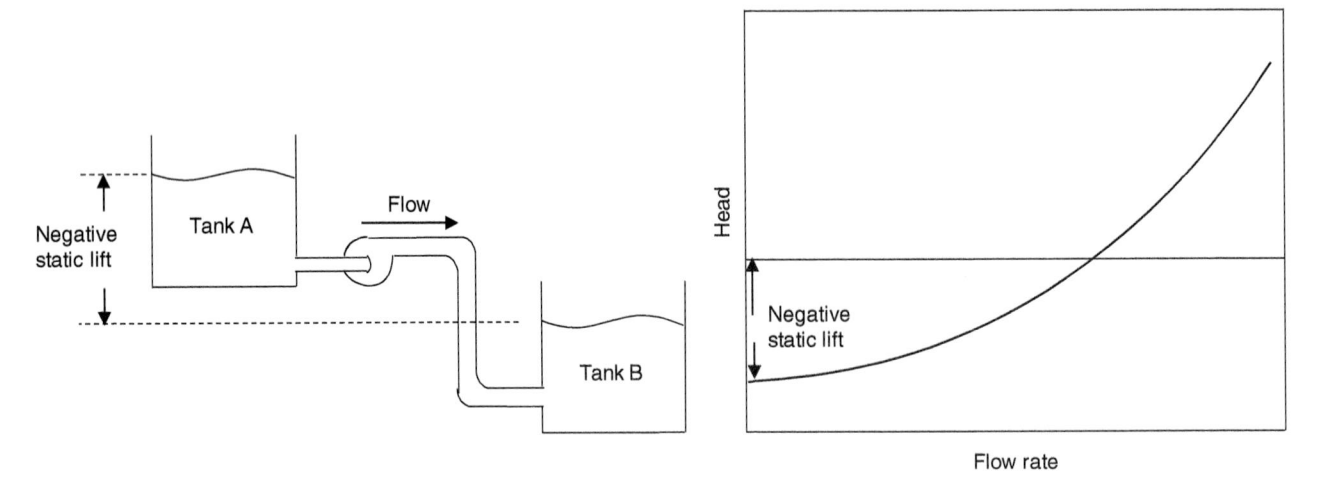

Figure 10.10 System curve for negative static lift.

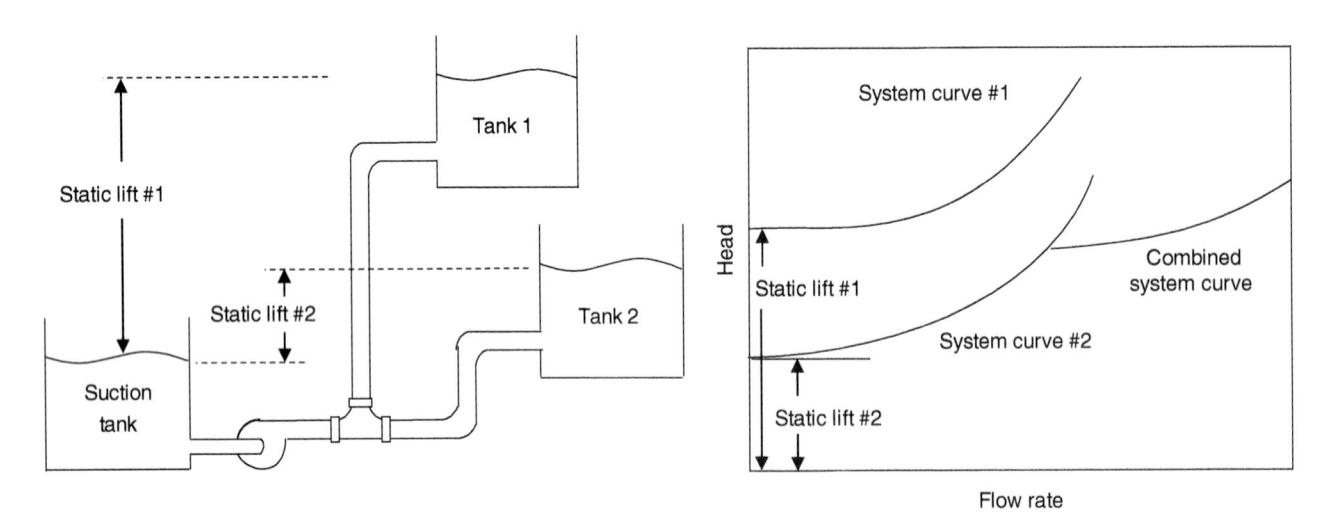

Figure 10.11 System curve for double discharges.

10.7 Pump Characteristics: Pump Curve

A pump curve indicates the pump capability to overcome the system head, which is different from the system curve which indicates the system requirement of pump head in terms of elevations and friction losses. Thus, while the system curve is a function of the process, the pump curve is the function of the pump. You may ask how a pump curve is generated and how to interpret it.

To answer this question, let us conduct a simple experiment with the pump suction at atmospheric condition: raise the discharge pipe end of the pump vertically until the flow stops. This means that the pump cannot raise the fluid higher than this point. Thus, we generate a zero flow point with the maximal pump head (H_T). For illustration, let us assume $H_{T,max} = 80$ ft. The maximum liquid height corresponds to the maximum discharge pressure as:

$$H_{T,max}\big|_{Q=0} = \frac{2.31 P_{D,max}}{SG} \quad (10.14)$$

At the zero flow when the pump is still running, the friction head is zero since there is no flow. Thus, the pump head is equal to the static head with zero flow. The pump head at zero flow is called shut-off head when the discharge valve is closed. Of course, a pump should not run under this condition continuously as the liquid would rapidly heat up to a temperature greater than what the seal can tolerate and the pump could be severely damaged.

If the discharge pipe is now cut at a slightly lower height, say 70 ft, there is a certain amount of flow out of the pipe, which can be measured and assumed to be

60 GPM for this example. Now, we have a second point at 60 GPM flow on the flow (Q)–head (H) diagram.

If we continue to cut the pipe at several descending heights and measure the flow rates, more Q–H points can be generated as shown in Figures 10.12. By connecting these points, a pump performance curve is obtained as in Figure 10.13. This may explain how a pump curve is generated. However, use of cutting the pipe here is just for a symbolic illustration. In reality, it relies on flow measuring and pressure reading devices.

Key pump characteristics can be summarized as below:

- Head and flow: The performance curve indicates the range of pump head that a pump is capable of providing over a range of flow rates. The process conditions dictate the pump head that the pump selected will operate at, while the pump curve determines the flow rate according to the head. If the head goes down, the flow rate goes up. If the head goes up, the flow rate goes down.
- Design point: The system curve defines the process requirement, which becomes the basis for choosing a pump. The pump selected must satisfy the process requirement of flow rate and head. Thus, by plotting the selected pump curve on top of the system curve, the intersection of these curves as shown in Figure 10.14 is known as the design point.
- Impeller size: Increasing the impeller diameter changes the performance capability. As the result, the pump curve moves upward. As an example, Figure 10.15 shows pump curves for pump impellers with different diameters under a given operating speed. See Figure 10.16 for more details and assume the process head is 2400 ft. If the pump has a 9½ inch

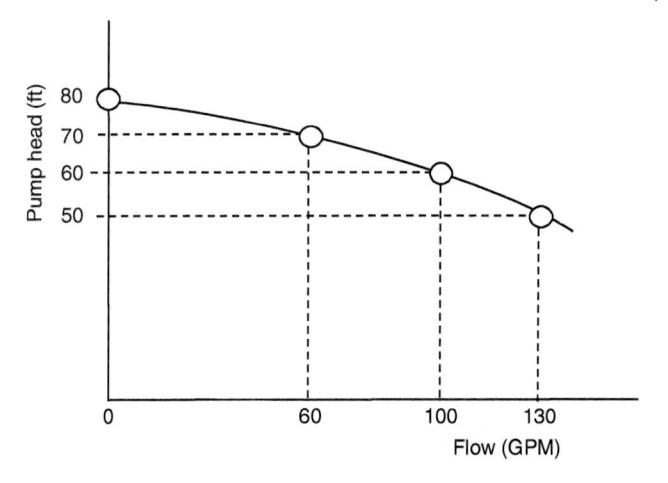

Figure 10.13 Pump curve for Figure 10.12.

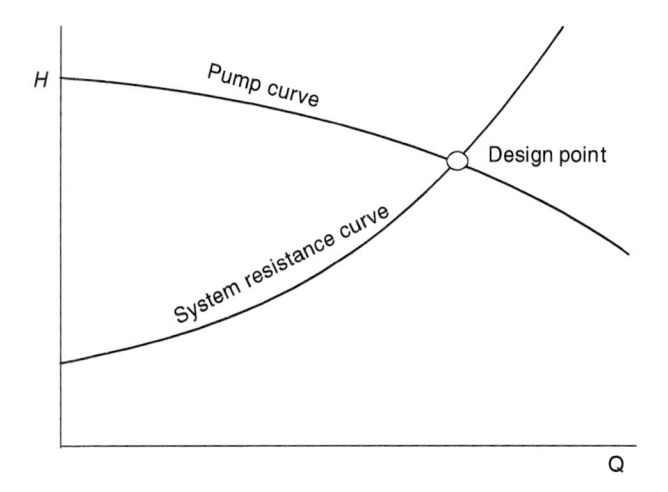

Figure 10.14 Pump normal operating point.

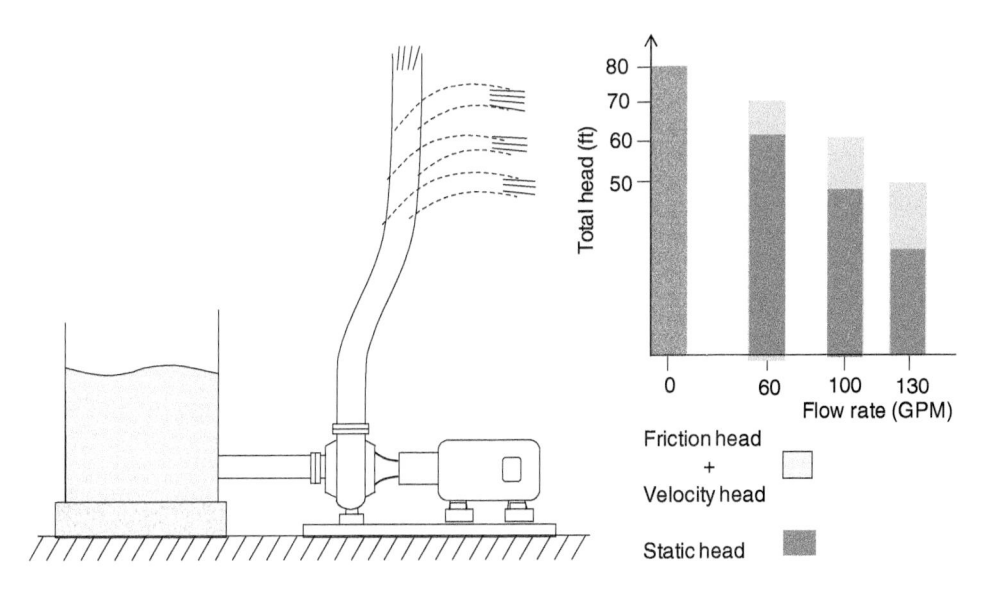

Figure 10.12 Pump head versus flow rate.

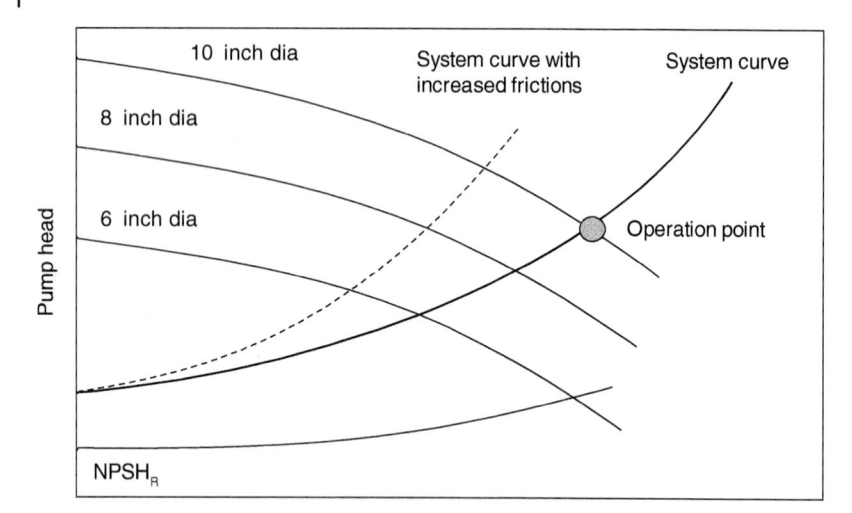

Figure 10.15 Pump curve and system curve could change.

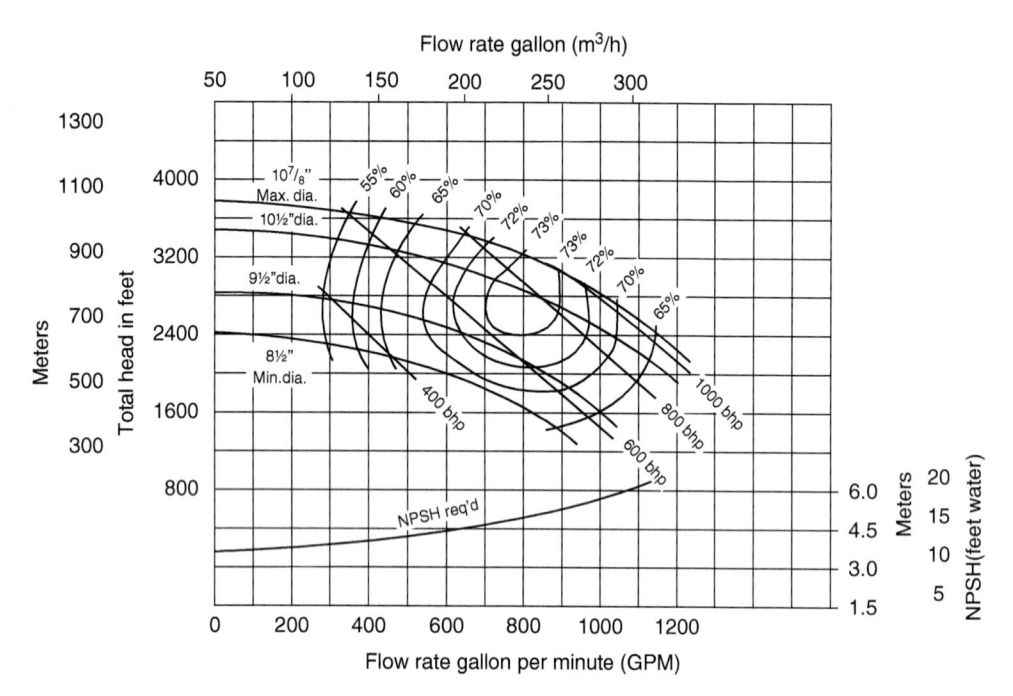

Figure 10.16 Pump curve.

impeller, the flow rate will be 600 GPM. If the pump has a 10½ inch impeller, the flow rate will be slightly over 1000 GPM.
- Pump speed: the pump curve indicates the performance at a certain speed (e.g. 3450 rpm, a common electric motor speed in 60 Hz countries). Increasing pump speed has similar effect as increasing impeller size.
- Brake power: The pump power in brake horsepower (BHP) can be calculated as

$$\text{BHP} = \frac{(\text{GPM} \times \text{SG}) H_T}{3960 \eta} \quad (10.15)$$

where the flow rate is in (GPM), total head H_T in (ft), and η is the pump efficiency in (%).

10.8 Best Efficiency Point (BEP)

The BEP indicates that the pump will operate most efficiently and reliably at the BEP flow. As per API 610, the rated flow (process condition) shall be 80–110% of the BEP. At the BEP, there will be minimal amount of vibration and noise. This is because at the BEP the impeller is balanced radially. At flows higher and lower than the flow at the BEP, there is a radial force on the impeller.

Of course, the pump can operate at other flow rates, higher or lower than the BEP flow.

As mentioned, increasing the impeller diameter and/or speed will raise the pump flow–head curve. Pump flow rate will vary with the suction and discharge pipe diameter and length as the system friction drop changes. A system with a long and narrow discharge pipe will lead to high friction loss and thus lower flow rate. In this case, the pump system curve will move upward. The pump is designed to produce a certain nominal flow rate for the piping system sized accordingly. The impeller size and its speed dictate the pump to deliver the nominal flow rate. To change the flow rate in operation, appropriate valves must be adjusted.

10.9 Pump Curves for Different Pump Arrangement

10.9.1 Series Arrangement

This arrangement may be required for higher head applications. Centrifugal pumps are connected in series if the discharge of the first pump is connected to the suction side and the second pump. Two similar pumps, in series, operate in the same manner as a two-stage centrifugal pump. Each of the pumps is putting energy into the pumping fluid, so the resultant head is the sum of the individual heads. If two of the same pumps are in series, the combined performance curve will have double the head of a single pump for a given flow rate (Figure 10.17). For two different pumps, the head will still be added together on the combined pump curve, but the curve will most likely have a piecewise discontinuity. At the same flow rate, the heads of the two pumps are added together.

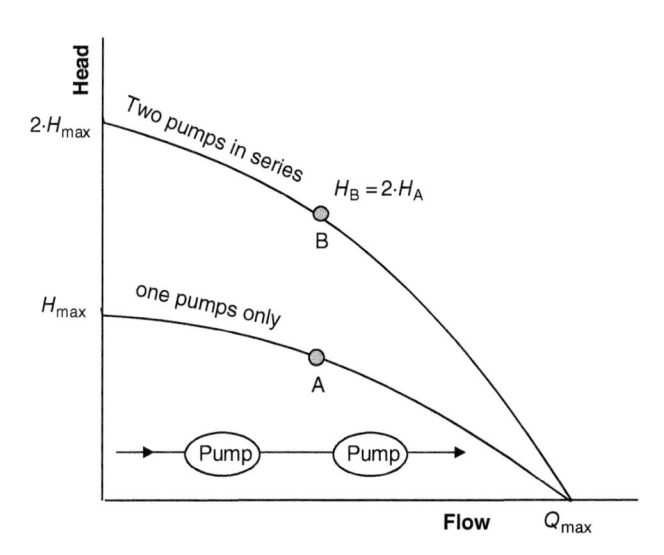

Figure 10.17 Pump curves for single and two pumps in series.

For example, if a single pump operating at 50 GPM at 70 ft of head and 3 BHP is put in series with an identical pump, the two pumps will provide 140 ft of head and require 6 BHP total. Each pump supplies the required flow at one-half the required head.

Some things to consider when you connect pumps in series:

- Both pumps must have the same width impeller. Otherwise, the difference in capacities could cause a cavitation problem if the first pump cannot supply enough liquid to the second pump.
- When pumps are operating in series, if either the low-pressure or the high-pressure booster pump fails, the remaining pump will operate at zero flow! Therefore, the second pump in series should be automatically shut down on low flow rate.
- Both pumps must run at the same speed (same reason).
- Be sure the casing of the second pump is strong enough to sustain the higher pressure. Higher strength material, ribbing, or extra bolting may be required.
- Be sure both pumps are filled with liquid during start-up and operation.
- Start the second pump after the first pump is running.

10.9.2 Parallel Arrangement

This arrangement may be selected for large flow rate. Pumps are operated in parallel when two pumps are connected to a common discharge line, and share the same suction conditions. The pump curves for single and two pumps in parallel are given in Figure 10.18. For example, a single pump operating at 50 GPM at 70 ft of head and 3 BHP is put in parallel with an identical pump. The result is a total flow of 100 GPM, requiring 6 BHP and operating at 70 ft of head. Each pump supplies one-half the required flow at the required head. For the same head the flow rates are added together.

Some things to consider when pumps are operated in parallel:

- Both pumps must produce the same head, which resulted from the same speed and the same diameter impeller.
- When pumps are operating in parallel, if the internal clearances in one pump deteriorate substantially more than the other, the stronger pump may force the other to operate below its minimum continuous capacity.
- Two pumps in parallel will deliver less than twice the flow rate of a single pump in the system because of the increased friction in the piping. If there is additional friction in the system from throttling, two pumps in parallel may deliver only slightly more than a single pump.

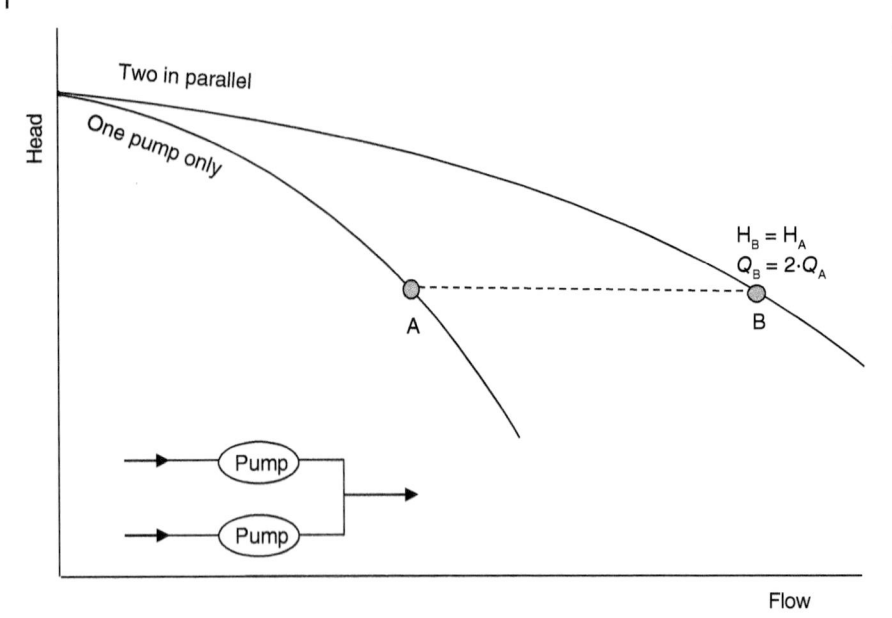

Figure 10.18 Pump curves for single and two pumps in parallel.

- Most plants read only total flow and cannot see the differences in individual pump performance.
- Parallel pumps are notorious for operating at different flows. Often a weaker pump is operating close to its shutoff point while a stronger pump is operating to the far right of its curve and running out of $NPSH_A$. This is why it is important to have similar curves which rise to shutoff.

10.10 NPSH

Cavitation occurs when there is presence of vapors in the impeller. As the fluid enters the pump and impeller, there is a pressure drop. There are entrance losses as the liquid enters the pump and friction losses in the pump nozzle. There are also friction and turbulence losses as the liquid enters the impeller. If this pressure drop causes the fluid to drop below its vapor pressure or bubble point, the fluid will start to boil and vapor bubbles will occur. When vapor bubbles occur, there is expansion. One cubic feet of water at room temperature will create 200 ft³ of vapor.

Pump cavitation is the formation and subsequent collapse of these vapor bubbles within the pump casing. When the fluid pressure rises as the fluid leaves the impeller, these vapor bubbles collapse. The liquid strikes the impeller and casing at the speed of sound. The noise generated from these collisions of vapor bubbles sounds like pumping marble stones. Cavitation could possibly stop flow altogether and damage the impeller. Thus, noise and capacity loss are the major indicators of cavitation.

Under cavitation, the first pump seal will be damaged due to high vibration. In propane service, running with too low a liquid level for a few hours will often damage the mechanical seal sufficiently to require taking the pump out of service. Passing vapor through the impeller causes rapid changes in the density of the fluid pumped. This uneven operation forces the impeller and shaft to shake, and the vibration is transmitted to the seal. A mechanical seal consists of a ring of soft carbon and a ring of hard metal pressed together. Their smooth, polished surfaces rotate past each other. When either surface is chipped or marred, the seal leaks. Continued operation of a cavitating pump will damage its bearing and eventually the impeller wear ring and shaft.

To avoid cavitation, the liquid pressure within the pump should never fall below the vapor pressure of the liquid at the pumping temperature. There must be enough pressure at the pump suction and thus not vaporize the fluid. This pressure available at the pump suction over the fluid vapor pressure is Net Positive Suction Head Available ($NPSH_A$), which is a function of the pumping system.

10.10.1 Calculation of $NPSH_A$

For better understanding of $NPSH_A$, consider a typical pump suction system shown in Figure 10.19 and $NPSH_A$ can be defined as

$$NPSH_A = P_{S,A} - P_{S,V} \qquad (10.16)$$

where

$$P_{S,A} = P_{S,SP} + \frac{H_{S,E} \times SG}{2.31} - \Delta P_{S,f}$$

$P_{S,A}$ = actual suction pressure (psia);

Figure 10.19 A typical pump suction system.

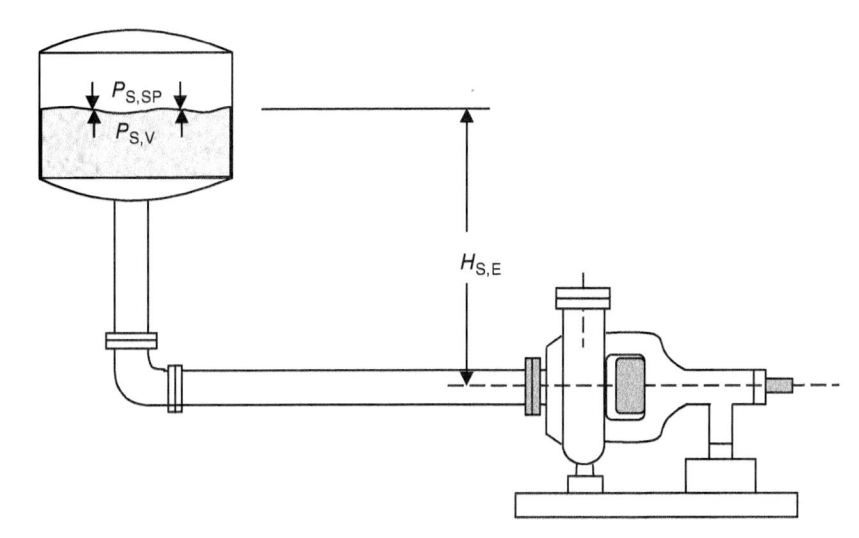

$P_{S,SP}$ = surface pressure at the liquid free surface in the suction (psia);
$P_{S,V}$ = vapor pressure of liquid under suction temperature (psia);
$\Delta P_{S,f}$ = pressure drop due to suction friction losses (psia).

If all pressures are converted to feet of liquid as shown in Figure 10.20, available NPSH can be calculated via

$$\text{NPSH}_A = (H_{S,SP} \pm H_{S,E} - H_{S,f}) - H_{S,V} \quad (10.17)$$

where

$H_{S,P} = 2.31 P_{S,SP}/\text{SG}$; absolute pressure at the liquid free surface converted to feet of liquid.
$H_{S,E}$ = suction head (takes "+") or lift (takes "−") with liquid specific gravity considered, in feed of liquid; make sure to use the lowest liquid level allowed in the tank.
$H_{S,V} = 2.31 P_{S,V}/\text{SG}$, vapor pressure of liquid (P_V) at pumping temperature converted to feet of liquid.
$H_{S,f}$ = friction loss through suction line, fitting, and entrance, in feet of liquid.

If the surface pressure is atmospheric pressure, 34 ft ($H_{S,P}$) is the value of atmospheric pressure at sea level. If the suction head ($H_{S,E}$) is 10 ft, NPSH$_A$ is 44 ft minus a small quantity of friction loss and vapor pressure based on Eq. (10.17). This NPSH$_A$ should be sufficient.

One should start to have concern when the NPSH$_A$ falls to within 4 ft of the NPSH$_R$. But how could this happen? This is possible if the pipe diameter is small and pipe length is long plus a lot of plugging which increases friction in the suction line. Also during start-up, the suction strainer is prone to plugging if the suction piping is not sufficiently cleaned.

Note that the above calculation does not include the velocity head, which is common for NPSH$_A$ as velocity head is relatively small. But velocity head is included in the NPSH$_R$ (required NPSH) curves provided by the manufacturer.

Example 10.3
Gasoline is stored in an open tank and the tank is piped to a centrifugal pump. The pump suction system is depicted in Figure 10.21. The pressure drop through the suction piping = 0.186 psi. Vapor pressure of gasoline at pumping temperature = 5.0 psia. SG of gasoline = 0.74. Calculate the available NPSH at a flow rate of 3000 GPM. If NPSH$_R$ is 22 ft, will the pump cavitate?

Solution

$$H_{S,SP} = 14.7 \times 2.31 / 0.74 = 45.89 \text{ ft};$$
$$H_{S,E} = 15 + 35 - 30 = 20 \text{ ft};$$
$$H_{S,f} = 0.186 \times 2.31 / 0.74 = 0.58 \text{ ft};$$
$$H_{S,V} = 5 \times 2.31 / 0.74 = 15.61 \text{ ft};$$
$$\begin{aligned}\text{NPSH}_A &= H_{S,SP} + H_{S,E} - H_{S,f} - H_{S,V} \\ &= 45.89 + 20 - 0.58 - 15.61 = 49.7 \text{ ft}.\end{aligned}$$

Thus, the pump will not cavitate as NPSH$_A$ = 49.70 ft is much larger than NPSH$_R$ = 22 ft.

10.10.2 NPSH Margin

Net Positive Suction Head Required (NPSH$_R$) is a characteristic of the pump and it is provided by pump vendor. As per API 610, the pump vendor can report NPSH required when there is a 3% loss of head due to cavitation. So if NPSH available is equal to NPSH required, there is cavitation. Therefore, there must be a margin of NPSH available over NPSH required. Typical margins are 4-ft for hydrocarbon liquids (including low SG) and 10-ft for boiling water.

During initial system design, one variable is suction vessel height. For example, the minimum liquid level in the suction vessel must be at a high-enough elevation to provide this 4-ft margin.

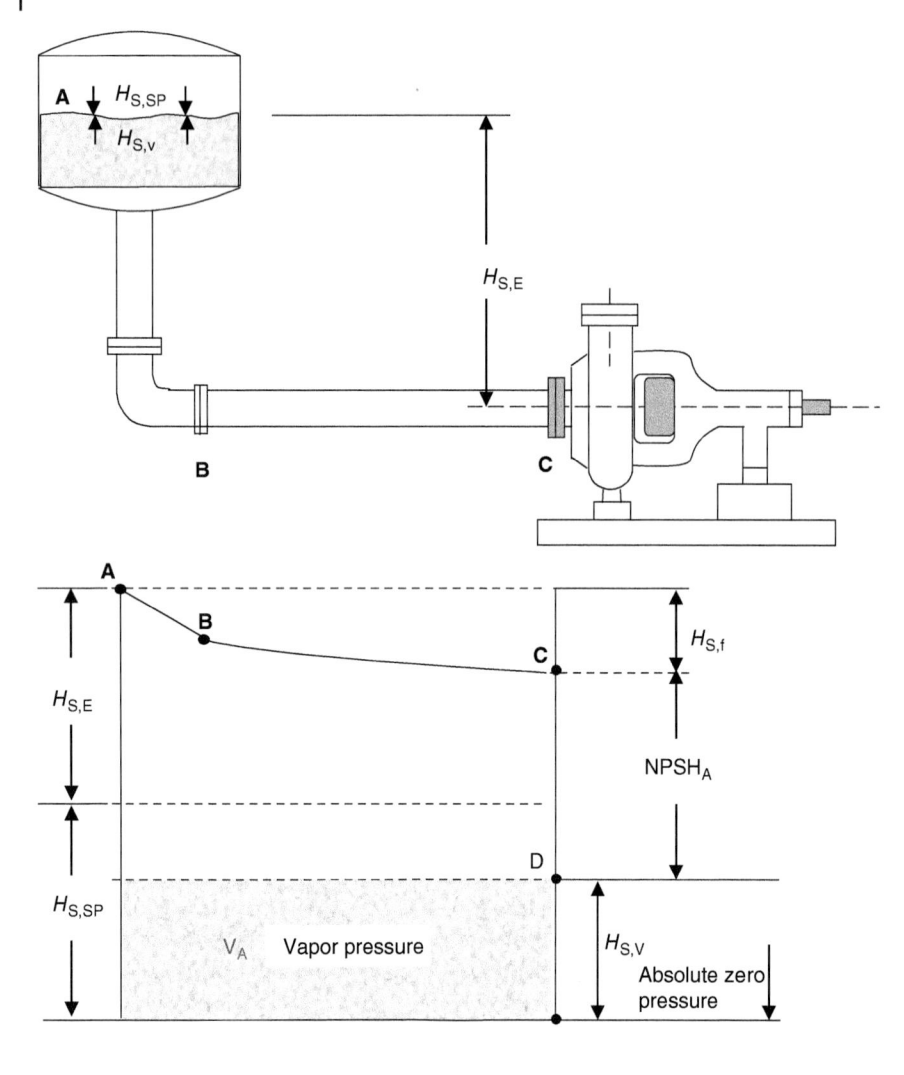

Figure 10.20 NPSH$_A$ expressed in feet for typical pump suction.

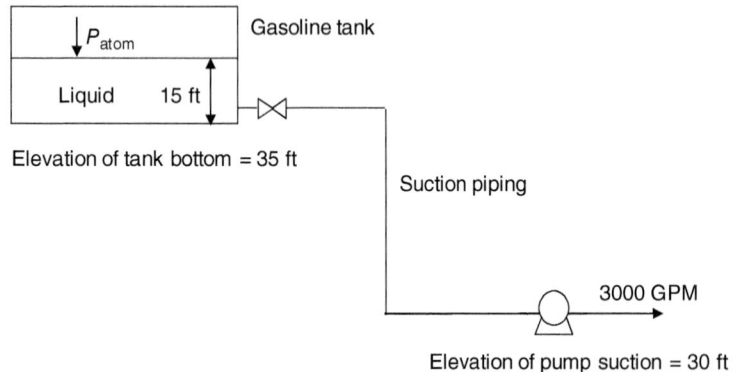

Figure 10.21 Pump suction for Example 10.3.

10.10.3 Measuring NPSH$_A$ for Existing Pumps

NPSH calculations must be conducted for every pump installation. It is also recommended to do the same for existing pumps. To know the true NPSHA for existing pumps, first install a compound gauge at the pump suction that can measure both vacuum pressures as well as positive gauge pressures. When the pump is running, the reading from this gauge will indicate the suction pressure. For given vapor pressure at the pump temperature, NPSHA can be calculated via Eq. (10.17) and the safety margin should be provided which depends on the company best practice. If this NPSHA value calculated is less than the pump's NPSHR provided by the manufacturer, this pump is under cavitation.

10.10.4 Potential Causes and Mitigation

There are various causes for a low NPSH and major causes are summarized below. If changes to the system are not adequate to increase $NPSH_A$, you may need to consult the pump manufacturer about reducing $NPSH_R$.

10.10.4.1 Lower P_S Due to Drop in Pressure at the Suction Nozzle

This is the basic cause when the fluid at pump suction is not available sufficiently above the vapor pressure of liquid at operating conditions.

10.10.4.2 Lower P_S Due to Low Density of the Liquid

Light hydrocarbon and vacuum services encounter low NPSH more frequently.

10.10.4.3 Lower P_S Due to Low Liquid Level

The problem may be a low liquid level in the suction side vessel. The level controller on the pump discharge line may have malfunctioned. To check this, blow out the taps on the vessel gauge glass and verify that there is a good level in the vessel.

10.10.4.4 Lower P_S Due to Increase in the Fluid Velocity at Pump Suction

Higher liquid velocity leads to lower suction pressure. As the flow rate increases, the fluid velocity increases and friction drop in the suction piping increases. At the same time, higher velocity results in higher friction losses in the piping and fittings in the pump inlet system. The above consequence is the lower pressure available at the pump suction and thus cavitation has a greater chance to occur.

10.10.4.5 Lower P_S Due to Plugged Suction Line

A suction-line restriction will also cause a pump to cavitate. To verify this, run the pump with its discharge valve pinched back just enough to suppress cavitation. Then, measure the pressures at the upstream vessel and the pump suction (use the same gauge) to calculate the pressure difference ΔP. Then, ΔP is converted to the liquid height as $2.31 \Delta P/SG$ of liquid. Finally, subtract the vertical distance between the 2 gauges from the liquid height to give the pump head. The resulting head at the pump suction should only be 1–2 ft less than the head of liquid in the vessel. If the pressure difference is quite a bit more, there is probably a plugged suction line.

10.10.4.6 Higher P_V Due to Increase in the Pumping Temperature

Vapor pressure is a function of temperature only. Increase in liquid temperature at the pump suction increases the vapor pressure of the liquid. It becomes more likely for operating pressure to fall below this vapor pressure. In some cases, a slight warming of the fluid at the pump suction could promote flashing.

10.10.4.7 Reduction of the Flow at Pump Suction

A certain minimum flow as indicated by the pump curves is required to keep the pump from running dry. If liquid flow falls below this limit, it has greater possibility of developing vapor within the pump and the likelihood of cavitation increases.

10.10.4.8 The Pump Is Not Selected Correctly

Every centrifugal pump has a certain requirement of positive suction head ($NPSH_R$). If the pump is not selected properly, $NPSH_A$ might fall below this $NPSH_R$ limit, causing cavitation.

According to Fernandez et al. (2002), increasing static head is the most viable approach. There are three basic methods to raise the static head:

- Lower the pump elevation. However, this could be proved to be less practical since pumps are typically located just above the ground level. Lowering the pump suction may require the suction nozzle to be below grade, which usually results in a more expensive pump.
- Raise the level of fluid in the suction tank. Application of this method depends on company operating policy.
- Increase the suction piping diameter and remove elbows
- Run both operating and spare pump in parallel (each pump will be at lower flow rate) resulting in lower NPSH required.

Reduction in friction losses through suction piping and fittings can also mitigate the risk of low NPSH. Reducing friction is more appealing in the existing plants where throughput is usually increased above the design throughput.

There are also other options available, which include using a larger but slower speed pump, a double-suction impeller, a larger impeller inlet area, etc.

10.11 Spillback

A "spillback" is a jargon for a partial flow recycle as a small percentage of a pump's discharge flow is routed back to the suction of the pump to ensure that the pump has a sufficient continuous flow. Spillback may be required for preventing cavitation and failures in bearing and seals if the process flow will fall below the pump minimum continuous flow for a prolonged duration. In some cases, spillback may be required to support off-design operation requirements, particularly start-up. A proper spillback system must be selected for a specific application.

There are three kinds of minimum flow limits. The objective of defining minimum flow limits is to prevent undue wear and tear in bearings and seals. In the real environment of a process plant, a pump is operated at just about any condition demanded by the situation at hand. Thus, these different pump minimum flows are used for different application purposes. In other words, for a specific application, the pump may be governed by a certain minimum flow limit.

The minimum continuous stable flow (MCSF) is the lowest flow at which the pump may operate without exceeding vibration limits imposed by API Standard 610 for the hydrocarbon process industry. It is the flow below which the pump should not be operated continuously. If the minimum continuous flow through a pump is insufficient, the resultant pump damage takes place over the long term and usually results in bearing or seal failure.

Generally speaking, single-stage, overhung process pumps with typically low power requirements are less susceptible to damage and cheaper to repair when subjected to minimum flow limits. Multistage are much more susceptible to damage when subjected to these situations and are much more costly to repair. Therefore, multistage pumps may require a spillback system and an automatic low flow shutdown of the driver. The automatic low flow shutdown removes energy input in the upset event and limits the duration/severity of dry running condition when the suction vessel liquid level is lost.

Most Sundyne pumps require spillback systems because of their limited turndown capability, high temperature rise, and drooping head curves. Specify a spillback anytime the pump turndown requirement is 60% or lower of the rated pump capacity.

10.12 Reliability Operating Envelope (ROE)

From the pump curve provided by manufacturer, we can determine the preferred operation ROE in the flow range of 70–120% of BEP as shown in Figure 10.22. Outside of the allowable operation region is unstable operation (Figure 10.22). When the flow is too high, the pump may suffer high-velocity cavitation where too large friction losses could cause too low suction pressure. On the other hand, when the flow is too low, the pump will suffer cavitation and failures in bearing and seals.

The bad-actor pump (more than one component failure per year) should be checked with ROE (Forsthoffer 2011).

To determine if a pump is operating within ROE, calculate the pump head required and measure the operating flow. Plot the flow-head point on the pump curve. If the bad-actor pump is operating outside the ROE, the equipment engineer must discuss with process engineer to determine the operating target ranges for flow, motor amps, control position, and temperature difference between the suction and discharge.

10.13 Pump Control

In operation, the pump head will change due to variations in throughput to meet processing objectives. There are two control options available for the pump to meet varied throughput (Figure 10.23). If the pump is driven by a steam turbine or variable speed motor, the driver speed can be adjusted to vary the pump head produced,

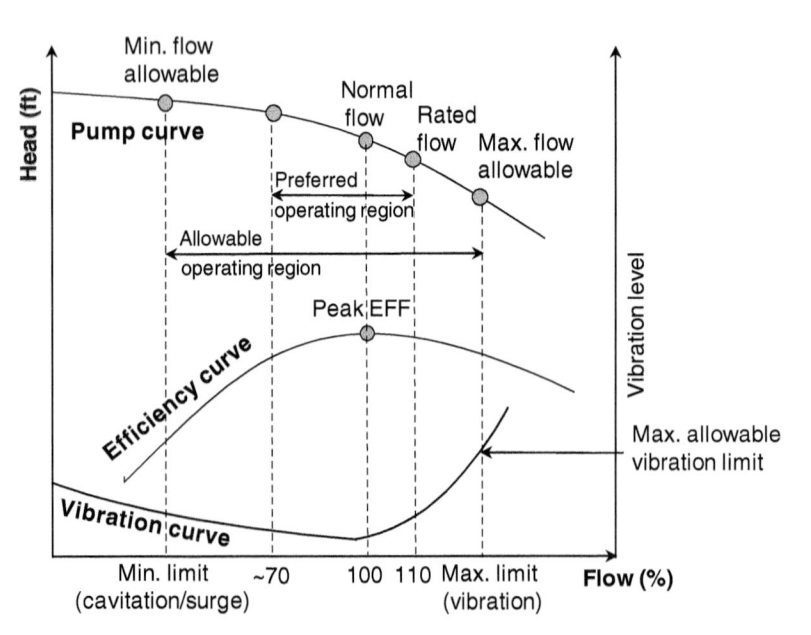

Figure 10.22 Reliability operating envelope.

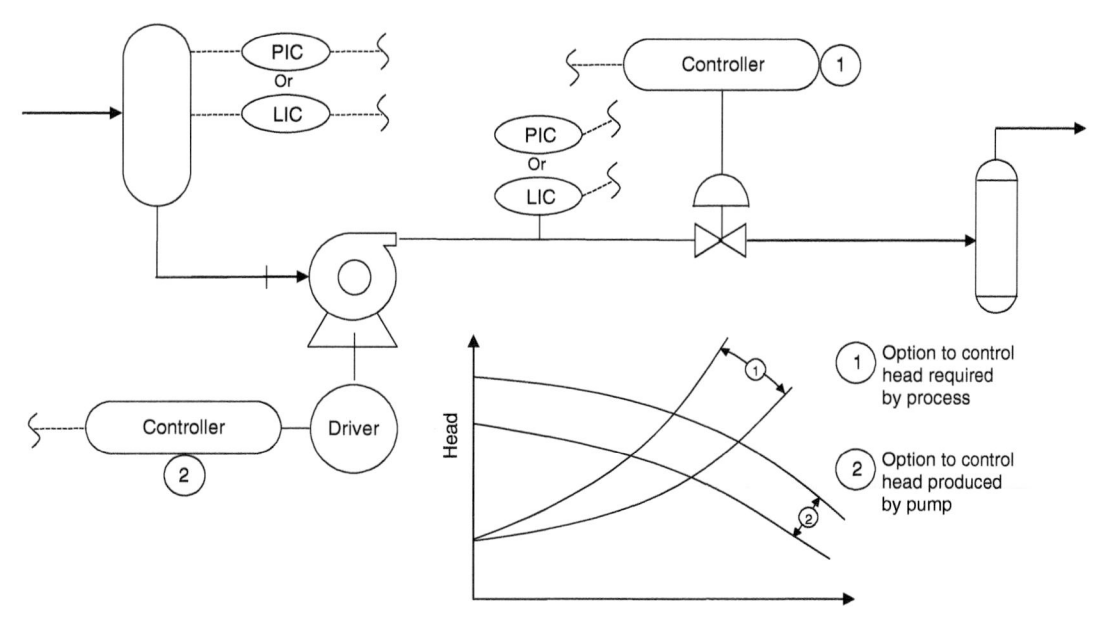

Figure 10.23 Two flow control options.

which is essentially to change the pump performance. The other option is to adjust discharge control valve to vary the pump head required, which is essentially to change the system characteristics (change the system friction losses). It can be harmful to restrict a pump's flow by putting a valve on the suction line. This can cause pump cavitation because the suction pressure will be reduced.

10.14 Pump Selection and Sizing

Both pump curves and the system curve must be evaluated to choose the right pump for a specific application. As a general guideline, the following steps are taken for selecting a centrifugal pump.

Step 1: Determine the process conditions
Correctly specified process conditions are essential for defining the operating requirements leading to proper pump selection. A process flow diagram for the pump system should be generated which covers all the components to be included in the system. The following factors must be considered in generating the pump system flow diagram.

- **Flow rate:** Flow rates including minimum, normal, and rated should be specified in the data sheet. Normal flow is to achieve a specific process operation while the rated flow is typically 10% over the normal flow depending on company practices to accommodate process variation and pump wear. The minimum flow rate for process turndown operation must be provided in order to establish if a flow bypass line is required in process design.
- **Liquid properties:** Viscosity, vapor pressure, and specific gravity are important parameters in achieving the required reliability of a pump. The viscosity affects pump performance. Since the performance of most centrifugal pumps is determined from water, the procedure developed by the Hydraulic Institute is adopted to correct the performance curves when pumping viscous fluids. Vapor pressure of the process liquid at the suction temperature is an important property when determining whether there is a sufficient NPSH. Specific gravity is the liquid property used to calculate the pump head required to overcome the resistance of the suction and discharge systems. The process engineer must specify not only the normal values for density and viscosity but also the maximal and minimal values that the pump may encounter in abnormal operations such as start-up, shutdown, turndown, and process upset. In addition, the engineer must specify the maximum operating temperature.

Step 2: Determine the total head required
The pump head at the design point (Figure 10.14) is determined based on the normal process conditions while rated pump head is based on consideration of design margin, which could be 10–25% to account for variations in physical properties and process conditions.

When operation is toward the end-of-run, pressure drops become much higher in heat exchangers and heaters due to fouling accumulation. This requires higher pump head, which is considered in determining rated discharge pressure.

Step 3: Select the pump

The pump is selected based on rated flow and head. Adequate knowledge of the process conditions and compositions is the most important aspect for optimizing pump selection. Process constraints must be considered such as operation flexibility, turndown, start-up, shutdown, etc.

Since centrifugal pumps are not normally custom designed, it is important to ensure that each vendor will provide quotes for similar pump configurations for the specific operating conditions. Often, there could be a group of pumps available for selection. This provides opportunity for good pump selection to achieve optimal trade-off between pricing, efficiency, and reliability. For given rated pump head and flow, the preliminary pump selection can be made based on the normal operating ranges (Fernandez et al. 2002).

Figure 10.16 shows that there are a number of different impeller diameters available for each pump. Selection of geometry and type are governed by the operating conditions, and properties and compositions of the liquid. Select an impeller that allows for future changes in the diameter. Pumps are rarely operated at their exact rated point. Therefore, the flow or head may need to be changed to increase the pump efficiency, or to accommodate changes in process requirements. API 610 requires that 5% higher head must be achieved with a larger impeller so the largest impeller for a casing size should not be selected.

The rated flow should be no greater than 10% to the right of the BEP. This will result in both rated and normal operation close to the BEP (see Figure 10.24). The pump selected must ensure reliability, which is discussed in more detail above.

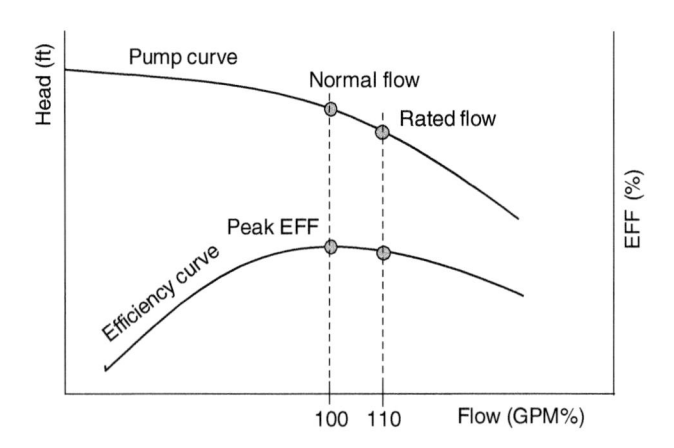

Figure 10.24 Optimal pump selection.

The next task is to match $NPSH_A$ and $NPSH_R$. $NPSH_A$ is calculated based on Eq. (10.17). It is prudent to incorporate a margin of safety for $NPSH_A$ above $NPSH_R$ to effectively prevent potential cavitation. The actual margin will vary from company to company. Some use the normal liquid level as the datum point, while others use the bottom of the vessel. Typical margins are 4-ft for hydrocarbon liquids (including low SG) and 10-ft for boiling water.

Step 4: Select the driver

When sizing a motor driver to fit an application, it is necessary to consider whether the pump will ever be required to operate at a flow higher than the duty point. The motor will need to be sized accordingly. If the pump may flow out to the end of the curve (if someone opens the restriction valve all the way, for

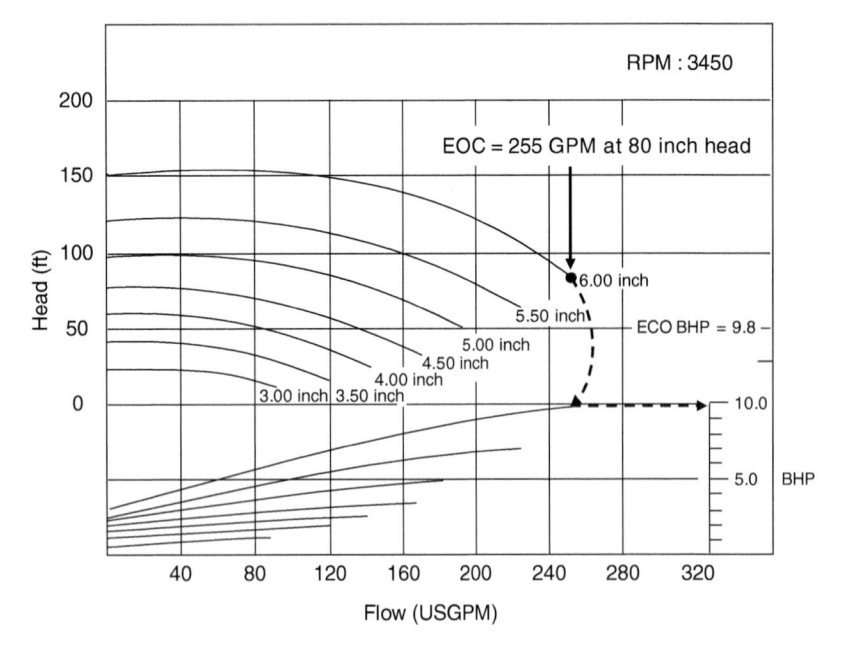

Figure 10.25 Pump curves with corresponding impeller diameters and BHP curves.

example), it is important that the motor does not become overloaded as a result. Therefore, it is normal practice to size the motor based on the end of curve (EOC) horsepower requirements. Figure 10.25 shows an example where a 7.5 hp motor would adequately power the pump at a duty point of 120 GPM at 150 ft. But notice that the *end of curve* BHP requires that a 10 hp motor be used. Note that at the bottom of the pump performance curve (Figure 10.25), BHP lines slope upward from left to right. These BHP lines are developed based on water with SG = 1. These lines correspond to the pump performance curves above them (the top performance curve corresponds to the top BHP line, and so on). These BHP lines indicate the amount of driver BHP required at different points of the performance curve.

Nomenclature

BHP   brake horsepower (hp)
BEP   best efficiency point
GPM   gallon per minute (gall/min)
H_E static head or elevation head (ft)
H_f friction head (ft)
H_{SP} surface pressure head (ft)
H_T pump head (ft)
H_v velocity head (ft)
H_V vapor pressure head (ft)
$NPSH_A$ Net Positive Suction Head Available (psi or ft)
$NPSH_R$ Net Positive Suction Head Required (psi or ft)
P_D pump discharge pressure (psi)
P_S pump suction pressure (psi)
P_V vapor pressure of liquid (psi)
Q flow rate (l/s)
SG   specific gravity, dimensionless
u fluid velocity (ft/s)
Z elevation (ft)

Greek Letters

η pump efficiency (%)

References

Fernandez, K., Pyzdrowski, B., Schiller, D., and Smith, M.B. (2002). Understand the basics of centrifugal pump operation, May issue, CEP, AICHE.

Forsthoffer, W.E. (2011). *Forsthoffer's Best Practice Handbook for Rotating Machinery*. Butterworth-Heinemann.

Part IV

Energy and Process Integration

11

Process Integration for Higher Efficiency and Low Cost

11.1 Introduction

In the nineteenth century, oil refineries processed crude oil primarily to recover kerosene for lanterns, and the refinery design was very simple. The invention of the automobile shifted the demand to gasoline and diesel, which remain the primary refined products today. Fuel production for automobiles required more conversion and hence increased plant complexity. In the 1990s, stringent environmental standards for low sulfur drove refineries to include more hydro-processing, leading to greater refinery complexity. In recent times, higher energy efficiency and lower greenhouse emissions requirements have resulted in designs incorporating more heat recovery with additional equipment and thus have further elevated the process complexity. This trend is illustrated by Figure 11.1.

The major challenge is how to achieve first-class energy efficiency process design with simplified process design and low capital costs. Recent work (Zhu et al. 2011) has pointed to the areas of process and equipment innovations, which play important roles to overcome this challenge. Implementing process and equipment innovations often result in combined benefits in process yields, throughput, energy efficiency, and reduced capital costs at the same time. Many of these areas include optimizing process flowsheet and condition as well as use of advanced equipment. The powerful concept and methodology for connecting these innovations to achieve high energy efficiency and simpler designs with low capital is the process integration.

11.2 Definition of Process Integration

To define the concept of process integration, we need the traditional design approach, which can be described by so-called "Onion Diagram" as shown in Figure 11.2 that provides an overall view of energy considerations throughout the traditional design procedure.

The design of a process complex starts from defining a design basis. This step consists of defining physical and chemical conditions for feeds, products, and utilities. The design then concentrates on the chemical reaction system. The reaction system is the core of a process complex where the conversion of feeds to products takes place. The goal of a reaction system design is to achieve a desirable product yield structure via selection of catalyst and design of reactors. As reaction effluents contain a large amount of heat at high temperature, the heat recovery of reaction effluent is a major consideration for process energy efficiency.

After the reaction, the reaction effluent goes through a separation system in order to separate desirable products from by-products and wastes. For separating multiple products, a separation system involving several columns are required. Heat recovery from products makes significant contributions to process energy efficiency. At the same time, there could be a large amount of excess heat available in fractionation columns where multiple products are made. It is essential that this excess heat is removed from pump around and used for process heating purpose.

Process heat recovery design comes after design of reaction and separation systems. The latter defines the basis for process energy demand, for reaction, fractionation, and separation, via selection of process conditions for reaction and separation. In the heat recovery design step, the goal is to minimize overall process energy use (fuel, steam, and power) for a given process energy demand. This is achieved by heat recovery between those process streams with heat available (such as reaction effluent, separation products, and column overhead vapor), and process streams with the need of heat (such as reaction feed, separation feed, and reboiling).

After the process heat recovery is done, the next design step is to determine the utility supply in terms of heating

---

Efficient Petrochemical Processes: Technology, Design and Operation, First Edition.
Frank (Xin X.) Zhu, James A. Johnson, David W. Ablin, and Gregory A. Ernst.
© 2020 John Wiley & Sons, Inc. Published 2020 by John Wiley & Sons, Inc.

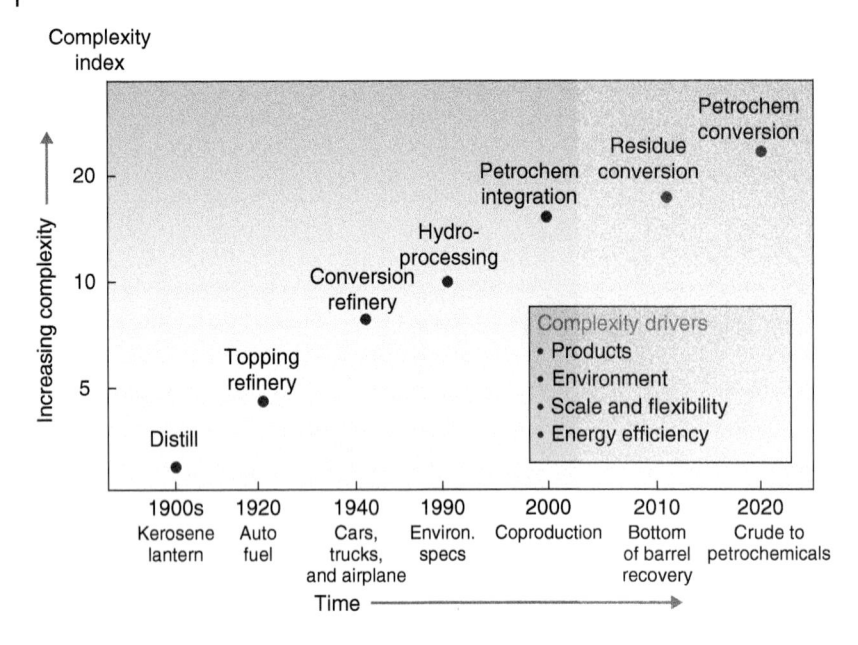

Figure 11.1 The trend of increased refinery complexity over time.

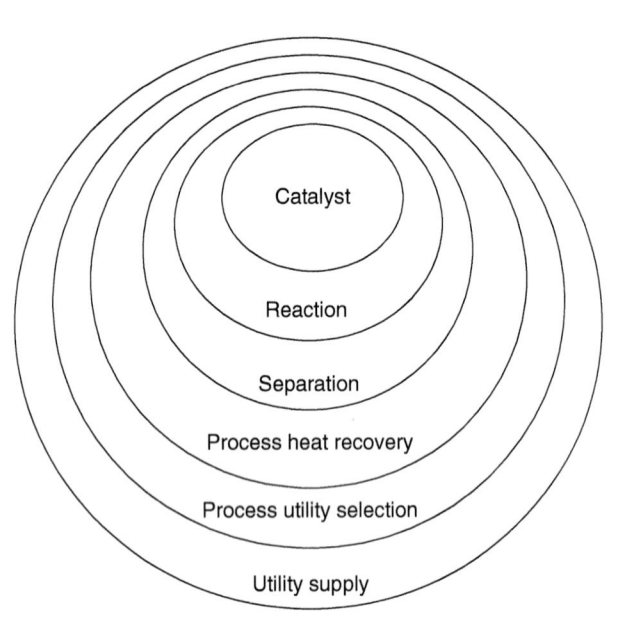

Figure 11.2 Sequential process design: traditional design approach.

and cooling, and power based on the needs and characteristics of process energy demand. In this step, the means of heat supply for reaction and separation system will be addressed. For example, a choice for the reboiling mechanism must be made for a separation column between a fired heater and steam heater. Similarly, a choice of process driver between steam turbine and motor will be determined. Selection is made based on operation considerations, reliability and safety limits, and capital cost. Selection of process utility supply defines the basis for design of a steam and power system.

The above steps complete the process design in the process battery limit. The last design step is to design utility system, which is mainly the steam and power system. The main design consideration of the steam and power system is technology selection in terms of combined cycle (gas turbine plus steam turbine) or steam ranking cycle (steam turbine) for power generation. At the same time, fuel selection, system configuration, and load optimization of steam and power system need to be determined. Furthermore, off-site utility demand should be addressed and this involves feed and product tank farm design with proper insulation and heating.

In short, the traditional design approach adopts a sequential design approach. In contrast, process integration methodology for process design takes a different approach in that process design aspects in the inner part of the onion diagram are allowed to change which may enhance the possibility of heat recovery and enable more energy savings in the utility system in the outer part of the onion.

Therefore, the discussions in this chapter focus on effects of process changes on energy usage. It can be found that the pinch analysis method (Linnhoff et al. 1982) is powerful in evaluating process changes in the early stage of design without waiting for completion of process design.

11.3 Composite Curves and Heat Integration

Identifying saving opportunity for process heat recovery should be the first step for process energy optimization and this can be accomplished by energy targeting based on composite curves.

11.3.1 Composite Curves

Composite curves were developed for heat recovery targeting (Linnhoff et al. 1982). The word "composite" reveals the basic concept behind the composite curves method: system view of the overall heat recovery system. The problem of assessing a complex heat recovery system involving multiple hot and cold streams is simplified as a problem of two composite streams (as shown in Figure 11.3). One hot composite stream (shown in black) represents all the hot process streams while one cold composite stream (shown in gray) represents all the cold process streams. In essence, the hot composite stream represents a single-process heat source while the cold composite stream a single-process heat sink. The composite curves can indicate current process heat recovery (Figure 11.3a) as well as the targeted heat recovery (Figure 11.3b). The difference between these two is a measure of the potential for heat recovery.

Generation of composite curves starts with identification of a representative base case for the process. The preliminary heat and mass balances for the base cases are then generated, to determine the critical process conditions such as reaction temperature and pressure, separation temperatures and pressures, conditions of feeds, products and recycle streams, etc. The stream data from this base case is then used to build the composite curves.

11.3.2 Basic Pinch Concepts

Composite curves can reveal very important insights for a heat recovery problem regarding process heat recovery, pinch point, and hot and cold utility targets, which can be visualized in Figure 11.4. The basic concepts are explained below.

1) *Minimum temperature approach ΔT_{min}*: For a feasible heat transfer between the hot and cold composite streams, a minimum temperature approach must be

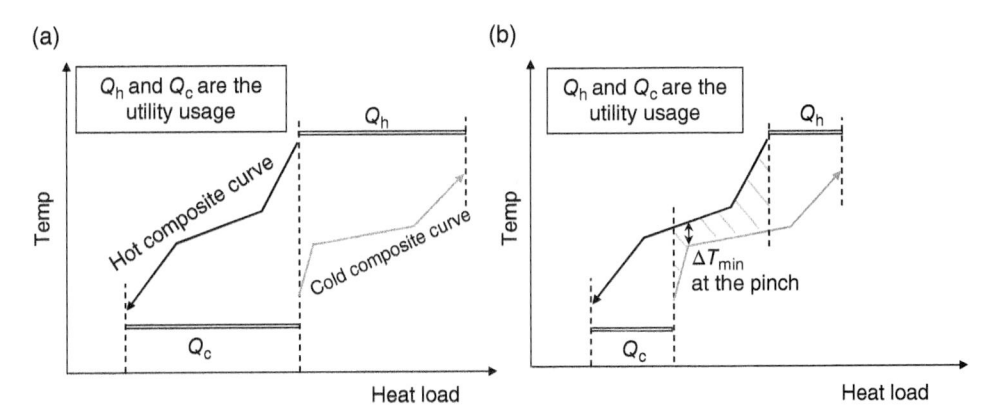

Figure 11.3 Composite curves: heat demand (gray) vs. heat availability (black) profiles. (a) No heat recovery case; (b) heat recovery (hatched area).

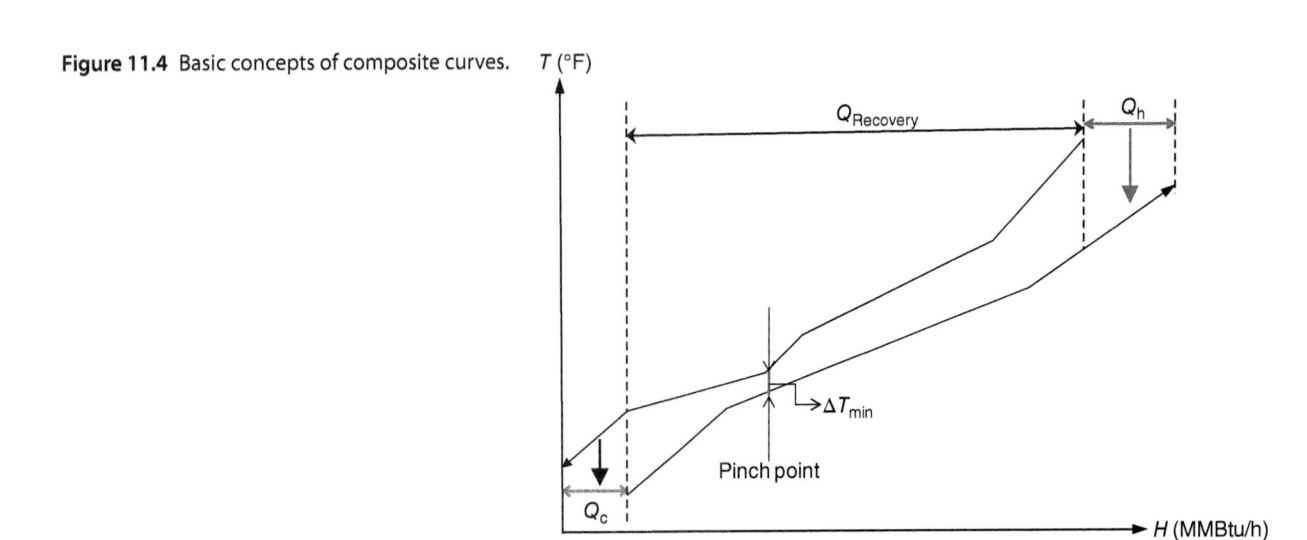

Figure 11.4 Basic concepts of composite curves.

specified, which corresponds to the closest temperature difference between the two composite curves on the T/H axis. This minimum temperature approach is termed as the network temperature approach and defined as ΔT_{min}.

2) *Process heat recovery*: The overlap between the hot and cold composite curves represents the maximal amount of heat recovery for a given ΔT_{min}. In other words, the heat available from the hot streams in the hot composite curve can be heat-exchanged with the cold streams in the cold composite curve in the overlap region to achieve maximal heat recovery.

3) *Hot and cold utility requirement*: The overshoot at the top of the cold composite represents the minimum amount of external heating (Q_h) while the overshoot at the bottom of the hot composite represents the minimum amount of external cooling (Q_c).

4) *Pinch point*: The location of ΔT_{min} is called the process pinch. In other words, the pinch point occurs at the minimum temperature difference indicated by ΔT_{min}. When the hot and cold composite curves move closer to ΔT_{min}, the heat recovery reaches the maximum and the hot and cold utility come to the minimum. Thus, the pinch point becomes the bottleneck for further reduction of hot and cold utility. Process changes must be made if further utility reduction is pursued.

11.3.3 Energy Use Targeting

By assuming a practical ΔT_{min}, the composite curves can indicate targets for both hot and cold utility duties. This task is called energy targeting.

The procedure of obtaining the composite curves and energy targets can be summarized. First, the base case is determined from which stream data are collected based on the heating and cooling requirements of the process streams in terms of temperatures and enthalpies. Next, the hot streams are plotted on temperature–enthalpy axes and then individual stream profiles are combined to give a hot composite curve. This step is repeated for the cold streams to generate the cold composite curve. Finally, the two composite curves are plotted together to obtain the composite curves (Figure 11.5a) for a given ΔT_{min} and thus the minimum hot and cold utility targets for the process can be determined.

The general trend is that a large ΔT_{min} corresponds to higher energy utility but lower heat transfer area and thus lower capital cost, and vice versa (Figure 11.5b). The calculations for capital cost for a heat recovery system can be seen in Section 11.3.5.

11.3.4 Pinch Design Rules

Once the pinch is identified, the overall heat recovery system can be divided into two separate systems: one above and one below the pinch, as shown in Figure 11.6a. The system above the pinch requires a heat input and is therefore a net heat sink. Below the pinch, the system rejects heat and so is a net heat source. When a heat recovery system design does not have cross-pinch heat transfer, i.e. from above to below the pinch, the design achieves the minimum hot and cold utility requirement under a given ΔT_{min}.

On the other hand, if cross-pinch (XP) heat transfer is allowed (Figure 11.6b), XP amount of heat is transferred from above to below the pinch. The system above the pinch, which was before, in heat balance with Q_{hmin}, now loses XP units of heat to the system below the pinch. To restore the heat balance, the hot utility must be increased by the same amount, that is, XP units. Below the pinch, XP units of heat are added to the system that had an excess of heat, therefore the cold utility requirement also increases by XP units. The consequence of a cross-pinch heat transfer (XP) is that both the hot and cold utility will increase by the cross-pinch duty (XP).

Based on the same principle, if external cooling is used for hot streams above the pinch, it increases the hot utility demand for the cold streams by the same amount. Similarly, external heating below the pinch increases the cold utility requirement by the same amount.

To summarize, there are three basic pinch golden rules that must be followed in order to achieve the minimum energy targets for a process:

1) Heat must not be transferred across the pinch.
2) There must be no external cooling above the pinch.
3) There must be no external heating below the pinch.

Breaking any of these rules will lead to cross-pinch heat transfer resulting in an increase in the energy requirement beyond the target.

11.3.5 Cost Targeting: Determine Optimal ΔT_{min}

The optimal ΔT_{min} is determined based on the trade-off between energy and capital such that the total cost for the heat recovery system design is at minimum. The total annual cost for a heat recovery system consists of two parts, namely the capital cost and the energy operating cost:

- The energy operating cost includes energy expenses for both hot and cold utilities which is billed regularly in $/year.
- The capital cost of the network includes surface area costs for all individual heat exchangers, water coolers, air coolers and refrigeration, fired heaters and steam heaters, as well as related costs including foundation, piping, instrumentation, control, etc. Thus, it is a total investment ($) required to build the entire heat transfer system.

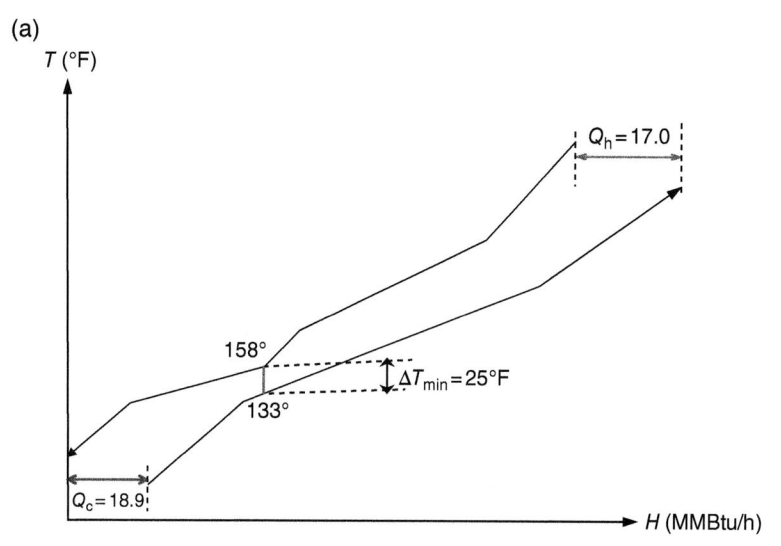

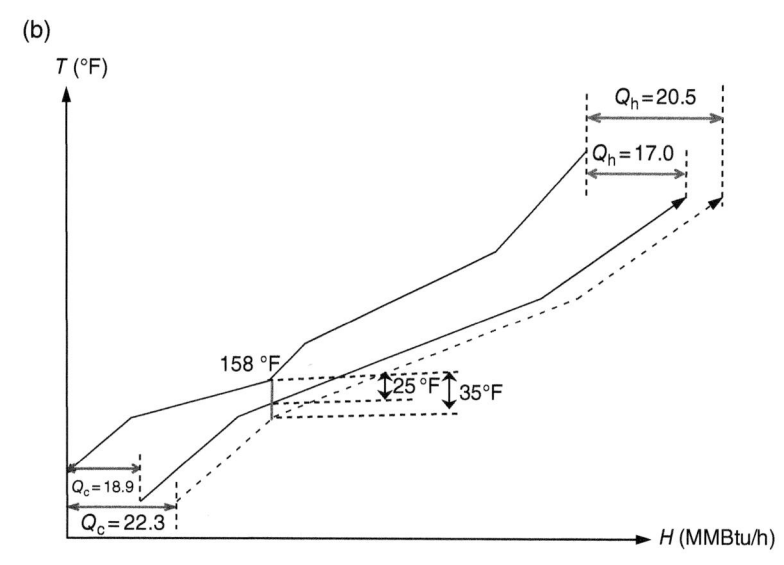

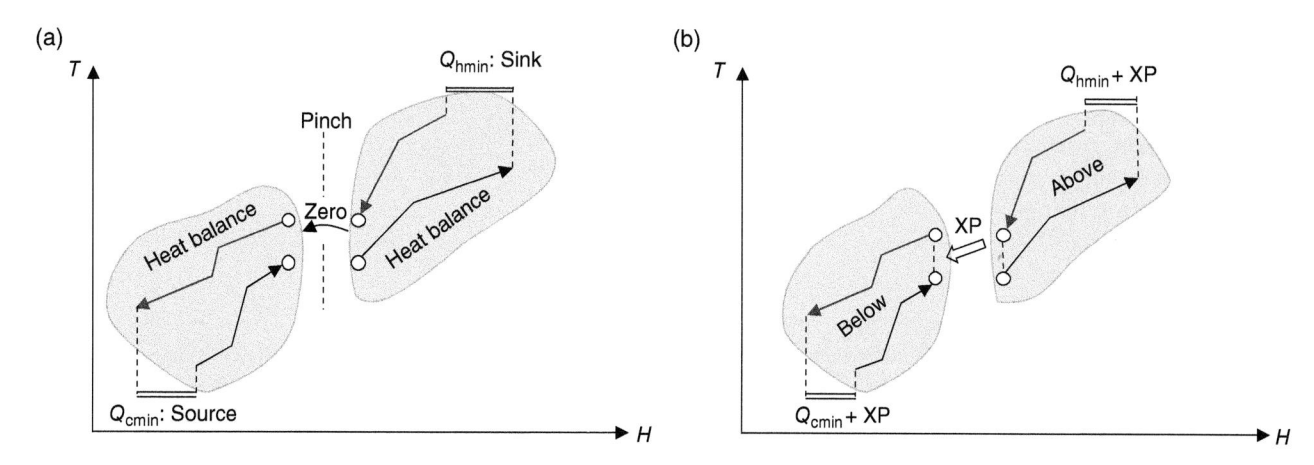

Figure 11.5 (a) Energy targets for a specified ΔT_{min}. (b) Energy targets for different ΔT_{min}.

Figure 11.6 Pinch principle: penalty of cross-pinch heat transfer. (a) Zero cross-pinch heat flow. (b) With cross-pinch heat flow.

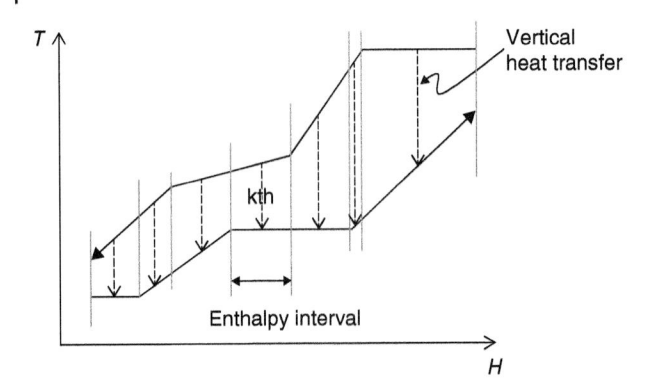

Figure 11.7 Calculation of surface area from the composite curves.

Heat exchanger capital cost is estimated based on exchanger surface area. The overall exchanger area can be directly calculated from the composite curves using the area model (Townsend and Linnhoff 1983a, b). To do this, utilities are added to composite curves to make heat balance between hot and cold composites. Then, the balanced composite curves are divided into several enthalpy intervals and each enthalpy interval must feature straight temperature profiles (Figure 11.7).

There could be several hot and cold streams within an enthalpy interval. For each heat exchange involving hot stream i and cold stream j in kth interval, the surface area and cost can be calculated via Eqs. (11.1) and (11.2), respectively, as follows:

$$A_{i,j,k} = \frac{Q_{i,j,k}}{U_{i,j,k} \text{LMTD}_{i,j,k}} \tag{11.1}$$

$$C_{i,j,k} = a_{i,j,k} + b_{i,j,k} \left(A_{i,j,k} \right)^{c_{i,j,k}}$$
$$= a_{i,j,k} + b_{i,j,k} \left(\frac{Q_{i,j,k}}{U_{i,j,k} \text{LMTD}_{i,j,k}} \right)^{c_{i,j,k}} \tag{11.2}$$

where

A = surface area (ft^2)
Q = heat load (MMBtu/h)
LMTD = logarithmic mean temperature difference (°F)
U = overall heat transfer coefficient [MMBtu/h/(ft^2 °F)]
C = exchanger cost ($)
a = fixed cost for exchanger ($)
b = surface area cost ($/ft^2)
c = economic scale factor, fraction

Thus, total surface area and purchase cost for all exchangers in kth interval can be calculated via Eqs. (11.3) and (11.4):

$$A_k = \sum_{i,j} A_{i,j,k} = \sum_{i,j} \frac{Q_{i,j,k}}{U_{i,j,k} \text{LMTD}_{i,j,k}} \tag{11.3}$$

$$C_k = \sum_{i,j} \left[a_{i,j,k} + b_{i,j,k} \left(\frac{Q_{i,j,k}}{U_{i,j,k} \text{LMTD}_{i,j,k}} \right)^{c_{i,j,k}} \right] \tag{11.4}$$

where

A_k = total surface area for kth interval (ft^2)
C_k = total exchanger cost for kth interval ($)

The overall surface area and capital cost for the network can be calculated via Eqs. (11.5) and (11.6). It is important to point out that the exchanger equipment costs must be converted to installed costs which include exchanger purchase cost, foundation, piping, instrumentation, control, erection, etc.

$$A_{\text{Network}} = \sum_k^{\text{intervals}} A_k = \sum_{i,j,k} \frac{Q_{i,j,k}}{U_{i,j,k} \text{LMTD}_{i,j,k}} \tag{11.5}$$

$$C_{\text{Network}}^{\text{Cap}} = \sum_k C_k + \sum_p C_p$$
$$= \sum_k \sum_{i,j} I_{i,j,k} \left[a_{i,j,k} + b_{i,j,k} \left(\frac{Q_{i,j,k}}{U_{i,j,k} \text{LMTD}_{i,j,k}} \right)^{c_{i,j,k}} \right]$$
$$+ \sum_p \left(a_p + b_p Q_p^{c_p} \right) \tag{11.6}$$

where

A_{Network} = total surface area (ft^2)
C_p = fired heater cost for heater p ($)
$C_{\text{Network}}^{\text{Cap}}$ = total installed cost ($)
$I_{i,j,k}$ = installation factor for exchanger between streams i and j in kth interval
Q_p = heat load for fired heat p (MMBtu/h)
a_p = fixed cost for fired heater ($)
b_p = heater duty cost ($/MMBtu)
c_p = economic scale factor, fraction

On the other hand, utility consumption and costs can be calculated for both hot and cold utilities, respectively, as

$$Q_{h,\text{Network}} = \sum_m Q_{h,m} \tag{11.7}$$

$$Q_{c,\text{Network}} = \sum_n Q_{c,n} \tag{11.8}$$

$$C_{\text{Network}}^Q = \sum_m c_{h,m} Q_{h,m} + \sum_n c_{c,n} Q_{c,n} \tag{11.9}$$

where

$Q_{h,m}$ = heat load for hot utility m (MMBtu/h)
$Q_{c,n}$ = heat load for cold utility n (MMBtu/h)
$c_{h,m}$ = cost for hot utility m ($/MMBtu)
$c_{c,n}$ = cost for cold utility n ($/MMBtu)
C_{Network}^Q = total utility cost ($/h)

The capital cost for each ΔT_{min} can be calculated based on Eq. (11.6) while the energy cost by Eq. (11.9). By calculating both energy and capital costs for different ΔT_{min}'s, two cost curves, namely the energy and capital cost curves for a range of ΔT_{min} can be obtained as shown in Figure 11.8.

The total annualized cost for the entire heat recovery system can then be defined as

$$C_{Network}^{Total} = F \times C_{Network}^{Cap} + K C_{Network}^{Q} \quad (11.10)$$

where

F = capital annualized factor (1/year)
K = time annualized factor = 24 h/day*operating days/year
$C_{Network}^{Total}$ = total annualized cost ($/year)

For different ΔT_{min}, it will be expected to have different capital cost and utility cost. When ΔT_{min} increases, capital costs drop as exchanger LMTD increases and thus surface area reduces. At the same time, the utility consumption raises. The impact of reduced ΔT_{min} is opposite: capital costs go up as LMTD reduces while utility consumption goes down. Thus, there is a trade-off between utility cost and capital cost as shown in Figure 11.8. This trade-off can be better visualized by plotting the total annualized cost versus ΔT_{min} on a graph in Figure 11.9. The optimal network approach, denoted as $\Delta T_{min,opt}$, is the one corresponding to the lowest total annualized cost. In many cases, there is a range of ΔT_{min} values in which total costs are similar and thus selection of $\Delta T_{min,opt}$ in this range should be made toward design simplicity.

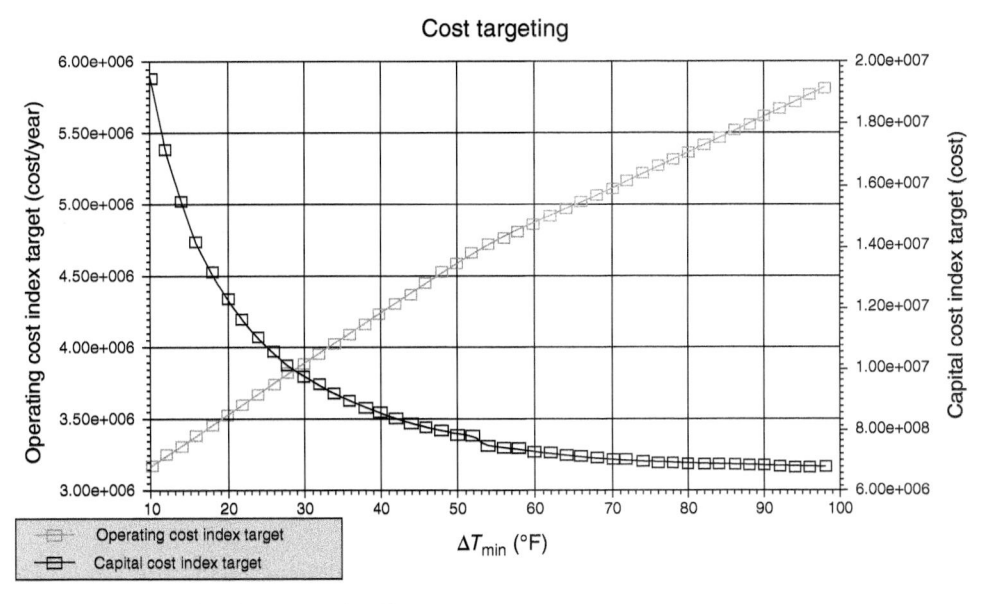

Figure 11.8 Capital and energy trade-off.

Figure 11.9 Cost targeting for determining $\Delta T_{min,opt}$.

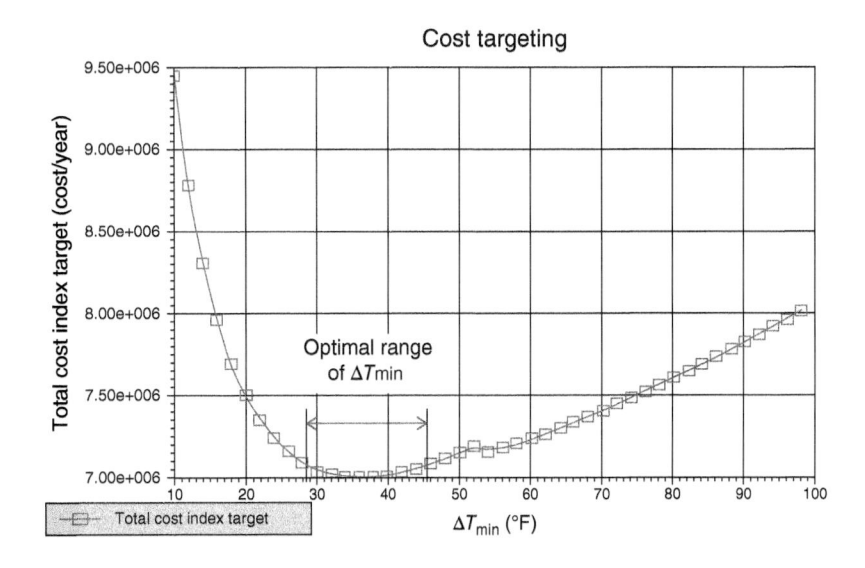

The significance of $\Delta T_{min,opt}$ is in setting the design basis for where the heat recovery design should start and what to expect for the utility and capital costs to result from the design. If the design comes up with much high costs than the targeted costs, a design evaluation must be conducted to find out why and figure out measured correction.

11.4 Grand Composite Curves (GCC)

The pinch analysis tool that can be used for assessing process changes is called the grand composite curve (GCC), which can be constructed based on the composite curves that were discussed above. The first step is to adjust the temperatures of the composite curves in Figure 11.10a to derive the shifted composite curves of Figure 11.10b. This involves increasing the cold composite temperature by ½ΔT_{min} and decreasing the hot composite temperature by ½ΔT_{min}.

Because of this temperature shift, the hot composite curve moves down vertically by ½ΔT_{min} while the cold composite curve moves up by ½ΔT_{min}. Thus, shifted hot and cold composite curves touch each other at the pinch (see Figure 11.10b). In doing so, the minimum approach (ΔT_{min}) condition is built in for the shifted composite curves, which makes the task easier for utility selection on GCC (this will become self-evident later).

The GCC is then constructed from the enthalpy (horizontal) differences between the shifted composite curves at different temperatures (shown by distance α in Figures 11.10b and c). The GCC provides the same overall energy target as the composite curves, i.e. targets are identical in Figure 11.10a and c. Furthermore, GCC represents the difference between the heat available from the hot streams and the heat required by the cold streams, relative to the pinch, at a given shifted temperature. Thus, the GCC is a plot of the net heat flow for any given shifted temperature, which can be used as the basis for assessing process changes and intermediate utility placement.

11.5 Appropriate Placement Principle for Process Changes

Let us see how GCCs are applied for process evaluation.

11.5.1 General Principle for Appropriate Placement

Assume there is a hot utility that can be used for process heating at any temperature levels. Where should we place it for process heating? Of course, we do not want to use it below the pinch according to the pinch golden rule (Chapter 10): do not use hot utility below the pinch. To be smart, we should consider minimizing its use since the hottest utility is the most expensive. If intermediate utilities are available, we should consider maximizing the use of the utility at lowest temperature first (e.g. low pressure steam) and then the second lowest temperature (e.g. medium pressure [MP] steam) and so on (e.g. high pressure [HP] steam) above the pinch prior to the hottest utility (e.g. furnace heating).

Similarly, the cooling utility at the highest temperature should be used first (e.g. air cooling) and then second highest temperature (e.g. cooling water [CW]) and so on (e.g. chilled water) below the pinch prior to the coldest utility (e.g. refrigeration).

The above discussions point to the general principle for *Appropriate Placement for process changes*, originally introduced by Townsend and Linnhoff (1983a, b). The penalty of violating this principle is that both the hot and cold utility requirements go up and the process no longer achieves its energy targets.

This general principle was developed for utility selection in terms of the correct levels and loads. But it is much less obvious that the principle also applies to process changes. For better illustration, application of this principle for utility selection will be discussed first and then discussions will cover unit operations such as reactors, separation columns, feed preheating, etc.

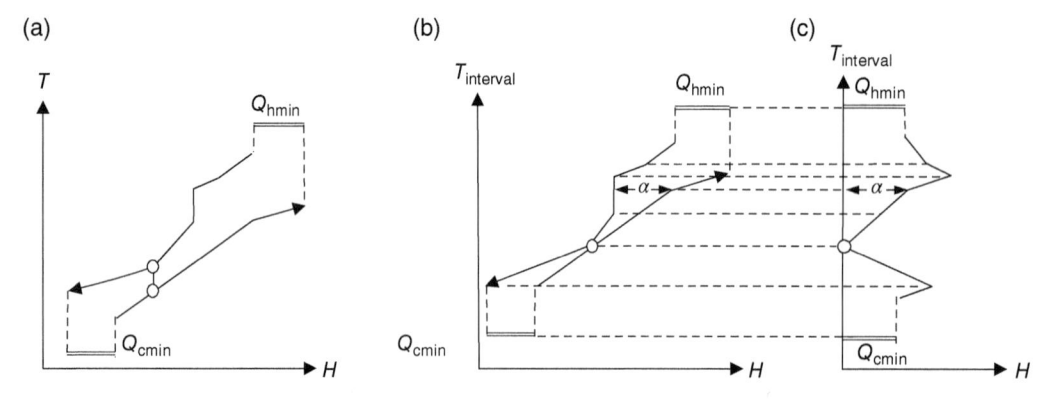

Figure 11.10 Construction of grand composite curve. (a) Composite curves, (b) shifted composite curves, and (c) grand composite curve.

11.5.2 Appropriate Placement for Utility

When multiple utility options are available, the question is which utility is to be selected to reduce overall utility costs. This involves setting appropriate loads for the various utility levels by maximizing cheaper utility loads prior to use of more expensive utilities. The GCC is an elegant tool for accomplishing this purpose.

Consider a process which requires heating and cooling. HP steam is sufficient for heating at any temperature levels and likewise, refrigeration is sufficient for cooling at any temperature levels. The simplest way of utility selection is to use HP steam everywhere for heating and refrigeration everywhere for cooling as explained in Figure 11.11a. However, this could be a very costly option as HP steam and refrigeration are expensive. However, there exist intermediate utilities for use. If MP steam and CW can be used, a GCC can be constructed as shown in Figure 11.11b. The target for MP steam is set by simply drawing a horizontal line at the MP steam temperature level starting from the vertical (shifted temperature) axis until it touches the GCC. Remember that the minimum approach temperature is built in when constructing GCC via shifting hot composite curves as explained previously. The remaining heating duty is then satisfied by the HP steam. This maximizes the use of MP steam before HP steam and therefore minimizes the total hot utility cost as MP steam is cheaper than HP steam. The additional benefit from using MP steam versus HP steam is that higher latent heat is available in MP steam, which reduces the MP steam rate to meet the same duty requirement. Similarly, maximal use of CW before refrigeration reduces the total cold utility costs. The points where the MP and CW levels touch the GCC are called the "Utility Pinches."

If the process requires furnace heating at high temperature, how can the furnace duty be reduced in design because furnace heating is more expensive? Figure 11.11c shows the possible design solution where the use of MP steam is maximized. In the temperature range is above the MP steam level, the heating duty must be supplied by the furnace flue gas. The flue gas flow rate is set as shown in Figure 11.11c by drawing a sloping line starting from the MP steam temperature to theoretical flame temperature (T_{TFT}).

The above discussions lead to the design guidelines for minimizing utility costs as follows:

- Minimize furnace heating or HP steam via maximizing the use of lower quality hot utility first.
- Minimize refrigeration or chilled water by maximizing the use of air and water cooling first.
- Maximizing generation of higher quality utility first.

11.5.3 Appropriate Placement for Reaction Process

Reaction integration implies appropriate heat integration of reaction effluent. The reactor integration can be evaluated by the process GCC which is constructed without the reaction effluent stream and then the reaction effluent stream is placed on top of GCC. The general Appropriate Placement principle states: The heat of reaction effluent should be released above the process pinch.

With guidance provided from the GCC, the reactor integration for new and existing process designs can be assessed. For existing processes, the process GCC is fixed, but reaction temperature might be adjusted to a small degree as well as integration of the reaction effluent can be modified by retrofitting the existing heat

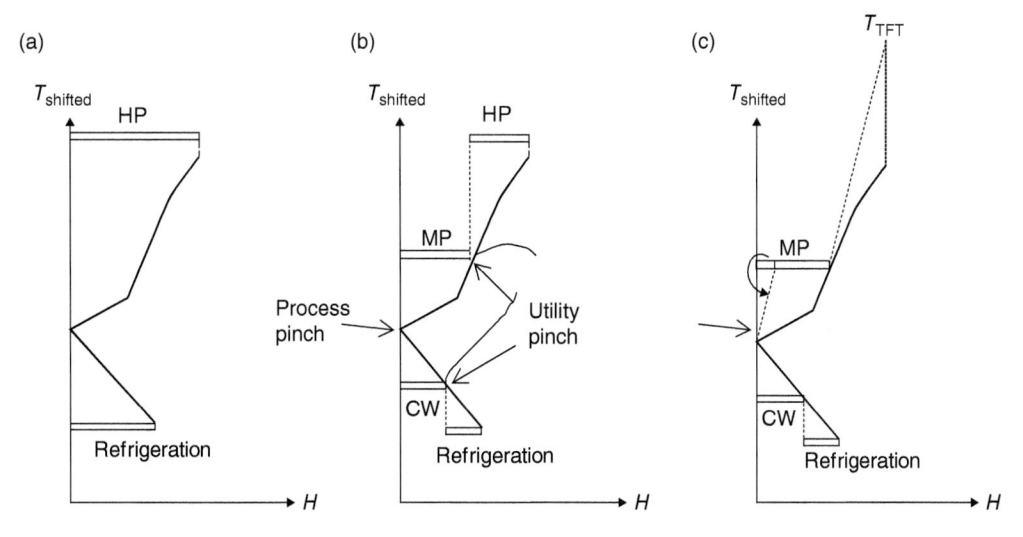

Figure 11.11 Selection of multiple utility. (a) Original grand composite curve. (b) Utility placement using MP and HP steam. (c) Utility placement using MP and fired heater.

exchange scheme. The general guideline is to maximize heat recovery of reaction effluent heat above the pinch.

For new design, if reaction effluent stream does not fit well with the background process (Figure 11.12a), the reaction conditions such as temperature may be required to vary. However, only small changes in reaction conditions may be tolerated because any significant change would impact on conversion and product yields, which usually outweighs energy costs. Thus, in grassroots design, there is little opportunity to change the desired reaction temperature which is determined based on yield. However, instead of changing the reaction temperature, can we modify the background process to have better reaction integration (Figure 11.12b)? This topic is discussed in more detail by Glavic et al. (1988).

11.5.4 Appropriate Placement for Distillation Column

There are several key opportunities for column integration which include reflux ratio improvement, pressure changes, feed preheating, side reboiling/condensing, and feed stage location. A pinch tool called column grand composite curve (CGCC, Dhole and Linnhoff 1993) was developed to provide aids for evaluation of these improvements.

11.5.4.1 The Column Grand Composite Curve (CGCC)

The CGCC can be constructed based on a converged column simulation as shown in Figure 11.13a. From the simulation, the column stagewise data are extracted and these data are then organized to generate the CGCC in Figure 11.13b. The stagewise data relate to "Ideal Column" design. For ideal column design the column requires infinite number of stages and infinite number of side reboilers and condensers as shown in Figure 11.13c, which represents minimum thermodynamic loss in the column. In this limiting condition, the energy can be supplied to the column along the temperature profile of the CGCC instead of supplying it at extreme reboiling and condensing temperatures. The CGCC is plotted in either T–H (T = temperature; H = enthalpy) or Stage-H diagrams. The pinch point on the CGCC is usually caused by the feed.

Similar to the GCC for utility selection for a process, the CGCC provides a thermal profile for evaluating heat integration ideas for a column such as side condensing and reboiling (Figure 11.13b). In a practical column, energy is supplied to the column at feasible reboiling and condensing temperatures.

11.5.4.2 Column Integration Against Background Process

Column integration implies heat exchange of the column heating/cooling duties against background process or the external utility available. The principles of appropriate placement of columns against a background process can be explained as below.

Let us look at Figure 11.14a where the reboiler receives heat above the pinch of the background process while the condenser rejects heat below the pinch. The background process is represented by its GCC. Therefore, this distillation column is working across the pinch. In this case, Figure 11.14a represents a case of no integration of the column against the background process. The column is therefore inappropriately placed as regards its integration with the background process.

Assume the pressure of the distillation column is raised and the condenser and reboiler temperatures can increase accordingly. As a result, the column can fit entirely above the pinch. This case represents a

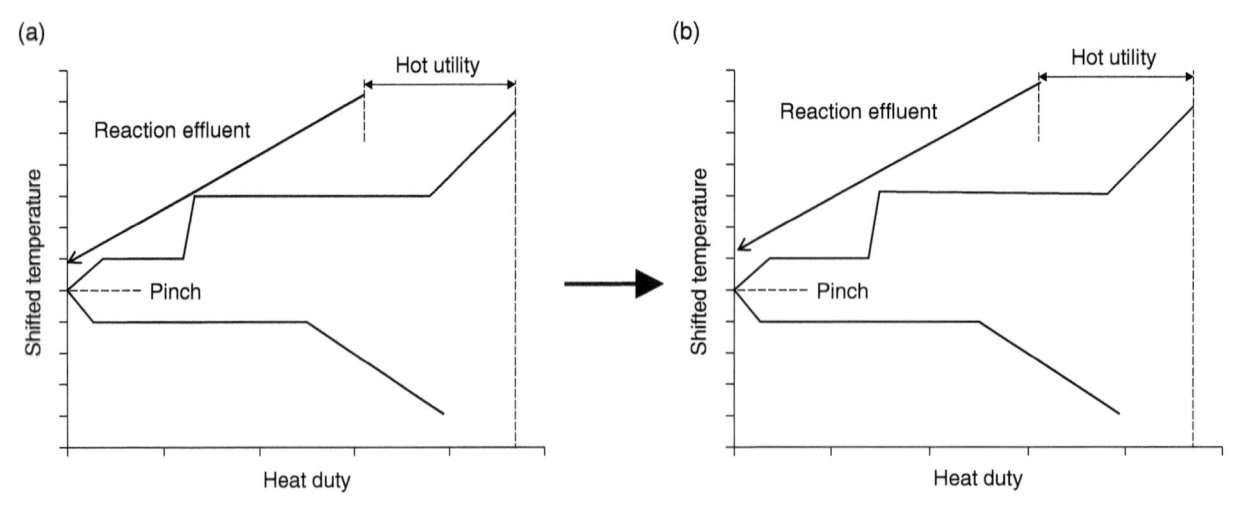

Figure 11.12 Reaction integration against process. (a) Poor reaction integration, (b) better reaction integration.

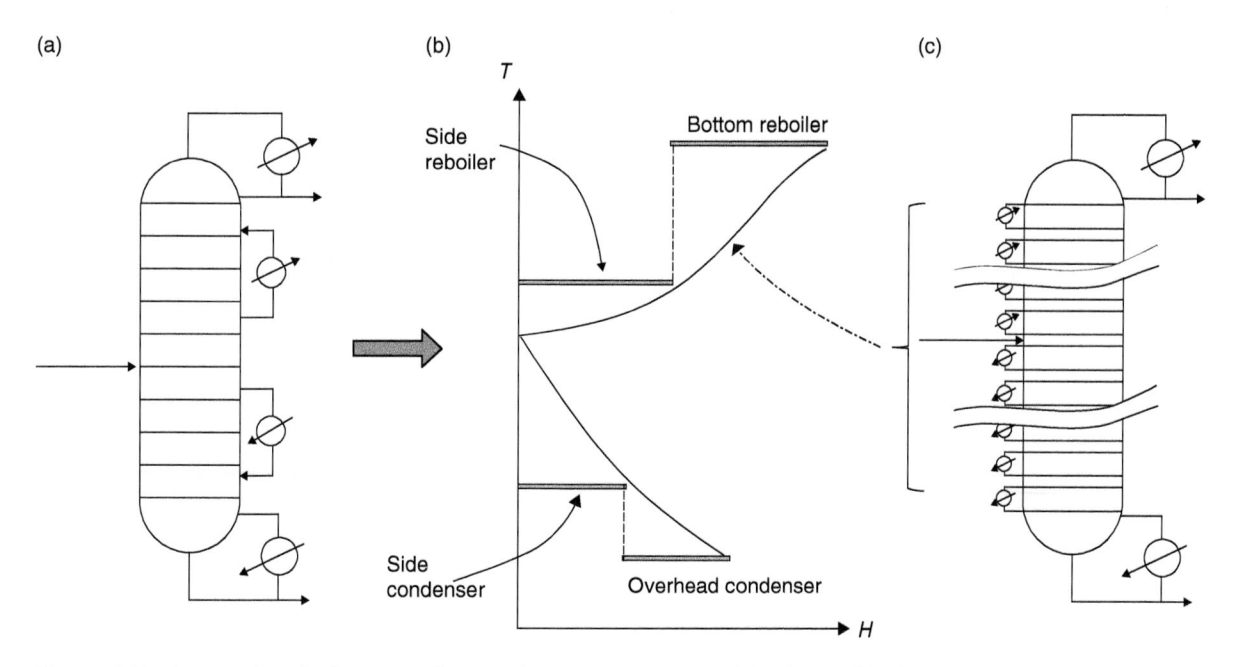

Figure 11.13 Construction of column grand composite curve. (a) Converged simulation, (b) column grand composite, (c) ideal column.

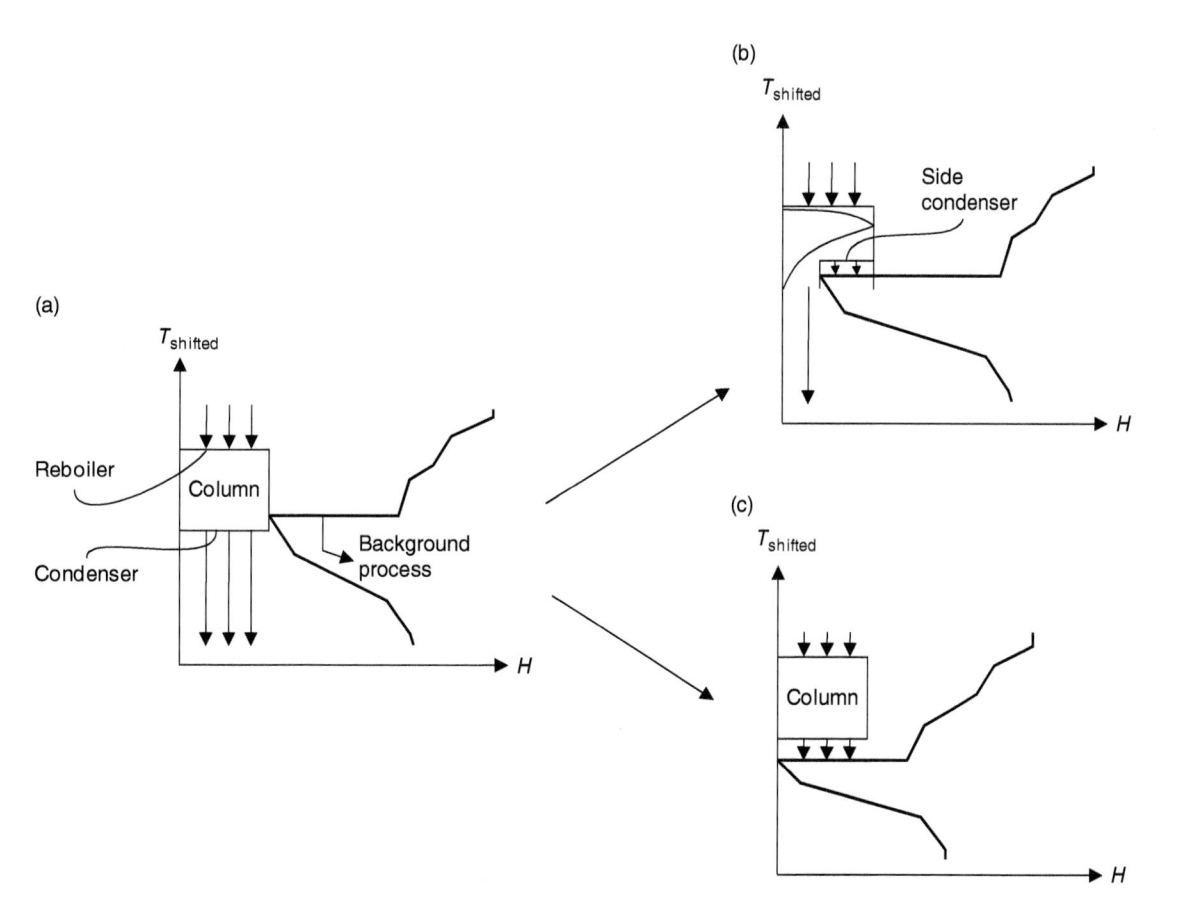

Figure 11.14 Column integration with process. (a) Inappropriate placement, (b) using column modification for integration, (c) appropriate placement.

complete integration between the column and the background process via the column condenser as shown in Figure 11.14c. The column is now on one side of the pinch (not across the pinch). The overall energy consumption (column plus background process) equals the energy consumption of the background process. In energy-wise, the column is running effectively for free. The column is therefore appropriately placed as regards its integration with the background process. Alternatively, lowering the column pressure so that its temperature drops will make the column fit below the pinch. Placing the column above or below the pinch is another application of the Appropriate Placement principle.

In practice, large changes to the operating pressure of the distillation column are rarely possible due to the process limits such as product specifications, capital costs, safety, or other considerations. However, there are other ways of reducing heat transferred across the pinch. One option is to install intermediate condenser so that it works at a higher temperature than the main condenser at the top of the column. Figure 11.14b shows the CGCC of the column. The CGCC indicates a potential for side condensing. The side condenser enables greater integration between the column and the background process. Compared to Figure 11.14a, the overall energy consumption (column plus background process) has been reduced due to the integration of the side condenser. Alternatively, use of intermediate reboiler or pump around can be considered.

In summary, the column is inappropriately placed if it is located across the pinch because the column has no heat integration with the background process. On the other hand, the column is appropriately placed if it is placed on one side of the pinch and can be integrated against the background process. Although appropriate column integration can provide substantial energy benefits, these benefits must be compared against associated capital investment and difficulties in operation. In some cases it is possible to integrate the columns indirectly via the utility system which may reduce operational difficulties.

11.5.4.3 Design Procedure for Column Integration

The design procedure for column integration is shown in Figure 11.15, which can be applied for new and revamp projects. Let us walk through the procedure as below.

11.5.4.3.1 Feed Stage Optimization

The feed stage location of the column is optimized first in the simulation prior to the start of the column thermal analysis since the feed stage may strongly interact with the other options for column improvements. This can be carried out by trying alternate feed stage locations in simulation and evaluating its impact on reboiling duty.

In principle, there could be several stages which can be used as feed stages. When the feed stage is too low, there is a big jump in temperature in the region below the feed stage since too much change in composition is happening than necessary. To get the composition change needed to meet bottoms spec, more reboiler duty is required leading to higher boilup and liquid and vapor traffic in the bottom section. Because of the higher flow rates, the bottom section will have a larger diameter. Having the feed too high does not have the dramatic change as having the feed too low.

Since the objective for feed stage optimization is to minimize energy use or reboiling duty for the separation without the need of additional trays, a plot of reboiler duty versus stage number can be obtained as Figure 11.16. The optimal feed stage should be in the flat region away from the steep change.

After the feed stage optimization is accomplished, the CGCC for the column is then obtained, which is used as the basis for the next step optimization.

11.5.4.3.2 Reflux Rate Optimization

The next step is to optimize reflux rate for the column. As shown in Figure 11.15b, the horizontal gap between the vertical axis and CGCC pinch point is the scope for reflux improvement. The CGCC will move closer to the vertical axis when the reflux ratio is reduced. The reflux rate optimization must be considered first prior to other thermal modifications since it results in direct heat load savings from the reboiler and the condenser. In an existing column the reflux can be improved by addition of stages or by improving the efficiency of the existing stages.

11.5.4.3.3 Feed Conditioning Optimization

After reflux improvement, the next step is to address feed preheating or cooling. In general, feed conditioning offers a more moderate temperature level than side condensing/reboiling. Also, feed conditioning is external to the column and is therefore easier to implement than side condensing and reboiling. Feed conditioning opportunity is identified by a "sharp change" in Stage-H (H: enthalpy) CGCC close to the feed point as shown in Figure 11.15c. The extent of the sharp change approximately indicates the scope for feed preheating. Successful feed preheating allows heat load to be shifted from reboiler temperature to the feed preheating temperature. Analogous procedure applies for feed precooling.

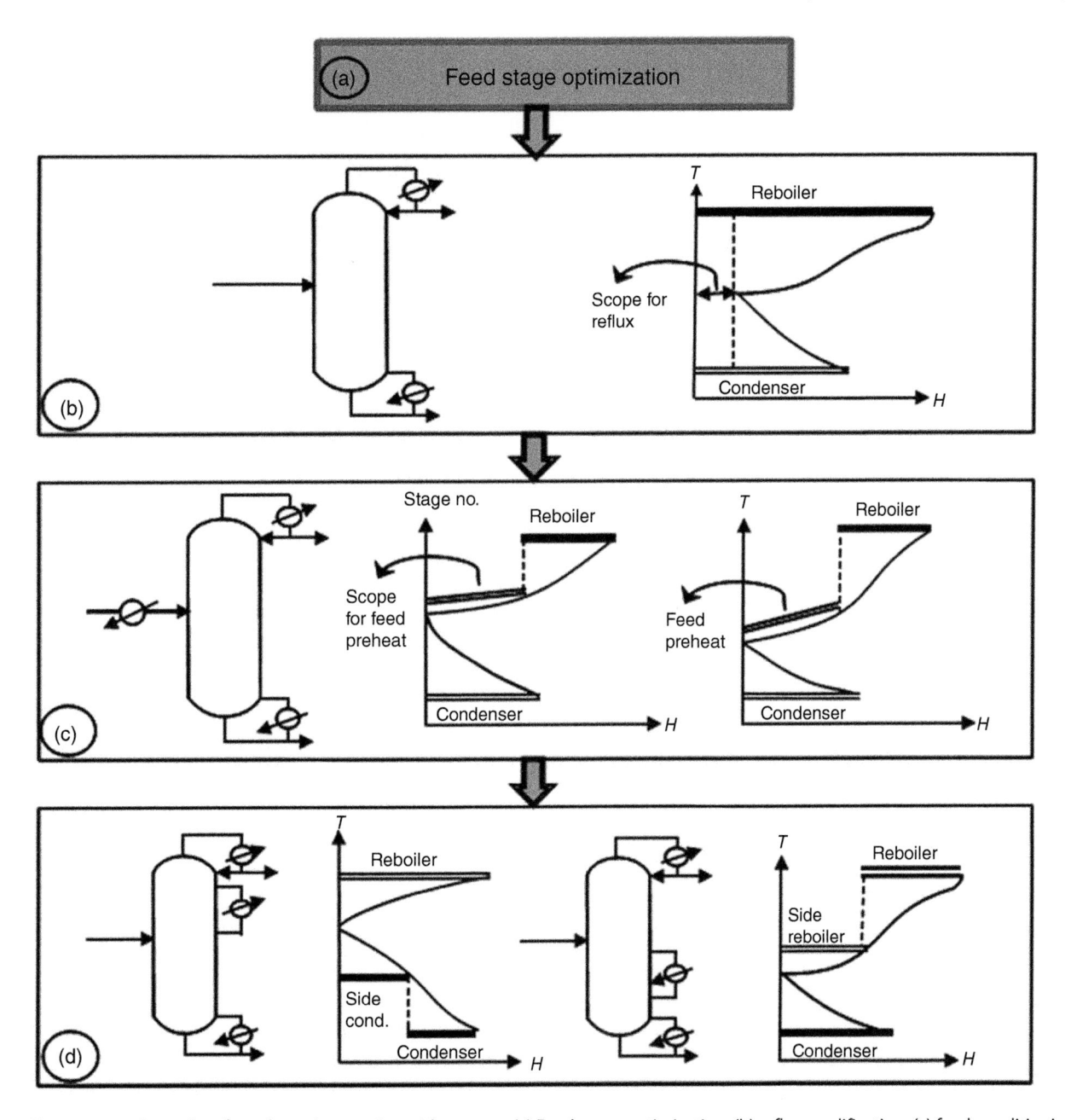

Figure 11.15 Procedure for column integration with process. (a) Feed stage optimization, (b) reflux modification, (c) feed conditioning, (d) side condensing/reboiling.

11.5.4.3.4 Side Condensing/Reboiling Optimization

Following the feed conditioning, side condensing/reboiling should be considered. Figure 11.15d describes CGCC's which show potential for side condensing and reboiling. An appropriate side reboiler allows heat load to be shifted from the bottom reboiling to a side reboiling without significant reflux penalty.

Another column integration option is column thermal coupling, which can be direct column integration or indirect column integration. This option will be discussed in detail in Section 11.7.2.

11.6 Systematic Approach for Process Integration

Traditional energy efficiency improvements have a narrow focus on energy recovery alone with little consideration of interactions with process flowsheeting, equipment design, and process conditions. As a result, energy-saving projects usually have limited economic benefits and thus have difficulty in competing with capacity and yield-related projects. In contrast, the process integration methodology as explained in Figure 11.17 takes a different approach in that energy optimization is closely integrated with changes to both process flowsheeting and conditions, as well as equipment design. The goal of this methodology is to achieve optimal process designs featuring maximum energy efficiency with lowest capital cost possible.

This methodology consists of four core components, namely process simulation development, equipment rating analysis, process integration analysis, and opportunity interaction optimization. This methodology has been applied to numerous design projects (Zhu 2014) with common features such as:

- Clearly defined needs, objectives, scope, and basis.
- Reduced capital investment and operating cost to achieve the objectives.
- Simplified process design and enhanced equipment performance.

The purpose of process simulation development is to represent current plant design for the base case, defined in terms of key operating parameters and their interactions. Thus, the simulation can provide the specifications for equipment rating assessment.

The key role of equipment rating analysis is to assess equipment performance and identify equipment spare capacity and limitations. Utilization of spare capacity can enable expansion up to 10–20% in general and accommodate improvement projects with low capital cost investment. When equipment reaches hard limitations – for

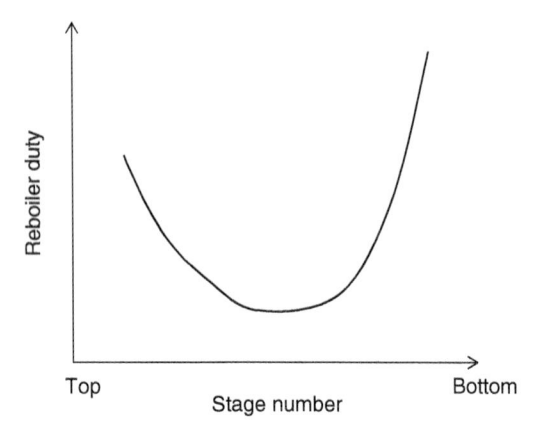

Figure 11.16 Feed stage optimization.

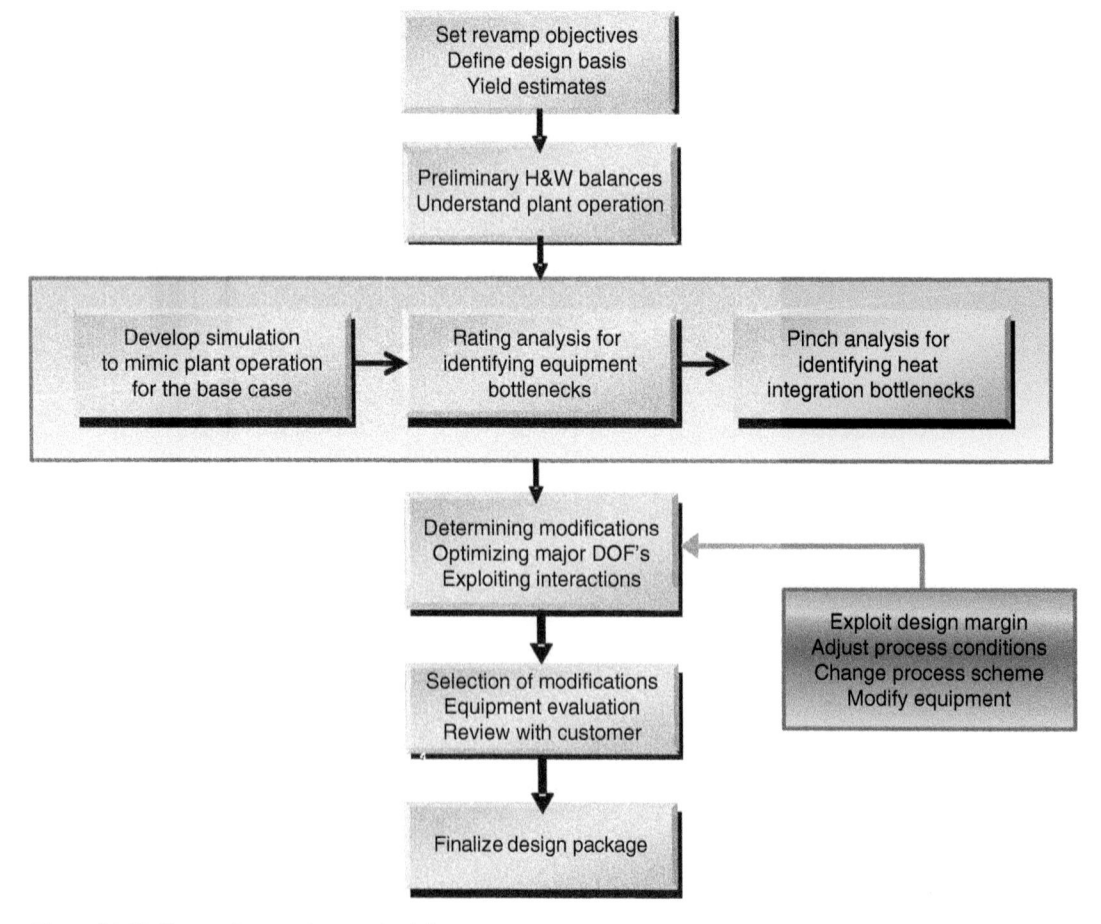

Figure 11.17 Process integration methodology.

example, a fractionation tower reaches its jet flood limit, or a compressor reaches its flow rate limit, or a furnace reaches its heat flux limit – it could be expensive to replace or install new equipment. The important part of a feasibility study is to find ways to overcome these constraints, which is accomplished in the next two steps.

The third step is to apply the process integration methods to exploit interactions and identify changes to process conditions, equipment, process redesign, and utility systems, with the purpose of shifting plant bottlenecks from more expensive to less expensive equipment. By capitalizing on interactions, it is possible to utilize equipment spare capacity and push equipment to true limits in order to avoid the need to replace existing equipment or install new equipment. This is a major feature of this process integration methodology.

A simple and effective example is fractionation tower feed preheat. A tower reboiler could reach a duty limit. With a tower feed preheater, the required reboiling duty is reduced. A column assessment may show the effects on separation with increased feed preheat and reduced reboiling at the bottom. If the effects are acceptable, this modification by adding a feed preheater could eliminate the need of installing a new reboiler, which is expensive.

The fourth step is integrated optimization, and the driver is to exploit interactions between equipment, process redesign, and heat integration. Making changes to process conditions provides a major degree of freedom to achieve this. One direct benefit of optimizing process conditions in this context is that spare capacity available in existing assets can be utilized. Process redesign provides another major degree of freedom as it can increase heat recovery and relax equipment limitations.

In summary, major changes to infrastructure and installation of key equipment such as a new reactor, main fractionation tower, and/or gas compressor could form a major capital cost component. In many cases, it is possible that the level of modification to major equipment could be reduced or even avoided by exploiting design margins for existing processes and optimizing degrees of freedom available in the existing design and equipment. It is the goal of the process integration methodology to achieve minimum operating and capital costs. Applying this integration approach can give results in three categories. Firstly, alternative options for each improvement idea will be provided. Secondly, any potential limitations, either in process conditions or equipment, will be flagged. Thirdly, solutions to overcome or relax these limits will be obtained by exploiting interactions between process conditions, equipment performance, process redesign, and heat integration.

The process integration principles outlined above provide the guidelines for process changes in general. By applying the process integration methodology, the design is no longer confined in a subsystem, from reaction system to separation system, heat exchanger network, and site heat and power systems. In many studies, the energy savings from process change analysis far outweigh those from heat recovery projects.

11.7 Applications of the Process Integration Methodology

Benzene and *para*-xylene are key chemical building blocks to produce synthetic films, fibers, and resins that comprise the household products we use every day. These chemicals are produced at very high purities, 99.9 wt.%, in an aromatics complex. Aromatics complexes are very energy intensive because the process involves very intense fractionations to produce these pure products. The block flow of a conventional aromatics complex is shown in Figure 11.18.

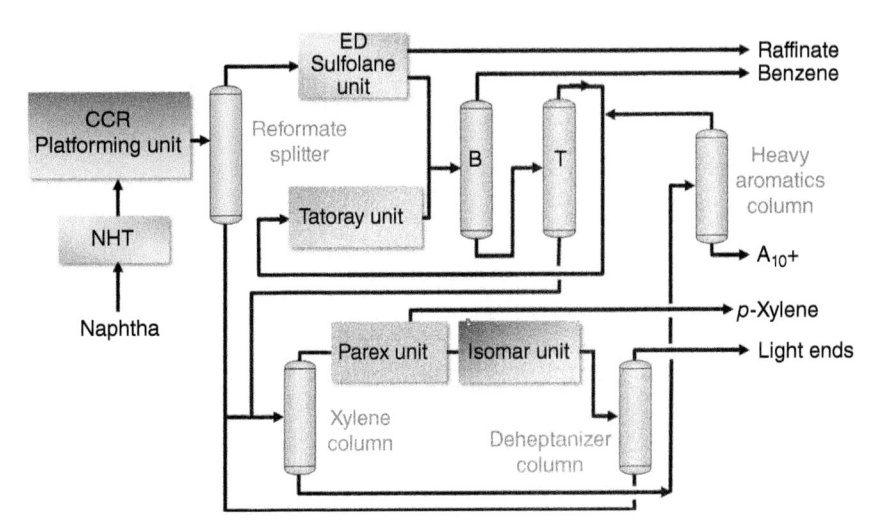

Figure 11.18 Block flow of a conventional aromatics complex.

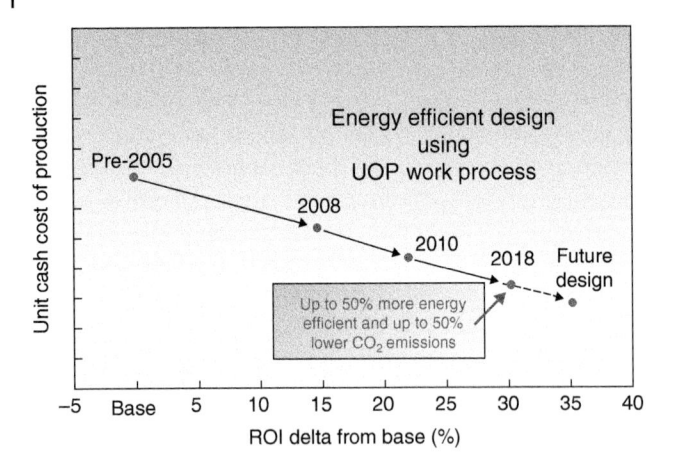

Figure 11.19 Energy improvements over time.

Much effort has been made to improve overall energy efficiency in the aromatics complex. Continuous innovation has led to process intensification and energy efficiency gains in the UOP Aromatics Complex (AC) design. Figure 11.19 shows a trend of energy efficiency improvement for the AC design. It can be seen that more than 50% energy use has been reduced from the base case. Some of the key advances are listed below with detailed discussions to follow:

1) Column split design of single xylene column with thermal coupling.
2) Column split design of extract column with thermal coupling.
3) Dividing wall column (DWC) for reformate splitter and benzene–toluene columns.
4) Use of light desorbent.
5) Heat pump for paraxylene column.
6) Indirect column heat integration.
7) Process–process heat integration.
8) Organic Rankine cycle (ORC) for low-temperature heat recovery.
9) Variable frequency driver on adsorbent chamber circulation pumps.

11.7.1 Column Split for Xylene Column with Thermal Coupling

In traditional SMB-based AC design, there is a single xylene column which receives three feed streams, namely the recycle stream from the Isomar deheptanizer, heavy reformate from the reformate splitter, and toluene from the toluene column. The separation objective of a xylene column is to separate A_8 component from A_8+. However, these three streams have very different compositions some of which contain high A_8 component while others much less. Mixing of these streams will undo the separation. Consequently, higher energy use is required for the xylene column.

To reduce the energy use, the concept of thermal coupling based on a single xylene column was considered. This single xylene column is elevated in pressure so that it can directly reboil the Parex Unit raffinate and extract columns (and other columns depending on the amount of condensing duty available). Large quantities of low temperature heat are rejected to the atmosphere via the raffinate and extract column condensers. Pressurizing the raffinate and extract columns would raise the temperature of the overheads of these columns sufficiently to be used to either directly or indirectly reboil other columns within the complex and thereby reduce the overall energy consumption within an aromatics complex. To do this, the xylene column pressure must be increased further if it is to continue to directly reboil the raffinate and extract columns. However, the extent to which the xylene column pressure can be elevated in pressure is limited due to the potentially detrimental effect of thermal degradation/coking of the xylene column bottoms material (primarily A_9+).

To overcome this limitation, the idea of column split for the single xylene column is developed. In other words, the single xylene column is split into two xylene columns, which are thermally coupled implying the overhead of one xylene column is used as the reboiling service for the other xylene column. The fundamental behind this idea is to segregate the A_8 containing streams from A_8+ streams for the ease of separation of desirable components.

More specially, to achieve the xylene segregation, the Isomar deheptanizer bottoms (containing high A_8) will be processed in a dedicated xylene column to improve utilization of overhead heat and avoid concerns of degradation that arise from excessive temperatures. At the same time, the bottom streams from the reformate splitter and toluene column will be processed in a different xylene column (Figure 11.20). The former xylene column will be pressurized to the extent that it can be used to directly reboil the latter xylene column as well as the extract column. The latter xylene column will also be pressurized to either directly or indirectly (via an intermediate heating medium such as MP steam) reboil other columns within the aromatics complex.

11.7.2 Column Split for Extract Column with Thermal Coupling

The extract column from the original flowscheme is also divided into two columns, namely extract columns no. 1 and no. 2. Extract fractionation is a large energy consumer in the aromatics complex and splitting the column allows for two columns to share the required extract duty. In addition, the columns operate as a low-pressure and a high-pressure extract column. This creates the

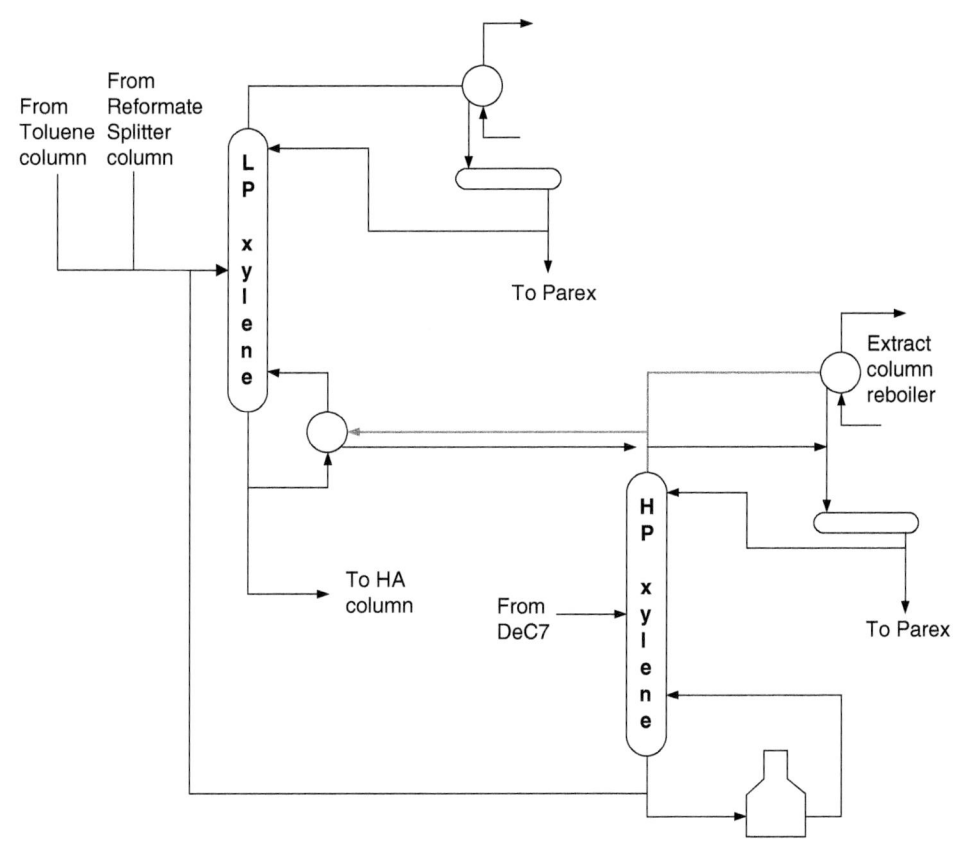

Figure 11.20 Column split for the xylene column and column integration.

opportunity of using the overhead of the higher pressure (higher temperature) extract column to provide the reboiling duty to the lower pressure extract column; therefore, cascading heat and delivering a large energy savings to the aromatics complex.

In the column split design, extract column no. 2 handles more of the toluene/xylene fractionation and therefore requires a higher percentage of the extract fractionation duty. The extract columns were optimized so that extract column no. 2 overhead reboiled extract column no. 1 as well as provided feed preheat to the raffinate column.

11.7.3 Use of Dividing Wall Columns (DWC)

It must be pointed out that there are two kinds of heat integration in general, one is latent heat integration for columns via thermal coupling and heat pump; the other is sensible heat integration via process–process heat exchange. Furthermore, DWC combining multiple simple column is a different kind of heat integration, which is based on column internal thermal coupling. Applications of these integration techniques to the aromatics complex are discussed below.

Another option of column integration is use of a DWC for combining multiple simple columns into one DWC. There are currently two applications of DWC in the aromatic complex, one is for reformate splitter and the other is for benzene–toluene separation.

11.7.3.1 Benzene–Toluene Fractionation Dividing Wall Column

The benzene–toluene (BT) fractionation unit in an aromatics complex is used to recover high-purity benzene and toluene. The conventional design for BT fractionation includes a sequence of two fractionation columns: a benzene column and a toluene column. High-purity benzene is recovered as an overhead product from the benzene column and toluene is recovered as an overhead product from the toluene column. C_8+ material from the bottom of toluene column is sent to the xylene fractionation section of the aromatics complex.

A DWC can be used in lieu of the conventional design. In the DWC, toluene is recovered as a side product while the benzene is recovered overhead, and the heavies are recovered as bottoms product. This design is applicable to cases where toluene is not a final product but rather an intermediate stream sent to other units for further processing (e.g. UOP Tatoray unit) since there is an expected loss of purity in toluene product compared to the conventional two-column design. The DWC is designed to satisfy the minimum energy requirements with optimal internal vapor and liquid flows while meeting required product specifications. The wall is positioned in the column to geometrically split the vapor to the desired ratio and the liquid is split externally and returned to each side of the wall (US Patent 6,551,465). The design of the column and trays are crucial for obtaining the optimal performance for the DWC and the required product specifications.

This DWC design (Figure 11.21) results in significant energy savings and capital cost savings in an aromatics complex. The estimated capital cost savings are around 26% and the estimated energy savings are 32% compared with the two-column B–T separation. The use of UOP proprietary MD trays in a BT DWC can further reduce the capital cost compared to conventional valve trays by reducing the tangent length and diameter of the DWC. In addition, the use of MD trays allows proper vapor distribution to be obtained to maintain the product specifications.

UOP MD trays have traditionally been used for reducing column diameter in grassroots designs, increasing the capacity of existing columns, and for the minimization of fouling/foaming. It is found that UOP MD trays are ideally suited for DWCs. The multiple downcomers and fewer receiving pans allow for the weir length to be maximized and the cross-sectional area to be fully utilized. The 90° rotation (tray-to-tray) results in unit cells that allow for accurate distribution across each tray. Regardless of where the dividing wall is located, optimum distribution can be achieved on both sides. MD tray designs can be optimized to equalize the pressure drop between the two sides of the wall to ensure proper vapor distribution. Although MD trays have lower efficiency compared with valve trays, they can be used at close tray spacing. This results in a shorter tangent height even if more trays are required.

11.7.3.2 Reformate Splitter Dividing Wall Column

The reformate splitter is a DWC that separates the fresh reformate feed into a benzene-rich overhead stream, a toluene-rich side-draw stream, and a xylene-rich bottoms stream (Figure 11.22). The dividing-wall application enables a side draw to produce negligible benzene content, which makes it ideal for gasoline blending. Toluene recovery to the side-draw stream can vary and will have a direct impact on the relative sizes of the overhead and side-draw product streams. Higher toluene recovery to the side draw will reduce the overhead product and thus reduce Sulfolane unit size and energy requirements. Xylenes in the side draw have a major effect on the size of the Tatoray unit (due to effects on conversion) and the overall yield of the complex (due to

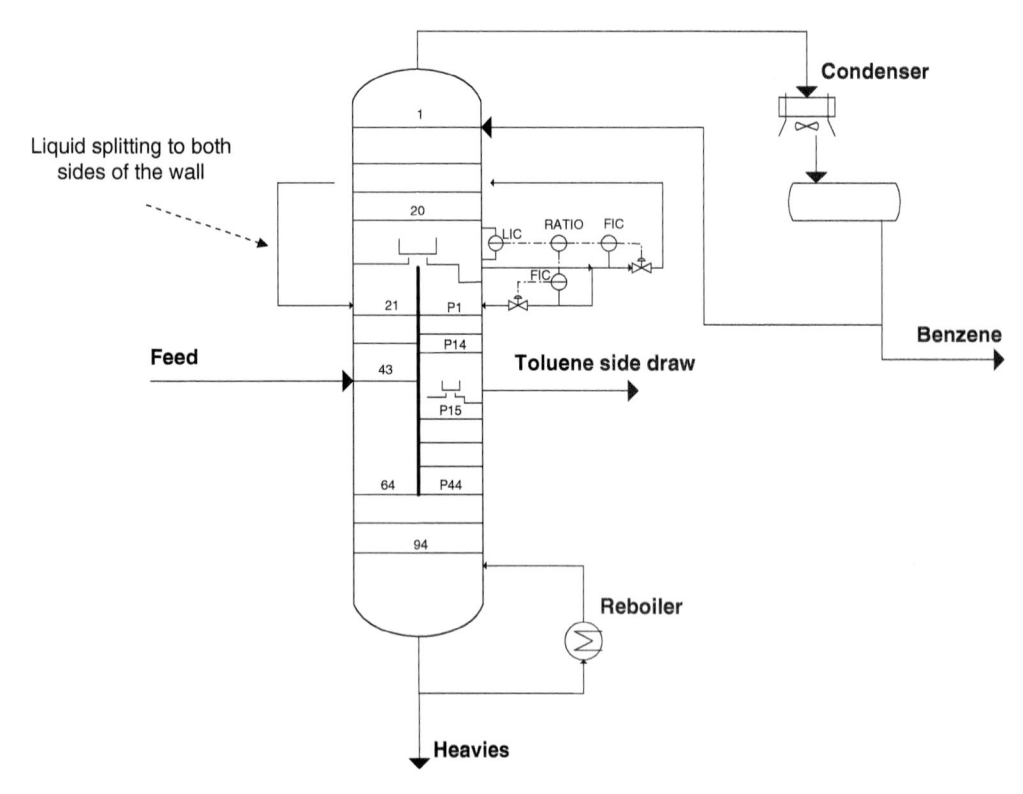

Figure 11.21 Benzene–toluene dividing wall column.

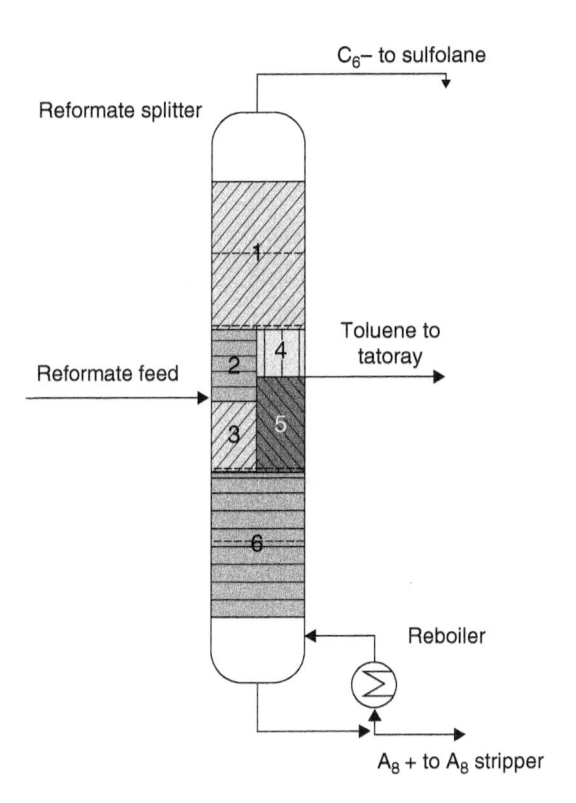

Figure 11.22 Reformate splitter stage layout.

xylene feed that were necessary to avoid heavy desorbent contamination were now relieved. This led to the combination of columns with resulting capital cost savings, as well as a dramatic reduction in energy requirements for xylene feed recovery. The energy savings outweigh by far the difference in desorbent recovery cost between light and heavy desorbent systems.

The toluene desorbent system also enabled less stringent desorbent recovery in the raffinate and extract columns. The heavy desorbent system uses a specialty chemical, typically *para*-diethylbenzene, which must be recovered and recycled within the adsorptive separation unit to avoid economic loss. When *p*-DEB leaves the separation unit, it will be converted to by-product and does not return, hence costly make-up desorbent is required. This is not the case with toluene desorbent at all!

Toluene is abundant in a typical reformate feed and it is intended to be converted into *para*-xylene or by-products. Thus, there is no need to restrict the movement of toluene and the column separations may be optimized accordingly. Toluene moves freely throughout the internal recycle streams of the complex without incurring any make-up costs. This enables further energy savings.

With light desorbent, there are four (4) distillation columns that can be eliminated from the previous generation heavy desorbent complex design. Moreover, there are capital cost savings afforded by eliminating two (2) dedicated storage tanks and an underground closed drain system with desorbent sump tank; these were needed to store make-up desorbent and preserve heavy desorbent inventory during shutdown periods. Finally, the cost of the desorbent inventory itself is lower, since the cost of reformate feed is far lower than a specialty chemical at very high purity. For a 1000 KMTA *para*-xylene complex, the inventory savings are nearly $8 MM US.

The continued innovation in aromatic complex design and aromatic complex adsorbents and catalysts drives both capital and utility cost reductions. It is the fundamentals of process change analysis and pinch analysis that offer paths forward to process intensification and maximum profitability.

ring loss in the Tatoray unit). The toluene concentration in the bottoms stream influences the energy required for separation of xylenes.

11.7.4 Use of Light Desorbent

The earliest use of selective adsorption for *para*-xylene production in the 1970s was based on a light desorbent (toluene) system. As unit capacities and minimum product purity requirements increased over time, the light desorbent system became less attractive and an alternate heavy desorbent system became the industry standard. The heavy desorbent systems brought a utility advantage compared to light desorbent systems because the recycle desorbent could be recovered from the raffinate and extract column feed streams without need to vaporize and lift the desorbent to the column overhead as distillate. However, the heavy desorbent system also required additional capital expense and other constraints that led to its ultimate obsolescence. Continuous innovation and improvements to the adsorbents eventually reversed the equation. With modern high-capacity adsorbent, the required desorbent-to-feed ratio decreased dramatically and the utility cost of light desorbent recovery diminished.

The light desorbent system enabled major changes to the fractionation throughout the complex. The stringent C_9 aromatic (Corradi et al. 2016) specifications on the

11.7.5 Heat Pump for Paraxylene Column

Heat pump application is part of low-temperature heat recovery because it lifts low temperature heat from the column overhead vapor up to a temperature sufficient to provide reboiling service for the column bottom. The paraxylene column heat pump is utilized to provide duty to the paraxylene column reboiler. The utility savings of using the heat pump is estimated to be between 5 and 10% over the original design depending on the relative prices of electric power vs. hot utilities, such as fuel and steam. The paraxylene column heat pump uses nitrogen for the buffer/flush gas required for the dry gas seals. Nitrogen as the cold dry gas is preferred by the dry gas seals. Non-condensable buildup is a concern, a purge line from the extract column no. 2 receiver to extract column no. 1 overhead condenser is required to address this concern. The use of the compressor discharge gas was also considered. However, this posed the concerns of extra equipment (dedicated seal gas heater) and condensation of the discharge gas (near dew point) in the dry gas seals.

11.7.6 Indirect Column Heat Integration

There are two types of column heat integration. The first type is direct thermal coupling, where the column overhead stream is used directly as reboiling for other columns. Xylene and extract columns' thermal couplings are good examples of direct column integration. Direct coupling yields the greatest energy efficiency and smallest energy footprint because the hot and cold process streams can operate at minimum approach temperatures. However, direct coupling introduces operating constraints and requires careful consideration to ensure process streams are available for heating and cooling when needed. Spare or redundant equipment systems may be needed to ensure expensive downtime is avoided.

The second type is indirect thermal coupling, where an intermediate stream is introduced to recover energy. The benefit of indirect coupling is the flexibility to operate individual process units without constraint, while still realizing the energy benefits afforded by direct coupling. A simple example of indirect coupling is the use of a boiler feed water (BFW) stream to recover heat from a hot process stream. The steam generated from the BFW by the hot stream is not constrained and the energy recovered as steam may now be used anywhere in any process unit. This flexibility is not free; however, as the use of the intermediate BFW stream means a greater approach temperature between the coupled hot and cold processes. The hot process stream requires a minimum temperature approach to the BFW, plus the generated steam requires a minimum temperature approach to the cold process stream. In addition, the indirect coupling requires an additional exchanger shell to transfer the heat fully from the hot process to the cold process.

Indirect coupling is not limited to process heat from distillation columns. Valuable heat recovery is also demonstrated in heater convection coils using a variety of process or utility streams. It is not uncommon to generate HP steam in reactor charge heater and reactor inter-heater convection coils. The generated HP steam is extremely versatile and may be used for thermal purposes, driving rotating equipment, or power generation. Figure 11.23 shows the indirect column heat integration via steam while Figure 11.20 presents a direct column heat integration (also called column thermal coupling).

11.7.7 Benefit of Column Integration

The column heat integration can be depicted by use of "Tetris-style" diagram, a schematic of the temperature vs. relative enthalpy of all the reboilers and condensers in the aromatics complex (Figure 11.24). The illustration by this diagram is useful in evaluating heat integration options by representing the reboilers and condensers in a stacked heat source and sink configuration with respect to relative enthalpies and temperatures. What can be observed is that in the top tier, both A_8 and raffinate columns are reboiled by fired heater. The overhead vapors of these two columns provide reboiling heat to other columns in the lower tier, with raffinate overhead vapor heat supplying the majority of the heat duty. While receiving reboiling from raffinate column overhead vapor, extract column no. 2 also provides reboiling service for extract column no. 1. This "double-cascade" of energy represents very intense column integration and significantly contributes to energy reduction.

11.7.8 Process–Process Stream Heat Integration

In addition to column integration, sensible heat integration is accomplished by optimal heat exchanger network design to maximize heat recovery and reduce utility usage further. Numerous services for sensible heat integration are added by applying fundamental principles of pinch analysis, most importantly avoiding wasteful cross-pinch heat transfer. Depending on the nature of the feedstocks to an aromatics complex, the sensible heat exchanger network will vary and a custom design is typical when striving for maximum energy efficiency.

The feed streams to a pressurized column is an attractive target for sensible heat transfer because the feed conditioning reduces the column reboiler duty. The raffinate column feed will recover heat from a variety of process sources, including waste heat from fired heater convection coils. Moreover, specialty exchangers such as plate exchangers may be installed to recover large quantities of sensible heat at close approach temperatures. Pressurized columns in aromatics complexes operate at higher temperatures and their reboilers are frequently fuel consumers. Any reduction in reboiler duty subsequently yields benefits in not only reduced utility consumption but also reduced column capital costs, reboiler heater CO_2 emissions, and complex flare header size.

Figure 11.23 Indirect column integration.

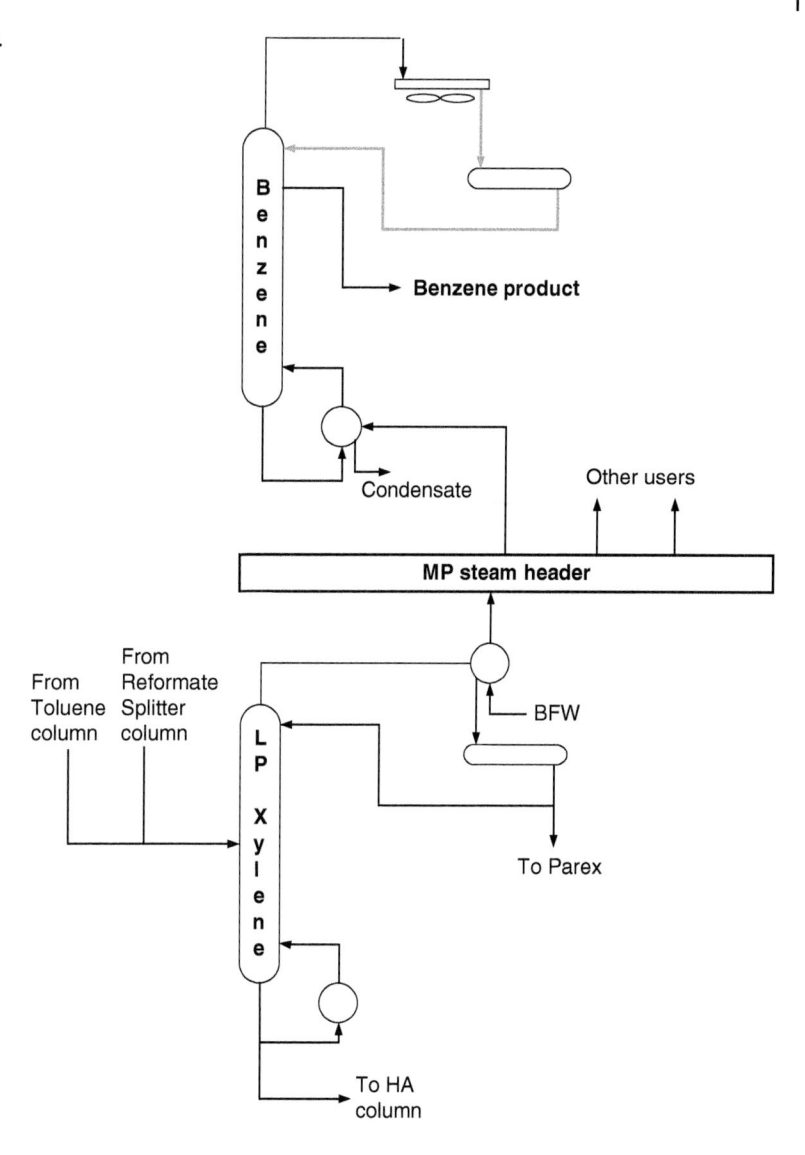

11.7.9 Power Recovery

Power recovery includes variable speed control to minimize wasteful spillback flows through compressors and pumps, while it also includes power recovery from large liquid pressure drop and from low-temperature heat rejection. In most chemical processes, large quantities of low temperature heat are rejected to the atmosphere from air-cooled exchangers in low pressure or atmospheric column condensing service. These condensers account for more than 50% energy loss. Without recovery of low temperature heat, further efforts to improve process energy efficiency become limited. Some improvements in this area are discussed here.

11.7.9.1 Organic Rankine Cycle for Low-Temperature Heat Recovery

The improvement idea discussed here is to utilize the column overhead heat to generate power using an ORC. An ORC is a power cycle that generates power using a low- or medium-temperature heat source and an organic fluid as a working fluid. Usually, the heat source temperature is less than 350 °F.

This idea aims to use low temperature heat rejected from column overheads, product rundown streams, and reactor effluent streams in an aromatic complex to evaporate an organic working fluid, feed the vapors of the working fluid to a turbine to drive a generator or other load, condense the exhaust vapors from the turbine, and pump the recycled condensed fluid back to the evaporator. The power generated with the ORC system can be used within the aromatics complex to run compressors and/or pumps or sent to a central power station. Suitable working fluids include nonflammable, low-toxicity chemical compounds with a boiling temperature range that are preferably 10–25 °F lower than the column overhead return temperature and a power cycle efficiency of

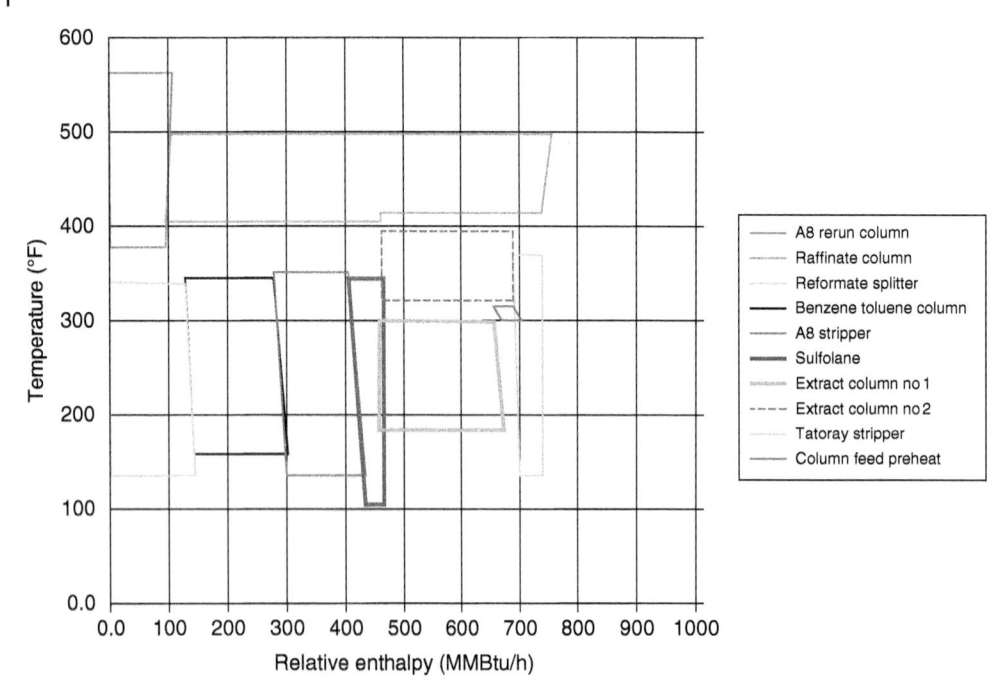

Figure 11.24 Teris-style T-H diagram for column integration.

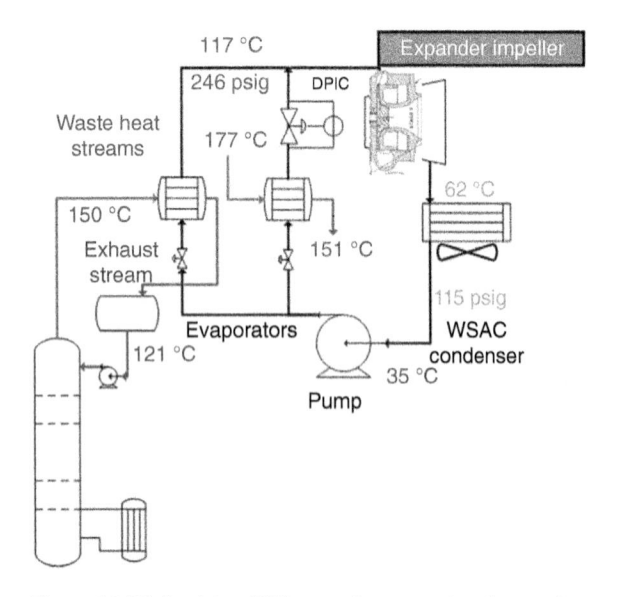

Figure 11.25 Applying ORC to a column overhead vapor for power generation.

10–20%. The ORC application is suitable for any distillation/fractionation/separation process if the overhead temperature satisfies the condition for vaporizing the working fluid (Figure 11.25). For example, the minimum column overhead temperature should be around 100 °C if a suitable refrigerant (e.g. Genetron 245fa) is used as a working fluid. UOP High Flux and High Cond tubes can be used to improve heat transfer in the evaporator and condenser, to reduce the temperature approach, and increase cycle efficiency and reduce plot space.

An example is provided for an existing aromatics complex producing 900 KMTA of *para*-xylene. The total energy requirement (fuel + steam) for this existing complex is around 300 MW. Heat loss through raffinate and extract column overheads alone is around 125 MW (42% of total energy requirement). The column overhead heat is typically rejected by an air cooler and the column receiver temperatures are around 120–140 °C.

Using an ORC with Genetron 245fa refrigerant commercially available by Honeywell as a working fluid, the net power benefit is between 12 and 13 MW depending on the cycle temperatures and pressures. The net power benefit includes all the efficiency factors and lost power (pump, air cooler fans, cooling tower pumps, etc.) depending on the cycle conditions. This corresponds to approximately 30% of the power consumed in the aromatics and naphtha complex. The power generated by the ORC can be sold or supplied to the compressors and pumps in the complex. Based on power pricing between $0.07 and $0.10/kWh, the energy cost saving is around $7–10 MM/year. If CO_2 credits are included at $30/MT CO_2 pricing, the cost saving could be between $8.5 and 12.5 MM/year. The installed capital cost for this system is around $1500/kW electricity generated and thus the installed cost of $18 MM is required. For 50% OSBL an investment cost around $27–28 MM may be required. Then, the payback period for this technology could be around 2–3 years depending on the power price and CO_2

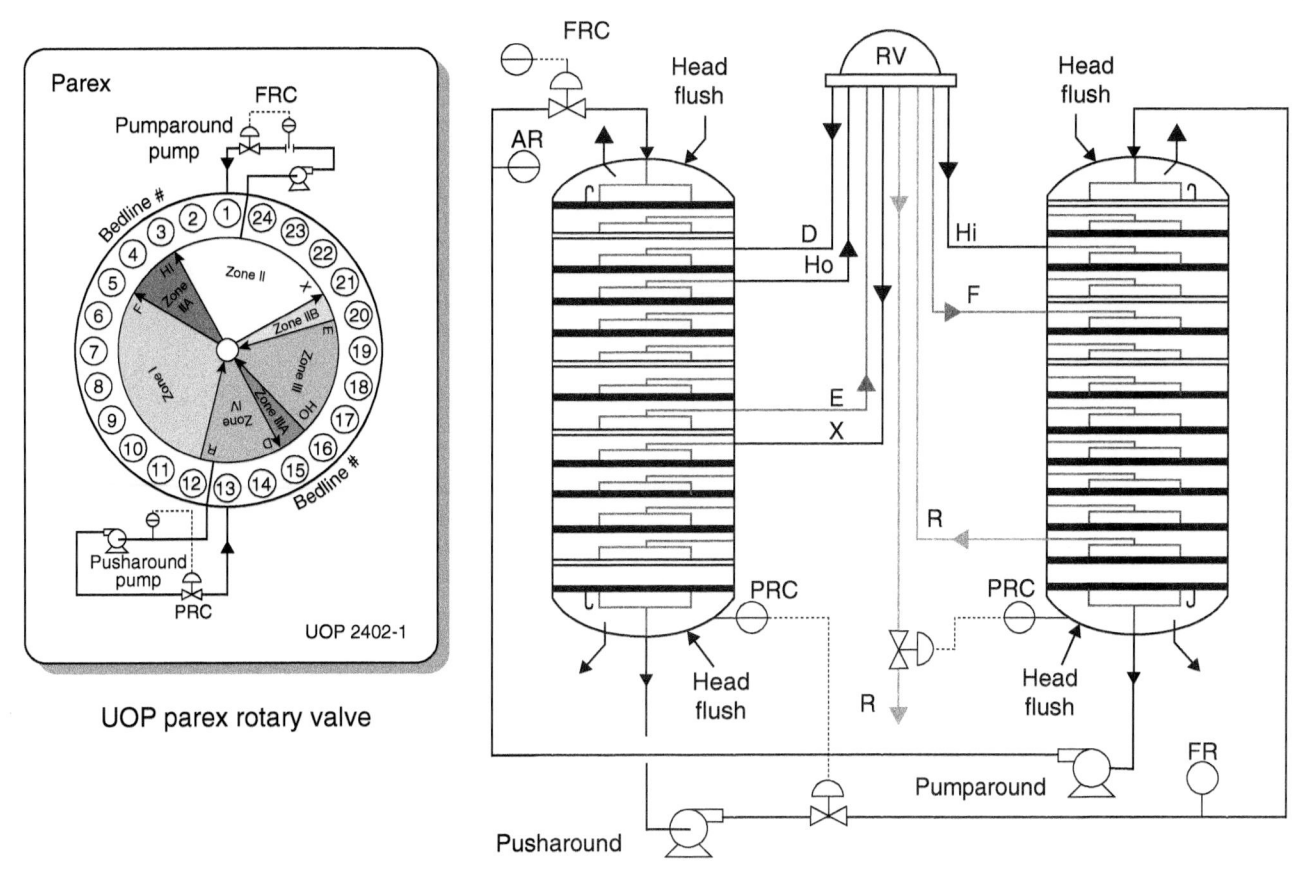

Figure 11.26 Circulation pumps in Parex chambers.

credits. Currently, the industry has not readily adopted ORC in general because the payback period is too high if without green energy credit. But eventually, regulatory will lead to capture of the low temperature energy via ORC, although it is not happening just yet. It is common to observe that this kind of low temperature heat is captured for implementing simple hot water recovery or of a mega-plant, a desal water stream.

Refineries and petrochemical plants are under pressure to reduce CO_2 emissions. Applying this ORC technology, low temperature heat can be utilized to generate power, which could qualify green energy or CO_2 credit in some countries, which helps to pay off the ORC investment.

11.7.9.2 Variable Frequency Driver on Adsorbent Chamber Circulation Pumps

The adsorbent chamber circulation pumps (Figure 11.26) are in a unique service. Unlike a typical pump with a steady-state flow, the chamber circulation pumps cycle through a multitude of flow rate targets. The pumps must support each individual zone flow rate for a prescribed length of time during each cycle of the rotary valve. As shown in Figure 11.28, there are large differences in zone flows. With a fixed speed machine, there is far greater head developed than needed for many of the zones. This is evident from the family of pump curves where much lower speeds are sufficient for the prescribed flow and head (Figure 11.27). Therefore, it was considered to install a variable frequency driver on the pumps (Frey et al. 2015). Given the special nature of this service where process performance is governed by sharp, square-wave flow transitions, a typical variable-frequency driver is insufficient. A proprietary control strategy is required to achieve the sharp flow transitions via both control valve and pump speed actions. With this strategy in place, an estimated 40–50% power savings can be realized compared to a fixed speed driver (Figure 11.28).

11.7.10 Process Integration Summary

To visualize the effect of overall process improvements, the composite curves are built for the overall aromatics complex as shown in Figure 11.29, which indicates more than 30% reduction in hot utility compared with the base case design. When counting for power recovery options, total energy saving is about 50% comparing with the base case. The improvement is very significant, which warrants more detailed feasibility assessment for any potential showstoppers, eventually leading to practical design.

260 | *11 Process Integration for Higher Efficiency and Low Cost*

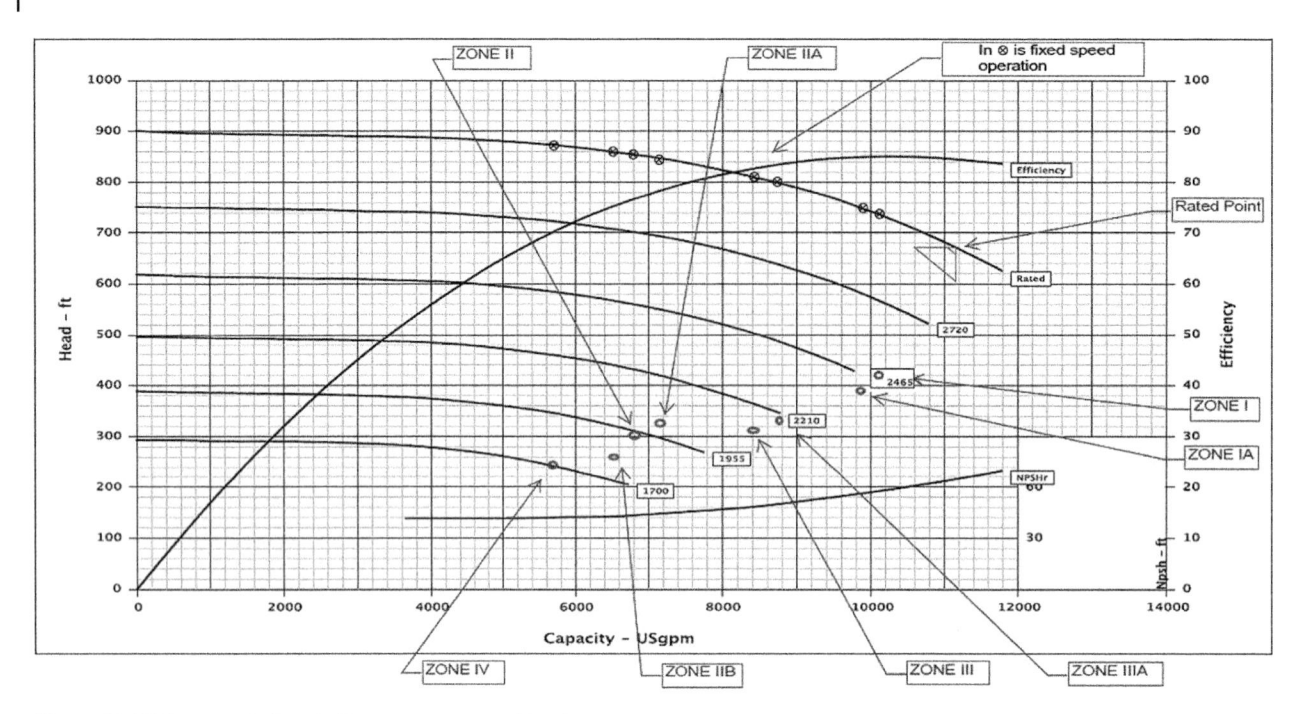

Figure 11.27 Large variations in Parex pump-around loads.

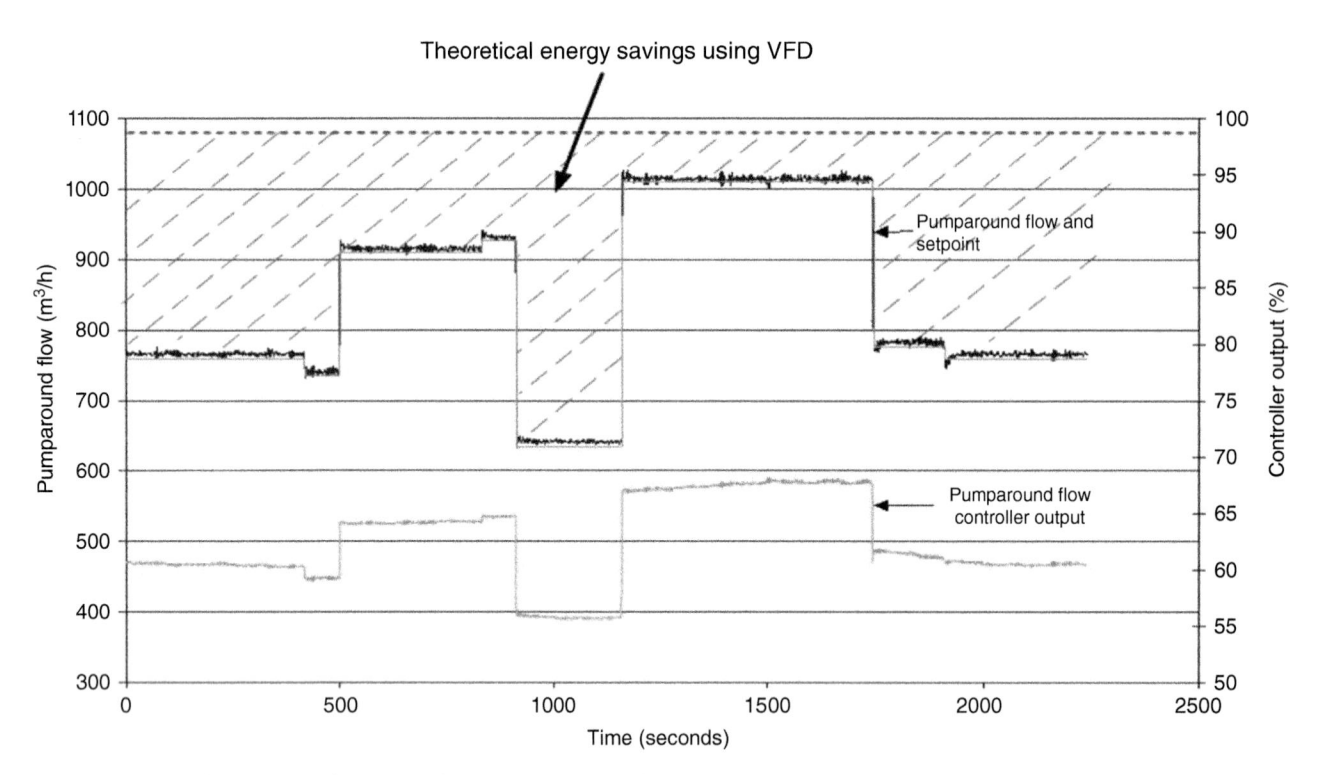

Figure 11.28 Benefit estimation for variable frequency driver vs. a fixed speed drive.

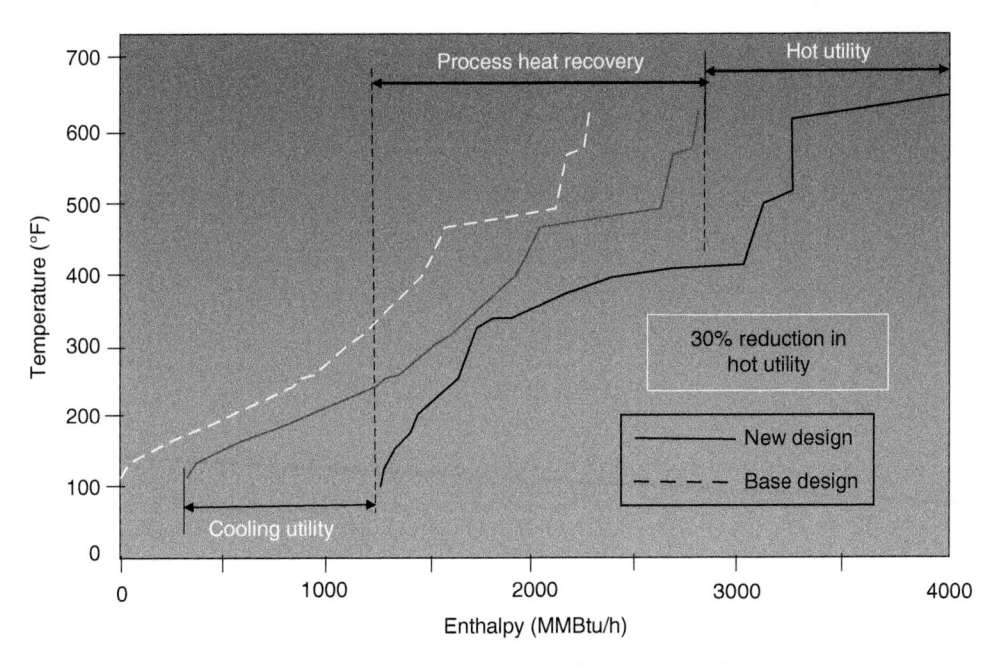

Figure 11.29 Composite curves for the aromatics complex for the new and base designs.

References

Corradi, J., Werba, G., and Gattupalli, R. (2016). Systems and methods for separating xylene isomers using selective adsorption. US Patent 9,302,955, 5 April 2016.

Dhole, V.R. and Linnhoff, B. (1993). Distillation column targets. *Computers and Chemical Engineering* 17 (5/6): 549–560.

Frey, S., Pettengill, L., and Van de Cotte, M. (2015). Energy efficiency in adsorptive separation. US Patent 9,085,499, 21 July 2015.

Glavic, P., Kravanja, Z., and Homsak, M. (1988). Heat integration of reactors; 1 Criteria for the placement of reactors into process flowsheet. *Chemical Engineering and Science* 43: 593.

Linnhoff, B., Townsend, D.W., Boland, D. et al. (1982). *User Guide on Process Integration for the Efficient Use of Energy*. Rugby: IChemE.

Townsend, D.W. and Linnhoff, B. (1983a). Heat and power networks in process design. Part 1: Criteria for placement of heat engines and heat pumps in process networks. *AIChE Journal* 29 (5): 742–748.

Townsend, D.W. and Linnhoff, B. (1983b). Heat and power networks in process design. Part II: Design procedure for equipment selection and process matching. *AIChE Journal* 29 (5): 748–771.

Zhu, X.X. (2014). *Energy and Process Optimization for the Process Industries*. Wiley/AIChE.

Zhu, X.X., Maher, G., and Werba, G. (2011). Spend money to make money. September Issue,. *Hydrocarbon Engineering*: 33–38.

12

Energy Benchmarking

12.1 Introduction

Energy benchmarking defines an intensity measure of process energy performance. It can be used to determine the baseline of energy performance to compare with peers and measure the effects by operation and process changes.

When you are given a task to improve energy performance for the plant or process unit, your immediate response would be: Where should I start? To answer this question, you need to determine both current energy use and an energy consumption target. Only then it is possible to establish the baseline and to know how well the process unit is doing by comparing current performance against a target. We refer to the exercise of establishing a baseline as benchmarking.

The most important result of energy benchmarking is the indication of energy intensity for individual processes. The energy intensity is then used with a performance target as defined by a corporate goal, or industrial peer performance, to determine the process energy performance in comparison with targeted performance. In general, benchmarking assessment can give several indications:

- The need for an overall energy optimization effort: If large gaps are available for a majority of the process units, this could imply there are many opportunities available and require coordinated effort across the plant. A dedicated energy team may need to be established to identify and capture the opportunities.
- Areas for focus: Process units with large performance gaps can be selected as focus areas. This allows us to effectively concentrate efforts on the areas with the greatest potential for improvement. Specialists may need to be assembled to form a project team for individual process units.
- Update targets: If all major process units are performing well relative to the targets, the plant may concentrate efforts on continuous improvements via monitoring and control.

12.2 Definition of Energy Intensity for a Process

Let us start with the specific question: How to define energy performance for a process? People might think of energy efficiency first. Although energy efficiency is a good measure as everyone knows what it is about, it does not relate energy use to process feed rate and yields and thus it is hard to connect the concept of energy efficiency to plant managers and engineers.

To overcome this shortcoming, the concept of energy intensity is adopted which connects process energy use and production activity. The energy intensity was originated from Schipper et al. (1992), who attempt to address energy intensity through historic energy use and economic activity in five nations: United States, Norway, Denmark, West Germany, and Japan. The concept of energy intensity allows them to better examine the trends that prevailed during both increasing and decreasing energy prices.

By definition, energy intensity (I) is described by

$$I = \frac{\text{Energy use}}{\text{Activity}} = \frac{E}{A} \qquad (12.1)$$

Total energy use (E) becomes the numerator while common measure of activity (A) is the denominator. For example, commonly used measures of activity are vehicle-miles for passenger cars in transportation, kWh of electricity produced in power industry, unit of production for the process industry, respectively.

Physical unit of production can be barrel per day or tons/h or m^3/h of total feed (or product). Thus, industrial energy intensity can be defined as

$$I = \frac{\text{Quantity of energy}}{\text{Quantity of feed or product}} \qquad (12.2)$$

Energy intensity defined in Eq. (12.2) directly connects energy use to production as it puts production as the basis (denominator). In this way, energy use is measured on the basis of production which is in the right direction of thought: a process is meant to produce products supported by energy. For a given process, energy intensity has a strong correlation with energy efficiency. Directionally, efficiency improvements in processes and equipment can contribute to observed changes in energy intensity.

Therefore, we can come to agree that energy intensity is a more general concept for measuring of process energy efficiency indirectly.

Before adopting the concept of energy intensity, the measure of activity must be defined by either feed rate or product rate. For plants with a single-most desirable product, the measure of activity should be product rate. For plants making multiple products, it is better to use feed rate as the measure of activity. The explanation is that a process may produce multiple products and some products are more desirable than other in terms of market value. Furthermore, some products require more energy to make than other. Thus, it could be very difficult to differentiate products for energy use. If we simply add all products together for the sum to appear in the denominator in Eq. (12.2), we will encounter with a problem which is the dissimilarity in product as discussed. However, if feed is used in the denominator, the dissimilarity problem is nonexistent for cases with single feed. The dissimilarity is much less a concern for multiple-feed cases than for multiple products because, in general, multiple feeds are more similar in compositions to each other than multiple products are similar to each other.

In the case of an aromatic complex, the desirable product is *para*-xylene (*p*X) and thus the above discussions lead us to define the process energy intensity on the product basis as

$$I_{process} = \frac{\text{Quantity of energy}}{\text{Quantity of product}} = \frac{E}{P} \quad (12.3)$$

It is straightforward to calculate the energy intensity for a process using Eq. (12.3), where E is the total net energy use and P is the total product produced from the process. Net energy use is the difference of total energy use and total energy generation. Process energy use mainly includes fuel fired in furnaces, steam consumed in column stripping and reboiling, as well as steam turbines as process drivers and electricity for motors. In addition, it also includes other energy usage such as boiler feed water generation, cooling water, and condensate return from steam heaters and turbines. Process energy generation mainly comes from process steam generation, while power generation comes from process pressure reduction.

12.3 The Concept of Fuel Equivalent (FE) for Steam and Power

There is an issue yet to be resolved for the energy intensity defined in Eq. (12.3). The energy use (E) for a process consists of fuel, steam, and electricity. They are nonadditive because they are different in energy forms and quality. However, if these energy forms can be traced back to fuel fired at the source of generation, which is the meaning of fuel equivalent (FE), they can be compared on the same basis, which is fuel. In other words, they can be added or subtracted after converted to FE. For simplicity of discussions, definitions of FE for different energy forms are given here while examples of FE calculations are provided below.

In general, FE can be defined as the amount of fuel fired (Q_{fuel}) at the source to make a certain amount of utility (G_i):

$$FE_i = \frac{Q_{fuel}}{G_i} \quad i \in (\text{fuel}, \text{steam}, \text{power}) \quad (12.4)$$

In most cases, Q_{fuel} is calculated based on the lower heating value of fuel. G_i is quantified in different units according to specifications in the market place, namely Btu/h for fuel, lb/h for steam, and kWh for power. Thus, specific FE factors can be developed as follows based on this general definition of FE.

12.3.1 FE Factors for Fuel

By default, fuel is the energy source. No matter what different fuels are used, tracing back to itself makes "fuel equivalent for fuel" equal to unity, i.e.

$$FE_{fuel} = \frac{Q_{fuel\,at\,source}}{G_{fuel}} \equiv 1 \text{ Btu}/\text{Btu} \quad (12.5)$$

12.3.2 FE Factors for Steam

A typical process plant has multiple steam headers, typically designated as high pressure, medium pressure, and low pressure. In some cases, very high pressure steam is generated in boilers, which is mainly used for power generation. For calculating FE of steam, a top-down approach is adopted starting from steam generators. The total FE for each steam header is the summation of all FE's entering the steam header via different steam flow paths, which include steam generated from on-purpose boilers and waste heat boilers, steam from turbine exhaust, steam from pressure letdown valves, etc. The FE for each steam header is the total FE divided by the amount of steam generated from this header, i.e.

$$\text{FE}_i = \frac{Q_{\text{fuel}}}{G_i} = \left.\frac{\text{Total FE consumed}}{\text{Total steam generated}}\right|_i$$
$$\text{header } i \in (\text{HP}, \text{MP}, \text{LP}) \text{ kBtu/lb}$$

(12.6)

12.3.3 FE Factors for Power

For power, FE_{power} is expressed as

$$\text{FE}_{\text{power}} = \frac{Q_{\text{fuel}}(\text{Btu/h})}{Q_{\text{power}}(\text{Btu/h})} = \frac{1}{\eta_{\text{cycle}}} \text{ Btu/Btu} \quad (12.7)$$

where η_{cycle} is the cycle efficiency of power generation and Q_{power} represents the amount of energy associated with power in unit of Btu/h.

By using the conversion factor of 1 kW = 3414 Btu/h, Eq. (12.7) is converted to

$$\text{FE}_{\text{power}} = \frac{1}{\eta_{\text{cycle}}} (\text{Btu/Btu}) \times 3414 \,(\text{Btu/kWh})$$
$$= \frac{3414}{\eta_{\text{cycle}}} \text{ Btu/kWh}$$

(12.8)

Equation (12.8) can be generally applied to different scenarios for power supply such as power import, on-site power generation from back pressure and condensing steam turbines as well as from gas turbines.

12.3.4 Energy Intensity Based on FE

By converting different energy forms to FE, process energy intensity in Eq. (12.3) can be revised to give

$$I_{\text{process}} = \frac{\text{FE}}{P} \text{ Btu/unit of product} \quad (12.9)$$

where FE is the total FE as summation of individual FE for different energy forms across the process battery limit and P is the most desirable product generated from the process.

Let us walk through calculation of process energy intensity via an example.

12.4 Calculate Energy Intensity for a Process

In industry, steam is measured in mass flow while fuel in volumetric flow and electricity in electrical current. To compare them on the same basis, all the energy use and generation need to be traced back to fuel fired at the source of energy generation in order to obtain FE, which is a cardinal rule for energy benchmarking calculations.

For energy benchmarking of a process unit, the important thing is to identify the main energy consumers and provide a reasonable estimate for missing data. Going overboard to collect miniature details and chasing utmost precision should be avoided. Doing so may be wasted effort because such fine details are most likely not needed in the benchmarking calculations and will not make meaningful impact on energy optimization.

Example 12.1
Calculate process energy intensity for the existing aromatics complex design where *para*-xylene is the desirable product. The complex consists of six process sections, namely aromatics extraction, benzene–toluene Fractionation, xylene fractionation, transalkylation, xylene isomerization, and *p*-xylene selective adsorption. The last row of the table shows energy use on FE basis. We will explain how FE usages are calculated below.

Table 12.1 gives an example for the relevant data needed at this stage for establishing the process energy balance and calculating energy performance. As a general guideline, the fuel produced from a process unit, in the forms of fuel gas, vent gas, and fuel oil, should not be included in the energy balance for the unit. The value of vent gas and liquid slop streams are kept separate from the energy balance because these streams may be processed further to valuable by-products instead of burned for heating value only.

Assumptions: Firstly, related FE factors need to be obtained as below and the basis for deriving these assumptions will be explained later. Assumed FE factors for this example are

- FE for purchased power = 9.09 MMBtu FE/MWe
- FE for MP steam = 1350 Btu FE/lb
- FE for LP steam = 1200 Btu FE/lb
- FE for condensate = 150 Btu FE/lb
- FE for BFW at 221 °F (105 °C) = 177 Btu FE/lb
- FE for CW at 90 °F (32 °C) = 5 Btu FE/lb
- FE for standard (std.) oil = 39.7 MMBtu FE/Metric Ton std. oil

Convert energy inputs and outputs to FE:

- FE for power usage = −20.5 MW × 9.09 MMBtu FE/MWh = −186.8 MMBtu FE/h
- FE for MP steam export = 5.2 klb/h × 1.35 MMBtu FE/klb = 7.0 MMBtu FE/h
- FE for LP steam usage = −26.4 klb/h × 1.2 MMBtu FE/klb = −31.6 MMBtu FE/h
- FE for BFW usage = −16.3 klb/h × 0.177 MMBtu FE/klb = −2.9 MMBtu FE/h
- FE for condensate return = 23.4 klb/h × 0.15 MMBtu FE/klb = 3.5 MMBtu FE/h

Table 12.1 Example data set for energy use.

Process sections (total pX product = 1.2 MMMT/year)	Electric	MP	LP	BFW	Condensate	Cooling water	Fuel
	Duty	150 psig	60 psig	at 221 °F	3 psia	at 90 °F	Fired duty
	MW	klb/h	klb/h	klb/h	klb/h	klb/h	MMBtu/h
ED Sulfolane	−0.4	0.0	−0.3	0.0	0.3	−2310.1	0.0
BT Frac	−0.6	0.0	−1.0	0.0	1.2	−68.3	0.0
Tatoray	−2.4	0.0	−1.6	0.0	1.8	−524.2	−35.4
Xylene Frac	−3.2	−13.1	−4.2	−0.5	0.0	−185.4	−190.3
Parex	−12.3	18.2	−17.4	−16.2	18.6	−714.4	−532.9
Isomar	−1.8	0.0	−1.9	0.4	1.5	−8.5	−74.0
Net total energy usage	**−20.5**	**5.2**	**−26.4**	**−16.3**	**23.4**	**−3811.1**	**−832.7**
Multiply FE factors	9.09 MMBtu/ MWe	1.35 MMBtu/ klb MP	1.2 MMBtu/ klb LP	0.177 MMBtu/ klb BFW	0.15 MMBtu/ klb Cond.	0.005 MMBtu/ klb CW	1.0 MMBtu/ MMBtu Fuel	**Grand total**
Total FE use (MMBtu/h)	**−186.8**	**7.0**	**−31.6**	**−2.9**	**3.5**	**−19.1**	**−832.7**	**−1062.5**

Note: A positive value indicates quantity produced. A negative value (−) indicates quantity consumed.

- FE for cooling water usage = −3811.1 klb/h × 0.005 MM Btu FE/klb = −19.1 MMBtu FE/h
- FE for fuel = −832.7 MMBtu/h × 1.0 MMBtu FE/MMBtu-fuel = −832.7 MMBtu FE/h
- Net total FE for the complex = summation of all the above = −1062.5 MMBtu FE/h

Let us define specific energy use the same as the energy intensity

$$\text{Specific energy} = \frac{\text{Net energy input}}{\text{Product rate}} \quad (12.10)$$

Assume on-stream utilization as 95%, thus hours on-stream per year is 8322 and Metric Ton p-Xylene production on-stream per year is equal to 1.2 MM MT/year/8322 = 144.2 MT/h.

Applying Eq. (12.10) yields

Specific energy use for the complex = 1062.5 MMBtu FE/h / 144.2 MT p−Xylene/h = 7.37 MMBtu FE/MT p−Xylene

Alternatively, specific energy use can be expressed based on standard oil equivalent (SOE) as below.

Specific energy usage for the complex = 1062.5 MMBtu FE/h / [39.7 MMBtu FE/MT std−oil × 1000 kg SOE/MT std−oil] / 144.2 MT p−Xylene/h = 185.6 kg SOE/MT p−Xylene

Specific energy use is a very insightful concept as it represents the *energy intensity of production* indicated by the amount of energy required for producing one unit of product. Let us see how to apply this concept that is discussed in Section 12.6.

Example 12.2

As an exercise, you may calculate the operational specific energy usage for the above process based on the FE data given above and the operation data are given in Table 12.2. For reference, specific energy usage for this example = 7.72 MMBtu/MT Product (equivalent to 194.3 kg SOE/MT Product). Use the same FE factors given in Table 12.1.

12.5 Fuel Equivalent for Steam and Power

In previous discussions, some assumptions of FE factors were made for power and steam. You may ask: What is the basis for making these assumptions? How to determine FE values for power and steam in your plant? The fundamental concept for energy benchmarking calculations is that different types of energy must be traced back to fuel fired at the source of energy generation in order to obtain FE. Let us consider the calculation of FE for power first.

12.5.1 FE Factors for Power (FE$_{power}$)

FE$_{power}$ is expressed as

$$FE_{power} = \frac{Q_{fuel}}{Q_{power}} = \frac{1}{\eta_{cycle}} \text{ Btu/Btu} \quad (12.11)$$

where η_{cycle} is the cycle efficiency of power generation and thus $\eta_{cycle} = Q_{power}/Q_{fuel}$ with Q_{power} (in Btu/h) representing the amount of heat associated with power with a conversion factor of 3414 Btu/kWh.

Table 12.2 Process energy use data for Example 12.2

Process sections (total pX product = 1.2 MM MT/year)	Electric Duty MW	MP 150 psig klb/h	LP 60 psig klb/h	BFW at 221 °F klb/h	Condensate 3 psia klb/h	Cooling water at 90 °F klb/h	Fuel Fired duty MMBtu/h
ED Sulfolane	−0.4	0.0	−0.3	0.0	0.3	−2310.1	0.0
BT Frac	−0.6	0.0	−1.0	0.0	1.2	−68.3	0.0
Tatoray	−2.9	0.0	−1.6	0.0	1.8	−524.2	−41.0
Xylene Frac	−3.7	−13.1	−4.2	−0.5	0.0	−185.4	−196.2
Parex	−12.0	18.2	−17.4	−16.2	18.6	−714.4	−544.1
Isomar	−1.9	0.0	−1.9	0.4	1.5	−8.5	−93.5
Net total energy usage	**−21.4**	**5.2**	**−26.4**	**−16.3**	**23.4**	**−3811.1**	**−874.8**

Note: A positive value indicates quantity produced. A negative value (−) indicates quantity consumed.

By using the conversion factor of $1\,kW = 3414\,Btu/h$, Eq. (12.11) can be converted to

$$FE_{power} = \frac{1}{\eta_{cycle}}(Btu/Btu) \times 3414\,(Btu/kWh)$$
$$= \frac{3414}{\eta_{cycle}}(Btu/kWh) \qquad (12.12)$$

Rearranging Eq. (12.12) leads to

$$FE_{power} = \frac{3414}{\eta_{cycle}} = \frac{3414}{Q_{power}/Q_{fuel}} = \frac{Q_{fuel}/Q_{power}}{3414}$$
$$= \frac{Q_{fuel}}{W}(Btu/kWh) \qquad (12.13)$$

where $W = (Q_{power}/3414)$ and W (in kW) represents the amount of power. By converting the unit of FE_{power} from Btu/Btu in Eq. (12.12) to Btu/kWh in Eq. (12.13), the expression of FE_{power} in Eq. (12.13) becomes exactly the same as that of heat rate for power generation. Let us look at three cases for applying Eq. (12.13) as below.

Case 1 Importing Power from Coal Power Plants

Average efficiency for today's coal-fired plants is 33% globally while pulverized coal combustion can reach efficiency of 45% (LHV, net) (IEA 2012). Thus, FE factors (FE_{power}^{ST}, MMBtu/MW) for purchased coal power are in the range of 7.58 (45% of power efficiency) and 10.34 (33%). For example, if assuming steam cycle efficiency is 37.56%, applying Eq. (12.12) yields

$$FE_{power} = \frac{3414}{\eta_{cycle}} = \frac{3414}{0.3756} = 9090\,(Btu/kWh) \qquad (12.14)$$

Note that 9090 Btu/kWh is the FE_{power} factor used in the previous assumption for power.

Case 2 On-Site Power Generation from Steam Turbines

For on-site power generation, usually heat rate is known and it should be used as FE_{power}. If unknown, a typical condensing steam turbine cycle efficiency of 30% could be used to yield

$$FE_{power} = \frac{3{,}414}{\eta_{cycle}} = \frac{3{,}414}{0.3} = 11{,}380\,(Btu/kWh) \qquad (12.15)$$

FE_{power} factors for back pressure steam turbines could be much higher than 11,380 Btu/kWh. What is the interpretation of a higher FE_{power} from on-site power generation than that of purchased power? The implication is that a commercial power plant can make power more efficient than a process plant if cogeneration is not involved. Does it mean that use of motor is more efficient than using on-site condensing turbine for process drivers? The answer is Yes. You may stretch out to think: The back pressure turbines could be even worse as process drivers. Is it true? The answer for this question relies on the steam balances. If the exhaust steam from the back pressure turbines is used for processes, the back pressure turbines have much high cogeneration efficiency (power plus steam).

Case 3 On-Site Power Generation from Combined Gas and Steam Turbines

When power is generated by a gas turbine (GT), GT exhaust is usually sent to heat recovery steam generator (HRSG) for steam generation. Steam is then used for further power generation via steam turbines. A configuration such as this is known as a GT–steam combined cycle.

The combined cycle efficiency can be expressed as

$$\eta_{CC} = \eta_{GT} + \eta_{ST} - \eta_{GT} \times \eta_{ST} \qquad (12.16)$$

By applying Eq. (12.12), FE factor for power generated from a combined cycle would be

$$FE_{power}^{CC} = \frac{3414}{\eta_{CC}} \qquad (12.17)$$

Suppose that a GT cycle has an efficiency of 42%, which is a representative value for gas turbines, and the steam turbine has an efficiency of 30%. The combined cycle efficiency (η_{CC}) is 59.4% based on Eq. (12.16) and FE factor is 5747 MMBtu/MW based on Eq. (12.17). In general, the combined cycle is much efficient in power generation than the steam cycle alone.

12.5.2 FE Factors for Steam, Condensate, and Water

Steam headers are the central collection points where steam enters each header from different sources and distributes to different sinks. The total FE for each steam header is the summation of all FE's entering the steam header via different flow paths. The FE for each steam header is the total FE divided by the amount of steam generated from this header, i.e.

$$FE_{Header\,i} = \left.\frac{\sum FE\ consumed}{\sum steam\ generated}\right|_{Header\,i} \qquad (12.18)$$
$$i = (HP, MP, LP)\ \ MMBtu/klb$$

A top-down approach is adopted for FE calculations. First, FE for HP steam is calculated and then cascading down in the order of pressure levels, FE's for other steam headers are determined. Let us look at the example below.

Example 12.3
Calculate the FE values for the steam headers in Figure 12.1.

Solution
To determine the FE for steam headers, the actual ways of producing steam must be identified which could have influenced the FE for the steam.

a) *FE for HP Steam*

There are two paths for making HP steam, namely boiler 1 with 75% thermal efficiency and boiler 2 with 85% thermal efficiency, respectively. The FE factors for both HP generation sources can be calculated as

$$FE_{HP, boiler\ 1} = \frac{Q_{B1}}{M_{B1}} = \frac{179}{108} = 1.66 \text{ MMBtu/klb}$$

$$FE_{HP, boiler\ 2} = \frac{Q_{B1}}{M_{B1}} = \frac{156}{108} = 1.44 \text{ MMBtu/klb}$$

The average FE for HP steam can be calculated as

$$FE_{HP} = \frac{Q_{B1} + Q_{B2}}{M_{B1} + M_{B2}} = \frac{179 + 156}{108 + 108} = 1.55 \text{ MMBtu/klb}$$

For evaluating a base case scenario, the average FE factor for HP steam should be used. In the case when opportunities for steam saving or extra steam use are explored, the generation-source-based FE factors must be considered. For this example, when capturing the steam saving opportunity, steam generation should be reduced from boiler 1, the less efficient boiler. On the other hand, when extra HP steam is required from processes, it should be generated from boiler 2, the more efficient boiler.

In general, high pressure steam is defined as steam produced from steam generators, mainly boilers. If using boiler feed water as the reference point, the FE of high pressure steam can be derived as

$$FE_{HP} = \frac{Q_{fuel}}{M_{HP}}\bigg|_{boiler\ i} = \frac{h_{HP} - h_{BFW}}{\eta_{boiler\ i}} \text{ kBtu/lb} \quad (12.19)$$

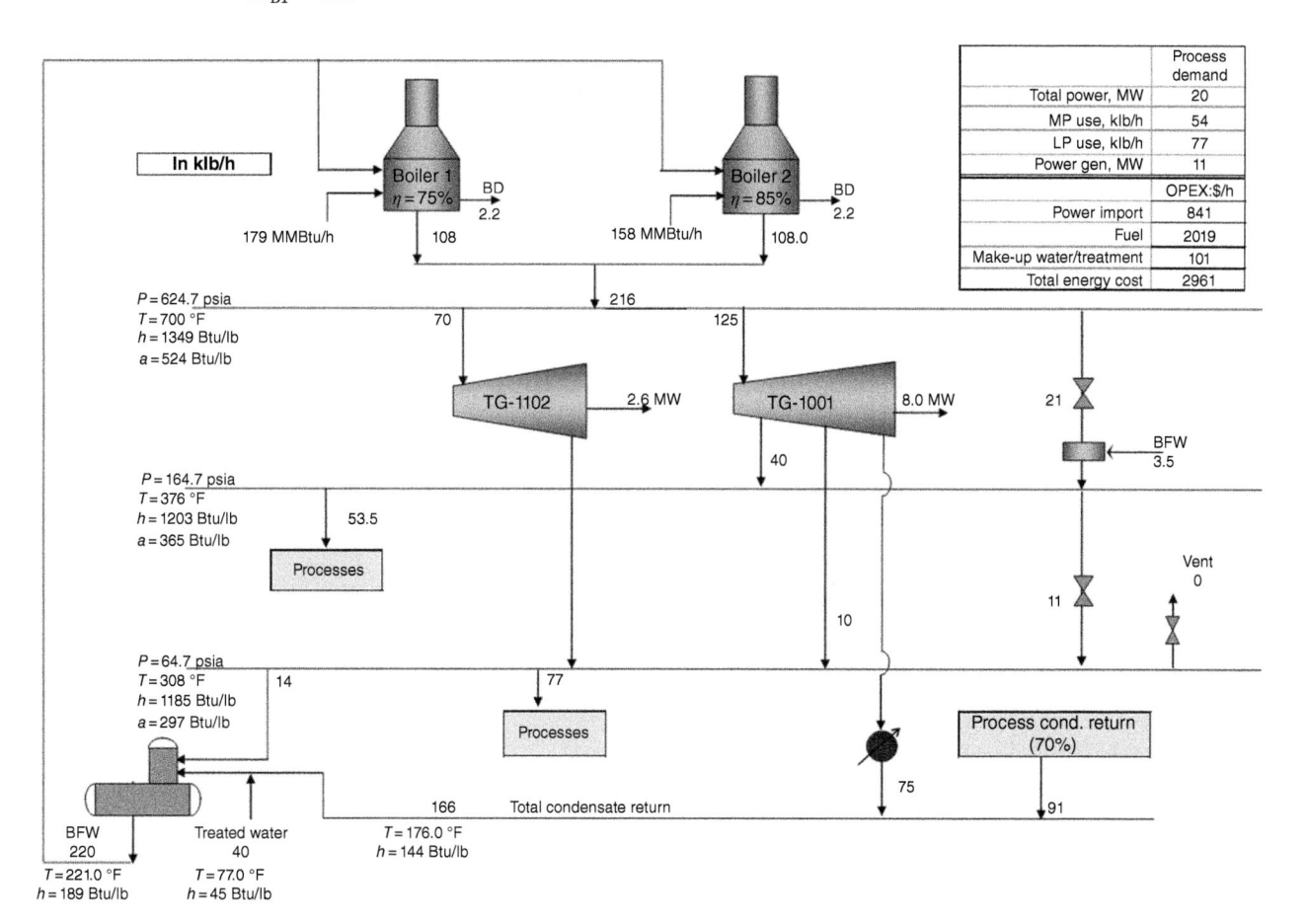

	Process demand
Total power, MW	20
MP use, klb/h	54
LP use, klb/h	77
Power gen, MW	11
	OPEX:$/h
Power import	841
Fuel	2019
Make-up water/treatment	101
Total energy cost	2961

Figure 12.1 The steam system for Example 12.1.

where h_{HP} and h_{BFW} are specific enthalpies for high pressure steam and boiler feed water while $\eta_{boiler\,i}$ is the boiler efficiency and M_{HP} is the amount of HP steam generated from the boiler.

In most cases, multiple boilers are used. In this case, Eq. (12.18) can be applied to derive the weighted average of FE for combined HP steam going to the HP header as

$$FE_{HP} = \frac{\sum_{i}^{Boilers} M_{i,HP} FE_{i,HP}}{\sum_{i}^{Boilers} M_{i,HP}}$$

$$= \frac{\sum_{i}^{Boilers} M_{i,HP} \times (h_{HP} - h_{BFW})/\eta_{boiler\,i}}{\sum_{i}^{Boilers} M_{i,HP}} \text{ kBtu/lb} \quad (12.20)$$

b) *FE for MP Steam*

Three paths of MP generation are identified as follows:

- Path 1: 40 klb/h of MP extraction from TG-1001 with specific steam rate $m_{HP\text{-}MP}$ at 35.6 klb/MWh. The FE for the MP steam exhaust can be calculated via

$$FE_{MP\text{-}steam} = FE_{HP\text{-}steam} - \frac{FE_{power}^{import}}{m_{HP\text{-}MP}} \quad (12.21)$$

The reason why FE factor for power import is used in Eq (12.21) is that power import is the marginal power source. In other words, if a steam turbine is replaced by motor, purchased power will be used. Assume FE factor for purchased power as 9.09 MMBtu/MWh, thus

$$FE_{MP\text{-}steam}^{P1} = 1.55 - \frac{9.09}{35.6} = 1.29 \text{ MMBtu/klb MP}$$

- Path 2: 21 klb/h of the letdown valve, $FE_{MP\text{-}steam}^{P2} = FE_{HP} = 1.55$ MMBtu/klb because a letdown is an adiabatic process and thus FE does not change through the letdown valve.
- Path 3: 3.5 klb/h of BFW addition for desuperheating, $FE_{MP\text{-}steam}^{P3} = FE_{BFW} = 177$ Btu/lb.

Thus, the average FE for the mixed MP steam can be calculated based on Eq. (12.18) as:

$$FE_{MP}^{av} = \frac{M_{MP}^{P1} \times FE_{MP}^{P1} + M_{MP}^{P2} \times FE_{MP}^{P2} + M_{MP}^{P3} \times FE_{MP}^{P3}}{M_{MP}^{P1} + M_{MP}^{P2} + M_{MP}^{P3}}$$

$$= \frac{40 \times 1.29 + 21 \times 1.55 + 3.5 \times 0.177}{40 + 21 + 3.5}$$

$$= 1.31 \text{ MMBtu/klb}$$

c) *FE for LP Steam*

There are three paths for making LP steam:
- Path 1: 70 klb/h from the TG-1002 turbine with a specific steam rate of 26.8 lb/kWh. The FE for the MP steam exhaust can be calculated via

$$FE_{LP\text{-}steam} = FE_{HP\text{-}steam} - \frac{FE_{power}^{import}}{m_{HP\text{-}LP}} \quad (12.22)$$

Assume FE factor for purchased power as 9.09 MMBtu/MWh, thus

$$FE_{LP\text{-}steam}^{P1} = 1.55 - \frac{9.09}{26.8} = 1.2 \text{ MMBtu/klb LP}$$

- Path 2: 10 klb/h from the TG-1001 LP extraction with a specific steam rate of 22.9 lb/kWh.

$$FE_{LP\text{-}steam}^{P2} = 1.55 - \frac{9.09}{22.9} = 1.15 \text{ MMBtu/klb LP}$$

- Path 3: 11 klb/h of the letdown valve, $FE_{LP\text{-}steam}^{P3} = FE_{MP} = 1.31$ MMBtu/klb because a letdown is an adiabatic process. BFW desuperheating is not needed for the LP steam in this case because the superheated fraction in LP steam is very small.

Thus, the average FE for the mixed LP steam is calculated based on Eq. (12.18):

$$FE_{LP}^{av} = \frac{70 FE_{LP}^{P1} + 10 FE_{LP}^{P2} + 11 FE_{LP}^{P3}}{(70 + 10 + 11)}$$

$$= \frac{(70 \times 1.21 + 10 \times 1.15 + 11 \times 1.31)}{91}$$

$$= 1.2 \text{ MMBtu/klb}$$

What about the FE for vented LP steam? In this case, FE_{LP} should also be calculated based on the path from which this vented LP steam is generated. This is because a certain amount of FE is consumed to make the LP steam no matter it is used or vented or not. For vented LP steam, the value is zero but FE is not.

d) *FE for Condensate*

Condensate temperature is similar to the deaerator temperature typically around 200 °F. The condensate FE_{Cond} can be determined by the difference of condensate temperature and raw water temperature (ambient). FE condensate is usually around 150 Btu/lb of condensate. Although FE condensate is small relative to steam, accumulated loss could be significant for a large amount of condensate loss. Also, condensate loss is costing due to extra chemicals required to treat make-up water.

e) *FE for BFW Water*

The energy required for providing boiler make-up water includes the heat content and the pump power used to elevate its pressure. The BFW heat content is the major portion of the BEW FE factor, which is determined by the difference of BFW temperature and raw water temperature (ambient).

The FE factor for BFW is calculated based on Eq. (12.23), which assumes that LP steam is used for BFW preheat. It must be pointed out that the FE of the pumping power is ignored as it is very low (~10 BTU/lb of BFW), even with the very high pump ΔP.

$$\text{FE}_{\text{BFW}} = \text{FE}_{\text{LP steam}} \times \frac{h_{\text{BFW}} - h_{\text{Ambient water}}}{h_{\text{LP steam}} - h_{\text{Ambient water}}} \quad (12.23)$$

The FE_{BFW} is in the range of 150–200 Btu/lb of BFW.

f) *FE for Cooling Water*

Energy for providing cooling water includes pump power and the fan power in running the cooling tower fans. The FE_{cw} is in the range of 2–10 Btu/gal of cooling water.

12.6 Energy Performance Index (EPI) Method for Energy Benchmarking

This FE calculation method is built on the concept of guideline energy performance (GEP), which is used as a benchmark against which actual energy performance (AEP) is compared. The rationale of using this concept as the basis for assessing process energy efficiency is revealed in the following.

Let a ratio of AEP and GEP be defined as below. This ratio shall be labeled the energy performance index (EPI):

$$\text{EPI} = \frac{\text{Actual energy performance}}{\text{Guideline energy performance}} = \frac{\text{AEP}}{\text{GEP}}$$
(12.24)

By definition, EPI represents the energy efficiency for the process unit on the basis of GEP. In this way, any improvements in operation, design, equipment, and technology upgrade can be measured using EPI. Application of the EIP method is discussed below.

Generally speaking, an EPI gap of less than 5% between AEP and GEP belongs to an operational gap. In other words, better operating practices and control could close this gap. An EPI gap of larger than 5% may require small-energy retrofit projects, which can feature a quick payback, for the gap to be closed. If the EPI gap is in the order of 10+%, it may require significant energy and process retrofit projects to close the gap.

12.6.1 Benchmarking: based on the Best-in-Operation Energy Performance (OEP)

By applying the method for calculating specific energy use, you can obtain a plot of specific energy versus time based on the historic data. This plot can pinpoint the best-in-operation energy performance (OEP) that your process unit has achieved at a time when there was institutionally dedicated effort for operation performance and with technical know-how available. You could confirm this by talking to engineers and operators who have worked in the plant during this period. As a result, you will be able to determine the OEP as the energy guideline performance (GEP) representing the best-in-plant performance.

Assume the specific energy use based on OEP is the same as the design performance of 185.6 kg SOE/MT *p*-Xylene as calculated in Example 12.1. With the actual energy use of 194.3 kg SOE/MT Product calculated from Example 12.2, EPI for the process unit can be calculated via

$$\text{EPI} = \frac{\text{AEP}}{\text{OEP}} \times 100\% = \frac{194.3}{185.6} \times 100\% = 105\% \quad (12.25)$$

Equation (12.25) indicates that AEP has a total energy consumption of 5% higher than the OEP, which indicates poor operation performance. Such a gap is significant, which should alert you to initiate investigations for root causes. The mere factor of determining EPI gives you an immediate indicator as to where your process unit stands in energy performance, so that you can quickly spot problematic areas.

At this point, we have a very good starting point. You know three essential facts: the energy intensity for your process unit, the performance target, and the gap against the target. Your mind may be racing with questions like: What has gone wrong with my process unit? How can the AEP be reduced to OEP for the process unit? These questions will be answered in Chapter 13 for key energy indicators.

12.6.2 Benchmarking: based on Industrial Peers' Energy Performance (PEP)

In industry, there are peer survey groups organized based on industrial sectors and process technology. Organizers for the survey groups send questionnaires to survey members to gather sample data on yearly basis and conduct performance calculations. Consequently, the peer performance results are shared among survey members. If your plant belongs to the survey group, you could obtain the best peer energy performance (PEP) via the representative in your organization. For certain large companies, there are community of practice (CoP) networks based on process technology. You should seek out the best PEP for your process unit via the CoP in your company.

Assume the specific energy for PEP is 176 kg SOE/MT Product for the example. Based on the actual energy use of 185.6 kg SOE/MT Product, we can calculate EPI for the process unit as

$$\text{EPI} = \frac{\text{AEP}}{\text{PEP}} \times 100\% = \frac{185.6}{176} \times 100\% = 105.5\% \quad (12.26)$$

This indicates that the current energy consumption in the plant is 5.5% higher than the best peer performance, which is significant and requires effort to get it down in order to stay in the same peer group. Usually, the survey group is divided into tiered performance structure such as first, second, third, and fourth quartiles. Based on the EPI calculated above, you can find out which performance quartile your process unit belongs to. This indicates where your process unit stands among your peers, which may motivate your management to consider to revamp the process or update the technology.

12.6.3 Benchmarking: based on the Best Technology Energy Performance (TEP)

With technology advancement in catalyst, equipment, process design, and control, process energy efficiency could improve. It is not difficult to gather the performance data for state-of-the-art technology. In some cases, the data are published in public by government offices and you could find them via web search. If not available in public, you can contact technology companies – they are often eager to provide the data to customers.

Assume the operation is improved for the example process and the energy use is reduced by 5% from 194.3 to 185.6 kg SOE/MT product. The plant management is interested to know the scope of further energy improvement by applying better process technology. Assume the TEP is 155 kg SOE/MT product. Thus, the EPI for the process unit can be calculated by

$$\text{EPI} = \frac{\text{AEP}}{\text{TEP}} \times 100\% = \frac{185.6}{155} \times 100\% = 120\% \quad (12.27)$$

Technology updates can improve the energy performance, further by 20% on top of 5% operation improvement; but technology improvement usually requires high capital costs and long implementation periods. Justification of technology improvement may mainly come from combined benefits from throughput increase, yield, and energy improvement.

12.7 Concluding Remarks

There are three fundamental concepts discussed in this chapter. The first one is the concept of converting all energy back to FE. This concept places all forms of energy on the same basis, i.e. fuel fired or FE. The second one is specific net energy, which describes the energy intensity for production. The third concept is GEP as the best alternative for comparison with actual performance.

In combination, these three concepts make it a much simpler yet effective approach for assessing the energy performance for a process unit and require minimal data. Therefore, the EPI method is designed for practical applications.

The strategy for achieving the target can involve changes to operating practice, new control strategy, process equipment modifications, or technology upgrade or combinations of the above. In general, closing the gap between average and the best potential performance of an individual unit involves operational and maintenance improvements. Eliminating the gap between an existing unit and its peers in industry often involves retrofit with modifications to operation and the process design. Reaching the state-of-the-art performance usually involves technology upgrade.

You may have questions during data extraction. Which data periods should be used as the basis for energy benchmarking? What data are more representative than other? How to prevent inefficient usage of time from data collection? Although the general guideline is to collect data that represent the most common operation, specific guidelines are provided below.

12.7.1 Criteria for Data Extraction

- Near-maximum feed rate or the most commonly used feed rate.
- Use middle-of-the-run historic data.
- Use 24 hour rolling averages based on hourly average data to smooth out fluctuation.
- One year of data could be a good representation; get rid of bad data by all means.

The reason for using middle-of-the-run historic data is because it represents an "average" operation performance. In contrast, both start-of-run (SOR) and end-of-run (EOR) represent two extreme operation modes and hence the data would give biased indications of energy use.

Annual data can cover changes in season and operation modes while monthly data can zoom into focus on a particular operation mode.

12.7.2 Calculation Accuracy for Energy Benchmarking

At this stage, you need to start focusing on important data. Make quick estimates for small consumption users as they usually do not have meters. Do not chase decimal point of precision as the key is getting the order of magnitude right. Some guidelines could be helpful to you:

- Ask instrumentation engineers to recheck critical meters to make sure they are functioning properly.
- Major consumptions need to be verified. Use design data for small consumptions if meters are not available.

- Corrections may be necessary to reflect the difference in temperature, pressure, and mass flow.
- Fill missing data by heat and mass balances.
- All forms of energy must be converted to FE. Adjustments may be necessary in order for the actual energy use to be on the same basis as guideline energy use.
- Specific energy can be on feed volume or mass basis depending on the norm used in the industry. Specific energy can also be on a product basis, which is the ratio of total net energy usage to a desirable product rate on either volume or mass basis. For a process involving both reaction and separation, use feed as the basis for calculating specific energy use. If a process only involves separation and makes single product, use product as the basis for calculating specific energy use.

References

IEA (2012). Technology Roadmap: High-Efficiency, Low-Emissions Coal-Fired Power Generation, December 4.

Schipper, L., Howarth, R.B., and Carlassare, E. (1992). Energy intensity, sector activity and structural changes in the Norwegian economy. *Energy* 17: 215–233.

13

Key Indicators and Targets

13.1 Introduction

If you ask operators and engineers how their plant is doing, they would tell you that the plant is under good control. Although often true, the process performance could become much better, both in terms of economics and operating efficiency. The root cause of this performance gap is the fact that there are many process variables and strong interactions among these variables in an operating unit. Engineers and operators are often unable to properly monitor key indicators, and may lack appropriate process optimization capabilities (Zhu 2017).

To address this, the engineer and operator need to know what key parameters to monitor, have a good understanding of their relationships, and target values for these parameters to achieve better energy utilization. Although process energy benchmarking in Chapter 12 gives a measure of process energy intensity for process units, the energy intensity does not provide indications of the root causes, nor which operating parameters should be adjusted to improve energy performance. To determine how well a process unit is doing, a system of performance metrics should be developed so that actual energy usage can be compared with a consumption target. Only then is it possible to conduct root cause analysis and take appropriate remedial actions. To accomplish this goal, the concept of key energy indicator (KEI) is introduced (Zhu and Martindale 2007), which is the foundation for systematic performance assessment and optimization.

The rationale of introducing KEIs is to seek answers to this critical question: "How can engineers characterize energy use in a process unit with an emphasis on major energy users in terms of their needs, and what are the justifications and practical methods to minimize energy use for these needs?" Application of the KEIs in reality follows a methodology based on three steps: defining key indicators, setting targets, and identifying actions to close gaps between the current indicator values and their targets. Using this methodology, a process unit can be described by a small number of key indicators to measure energy performance, which can be further developed based on process knowledge and experience. Application of key indicators will allow focus on higher priority issues and avoid falling into a trap of details.

13.2 Key Indicators Represent Operation Opportunities

The intention of defining key indicators is to describe the process and energy performance with a small number of operating parameters. A key indicator can be simply an operation parameter. Some examples of key indicators are product rates, a column reflux ratio, column overflash, spillback of a pump, heat exchanger U-value, and so on. The parameter identified as a key indicator is important due to its significant effect on process and energy performance.

In defining key indicators, one needs to understand the strong interactions between process throughput, yields, and energy use. In the traditional view, energy use is regarded as a supporting role. Any amount of energy use requested from processes is supposed to be satisfied without question and challenging. This philosophy loses sight of synergetic opportunities available for optimizing energy use for higher throughput and better yields. The following discussions will provide insights into these kinds of opportunities, which form the basis for defining key indicators for your process units to capture specific opportunities.

13.2.1 Reaction and Separation Optimization

Optimizing energy use in reaction and separation systems could lead to significant energy saving because both reactions and product separation consume the majority of overall energy use in a complex. Much effort is commonly put into reducing energy losses incurred in heat exchangers, furnaces, steam leaks, insulations, etc.; but

Efficient Petrochemical Processes: Technology, Design and Operation, First Edition.
Frank (Xin X.) Zhu, James A. Johnson, David W. Ablin, and Gregory A. Ernst.
© 2020 John Wiley & Sons, Inc. Published 2020 by John Wiley & Sons, Inc.

until recently little effort has been spent on minimizing energy use for reactions and separations, which are the heart of the processes. Very often, energy demands in these systems are considered as "must meet," with expectation of no challenges from engineers and operators. However, in reality, there is a large scope in minimizing energy use in these areas.

Reaction condition optimization considers reaction severity in terms of temperature and pressure profiles in accordance with catalyst performance over the entire run length. Optimizing reaction conditions, selecting better catalysts, and maintaining catalyst performance in operation have significant effects on both yields and energy efficiency. Consider reaction temperature as an example. In the catalyst cycle, the catalyst performance deteriorates which affects the reaction conversion and desired product yields. To compensate, the reaction temperature may be increased. However, more severe reaction conditions require more heat from hot utilities such as fired heaters. Further, more severe conditions produce more desirable products as well as undesirable by-products. The question is how to determine the optimal reaction temperature which is a function of reaction conversion, production rate, and energy use.

The primary goal of separation optimization is to achieve product recovery and quality with minimum energy use. Consider the reflux-to-feed ratio for a given separation. The reflux-to-feed is measured as the reflux rate divided by the feed rate, expressed as a simple ratio. Higher reflux-to-feed ratios will typically allow a column to achieve a better separation of heavy and light components; however, this will require higher heat input. Higher heat input will allow for light key components to be rejected from the bottoms stream, and the higher reflux rate can better reject the heavy key components from the column overhead. Excessively high reflux-to-feed ratios can cause column flooding. Operating at a lower reflux ratio will allow for higher column throughput, as the column may be farther from its flood point, and provide margin in the heat input requirement from the reboilers. Columns will be designed for an expected reflux-to-feed ratio, but this can be optimized when conditions of the column change, and as the column product qualities can be adjusted to optimize complex performance. The optimization of a particular column also needs to be balanced with the heat source providing heat to the reboiler, and the heat sinks that provide condensation capability in the column overhead. Columns may also be designed with some margin from the expected operating conditions.

Minimizing process recycle is another optimization example in separation. Another common observation in a process plant is that unconverted streams or off-specification streams are recycled back to the front of a process. Recycle streams undo separation. Minimizing recycle presents a significant opportunity for process plants to smartly reduce energy consumption. However, minimizing recycle requires proper operation of fractionation columns across the entire plant.

13.2.2 Heat Exchanger Fouling Mitigation

Fouling mitigation represents large opportunity for increased heat recovery in operation. Fouling occurs in many heat exchange services and it reduces heat transfer duty significantly. Effective fouling mitigation can save substantial amounts of energy. The overall heat transfer coefficient, or the U-value, is the single-most important parameter for fouling monitoring for a stand-alone heat exchanger. However, for a complex exchanger network involving many exchangers, determining which exchangers should be selected for cleaning and the frequency of cleaning is not trivial. Selection of the most fouled heat exchanger for cleaning could lead to a suboptimal solution. The primary objective of heat exchanger fouling mitigation should be to minimize energy use. It is possible to select the exchanger for cleaning which may not be the most fouled, but it could yield the greatest reduction in energy use than cleaning the one that is most fouled. The streams in an aromatics complex are typically quite clean, though there are still a number of services where fouling can be seen more frequently than others. For example, in the aromatics extraction unit if there is a solvent degradation problem, or at locations where outside feeds that may be contaminated with oxygen or acidic compounds are being brought into the unit. Methods for fouling mitigation are discussed in detail in Chapter 17.

13.2.3 Furnace Operation Optimization

Fired heaters provide heat for both reaction and separation. Reliability is the major concern for furnace operation, with heat flux and tube wall temperature (TWT) as the most important reliability parameters for large heaters and small heaters, respectively. Increasing either heat flux or TWT can increase furnace efficiency. When operating heat flux is much lower than the maximum limit, this is an indication that the furnace being underutilized, and thus presents an opportunity for increased feed rate. Increasing feed rate is a win–win adjustment since more feed also results in reduced energy intensity. Practically speaking however, increasing feed rate may not always be possible.

Besides reliability, efficient furnace operation is a major part of an energy management system. Oxygen content (O_2%) in the flue gas and furnace stack temperature are the two key operating parameters. Correct measurement of O_2% in the flue gas is the first step, while good control of air intake is the necessary action to achieve low excess O_2%. Proper maintenance is essential for burners to function properly and to eliminate combustion zone air leaks. However, too low O_2 content could promote uneven distribution of the combustion flame, damage furnace tubes, and cause poorer reliability. Typically, 3% oxygen is the industrial average excess oxygen, although furnaces with state-of-the-art burners and control systems may achieve less than 3%. When reducing air intake to minimize O_2 content in operation, it is imperative to minimize air leaks and have proper oxygen measurement.

Reducing the stack temperature could yield greater reduction in combustion fuel. The limit for the stack temperature is frequently set by the sulfur dew point. For furnaces with a convection section, the stack temperature could be reduced by adding an "economizer" service into the existing convection section. Economizers are typically water heating services for boiler feed water preheat or saturated steam generation. For furnaces without a convection section, installing a convection section could certainly help complex operating efficiency. However, these options need to go through a thorough feasibility evaluation of foundation strength, along with thermodynamic and hydraulic considerations. Preheating combustion air is another opportunity in reducing stack temperature.

13.2.4 Rotating Equipment Operation

Rotating equipment used in industrial processes includes pumps and compressors. For pumps and compressors with fixed speed motors, minimizing spillbacks could result in significant power saving. Optimizing steam turbine operation and maintenance could save steam. Selection of process drivers, namely motors vs. steam turbines, could save energy cost.

13.2.5 Minimizing Steam Letdown Flows

Steam letdown valves are used to give steam supply flexibility and temperature control. However, letdown steam represents lost opportunity for power generation. Steam balance optimization could minimize the letdown steam flow and hence reduce the loss in power-generating potential.

13.2.6 Turndown Operation

Poor turndown operation implies that when the feed rate reduces, energy use does not reduce accordingly. For example, air intake for furnaces should reduce accordingly when the feed rate drops. Heat input to separation columns should be reduced based on reflux-to-feed ratio instead of keeping a fixed heat input and maintaining the key component splits. Recording these heat input values for future use is useful in aromatics units, as there tends to be minimal variation in column feed compositions. Some of the hand valves for steam turbines can be partially closed when the feed rate drops significantly, so that the governor can be kept wide open to minimize steam rate. Proper turndown operation could generate significant energy saving for plants operating under large feed rate variations.

13.3 Defining Key Indicators

The above discussions reveal the fact that major improvement opportunities can be captured by paying attention to key operating parameters. This is the foundation of introducing key indicators. The method for determining key indicators follows a basic thought process: understand the process objectives → understand energy needs → develop measures for the needs → define the key indicators representing the measures. As an example of how to define key indicators for a process, let us consider an ethylbenzene dealkylation-type of xylene isomerization unit in an aromatics complex.

The flow scheme as shown in Figure 13.1 is a commonly used xylene isomerization unit (such as the UOP Isomar Process), converting a raffinate stream from an upstream *para*-xylene separation unit, which is depleted in *para*-xylene, to a product that is a near-equilibrium mixture of xylene isomers. In the most common style of this process unit, ethylbenzene in the feed is converted to benzene and ethane.

To reduce the charge heater duty, the reactor effluent stream is heat-exchanged against both feed and recycle gas to recover process heat in a combined feed exchanger (CFE). Afterward, reactor effluent is sent to a product separator where the non-condensable vapors that remain after the stream passes through the CFE and condenser are rejected to a recycle gas stream, and liquid is sent onward to fractionation. In more recent designs, two separators may be used – the first being a hot separator that is placed upstream of the condenser, and the second is a cold separator placed downstream of the condenser. This reduces the duty requirement of the condenser, as already condensed liquid from the CFE does not need to pass through the condenser itself. The fractionator, a deheptanizer column in this case, is partially condensing. Off-gases are vented, as the feed to the column will still contain dissolved light gases such as hydrogen and ethane (and small amounts of other light gases that will have entered with the reactor section make-up gas such as methane and propane). The overhead goes to a light aromatics recovery system, first typically a stripper or stabilizer column to remove light non-aromatics and any remaining dissolved light gases, and then an aromatics extraction unit, with benzene as

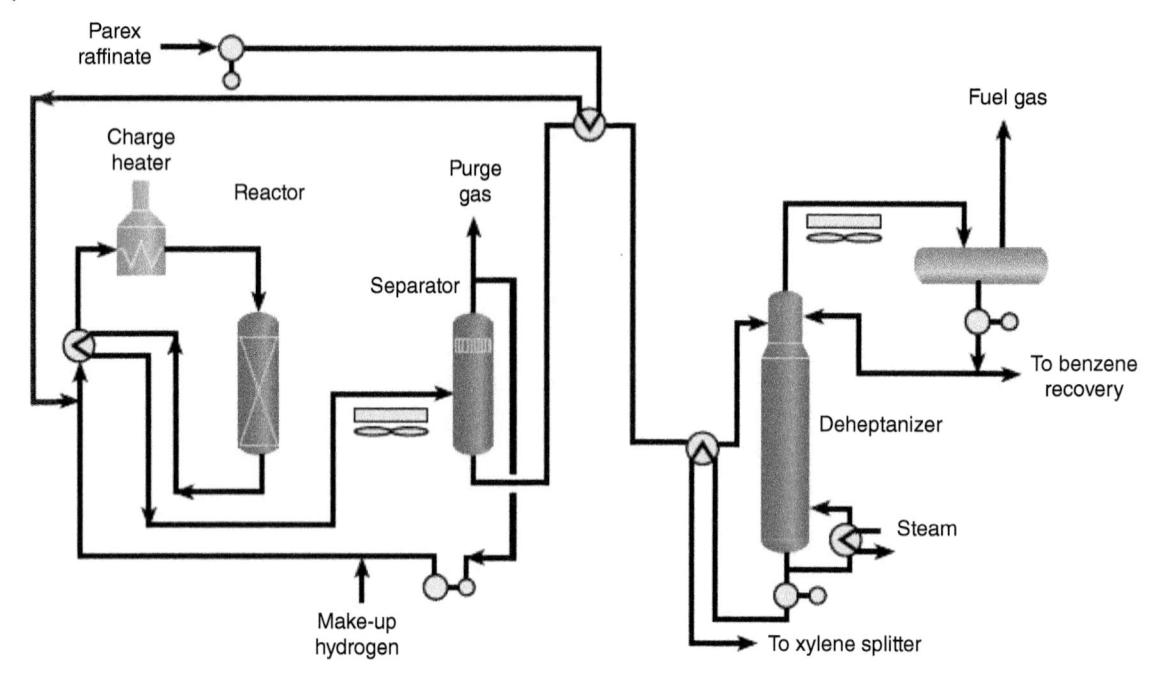

Figure 13.1 Typical xylene isomerization unit, example from a UOP aromatics complex.

the main product from a downstream benzene–toluene fractionation unit. The fractionation column bottoms liquid is recycled back to the xylene fractionation unit upstream of the *para*-xylene separation unit.

13.3.1 Simplifying the Problem

To simplify the overall task of defining key indicators for the whole unit, the unit can be divided into two main sections, namely reactor section and fractionation section. The goal is to define a set of key indicators for each section, which can be used for monitoring and optimization.

13.3.2 Developing Key Indicators for the Reaction Section

13.3.2.1 Understanding the Process

In this example, the primary process objective is actually twofold – conversion of feedstocks to an equilibrium mixture of xylenes, and conversion of most of the ethylbenzene in the feed to benzene and ethane. The equilibrium mixture of product xylenes will typically contain around 23.5–24.0 wt.% *para*-xylene, compared to a feed that contains less than 1 wt.%. The ethylbenzene conversion is typically around 60–80 wt.%, with the feed ethylbenzene content typically around 2–15 wt.%. The target ethylbenzene conversion and concentration in the feed are heavily dependent on the complex processing objectives and feed sources, respectively. The reactor contains a catalyst that is intended to maximize production of the desired products, with minimal side reactions. The energy efficiency for this unit will largely depend on how effectively the reaction effluent heat is recovered, and how well the side reactions can be minimized.

13.3.2.2 Understanding the Energy Needs

The following items are identified as major energy users in the reactor circuit and their distinct roles and significances will be discussed.

The feed heater is used to increase the feed temperature and control the reactor inlet temperature. Although the heater efficiency depends on how it is designed and operated, the heater duty is determined by feed preheating. For a given feed preheat, the heater duty is mainly a function of the heat of reaction and heat recovery. Xylene isomerization processes typically have a low exotherm (5–10 °C is typical). A process engineer can determine the ways to maintain process heat recovery, heater efficiency, and heat flux.

The compressor for recycle gas is a large power user. The role of a recycle gas compressor is to provide the required amount of hydrogen to the reaction, and to promote catalyst stability and selectivity in the presence of hydrogen. The recycle compression work depends on the gas flow and its molecular weight. The gas flow depends on the hydrogen purity in the recycle gas and pressure of the reactor section. The role of a process engineer is to optimize the recycle gas rate, purity, and hydrogen-to-hydrocarbon (H_2/HC) ratio. The optimization may indicate to operate the compressor at the lowest possible recycle gas rate, purity, and H_2/HC that will still provide good catalyst stability and selectivity.

13.3.2.3 Effective Measures for the Energy Needs

Based on the above understanding of the energy needs, we can go one step further to develop efficiency measures in providing these needs.

- Reactor activity: Operating at a higher reactor temperature will result in higher ethylbenzene conversion. As the catalyst deactivates, the temperature requirement to maintain ethylbenzene conversion will increase. The xylene isomerization function is relatively constant over the typical temperature window and life of the catalyst, though there is a minor shift in the equilibrium concentration as temperatures change. Operating at higher severity will result in shorter catalyst life, higher hydrogen consumption, and higher side reactions. Ethylbenzene conversion is the standard measure of severity for these units.
- Heater reliability: Heat flux (for large heaters) or TWT (for small heaters) is the key reliability parameter for a heater. When heater operation is higher than the limit of heat flux or TWT, a heater is under risk of reliability failure because the tube life is shortened at higher fluxes. On the other hand, operating a heater at much lower than the limit makes the heater underutilized, which represents an opportunity for increased feed rate or higher process severity. Beside these two operating limits, flame impingement is another reliability measure which is usually caused by too low O_2 content. The combustion flame becomes longer with too little O_2 and could reach the tube, posing a serious reliability risk.
- Heater efficiency: Efficiency can be affected by the excess O_2 content, or extra air for combustion and a high stack temperature. Inappropriate O_2 content could be caused by lack of control, air leaks, and poor burner performance. A high stack temperature corresponds to high heat loss in the flue gas. A high heater approach temperature, defined as the temperature difference between flue gas to the stack and heater feed inlet, could be caused by heater fouling in operation and by poor or outdated heater design. Xylene isomerization is generally considered a non-fouling service, as such heater fouling here is unlikely.
- Feed exchange: The feed is mainly heated by the reactor effluent in a combined feed exchanger before the charge heater. The hot-end approach temperature on combined feed exchanger is a good indication of heat recovery performance by the feed preheating system.
- Reactor product condenser inlet temperature: After the reactor effluent transfers its heat to the feed and recycle gas, it goes to an air condenser. Thus, the condenser inlet temperature on the reactor effluent side is a good indication of how effectively the reactor effluent heat has been recovered.
- Hydrogen-to-hydrocarbon ratio (H_2/HC): This ratio defines the amount of hydrogen in the recycle gas compared to the amount of hydrocarbon in the feed, on a molar basis. Too high of a H_2/HC ratio could cause greater power usage due to the higher recycle gas rate, while too low of a ratio could impact yield and shorten the catalyst life. The allowable H_2/HC for xylene isomerization reactors is getting progressively lower throughout the industry, as catalyst manufacturers are able to design more stable catalysts. This reduces the required compressor power requirement for newer units. The lower hydrogen requirement of new catalysts is particularly helpful for a unit when higher throughput is desired, and the compressor is thought to be a key equipment limitation. Though it is also important to consider the impact of operating at a lower H2/HC value on other related equipment, such as the combined feed exchanger.

13.3.2.4 Developing Key Indicators for the Energy Needs

Through the above exercise of simplifying the problem and developing understanding of major energy needs and measures of efficiency in providing the needs, we can define the following key indicators for the reaction section:

- Ethylbenzene conversion
- H_2/HC ratio
- CFE hot-end approach temperature
- Product condenser inlet temperature

Specific indicators for heaters could include:

- Heater O_2 content
- Heater stack temperature
- Heat flux
- Flame impingement

13.3.3 Developing Key Indicators for the Product Fractionation Section

13.3.3.1 Understand Process Characteristics

The deheptanizer is intended to remove toluene and lighter components from the bottoms product of the column. The bottoms product should be primarily C_8 aromatics and heavier, which will be further reprocessed in a column upstream of the *para*-xylene separation unit to remove any heavies which may have been generated. It is okay to let small amounts of toluene in the C_8A+ fraction, but most of it should be recovered overhead as it will increase the load on the *para*-xylene separation unit.

The overhead aromatics product, mostly benzene and toluene, is typically taken as a distillate product. This product should be mostly devoid of C_8 aromatics. This is important to maintain proper function of downstream units, in particular the aromatics extraction unit. For sake of storage tanks, the RVP of this stream needs to be controlled. As such, a stabilizer column is often used as well, though it is not always within the scope of the xylene isomerization unit itself.

It is essential to avoid flooding and/or slumping this column, which would severely impair its separation capability and thus energy efficiency. Fractionation efficiency can be monitored by the column internal vapor-to-liquid ratio (V/L). The ideal V/L can be achieved jointly by optimizing the reboiler heat input and overhead composition control scheme. This should be done to maintain good vapor–liquid contact. Operating at slightly below the flood point is a good practice, though this is not always straightforward to measure. An optimal reflux-to-feed ratio should be determined, typically the lowest possible that achieves the desired splits. This will generally reduce heat input per amount of feed processed.

Column pressure also has an impact on fractionation efficiency. While lower pressure can allow for better separation of the key components, reducing column pressure will bring the column closer to its flood point. Higher pressure will increase the column temperatures, reducing the LMTD in the reboiler, making heat exchange more difficult. There may also be considerations for the destination of the column off-gas that set a minimum allowable column pressure.

13.3.3.2 Understand the Energy Needs

Heat input is provided to the deheptanizer column to facilitate separation of benzene and toluene overhead, from the main bottoms product of C_8+ aromatics. The main role of the process engineer is to determine the appropriate heat input and R/F ratio that will allow for good rejection of C_8 aromatics from the overhead, and toluene from the bottoms product. The overhead product is typically controlled by a temperature controller tied to a column tray temperature. Increasing or decreasing this temperature will increase or decrease the amount of xylenes allowed into the distillate product, respectively. Higher heat input will generally result in a better separation, but the separation typically does not need to be perfect. Further, very high input can result in column flooding.

If an additional stabilizer exists, its purpose is to remove light non-aromatics from the light aromatics product stream from the deheptanizer overhead. This is frequently required to suit vapor pressure requirements for downstream destinations of this stream, or to reduce the likelihood of foaming in a downstream extraction unit.

13.3.3.3 Effective Measure for the Energy Needs

Fractionation efficiency can be determined by the minimum R/F ratio that allows for the target values of the key component splits in the deheptanizer distillate and bottoms liquid product streams.

Heat input for each column should be measured as heat transferred in the reboilers for the deheptanizer and stabilizer (if applicable) columns. It is easier to strictly measure the flow rate of the reboiler heating medium, for instance, a condensate return flow rate, but this is not always a suitable proxy for heat input. The conditions of the heating medium (i.e. temperature and pressure) may be different, changing the heat available for exchange. For this reason, the actual heat input should be measured based on inlet and outlet conditions of the hot and cold sides of the reboiler exchangers.

13.3.3.4 Developing Key Indicators for the Energy Needs in the Main Fractionation System

Based on the understanding of major energy needs and measures of efficiency in providing the needs, we can define the following key indicators for the fractionation section:

- Fractionator column reflux ratio(s)
- Reboiler heat input(s)
- Column pressure(s)

13.3.4 Remarks for the Key Indicators Developed

For a typical xylene isomerization unit, there could be several thousands of data points measured and collected. The key indicators listed above only account for a very small fraction of the overall data available, but capture the key performance parameters which ultimately reflect directly on operating costs. If these indicators can be monitored and optimized, the unit can operate closer to optimal performance.

In many cases, parameters related to feed and product yields are measured and controlled using basic control systems or advanced process control (APC) systems. However, process indicators are typically not integrated with energy use. Furthermore, many energy parameters are not even measured, much less recorded. By identifying the key process and energy indicators and optimizing them together, the optimization does not only reduce energy cost but may also allow for an increase in throughput, improve product quality, and minimize product specification giveaway.

13.4 Set Up Targets for Key Indicators

To improve from current performance, targets must be established for the key indicators. These targets provide standards against which existing facilities are measured and equipment improvements are evaluated. The difference between a target and the current performance for each key indicator defines the performance gap. Each performance gap should be associated with dollar value which represents opportunity to be captured. Each indicator is correlated to a number of parameters, including process and equipment conditions together with equipment limits. In this way, energy optimization is connected with process conditions and constraints.

How would one make the concept of KEIs work for a process unit or the complex as a whole? Consider the example of the deheptanizer for the xylene isomerization unit previously discussed.

13.4.1 Problem

The deheptanizer column sketch is shown in Figure 13.2. The reboiler at the bottom of the tower is to provide sufficient vapor flow on the trays for separating toluene and lighter components from the C_8 aromatic and heavier components in the feed. Most of the toluene and lighter boiling components will be withdrawn at the overhead, while the C_8 aromatics and heavier components leave at the bottom. A certain amount of toluene in the bottoms product is allowed as a component of energy optimization for the complex. Too much toluene will limit the *para*-xylene separation unit, but trying to remove too much will require more heat input to the deheptanizer than it is worth. Reboiling duty is the main variable in controlling the toluene amount in the bottoms product. In other words, the reboiling duty must increase when toluene amount in the bottom exceeds the target. The operating objective is to minimize the reboiling duty while achieving the maximum amount of toluene that the downstream units can process. This typically optimizes to less than 1 wt.% in the bottoms product.

13.4.2 Rationale

The task at hand is to develop a relationship between reboiling duty and toluene in the deheptanizer bottoms product so that the toluene content in weight percent can be controlled by adjusting the reboiling duty. However, other operating parameters also affect the reboiling duty, which include column feed conditions, feed enthalpy, and tower conditions. If a correlation of reboiling duty against the above influencing parameters could be generated, reboiling duty can be adjusted according to any of the changes in the related parameters and thus avoid the need of trial and error.

13.4.3 Solution

There are several ways to develop such a correlation. The simplest way is by use of a data historian. This method can be applied if three conditions are met:

1) the related parameters are measured and data available in the historian
2) the measured data reflects the operation at the time the toluene content was measured
3) the historian data covers all possible operating scenarios. After all, online data is the true representation of real simulation!

Development of a correlation using the historian data can be conducted readily in a spreadsheet using regression techniques. After gathering the data from the historian, multiple-variable regression can be applied to develop such a correlation. The overall correlation coefficient must be higher than 85% for sufficient regression fidelity.

The second option is to use the step-test method usually used for developing a parametric relation for control systems. By making a small step change to the independent variable of interest, a response from the dependent control variable (in this case, reboiling duty) can be recorded after reaching the steady-state condition. This response can be called an energy response. Finally, the regression method is applied to derive the correlation of reboiling duty against all related variables.

In many cases, the conditions above for using the data historian are difficult to satisfy. Also, it could be labor intensive and inconvenient in operation to adopt the step-test method. Thus, the most common method is to use the simulation method for developing

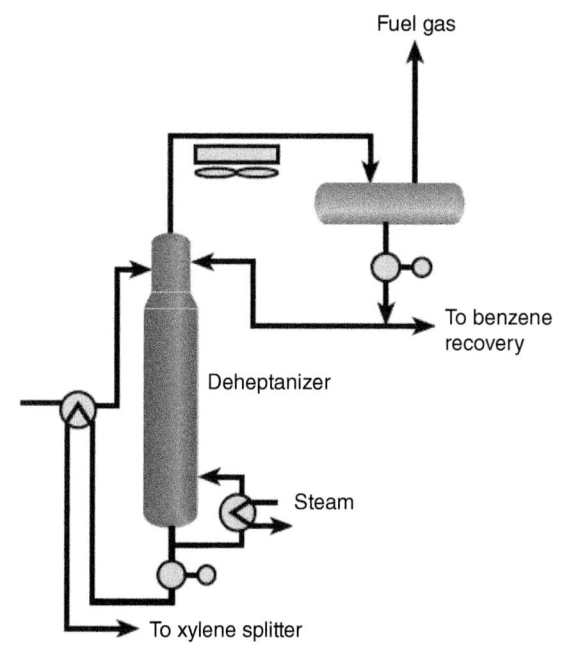

Figure 13.2 Deheptanizer column in a xylene isomerization unit.

relationship correlations. To do this, a simulation model for the tower can be developed readily based on the feed (rate and compositions) and tower conditions (temperature, pressure, and theoretical trays), with product specifications (toluene weight percent in the bottom and C_8 aromatic weight percent in the overhead) established as set points in the simulation. Operating parameters such as reflux rate and reboiling duty can be adjusted to meet the product specifications. The simulation model is verified and revised against high-quality performance test data.

For evaluating the effect of individual parameters, simulation cases can be developed by prespecifying the values for independent variables of interest, and the energy response (reboiling duty) will be recorded automatically. For example, to evaluate the effect of feed preheating, the UA value of the feed preheating exchanger is varied with prespecified values. During simulation runs, the feed temperature before the tower will change according to the UA values, which will cause the reboiling duty to vary automatically in the simulation. A set of four curves for such one-to-one relationships can be obtained as shown in Figure 13.3a–d. Brief explanations are given below for each figure.

Reducing the column overhead pressure will reduce reboiling duty with the trend as shown in Figure 13.3a. Toluene in the bottom is the specification which the ultimate C_8 aromatic feed to the *para*-xylene separation unit should meet. But, too low toluene content is not necessary; there are diminishing returns for sake of the *para*-xylene separation unit performance compared to the cost of extra reboiling duty due to the steeper part of the curve, shown in Figure 13.3b. C_8 aromatics in the overhead product, shown in Figure 13.3c, is the indication of C_8 aromatics lost to the aromatics extraction unit, which should be avoided. This loss may affect the operation of the extraction unit, and also results in unnecessary recycle of C_8 aromatics through the complex. Adding feed preheat also reduces the reboiling duty, but raises condensing duty (Figure 13.3d).

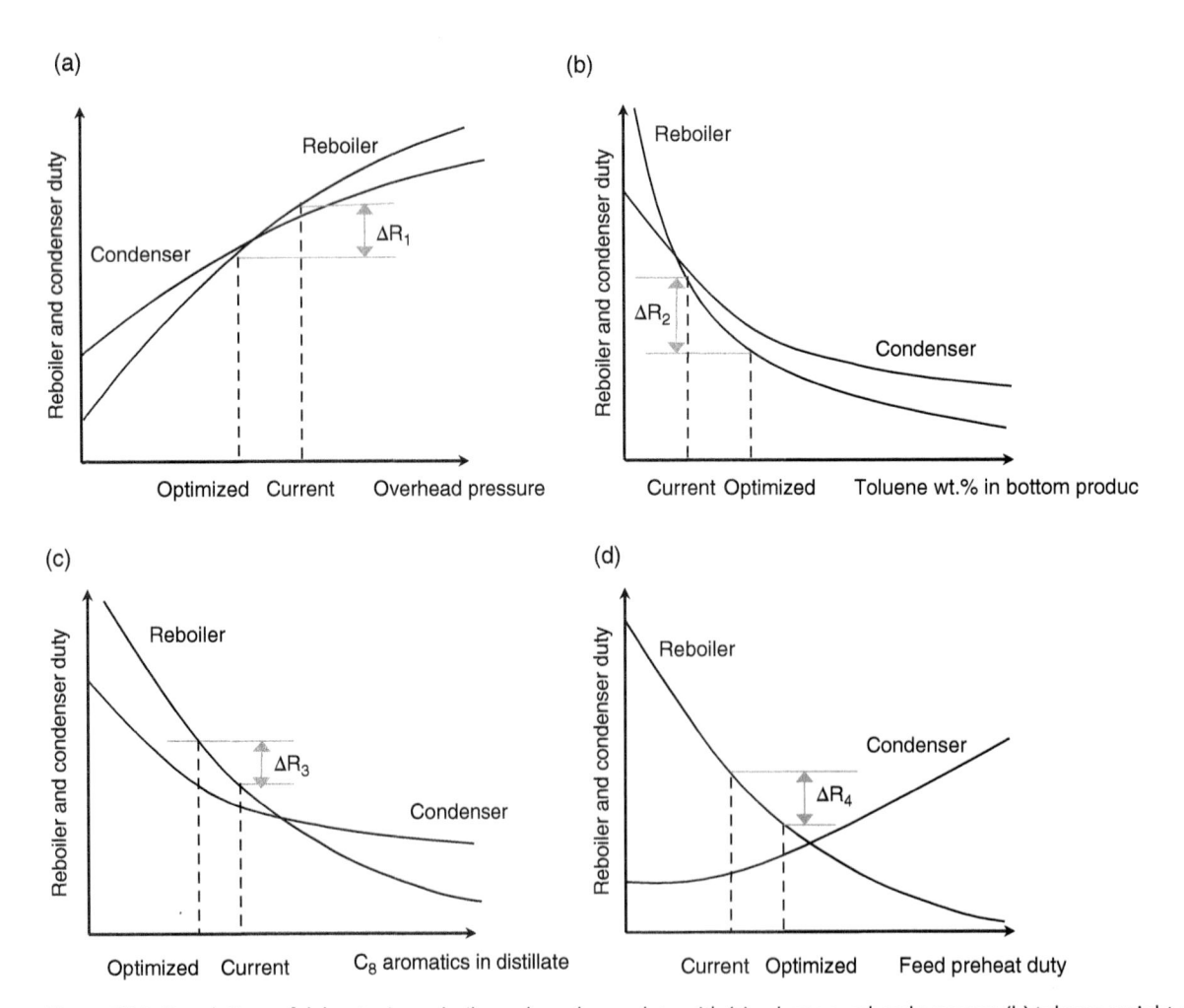

Figure 13.3 Correlations of deheptanizer reboiler and condenser duty with (a) column overhead pressure, (b) toluene weight percent in the bottom product (c) C_8 aromatics in the distillate, and (d) feed preheat duty.

One must be aware of the capacity limit of the existing condenser when increasing feed preheat, as this will require extra condensing. On the other hand, when the feed toluene concentration changes, the reboiling duty will need to be adjusted accordingly. There is no control for the feed quality for the deheptanizer operation because the feed quality is the consequence of side reactions in the xylene isomerization unit. For the other four parameters above, operators can make changes to the column overhead pressure, toluene content in the bottoms product, C_8 aromatics in the overhead product, and feed preheating. Optimizing these parameters could give around 5% reduction in reboiling duty, compared to operating based on experience and meeting target key component splits. This can be very significant, even more so if this methodology is applied to every column in a complex.

Reboiling and condensing duty can be described based on the relationship with individual parameters as shown in Figure 13.3a–d. If assuming a polynomial form of correlations with order three, the equation becomes:

$$R_i = a + bx_i + cx_i^2 + dx_i^3$$
$$x_i = (\text{Toluene\% in bottom, } C_8A\% \text{ in ovhd,}$$
$$\text{preheat, ovhd pressure}) \quad (13.1)$$

where

x_i is the value of the particular operating parameters
R_i is the reboiling duty calculated for a given x_i value

The incremental effect (ΔR_i) from each individual parameter (Δx_i) can be determined as

$$\Delta R_i = R_{i,new} - R_{i,\text{base}} = a + b(x_{i,new} - x_{i,\text{base}})$$
$$+ c(x_{i,new} - x_{i,\text{base}})^2 + d(x_{i,new} - x_{i,\text{base}})^3 \quad (13.2)$$

If there are no interactions among these four operating parameters, the total effect of changes in these parameters on the required reboiling duty would be the simple summation of the individual effects:

$$\Delta R = R_{new} - R_{\text{base}} = \sum_{i=1}^{4} \Delta R_i \quad (13.3)$$

However, in many cases, there could be strong interactions among operating parameters. If this were the case, two or more parameters could appear together in one term and the bilinear ($x_1 \cdot x_2$) is the simplest form of interaction. To develop a relationship of parameters with interactions, several parameters need to vary at the same time in the plant test or simulation, and the effect on reboiling duty can be seen as the statistically significant result of the interaction parameter. A set of data with changes to the operating parameters and the energy response can be obtained, and regression can be applied to derive a correlation that accounts for these interactions.

When dealing with correlations involving multiple variables, economic sensitivity analysis is essential to determine the most influential parameters. For example, the reboiling duty requirement is very sensitive to feed preheat and column overhead pressure, more so than other operating parameters for the deheptanizer. Getting the most sensitive parameters right in operation can get the greatest economic and technical response.

The correlation developed can be implemented into the control system so that reboiling duty can be controlled automatically to achieve the minimum energy requirement at all times. Alternatively, the correlation can be used as a supervisory tool. Whenever a variation is expected, adjustments to operating parameters should be made to optimize the reboiling duty. This reboiling duty is the minimum with major independent variables considered, and is the target for the conditions at hand. This target and operating cost value for closing the gap must be communicated with board operators in each shift so that actions will be taken for achieving the targets, and the actual cost savings can give operators a sense of pride in their direct contribution.

13.5 Economic Evaluation for Key Indicators

Operation variability is a major cause of operation inefficiency. In general, there are two kinds of variability which can be observed in reality: inconsistent operation, and consistent but nonoptimal operation. Figure 13.4 represents the operating data of a reboiler heat input as measured by condensate rate in the deheptanizer in a xylene isomerization unit. In Figure 13.4a, the heat input data points appear to be randomly scattered showing an example of inconsistent operation. This is usually caused by either poor control strategy or different operating philosophies used by operators for running the column. In contrast, Figure 13.4b shows a consistent operation, but one which is nonoptimal. In this case, a consistent operating strategy was adopted, but it was far away from the target for adjusting reboiling duty against column feed rate. The target operation represents the minimum reboiling duty required to achieve product specification.

The variability of any operating parameter occurs due to various reasons. The question is how to identify variability in operation and the economic value of minimizing this variability.

Variability assessment starts with simple statistical analysis of operating data. For example, the operating data for C_8 aromatics in the deheptanizer column overhead product under normal conditions can be extracted from the historian as shown in Figure 13.5a, with the specification limit provided. To understand

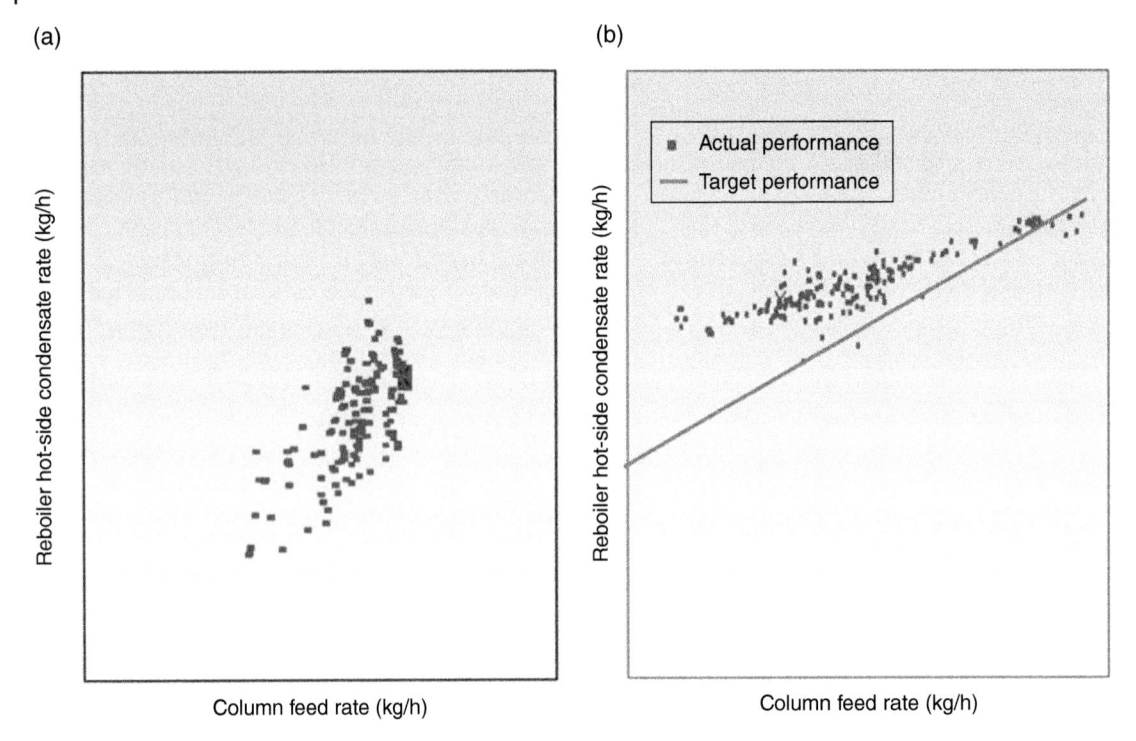

Figure 13.4 Two common operating patterns: (a) inconsistent operation; (b) consistent operation but nonoptimal.

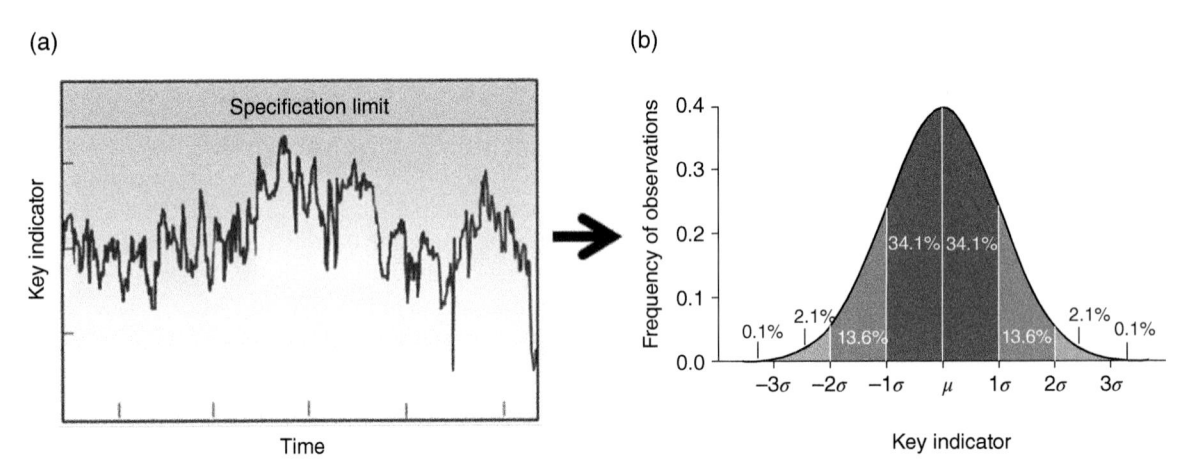

Figure 13.5 Operating data: (a) historian; (b) frequency distribution.

the variability, data in Figure 13.5a are converted to a normal distribution curve, which represents frequency of observations as shown in Figure 13.5b. Frequently, the operating data will exhibit a normal distribution.

Two parameters describe the normal distribution, namely the mean or average (μ), and the variance or variability (σ). σ defines the shape of the normal distribution. The larger the σ value, the broader the curve peak. Conversely, the smaller the σ value, the thinner the curve peak. μ and σ can be calculated by:

$$\mu = \frac{\Sigma x}{N} \tag{13.4}$$

$$\sigma = \frac{\Sigma(x-\mu)^2}{N} \tag{13.5}$$

where

x is the value of the key indicator obtained from the historian
N is the number of sample data points for the key indicator

Refer to the example discussed in White (2012) as shown in Figure 13.5. There are two shortcomings in performance shown in Figure 13.6a. The first is the large variability, and the second is too much conservatism from reaching the specification limit. If the operation or control strategy improves, the variability could be

Figure 13.6 Operation performance: (a) poor control in the current operation; (b) better control in reduced variability; (c) increased profit by changing target. *Source:* White (2012), reproduced with permission by AIChE.

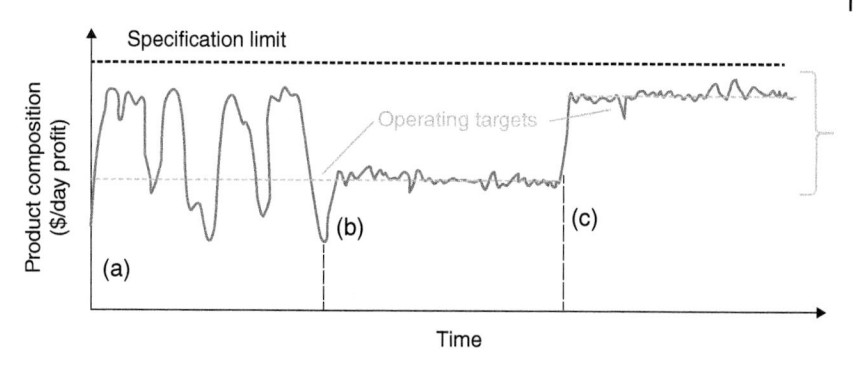

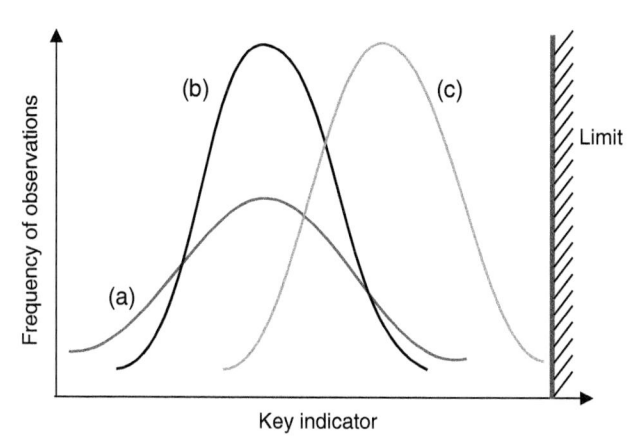

Figure 13.7 Convert time series data in Figure 13.6 into normal distribution curves: (a) current; (b) reduced variability; (c) increased profit.

minimized to achieve more consistent operation (Figure 13.6b) but performance still remains far away from the specification limit. The limit is usually set by a physical limit such as product purity specification, maximum temperature or pressure, maximum valve opening, maximum vapor loading in a separation column, maximum space velocity in a reactor, and so on. The operation can be improved further (Figure 13.6c) by moving the average closer to the limit by adopting a better control strategy. Time series data in Figure 13.6 can be converted to normal distribution curves as shown in Figure 13.7.

The economic value, V_i, for a given frequency that a particular key indicator is observed is given by

$$V_i = C_i \cdot x_i \cdot f_i \quad (13.6)$$

where

C_i is the economic value for the key indicator
x_i is the numerical value of the key indicator of interest
f_i is the frequency of observations for the key indicator

For example, C_8 aromatic content in the deheptanizer overhead product represents the high-value component C_8 aromatics lost in the deheptanizer overhead, representing an additional cost for reprocessing. The key indicator could be defined as the difference of actual C_8 aromatics in the deheptanizer overhead and the C_8 aromatics operating target. As an example, if the distillate rate produced from the column is 1000 MT per day with C_8 aromatics at 10% higher than the specification, or $x_i = 10\%$ with frequency of occurrence as $f_i = 30\%$, and if the cost of reprocessing these C_8 aromatics is $10/MT, the economic value to avoid this occurrence is

$$\begin{aligned} V_i &= C_i \cdot x_i \cdot f_i \\ &= \$10 / \text{MT} \times 10\% \times 1000 \text{MT} / \text{day} \times 30\% \\ &= \$300 / \text{day} \end{aligned} \quad (13.7)$$

Similarly, for other occurrences, C_8 aromatics in the distillate could be lower or higher than 10%, and the economic values can be calculated accordingly.

In this way, the normal distribution curve can be converted to an economical curve as shown in Figure 13.8. The conversion is calculated to an economic difference.

With the statistic-based economic evaluation method mentioned above, improved operation can be quantified with economic values based on statistical distribution of operating data. The current operation with large variance, such as the case shown in Figure 13.9a, is improved by more consistent operation and/or control strategy to reduce variability as shown in Figure 13.9b, while optimized operation shown in Figure 13.9c utilizes the potential capability available in the process and equipment and pushes the economic value even higher.

13.6 Application 1: Implementing Key Indicators into an "Energy Dashboard"

The concept of KEIs and targets can be readily implemented into an energy dashboard, which can quickly show the performance gaps between current performance and targets on a computer screen. The magnitude of a gap indicates the severity of deviations and forms the basis to assign a "traffic light" or similar indication for each KEI – i.e. a green light indicates the current

Figure 13.8 Converting the normal distribution curve to an economic curve.

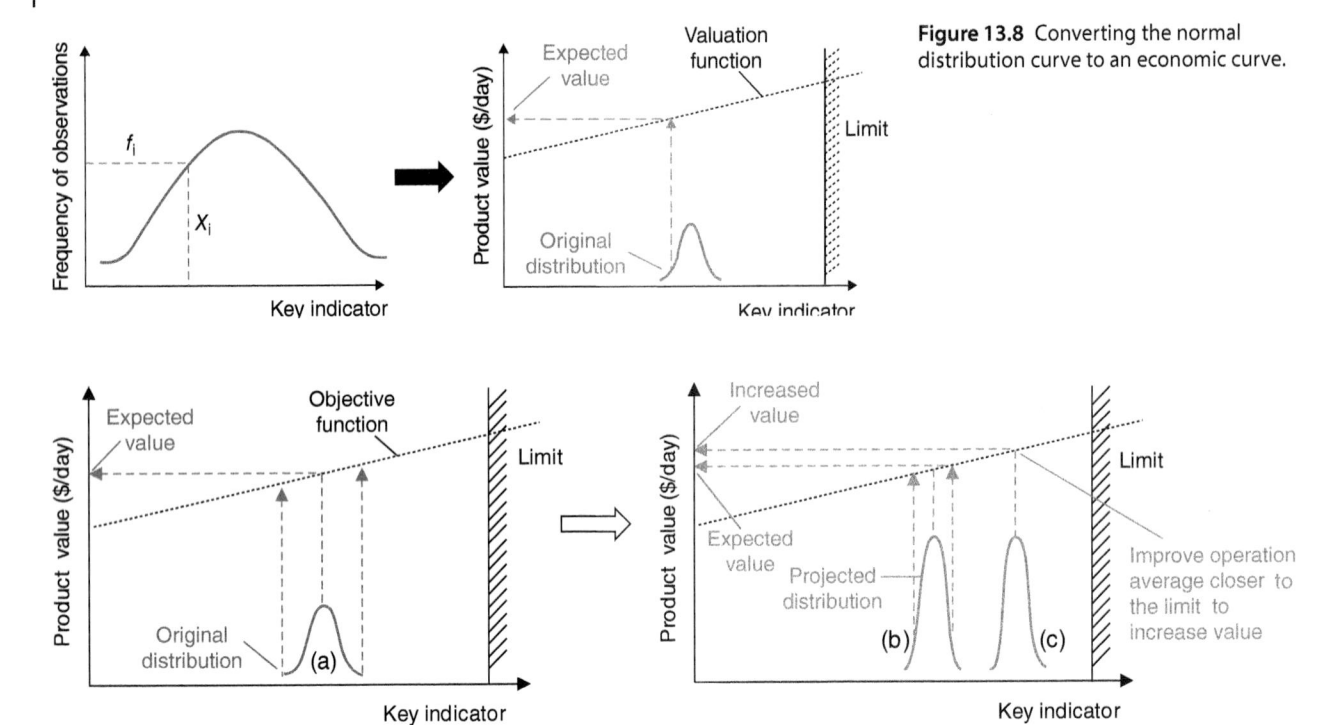

Figure 13.9 Economic curves generated based on normal distributions. Distribution (a) shows the original distribution with a relatively large variability in the key indicator resulting in lower product value per day. Distribution (b) shows an improved, tighter distribution with reduced variability of the key indicator, allowing for optimization to distribution (c) where the key indicator has the similar low variability but has been optimized to provide higher product value per day. *Source:* White (2012), reprint with permission by AIChE.

performance is acceptable as it is within the target range; a yellow light is a warning sign indicating that a gap occurs and requires attentions; or a red light is an alarm urging for taking actions at the earliest time possible. An example tool of monitoring key indicators is Honeywell's Energy Dashboard (Sheehan and Zhu 2009), and has been further improved upon in Honeywell's Connected Plant applications. Tools such as these could be tremendously valuable to operators and engineers, as they indicate the most important variables to watch, helping to suggest which parameters should be adjusted and when.

KEIs applied to a system can be defined in a hierarchical structure, from the overall complex, to each process unit, down to major equipment, and finally to individual operating parameters. The sum of all incentives (the opportunity gap between current performance and optimal targets) from all pieces of equipment represents the total opportunity for the entire process and the overall site. This hierarchical structure allows engineers to drill down from overall performance to specific parameters and thus identify specific actions. Listed below is an example of what can be visualized in the Energy Dashboard.

- **Overall Site View** shows the site-wide energy consumption and emissions versus overall targets. On the same screen, the overall site view shows the relative amount of energy consumption and emissions from each process unit. A traffic light color is assigned to indicate which processes are furthest away from the targets.
- **Process Unit View** indicates the process performance which can be measured by 10–20 KEIs. These KEIs are developed from a combination of design, process simulation, and historical data. These predicated energy targets are automatically adjusted to reflect current operating conditions such as feed rate and compositions, operating mode, product yields, quality, etc. Color coding is assigned to each KEI, which could suggest the need to drill down in the next level of key indicators to identify root causes and logical actions.
- **Equipment View** shows equipment performance via several key operating parameters with indications of current values versus their corresponding targets. The operators may decide to perform more detailed investigations for the root causes if the gap between these values is large.
- **Deviation Trends View** allows operators to review the time periods during which the KEIs deviated significantly from the targets, and to determine the major causes of the deviation. By building up a history of causes, operators are able to look back over time and see the most common causes of deviations. This can

lead to recommendations about remedial actions for improving equipment performance, and hence, overall process performance.

For each key indicator, a target is established as the basis to compare with current performance. The difference between the target and the current performance for each key indicator defines the performance gap. Different gap levels indicate the severity and level of urgency for actions.

Gap analysis is then used to identify root causes. Potential causes may include inefficient process operation, insufficient maintenance, inadequate operating procedures, poor operating practices, poor process control, inefficient energy system design/heat integration, and outdated technology. Gap analyses are translated into specific corrective actions to achieve targets via either manual adjustments or by automatic control systems. Finally, the results are tracked so the improvements and benefits achieved can be quantified.

13.7 Application 2: Implementing Key Indicators to Controllers

Many opportunities for energy improvement can be achieved directly by adjusting the control set points of key variables. In some cases, these opportunities for energy improvement may be automated by incorporating these key variables into an APC system, if the investment for such a system can be justified by the value to be captured.

Multivariable, predictive control, and optimization applications have been commonly applied in the process industry. The ability to take models derived from process data and simulations, and configure the models in a highly flexible manner, allows the engineers to design controllers that can be suitable for multiple purposes. The same controller can be used to maximize throughput, maximize yields, and/or minimize energy consumption just by changing the cost factors in the objective function. This APC environment is suitable for incorporating energy strategies into overall operating objectives. In fact, adding energy operating costs into an existing objective function and inserting related KEIs with corresponding correlations and operating limits is appropriate. In this manner, minimizing energy cost will not be accomplished at the expense of the most valuable product yields.

There are many energy-saving opportunities that can be incorporated into APC applications, such as:

- Furnace pass balancing and excess O_2 control
- Reaction conversion control
- Reaction yield maximization
- Column feed preheating maximization
- Separation column reboiler duty control
- Recycle stream minimization

13.8 It Is Worth the Effort

As demonstrated above, the concept of key indicators and targets can play an important role for process and energy integrated optimization. Process optimization without taking energy use into account will lead to high energy costs, while energy optimization without fully addressing process requirements will cause penalties in processing capacity, product quality, and yields. With an appropriate work process fitting into the existing technical management system, the concept of key indicators and targets can become the cornerstone of energy management.

Due to this significance, developing key indicators and targets should become a corporate-backed effort, as support from management is critical. First of all, significant effort is required in developing technical targets for key indicators. Setting up targets requires modeling of major equipment. Once a base case of operation is defined and simulated, operation variations can be simulated and correlations can be developed for key indicators in relation with other operating parameters. Using the deheptanizer column as an example, the reboiling duty requirement is the key indicator which can be affected by feed rate and composition, overhead pressure, feed preheat, and desired key component splits. Regression analysis of simulation results may be required to develop the correlation for the reboiling duty and these operating parameters. Whenever operating conditions change, the correlation could be applied to determine the minimum reboiling duty under the new conditions.

The engineering effort required to develop a system of simulations for setting up technical targets for key indicators at different operating conditions is appreciable. It may take 1–2 man-years to develop a system of key indicators and targets for processes and major equipment. In one example of a large refinery complex, two man-years were required to develop such a target-based system. Implementing this target system in day-to-day operation is essential as is integrating the target system into existing technical management system. Confidence in the system will grow once people can actually observe the operating cost savings made from applying such a system. The benefits will pay for the investment alongside the development. The lessons learned and experience gained will then likely be disseminated to other process units.

References

Sheehan, B. and Zhu, X.X. (2009). The first step in energy optimization. *Hydrocarbon Engineering Journal*: 25–29.

White, D.C. (2012). Optimize energy use in distillation. *Chemical Engineering Progress* 108 (3): 35–41.

Zhu, X.X. (2017). Key indicators and targets. In: *Hydroprocessing for Clean Energy: Design, Operation, and Optimization* (ed. X.X. Zhu), 447–467. Wiley.

Zhu, X.X. and Martindale, D. (2007). Energy optimization to enhance both new projects and return on existing assets. *Proceedings of the 105th NPRA (National Petroleum Refining Association) Annual Meeting, Texas, USA* (March).

14

Distillation System Optimization

14.1 Introduction

Distillation is the core of a process unit and also of the major energy users. Design and operation of distillation and separation columns involve trade-off between energy use and product recovery. When energy usage is less than design, product recovery and quality may suffer. On the other hand, when energy is more than design, product quality is better than the specification, which is called product spec giveaway. In abnormal operations, little product recovery can be achieved regardless of how much energy is used.

However, reducing energy usage in a distillation system is not straightforward. This is because a distillation system involves many operating parameters including those within and outside the process battery limit. In particular, variations in conditions of feed and products as well as prices of feeds and products add much complexity to the economic operation of the process. This feature leads to strong dynamic behaviors of operating parameters. Furthermore, most of the parameters interact in a nonlinear manner and have numerous constraints on their operation, which further complicate the task of energy optimization. If some of the constraints can be relaxed, this could improve operating margin significantly.

The performance assessment was discussed in Chapter 6 while this chapter focuses on economic operation within the feasible operation region.

14.2 Tower Optimization Basics

Tower optimization is a difficult task as product pricing and unit constraints often change daily or weekly, but changing unit operating philosophy and addressing hardware constraints can take months to accomplish. Even after the steps to improving optimal performance have been identified and implemented, if the desire to improve is removed, operation tends to return to the older, more comfortable routine, or constraints in other areas often prevent operation in the most profitable mode. Thus, it is highly recommended that for a complex system, performance optimization should be implemented in advanced process control (APC) which can improve tower operation to the most economic mode on a regular and consistent basis and in an automatic manner.

To establish tower optimization, key operating parameters must be defined and correlations must be developed to understand the relationships between key parameters. Finally, optimization objective function must be developed to determine the optimal set points for the key parameters. The optimization can be conducted in two ways; one is semi-manual based while the other is APC based. With the semi-manual-based approach, operating parameters are manually adjusted while optimization is done in an off-line (online) manner. In contrast, with the APC approach, operating parameters are automatically adjusted while optimization is done online. But, both the methods adopt the common ground of optimization: using an objective function to derive the optimal set points for key parameters based on economic trade-offs and correlations to represent relationships between key parameters and constraints to define process and equipment limits. Noticeably, the optimization pushes operating limits in obtaining the optimal solution and relaxation of sensitive constraints or limits could generate significant benefits.

14.2.1 What to Watch: Key Operating Parameters

As discussed extensively in Chapter 13, it is important to define major operating parameters or key indicators as they can describe the process and energy performance. A key indicator can be simply an operation parameter like desirable product rate, column overhead reflux ratio, column overflash, column temperature and pressure, etc. By the name of key indicator, the parameter identified is important and has a significant effect on process and energy performance.

Although primary operating parameters affecting both fractionation and energy use are tower specific, common

Efficient Petrochemical Processes: Technology, Design and Operation, First Edition.
Frank (Xin X.) Zhu, James A. Johnson, David W. Ablin, and Gregory A. Ernst.
© 2020 John Wiley & Sons, Inc. Published 2020 by John Wiley & Sons, Inc.

operating parameters can be identified which are discussed below.

14.2.1.1 Reflux Ratio

Reflux ratio is defined as the ratio of reflux rate to distillate rate (R/D) or the reflux rate to feed rate (R/F). In essence, a reflux rate is to set a tower top temperature required for making the distillate (overhead product) to meet specification. Reflux is generated by energy either via tower reboiler or feed heater. Lower reflux rate saves energy but too low reflux rate could affect product quality. On the other hand, too high reflux rate could be wasting of energy if product quality is better than the specifications already. In this case, the quality that is better than the specification is given away for free because there is no credit in pricing for the extra better quality.

Optimal reflux rate in operation depends on the operating margin which is defined as the difference of product sales minus feed cost and energy cost. When energy cost is too high, it could drive the operation toward lower reflux rate and vice versa for the case of lower energy cost.

In tower design, the reflux ratio is determined based on the trade-off between operating cost in reboiler and capital cost for the tower. In other words, use of more separation stages requires less reflux rate and in turn less reboiling energy but at the expense of additional capital cost. The minimum reflux ratio is calculated based on Underwood (1948). A tower requires an infinite number of stages to achieve the minimum reflux ratio. To make tower feasible in operation and affordable in cost, a reflux ratio larger than the minimum is used. Typical reflux ratio is 1.1–1.3 of the minimum reflux ratio. With a high reflux ratio, the number of theoretical stages is lower resulting in lower capital cost for a tower but at the expense of higher reboiler duty and vice versa. The optimal reflux ratio in tower design is determined based on the minimum total cost.

14.2.1.2 Overflash

Overflash is defined as the ratio of internal reflux at the feed vaporization zone and the feed rate. By definition, overflash represents the percentage of feed vaporized more than the amount of products drawn from above the feed tray. Overflash is a function of reflux rate, feed temperature, and tower pressure.

Overflash is generated from the overhead reflux rate. Thus, it can be said that overflash is generated by a reboiler or feed heater. Overflash is an indication of reflux rate sufficiency for proper separation throughout the tower. A small overflash implies less reboiling duty and thus saves energy, but it could negatively affect the fractionation efficiency and hence product quality and vice versa. Therefore, overflash connects fractionation efficiency and energy efficiency for a tower. A tower could be making poor product quality even with high overflash when the tower is operated under abnormal operation such as flooding or dumping.

Overflash is typically controlled between 2 and 3%. An operation policy focusing on throughput would operate a tower at very low reflux rate; it is not uncommon to observe that a tower is operated at close to 1% overflash. This low reflux operation could be beneficial if the tower produces intermediate products which will be processed further via downstream reaction and separation processes. In this case, this operation could lead to energy efficiency as well as high economic margin.

14.2.1.3 Pressure

Lower pressure typically saves energy. This is because the lower the tower pressure the less heat required for liquid to vaporize and thus less energy required. This results in better fractionation as it is easier for vapor to penetrate into liquid on the tray deck.

The condenser pressure controls the tower pressure and thus the feed tray pressure. There is a pressure valve in the overhead which can be used to control the tower pressure. The lower limit of the tower pressure is defined by the column overhead condensing duty, net gas compressor capacity, and column flood condition. During extended turndown periods, reducing pressure up against an equipment limit can improve efficiency. Many of the new APC systems use pressure control to save energy.

Heat exchanger fouling in overhead condensers could cause higher pressure drop or lower heat transfer and thus result in high tower pressure. On the other hand, higher reflux rate could lead to high pressure drop in the overhead loop causing high tower pressure.

14.2.1.4 Feed Temperature

A hotter feed can increase feed vaporization and thus reduce reboiling duty. However, higher temperature feed could cause too much vapor resulting in rectification section flooding. For a given tower, the optimal feed temperature corresponds to the lowest reboiling duty while the tower can meet product specifications.

14.2.1.5 Stripping Steam

Some towers may have stripping steam in the feed zone. Stripping steam reduces flash zone pressure and provides partial heat to the feed and thus helps to increase the lift of light components from the bottom product. Stripping steam for a fractionation tower is controlled based on the lift while stripping steam for a stripper is controlled based on stripper product specification. Be aware of too much stripping steam as it could lead to high energy cost and also cause vapor loading limitations in the overhead system.

14.2.1.6 Pump Around

Many fractionation towers have pump arounds to remove excess heat in the key sections of the tower. The effect of increasing pump around rate is reduced internal reflux rate in the trays above the pump around but increased internal reflux rate below the pump around. Thus, change in pump around duty affects fractionation. On the other hand, pump around rates and return temperature have effects on heat recovery via the heat exchanger network. It is not straightforward in optimizing pump around duties and temperatures since the effects on both fractionation and heat recovery can only be assessed in a simulation model. An APC application incorporated with process simulation should be able to handle this optimization.

14.2.1.7 Overhead Temperature

In hot weather, tower overhead fin fan condenser could be limited and thus the tower top temperature can go up. As the result, valuable components could be vaporized into overhead vapor leading to yield loss. There are a number of ways to reduce the overhead temperature such as increasing cooling water rate, turning on spare overhead fan for air cooler, and increasing reflux rate. On the other hand, when overhead temperature is too cold, salt condensation in the condenser could occur and cause corrosion.

14.2.2 What Effects to Know: Parameter Relationship

How do operating parameters relate to each other? Which parameters are more sensitive to fractionation and energy use? What is the impact of changing one parameter to another? Understanding these could provide insights and guidelines for operational improvements. The objective of developing key indicators is to understand the strong interactions between process throughput, yields, and energy use so that the trade-off among them can be optimized with the objective of maximizing operating margin. In the traditional view, energy use is regarded as a supporting role. Any amount of energy use requested from processes is supposed to be satisfied without question and challenging. This philosophy loses sight of synergetic opportunities available for optimizing energy use for more throughput and better yields.

In developing correlations, one needs to connect energy with product yields and quality. One such example is discussed in detail in Chapter 13. The correlations can be applied for operation optimization. For automatic control, the correlations can be implemented into an APC system, which determines the set points for primary or independent operating parameters. For manual control, operating targets for primary parameters can be obtained based on the correlations.

A process simulation could be a very good vehicle in developing correlations of primary parameters. To do this, a simulation model for the tower can be developed readily based on the feed conditions (rate and compositions) and tower conditions (temperature, pressure, and theoretical trays) with product specifications established as set points in simulation. Operating parameters such as reflux rate and reboiling duty can be adjusted to meet product specifications. The simulation model is verified and revised against performance test data based on clean conditions. Different operating cases can be generated in simulation and simulation results can be transferred to a spreadsheet with relationship between dependent and independent variables. Then, the regression method is applied to derive the correlations.

When dealing with correlations involving multiple variables, an economic sensitivity analysis is essential to determine the most influential parameters on process economics. For example, in a debutanizer, feed preheat and reflux drum pressure are very sensitive to reboiling duty more than any other operating parameters. Getting the most sensitive parameters right in operation can get the greatest bang.

14.2.3 What to Change: Parameter Optimization

A tower is built to make separation of products. Therefore, tower optimization is to maximize operating margin and minimize energy usage. This processing goal can be described mathematically in an objective function with the parameters in the objective function connected to other processing parameters. All these parameters are defined as constraints in two forms: inequality equations (larger and smaller than) which are used for describing operating minimum and maximum operating limits. Therefore, these constraints form a feasible operating region in which the objective function is constrained within during optimization. The objective function plus the set of constraints form an optimization model. The results of solving this model yield the values for a set of operating parameters which can be adjusted in operation to achieve the maximal operating margin defined in the objective function.

A generic form of a process optimization model is provided as

$$\text{Objective function: Maximize } Z = \sum_i^m c_i P_i - \sum_j^n c_j P_j - \sum_z^p c_z Q_z$$

$$\text{Subject to: } f_k(X_c, X_m, P, F, Q) = 0; \quad k = 1, 2, \ldots, q$$

$$X_{c,\min} \leq f_l(X_c) \leq X_{c,\max}; \quad l = 1, 2, \ldots, u$$

$$X_{m,\min} \leq f_p(X_m) \leq X_{m,\max}; \quad p = 1, 2, \ldots, w$$

where

c_i are the unit prices of products
c_j are the unit prices of feeds
c_z are the unit costs of energy including steam, fuel, and power
F's and P's are the mass flows of feed and products
Q's are the amount of energy
Essentially, the objective function Z represents the upgraded value from feed to products at the expense of energy in terms of fuel, steam, and power
X's are the operating parameters with X_c's the independent or control variables and X_m's the dependent or manipulated variables for the tower
$f(X_c, X_m, P, F, Q)$ are the relationship constraints.

Maximizing the objective function Z under these correlations with proper limits of the operating parameters will change the related parameters to the values that economic value Z will be maximal. In this case, the operating parameters achieve optimal values which can be used either as set points for the close-loop or open-loop control.

In building this optimization model, the most important thing is to include all the major operating parameters which affect operating margin. Then, correlations are developed to describe the relationship among these major parameters and between these major parameters and other operating parameters. When defining operating limits, it is very important to distinguish soft and hard constraints. Hard constraints refer to mechanical performance limits, for example, the tower tray flood limit, or the compressor flow rate limit, or the furnace heat flux limit and so on. While making sure that hard constraints must be satisfied, relaxing soft constraints could play a significant role in improving operating margin.

14.2.4 Relax Soft Constraints to Improve Margin

Finding ways to relax plant limitations is one of the most important tasks in improving operating margin. Mathematically, relaxing constraints will lead to a large operating region and push the objective function toward the edge of the enlarged region. But the bottom line is: What can be done in reality for relaxing plant limitations? Equipment rating analysis could be a very effective way to identify equipment spare capacity and limitations. Utilization of spare capacity can allow capacity expansion up to 15–20% in general and accommodate improvement projects without or with little capital cost. The important part of a feasibility study (Chapter 5) is to find ways to overcome soft constraints; at some times, hard constraints, if they can be overcome using cheap options.

Three general limitations for equipment are pressure, temperature, and metallurgy as each equipment is designed with the limits for these parameters. If it is identified that the equipment will operate at higher pressure than the design limit, operating pressure needs to be reduced if possible. Otherwise, it is necessary to replace it which comes with a cost and it is usually expensive. This is similar to the case when operating temperature will exceed the design limit. These constraints could be resolved if process conditions could be changed. However, if metallurgy is found less than required, this could be a major hard constraint and it needs to be flagged out for metallurgist's attention as earlier as possible. In some cases, the equipment could still be usable if it is agreed by metallurgist and the plant takes actions for routing inspection.

For fired heaters, the limitations could be heat flux or tube wall temperature (TWT). The former is applied to heaters with pressure drop larger than 20 psi, which is the most common. The latter is for low pressure drop heaters. When the heater duty must be increased to handle duty much larger than design, the existing heater may be insufficient in meeting the limits. Installing a new heater could be very expensive. The most effective way to avoid this is to increase feed preheating via process heat recovery, adding a feed preheater or adding tube surface area into the heater, or installing a new one.

For separation/fractionation columns, the major limitation is the tray loadings. Typically, the column is design with 80–85% jet flooding. In operation, the column can run harder and flooding limit can be relaxed to 90% or even 95%. It is possible to push extra throughput at the column flooding limit if product cut point specifications can be relaxed. If higher throughput is desired in revamp, more efficient and higher capacity trays could be used to replace the old ones partially or completely. Although it is costly, it is cheaper than installing a new column.

For the compressor, when it reaches the capacity limit in flow or head, the direct solution is reducing the gas flow. The alternative solutions include increasing speed by gear change or adding wheels to rotor or adding a booster compressor. Similarly for pumps, the major constraint is insufficient capacity in flow or head. The low-cost solution is replacing impeller with a larger size. The alternative solution is to add a booster pump.

For heat exchangers, the constraint is insufficient surface area. In resolving this constraint, adding surface area via more tube counts or tube enhancement to the existing exchanger could help; but area addition could only be up to typically 20%. Beyond this, adding a new shell in parallel may be required, which could reduce ΔP, but it can cause problems in flow distribution.

The above constraints occur in the internal system battery limit (ISBL). However, it is important to identify outside system battery limit (OSBL) as well such as offsite utility, layout, piping, substations, etc. For example, a revamp project requires additional HP steam for process use but existing steam system may reach the boiler capacity limit of HP steam generation. Installing a new boiler could kill the economics of a revamp project. Another revamp case could require installing a new exchanger, which is well justified from ISBL conditions; but the piping to connect two process streams in the exchanger is too long in distance, which becomes cost prohibitive. A recent revamp project determined the great benefit of installing a new motor of 5 MW in replacement of the steam turbine to run recycle gas compressor in a reforming unit. The benefit of this project was due to the facts that the turbine has a low power generation efficiency and the electrify price is very low. But the project was deemed infeasible because the electricity requirement of the new motor is beyond the capacity of the nearby substation. Installing a new substation could cost multiple million dollars. Numerous ISBL and OSBL limitations could occur which are not listed here in detail.

14.3 Energy Optimization for Distillation System

Operation of separation columns involves trade-off between energy use and product recovery or purity. When energy usage is lower than the requirement, product recovery and purity suffer. On the other hand, when energy is more than the requirement, product purity and yields are improved. The optimal trade-off determines the operating target. The operating parameters involved in the trade-off include reflux ratio, feed temperature, column temperature, reboiling duty, column pressure, and so on. The questions are: How to obtain the energy target for a fractionation column and how to achieve the most economic operation of the column?

Let us look at an example that consists of a debutanizer column (Figure 14.1) for which White (2012) gave excellent explanations. This example gives perspective and guidelines in principle for optimizing a separation column and this example is reproduced here with permission from AIChE.

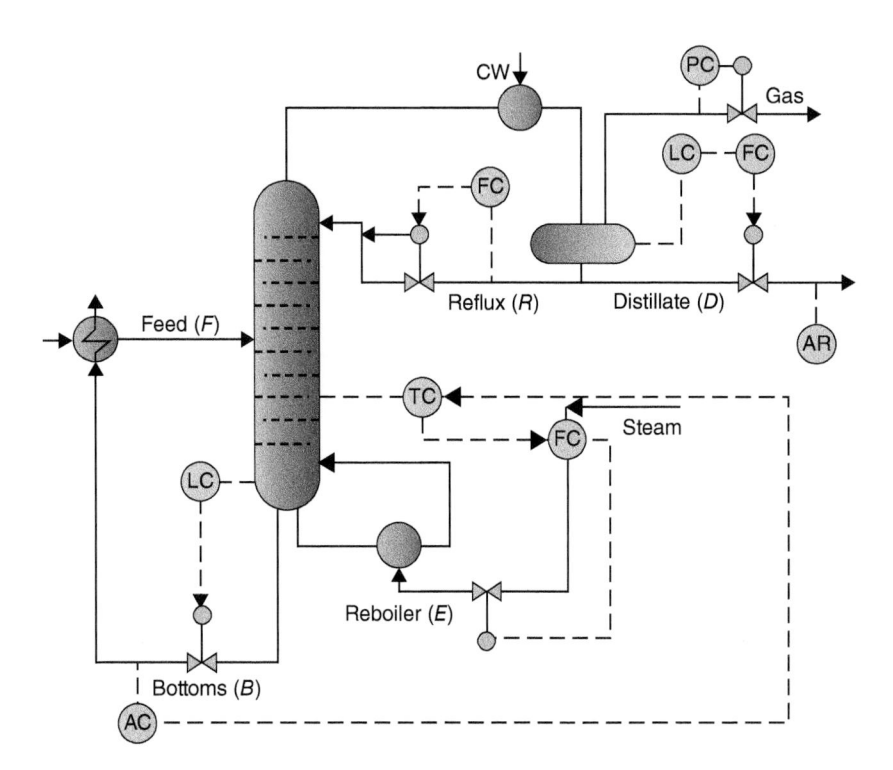

The feed and product specifications and prices for this example are listed in Table 14.1. Both products have tiered prices: on-specification product is priced much higher than the one that does not achieve specification. If the top product, butane, achieves the specification, i.e. less than 3% C_5, it is sent to the downstream unit for further processing leading to eventual sales. Off-spec butane will be used as fuel which has a low value than selling as a product. Similarly, the bottom product, pentane, if achieving specification, is used for making a high-value product. Otherwise, it will be sent to the tank for reprocessing. In operation, the operator changes column temperature and reflux–to-feed ratio to achieve the product specification.

14.3.1 Develop Economic Value Function

The objective function (economic value) representing the economic operating margin for the column is shown in Figure 14.2, which is defined as the difference between the value of products (top butane product and bottom pentane

Figure 14.1 Debutanizer example: energy optimization based on reflux ratio. *Source:* White (2012), reprint with permission by AIChE.

Table 14.1 Product specifications and prices.

Stream	Composition/specification	Value
Feed, 20 000 barrels per day	25% C_3 25% nC_4 25% nC_5 25% nC_6	$60/barrels
Bottoms product = C_5	≤5% C_4	$80/barrels
	>5% C_4	$60/barrels
Top product = C_4	≤3% C_5	$60/barrels
	>3% C_5	$40/barrels
Steam		$15/MBtu

Source: White (2012), reprint with permission by AIChE.

product) and costs of feed and energy. The value function features two discontinuities. The first, which occurs when the composition of the bottom product is about 1% butane, corresponds to a change in the top product from off-spec to on-spec. The second discontinuity occurs when the bottom product becomes off-spec at 5% butane.

14.3.2 Setting Operating Targets with Column Bottom Temperature

To choose the bottom temperature target, first assume that the reflux rate is fixed, and that the bottom product is on-spec but the top product is off-spec because of its high pentane content. This would correspond to a very hot bottom temperature. When the bottom temperature is slowly reduced, the amount of bottom product increases but the percentage of butane in the bottom also increases simultaneously. As the amount of pentane product increases, the total product value improves. The middle line in Figure 14.2 represents the above operation.

Normally, one would select a temperature target such that the bottom composition is as close to the specification limit as possible. There will always be some variability in the control performance due to external disturbances and limitations on loop control action. If composition control is poor and has a high variance, the observed composition probability distribution could look like a normal distribution in relation to operating margin as shown in Figure 14.3.

The product composition target is the mean value of observed composition distribution as shown in Figure 14.3. The mean value of the operating margin in Figure 14.3 is calculated based on the weighted average composition of the observed distribution – i.e. the percentage at each composition is multiplied by the margin value at that composition to determine the overall value.

Figure 14.3 shows that part of the column operation is the bottom product being off-spec. The mean product value does not correspond to the value at the mean of product composition, which is also the operating target. This is because of the nonsymmetrical nature of the operating margin and low value of off-spec products.

After reducing the variability through improved control-valve performance and reduced measurement error, the new mean value of the operating margin increases at the same operating target or bottom composition target (Figure 14.4). It can be seen that reduced variability results in increase in the mean value of operating margin.

14.3.3 Setting Operating Targets with Column Reflux Ratio

The above discussions involved constant reflux ratio. Next, consider the situation where the reflux rate is varied and the bottom temperature is constant. Fundamentally, the reflux rate provides internal reflux needed for

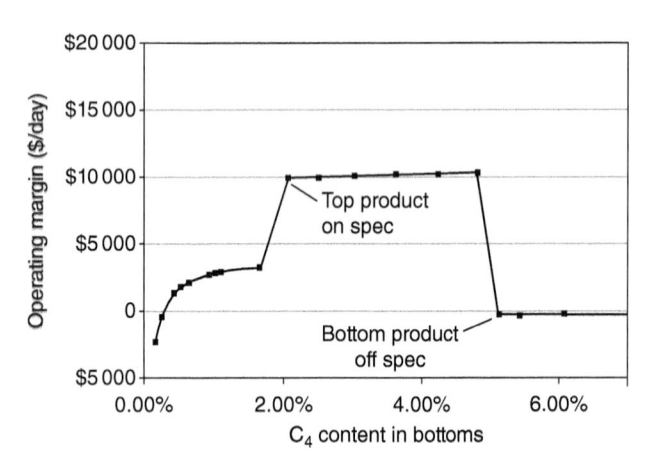

Figure 14.2 Operating margin as a function of the bottom composition. *Source:* White (2012), reprint with permission by AIChE.

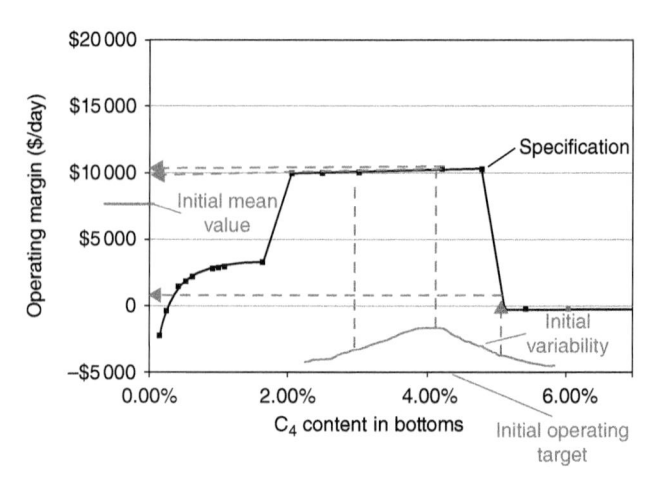

Figure 14.3 Observed composition normal distribution vs. operating margin. *Source:* White (2012), reprint with permission by AIChE.

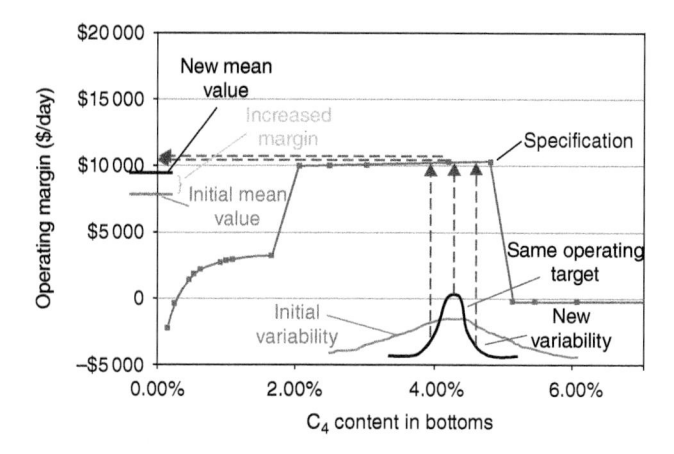

Figure 14.4 Improved composition normal distribution versus operating margin. *Source:* White (2012), reprint with permission by AIChE.

separation in the tower and it is generated by either feed heater or reboiler. Thus, lower reflux rate saves energy but too low reflux rate could affect product quality. On the other hand, high reflux rate could improve production of more valued product.

In tower design, the reflux ratio is determined based on the trade-off between operating cost in reboiler and capital cost for the tower. In other words, use of more separation stages requires less reflux rate and thus less reboiling energy but at the expense of additional capital cost. The minimum reflux ratio is calculated based on the work of Underwood (1948). A tower requires an infinite number of stages to achieve the minimum reflux ratio. To make separation feasible and at the same time a tower affordable, a reflux ratio larger than the minimum is used. Typical design reflux ratios are 1.1–1.3 of the minimum reflux ratio. With a high reflux ratio, the number of theoretical stages is lower resulting in lower capital cost for a tower but at the expense of higher reboiler

duty and vice versa. The optimal reflux ratio in tower design is determined based on the minimum total cost.

However, in operation, the optimal reflux ratio is determined based on the trade-off product value and energy cost. When the reflux ratio increases, the separation improves at the expense of increased reboiling duty (Figure 14.5). As the result, the top product rate decreases while the bottom product rate increases. As shown in Figure 14.5, the cost of reboiling duty presents a linear relationship with reflux ratio but the product rate is nonlinear and presents a different trend as reboiling duty.

Figure 14.6 shows the operating margin for different energy prices, assuming constant product prices. The optimum reflux rate depends on the price of energy. At a high energy price, the optimum reflux rate is at the minimum value which allows the column to maintain the top product in specification. At the lower energy prices, the optimum reflux rate increases.

The conclusion is that operating targets should be a function of energy costs rather than a fixed number even with fixed composition limits. It is common to observe separation columns operating at reflux rate that are 50% higher than the optimum. For the debutanizer column operation discussed here, such an operation could cost operating margin in excess of $500 000 per annum.

14.3.4 Setting Operating Pressure

It is generally known that reducing the operating pressure of separation columns reduces energy consumption. This is because the lower the tower pressure the less heat required for liquid to vaporize (thus less energy required) and the easier for vapor to penetrate into liquid on the tray deck (thus better separation). Yet, many columns are operated well above their potential minimum pressure. One may ask: If benefit of reducing pressure is well

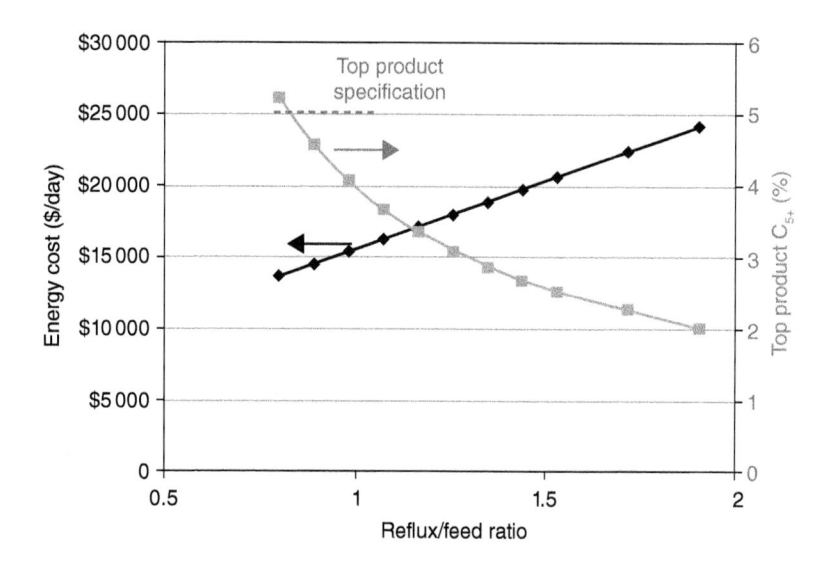

Figure 14.5 Energy separation trade-off: energy cost increases linearly as reflux rate while the top product quality improves. *Source:* White (2012), reprint with permission by AIChE.

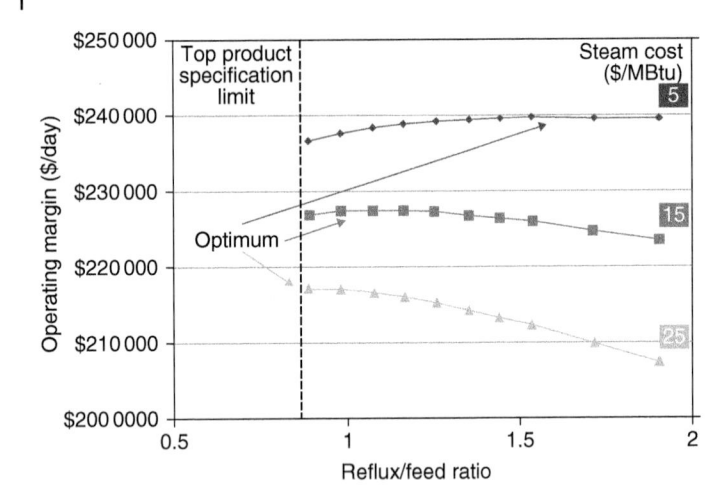

Figure 14.6 Optimum reflux rate depends on energy price. *Source:* White (2012), reprint with permission by AIChE.

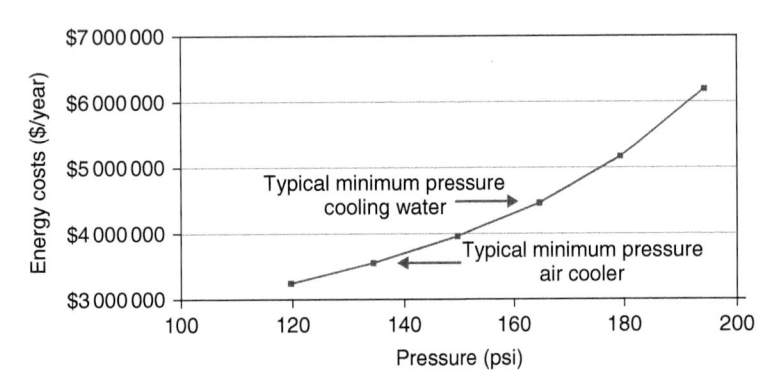

Figure 14.7 Pressure has significant effect on energy cost. *Source:* White (2012), reprint with permission by AIChE.

known, why is it not widely implemented? There appears to be three primary reasons for this.

First, changing column pressure requires simultaneously changing the bottom temperature set point to hold the product composition at their targets. This is difficult to do manually – advance composition control is required.

Second, changes in column pressure have other impacts such as changes in the off-gas rate, the amount of reboiler duty, and hydraulic profile of the plant. In the case of partial condensation, pressure control can interact with the overhead receiver level. While these effects are real, their magnitude is sometimes exaggerated and cited as reasons for not making any changes.

Finally, plant personnel frequently do not agree on the amount of operating margin required to handle major disturbances. For instance, questions often arise about the dynamic response of an air-cooled condenser to a rainstorm and the ability of the overall control system to handle such conditions. A well-designed overall control system for the column can compensate for such disturbances.

The condenser pressure controls the tower pressure and thus the feed tray pressure. There is a pressure valve in the overhead which can be used to control the tower pressure. The lower limit of the tower pressure is column overhead condensing duty, net gas compressor capacity, and column flood condition. Many of the new APC systems are using pressure control to save energy. During extended turndown periods, reducing pressure up against an equipment limit can improve efficiency (Figure 14.7).

14.4 Overall Process Optimization

This example comes from Loe and Pults (2001) and is reproduced, with permission from AIChE, for the purpose of explaining how a tower could be optimized. This example provides a case history of how operational improvements for a single deisopentanizer (DIP) fractionation tower are identified, implemented, and sustained. To do so, the current operation is simulated and assessed. Then, improvement opportunities are identified and the limiting factors are determined. Optimal solutions are obtained by optimizing tower ISBL conditions and OSBL conditions. The improvements on this single tower have generated over $500 000 as compared with historical operation over the first six months.

14.4.1 Basis

The DIP tower, shown in Figure 14.8, processes light straight-run naphtha from two sources. The LPG fractionation unit debutanizer bottoms consist of primarily

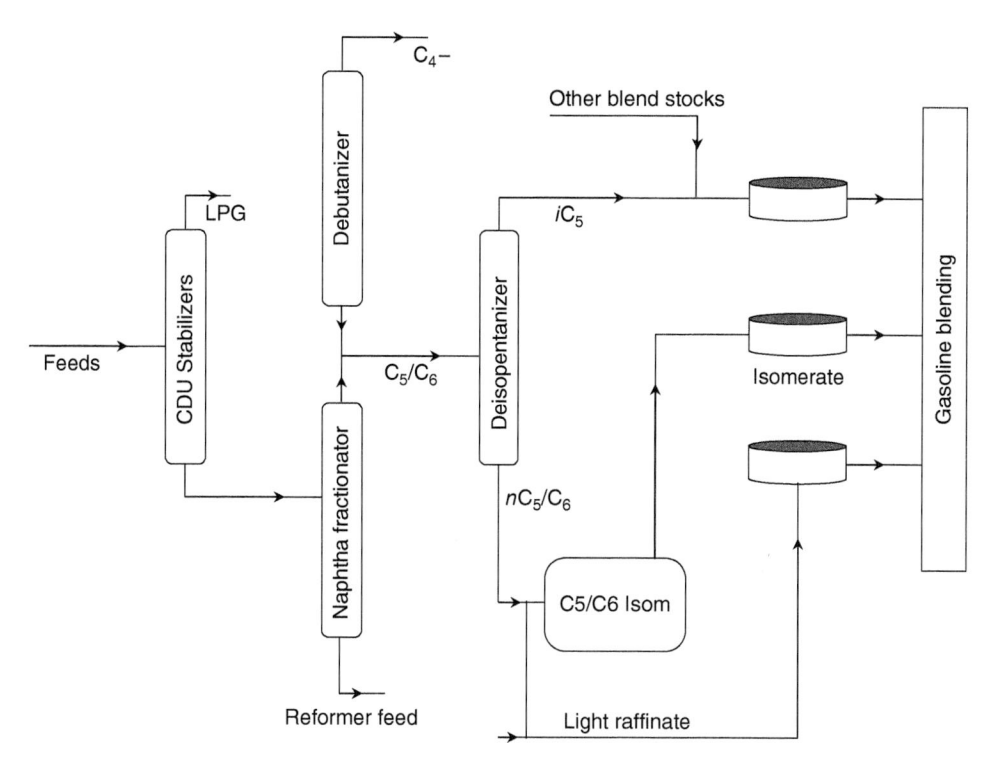

Figure 14.8 Deisopentanizer flow scheme. *Source:* Loe and Pults (2001), reprint with permission by AIChE.

iso and normal pentane (iC_5 and nC_5), with small fractions of butane and C_6+ components. The overhead from the naphtha fractionator tower contains mostly C_5 and C_6 paraffin compounds, with some benzene and C_6 naphthenes and a small amount of butane. The combined feed to the DIP is typically in the range of 9000–15000 barrels per day.

The DIP overhead product is normally rich in isopentane, and is routed to gasoline blending along with other high octane, low Reid vapor pressure (RVP) gasoline components. The DIP bottoms, which are rich in nC_5 and C_6 paraffins, are routed to the light naphtha isomerization unit, along with light raffinate from the aromatics extraction unit.

The DIP tower has 50 trays in comparison with 70+ trays used in a typical DIP. As a consequence, the DIP tower often has a difficult time making a good split between isopentane and normal pentane components.

14.4.2 Current Operation Assessment

Historically, the tower had been operated with a target of 10% nC_5 in the overhead, and 20–30% iC_5 in the bottom product. The tower was reported to be limited by reboiler or condenser duty. One of the two steam reboilers had been out of service for some time, and 20–30% of the condenser fin fan motors were not operating and in need of repair. The DIP equipment had not been a maintenance priority, in part because no economic penalty had been calculated for having a reboiler or condenser out of service. There was also a concern that the tower could flood if both reboilers were placed in service.

The DIP process control was accomplished with a distributed control system (DCS) equipped with an advanced process control (APC) algorithm. The controller was set to target 10% nC_5 in the overhead and 10% iC_5 in the product, and would increase reboiler steam and reflux rate until reaching the maximum limits for these flows. The tower pressure was also controlled within a specified range by the APC, and this could indirectly limit the reboiler duty as well, if the tower pressure increased beyond its set maximum. Inferential estimates for product iC_5 and nC_5 qualities were calculated based on tower temperatures and pressures, and a bias for these values was continually updated based on daily lab data.

To establish the historic performance, the operating and laboratory data were collected from the past year, eliminating periods of known equipment failure or poor unit volume balance. The data showed an average of 19% nC_5 in the DIP overhead, and 27% iC_5 in the bottoms product, with a wide variation in the product quality, as shown in Figure 14.9. An average of 18% of normal butane was observed in the overhead product, indicating poor debutanization in the upstream fractionation towers.

The process design for the tower showed an available reboiler duty of 51 000 lb/h of steam, but the average for the data collected showed only 27 000 lb/h. The data show an erratic variation in reboiler duty.

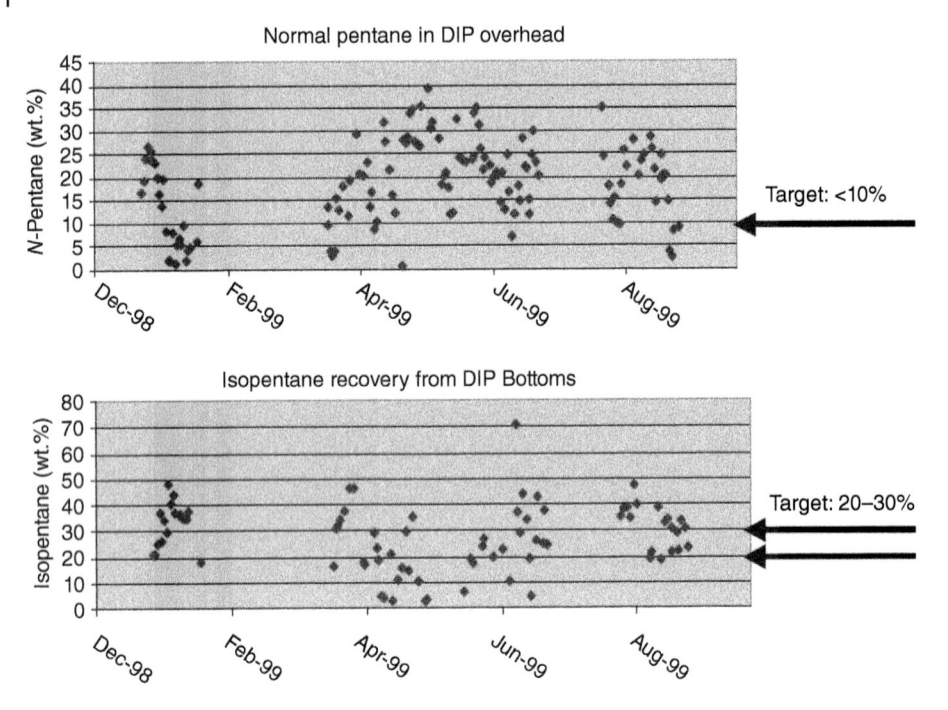

Figure 14.9 Variation of DIP performance. *Source:* Loe and Pults (2001), reprint with permission by AIChE.

14.4.3 Simulation

Using the averaged process and lab data, a simulation model for the DIP was developed. The feed rate and composition to the tower were fixed, as well as the reflux rate, tower pressures, and overhead rate. The model results for reboiler duty, tower temperatures, and compositions compared favorably with the unit operating data, as shown in Table 14.2.

The calibrated model was used to simulate DIP performance at the design reboiler duty, to determine if available condenser duty and tower tray capacity would be adequate for this operation. As shown in Table 14.3, the predicted condenser duty for this operation was only slightly above the design value, and tower tray parameters indicated that flooding was unlikely. Also the separation of iC_5 and nC_5 improved dramatically vs. the historical operation as would be expected with the increased tower traffic.

14.4.4 Define the Objective Function

The profitability of the DIP column was determined based on the value of separating iC_5 for direct blending to gasoline and nC_5 to be used as feed for the C_5/C_6 isomerization unit, less the utility and downstream unit opportunity costs incurred to do so. Lighter feed components, such as *n*-butane, were assumed to always be fractionated into the DIP overhead, and components heavier than nC_5 were assumed to always be found in the DIP bottoms stream. Thus, only the disposition of iC_5 and nC_5 components was considered in the profitability calculation.

Therefore, the objective function for optimizing the DIP tower is defined as below:

Table 14.2 Simulation results vs DIP operating data.

Result	Data	Model
Reboiler duty (MMBtu/h)	23.6	26.7
Top temperature (°F)	143	153
Bottom temperature (°F)	179	183
nC_5 in overhead (wt.%)	18%	16%
iC_5 in bottoms (wt.%)	31%	30%

Source: Loe and Pults (2001), reprint with permission by AIChE.

Table 14.3 Simulation results at designed reboiler duty.

Result	Design	Model
Reboiler duty (MMBtu/h)	48.8	48.8
Condenser duty (MMBtu/h)	45.0	48.1
Max jet flood	85%	74%
Max downcomer backup	50%	37%
nC_5 in overhead (wt.%)	NA	5%
iC_5 in bottoms (wt.%)	NA	18%

Source: Loe and Pults (2001), reprint with permission by AIChE.

DIP upgrade value = Overhead value + Bottoms value
− Feed value − Reboiler steam cost
− Isom operating cost
− Isom capacity penalty

The DIP feed and overhead values are calculated as the gasoline blending value of the iC_5 and nC_5 in this stream, with corrections for road octane and RVP of these components vs. those of conventional regular gasoline. The DIP bottoms stream is normally processed at the isomerization unit, where 75% of the exiting C_5's are assumed to be iC_5. After this equilibrium conversion, the value of the resulting iC_5/nC_5 stream is calculated at gasoline blending value as described for feed and overhead above. The reboiler steam cost is calculated assuming a 70% generation efficiency from refinery fuel gas, and the isomerization unit operating cost (for fuel, power, and catalyst) was taken to be the same value per barrel as used in the refinery planning model.

During some periods, the refinery isomerization unit has more feed available than can be processed. If additional DIP bottoms are produced, less capacity is available to process light raffinate from the aromatics extraction unit. The Isom capacity penalty, or the opportunity cost for processing additional DIP tower bottoms at this unit, was therefore estimated by evaluating the octane upgrade of light raffinate.

14.4.5 Off-Line Optimization Results

Once the economic evaluation criteria were determined, it was implemented in the simulation model. Numerous case studies were conducted via process simulation to determine the optimum operating point for the DIP tower under different scenarios. It quickly became apparent that in nearly all economic and operating situations, maximizing the DIP reboiler duty up to the maximum limit gave the highest profitability.

For subsequent case studies, the simulation was completed with maximum reboiler duty, and the tower pressure and nC_5 content of the overhead product were also fixed. These constraints completely specified the tower operating conditions.

The DIP profitability was first examined for scenarios where the isomerization unit has available capacity. The DIP feed rate and overhead nC_5 content were varied, and profitability calculated, as shown in Figure 14.10. This analysis showed that the optimum target was around 5% of nC_5 in the overhead, regardless of the tower feed rate.

Profitability was then examined assuming the isomerization unit was at its maximum charge rate, and additional production of DIP bottoms would result in bypassing of light raffinate around the Isom, direct to gasoline blending. The cost of losing the light raffinate octane upgrade can vary between $2 and $5/barrels, and so simulation cases were completed for both of these scenarios as shown in Figure 14.11. In these cases, the optimum nC_5 in DIP overhead target is dependent on the charge rate to the tower. At low charge rates, the available reboiler duty is sufficient to obtain good separation between iC_5 and nC_5 components, so that minimal iC_5 is lost into the DIP bottoms when targeting 5% nC_5 in the overhead. At higher charge rates, more iC_5 is lost to the bottoms stream, and it is more profitable to increase the overhead nC_5 target, reducing the DIP bottoms rate to the Isom unit and allowing additional raffinate upgrading. Thus, the optimal nC_5 in overhead target varies between 5 and 20%, depending on the DIP charge rate and the value of light raffinate upgrading.

Based on a comparison of the optimal tower operation as determined above, and the historical performance, an incentive of around $1.5 million per year was identified to improve DIP fractionation.

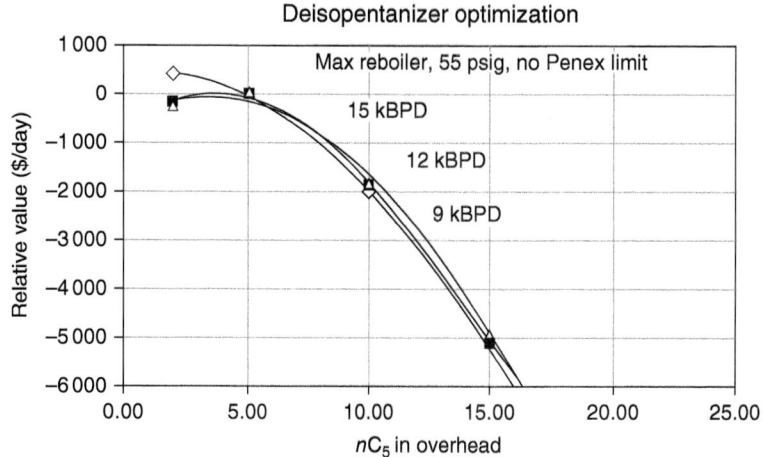

Figure 14.10 Optimization without Isom capacity constraint. *Source:* Loe and Pults (2001), reprint with permission by AIChE.

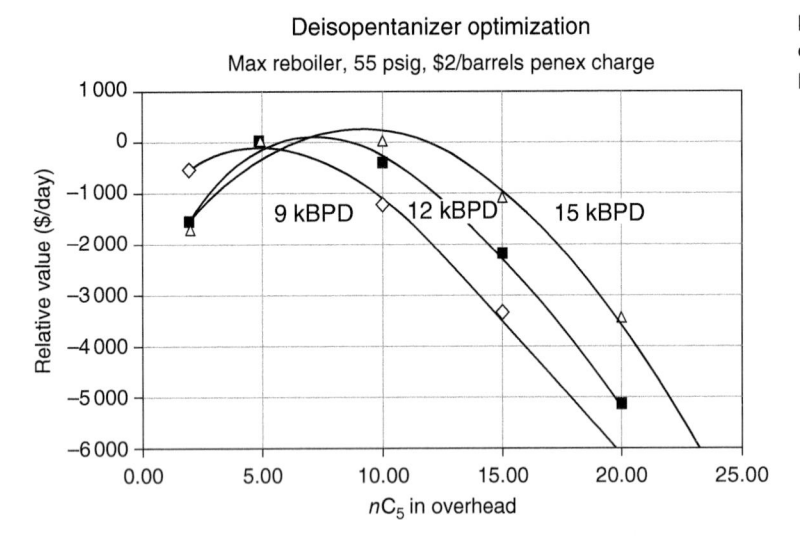

Figure 14.11 Optimization with Isom capacity constraint. *Source:* Loe and Pults (2001), reprint with permission by AIChE.

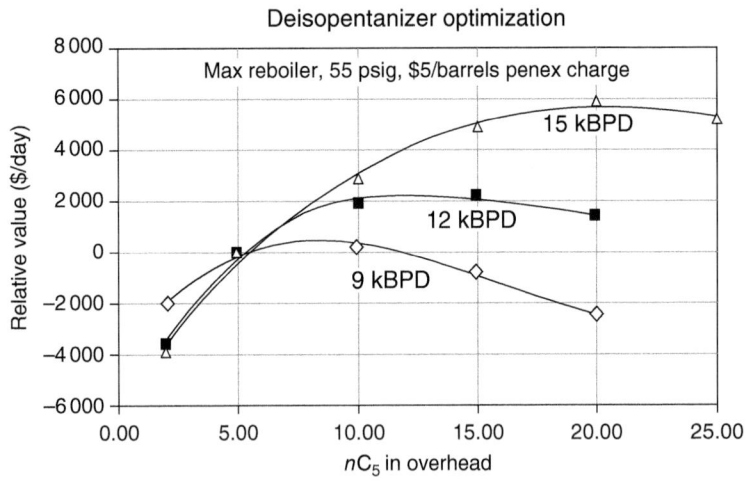

14.4.6 Optimization Implementation

In order to realize the benefit indicated by the optimization results above, several unit hardware, process control, and operating philosophy changes were needed. First, it was clear that both reboilers would be required, so the spare bundle was leak-tested and returned to service by operations. Several fin fan motors were also quickly repaired to ensure that design condenser duty was available and over-pressurization of the tower would not limit the reboiler duty that could be applied.

Secondly, the optimum target for the DIP overhead nC_5 was implemented into the APC system. The APC controller on the DIP DCS system was reconfigured to operate at this nC_5 target, while maximizing the reboiler duty as limited by the high limit on tower pressure. This APC system allowed the DIP operation to be maintained at an economic optimum, accounting for the isomerization unit capacity and the economics of the day.

Thirdly, communication of the new operating philosophy was also critical in improving DIP performance. Operators and unit supervisors were trained on the importance of always maximizing the reboiler duty and setting the nC_5 in overhead target. The iC_5 content of the bottoms stream was still measured by lab and inferred analysis, but this was no longer a tower control variable. The reboiler duty and overhead nC_5 were tracked on a daily basis, and performance for these key performance indicators (KPIs) were discussed at weekly operations and planning meetings.

14.4.7 Online Optimization Results

The economic benefit of improving DIP fractionation was tracked on a monthly average basis shortly after implementation of the new optimization strategy. Economic performance versus the baseline operation is shown in Figure 14.12 using actual monthly averaged economics and unit operating and lab data. Monthly benefits of over $100 000 were achieved in several cases during the summer months, when octane values were at their highest

Figure 14.12 DIP economic improvements. Source: Loe and Pults (2001), reprint with permission by AIChE.

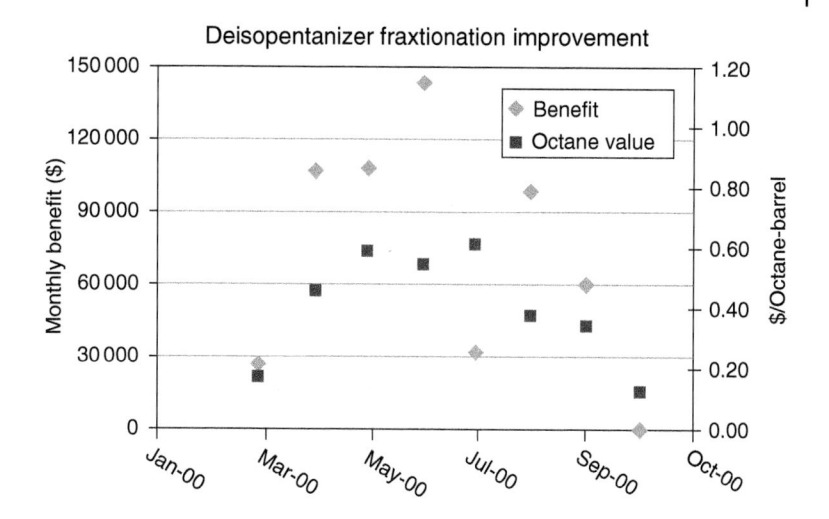

level. As octane values dipped during spring and fall months, benefits from the improved DIP fractionation dropped off as well.

A significant drop in benefits can be seen for July. This was due to poor operation of the DIP tower, caused by a high butane content in the tower feed from the crude unit stabilizers. The C_4's caused the DIP tower pressure to increase up to its safe operating limit, and the reboiler duty was cut back to avoid over-pressuring the tower. This resulted in a reduction in fractionation efficiency and profitability during part of July, and represented one of the challenges encountered in sustaining the improved DIP performance.

14.4.8 Sustaining Benefits

Without ongoing attention, optimization improvements from initiatives such as that on the DIP tower tend to fade over time, for a multitude of reasons. Challenges to the new level of performance must be tackled as they arise, whether they result from hardware or control problems, misunderstandings, or operating changes in other parts of the refinery. Tracking the economic benefits of the initiative is a critical element of sustaining the change, since knowing the lost profits associated with a loss in performance helps to set work priorities within the refinery. Several problems occurred during the first six months of improved DIP operation which reduced the profit derived from this tower, and these issues were quickly addressed to sustain the improvement.

During the summer months, a change in routing of a portion of the refinery condensate resulted in a hydraulic constraint on the amount of condensate that could be removed from the DIP reboilers into the condensate header. This caused reboiler duty reduction leading to fractionation efficiency drop. The economic calculations clearly showed that the benefit derived from the additional reboiler duty was much higher than the value of recovering the condensate. For this reason, the condensate was safely spilled into the sewer until normal condensate header operation was restored and DIP fractionation was maintained.

A second challenge occurred when it was noticed that the DIP reflux drum temperature had increased above 140 °F, which was higher than the recommended rundown limit to tankage. Initially, reboiler duty was decreased to ensure safe operation. However, in meeting with the tank farm operators about the problem, it was found that the DIP overhead mixed with several other much larger streams before entering a gasoline blending tank. Calculations showed that the effect of the higher DIP rundown temperature on the tank temperature was minimal and DIP reboiler duty was again increased.

Another problem in maintaining reboiler duty was identified when the DIP tower pressure increased due to butanes in the feed, as mentioned above. It was found, however, that the maximum tower pressure was set well below the vessel design pressure, and so the safe operating limit of the tower could be increased after an appropriate management of change (MOC) review. Simulation modeling showed that operation at the higher pressure did not significantly impact the tower profitability if the reboiler duty could be maintained near the maximum. The pressure limit was increased and performance of the tower again improved.

A fourth reduction in DIP profitability was noticed when the APC controller began cutting reboiler duty, for no apparent reason. Further investigation showed that the inferred nC_5 content of the overhead stream had been deviating from the lab value for a few days, because of a computer glitch. This was quickly corrected and DIP operation returned to normal.

Although all of these obstacles reduced profitability for a short period, timely identification and resolution of the constraints averted potentially long periods of underperformance.

14.5 Concluding Remarks

The profit improvement process should pass through several phases. First, current performance must be assessed, evaluating upstream and downstream constraints and unit equipment limitations. Understanding of how the unit operation can be optimized can then be gained by use of an appropriate process model and refinery economics, and a new operating strategy is then developed to improve profitability. Implementation of this strategy requires good communication of its benefits throughout the plant. Upgrading or repair of process equipment may be needed to allow operation under desired conditions. Tracking of key operating parameters as well as economic performance of the unit and distributing these results within the plant are essential to sustaining the improvement, as this process flags deterioration of profitability. The profit improvement process requires input from technical, operations, safety, environmental, and economics groups within the plant.

References

Loe, B. and Pults, J. (2001). Implementing and sustaining process optimization improvements on a de-isopentanizer tower. AIChE Spring Meeting, April, Houston.

Underwood, A.J.V. (1948). Fractional distillation of multi-component mixtures. *Chemical Engineering Progress* 44: 603.

White, D.C. (2012). Optimizing energy use in distillation. *Chemical Engineering Process* 44: 35–41.

15

Fractionation and Separation Theory and Practices

15.1 Introduction

Separation technologies are commonly used in most chemical processes for separating mixtures into its constituent elements with the tasks such as removal of impurities from raw materials, segregation of individual products, purification of products, and recycling of solvents and unconverted reactants. This chapter will provide a brief overview of the most common separation technologies that are available.

As with all design choices, the process engineer needs to be certain that the design will work. This requires a good understanding of the separation technology and the fluid properties of the stream to be separated. Distillation is the most common separation technology which results in many good sources for the needed knowledge. Knowledge about other separation technologies is much more limited.

Each technology has key thermodynamic properties that allow the separation of the components. For distillation, this is relative volatility. For crystallization, this is the freeze point. For adsorption, this is affinity. The thermodynamic properties needed for distillation designs are frequently readily available. The thermodynamic properties for non-distillation designs are frequently more difficult to obtain.

Most engineering departments have sufficient experience to design a broad range of distillation separations. A much smaller number of departments will have sufficient experience to allow design of non-distillation separations. In all cases, the process engineer needs to be comfortable discussing the design issues that each separation technology has. The process engineer needs to select the separation technology that provides the best capital and operational costs. Distillation is the most frequent answer but there are many separations that are better done with other technology. The most difficult separations may require the combination of several technologies, such as membrane technology and distillation technology.

The time allowed for design work is almost always an important issue. When identical or similar separations have been done in the past, there will likely be large time benefits to reusing the previous technology and design methods. Care is needed here to allow some time for innovation that may result in a better choice of technology.

In the aromatics complex, and most chemical processes, distillation is the most common technology adopted, which will be the focus of the discussions in this chapter. At the same time, simulated moving bed (SMB) and crystallization will also be discussed in detail as these two technologies lay down the foundation for two completely different kinds of aromatic complex offerings available in the market. Furthermore, extractive distillation as an alternative to both SMB and crystallization will be briefly discussed. To provide a better understanding for SMB technology, adsorption as the cornerstone for SMB will be discussed in more detail.

15.2 Separation Technology Overview

Separating a feed stream into two or more product streams can require the use of substantial energy and the addition of non-feed material (Figure 15.1). The term "feed stream" refers to the main stream that is to be separated into two or more products. The term "non-feed stream" refers to another stream that is only added to the separation process to facilitate the separation of the "feed stream." Design effort is always needed to balance energy requirements versus capital cost. The addition of a non-feed material may either be required or may just be advantageous. When the non-feed material is used, the issue of its presence in the products needs to be addressed. Removal of the non-feed material is normally desirable either due to its presence being unacceptable or due to a desire to reuse this material.

Efficient Petrochemical Processes: Technology, Design and Operation, First Edition.
Frank (Xin X.) Zhu, James A. Johnson, David W. Ablin, and Gregory A. Ernst.
© 2020 John Wiley & Sons, Inc. Published 2020 by John Wiley & Sons, Inc.

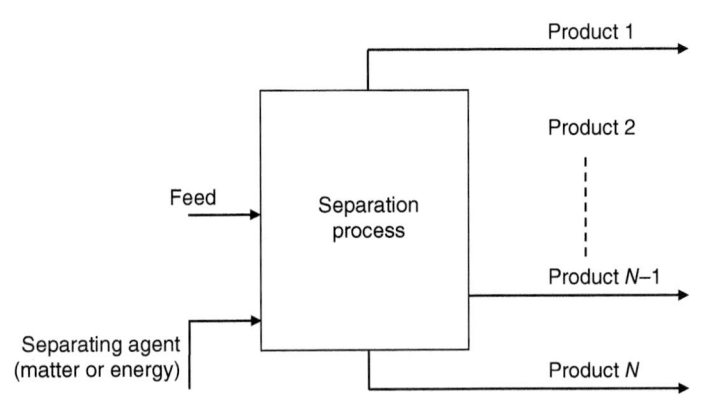

Figure 15.1 Concept sketch for separation process.

The following types of separation technology will be briefly reviewed below while the detailed discussions will be given later in this chapter.

- Single-stage separation
- Distillation
- Liquid–liquid extraction
- Adsorption
- SMB chromatography
- Crystallization
- Membrane

15.2.1 Single-Stage Separation

It is where the feed stream is separated into a vapor stream, a light liquid-phase stream, and a heavy liquid-phase stream, as appropriate. The simplest fluid-phase equilibrium separation is a vessel/process that will separate a stream into its various phases, vapor/light liquid/heavy liquid. Here, phase separation relies on mechanical means based on gravity, centrifugal force, or an electric and/or magnetic field. A phase separation operation is frequently needed prior to more complex separation technologies, such as adsorption, that normally requires a single-phase feed. However, distillation columns can normally accept multiphase feeds.

15.2.2 Distillation

It is the separation of a liquid mixture into its components on the basis of differences in relative volatility. Distillation requires a source of liquid to the top stage and a source of vapor to the bottom stage. The top liquid may be obtained by condensing some of the vapor rising from the top stage. Other sources could be a feed stream or a non-feed stream. Some of this stream will be stripped into the vapor from the top stage.

The vapor to the bottom stage may be obtained by vaporizing some the liquid descending from that stage. Other sources could be a feed stream or a non-feed stream. Some of this stream will be absorbed into the liquid from the bottom stage.

There are some variations to distillation, which include extractive distillation, absorbers, and strippers. Extractive distillation units add extraction material to improve the relative volatility of the key components. Absorbers may add a non-feed absorption liquid to remove desired relative heavy components from a vapor feed stream. Strippers may add a (non-feed) stripping vapor to remove desired light components from a liquid feed stream. In each of these cases, some of the added material ends up in each of the product streams. Frequently, additional separation columns are required to recover and reuse this material.

15.2.3 Liquid–Liquid Extraction

It is where a descending heavy liquid stream is repeatedly contacted with a rising light liquid stream. This is very similar to distillation but involves the use of two liquid phases. A heavy liquid phase is fed onto the top stage and a light liquid phase is fed into the bottom stage. Both flooding calculations and efficiency calculations are much more difficult for extraction as compared to distillation.

15.2.4 Adsorption

It uses particles to preferentially adsorb some components from a feed stream. A bed of solid particles is used to adsorb the material from a feed stream. Non-regenerated adsorbers can be used to adsorb very small amounts of material out of a feed stream. This is true as the adsorbent is replaced rather than be regenerated.

Regenerated adsorbers are much more common than non-regenerated adsorbers. The regeneration is sometimes done by lowering the pressure or sweeping a regenerant material over the adsorbent. The regenerant is frequently heated to assist with regeneration. It is common that multiple bed configuration will be used so the one or more beds can be adsorbing while one of the beds are being regenerated. In cases where large amounts of high-purity products are needed, SMB design can be used.

15.2.5 Simulated Moving Bed Chromatography

This is a complex form of adsorption process where liquid mixture is separated using a SMB with packed adsorbent that preferentially adsorbs some of the feed components. A non-feed adsorbent liquid is used to remove the feed components that are absorbed. The adsorbent material is typically a zeolite. The preferential adsorption of some of the feed components can either be the result of the molecular sizes (e.g. UOP Molex process) or the result of chemical bonding or affinity (e.g. UOP Parex process).

15.2.6 Crystallization

Crystallization is a chemical solid–liquid separation technique, in which mass transfer of a solute from the liquid solution to a pure solid crystalline phase occurs. Crystallization occurs in two major steps. The first is nucleation, the appearance of a crystalline phase from either a super-cooled liquid or a supersaturated solvent. The second step is known as crystal growth, which is the increase in the size of particles and leads to a crystal state. An important feature of this step is that loose particles form layers at the crystal's surface and lodge themselves into open inconsistencies such as pores, cracks, etc. Attributes of the resulting crystal depend largely on factors such as temperature, air pressure, and in the case of liquid crystals, time of fluid evaporation.

15.2.7 Membrane

Gas separation membranes work according to the principle of selective permeation through the membrane surface. The permeation rate of each gas depends on its solubility in the membrane material and on the diffusion rate of the gas.

Gases with high solubility and small molecules pass through the membrane very quickly. Less soluble gases with larger molecules take more time to permeate the membrane. In addition, different membrane materials separate differently. The driving force needed to separate gases is achieved by means of a partial pressure gradient.

The driving force for a gas to permeate through a membrane is the partial pressure difference; in other words, the partial gradient between the inside of the hollow fiber (retentate side) and the outside of the hollow fiber (permeate side). The greater the difference, the more gas permeates through the membrane. For example, if carbon dioxide and methane are being separated, as is the case with biogas upgrading, carbon dioxide permeates through the membrane very quickly while the methane tends to be held back.

15.3 Distillation Basics

Distillation is the common separation technology used in most chemical processes due to the maturity of the technology, significant experience in the design and operation of distillation columns. Distillation is converting multicomponent streams into desirable products based on the difference in boiling points between key components. Improving energy utilization, reducing capital costs, and enhancing operational flexibility are spurring increasing attention to distillation column optimization during design and operation.

15.3.1 Difficulty of Separation

The difficulty of separation for a single column can be determined by considering the desired degree of separation and the intrinsic difficulty of separating the key components. It is common to define a light-key (LK) component and a heavy-key (HK) component when designing a distillation column. In many cases the choice of these components is clear. The LK is the lightest of the components that are desired to go to the overhead. The HK is the heaviest of the components that are desired to go to the bottoms. In the cases where the choice of these components is not clear, it may be necessary to consider several different pairs.

As shown in Figure 15.2, the separation column involves a feed, overhead product D which consists of most light components and small amount of heavy components, and bottom product B which consists of

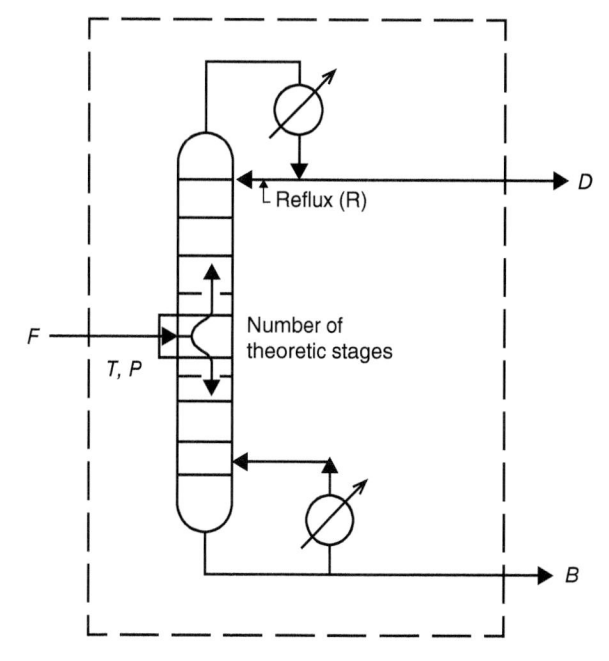

Figure 15.2 Simple distillation column.

most heavy components and small amount of heavy components. The overall difficulty of separating two components in a column includes the relative volatility of those components and the degree of separation that is required. The relative volatility is simply the ratio of the vapor–liquid equilibrium K values for the key components. This value does vary from top to bottom in the column. The degree of separation can be represented by the separation factor which is defined as the ratio of the key component in the distillate divided by the ratio of the key components in the bottoms (King 1971). The difficulty of separation can be defined as relative volatility, α, for a given temperature and pressure as

$$\alpha_{LK,HK} = \frac{y_{LK}/x_{LK}}{y_{HK}/x_{HK}} = \frac{K_{LK}}{K_{HK}} \quad (15.1)$$

where

α = relative volatility of the more volatile component (LK) to the less volatile component (HK)
D = distillate product; B = bottom product
y_{LK} = vapor–liquid equilibrium concentration of LK component in the vapor phase
x_{LK} = vapor–liquid equilibrium concentration of HK component in the liquid phase
y_{HK} = vapor–liquid equilibrium concentration of HK component in the vapor phase
x_{HK} = vapor–liquid equilibrium concentration of HK component in the liquid phase
K = vapor–liquid distribution ratio (y/x)
K_{LK} = K-value of LK; K_{HK} = K-value of HK.

As a separation becomes more difficult, the column required becomes larger and more expensive. The height of the column is strongly related to the number of theoretical stages that are needed. The Fenske equation (1932) allows the calculation of the minimum number of stages as

$$N_{min} = \frac{\log\frac{(LK_D/HK_D)}{(LK_B/HK_B)}}{\log \alpha_{ave}} \quad (15.2)$$

where

$$\alpha_{ave} = \sqrt{\alpha_D \cdot \alpha_B} \quad (15.3)$$

where

N_{min} = the minimum number of theoretical plates required at total reflux
LK_D (HK_D) = the mole fraction of more (less) volatile component in the overhead distillate
LK_B (HK_B) = the mole fraction of more (less) volatile component in the bottoms

α_{ave} = the average relative volatility of the more volatile component to the less volatile component
α_D = the relative volatility of LK to HK at the top of column;
α_B = the relative volatility of LK to HK at the bottom of column.

The LK and HK component concentration should be known or estimated based on the separation requirements. The α value is normally obtained from a simulation program. The minimum number of stages is based on using infinite reflux. A typical design will use 2.5–3.0 times the minimum number of theoretic stages. Equation (15.2) can be applied for multicomponent systems and more details can be seen in the work of Wankat (2006).

For a perfect separation between the key components while using an infinite number of stages, the minimum reflux-to-feed ratio, R_{min}/F, can be calculated with the following equation:

$$\frac{R_{min}}{F} = \frac{1}{1-\alpha} \quad (15.4)$$

The actual required reflux will vary due to the actual desired separation and the actual number of stages. As a result, the above is of little value other than demonstrating that the reflux requirement is a strong function $1/(1-\alpha)$. It is rare to have a commercial distillation separation for systems that have an α value less than 1.05 (Perry and Green 1997).

15.3.2 Selection of Operating Pressure

The selection of the operating pressure is normally based by reviewing its effect on the condenser and reboiler, as well as separation difficulty. Separation difficulty is almost always better with lower pressure as most relative volatilities are larger at lower pressures. This benefit would make vacuum operation of great interest except that there are several reasons that cause column operation below atmospheric pressure to be expensive. As such, vacuum operation is only considered when it is needed to keep the bottom temperature below some maximum value. A maximum bottom temperature may be required to limit the amount of cracking or polymerization. Column operation slightly above atmospheric pressure is sometimes needed to supply improved pressure control or to insure column isolation from non-feed material. Column operation at higher pressures is frequently needed to allow condenser operation. When a liquid distillate is desired, the pressure needs to be high enough to provide a bubble point temperature that will allow the use of the desired cooling medium. Below is some discussion on several possible cooling mediums.

- Air is frequently the first choice that should be considered. The generation of hot air is normally much less of a problem than the generation of hot/warm water. However, air cooling can be a problem as it may require a substantial footprint. The possible process temperature will normally be slightly hotter than what is obtainable with recirculating cooling water.
- Recirculating cooling water normally is the lowest cost choice for the condenser and requires a small footprint. It normally can obtain a lower process temperature than available via air cooling. The cooling water is normally supplied from and returned to a cooling tower. The warm returning water is cooled by direct contact with air with most of the cooling being supplied by the evaporation of some of the water. Chemical addition is required to prevent fouling of the cooling water and exchangers. A water purge is needed to avoid chemical buildup in the cooling water. While the cost of the condenser is relatively low, the cost of the cooling tower-related equipment is of significance.
- Once-through cooling water can be sourced and returned to a local stream or ocean. Some pretreatment of the water is needed to prevent fouling and corrosion problems. In some cases a significantly lower process temperature may be obtainable with this choice.
- Steam generation is possible when the condenser outlet temperature is very high. Using the condenser energy to generate steam produces a lot of value provided that the steam is needed as a heat source.
- Exchange with another process stream may have significant economic benefit. A common service is to have the condenser energy supply the reboiler duty for another column. However, this type of heat integration does result in possible operation problems. The two columns are linked, making start-up a special consideration. In addition, neither column can be operated in normal mode without the other column. For example, it would be very difficult to have maintenance in progress on one column while the other is in operation.

In the case where a two-phase distillate is acceptable, it is common that maximizing the liquid portion is still desirable. This means that a high operating pressure is desirable as is a lower condenser outlet temperature.

In the case where the distillate is desired as a vapor, the condenser still needs to have an outlet temperature that is the desired distillate's dew point temperature and that is low enough to allow the use of the desired cooling medium. A higher pressure will result in a higher dew point temperature.

The reboiler consideration is much the same as the condenser except that the push is normally to a lower pressure which yields a lower bottoms temperature. A low bottoms temperature may be required to avoid heat damage to the bottoms product (cracking or coking) or a low bottoms temperature may be needed to allow the desired heating medium to function. Below is some discussion on several possible heating mediums.

- Steam is a very common reboiler heating medium. Frequently, steam is available at several different pressure levels. The cost of generating steam increases with the pressure level of that steam. Hence, it is normally desirable to use the lowest of steam that will work.
- Some locations may have circulating hot oil available. It is possible that the hot oil may be significantly hotter than available steam. Once the hot oil has been used (cooled), it is normally reheated via a fired heater. The advantage of hot oil over direct use of a fired heater is that a single fired heater may be used to supply heat for many services. Fired heaters are a safety concern and an environmental concern. Having one large fired heater rather than several smaller fired heaters allows easier management of these issues.
- The use of a fired heater is typically the last resort of reboiler heat input. However, its use is by no means unusual.
- Reboiler energy can be supplied by an available process stream with sufficient temperature and energy content. The use of a column overhead from a nearby column will provide a large economic benefit. In most cases the column overhead material needs to be a narrow boiling range material so that the dew point and bubble point are relatively close.

In overall, the column pressure is selected based on review of the condenser and reboiler operations.

15.3.3 Types of Reboiler Configurations

In general, there are two common types of reboiler configurations, namely recirculating and once-through reboilers. Recirculating reboilers (Figure 15.3) use the bottom product material to feed to the reboiler. The bottoms material is bubble point liquid with its separation requirement setting its composition and hence its temperature. The feed to the reboiler has this temperature. The outlet temperature of the reboiler will be higher and is set by the flow rate to the reboiler and the duty of the reboiler. A thermosiphon reboiler and most fired heater reboilers have this configuration.

Once-through reboilers (Figure 15.4) have the bottom product as the liquid that is leaving the reboiler. The reboiler outlet temperature is the bottom product temperature. The reboiler inlet material is the sum of the bottom product material and the vapor that leaves the reboiler. The reboiler inlet material is at its bubble point and will be cooler than the reboiler outlet material. This

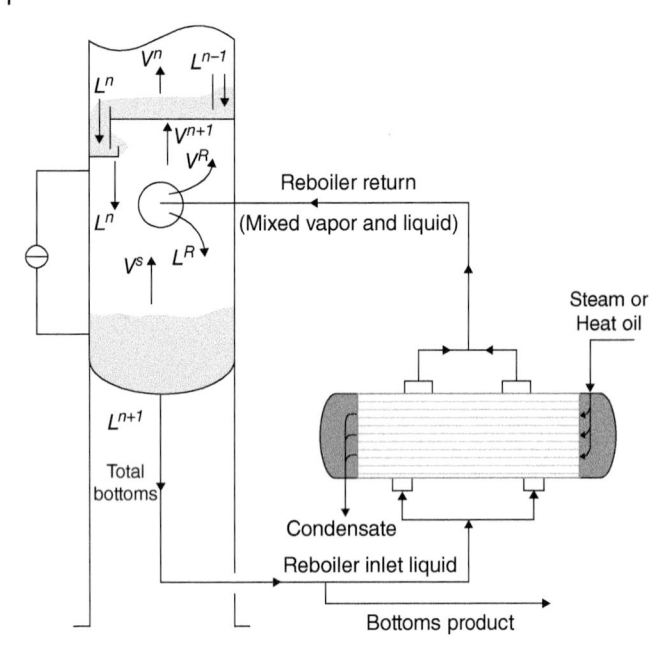

Figure 15.3 Recirculating thermosiphon reboiler.

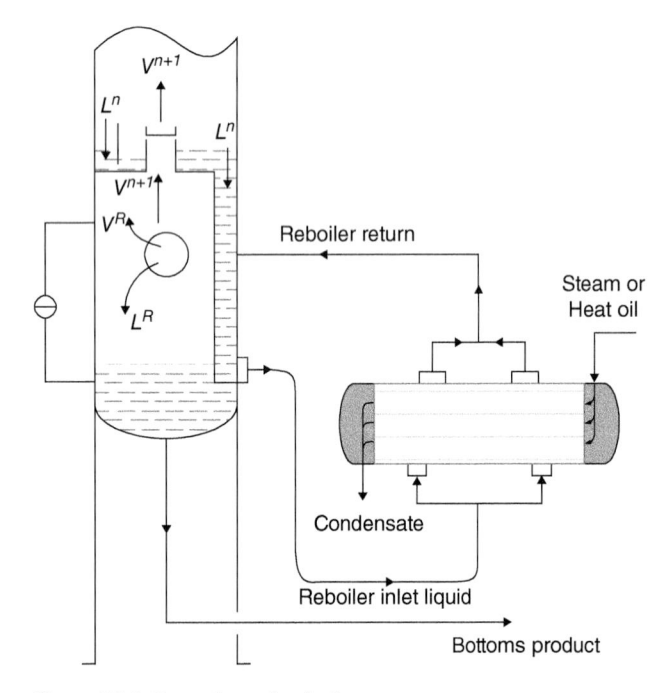

Figure 15.4 Once-through reboiler.

configuration has colder cold-side temperatures when compared to a recirculating reboiler. The amount of this temperature difference is a function of the boiling range of the bottoms material. The larger the boiling range the more the temperature difference will be. This temperature advantage may allow for the use of a less expensive heating medium. For example, perhaps a steam reboiler can be used rather than a fired heater reboiler. Perhaps medium pressure steam can be used rather than high-pressure steam. This benefit may justify this configuration over a recirculating reboiler configuration. Note that the cost and complexity of a once-through reboiler is higher than a recirculating reboiler. Once-through reboilers include stab-in and kettles. It is also possible to have other column bottoms internals that will provide this configuration.

There is a rarely used third type of reboiler configuration that is referred to as a preferential once-through reboiler. This is a modification of a once-through reboiler scheme. The bottom product is the same composition as the liquid that is leaving the reboiler; however, some reboiler outlet liquid is recycled back to the reboiler inlet. This recycled liquid allows the feed to the reboiler to be kept constant despite possible variations in the bottom rate. Constant flow control of the feed is normally considered critical for fired heater reboilers.

15.3.4 Optimization of Design

Optimization of a distillation column is almost always done by use of a simulator that uses an equilibrium-stage model to produce a full heat and weight balance for the column. When possible, obtaining specifications for a reference job will be of great value. The reference job should include a column that is like the new column with regard to key components and the desired separation. Having a reference job provides an easy source of good variable estimates that are required for a column simulation. To optimize a design, specifications for the overhead and bottoms products are required, along with the feed composition and condition, the total number of stages, the pressure of each stage, and the feed stage location. Estimated values for the reflux rate and the temperature profile are also typically required to assist with attaining a converged solution. In cases where good estimated values are not provided, the simulator may not be able to produce a converge simulation. An easier convergence is normally obtained when flow specifications are used rather than composition specifications. Instead of using distillate and bottoms composition, specification use a distillate flow rate and a reflux flow rate specification. The results can be reviewed, and the specified values can be adjusted to obtain approximate desired composition values. This converged simulation can be used to supply improved estimates for a simulation with the desired composition specifications. Note that for any given separation there is a minimum number of stages required. If there is no convergence for a column, consider the possibility that the specified stage count is close to or less than the minimum number of stages.

For this discussion the optimization will be limited to total stage count, feed stage location, and feed enthalpy. Each of these three variables does affect the optimum value of the other two variables so there is normally a little bit of a random walk as the optimum of each one is determined. Note that in most cases it is best to have the simulator use theoretical stage rather than actual trays. Most simulators allow for the use of trays with an assigned efficiency for each stage. Generally, this choice adds another layer of complexity/ambiguity without adding enough value. Once a design is complete, the needed stages should be converted to needed trays or packing height.

For a given number of stages there will be a feed stage location that supplies the lowest reflux rate/reboiler duty. This is clearly the best feed stage location for that total stage count design (see Figure 15.5).

As stages are added to either the stripping or rectifying location, there will be a reduction in the required reflux rate/reboiler duty (Figure 15.6). The value of the reduced reflux/reboiler duty needs to be considered against the cost of the added stages. The cost of a stage is normally considered to be small so a reduction in reflux/reboiler duty of 0.8% is considered clear justification for the addition of that stage. For many situations a value of 0.4% or lower can be justified. Having a lot of stages has a second benefit in addition to the lower utility cost. This benefit is a design that is much more tolerant to an unexpected tray or packing efficiency problem. While conservative values for a tray or packing efficiencies should always be used, surprises are, unfortunately, still possible. Column design procedures normally have implicit overage that will allow operation with an additional 10% reflux. By designing a column well away from the minimum stages, the implicit reflux overage will cover some shortfall in the design efficiency.

The last optimization variable, discussed here, is the feed enthalpy. Again, an equilibrium stage simulation of the column is almost always used to determine a good feed enthalpy value. The cost of adding feed enthalpy and the value that results are normally complex/nonobvious. When enthalpy is added to the feed, the required reboiler duty will decrease but by less than the enthalpy that is added to the feed, when the separation requirements for the column are held constant. When comparing two cases, the change in the reboiler duty divided by the change in the feed enthalpy can be considered as a preheat efficiency (Figure 15.7). As more enthalpy is added to the feed, the preheat efficiency will decrease. Determination of the preheat efficiency is not normally of great value but an appreciation of this behavior is of value. Below are several sample cases/discussion points.

- Column feeds are frequently relatively cold and can be an excellent place to recover enthalpy in other streams that would otherwise be sent to a cooler or have some other low value use. In these cases, it is likely that a

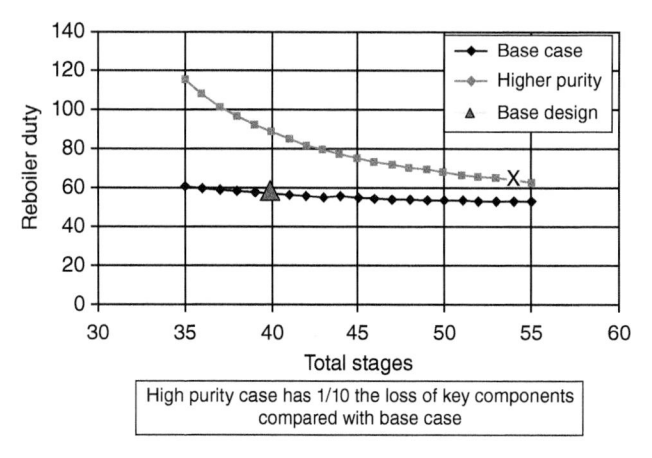

Figure 15.6 Stages vs. reboiler duty.

Stage	R/D
15	1.485
16	1.472
17	1.454
18	1.470
19	1.480
20	1.499
21	1.525

Figure 15.5 Feed stage selection.

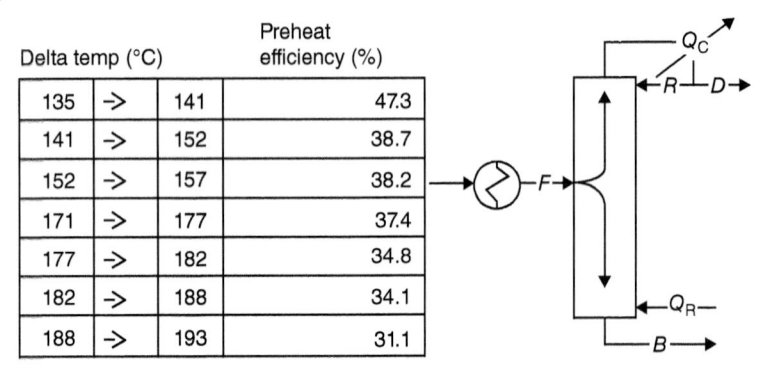

Figure 15.7 Feed preheat optimization vs. preheat efficiency.

Delta temp (°C)			Preheat efficiency (%)
135	->	141	47.3
141	->	152	38.7
152	->	157	38.2
171	->	177	37.4
177	->	182	34.8
182	->	188	34.1
188	->	193	31.1

Preheat efficiency = change in reboiler duty/change in feed enthalpy

feed exchanger with a close temperature approach can be justified. Consider a column that has a cold feed and a bottoms stream that will be cooled and sent to storage. A feed-bottoms exchanger will allow some of the bottoms stream enthalpy to be moved to the feed stream where it will be better used.

- In some cases the column feed may be available with a very high enthalpy. Here, it may be advantageous to remove some of this enthalpy rather than allow it to enter the column. Consider a two-column arrangement. The first column is high pressure with a cold feed. The bottoms of the first column are the feed to the second column. The second column pressure is at a much lower pressure than the first column. If no enthalpy is removed from the feed to the second column, the feed will have a very high enthalpy resulting in high fraction of vaporization. It is common that first-column bottoms are cooled by heat exchange with the feed to the first column. Properly sizing this exchange will result in minimizing the cost of reboiling the two columns.
- A good understanding about the effect of feed enthalpy will allow better column design. As stated, adding feed enthalpy will decrease the needed reboiler duty and will increase the condenser duty, when the separation is held constant. The lower reboiler duty results in worse fractionation in the stripping section due to a V/L ratio that is farther away from 1.0 than it was. (The best fractionation for any column section is always attained under total reflux conditions, V/L is 1.0.) As a constant separation is desired, a higher reflux rate is required. The V/L in the rectification section moves closer to 1.0 in order to supply the reduced fractionation that occurred in the stripping section. The required diameters of the stripping and rectification sections are loosely related to the reboiler duty and the condenser duty, respectively. There is no right ratio of these two diameters but there are important implications that should be considered by the design engineer. When the rectification diameter is much smaller than the stripping section diameter, feed enthalpy and reboiler duty are likely critical to the stripper section diameter and the condenser duty. Small variations in the reboiler duty/feed enthalpy or in the differences between them can result in rectification section problems. Good control of the feed enthalpy and reboiler duty are then very important. In most designs, the reboiler duty is well controlled. This is not always the case for feed enthalpy.

Having the rectification section diameter larger than the stripping section diameter is not common but not rare either. The mechanical support for the rectification can be expensive. Some consideration to alternate designs, with a lower feed enthalpy, should be reviewed. However, extraordinary effort is not needed as there are many cases where a reverse-column swage is appropriate.

15.3.5 Side Products

Occasionally, there is a desire to produce a side-draw product that may be located either above or below the feed stage. A key point of understanding about these configurations is that this product will always have significant content of either the overhead product or the bottoms product. When the side draw is located above the feed, there is no way to avoid significant content of the overhead material as it must go past this side-draw location. Similarly, when the side draw is located below the feed, there is no way to avoid significant content of the bottoms material as it must go pass this side-draw location. Increasing the stages or reflux will have only a minor effect on these contaminations.

In most cases a liquid side draw is produced as there tends to be less issues with moving and processing a liquid stream as compared to a vapor stream. When the side draw is above the feed, there is the additional advantage of having a full stage of separation with overhead product going by this point as a vapor. In order to gain a similar advantage when the side draw is below the feed, the side draw needs to be a vapor. A full stage of separation is then available as the bottoms product needs to go by this point as a liquid. The benefit of a single stage of separation can normally only be of significance when the relative volatility of the key components is large. Consider the case when the column feed contains a very small quantity of very heavy polymer and the side draw needs to be located below the feed. Here, the side draw may be the major heavy product and is located below the feed to allow good separation between the overhead product and the side draw. Most of the heavy polymer, that is in the feed, can be routed past the side-draw location by having a vapor draw.

The next level for improved purity of the side-draw material can be obtained by the addition of a side column. If the side-draw material is a liquid from above the feed, it can be routed to the top of a stripper with the stripper overhead vapor returning to the main column. If the side-draw material is a vapor from below the feed, it can be routed to the bottom of a rectifier with rectifier bottoms returning to the main column. The use of side columns will certainly reduce the contamination of the side product but the cost and effectiveness may not be sufficient to meet the desired targets. The use of two columns in series or the use of a dividing wall column (DWC) may be appropriate.

15.4 Advanced Distillation Topics

15.4.1 Heavy Oil Distillation

Heavy oil distillation has a bottoms product that is too heavy to allow the use of a reboiler. The problem that prevents the use of the reboiler is that the reboiler outlet temperature would result in unacceptable coking or cracking or polymerization. Heavy oil distillation columns address this issue by using a feed that is as hot as possible without causing the temperature-related issues mentioned above. Most frequently there is a fired heater on the feed with the outlet temperature controlling the heat input (Figure 15.8). This normally results in a feed that is highly vaporized and causes difficulty with the feed introduction to the column. Special internals with significant spacing above and below the feed nozzle are the norm. Any liquid that enters the column through the feed nozzle is joined with liquid that comes from above the feed nozzle location. This combined liquid stream is frequently routed to a steam stripping section below the feed nozzle. The liquid to the stripping section contains light material that is desirable to recover as an upper product. The use of steam provides stripping vapors that did not require increasing the liquid temperature. The steam is a non-feed stream that proceeds up the column with its presence, in all products, needing to be considered.

As the vapor proceeds up the column, from the feed point, it needs to be contacted with descending liquid in order to knock down heavy components that are not desired in the next product draw location above. The V/L ratio sets the heavy removal from the rising vapor, with a low value (closer to 1.0) providing better removal. The negative side of a low V/L ratio is that more light material is forced into the column bottoms. The loss of light material into the bottoms can only be reduced by increasing the feed enthalpy or increasing the bottom steam stripping rate.

Between adjacent draw points the V/L ratio tends to decrease going up the column due to the temperature decrease. This effect is frequently strong enough that excess heat can be removed from the column at the side-draw locations. This is done via a pump around. Liquid is withdrawn from the column and is used as a hot oil to provide energy to an appropriate service. The cooled liquid is then returned to the column. The amount of

Figure 15.8 A heavy oil configuration of a distillation column.

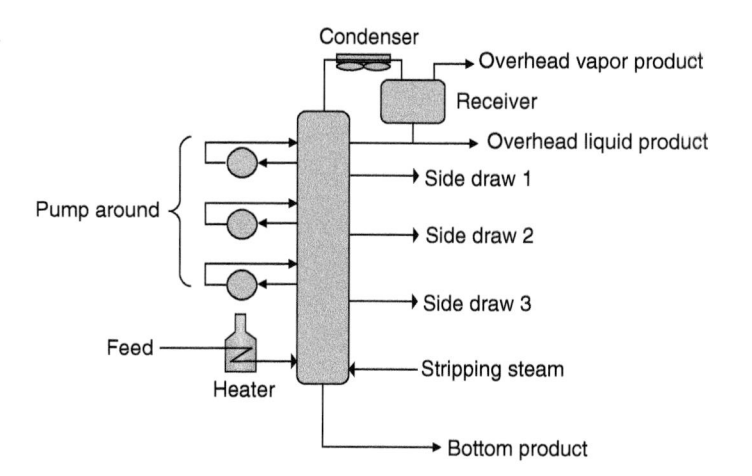

heat that is removed at a pump-around location changes the V/L above that location and therefore effects the heavies that will be allowed to rise to the next product above that point.

The amount of light material in each side draw cannot be significantly affected by changes in the V/L ratio. As discussed before, side-draw products that are located above the feed point always have significant content of light material that is product above the feed point. To address this issue a stripper column is commonly located on the side product streams. Stripping vapor for these columns can either be provided by a reboiler or steam injection.

15.4.2 Dividing Wall Column

A DWC is a special type of distillation column containing a vertical partition wall inside the column, allowing three or more products to be produced using a single vessel. This type of distillation column can not only reduce both the capital and energy costs of fractionation systems comparing with simple columns but also produce higher purity products than the products from a simple side-draw column. The process integration theory via the grand composite curve can be applied to DWCs.

15.4.2.1 DWC Fundamentals

Consider a mixture consisting of three components, A, B, and C, where A is the lightest and C is the heaviest. Figure 15.9a shows how a direct sequence of two distillation columns can be used for this separation. For some mixtures, for instance when B is the major component and the split between A and B is roughly as easy as the split between B and C, this configuration has an inherent thermal inefficiency, as illustrated in Figure 15.9b for a generic example. In the first column, the concentration of B builds to a maximum value at a tray near the bottom of the column. On trays below this point, the amount of the heaviest component, C, will continue to increase, diluting B so that its concentration profile will now decrease on each additional tray toward the bottom of the column. Energy has been used to separate B to a maximum purity on an intermediate tray in the first column, but because the B has not been removed at this point, it is remixed and diluted to the concentration at which it is removed in the bottoms. This remixing effect creates a thermal inefficiency.

Figure 15.10 shows a configuration that eliminates this remixing problem. This prefractionator arrangement performs a sharp split between A and C in the first column, while allowing B to distribute between the overhead and bottoms streams. All of the A and some of the B are removed in the overhead of the smaller prefractionation column, while all of the C and the remaining B are removed in the bottoms of the prefractionation column. The upper portion of the second column then separates A from B, while the lower portion separates B and C. The fraction of B separated in the overhead of the prefractionator can be specified in order to eliminate any of the remixing seen in the direct sequence of Figure 15.10a, leading to a significant energy saving, as shown in Figure 15.10b. This saving can be about 30% for a typical design, but can reach 50 or 60% for unconventional designs.

The prefractionator arrangement (Figure 15.11a) can be thermally integrated, creating what is known as the Petlyuk arrangement (Figure 15.11b). Vapor and liquid streams from the second (main) column are used to provide vapor and liquid traffic in the prefractionator. This system has only one condenser and one reboiler, and both are attached to the second column. Because the Petlyuk arrangement has fewer major equipment items than the conventional two-column sequence, the total capital costs may be reduced. The prefractionation column can then be integrated into the same shell as the main column, as shown in Figure 15.11c, forming the DWC. Assuming that heat transfer across the dividing wall is negligible, a DWC is thermodynamically equivalent to a Petlyuk arrangement. When compared to a conventional two-column system, reboiling duty saving and capital cost saving of up to 30% is typical.

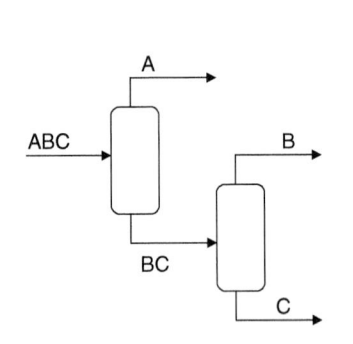

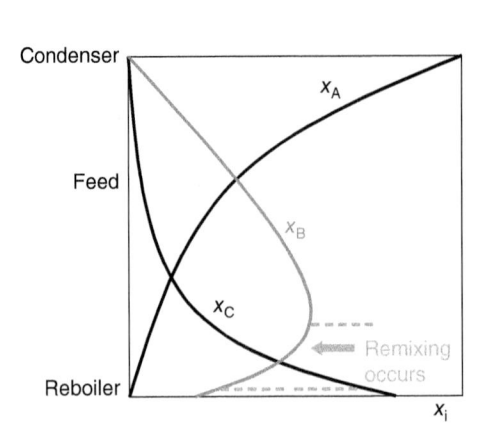

Figure 15.9 Thermal inefficiency in direct sequence. (a) Direct distillation sequence. (b) Component profiles for the columns.

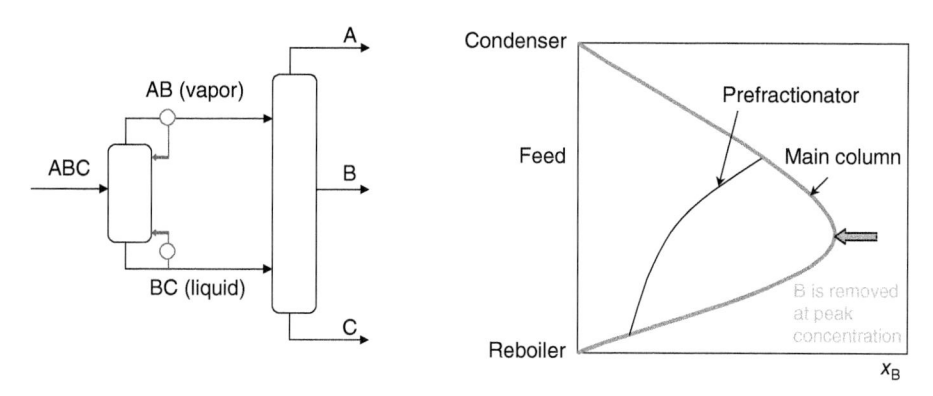

Figure 15.10 Thermal efficiency for prefractionator arrangement. (a) Prefractionator arrangement. (b) Component profiles for the columns.

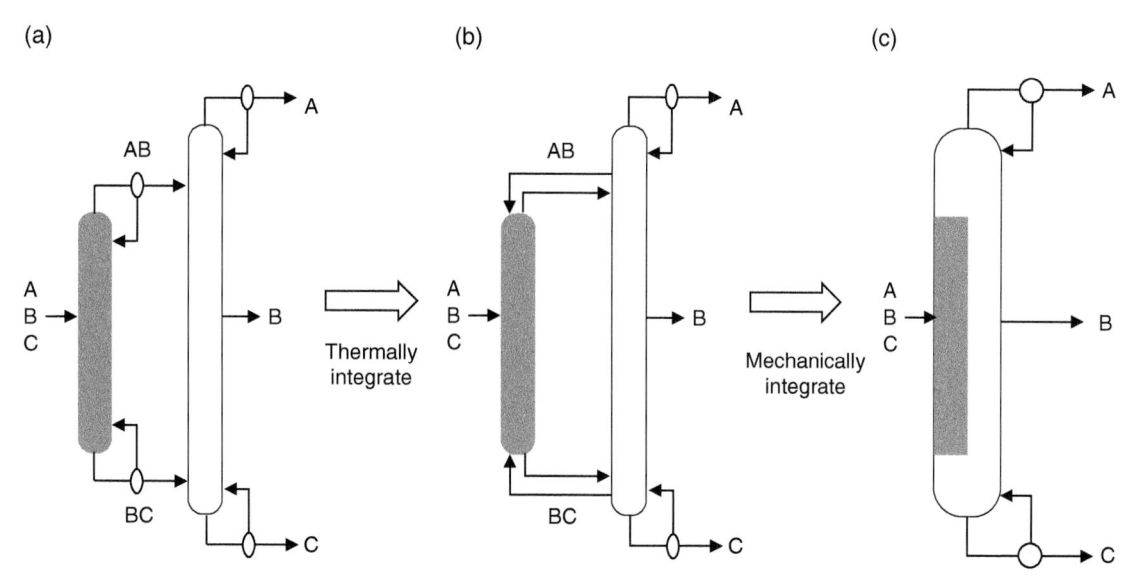

Figure 15.11 From prefractionation to dividing wall. (a) Prefractionator. (b) Thermally coupled columns (Petyluk column). (c) Dividing wall column.

15.4.2.2 Guidelines for Using DWC Technology

It seems DWC has significant advantages over simple distillation columns. Can it be applied to replace simple columns when three products are produced? The answer is no as there are a certain conditions for applying DWC. When evaluating the possibility of using a DWC, it is important to consider the properties and composition of the separating material as well as the product specifications. The following guidelines (G) (Zsai et al. 2016) describe situations when a DWC may be beneficial.

- Guideline for feed composition (G_1): Neither components A nor C should dominate, although they do not necessarily have to be equal.
- Guideline for relative volatility (G_2): The relative volatility ratios of A/B and B/C should be comparable.
- Guideline for pressure (G_3): The corresponding two-column system should be at similar pressures.
- Guideline for material construction (G_4): The corresponding two-column system should not already be at mechanical design limits or require distinct metallurgies with vastly different costs.

As with any heretic rules, there are exceptions, and in certain situations the rules may contradict each other, but these guidelines can be useful when identifying applications in which a DWC may be better than other distillation options. Two case studies will be discussed below to show how these guidelines are applied and what benefits could be achieved by use of DWC.

15.4.2.3 Application of Dividing Wall Column

Naphtha isomerization processes typically use distillation columns for both prefractionation of the feed and post fractionation of the reactor products. Depending on a refiner's gasoline blending requirements, C_5–C_6 isomerization unit prefractionation schemes can include depentanizers (DP) and deisopentanizers (DIP). The DP column is designed to produce a net overhead product that is 95 mol.% isopentane per nC_5 and iC_5 components and recover 67.4% of the total isopentane of the feed. The side-draw product is designed to recover 94% of the n-pentane in the feed. The remaining bottoms stream contains C_6+ material. These tower columns are large in dimension and consume a significant amount of utilities.

Currently, the DP and DIP towers can be combined into a single three-cut tower (Figure 15.12) with side draw and while there are savings in plot space and capital when using a single tower, total utility requirements are similar to the two-column arrangement. Product purities are also reduced when going to a single tower with side draw. There is consideration to find a more economical solution for cases where we want to produce separate iC_5, nC_5, and C_6-rich streams for isomerization complexes.

The sidecut column in Figure 15.12 is designed with 72 theoretical stages or 80 real stages with a thermosiphon reboiler as per the column specifications above. To achieve the purity specification for the n-pentane, the resulting reflux to feed (R/F) must be high at 2.96 with a required reboiler duty of 248 MMBTU/h. To reduce the reboiling duty, the number of real stages is increased to 107 with reboiling duty reduced to 209 MMBtu/h. Beyond this, any further addition of stages generates very small reduction in reboiling duty.

Potential application of DWC for this case is verified based on the guidelines provided above. The sidecut n-pentane purity remains the same as the base case which satisfies G_1. Although the sidecut B is not in excess, it has 25 mol.% which is not a small quantity. Thus, G_2 is loosely satisfied. Furthermore, the most important guideline is G_3 which is satisfied with α (iC_5/nC_5) = 1.15–1.21 while α (nC_5/C_6) = 1.4–1.5 across the pressure range of the distillation. This implies the separation of components B and C in the bottom section is easier than A and B, which is the fundamental reason why a DWC can achieve significant reduction in reboiling duty.

The DWC concept is considered to maximize the recovery of iC_5 in the overhead product while minimizing reboiler duty (King et al. 2014). The advantage of utilizing a DWC for this application is that the LK components (iC_5) would need to boil up over the wall and condense back down to the side-draw location. Thus, this allows more stages of separation to ensure a minimal amount of iC_5 is in the side draw while keeping the fractionation utility minimal.

Therefore, a dividing wall column (DWC) (Figure 15.13) to combine the DIP and DP columns instead of combining them into a simple column with side draw (Figure 15.12). The DWC was designed to the specifications outlined above and optimized toward reducing reboiler duty via optimizing feed tray location, location of the top of the dividing wall, location of the bottom of the dividing wall, liquid reflux to each side of the wall, and vapor split to each side of the wall.

For a recent C_5–C_6 isomerization unit design, we found that using a DWC instead of a simple column with side draw reduced capital cost by $4.6 MM USD (32%)

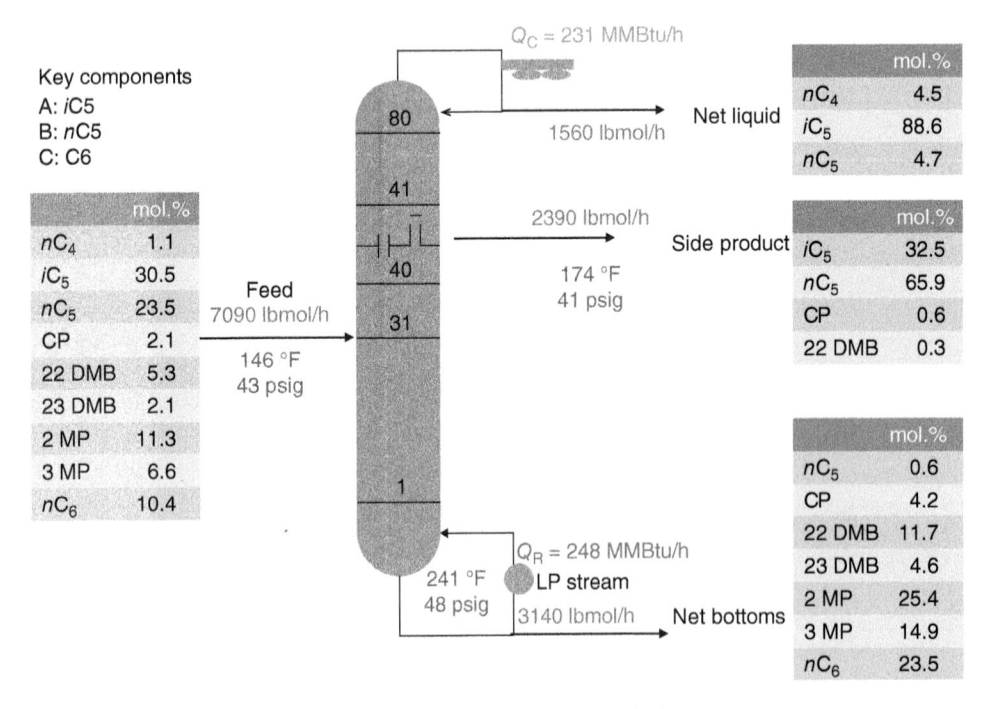

Figure 15.12 Current design: single-column depentanizer with side draw.

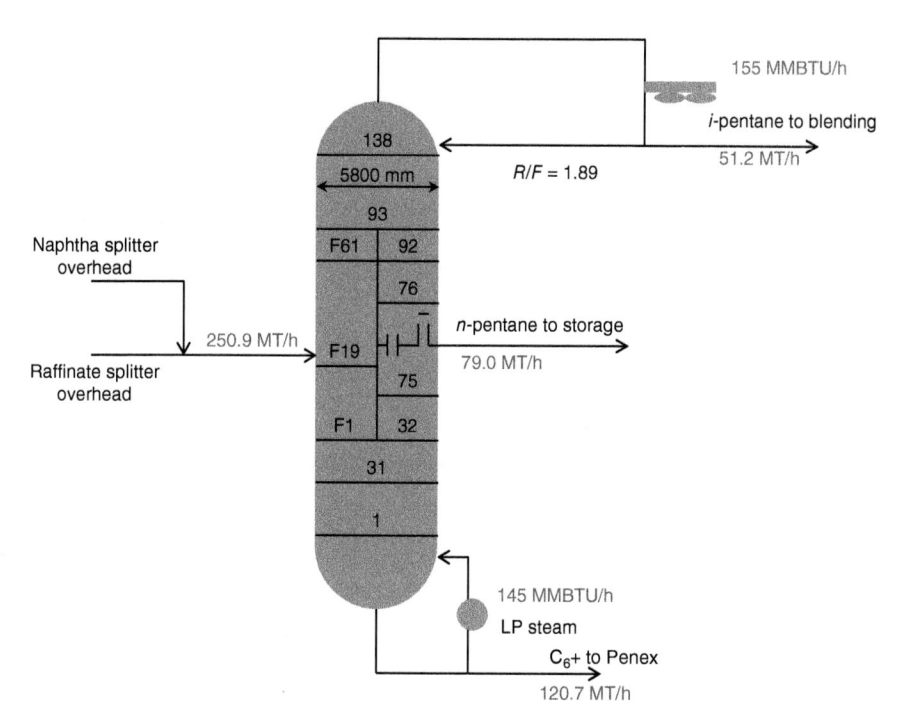

Figure 15.13 Dividing wall column for the depentanizer.

and more importantly, the required reboiler duty was reduced by 30% resulting in a savings of $2.6 million USD in operating costs per annum for a 2000 KMTA capacity unit.

15.4.3 Choice of Column Internals

15.4.3.1 Crossflow Trays

These are the most common type of column internals. Here, the trays consist of downcomers and bubbling areas. The downcomers collect the liquid from the tray above and move the liquid to the tray below. The size of the downcomers needs to be large enough to disengage the vapor out of the descending liquid. The bubbling area has horizontal flow of liquid that proceeds from a downcomer outlet to a downcomer inlet. Perforations in the bubbling area allow vapor to pass through the floor of the tray and generate a froth/spray mixture. The bubbling area needs to be large enough to allow the liquid to disengage from the rising vapor prior to the vapor entering the tray above. Crossflow trays are the most common type of internals because they are inexpensive, are relatively easy to design, and are tolerant to design and operational problems.

The bubbling area may contain round hole, fixed valves, floating valves, or bubble caps. Trays that have round hole are referred to as sieve trays. Fixed valve trays have a housing above the perforation that deflects the vertical vapor flow to a horizontal vapor flow. The housing may be formed out of the tray floor or may be a separate piece of metal attached to the tray floor. In either case this housing cannot move, hence the name fixed valve. Fixed valve trays have an advantage over sieve trays as the horizontal flow of vapor helps prevent solids settling onto the tray floor and fouling the tray. The horizontal flow reduces the possibility of weeping of liquid through the tray floor and also results in a more uniform vertical flow which may yield a better capacity. Floating valves are like fixed valve but with a movable cover over each perforation. This creates a variable open area through which the vapor flows. The intent is that the variable area will decrease the likelihood of weeping when the vapor flow rate is low. This works well for single-pass tray designs.

For multipass trays there is concern that floating valves may allow maldistribution between the passes. In recent years the use of fixed valve trays has gained popularity but the use of sieve and floating valves are still common. Bubble caps have vapor risers that are attached to the tray. The riser is covered with a cap that forces the vapor to reverse flow and then exit into the bubbling area in a horizontal direction. Bubble cap tray designs were very popular in the distant past; however, they are much more expensive and have a lower capacity than a valve of sieve tray. This combination means that these trays are rarely used for a current design. There are two advantages to bubble caps. The riser means that it is very difficult for this tray to ever weep. Also, the riser allows the accumulation of significate debris which can be of some benefit in certain cases.

A subset of crossflow trays are sometimes referred to as "high capacity trays." These trays typically have a combination of several special features for the downcomer and bubbling area design. Some such features are heavily sloped downcomers, truncated downcomers, multichord or arched downcomers, small perforations, short flow path lengths, and small tray spacings. Many of these may reduce the operating range of the column and sometimes the tray efficiency.

15.4.3.2 Packing

The use of packing in columns has several negative features that make their use much less common than trays. The cost of packing is significantly more than trays. For difficult separations, a well-designed distributor with careful installation is required. Even when this occurs, the presence of solids can foul the distributor and yield poor distribution. However, there are several advantages of packing that can justify its use. Packing will almost always have a much lower pressure drop per theoretical stage than trays. When a vacuum column operation is required, it is common that packing is the best choice for the column internals. Trays should still be considered when only a small amount of vacuum is needed.

The two most common types of packing are referred to as random and structured. Random packing is normally less expensive than structured but typically has a high pressure drop per theoretical stage. Structured packing has frequently shown erratic behavior at high liquid flux rate, say above $20\,GPM/ft^2$. As such, it should not be used in these cases.

15.4.3.3 Super System Limit Devices

Both crossflow trays and packing rely on gravity to allow the liquid to move down the column while the pressure gradient is moving the vapor up the column. The highest capacity trays and packings have been used to define a system limit which is the maximum flux rate for any set of vapor and liquid flows. There are a small number of devices that can exceed the system limit. This is done by pairing a concurrent vapor/liquid contact step with a separation step that uses centrifugal or impact technology. Cost and operation concerns have limited the use of such device to only a small number of services.

15.4.4 Limitations with Distillation

Although distillation is the most common separation method in the refining and petrochemical processes due to the maturity of the technology and good experience in designing the distillation column and operating it, there are some limitations associated with distillation technology for which other separation methods may be considered as alternatives.

15.4.4.1 Separation Methods for Low Relative Volatility

Distillation is not effective to separate key components with very low relative volatility as it requires a very large column with too many stages leading to prohibitive capital cost. When relative volatility of key components is less than 1.05, other separation technology should be considered, such as extractive distillation, crystallization, membrane and adsorption, which are discussed below.

15.4.4.2 Separation Methods for Low Concentration

If only a small amount of one component is to be removed from a large amount of other components, changing the phase of the latter components should be avoided. In such a case, extractive distillation or selective adsorption are the preferred techniques to remove the minor component. The use of a non-feed vapor stream is sometimes used to remove small amounts of volatile chemicals from aqueous systems. It is particularly good for components that are immiscible with water and tend to have extremely high volatilities such as benzene, toluene, and other hydrocarbons. Stripping of low volatility components that are completely miscible with water is generally uneconomical because of the large amounts of stripping gas or steam that are required.

15.5 Adsorption

Adsorption is selectively concentrating a certain component(s) from a gas or liquid at the surface of a microporous solid. The mixture of adsorbed components is called the adsorbate while the solid is the adsorbent. Similar to absorption, adsorption is used to remove species in low concentration. The difference is that solid adsorbent is utilized in adsorption versus liquid solvent used in absorption. The advantage of adsorption is that sharper separation can be achieved than absorption as little cross contamination occurs. Adsorption is a surface-based process while absorption involves the whole volume of the material. The term sorption encompasses both processes, while desorption is the reverse of it.

Adsorption exploits the difference in van der Waals force (or affinity) of the key component to adsorbent in the form of solid particles. The adsorbate can be released (desorbed) by raising the temperature or reducing partial pressure of component in gas phase because the interaction energy is weak. Other properties that may be successfully exploited in the separation of isomer mixtures are the difference in kinetic diameter and complexation behavior resulting from the difference in isomer structures. In the case of p-xylene separation, this is used in a large-scale continuous adsorption process where shape-selective zeolites are used as adsorbents to achieve full separation. One of the properties they use is the relative linearity of the *p*-xylene molecule compared to the other xylene isomers. As a result, the *p*-xylene molecules fit better, the kinetic diameter is smaller, into the zeolite pores and an increase in the separation factor to values exceeding three is obtained.

There are gas and liquid adsorption. For gas adsorption, the key physical properties for separation are the molar volume and vapor pressure (or boiling points). For molecular sieve-based adsorbents, the key physical property for gas separation is the molecular size of species to be separated. For liquid adsorption, the key property difference for separation is the molar volume and the solute solubility in the liquid phase.

Selectivity in adsorption is controlled by molecular sieving based on molecular size or adsorption equilibrium based on affinity. When solutes differ significantly in molecular size and/or shape, zeolites and carbon molecular-sieve adsorbents can be used to advantage. These adsorbents have very narrow pore-size distributions that are capable of very sharp separations based on differences in the molecular kinetic diameter. Adsorbents made of activated alumina, activated carbon, and silica gel separate by differences in adsorption equilibria, which must be determined experimentally.

Adsorbents are highly porous and have large surface areas per gram of adsorbent. The adsorbent particles are commonly packed in a column. In general, although the particles will be of different sizes and shapes, they are described by an average interparticle porosity. In a poorly packed bed surface area may vary considerably in different parts of the column. This can lead to poor flow distribution or channeling and will decrease the separation. Since the particles are porous, each particle has an interparticle (within the particle) porosity. Approximately 2% of the surface area is on the outer surface of the packing; thus, most of the capacity is inside the particles.

Molecular sieve zeolites differ from both these pictures. The zeolite crystals form a porous three-dimensional array and have a highly interconnected, regular network of channels and cavities of very specific sizes. Thus, the crystal geometry is well defined. Commercial zeolite adsorbents are pelleted agglomerates of zeolite crystals and binders. The binders have large pores and relatively little sorption capacity compared to the zeolite crystals.

Adsorption is the cornerstone of SMB technology which will be discussed below. In addition, it is also widely used in industrial applications such as heterogeneous catalysts, activated charcoal, capturing and using waste heat to provide cold water for air conditioning and other process requirements (adsorption chillers), synthetic resins, increasing storage capacity of carbide-derived carbons, and water purification.

15.6 Simulated Moving Bed (SMB)

Separations of isomers such as *p*-xylene (*p*X) from the *o*-xylene (*o*X) and *m*-xylene (*m*X) involve very similar boiling points. As shown in Table 15.1, the boiling points of *p*-xylene, *m*-xylene, and ethylbenzene differ only marginally. With a separation factor of only 1.02, ordinary distillation would require about 1000 stages and a reflux ratio of more than 100.

However, the property difference that stands out is kinetic diameter, which results from the difference in isomer molecular structures. One of the properties exploited is the relative linearity of the *p*-xylene molecule compared to the other xylene isomers. In other words, the *p*-xylene molecules have smaller kinetic diameter than other isomers. If zeolite with good shape selectivity is used, *p*-xylene molecules can fit better into the zeolite pores than other xylene isomers. This understanding leads to development of shape-selective zeolites as adsorbents.

When adsorbents are saturated with adsorbed species, it must be regenerated, which makes adsorption often batchwise. Because of practical difficulties with a continuously moving bed, SMB technique was developed for achieving full separation, which is the cornerstone of

Table 15.1 Properties of xylenes and ethylbenzene.

Property	o-xylene	m-xylene	p-xylene	Ethylbenzene
Molecular weight	106.2	106.2	106.2	106.2
Boiling point (°C)	144.2	139.5	138.7	136.5
Melting point (°C)	−25	−47.7	13.5	−94.7
Dipole moment (debye)	0.21	0.10	0.00	0.20
Polarizability ($10^{-31}/m^3$)	141	142	142	135
Molecular volume (cm^3/mol)	121	123	124	123
Kinetic diameter (Å)	6.8	6.8	5.8	6.3

UOP large scale of continuous adsorption process for aromatics complex.

This SMB-based process reduces energy use significantly comparing with alternative xylene separation technology such as conventional distillation, extractive distillation, etc. SMB is a continuous process and it simulates the countercurrent flow of a liquid feed over a solid bed of adsorbent. To appreciate the beauty of the SMB technology, we need to understand the concept of moving bed first.

15.6.1 The Concept of Moving Bed

The rationale behind the concept of moving bed is to ensure a plug flow of liquid against adsorbent solids to minimize backmixing. The countercurrent flow in a moving bed can be explained below by means of Figure 15.14.

- **Zone I (adsorption of A):** It is between the point of feed injection and raffinate withdraw. As the feed flows down through Zone 3, countercurrent to the solid adsorbent flowing upward, component A is selectively adsorbed from the feed into the pores of the adsorbent. At the same time, the component A displaces the desorbent, component D, and drives it out of the pores of the adsorbent.
- **Zone II (desorption of B):** It is between the point of feed injection and extract withdraw. At the fresh feed point, the upward flowing solid adsorbent contains the quantity of component A that was adsorbed in Zone 3. However, the pores will also contain a large amount of component B, because the adsorbent has just been in contact with fresh feed. The liquid entering the top of Zone 2 contains components A and D without B. Thus, component B is gradually displaced from the pores by A and as the adsorbent moves up through Zone 2. At the top of Zone 2, the pores of the adsorbent contain only A and D.
- **Zone III (desorption of A):** It is between the point of desorbent injection and extract withdraw. The adsorbent entering Zone 1 carries only A and D. The liquid entering the top of Zone 1 consists of pure D. As the liquid stream flows downward, component A in the pores is displaced by D. A portion of the liquid leaving the bottom of Zone 1 is withdrawn as extract and the remainder containing A and D flows down into Zone 2 as reflux.
- **Zone IV (isolation zone or partial desorption of D):** The main purpose of Zone 4 is to segregate the feed components in Zone 3 from extract in Zone 1. At the top of Zone 1, the adsorbent pores are completely filled with D. The liquid entering the top of Zone 4 consists of B and D. By properly regulating the flow rate of Zone 4, it is possible to prevent the flow of component B into Zone 1 and avoid contamination of the extract.

The desorbent liquid is selected so as to have boiling point significantly from those of the feed components. In addition, the desorbent must be capable of displacing the feed component from the pores of the adsor-

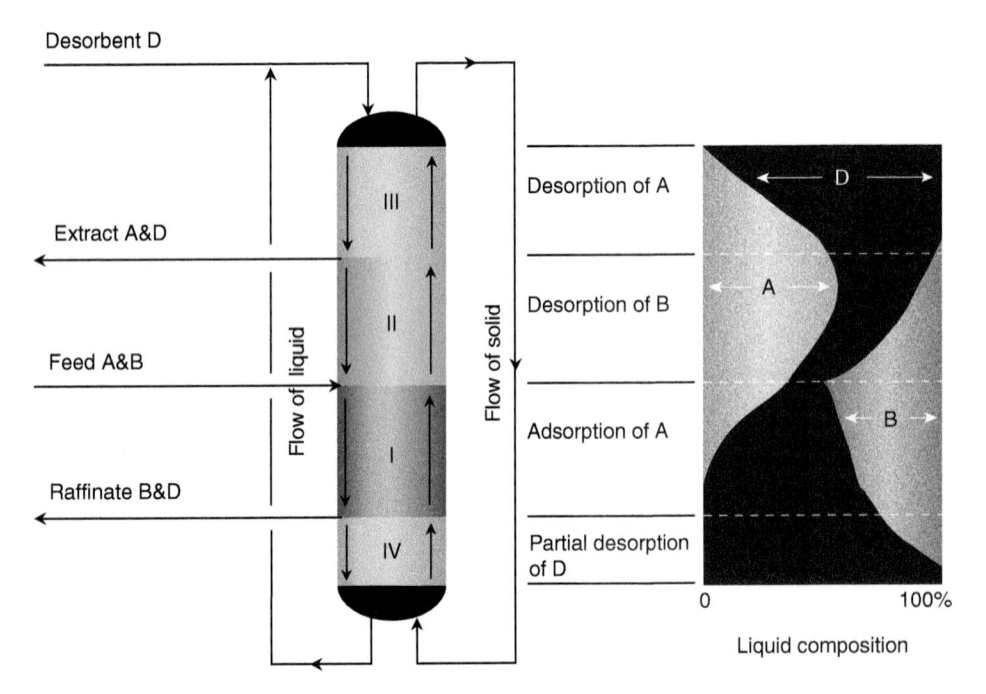

Figure 15.14 Moving bed analogy used for illustration.

bent. Conversely, it must also be possible for the feed components to displace the desorbent from the pores of the adsorbent. Thus, the desorbent must be chosen so as to be able to compete with the feed components for any available active pore space in the solid adsorbent, on the basis of concentration gradients.

However, when adsorbents are saturated with adsorbed species, it must be regenerated, which makes adsorption often batchwise. It is very difficult to move solid beds of adsorbent as each adsorbent bed could be up to 7 m in diameter. Maintaining a plug flow over such a large diameter would be virtually impossible.

15.6.2 The Concept of Simulated Moving Bed

In the SMB technology, instead of moving the bed, the feed inlet, the solvent or eluent inlet, and the desired product exit and undesired product exit positions are moved continuously, which is accomplished by a unique rotary valve (RV) (see details below), giving the impression of a moving bed, with continuous flow of solid particles and continuous flow of liquid in the opposite direction of the solid particles, which is the reason why the technology is named as SMB. In other words, in a SMB column, the countercurrent flow of liquid feed and solid adsorbent is accomplished without physical moment of the solid. Instead, countercurrent flow is simulated by periodically changing the points of liquid injection and withdraw along a stationary bed of solid adsorbent. In this "simulated moving bed" column, the concentration profile changes moving down the column with adsorbent beds (Figure 15.15). This movement of the liquid streams is accomplished by a RV, which is discussed below.

15.6.3 Rotary Valve

In the SMB technique, the feed inlet, the solvent or eluent inlet, and the desired product exit and undesired product exit positions are moved continuously, giving the impression of a moving bed, with continuous flow of solid particles and continuous flow of liquid in the opposite direction of the solid particles. This difficult material transportation task is achieved by a complex valve arrangement. The valve and piping arrangements and the predetermined control of these material flows allow switching at regular intervals the sample entry in one direction, the solvent entry in the same direction but at a different location in the continuous loop, while changing the fast product and slow product takeoff positions to also move in the same direction, but at different relative locations within the loop. 144 two-way valves could be used that sequence and direct the flows of feed, desorbent, and extract in and out of the multiple adsorbent beds. In contrast, UOP's Parex process uses one Rotary-Valve™ that can function the same as that of 144 two-way valves. The UOP RV is a hydraulically driven, rotating plate device that replaces the need for multiple separately controlled valves. The UOP RV has the major features as below:

- Process Performance – The RV allows maximum product purities and recoveries by optimal flow control throughout the unit, and due to optimal piping layouts with a single valve, minimizing liquid inventories outside the adsorbent beds.
- Process Reliability and On-steam Factor. The UOP RV provides greater overall reliability relative to individual valves and shorter time to on-spec product following unit restarts.

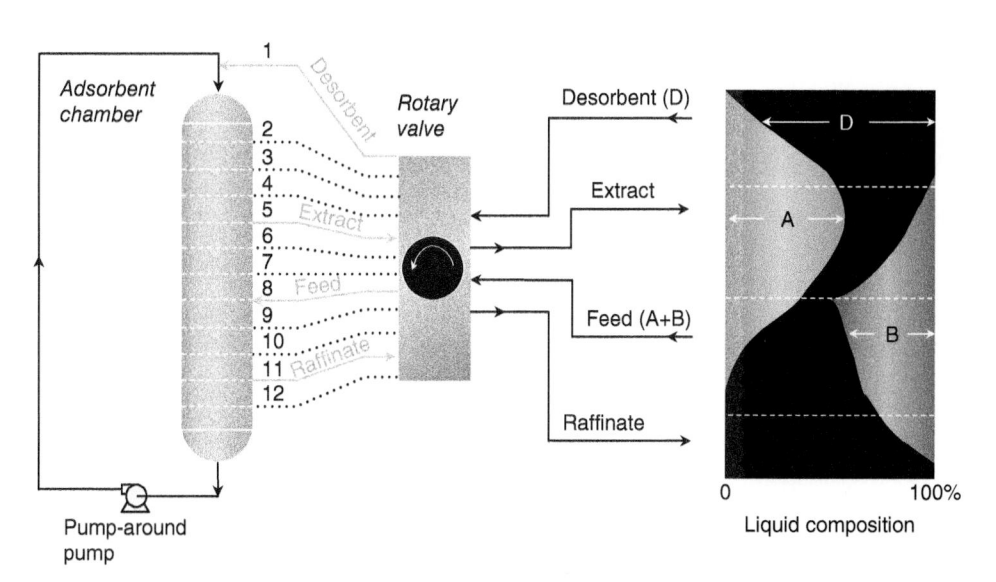

Figure 15.15 Simulated moving bed design.

- Lower Operating Costs – The UOP RV results in lower maintenance costs relative to banks of individual valves, and due to the process efficiency of the close-coupled RV and adsorbent chambers, lower desorbent circulation rates, and resultant utility consumption.

15.7 Crystallization

Crystallization is the process by which a solid is formed, where the atoms or molecules are highly organized into a structure known as a crystal. Some of the ways by which crystals form are precipitating from a solution, melting, or more rarely deposition directly from a gas. Attributes of the resulting crystal depend largely on physical properties such as temperature, air pressure, and in the case of liquid crystals, time of fluid evaporation.

Crystallization occurs in two major steps. The first is nucleation, the appearance of a crystalline phase from either a super-cooled liquid or a supersaturated solvent. The second step is known as crystal growth, which is the increase in the size of particles and leads to a crystal state. An important feature of this step is that loose particles form layers at the crystal's surface that lodge themselves into open inconsistencies such as pores, cracks, etc.

The majority of minerals and organic molecules crystallize easily, and the resulting crystals are generally of good quality, i.e. without visible defects. However, larger biochemical particles, like proteins, are often difficult to crystallize. The ease with which molecules will crystallize strongly depends on the intensity of either atomic forces (in the case of mineral substances), intermolecular forces (organic and biochemical substances), or intramolecular forces (biochemical substances).

Crystallization is also a chemical solid–liquid separation technique, in which mass transfer of a solute from the liquid solution to a pure solid crystalline phase occurs. In chemical engineering, crystallization occurs in a crystallizer. Crystallization is therefore related to precipitation, although the result is not amorphous or disordered, but a crystal.

Since there is no solvent, melt crystallization has the advantage that no solvent removal and recovery is required and contamination by the solvent is impossible. However, there is also no way to influence the melt properties (viscosity and diffusivity) and the chemicals being purified must be stable at the melting point.

The design of crystallization equipment is quite difficult. Knowledge of the phase equilibrium and physical properties does not allow one to predict the behavior of the crystallization process. Bench-scale experiments on the actual stream always are required, and equipment vendors usually require a pilot-plant test in small commercial equipment before designing the equipment. Even with such testing, operating adjustments usually must be made on the actual commercial installation before the equipment will operate to yield an acceptable product. Even with such testing, there are occasional failures at the commercial scale.

An example of crystallization is the separation of xylene isomers and ethylbenzene. Xylenes are obtained commercially from the mixed hydrocarbon stream manufactured in naphtha reforming units in oil refineries. p-Xylene is the desired product for the manufacture of phthalic acid and dimethyl terephthalate. As shown in Table 15.1, the boiling points of p-xylene, m-xylene, and ethylbenzene differ only marginally. With a separation factor of only 1.02, ordinary distillation would require about 1000 stages and a reflux ratio of more than 100. As the chemical nature of all four constituents is almost the same, the addition of an extractive solvent will act approximately the same on all isomers. The only macroscopic property that differs considerably is the melting point, allowing selective recovery of the p-xylene from the mixture by crystallization. For crystallization, the separation factor is nearly infinity due to the additional advantage of shape selectivity in crystal growth. The resulting flowsheet for industrial p-xylene recovery is shown in Figure 15.16.

15.8 Liquid–Liquid Extraction

Solvent extraction for liquid–liquid separation is a major alternative to distillation for more concentrated feeds and is appealing for separation of diluted species. The temperature of operation is typically ambient. The property that liquid–liquid extraction exploits is solubility. In the extraction, one key component is more attracted to the solvent than the other. The products are the extract (the liquid absorbed by the solvent) and raffinate (the liquid left behind). The solvent-to-feed ratio is reasonably high as it affects the concentration in the extract. The solvent regeneration is usually done by distillation. The combination of extraction and distillation is called extractive distillation.

It is a phase-equilibrium process, just like distillation. To obtain the components of the original mixture in the specified purity, it is necessary to separate the extract stream by recovering the extracted component from the extracting solvent. Usually, it is also necessary to recover the small amount of solvent dissolved in the raffinate stream to prevent either excessive solvent make-up costs or contamination of the raffinate product. Frequently, these auxiliary separations are accomplished by distillation. Therefore, an extraction process virtually always requires multiple processing steps to make acceptable products.

Solvent extraction has a great flexibility as wide range of solvents are available. The type of solvent to be chosen depends on the properties of solute (the species to be

Figure 15.16 Schematic of p-xylene production by crystallization and isomerization.

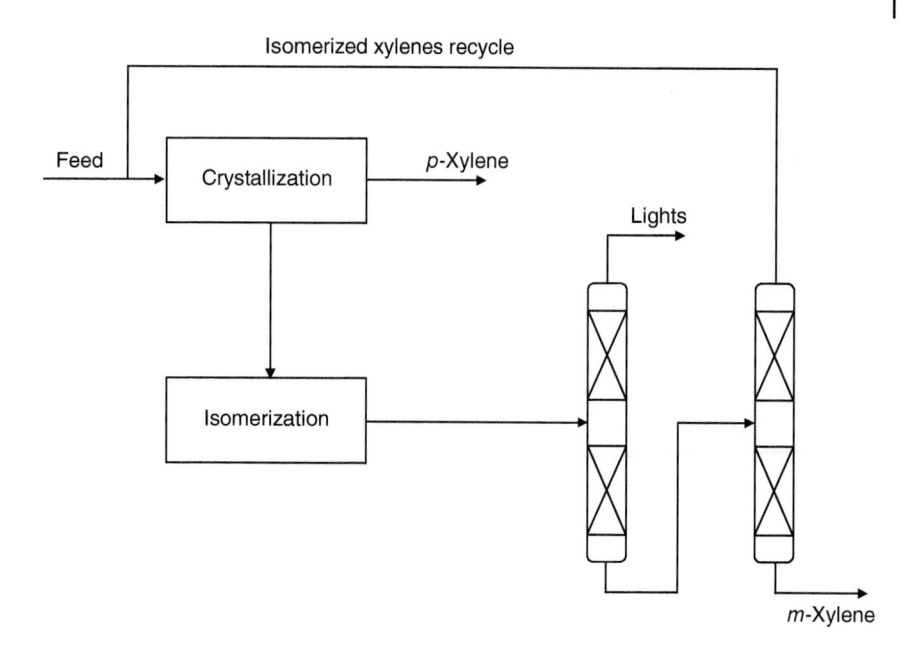

extracted). The design of extraction processes requires a knowledge of the phase–equilibrium relationships between the components to be separated and the extraction solvents to be used. With this information and a knowledge of the pertinent physical properties (densities, interfacial tension, viscosities, etc.), one can project a preliminary design that is reliable enough to determine the choice of solvents and operating conditions to be used and the type of extraction equipment to use. However, the designs generated in this manner are not nearly as reliable as those for distillation. Consequently, small-scale testing on the type of equipment to be used in the plant is required.

15.9 Extractive Distillation

Extractive distillation is effective in separating components with close boiling points or components forming azeotropes. In extractive distillation, an extra miscible component (called solvent) is added and the solvent alters the relative boiling points or volatility of original compositions making the separation possible (Figure 15.17). The key physical property difference that extractive distillation can exploit is relative solubility of two key components to the solvent used. In general, the solvent is the least volatile and thus is added near the top of the column.

Originally, extractive distillation was limited to two components' separation and the recent advances in solvent technology makes multicomponent separation possible. There are many commercial applications for extractive distillation. For example, butane and butene separation take place with furfural–water mixture as the solvent. In petroleum processing, aromatic hydrocarbons such as benzene, toluene, and xylenes (BTX) are

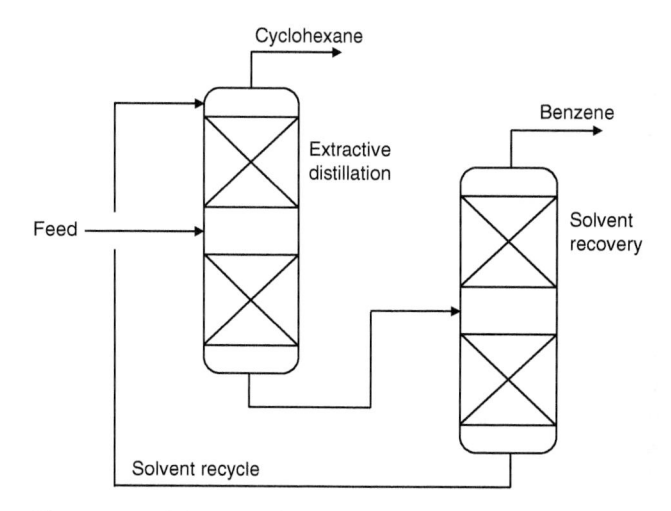

Figure 15.17 Schematic of an extractive distillation process.

separated by extraction with a solvent as Sulfolane. The mixture of Sulfolane and aromatics is sent to a distillation column, where the Sulfolane is the bottom product and is recycled back to the extractor. The BTX process of GTC Technology Corp. uses extractive distillation to replace conventional liquid–liquid extraction to separate BTX from catalytic reformate. This led to CAPEX and OPEX saving by 25 and 15%, respectively.

Other examples of applying extractive distillation include separation of high-purity cyclohexane from hydrocarbons, benzene and toluene from nonaromatics, methyl acetate from methanol, propylene from propane, and 1-butene from 1,3-butadiene.

Let us consider the separation of benzene from cyclohexane and another example is 1-butene to be separated from isobutane. Table 15.2 shows the properties of

Table 15.2 Properties of benzene, cyclohexene, 1-butene, and isobutane.

Property	Benzene	Cyclohexene	1-Butene	Isobutane
Molecular weight	78.1	84.2	56.1	58.1
Boiling point (°C)	80.1	80.7	−6.2	−11.7
Melting point (°C)	5.6	6.5	−185.3	−159.5
Dipole moment (debye)	0	0.3	0.3	0.1
Polarizability ($10^{-31}/m^3$)	103	110	80	82

these components. This table illustrates that the difference in boiling points between cyclohexane and benzene is so small that results in very low relative volatility of 1.01, which is far too small to apply traditional distillation method. In the second example, although the boiling point difference for the 1-butene/isobutane system is relatively larger, nonideal behavior of the liquid phase is responsible for obtaining only very low relative volatilities.

In these cases the molecular difference in chemical structure can be exploited to increase the relative volatility to a sufficiently high level. This can be done by the addition of a polar solvent (n-methylpyrrolidone, Sulfolane, and furfural) that increases the volatility of the nonaromatic components than the volatility of the aromatic components. So, introducing a polar solvent in the top of a distillation tower will preferentially force the cyclohexane to the vapor phase and thereby facilitate separation. In most of these low relative volatility situations, extractive distillation is economically favored. A second distillation column is required to separate the benzene from the extractive solvent. The resulting flowsheet is indeed more complex than a distillation flowsheet, but the mass transfer rate and the phase disengagement ease of distillation are also present, making them economically favored in a number of low relative volatility situations.

15.10 Membranes

A membrane is a selective barrier, which allows certain species to pass through more easily than other species, and thus results in separation to occur. The separation is realized by selectively permeating components. Some components permeate through a polymeric membrane more easily than other. The components with high sorption and high diffusivity in the polymer membrane will preferably permeate through the membrane while other components will be kept at the feed side of the membrane. In general, membrane separation does not give sharp separation as what adsorption does. The key property that membrane exploits is diffusivity which depends on molecular volume and solubility. The driving force for membrane can be concentration difference, pressure drop, or electrical field.

Membrane is used when targeted species is difficult to condense, which provides moderate purity and recovery. Selecting a membrane is the trade-off between selectivity and permeation. Usually, the grater the selectivity the less permeation rate and the overall capacity required.

The degree of selectivity of a membrane depends on the membrane pore size. Depending on the pore size, they can be classified as microfiltration (MF), ultrafiltration (UF), nanofiltration (NF), and reverse osmosis (RO) membranes. Membranes can also be of various thicknesses, with homogeneous or heterogeneous structure. Membranes can be neutral or charged, and particle transport can be active or passive. The latter can be facilitated by pressure, concentration, and chemical or electrical gradients of the membrane process. Membranes can be generally classified into synthetic membranes and biological membranes.

Membranes are manufactured from natural filberts and synthetic polymers, but also from ceramics and metals. Membranes are fabricated into flat and thin sheets, tubes, hollow fibers, or spiral-wound sheets and incorporated into commercial modules or cartridges.

Facilitated transport membrane (FTM) technology represents a step-change functional intensification of membrane technology, offering potentially much higher flux and selectivity. One example of FTM technology is separating propane and propylene. Propylene and propane are among the light hydrocarbons produced by thermal and catalytic cracking of heavy petroleum fractions. Although propylene and propane have close boiling points, they are traditionally separated by distillation. Because distillation requires more than 120 trays and high energy costs due to considerable reflux and boil up flow rates compared to the feed flow, great attention has been given to the possible replacement of distillation with a more economical and less energy-intensive option.

Based on the given properties of both species shown in Table 15.3, the fact that the relative volatility between these two components is very close to one reveals the sole reason why the separation in a distillation is very difficult as it requires a very tall and large column with lots of energy. Therefore, it is clear that distillation alone is not effective for this separation task.

One property difference that stands out calling for attention is the dipole number. Propylene has the

Table 15.3 Molecular properties of propylene and propane.

Property	Propylene	Propane
Molecular weight	42.1	44.1
van der Waals volume (m^3/mol)	34.08	37.57
Dipole moment (debyes)	0.4	0
Normal melting point (°F)	−301.4	−306.4
Normal boiling point (°F)	−53.8	−43.6

asymmetric location of the double bond leading to a higher dipole number than propane. This makes propylene a weak polar compound. Several separation methods are able to explore this property difference, which include extractive distillation, pressure swing adsorption, thermal swing adsorption, and membrane. FTM technologies are an attractive alternative for this type of olefin/paraffin separation, even though their use on large scale is limited by the harsh conditions of hydrocarbon-rich environment under pressure which strongly reduce membrane lifetime (Faiz and Li 2012).

It was found that facilitated transport (FT) membrane using impregnated silver nitrate can exploit the polar property difference identified by molecular analysis (Liu and Karns 2017), This is because FT membrane using impregnated silver nitrate features a high affinity for propylene so that it can be carried more readily and selectively through membrane (Li and Calo 1985; Teramoto et al. 1989). FT membrane separation is based on solution diffusion as well as FT, i.e. reversible olefin complexation through π bonds with metal cations in a polymer membrane where olefins are capable of forming reversible chemical bonds with transition metal ions incorporated into the membrane due to the specific interaction between the olefin's hybrid molecular orbitals and the metal's atomic orbitals.

However, FT membrane alone is limited in propylene recovery. Thus, a hybrid system consisting of FT membrane and a distillation tower was considered (Marzouk Benali and Aydin 2010). The simulation result indicates the hybrid system with FT membrane and distillation can increase propylene recovery from 80% (distillation only) to 95% (hybrid system). Many million dollars from propylene recovery can be obtained with relatively small capital investment.

This example shows that understanding of molecular properties provides insights for the most promising separation methods to use for the task in hand. There are other potential hybrid systems that could achieve the separation objectives with lower capital and operating costs than distillation alone. Such integration may lead to improved separation processes with reduced capital and operating costs. It may also be possible to achieve the extent of separations that cannot normally be achieved by any one of the single technique. Discussions of various separation methods can be seen in (De Haan and Bosch 2013; Seader et al. 2011).

15.11 Selecting a Separation Method

When selecting and designing a separation unit there is always a need to balance the capital cost of the unit versus the efficiency (energy consumption). Ultimately, the separation process having the lowest total cost (operating and capital) with low maintenance and high reliability is selected. On top of this need is a requirement to produce a working design in a reasonable amount of time. Some of the key factors influencing selection of separation methods are discussed below while more details can be seen in Seader et al. (2011).

15.11.1 Feed and Product Conditions

The important factors for feed conditions are composition and flow rate, because the other conditions (temperature, pressure, and phase) can be altered to fit a particular operation. However, feed vaporization, condensation of a vapor feed, or compression of a vapor feed can add significant energy costs to chemical processes.

The important product conditions are purities because the other conditions listed can be altered by energy transfer after the separation is achieved. When a very pure product is required, large differences in volatility or solubility or significant numbers of stages are needed for chemicals in commerce. Accurate molecular and bulk thermodynamic and transport properties are also required.

15.11.2 Operation Feasibility

The separation technology to be considered must be able to produce products with desirable quality under requirements of varied process conditions. This feasibility criterion is commonly used to make a first cut between separation methods that may work and certainly will not work. Often the question of process feasibility will have to do with the extreme processing conditions. Although the dividing lines are not easy to draw, the general guideline is that a process which requires very high or very low pressures or temperatures will always suffer in comparison with one that does not require extreme conditions.

15.11.3 Design Reliability

Design reliability relates to the amount of pilot-plant tests and scale-up that must be done before a suitable commercial-scale design is produced. It is therefore not

Table 15.4 Feature of the common separation operations.

Decreasing "ease of scale-up"	Ease of staging	Need for parallel units
Distillation	Easy	No need
Absorption	Easy	No need
Extraction	Easy	No need
Membrane	Easy	Almost always
Adsorption	Easy	Only for regeneration
Crystallization	Not easy	Sometimes

surprising that those separation operations that are well understood and can be readily designed from first principles and simulation and can be easily scaled up are favored in an industrial environment.

Table 15.4 ranks the more common separation operations according to ease of scale-up. Operations ranked near the top are frequently designed without the need for any laboratory data or pilot-plant tests. Operations near the middle usually require laboratory data, while operations near the bottom require pilot-plant testing on actual feed mixtures. Also included in the table is an indication of the ease of providing multiple stages and to what extent parallel units may be required to handle high capacities. Single-stage operations are utilized only when the separation factor is very large or only a rough or partial separation is needed. When higher product purities are required, either a large difference in certain properties must exist or efficient countercurrent-flow cascades of many contacting stages must be provided. Operations based on a barrier are generally more expensive to stage than those based on the use of a solid agent or the creation or addition of a second phase. Some operations are limited to a maximum size. For capacities requiring a larger size, parallel units must be provided.

15.11.4 Selection Heuristics

There are rules of thumbs generated based on experience over time and applied by engineering departments in industrial companies. Some of them are listed below.

- If the relative volatility of key components is greater than 1.2, distillation separation is almost always the best choice. If the relative volatility is between 1.05 and 1.2, distillation separation is still likely the best choice, but other methods should be considered. If the relative volatility is less than 1.05, other separation methods should be seriously considered.
- Avoid vacuum distillations.
- Favor known techniques, such as distillation, filtration, and extraction.
- Separate corrosive and reactive components first.
- Perform challenging separations last.
- Separate the component with highest content first.
- Favor separations where product flows are equal (e.g. 50/50).

References

De Haan, A. and Bosch, H. (2013). *Industrial Separation Processes–Fundamentals*. De Gruyter.

Faiz, R. and Li, K. (2012). Polymeric membranes for light olefin/paraffin separation. *Desalination* 287: 82–97.

Fenske, M.R. (1932). Fractionation of straight-run Pennsylvanian gasoline. *Industrial and Engineering Chemistry Research* 24: 482.

King, C.J. (1971). *Separation Processes*. New York: McGraw-Hill.

King, T., Schetler, D., and Zhu, X.X. (2014). Use of dividing wall column to replace a single 3-cut depentanizer column in C5-C6 isomerization process. US Patent application H0043305-01-8249, April, 2014.

Li, N.N. and Calo, J.M. (eds.) (1985). *Recent Developments in Separation Science*, vol. IX. CRC Press.

Liu, C.Q. and Karns, N.K. (2017). Stable facilitated transport membranes for olefin/paraffin separations. US Patent US 20180001268. Honeywell UOP.

Marzouk Benali, M. and Aydin, B. (2010). Ethan/ethylene and propane/propylene separation in hybrid membrane distillation systems: optimization and economic analysis. *Separation and Purification Technology* 73 (3): 377–390.

Perry, R.H. and Green, D.W. (eds.) (1997). *Perry's Chemical Engineers' Handbook*, 7e. McGraw-Hill.

Seader, J.D., Henley, E.J., and Roper, D.K. (2011). *Separation Process Principles*. Wiley.

Teramoto, M., Matsuyama, H., Yamashiro, T., and Okamoto, S. (1989). Separation of ethylene from ethane by a flowing liquid membrane using silver nitrate as a carrier. *Journal of Membrane Science* 45 (1–2): 115–136.

Wankat, P.C. (2006). *Separation in Chemical Engineering: Equilibrium Staged Separations*, 2e. Upper Saddle River: Prentice Hall.

Zsai, R., Zhu, X.X., and Steacy, P. (2016). Dividing-wall column screening guidelines and applications. AIChE Annual Meeting, November, San Francisco.

16

Reaction Engineering Overview

16.1 Introduction

Thermodynamics and kinetics are two factors that affect reaction rates. Chemical kinetics is concerned with the rate of a chemical reaction while thermodynamics determines the extent or equilibrium to which the reaction can go. A catalyst helps to speed up the reaction rate while reactor design determines the appropriate reactors to match the type of reaction and catalyst in order to achieve the desired conversion and product yield.

16.2 Reaction Basics

16.2.1 Reaction Rate Law

A rate law is an expression which relates the rate of a reaction to the rate constant and the concentrations of the reactants. Assume a reaction with reactants [A] and [B] making product [P] takes the form as

$$A + B \rightarrow P \tag{16.1}$$

The reaction rate law could be expressed as:

$$-r_A = k_A C_A C_B \tag{16.2}$$

r_A indicates the disappearance of component [A] with the concentration reducing and thus r_A takes negative sign. A similar expression can be made for disappearance of component [B]. For product [P], which increases its amount over time, the rate law takes the positive sign as

$$r_P = k_P C_A C_B \tag{16.3}$$

where k is reaction constant and it is expressed as Arrhenius equation as

$$k = k_o e^{-E_a/RT} \tag{16.4}$$

where

T is temperature in Kelvin,
E_a is Arrhenius activation energy in joules, and
R is the universal gas constant.

The unit of the reaction constant k depends on the overall order of the reaction. k_o is frequency constant or pre-exponential factor that has the same unit as that of k.

The rate constant, k, is a proportionality constant for a given reaction. As denoted in Eq. (16.4), the rate constant k is a function of activation energy (E_a) and temperature (T). Use of a catalyst in a reaction reduces E_a and thus increases k for speeding up a reaction as it enables a reaction starting at lower temperature.

If a chemical reaction is to occur in two different temperatures, it would be observed that the reaction occurring at a higher temperature would have a higher rate. This is because with a high temperature, the kinetic energy of the reactants increases, allowing far more collisions between the molecules. This, therefore, allows for products to be formed faster.

16.2.2 Arrhenius Activation Energy (E_a)

The net Arrhenius activation energy term from Arrhenius equation (16.4) indicates the threshold barrier that a reaction must overcome via raising the reactants' temperature. After climbing over the threshold, a reaction starts and proceeds to make products while going down the hill as the products have the lower energy state than reactants (Figure 16.1). However, most chemical reactions that take place in reactor vessels are like the hydrocarbon combustion: the activation energy is too high for the reactions to proceed significantly at ambient temperature.

16.2.3 Reaction Catalyst

To increase the rate of reaction, the threshold barrier, or activation energy must be lowered so that more collision among reactants can occur. An alternative route with lower threshold is created by adding a catalyst to reaction (Figure 16.2). A catalyst is a substance (usually in solid phase) that increases the rate of a chemical reaction. During reaction, the catalyst is not consumed although the activity may be reduced due to deactivation

---

Efficient Petrochemical Processes: Technology, Design and Operation, First Edition.
Frank (Xin X.) Zhu, James A. Johnson, David W. Ablin, and Gregory A. Ernst.
© 2020 John Wiley & Sons, Inc. Published 2020 by John Wiley & Sons, Inc.

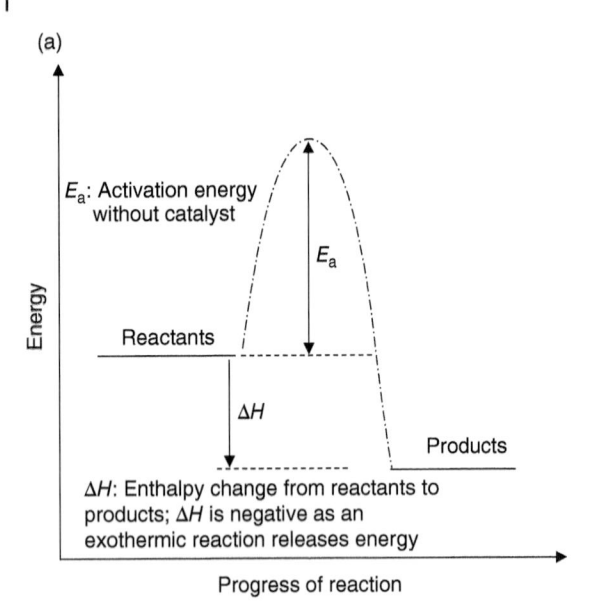

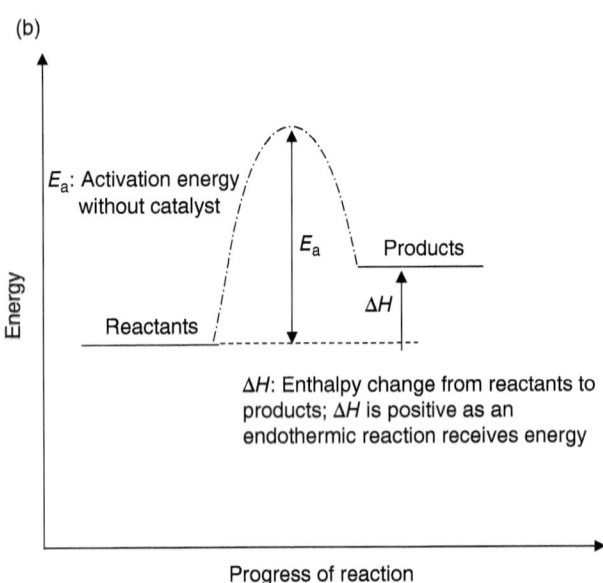

Figure 16.1 Activation energy without catalyst. (a) For exothermic reactions. (b) For endothermic reactions.

over time. Relatively small quantities of catalysts can catalyze relatively large masses of reactants and thus can be used repeatedly in certain period. Catalysts are developed to suit the needs of specific reactions.

It should be emphasized that while the catalyst lowers the activation energy, it does not change the energies of the original reactants or products. Rather, only the activation energy for the catalytic route is lowered.

16.2.4 Order of Reaction

Another important aspect of the reaction rate law is the reaction order. The reaction rate for a given reaction is a crucial tool that enables us to calculate the specific order of a reaction. The order of a reaction is important in that it enables us to classify specific chemical reactions easily and efficiently. Reaction order can be calculated from the rate law by adding the exponential values of the concentrations of the reactants.

Here is an example of how you can look at this. If the reaction orders with respect to [A] and [B] are $n = 2$ and $m = 1$, respectively, which basically means that the concentration of reactant A is decreasing by a factor of 2 while the concentration of [B] is decreasing by a factor of 1 during reaction. The overall reaction order is equal to 3 ($= n + m$).

It is important to note that the reaction order can be determined from experimental data as part of a rate law equation. If the correlated reaction order is the same as the stoichiometric coefficients in the chemical equation, the reaction is called an elementary reaction. Otherwise, the reaction is a nonelementary reaction.

16.2.5 Reactor Design

The objective of reactor design is to match the type of reaction and catalyst. Selection of reactor type and proper reactor design are very important as they influence not only the rate of reaction for product and by-product formation but also the capital costs of the reactor and downstream separation system. For example, if the by-product can be minimized, the downstream separators will be reduced in size.

16.3 Reaction Kinetic Modeling Basics

Reaction kinetics is determined via experiment based on the type of reaction and reaction mechanism.

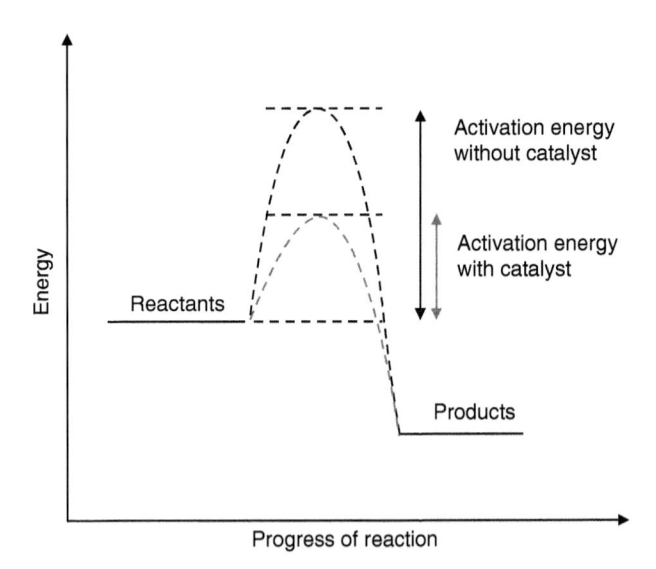

Figure 16.2 Activation energy with catalyst in exothermic reaction.

16.3.1 Elementary Reaction Rate Law

It is an elementary reaction if the order of reaction for each species in the rate expression is the same as the stoichiometric coefficients. Consider a single reaction with stoichiometric equation as

$$O_2 + 2NO \rightarrow 2NO_2 \tag{16.5}$$

The rate expression derived from experiment data is

$$-r_{NO} = k_{NO} C_{NO}^2 C_{O_2} \tag{16.6}$$

Since the order of reaction for each species matches with the stoichiometric coefficient of each species, it is an elementary reaction.

16.3.2 Reversible Reaction

The net reaction rate is equal to the difference between the rate of formation in the forward reaction and the rate of formation in the reverse reaction:

$$r_{net} = r_{forward} - r_{reverse} \tag{16.7}$$

At equilibrium, the rate of forward reaction is equal to that of reverse reaction, i.e. $r_{forward} = r_{reverse}$ or $r_{net} = 0$ and the rate law reduces to an equation that is thermodynamically consistent with the equilibrium constant for the reaction.

16.3.3 Nonelementary Reaction Rate Law

When there is no direct correspondence between the order of reaction for each species and stoichiometry, then the reaction is nonelementary. For example, consider the reaction as

$$A + B \rightarrow C + D \tag{16.8}$$

The rate expression derived from experiment data is

$$-r_A = k C_A^2 C_B \tag{16.9}$$

Thus, it is a nonelementary reaction because the reaction is second order for species A and first order for species B, which does not have direct correspondence with the reaction stoichiometry. Nonelementary reaction often involves the presence of intermediate reaction steps and intermediate products. To derive reaction rate law for a nonelementary reaction, the reaction mechanism containing intermediate reaction steps and products needs to be devised.

16.3.4 Steady-State Approximation

When a reaction involves one or more intermediates, the concentration of one of the intermediates remains constant at a certain stage of the reaction. Thus, the system has reached a steady state on which the method for deriving the rate law is called steady-state approximation. The method is based on the assumption that one intermediate in the reaction mechanism is consumed as quickly as it is generated. Its concentration remains the same in the duration of the reaction.

16.3.5 Reaction Mechanism

A reaction mechanism describes all intermediate steps from reactants to products. An intermediate is a species that is neither a reactant nor a product. Devising a reaction mechanism requires a broad understanding of the properties of reactants and products, and this is usually done by experienced chemists. The steady-state approximation is a technique for deriving a rate law from the proposed mechanism.

Let us consider the following example:

$$2N_2O_5 \rightarrow 4NO_2 + O_2 \tag{16.10}$$

The mechanism for this reaction is devised based on elementary reaction steps as below:

- Step 1: Let k_f and k_r be forward and reverse rate constants

$$N_2O_5 \leftrightarrow NO_2 + NO_3 \tag{16.11}$$

- Step 2:

$$NO_3 + NO_2 \rightarrow NO + NO_2 + O_2 \quad (\text{rate constant } k_2) \tag{16.12}$$

- Step 3:

$$NO_3 + NO \rightarrow 2NO_2 \quad (\text{rate constant } k_3) \tag{16.13}$$

NO and NO_3 are intermediate products and we try to replace them with final products.

The production rate of NO in Step 2 and consumption rate of NO in Step 3 are

$$r_{NO} = k_2 [NO_3][NO_2] \tag{16.14}$$

$$-r_{NO} = k_3 [NO_3][NO] \tag{16.15}$$

where the brackets indicate the concentration of a component.

Based on the steady-state assumption, the rate of production of an intermediate is equal to the rate of its consumption. Thus, we have

$$k_2 [NO_3][NO_2] = k_3 [NO_3][NO] \tag{16.16}$$

Solving Eq. (16.16) for the concentration of NO ([NO]) gives

$$[NO] = \frac{k_2 [NO_2]}{k_3} \tag{16.17}$$

Now, let us turn to the other intermediate, i.e. NO_3. The rate law for both production and consumption of NO_3 from Steps 1–3 can be derived as

$$r_{NO_3} = k_f [N_2O_5] \tag{16.18}$$

$$-r_{NO_3} = k_2 [NO_3][NO_2] + k_3 [NO_3][NO] + k_r [NO_3][NO_2] \tag{16.19}$$

Applying the steady-state assumption gives

$$k_f [N_2O_5] = k_2 [NO_3][NO_2] + k_3 [NO_3][NO] + k_r [NO_3][NO_2] \tag{16.20}$$

Solving Eq. (16.20) for the concentration of NO_3 ($[NO_3]$) gives

$$[NO_3] = \frac{k_f [N_2O_5]}{k_2 [NO_2] + k_3 [NO] + k_r [NO_2]} \tag{16.21}$$

Take a pause to review the three reaction steps: (i) Step 1 is an equilibrium reaction and thus cannot give a rate expression or $r_{NO} = 0$ under equilibrium. (ii) Step 2 gives the production of both intermediate and final products. The former leads to further production of NO_2 in Step 3. From Step 2, the production rate for O_2 is

$$\frac{d[O_2]}{dt} = k_2 [NO_3][NO_2] \tag{16.22}$$

Integrating [NO] in Eq. (16.17) and $[NO_3]$ in Eq. (16.21) with Eq. (16.22) yields

$$\frac{d[O_2]}{dt} = \frac{k_2 k_f}{k_r + 2k_2} [N_2O_5] = k_0 [N_2O_5] \tag{16.23}$$

where

$$k_0 = \frac{k_2 k_f}{k_r + 2k_2} \tag{16.24}$$

The reaction rate for product O_2 in Eq. (16.23) agrees with the experimental results, which means the reaction mechanism assumed as above is correct. Otherwise, a different reaction mechanism needs to be assumed and verified against experiment data until they both match.

16.4 Rate Equation Based on Surface Kinetics

In many reactions, a catalyst is used to speed up a reaction or it may slow a reaction (negative catalyst). Most catalysts are solid with porous structure. Reactants must interact with the catalyst for reaction to occur and thus three major steps occur successfully (Levenspiel 1999).

Step 1 (Adsorption): A molecule is absorbed onto the surface and is attached to an active site for reaction.
Step 2 (Reaction): It reacts with another molecule on the active site.
Step 3 (Desorption): Products are desorbed from the surface and then frees the site.

All species of molecules in the above steps are assumed to be in equilibrium.

Rate expression for catalytic reaction is of the form:

$$\text{Reaction rate} = \frac{(\text{Kinetic term})(\text{Concentration terms})}{(\text{Resistance terms})} \tag{16.25}$$

Consider the following reaction:

$A + B \leftrightarrow R + S$, based on reaction equilibrium constant K

The rate equation when adsorption of A controls in Step 1 is

$$-r_A = \frac{k(p_A - p_R p_S / K p_B)}{\left(1 + K_A p_R p_S / K p_B + K_B p_B + K_R p_R + K_S p_S\right)^2} \tag{16.26}$$

When reaction between adjacent site-attached molecules of A and B controls (Step 2), the rate equation is

$$-r_A = \frac{k(P_A P_B - P_R P_S / K)}{\left(1 + K_A P_A + K_B P_B + K_R P_R + K_S P_S\right)^2} \tag{16.27}$$

where P_A, P_B, P_R, and P_S are partial pressure of gaseous species A, B, R, and S, respectively. K_A, K_B, K_R, and K_S are adsorption equilibrium constants.

Whereas when desorption of R controls (Step 3), the rate equation becomes

$$-r_A = \frac{k(P_A P_B / P_S - P_R / K)}{\left(1 + K_A P_A + K_B P_B + K \times K_R P_A P_B / P_S + K_S P_S\right)} \tag{16.28}$$

The above shows that each controlling reaction has its own resistance and thus reaction rate equation based on surface kinetics involving reaction equilibrium constant K and several adsorption equilibrium constants.

The above reaction kinetics modeling does not take into account the effects of internal diffusion resistance on the reaction rate. To measure how much the reaction rate is lowered because of diffusion resistance, the effectiveness factor, η, is defined as

$$\eta = \frac{\text{Actual reaction rate}}{\text{Rate without diffusion resistance}}$$
$$= \frac{r_{A,\text{with diffusion}}}{r_{A,\text{without diffusion resistance}}} \tag{16.29}$$

or

$$r_{A,\text{with diffusion}} = \eta \cdot r_{A,\text{without diffusion resistance}} \quad (16.30)$$

Effectiveness factor η is determined based on a parameter called Thiele Modulus, M_T. M_T is a very important parameter for describing the effects of internal pore diffusion on the reaction rate in the porous catalyst pellets with no mass transfer limitations. The value of M_T is used in determining the effectiveness factor for the catalyst.

M_T expression is given as below and details of derivation can be seen in Levenspiel (1999): for first-order reversible reactions:

$$M_T = L\sqrt{\frac{k}{D_e X_A}} \quad (16.31)$$

for nth-order reversible reactions:

$$M_T = L\sqrt{\frac{(n+1)k C_{A_s}^{n-1}}{2 D_e}} \quad (16.32)$$

where

D_e is effective diffusion coefficient of fluid within catalyst pores,
L is the characterized size of catalyst particle,
X_A is the conversion of component A, and
C_{A_s} is the concentration of species A on the surface of the catalyst particle.

Figure 16.3 provides the definition of C_{A_s} and L for the pore length with a cylinder shape.

As indicated in Eqs. (16.31) and (16.32), the two parameters for M_T are diffusion coefficient (D_e) and catalyst size (L). Let us pay attention to effective diffusivity D_e first. The effective diffusivity accounts for the facts that: (i) not all of the area is available (i.e. the area occupied by solids) for the molecules to diffuse; (ii) the internal paths are tortuous; and (iii) the pores are of varying cross-sectional areas. These facts can be described graphically by Figure 16.4. Thus, D_e is defined below to address these facts:

$$D_e = \frac{D'_{AB} \phi_p \sigma_c}{\tau} \quad (16.33)$$

where

τ = Tortuosity
$= \dfrac{\text{Actual distance a molecule travels between two points}}{\text{Shortest distance between two points}}$
$\quad (16.34)$

ϕ_p = pellet porosity = $\dfrac{\text{Volume of void space}}{\text{Total volumn (voids and solids)}}$
$\quad (16.35)$

σ_c = Constriction factor $\quad (16.36)$

The constriction factor σ_c accounts for the variation in the cross-sectional area that is normal to diffusion. It is a

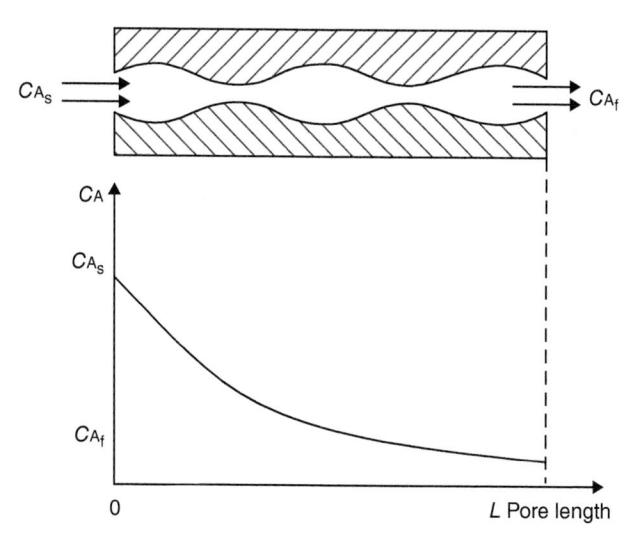

Figure 16.3 Concentration profile within a cylinder pore.

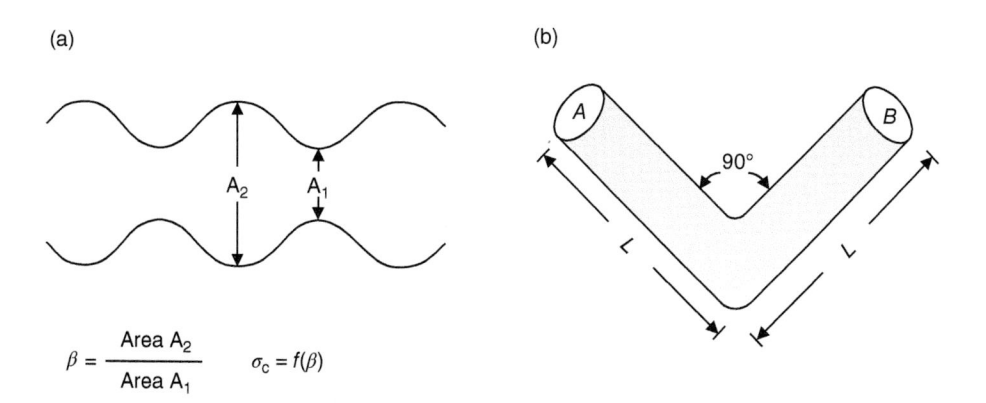

Figure 16.4 Catalyst geometry: (a) pore constriction; (b) pore tortuosity.

function of the ratio (β) of maximum and minimum pore areas (Figure 16.4a).

The second parameter is characteristic size L. To obtain the effective distance penetrated by reactants to reach all internal surfaces, the characteristic size of catalysts, L, is defined as

$$L = \frac{\text{Volume of particle}}{\text{Exterior surface avaiable for reactant penetration}}$$

e.g. $L = \delta/2$ for flat plates; $L = R/2$ for cylinders;

$$L = R/3 \text{ for spheres} \tag{16.37}$$

To find how the pore resistance affects the reaction rate, D_e and L are calculated first from Eqs. (16.33) and (16.37), and then M_T can be determined based on Eq. (16.31) or (16.32). The η–M_T charts in Figure 16.5 are used to determine η and finally calculate the actual reaction rate based on Eq. (16.30).

The η–M_T charts show that for $M_T < 0.4$, η is close to 1 and the concentration of the reactant does not reduce appreciably within the pore; thus, pore diffusion resistance has a negligible effect on the reaction rate. Small M_T implies a short pore and rapid diffusion rate. For large M_T, $\eta \ll 1$, and the reactant concentration drops rapidly to zero within the pore; hence, diffusion affects the reaction rate significantly. This is a regime of strong pore resistance.

16.5 Limitations in Catalytic Reaction

A catalytic reaction can be external diffusion limited or reaction kinetics limited or internal pore diffusion limited. Two parameters can be used to describe the external diffusion and reaction rate. The former can be described by external diffusion coefficient k_m while the latter by reaction constant k_0. The following discussions will explain how and why. Figure 16.6 describes the external diffusion near the surface of catalyst where δ is the thickness of the fluid film on the catalyst surface and the concentration gradient is described by the difference between $[A]_b$ (bulk flow concentration) and $[A]_s$ (concentration on the surface).

When reactants reach the surface of the catalyst pellet in Step 1 defined above, fluid transfers to the surface via diffusion based on concentration gradient. This implies that the flux (W_A) is equal to the rate of reaction (or disappearance of the reactants) on the surface, which can be expressed as

$$W_A = -r_A \tag{16.38}$$

or

$$k_m \left([A]_b - [A]_s \right) = k_0 [A]_s \tag{16.39}$$

$$[A]_s = \frac{k_m}{k_m + k_0} [A]_b \tag{16.40}$$

$$-r_A = k_0 [A]_s = \frac{k_0 k_m}{k_m + k_0} [A]_b \tag{16.41}$$

where

$$k_m = \frac{D_{AB}}{\delta} \tag{16.42}$$

$$k_m \propto v^{1/2} \tag{16.43}$$

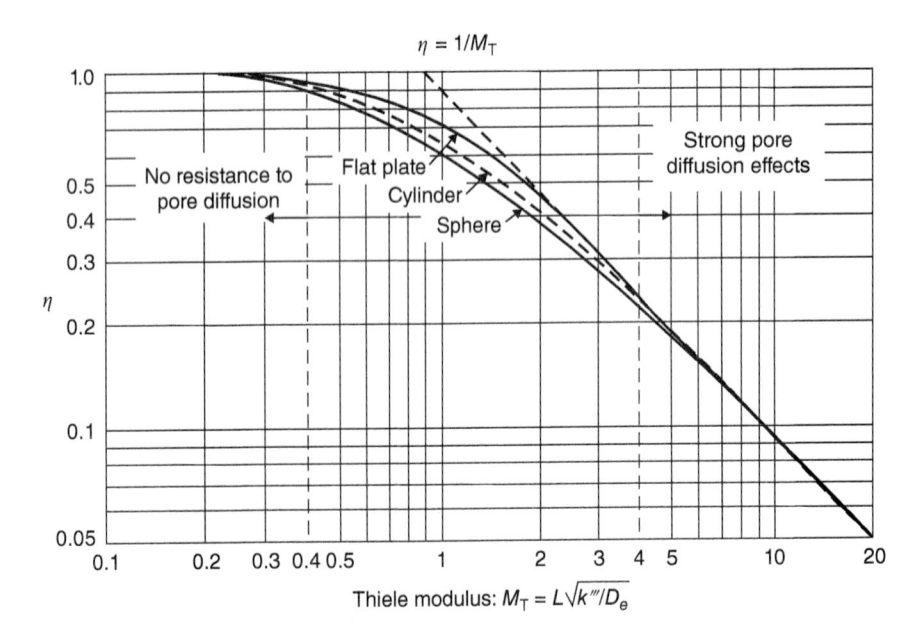

Figure 16.5 Effectiveness factor η versus M_T for porous particle with various shapes (Levenspiel 1999).

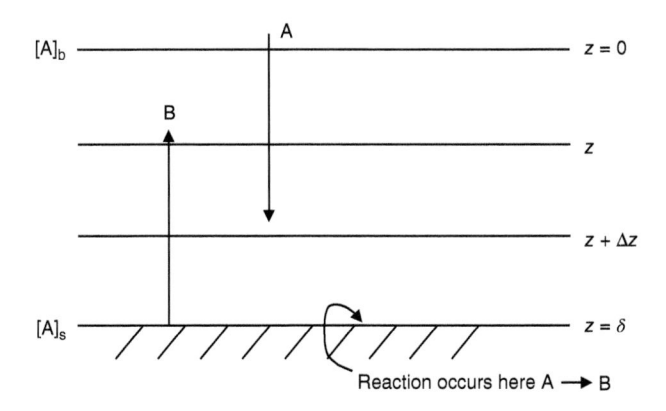

Figure 16.6 Concentration profile in stagnant film model.

where

D_{AB} is the diffusion coefficient,
δ is the film thickness with concentration gradient, and
v is the velocity of the bulk fluid.

Clearly, k_m increases with fluid velocity (v) while k_0 is independent of fluid velocity. Comparing mass transfer constant k_m with reaction rate constant k_0, two limiting cases can be defined as below. It must be pointed out that the overall rate of a chemical reaction is determined by the limiting step and thus called "rate determining" step.

16.5.1 External Diffusion Limitation

Case 1: Low velocity leading to low external diffusion. In this case, $k_m \ll k_0$, thus Eq. (16.41) reduces to

$$-r_A = k_m [A]_b \tag{16.44}$$

Equation (16.44) implies a rapid reaction on the surface and thus the overall reaction rate is limited or controlled by diffusion, thus the external diffusion is reaction rate-limiting.

Remarks: External diffusion-controlled reactions are the reactions that occur so quickly. As quickly as the reactants encounter each other, they react. The process of chemical reaction can be considered as involving the diffusion of reactants until they encounter each other in the right stoichiometry and form an activated complex which can form the product species. In the external diffusion-controlled reactions, the formation of products from the activated sites is much faster than the mass transfer rate of reactants.

Diffusion control is rare in the gas phase, where rates of diffusion of molecules are generally very high. Diffusion control is more likely in solution where diffusion of reactants is slower due to the greater number of collisions with solvent molecules. Reactions where the products form rapidly are most likely to be limited by diffusion control. Heterogeneous reactions are candidates for diffusion control.

One classical test for external diffusion control is to observe whether the rate of reaction is affected by stirring or agitation; if so, then the reaction is almost certainly external diffusion controlled under those conditions. The external diffusion control can be overcome by increasing reactant's velocity, enhancing the mixing, and reducing catalyst pellet size.

16.5.2 Surface Reaction Limitation

Case 2: High velocity, $k_m \gg k_0$. In this case, Eq. (16.41) reduces to

$$-r_A = k_0 [A]_b \tag{16.45}$$

Equation (16.45) implies a slow surface reaction as a rate-limiting case. To overcome this limiting case, we could increase the temperature and concentrations to increase the reaction rate.

The above discussions define two limiting cases, namely external diffusion limited and surface reaction limited. If an activation energy is in the range of 8–24 kJ/mol, which implies fast reaction, chances are that the reaction is strongly diffusion limited. When activation energy is around 200 kJ/mol, which implies slow reaction, the chances are that the reaction could be surface reaction limited.

16.5.3 Internal Pore Diffusion Limitation

The third limiting case is internal pore diffusion limited. The pore structure is defined by the catalyst morphology (size, porosity, tortuosity, and constriction). Thiele Modulus, M_T, is used to describe the relationship between diffusion and reaction rate for porous catalyst pellets with no mass transfer limitations. As mentioned above, for large M_T, $\eta \ll 1$, and reactant concentration drops rapidly to zero within the pore; hence, internal diffusion affects the reaction rate significantly.

16.5.4 Mitigating Limitations

For design of a new reactor or revamp of existing reactor using solid catalyst, the limitation (or controlling) case for the reaction rate must be determined first. If it is diffusion limiting, the change to the physics of the process (fluid velocity, mixing, catalyst size, and pore structure) could overcome the limitation and thus increase the reaction rate. If surface reaction (or kinetics at the catalyst site) is limiting, then the physical conditions of the process (temperature and concentration) together with chemistry (kinetics and catalyst receipt) may be altered.

Three limiting cases together with some of the ways to mitigate the limitations are listed in Table 16.1. For external diffusion-limited reactions, the reaction rate can be

Table 16.1 Limitation types and mitigation handles.

Type of limitation	Reaction rate increases with				
	Velocity	Pellet size	Pellet pore structure	Temperature	Concentration
External diffusion	$v^{1/2}$	$d^{-3/2}$	Independent	Linear	Independent
Internal pore diffusion	Independent	d^{-1}	$L^{-1}\phi^{1/2}\sigma^{1/2}$	Exponential	Independent
Surface reaction	Independent	Independent	Independent	Exponential	Power

altered by changing the fluid velocity, catalyst particle size, and temperature with different impacts. This is because the reaction rate is proportional to square root of fluid velocity, inversely proportional to the particle diameter to the three-halves power (strong impact), and approximately linear with temperature (week impact). In other words, if it is the external diffusion, the reaction rate can be increased by increasing the fluid velocity (strong impact), decreasing the particle size (strong impact), and the temperature (week impact).

It is internal diffusion limiting if the rate of reaction increases via reducing particle diameter (strong impact); making the pore short (L^{-1}), straight ($\sigma^{1/2}$), and porous ($\phi^{1/2}$) (strong impact); and increasing temperature (week impact).

It is reaction kinetic limiting if increasing temperature (strong impact) and reactant concentration (strong impact) as well as improving catalyst recipe (strong impact) will increase the rate of reaction. It should be noted that the exponential temperature dependence for diffusion limitation is usually not as strong a function of temperature as is the dependence for surface reaction limitation.

It should be noted that the impact of temperature change depends on the reaction mechanism. For simultaneous formation of product (p) and by-product (b), if $E_b > E_p$, then the by-product formation will increase faster relative to product formation when increasing temperature, thus reducing product quality.

16.5.5 Important Parameters of Limiting Reaction

Many factors influence rates of chemical reactions, some of which are discussed above. The key factors are summarized below.

16.5.5.1 Nature of Reactions
Depending upon what substances are reacting, the reaction rate varies. Acid–base reactions, the formation of salts, and ion exchange are fast reactions. When covalent bond formation takes place between the molecules and when large molecules are formed, the reactions tend to be very slow. Nature and strength of bonds in reactant molecules greatly influence the rate of its transformation into products.

16.5.5.2 Physical state
The physical state (solid, liquid, or gas) of a reactant is also an important factor for the reaction rate. When reactants are in the same phase (called homogeneous reaction), as in aqueous solution, thermal motion brings them into contact. However, when they are in different phases (called heterogeneous reaction), the reaction is limited to the interface between the reactants. Reaction can occur only at their area of contact; in the case of a liquid and a gas, at the surface of the liquid.

16.5.5.3 Concentration
Rate laws are expressions of rates in terms of concentrations of reactants. The reactions occur due to collisions of reactant species. The frequency with which the molecules or ions collide depends upon their concentrations. The more crowded the molecules are, the more likely they are to collide and react with one another. Thus, an increase in the concentrations of the reactants will usually result in the corresponding increase in the reaction rate, while a decrease in the concentrations will usually have a reverse effect. For example, combustion will occur more rapidly in pure oxygen than in air (21% oxygen). The actual rate equation for a given reaction is determined experimentally and provides information about the reaction mechanism.

16.5.5.4 Temperature
Temperature usually has a major effect on the rate of a chemical reaction. Usually, the higher the temperature, the faster the reaction. Molecules at a higher temperature have more thermal energy. Although collision frequency is greater at higher temperatures, this alone contributes only a very small proportion to the increase in rate of reaction. Much more important is the fact that the proportion of reactant molecules with sufficient energy to react so that actual energy is greater than activation energy: $E > E_a$.

16.5.5.5 Mass Transfer
Diffusive mixing is strongly related to the reaction because reactions involving multiple reactants cannot

occur without the reactants being contacted intimately at a molecular level. For the reaction to occur, the pure reactants need to be homogenized at the molecular scale so that molecules can collide. If the mixing and diffusion is fast enough, the intrinsic chemical kinetics governs the rate of production of new species. This requires a reduction of scale and of differences in concentration, which is the very definition of mixing as it pertains to chemical reactions.

In two known classes of reaction, i.e. consecutive and parallel reactions, the progress of the reaction depends heavily on how quickly the reactants are brought together. These reactions consist of two or more competitive reactions either occurring in parallel, where two or more reactions involving the same reactants take place at the same time, or in a consecutive sequence, where the desired product of one of the reactions participates in a second undesired reaction with the original reactants. Both types of reaction schemes can involve considerable production of unwanted by-product despite the desired reaction being much faster than the undesired reaction. The yield of desired product from these coupled reactions depends on how fast the reactants are brought together. Recent experimental results have suggested that the mixing effect may depend strongly on the stoichiometry of the reactions.

16.5.5.6 Catalysts

Catalysts play important roles in chemical reactions. As what mentioned previously, catalysts can increase the reaction rate via reducing the activation energy required. At a molecular level, catalysts can increase the success rate of the collisions between reactant molecules although they do not increase the frequency of collisions.

Catalysts generally work by either changing the structure of the reactants or by bonding to reactants in such a way as to cause them to combine, react, and release products or energy. This enables collisions between reactants with less kinetic energy than would otherwise be necessary for a reaction to occur to result in chemical changes (chemical bonds broken and new bonds formed). Therefore, a larger number of individual chemical changes takes place per unit of time, i.e. the overall rate of the reaction is increased.

Solid catalysts are the most popular type because they can be easily separated from reactant liquids.

Performance of catalysts can be improved by changing the materials (base material, promoter, and carrier) and physical structure (size, shape, and internal pore structure) to maximize the rate of mass transfer rate or diffusion rate. For solid-phase catalysts, only those particles that are present at the surface can be involved in a reaction. Crushing a solid into smaller parts and making the pores short in length and straight in passage means that more particles are present at the surface and reactants can reach these surfaces. Thus, the frequency of collisions between the reactant and the surface increases, and thus reaction occurs rapidly. For example, finely divided aluminum confined in a shell explodes violently. If larger pieces of aluminum are used, the reaction is slower and sparks are observed as pieces of burning metal.

16.6 Reactor Types

16.6.1 General Classification

16.6.1.1 Homogeneous and Heterogeneous

In homogeneous reactors, only one phase (a gas or a liquid) is present. If more than one reactant is involved, mixing the reactants is the way of starting off the reaction.

In heterogeneous reactors, multiple phases are present, and examples are gas–liquid, gas–solid, liquid–solid, and liquid–liquid reactions. Usually a solid catalyst is present. For example, gas–solid catalytic reactors particularly form an important class of heterogeneous reaction systems. Generally, heterogeneous reactors have a greater variety of configurations and contacting patterns than homogeneous ones.

16.6.1.2 Continuous Flow Reactors

There are two basic types of continuous flow reactors, namely stirred-tank reactor and tubular-flow (or plug-flow) reactor. In the tubular-flow reactor the aim is to pass the reactants along a tube so that there is as little intermingling between the reactants entering the tube. The tubular reactor is the natural choice for gas-phase reactions although sometimes used for liquid-phase reactions. Usually, a catalyst is used and temperature is raised to obtain high reaction rate in which case a relative small tubular reactor is sufficient to produce a high volumetric flow rate of gas.

The continuous stirred-tank reactor (CSTR) is also called backmix reactor, is by its nature well suited to liquid-phase reactions. In a CSTR, an agitator is used deliberately to disperse the reactants thoroughly into the reaction mixture immediately as they enter the tank in order to obtain uniform mixture with same composition everywhere within the reactor. The product stream is drawn off continuously with the same composition as the contents of the tank. A series of several CSTR's could approach a tubular reactor in terms of reaction rate versus concentration.

16.6.1.3 Semi-Batch Reactors

A semi-batch reactor is shown in Figure 16.7. It is essentially a batch reactor, and at the start of a batch it is charged with the first reactant (A). But the second

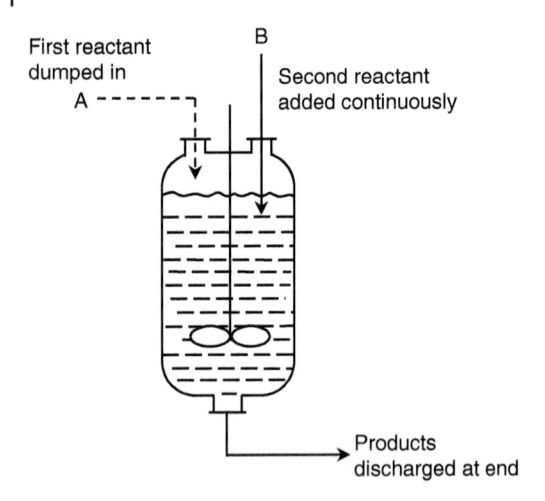

Figure 16.7 Reactant contacting scheme in semi-batch reactor.

reactant (B) is continuously added over the period of the reaction. This is the natural and obvious way to carry out many reactions. For example, if a liquid must be treated with a gas, perhaps in a chlorination or hydrogenation reaction, the gas is normally far too voluminous to be charged all at once to the reactor; instead, it is fed continuously at the rate at which it is used up in the reaction. Another case is where the reaction is too violent if both reactants are mixed suddenly together. Organic nitration, for example, can be conveniently controlled by regulating the rate of addition of the nitrating acid. The maximum rate of addition of the second reactant in such a case will be determined by the rate of heat transfer. In summary, a semi-batch reactor may be chosen for the following scenarios: (i) react a gas with a liquid; (ii) control a highly exothermic reaction, and (iii) improve product yield in suitable circumstances.

16.6.2 Practical Types of Reactors

With general classifications of reactors discussed above, the practical type of reactors is discussed below with their own advantages and disadvantages for selection.

16.6.2.1 Fixed Bed Reactors

The catalyst may have multiple configuration including: one large bed, several horizontal beds, several parallel packed tubes, and multiple beds in their own shells. The various configurations may be adapted depending on the need to maintain temperature control within the system. The flow of a fixed bed reactor is typically downward.

16.6.2.2 Trickle Bed Reactors

A trickle bed reactor is a fixed bed where liquid flows without filling the spaces between particles. Like with the fixed bed reactors, the liquid typically flows downward. At the same time, gas is flowing upward.

This reactor is often utilized to handle feeds with extremely high boiling points.

16.6.2.3 Moving Bed Reactors

A moving bed reactor has a fluid phase that passes up through a packed bed. Solid is fed into the top of the reactor and moves down and it is removed at the bottom. Moving bed reactors require special control valves to maintain close control of the solids.

16.6.2.4 Fluidized Bed Reactors

A fluidized bed reactor suspends small particles of catalyst by the upward motion of the fluid to be reacted. The fluid is typically a gas with a flow rate high enough to mix the particles without carrying them out of the reactor. The particles are much smaller than those for the above reactors, typically on the scale of 10–300 μm. One key advantage of using a fluidized bed reactor is the ability to achieve a highly uniform temperature in the reactor. In addition, for cases with catalysts deactivated rapidly, they can be treated and regenerated easily in fluidized bed, which offers overwhelming advantages over fixed beds design.

16.6.2.5 Slurry Reactors

A slurry reactor contains the catalyst in a powdered or granular form. This reactor is typically used when one reactant is a gas and the other a liquid while the catalyst is a solid. The reactant gas is put through the liquid and dissolved. It then diffuses onto the catalyst surface. Slurry reactors can use very fine particles and this can lead to problems of separation of catalyst from the liquid. Trickle bed reactors do not have this problem and this is a big advantage of them. Unfortunately, these large particles in trickle bed means much lower reaction rate. Overall, the trickle bed is simpler, the slurry reactors usually have a high reaction rate, and the fluidized bed is somewhat in-between.

There are additional features for above reactors that are discussed below.

Flow contacting: Fixed bed reactor approaches plug flow. In contrast, fluidized bed reactor has a very complex flow pattern which is unsatisfactory from the standpoint of effective contacting and requires more catalyst for high gas conversion, and greatly depresses the amount of intermediates which can be formed in series of reactions. Hence, if efficient contacting in a reactor is of primary importance, the fixed bed is favored.

Temperature control: The shortcomings of fixed beds is that they cannot use very small size catalysts because of plugging and high pressure drop. Effective temperature control of large fixed beds can be difficult because such systems are characterized with low heat conductivity. Thus, for highly exothermic reactions, hot spots are likely to develop which may ruin the catalyst. Fluidized beds can use small size catalysts. Thus, for very fast reactions for which external and internal diffusions may limit the reaction rate, the fluidized bed with its rapid gas–solid contacting and small particles will allow a more effective use of catalyst and isothermal operation.

- **Catalyst regeneration:** For cases with catalysts deactivated rapidly, they can be treated and regenerated easily in fluidized bed or moving bed or tubular reactors, which offers overwhelming advantages over fixed beds design.

16.7 Reactor Design

We shall see that reactor design involves all the basic principles of chemical engineering with the addition of chemical kinetics. Mass transfer, heat transfer, and fluid flow are all concerned and complications arise when, as so often is the case, interaction occurs between these transfer processes and the reaction itself. In designing a reactor, it is essential to weigh up all the various factors involved and, by an exercise of judgment, to place them in their proper order of importance. The theory of reactor design is being extended rapidly and more precise methods for detailed design and optimization are being evolved. However, if the final design is to be successful, the major decisions taken at the outset must be correct. Initially, a careful appraisal of the basic role and functioning of the reactor is required and at this stage the application of a little chemical engineering common sense may be invaluable.

16.7.1 Objective

Reactor design basically concerns with selection of the right type of reactor and the size as well as what method of operation we should employ for a given conversation.

In any process where there is a chemical change taking place, however, the chemical reactor is at the heart of the plant because its performance is usually the most important in the design of the whole plant.

When a new chemical process is being developed, at least some indication of the performance of the reactor is needed before any economic assessment can be made. As the project develops and its economic viability becomes established, further work is carried out on the various chemical engineering operations involved. Thus, when the stage of designing the reactor is reached, the project will already have acquired a definite form.

Among the major decisions which will have been taken is the rate of production of the desired product. This will have been determined from a market forecast of the demand for the product in relation to its estimated selling price. The reactants to be used to make the product(s) and their chemical purity will have been established. The basic chemistry of the process will almost certainly have been investigated, and information about the composition of the products from the reaction, including any by-products, should be available. On the other hand, a reactor may have to be designed as part of a modification to an existing process. Because the new reactor has then to tie in with existing units, its duties can be even more clearly specified than when the whole process is new. Naturally, in practice, detailed knowledge about the performance of the existing reactor would be incorporated into the design of the new one. As a general statement of the basic objectives in designing a reactor, we can say therefore that the aim is to produce a specified product at a given rate from known reactants.

After a reactor type is selected, the engineer will determine the process conditions including temperature, pressure, feed rate, and compositions at the reactor inlet, which form the basis for determining the reactor conversion, selectivity, and yield. This is probably the most critical step in reactor design.

16.7.2 Temperature and Equilibrium Constant

Most reactions are reversible, which is governed by the chemical equilibrium, through the equilibrium constant, determining the limit (how far) that a reaction can possibly proceed given sufficient time. On the other hand, the reaction kinetics determines at what rate (how fast) that the reaction will approach the equilibrium limit. In general, if the equilibrium constant is very large, the reaction is irreversible. Otherwise, the reaction is reversible. The equilibrium constant is a function of temperature.

In deciding process conditions, the two principles of thermodynamic equilibrium and kinetics need to be considered together; indeed, a complete rate equation for a reversible reaction will include the equilibrium constant or its equivalent but complete rate equations are not always available to the engineer. The first question to ask is: In what temperature range will the chemical reaction take place at a reasonable rate? The next step is to calculate values of the equilibrium constant K in this temperature range as

$$\frac{d(\ln K)}{dT} = \frac{\Delta H}{RT^2} \qquad (16.46)$$

where

ΔH is heat of reaction;
T is temperature and
R is ideal gas constant.

16.7.3 Pressure, Reaction Conversion, and Selectivity

For a given composition of reactant mixture and reaction time, reaction conversion solely depends on pressure with the mirror similarity with equilibrium constant that depends on temperature only. Thus, for a reversible reaction, pressure condition needs to be optimized to obtain an acceptable reaction conversion. Temperature, concentration, and catalyst pore size influence selectivity which were discussed in section 16.5.5.

16.7.4 Reaction Time and Reactor Size

Once reaction conversion is determined, reaction time and then reactor volume can be calculated based on the mass balance. Using a plug flow reactor as an example (Figure 16.8), the material balance for the finite element can be expressed as

$$F_A = (F_A + dF_A) + (-r_A)dV \quad (16.47)$$

F_{A_0} is the molar flow rate of component A at the reactor inlet. Amid the plug flow reactor (PFR), we have

$$F_A = F_{A_0}(1 - X_A) \quad (16.48)$$

Thus,

$$dF_A = d[F_{A_0}(1 - X_A)] = -F_{A_0} dX_A \quad (16.49)$$

With the replacement of dF_A of Eq. (16.49) into Eq. (16.47), we obtain

$$F_{A_0} dX_A = (-r_A)dV \quad (16.50)$$

Thus,

$$\frac{dV}{F_{A_0}} = \frac{dX_A}{-r_A} \quad (16.51)$$

Integrating Eq. (16.50) for the whole length of the reactor yields

$$\int_0^V \frac{dV}{F_{A_0}} = \int_0^{X_{Af}} \frac{dX_A}{-r_A} \quad (16.52)$$

Integrating Eq. (16.52) for V gives

$$\frac{V}{F_{A_0}} = \int_0^{X_{Af}} \frac{dX_A}{-r_A} \quad (16.53)$$

where X_{A_f} is the required conversion of A which can be achieved by the reactor volume V.

Equation (16.53) can be represented graphically as Figure 16.9.

The reaction time can be calculated as

$$\tau = \frac{C_{A_0} V}{F_{A_0}} \quad \text{or} \quad V = \frac{F_{A_0} \tau}{C_{A_0}} \quad (16.54)$$

$$\tau = C_{A_0} \int_0^{X_{Af}} \frac{dX_A}{-r_A} \quad (16.55)$$

One could use Eq. (16.55) to calculate the reaction time first and then use Eq. (16.54) to calculate the reactor volume for a PFR. Reactor volume and time for other types of reactors can be calculated according the mass balance and rate law equation. Details can be seen in Levenspiel (1999).

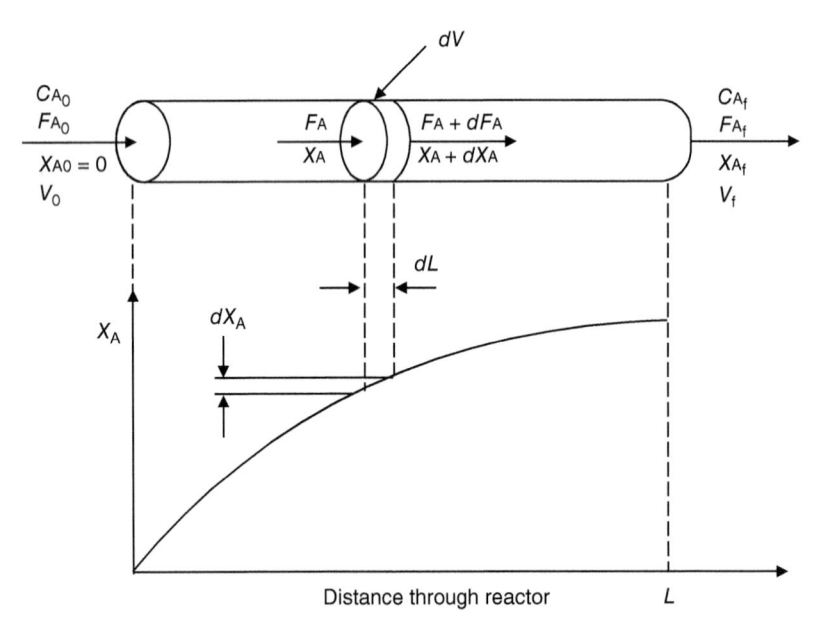

Figure 16.8 Plug flow reactor.

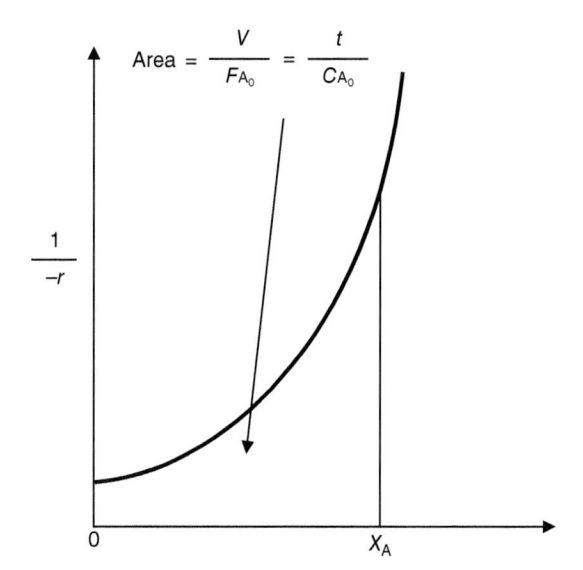

Figure 16.9 Reaction time and volume. Area = $\frac{V}{F_{A_0}} = \frac{t}{C_{A_0}}$, axes $\frac{1}{-r}$ vs X_A.

It must be noted that the volume estimated is only the active reacting volume, and the reactor layout must also consider the additional spaces required for supporting devices, which are discussed further below.

16.7.5 Determine the Rate-Limiting Step

A catalytic reaction can be surface reaction limited or mass transfer (diffusion) limited. The consequence from either case is reduced reaction rate. During the reactor design stage, the controlling case for the reaction rate must be determined first. If it is diffusion limiting, the change to the physics of the process (feed concentration, feed rate, mixing, catalyst size, and pore diffusion) could overcome the limitation and thus increase reaction rate. If it is surface reaction limited, the physical conditions of the process (temperature and concentration) together with chemistry (kinetics and catalyst receipt) may be altered. It is important to identify the true limiting step and understand the root causes of the limiting step to find ways to overcome it. The following parameters are discussed for identifying the rate-limiting step (Towler and Sinnott 2013).

- **Reaction mechanism:** In a reaction involving multiple reaction steps, the step with the slowest reaction rate is called the rate-limiting step which governs the overall reaction rate. The rate-limiting step can be determined by comparing the experimental rate law with mechanism-based rate law. The correct rate-determining step can be identified by comparing the rate law derived from the reaction mechanism with the rate law determined from experiment data. The elementary reaction step that has the same (almost the same) rate law as the experimental law is the rate-limiting in the overall reaction. An example is given for explanation of the rate-limiting step.
- **Mass transfer:** Often, the reaction rate is affected by the rate of mass transfer, particularly diffusion rate for porous catalysts. The η–M_T method as explained previously can be used to identify if diffusion is the rate-limiting case. If it is the case, catalyst morphology may be altered in terms of catalyst particle size and internal pore structure. In other words, diffusion rate could be improved by reducing particle diameter and making the pore short and straight.
- **Heat transfer rate:** For an endothermic reaction, the temperature of the reaction mixture will decrease significantly and the heat of reaction may become the governing factor in reactor design. If the rate of reaction is limited by the rate of heat addition, the reactor may be designed as a fired heater or heat exchanger. Alternatively, multiple reactors with intermediate heating may be designed to bring the intermediate reaction temperature up to a desired level.
- **Process conditions:** For an exothermic and fast reaction, we may feed the reaction with low concentration of a certain component(s) to "starve" the reaction to achieve optimal yield. At the same time, temperature is reduced to slow down the reaction. For an endothermic and slow reaction, high temperature may be used.
- **Mixing:** The time taken for mixing the reactants could be the limiting step for a very fast reaction. Thus, to overcome this mixing limit, the mixing rate must be fast enough to obtain the desired concentration profile for the reaction.

As an example of identifying rate-limiting reaction step, consider the gas-phase reaction: $NO_2 + CO \rightarrow NO + CO_2$. If this reaction occurred in a single step, its reaction rate (r) would be $r = k[NO_2][CO]$, where k is the reaction rate constant and square brackets indicate a molar concentration for reactants.

In fact, however, the observed reaction rate is second order in NO_2 and zero order in CO, with rate equation $r = k[NO_2]^2$ determined from experiment. This suggests that the rate is determined by a step in which two NO_2 molecules react, with the CO molecule entering at another (faster) step. A possible mechanism in two elementary steps which explains the rate equation is:

Step 1: $NO_2 + NO_2 \rightarrow NO + NO_3$ (slow step, rate-determining)
Step 2: $NO_3 + CO \rightarrow NO_2 + CO_2$ (fast step)

The rate equation for the first step is $r_1 = k[NO_2]^2$, which is the same as the rate equation for the overall reaction determined from the experiment. This implies that the first step is rate-limiting as its rate of reaction governs the overall reaction rate.

16.7.6 Reactor Design Considerations

Based on the reaction kinetics and equilibrium determined together with product specifications, the most important aspects must be determined including reactor type, size, and conditions as well as products and by-products from the reactor, which are described below (Richardson and Peacock 1994).

16.7.6.1 The Overall Size of the Reactor

This concerns about configuration of the reactors, the geometry of the reactor, and the dimensions and structures. These aspects define the basis for the capital cost of the reaction system. The volume estimated from above discussions is only the active reaction or catalyst volume while the reactor layout must also consider the following factors that may add to the volume required for the reactor vessel such as additional space needed for any internal heat transfer devices and for vapor–liquid distribution, spargers, vapor–liquid segregation, or redistribution, etc. A stirred-tank reactor should not be designed to operate more than 90% filled, and 65–75% is a better design guideline (Towler and Sinnott 2013).

16.7.6.2 Products of the Reactor

The exact composition and physical condition of the products must lie within the limits set in the original specification of the process.

16.7.6.3 By-products

Before taking up the design of reactors in detail, let us first consider the very important question of whether any by-products are formed in the reaction. Obviously, consumption of reactants to produce perhaps unsalable, by-products is wasteful and will directly affect the operating costs of the process. Apart from this, however, the nature of any by-products formed and their amounts must be known so that a plant for separating and purifying the products from the reaction may be correctly designed. The appearance of unforeseen by-products on start-up of a full-scale plant can be utterly disastrous. Economically, although the cost of the reactor may sometimes not appear to be greatly compared with separation systems, it is the composition of the mixture of products from the reactor which determines the capital and operating costs of the separation processes.

16.7.6.4 The Physical Condition of the Reactor

The basic processing conditions in terms of feed rate, pressure, temperature, and compositions of the reactants must be decided, if not already specified as part of the original process design. The temperatures prevailing within the reactor and any provision must be made for heat transfer. The operating pressure and any pressure drop associated with the flow of the reaction mixture must be determined.

The choice of temperature, pressure, reactant rates, and compositions at the inlet to the reactor is closely bound up with the basic design of the reactor. In arriving at specifications for these quantities, the engineer is guided by knowledge available on the fundamental physical chemistry of the reaction. Usually, the engineer will also have results of laboratory experiments giving the fraction of the reactants converted and the products formed under various conditions. Sometimes the engineer may have the benefit of highly detailed information on the performance of the process from a pilot plant, or even a large-scale plant. Although such direct experience of reactor conditions may be invaluable in particular cases, we shall here be concerned primarily with design methods based upon fundamental physical–chemical principles.

Optimization to reduce the reactor cost is usually a waste of time because the cost of the reactor is typically a relatively small fraction of the total capital cost; however, if the reactor validation experiments showed the presence of unexpected components, or showed different selectivity than were found at smaller scale, then it will be necessary to reevaluate the overall process optimization and confirm that the target conversion, yields, and selectivity still apply (Towler and Sinnott 2013).

Ultimately, the final choice of the temperature, pressure, reactant ratio, conversion, and selectivity at which the reactor will operate depends on an assessment of the overall economics of the process. This will consider the cost of the reactants, the cost of separating the products, and the costs associated with any recycle streams. It should include all the various operating costs and capital costs of the reactor and plant. During making this economic assessment, a whole series of calculations of operating conditions, final conversion, and reactor size may be performed with the aid of a computer. Each of these sets of conditions may be technically feasible, but the one selected will be the one that gives the maximum profitability for the project.

16.7.7 General Guidelines

Two key parameters, namely concentration and temperature, are available for optimizing desired product yields (Worstell 2001).

When there are two or more reactants, the type of reactors depends on concentrations of the feed materials, which can be obtained by certain components in excess, and by using the correct contacting patterns with the benefit of improving the yield of the desired product. Figure 16.10 provides the operating methods for two reactant fluids. When the concentration of B (C_B) is low

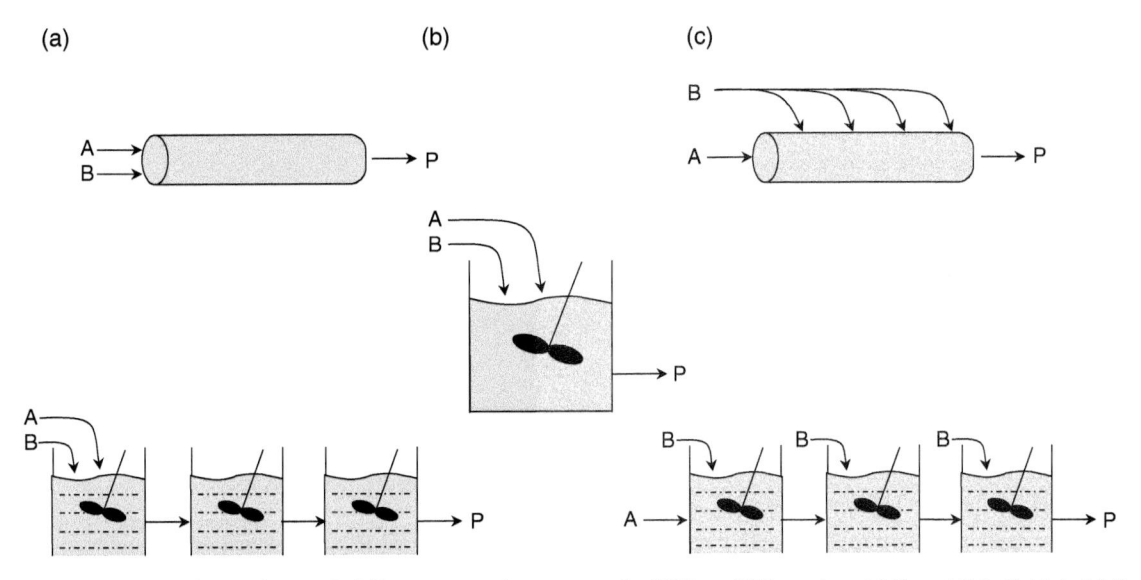

Figure 16.10 Feed procedure with different contacting patterns for CSTR and PFR reactors. (a) C_A and C_B both high. (b) C_A and C_B both low. (c) C_A high and C_B low.

Table 16.2 Reaction type, preferred reactor type, and feeding procedure.

Reactor type	Generalized reaction mechanism	Rate equation	Preferred feed procedure	Preferred reactor type
Simultaneous	$R_1 + R_2 \to P$ $R_1 + R_2 \to B$	$r_1 = k_1(R_1)^{1.5}R_2$ $r_2 = k_2 R_1 (R_2)^{0.5}$	Keep R_1 and R_2 concentrations as high as possible; feed all of R_1 and R_2 at inlet	Series CSTR, plug-flow batch
Simultaneous	$R_1 + R_2 \to P$ $R_1 + R_2 \to B$	$r_1 = k_1 R_1 (R_2)^{0.5}$ $r_2 = k_2 (R_1)^{1.5} R_2$	Keep R_1 and R_2 concentrations as low as possible; feed all of R_1 and R_2 continuously	Series CSTR
Simultaneous	$R_1 + R_2 \to P$ $R_1 + R_2 \to B$	$r_1 = k_1(R_1)^{1.5}R_2$ $r_2 = k_2 R_1 R_2$	Keep R_1 concentrations high to favor product formation; keep R_2 concentrations high to maximize reaction rate; feed all of R_1 and R_2 at inlet	Series CSTR, plug-flow batch
Simultaneous	$R_1 + R_2 \to P$ $R_1 + R_2 \to B$	$r_1 = k_1 R_1 R_2$ $r_2 = k_2 (R_1)^{1.5} R_2$	Keep R_1 concentrations low to favor product formation; keep R_2 concentrations high to maximize reaction rate; stage R_1 feed	Plug-flow; series CSTR
Parallel competitive	$R_1 + R_2 \to P$ $R_1 + R_1 \to B$	$r_1 = k_1 R_1 R_2$ $r_2 = k_2 (R_1)^2$	Keep R_1 concentrations low and R_2 concentrations high	Semi-batch, Staged plug flow
Parallel competitive	$R_1 + R_2 \to P$ $R_1 + X \to B$	$r_1 = k_1 R_1 R_2$ $r_2 = k_2 R_1 X$	Keep R_1 concentrations low to minimize by-product formation; keep R_2 concentrations high to maximize reaction rate	Semi-batch, Staged plug flow
Consecutive	$R_1 + R_2 \to P$ $P + R_2 \to B$	$r_1 = k_1 R_1 R_2$ $r_2 = k_2 (P)^{1.5} R_2$	Keep R_2 concentration low	Semi-batch, Staged plug flow

B, by-product; P, product.

relative to A, the feed of B is divided between several points in the tabular flow regime (Figure 16.10c) and the stirred-tank regime (Figure 16.10c) if the desired reaction is favored by a low C_B. These are known as cross-flow reactions.

Table 16.2 presents methods for maximizing product formation by optimizing the feed procedure and selecting reactor type in simplified form for homogenous processes. A detailed reaction mechanism can be developed using the appropriate scheme. The order of the reaction can be established by developing a reaction mechanism which is tested using laboratory experiments. Then, the appropriate type of reactors, design of each individual reactors, and their configuration can be made in the design phase. At the same time, downstream separation systems are designed with the overall objective to maximize desired products and minimize by-products while minimizing capital and operating costs for the process. Similar information can be used to optimize exiting reactor performance to reduce the operating costs or maximize the throughput of separation system.

Managing reactor feed and utilizing reactor geometry are the best methods for controlling concentration. Consider the simultaneous formation of product and by-product:

$$R_1 + R_2 \rightarrow P \tag{16.56}$$

$$R_1 + R_2 \rightarrow B \tag{16.57}$$

where

R_1 and R_2 are the reactants,
P and B are product and by-product, respectively.

Thus, the reaction rates (r_1, r_2) for P and B can be expressed as

$$r_1 = k_1 R_1^p R_2 \tag{16.58}$$

$$r_2 = k_2 R_1^b R_2 \tag{16.59}$$

$$r_1 / r_2 = (k_1 / k_2) R_1^{p-b} \tag{16.60}$$

p and b are the reaction orders for the formation of product and by-product, respectively. To achieve the maximum amount of desired product, we can take following steps:

- If $p > b$, then keep reactant concentration higher for maximum product concentration. In other words, keeping high R_1 in the feed favors formation of product.
- However, if $b > p$, keeping reactant concentration low for maximum product production. In other words, keeping low R_1 in the feed favors formation of product.
- In addition, if $b > p$, keeping high R_2 in the feed favors product formation. The choice in this case would be a semi-batch reactor.
- For $p > b$, R_2 is the continuous feed, for $b > p$, R_1 is the continuous feed.
- For $p = b$, change in reactant concentration will not affect the product then, because rate constant k_1 and k_2 are different at different temperature, we can keep our temperature such that the desired product will be high or use of catalyst would be an option which are selective in nature.

For consecutive reaction, for example,

$$R_1 + R_2 \rightarrow P + X \rightarrow B \tag{16.61}$$

where X is containment or reactant, by-product formation is minimized when the concentration of P in the reactor is maintained low. The preferred reactor type for Eq. (16.61) is semi-batch or staged plug flow reactors as shown in Table 16.2. Table 16.2 provides methods for maximizing product formation in homogeneous reaction by controlling concentration via feed procedure and reactor geometry.

Managing reactor temperature can also optimize product yields. For example, for simultaneous product and by-product formation, the instantaneous selectivity is

$$dP/dB = (A_P / B_B) e^{[(E_B - E_P)/RT]} \tag{16.62}$$

where A_P and A_B are the Arrhenius pre-exponential factors and E_P and E_B are the activation energy for product and by-product formation, respectively. If $E_P > E_B$, then a high reactor temperature can maximize product formation; if $E_P < E_B$, then a low reactor temperature will maximize product formation. Table 16.3 provides temperature optimization guidelines for product formation in homogeneous processes.

Optimizing a solid catalyst in terms of catalyst receipt and reactor design with catalyst to increase the conversion rate may change by-product production. For simultaneous reactions involving a common reactant, selectivity varies with reactant concentration. For a porous solid catalyst, if reactivity increases with decreasing reactant concentration, then reducing average pore diameter will raise product formation relative to by-product. If reactivity increases with increasing reactant concentration, the enlarging of the average pore diameter will improve selectivity.

For consecutive reactions, selectivity reduces with decreasing average pore diameter. On the other hand, selectivity increases with increasing average pore diameter. For parallel reactions as mentioned in Eq. (16.6), the rate constant ratio becomes $(k_P/k_B)^{0.5}$ for small pores. Thus, selectivity decreases as the average pore diameter reduces.

In some cases, it might not need to inform customers for physical changes to increase conversion as long as the product purity remains the same. However, it must inform customers for any chemical changes to a catalyst. In other cases, it requires catalyst vendors to inform customers for any changes to a catalyst.

16.8 Hybrid Reaction and Separation

An important class of reaction enhancement is combined reaction and separation. The objective is to enhance both conversion and selectivity against a conventional reactor on stand-alone via selective increasing reactants and/or selective removing products to overcome thermodynamic limits, which can be expressed as

Table 16.3 Temperature optimization guidelines.

Reactor type	Generalized reaction mechanism	Instantaneous yields	Preferred operating temperature
Simultaneous	$R_1 + R_2 \to P$ $R_1 + R_2 \to B$	$dP/dB = (A_P/A_B) \times \exp[(E_B - E_P)/RT]$	To maximize P, use high temperature when $E_P > E_B$. If $E_P < E_B$, then use low temperature to maximize P
Parallel competitive	$R_1 + R_2 \to P$ $R_1 + X \to B$	$dP/dB = (A_P/A_B) \times \exp[(E_B - E_P)/RT]$	To maximize P, use high temperature when $E_P > E_B$. If $E_P < E_B$, then use low temperature to maximize P
Consecutive	$R_1 + R_2 \to P$ $R_1 + X \to B$	$dP/dB = (A_P/A_B) \times \exp[(E_B - E_P)/RT]$	To maximize P, use high temperature when $E_P > E_B$. If $E_P < E_B$, then use low temperature to maximize P
Branched consecutive	$R_1 + R_2 \to P$ $P + X_1 \to B_1$ $P + X_2 \to B_2$	$dP/(dB_1 + dB_2) = A_P \times \exp(-E_P/RT) +$ $[A_{B_1} \times \exp(-E_{B_1}/RT) + A \times \exp(-E_{B_{1,2}}/RT)]$	High temperature favors P over B_1, but low temperature favors P over B_2 when $E_{B_2} > E_P > E_{B_1}$; optimum formation of P requires an intermediate temperature

B, by-product; P, product.

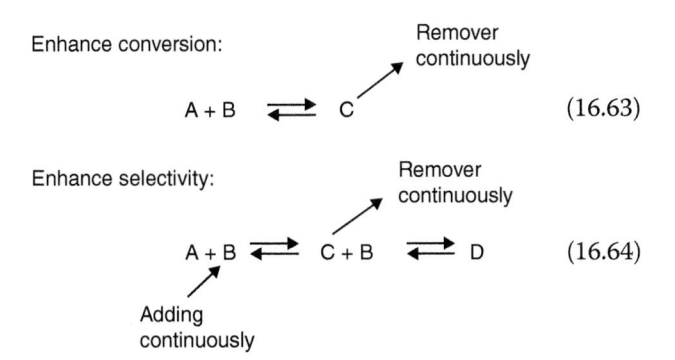

Enhance conversion:

$$A + B \rightleftarrows C \xrightarrow{\text{Remover continuously}} \quad (16.63)$$

Enhance selectivity:

$$\underset{\text{Adding continuously}}{A + B} \rightleftarrows C + B \rightleftarrows D \xrightarrow{\text{Remover continuously}} \quad (16.64)$$

Equations (16.63) and (16.64) conceptually illustrate that it is possible to shift the equilibrium via removing reaction product(s) continuously. Also, it shows that the selectivity can be enhanced by optimizing the concentration profile of one of the reactants or by selectively removing a certain product.

To enhance conversion and/or selectivity, hybrid reaction and separation includes reactive distillation, reactive membrane separation, reactive extraction, reactive absorption, and reactive adsorption. In membrane reactors, reaction and separation via the membranes are coupled. When the membrane is selective, both conversion and selectivity can exceed those of a conventional reactor. Reactive extraction, which combines reaction with liquid–liquid extraction, can be used to separate waste-by-products that are hard to separate with conventional techniques. Separation involving supercritical liquids are especially promising. One potential application of reactive extraction is deep desulfurization of fuels using oxidation of sulfur compounds.

Reactive adsorption, in which reaction and adsorption are coupled at a solid surface, is already used commercially for deep desulfurization, in the Phillips S Zorb process. Reactors based on reactive adsorption, such as the gas–solid–solid trickle bed reactor have great potential for improving process economics. Reactive adsorption has been used for a long time in bulk chemicals production and has potential in the gas treatment and separation of light olefins and paraffins.

Detailed discussions of hybrid separation is beyond the scope of this book. But it is worth to describe the context using reaction distillation as an example. In reactive distillation, chemical conversion and distillation are carried out in one vessel. The process of reactive distillation is shown schematically in Figure 16.11. A good example is shown in Figure 16.12 for production of ethylbenzene. The incentive for catalytic distillation in this case is to keep the ethylene/benzene ratio low, thereby slowing down the oligomerization of ethylene and limiting the amounts of polyethlyenes formed. In addition, the heat of reaction is now used directly integrated with distillation.

16.9 Catalyst Deactivation Root Causes and Modeling

Catalyst deactivation is that they lose activity and selectivity over time. The plant has to regenerate or replace the catalysts in a reactor. It is important to understand the deactivation mechanism.

There are three principal causes of deactivation:

1) The gradual buildup of reaction by-products, such as coking tar, that obstruct the active pores and eventually lead to the extinction of the catalyst.

   Catalyst poisoning. Poisoning is provoked by substances contained in the feed being processed, even when they exist in small, and sometimes undetectable, concentrations.

   During their passage through the reactor, these substances are adsorbed on active sites and the cumulative effect of this preferential adsorption eventually leads to a significant reduction of the activity of the catalytic mass.

2) The slow structural transformation of the catalytic species or the agglomeration and growth of micro-crystallites leading to a gradual decrease in the number of active sites per unit surface or – which amounts to the same thing – to a decrease of the active surface area. This is possibly the most harmful of the three causes of deactivation.

To overcome catalyst poisoning, the substance responsible for poisoning must be identified and the feed treated to eliminate the unwanted compounds, or a catalyst better resistant to poisoning can be used, or a portion of the catalytic bed located ahead of the actual bed can be sacrificed.

Structural transformation must be attacked at the root, for it is during the development of catalysts that the means to prevent transformation of the crystalline phases or recrystallization are tested. When designing reactors, engineers must obtain the maximum benefit from commercially available catalysts, whatever their advantages and disadvantages. Together with the equations describing the chemical kinetics of the process, at least one deactivation equation must be taken into consideration, regardless of the ultimate causes of deactivation.

Deactivation of the catalyst can be described via the use of kinetic equations and the simplest and most general version is proposed by Levenspiel (1999). It is based on empirical laws without concern for any of the underlying mechanisms. It turns out that deactivation processes, which most frequently cannot be analyzed because of their complexity, can generally be represented by relatively simple kinetic equations.

If the initial rate of a process can be expressed by:

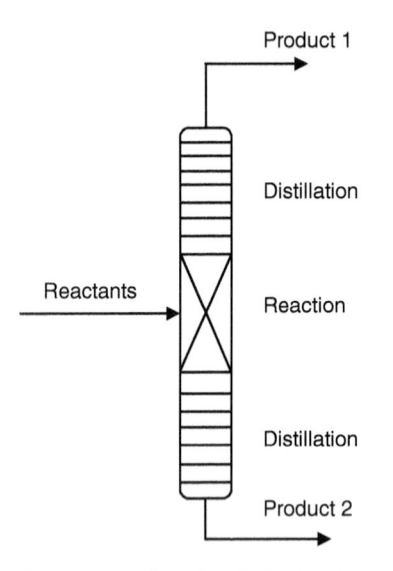

$$r_A = kC_A^n \tag{16.65}$$

Figure 16.11 Reactive distillation (or catalytic distillation).

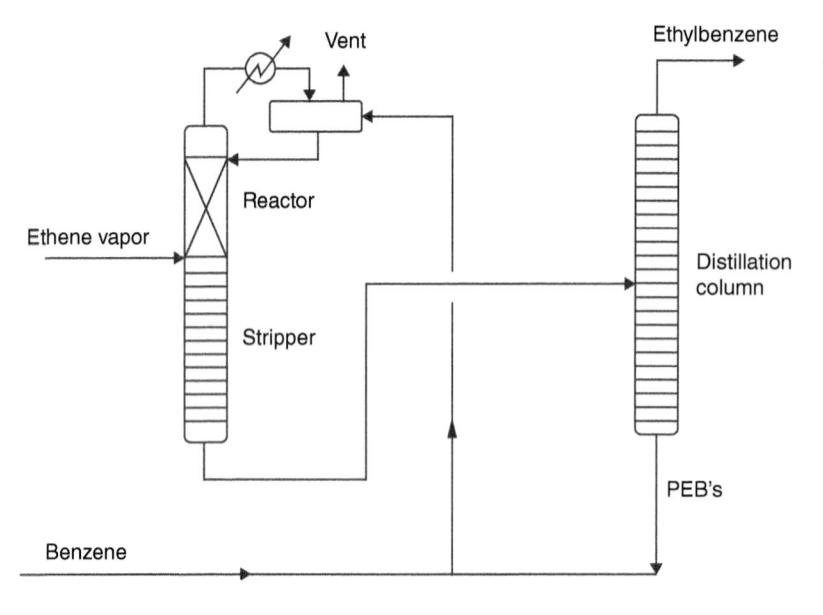

Figure 16.12 Reactive distillation for ethylbenzene production.

where the rate at time t, all other conditions being equal, can be written:

$$r_A = \alpha \cdot k C_A^n \tag{16.66}$$

where α is defined as a relative activity.

Different rate laws may be used, for example:

$$r_A = \frac{k C_A^n}{1 + k_d t} \quad \text{(hyperbolic form)} \tag{16.67}$$

or

$$r_A = k C_A^n e^{-k_d t} \quad \text{(exponential form)} \tag{16.68}$$

or

$$r_A = k C_A^n (1 + k_d t)^p \quad \text{(power law form; } p < 0 \text{)} \tag{16.69}$$

The variation of α with time ($d\alpha/dt$) is a function of concentration, temperature, and the activity α at time t, according to a general expression:

$$-\frac{d\alpha}{dt} = k_d \sum_i C_i^m \alpha^d \tag{16.70}$$

where k_d is the deactivation rate constant; ΣC_i^m the cumulative concentration of one or more compounds.

The most frequently found values for d, m, and the number of compounds included in ΣC_i^m are:

- d is between 1 and 3, and generally 1
- m is between 0 and 2, and generally 1
- The number of compounds included in ΣC_i^m is between 0 and 2, and is generally 1.

References

Levenspiel, O. (1999). *Chemical Reaction Engineering*, 3e. New York: Wiley.

Richardson, J.F. and Peacock, D.G. (1994). *Coulson and Richardson's Chemical Engineering Series, Volume 3 - Chemical and Biochemical Reactors and Process Control*, 3e. Elsevier.

Towler, G. and Sinnott, R.K. (2013). *Chemical Engineering Design: Principles, Practice and Economics of Plant and Process Design*, 2e. Elsevier.

Worstell, J. (2001). Don't act like a novice about reaction engineering, 68–72, CEP, March.

Part V

Operational Guidelines and Troubleshooting

17

Common Operating Issues

17.1 Introduction

The best way to address operating issues in a modern petrochemical complex is to avoid them in the first place through implementing a proper design, performing appropriate maintenance, and operating according to technology licensor and vendor guidelines. This starts before the complex is built, with both licensor and construction contractor selection. When selecting a technology licensor, their successful experience of designing reliable complexes must be considered. The same is true for the company or companies selected for detailed design and construction (Yuh 2017).

Technology licensors may offer training specific to their design for the plant operators and engineers. Training should address items such as design considerations, how to properly operate and maintain the process units, commissioning and shutdown procedures, and how to create unit-specific emergency procedures. It should also address common operating issues from the licensor's experience and what troubleshooting techniques will be useful to the operators and engineers. It is critical that operators and engineers are properly trained, to ensure long-term reliability of the complex.

After construction, it is important to maintain the equipment and instrumentation in the complex by performing appropriate routine maintenance, as recommended by the equipment manufacturers and technology licensor. Most equipment and instrumentation failures should be able to be avoided by keeping these parts properly maintained or replaced (as appropriate) at recommended intervals. Many unplanned outages occur as a result of equipment failures, due to either old or poorly maintained equipment breaking during operation. This is especially the case for equipment with moving parts, in particular rotating equipment and instrumentation (items such as control valves and turbine meters). Additionally, suboptimal operation, or operation whereby production rate is limited for an extended period of time, may occur if a piece of equipment becomes a bottleneck due to a partial failure.

So, proper design and maintenance ought to address a significant portion of operating problems before they occur by avoiding them outright. Nevertheless, operating problems can still occur. These may be due to how the complex is commissioned and started up (or restarted after a temporary shutdown), how it is shut down, or how individual process units are run during normal operation, often due to the human element. Just like cases that may arise out of design issues or improper maintenance, operational errors may result in immediate shutdowns, or they may result in prolonged operation that is inefficient or at lower than desired production rate.

When discussing suboptimal performance, this generally refers to running the complex such that a target production rate can be met, but a higher amount energy input is required to achieve this production rate. If the operating problem results in a loss of production, this is often considered to be more significant as the financial impact of producing less product will be far more costly than the incremental amount of higher utilities consumption. Loss of production can happen because of two general causes: either by having to run the unit at a lower feed rate due to an internal bottleneck, or due to an internal yield loss of desired product from conversion to side product, or product loss through other outlets such as vents or slop streams.

For an aromatics complex, the desired products are typically purified single components, in particular benzene and *para*-xylene. Modern aromatics complexes are often designed to produce other monocyclic aromatic compounds, though typically at lower rates. Toluene, *meta*-xylene, and *ortho*-xylene are the most common. Certain intermediates such as a purified mixed xylene stream, or a C_9 or heaver aromatics stream may be produced and sold to other companies that have a specific need for these as feeds to their complexes. There will typically be a heavy aromatics stream as a side product of an aromatics complex. In general, the quantity of this stream is minimized as much as possible, and may be sold as a solvent or lubricant, blended with fuel oil or

---

Efficient Petrochemical Processes: Technology, Design and Operation, First Edition.
Frank (Xin X.) Zhu, James A. Johnson, David W. Ablin, and Gregory A. Ernst.
© 2020 John Wiley & Sons, Inc. Published 2020 by John Wiley & Sons, Inc.

motor fuels, or sent to a different process unit for additional conversion.

The discussion to follow in this chapter will deal primarily with operation problems as they are related to complexes producing benzene and *para*-xylene as their primary products. Aromatic derivatives will not be addressed as this would require a detailed discussion of an entirely different type of complex; nor will the other less common products and by-products (such as toluene, *ortho*-xylene, or heavy aromatic solvents), as their scope is far more limited on a global production scale.

Data collection and subsequent analysis is a critical aspect of maintaining the optimal performance of an aromatics complex (or any chemical complex, for that matter). There are many ways to look unit performance, but usually this will focus on certain key performance indicators that track product and other process stream qualities, conversion (where applicable), and yield (or recovery). At the next level, this analysis should look at certain measures of utilities required to process the feed and make the desired product. Each process unit will have its specific key performance indicators, as will the complex as a whole, and subsections of any particular unit (such as specific columns or pieces of equipment like heaters and compressors). The discussions that follow are intended to highlight operating variable adjustments and operations that should be considered to improve and optimize the key performance indicators.

17.2 Start-up Considerations

For certain technologies where a catalyst is used for conversion, specific start-up procedures may be required to condition a catalyst before reaching normal operation. This is essentially saving the final manufacturing step to be performed in-situ, after the catalyst has been loaded into the reactor. The reason for this is that the conditions required for final conditioning are such that they may be difficult to achieve or control in a manufacturing facility, or are simply safer to do in the process plant setting. These conditions likely will require recirculation of hydrogen gas, similar to what will be used during normal operation, and may also require introduction of hydrocarbon feed. Once this conditioning step is completed, the catalyst should be in its appropriate chemical form to perform its intended reactions, while minimizing undesirable side reactions. These are typically required when the catalyst has a metal function, which is often the case for modern transalkylation and isomerization catalysts. The metals impregnated on these catalysts typically provide a hydrogenation function, but if the metal is in the improper state or is not well dispersed, it can result in poor catalyst activity along with excessive undesired side reactions.

Some common initial conditioning steps include:

- Reduction with hydrogen gas
- Sulfiding using a sulfur compound
- Attenuation by temporary operation at high severity with a hydrocarbon feed

17.2.1 Catalyst Reduction

In the case where reduction is required, it is because the pure metal is the active form of the metal on the catalyst, and the metal may have been provided in the oxidized state. Catalysts are often provided with the metal initially in an oxidized state, as one of the final manufacturing steps frequently occurs at high temperature in dry air.

Certain metals, such as platinum, are relatively mobile in the oxidized state. Metal atom mobility becomes a problem when the catalyst is exposed to high-temperature environments in the presence of certain compounds. Water and carbon monoxide are two known examples. If the catalyst is exposed to relatively high amounts of these compounds prior to reduction, while the metal is mobile, the metal atoms will tend to agglomerate into larger cluster. This agglomeration results in low availability of metal active sites, thus suppressing the catalyst metal function.

The hydrogenation function provided by the metal in these catalysts typically provides several functions. First, if the catalyst acid function cleaves an alkyl group from an aromatic, the metal function in a hydrogen atmosphere saturates the radical that is formed. Without this saturation function, the alkyl radical that is formed may reattach to an aromatic ring, or polymerize as heavies or coke. The ultimate result of this is that the catalyst will be seen to have lower than desired activity, and poorer stability. There will also likely be excessive side reactions, as these alkyl radicals can recombine with aromatic rings to form undesirable aromatic products. While that may not translate as an aromatic ring loss where the ring cannot be reclaimed to useful product within the complex, it may cause excessive recirculation through the complex due to poor conversion per pass. This will result in higher utilities consumption per amount of product.

As such, it is critical to ensure that the licensor's recommendations are followed explicitly when the catalyst calls for reduction immediately upon start-up. Properly reduced catalyst will provide the best possible catalyst activity, highest yield to desired product, and longest catalyst life.

17.2.2 Catalyst Sulfiding

For certain catalysts, the metal-sulfide is the active form of the metal function for the catalyst. While sulfiding may be done in certain manufacturing facilities, process plants are frequently better setup to handle the hazards inherent to sulfiding, primarily the presence of hydrogen sulfide (H_2S) gas.

During catalyst sulfiding, a sulfiding agent is typically added to the liquid feed immediately upon feed cut-in, as a part of the initial start-up procedure. Common sulfiding agents include dimethyl disulfide (DMDS) or certain olefin sulfide compounds. When it decomposes, the sulfiding agent generates H_2S, which at the appropriate conditions will sulfide the catalyst metal. The amount of H_2S present will typically be greater than the amount required by stoichiometry of the metal sulfiding reaction. This means that some H_2S will need to be vented. In a process plant, this can typically be handled by dilution in the fuel gas header. Such capabilities may not exist in a manufacturing facility.

When the metal-sulfide is the active form of the catalyst metal, it is typically to both attenuate the metal function, and to promote the desired reactions. While this may not be the case for all catalysts that require sulfiding, typically a slight over-sulfiding is not problematic for the catalyst (this should be confirmed with the catalyst supplier). Catalysts that require sulfiding tend to be more robust to the presence of excess sulfur, though if present during normal operation it may result in temporary activity suppression. Under-sulfiding, on the other hand, would likely result in overactivity and excessive side reactions. So it is usually better to err on the side of over-injection of sulfiding compound if this step is required – but again this point should be confirmed with the catalyst supplier. In some cases, under-sulfiding is able to be fixed after start-up by additional injection of sulfiding compound.

17.2.3 Catalyst Attenuation

Some catalyst may require an initial period of operating at a higher-than-normal severity target. If this is done, the purpose is to attenuate the catalyst by deactivating a small amount of the most easily available active sites. Depending on the catalyst, this step may last as short as one day, to as long as several weeks.

In order to operate at higher severity, there are a number of possible operation adjustments that can be made. For instance, the unit can be run at a higher temperature to increase activity. Or, the hydrogen-to-hydrocarbon (H_2/HC) ratio may be reduced to lower the catalyst stability and encourage coke deposition. The actual operational adjustments will vary according to the specific requirement for a given catalyst, and are determined by the catalyst supplier.

The ultimate goal of this high severity operation, regardless of how it is accomplished, is to attenuate catalyst activity to reduce the likelihood of undesired side reactions. Some examples of undesired side reactions resulting from overactive xylene isomerization catalyst are aromatics saturation, cracking, and transalkylation.

For transalkylation catalysts, some examples are aromatics saturation, cracking, and heavies formation.

17.3 Methyl Group and Phenyl Ring Losses

Production loss of *para*-xylene and benzene will be addressed primarily by a discussion of phenyl (benzene ring) losses and methyl group (those that are fed to the complex attached to phenyl rings) losses.

A fixed amount of phenyl rings and methyl groups are present in the feed, and the ratio of their presence in the feed can be characterized by a methyl-to-phenyl ratio, or *M/P*. This calculation takes the ratio of the total amount of methyl groups and phenyl rings, on a molar basis.

$$\text{Methyl to phenyl ratio} = \left(\frac{M_{\text{methyl}}}{M_{\text{phenyl}}} \right) \quad (17.1)$$

where

- M_{methyl} is the total moles of methyl groups bound to aromatic rings in the feed.
- M_{phenyl} is the total moles of aromatic rings in the feed.

As an example, pure benzene has an *M/P* ratio of 0.0, toluene is 1.0, xylene is 2.0, tri-methylbenzene is 3.0, and so on. Ethylbenzene (EB) will have an *M/P* ratio of 0.0, similar to benzene, due to the lack of methyl groups. A molecule such as methyl-ethylbenzene will have an *M/P* ratio of 1.0, similar to toluene, as the ethyl group is again not counted.

The *M/P* ratio discussion becomes more relevant when considering mixtures of molecules, and this will help to understand the theoretical maximum amount of *para*-xylene and benzene that can be produced. The typical operating philosophy of an aromatics complex is to make as much *para*-xylene as possible, and preserve the rest of the phenyl rings to ultimately produce benzene. The amount of methyl groups available in the complex feed will define the maximum amount of *para*-xylene that can be made. Consider the following example:

A hypothetical complex feed that contains 10% benzene, 30% toluene, 50% xylene, and 10% tri-methylbenzene (on a molar basis) has 16 mol of methyl groups to 10 mol of phenyl rings. This results in an *M/P* ratio of 1.6. The theoretical maximum amount of xylenes that could be made is to have all methyl groups rearranged to be bound on phenyl rings as *para*-xylene. If we assume 10 mol of feed molecules, 8 of the phenyl ring moles could take up the 16 mol of methyl groups. Therefore, 8 mol of *para*-xylene could be produced, leaving 2 mol of phenyl rings available as benzene product (Figure 17.1).

The above example assumes that there are no process units in the complex that can generate additional phenyl

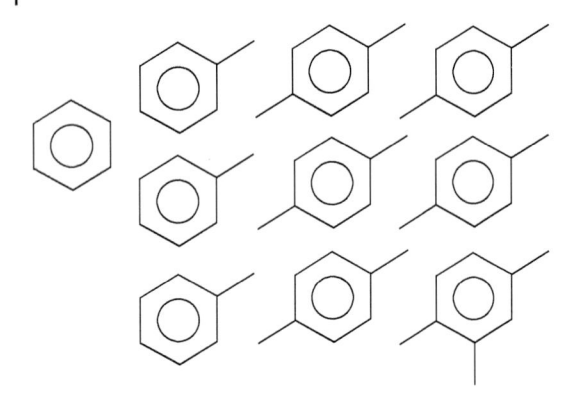

Figure 17.1 Hypothetical feed mixture with one benzene molecule, three toluene molecules, five xylene molecules, and one trimethylbenzene molecule.

rings (similar to what is done in a catalytic reforming unit), or process units that can add methyl groups (methylate) to phenyl rings. While it is fair to assume the first (most complexes are feed by reformate from a catalytic reforming unit, and this is understood to be upstream of the complex rather than within the complex), the second is not necessarily a given. At the time of this publishing, toluene methylation is gaining acceptance in aromatics complexes worldwide, though it is still evolving and has not become commonplace. Additionally, certain xylene isomerization technologies are able to convert ethyl groups to methyl groups by converting ethylbenzene to xylene, via a naphthene intermediate. However, this technology is becoming less popular due to the higher production rates capable by using the more common xylene isomerization technologies that convert ethylbenzene to benzene and ethane.

From the above evaluation, it becomes apparent that it is critical to preserve all available methyl groups on phenyl rings to maximize *para*-xylene production. If demethylation occurs (removal of a methyl group from a phenyl ring), that methyl group can be considered lost. It will ultimately be vented to a gas stream becoming fuel gas, a far less valuable destination than if it were coupled as part of a *para*-xylene molecule. Similarly, if a phenyl ring is destroyed by cracking or saturation side reactions, it is not possible to reform that molecule back into a phenyl ring and will be lost from the overall complex yield. In some cases, significant amounts of phenyl rings may be lost by venting or by improper fractionation control. Once these components are vented or lost to a waste stream, they typically cannot be reclaimed.

In order to understand how much yield loss may be occurring due to conversion or fractionation losses, proper account of methyl and phenyl groups must be done. This relies on maintaining high-quality data collection, and having well-calibrated flow meters that can be used to do mass balance calculations on individual components. In addition to this, the laboratory equipment and personnel must be capable of determining the composition of feed, intermediate, and product streams with high precision and accuracy. More will be discussed on this in Section 17.8.

17.4 Limiting Aromatics Losses

Aromatics complexes should limit aromatic ring losses to whatever extent is possible based on the available catalysts at the time of design. As technologies advance, waste streams are becoming smaller when compared on a total feed basis. Waste streams are mainly light ends in the range of fuel gas, or heavies in the range of fuel oil. These are typically valued for their fuel equivalent, though they may occasionally be blended with other product streams of higher value (if, for example, the complex is integrated with a refinery or other specialty chemicals manufacturing). Other waste streams include lower value products such as light non-aromatics that may be blended into a gasoline product stream.

Some of the waste product streams are directly a result of components in the feed that cannot contribute to a desired aromatics product, such as nonaromatics that were not converted in an upstream catalyst reforming unit. True losses in the complex itself, however, are a result of either fractionation losses or undesired reactions in the catalytic units in the complex. Limiting these losses should be a key objective of a plant engineer.

17.4.1 Olefin Removal in an Aromatics Complex

Olefins are often considered poisons in aromatics process units, and the final products of aromatics complexes are typically required to have low bromine indices in order be sold on the open market. The bromine index is a measure of the amount of bromine-reactive molecules in a mixture. Olefins are reactive with bromine, so in a laboratory, the test used to identify the quantity of olefinic species is based on how much of a bromine-containing solution reacts with a given mass of hydrocarbon sample (Figure 17.2).

Figure 17.2 Bromine reacting with an olefin as in the bromine index laboratory method.

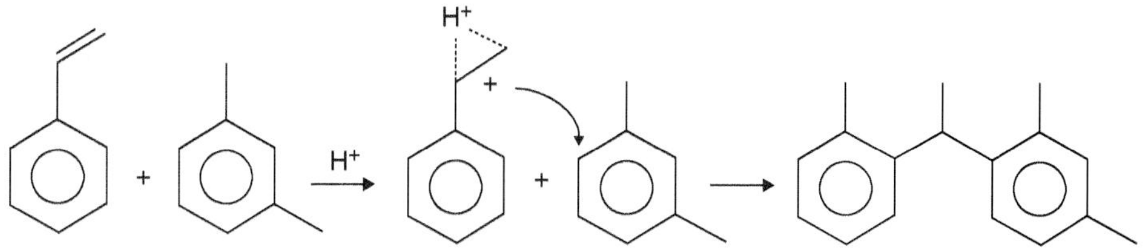

Figure 17.3 Formation of alkylate across clay acid site to remove olefins.

Most aromatics complexes include olefin removal as one of their process units. The most common way to do this is by clay treating. The clay loaded into clay treaters functions by acidic active sites causing olefins to react with other hydrocarbons, creating a heavier alkylate of the two combine molecules. Clay also does have capacity for olefin adsorption. One of the most common olefinic species in the aromatic complex feed is styrene (IUPAC name ethenylbenzene). When it reacts across a clay treater, it will most likely form a biphenyl heavy molecule, resulting in the loss of two aromatic rings (Figure 17.3).

From an operational standpoint, there is not much that can be done about this aromatic ring loss in clay treaters. The temperature may be reduced so that the clay treater operates more in an olefin-adsorption mode, but this will significantly reduce the life of the clay (typical life is on the order of several months between dumping and reloading). Another option is to run close to the maximum allowable bromine index, if adjusting the clay treater temperature allows that flexibility. But the impact on yield in doing either will be minimal.

Some licensors offer olefin reduction technologies that selectively hydrogenate olefins, rather than removing them by oligomerization. One example is the ORP Process Unit offered by Honeywell UOP. The benefit of this type of unit is that allows for olefinic aromatics to not be lost as a heavies product, instead the olefin is saturated, while the aromatic ring remains usable to be processed and ultimately be converted to one of the complex's intended products.

17.4.2 Fractionation and Separation Losses

A frequently overlooked source of losses is that which occurs in the fractionation section. Attention is often focused on reactor section losses that occur by converting feed to by-products, but this can easily be outweighed by loss of valuable components in fractionation columns and separator vessels.

17.4.2.1 Vent Losses

A common path of nonreactive aromatics losses is through product separator and column overhead receiver vents. Both of these types of vessels are essentially a single stage of separation. The temperature is typically relatively low compared to the vaporization temperature of the lightest aromatics at the given operating pressure. Still, some of the lighter aromatics might leave in these vents if the vessel conditions are keeping appreciable aromatics in the exiting vapor. This can be particularly problematic if there is poor disengagement of the vapor–liquid mixture entering the vessel, if the temperature is not low enough, or if the pressure is lower than expected.

Disengagement of the vapor–liquid mixture should be addressed during the design phase. The vessel and inlet nozzle should be constructed to allow separation over a reasonable range of operation. If it is critical to keep liquid droplets out of the vent gas, a mesh blanket will likely be included at the top of the vessel. This is typically a metal fabric mesh pad that entrains liquid droplets from the exiting gas, and allows them to drop back to the liquid level below.

From an operations standpoint, the pressure and temperature are the main handles available to keep as aromatics from exiting with the vent gas. If the temperature is higher than normal, then the vapor pressure of the light aromatics may increase to the point where aromatics leaving the vessel as a vapor becomes problematic. This may occur more frequently during the summer months, where condensing services become less effective, particularly for units where capacity is being pushed to its limits. If this is not purely due to operating a very high throughput, there may be problems with the condensers. For instance, if the condensing service is provided by fin fans, check to ensure that all of the fin fan motors are functioning properly. Also, check to ensure that the fan speed and blade pitches are appropriate, if they are adjustable. Of course, the tube fins should be visually checked to ensure they are clear of debris, and if not, blown clean with air or water.

Alternatively, if the pressure is lower than expected, it may approach the vapor pressure of the aromatics at their current temperature. This may occur due to too much pressure drop through a circuit, or simply because a pressure set point has been set inappropriately. There may also be a pressure measurement error, resulting in poor pressure control of the vessel in question.

17.4.2.2 Losses to Distillate Liquid Product

The main product streams from an aromatics complex require very high purities, targeting between 99.5 and 99.95 wt.% purity, depending on the specific compound and market. *para*-Xylene, for example, is typically traded at 99.70 wt.% purity, leaving only 3000 wt-ppm of possible impurities. The vast majority of these impurities are co-boiling molecules, so there is very small allowable margin for allowing heavier or lighter components into the product streams. As such, the fractionation targets can be very tight. If fractionation columns are not properly controlled, the aromatics product purities may be maintained but at the expense of component recoveries.

Aromatics complexes will have fractionation columns with stripping functions and rerun functions. Usually those columns that operate as strippers will send lighter aromatics to extraction in order to recover benzene and other light aromatics. This is typically not a pathway for loss of aromatics from the complex, however (except for those that are inadvertently lost to a raffinate stream). Heavier aromatics sent to this part of a complex will generally be recycled back into the complex for reprocessing. So, it may be a source of inefficiency, but typically not a direct source of loss. (As a side note, in cases where the heavier aromatics bring with them significant amounts of co-boiling non-aromatics, this can have negatively impact the extraction unit extract purity.)

17.4.2.3 Losses to Bottoms Liquid Product

There will typically be one or more columns with a rerun function to reject heavy hydrocarbons. A heavy aromatics column will remove the heaviest hydrocarbons as a bottoms product, essentially a waste stream usually used as fuel oil or sent elsewhere for further processing or blending. The overhead product of this column is typically feed for a transalkylation unit, and tends to rich with methyl groups. These components tend to convert very favorably in the transalkylation unit toward C_8 aromatics.

This stream may also contain small amounts of naphthalene and substituted naphthalenes. From a production standpoint, it is best to send all of the aromatics from this column as an overhead product that the transalkylation unit can handle. The reason is that the aromatic components in this stream will typically convert at a high percentage toward *para*-xylene. So, it is critical to let as little of these valuable components end up in the bottoms waste stream as is reasonably possible. Naphthalene is often considered a coke precursor, so catalyst manufacturers may have specifications on how much of this component is allowed into the transalkylation unit feed for sake of catalyst stability. But as the transalkylation units are equilibrium reactions (with the exception of any dealkylation function), the presence of small amounts of naphthalene in the feed may inhibit their formation across the reactor. Since naphthalenes will be formed at the expense of monocyclic aromatics, keeping a small amount of naphthalene in the feed will reduce the loss of these aromatics, ultimately improving production by a small amount. The specification on feed naphthalene should be confirmed with the catalyst manufacturer.

17.4.3 Extraction Losses

Aromatics complexes will typically include an aromatics extraction unit. This unit uses a solvent to extract aromatics from a feed stream containing both aromatic and nonaromatic (saturates) hydrocarbons (Jeanneret 1996a). These units operate under the principle that the solvent has high affinity for aromatic compounds, and they are extracted from the hydrocarbon solution. Non-aromatic compounds are rejected to a raffinate stream. If the solvent is overloaded or the conditions are such that less aromatics can be extracted, then the excess aromatics will be rejected along with the non-aromatics to the raffinate stream.

17.4.3.1 Common Variables Affecting Aromatic Recovery

There are a variety of aromatics extraction unit designs, depending on the licensor. There are also a number of solvents that may be used, depending on the particular design. In general, however, there are a few operating variables common to most extraction unit designs that can impact aromatics recovery. The primary operating variable is a solvent-to-feed ratio. Over a reasonable operating range, a certain amount of solvent flow is required to extract the aromatics from the feed. If the solvent-to-feed ratio is too low, loss of aromatics to the raffinate stream can be expected. Specific to each licensor's design, the unit will be sensitive to heat input and reflux-to-feed ratios, as well as allowable hydrocarbon and water loadings of the solvent for the unit to perform optimally.

17.4.3.2 Feed Composition

Feed composition can have an impact on the recovery of aromatics, particularly units where only two columns, an extractive distillation and recovery column, are used. A feed that contains only one aromatic species (i.e. of either benzene, toluene, or C_8 aromatics) is different from a feed that contains 2 or 3. The relative amounts of aromatics and nonaromatics are also significant. Wider boiling feeds are generally more difficult to process than more narrow boiling feeds. More aromatic feeds are generally easier to process than less aromatic feeds.

When processing a feed containing benzene and toluene, the extractive distillation column must vaporize nonaromatics that co-boil with toluene and the solvent flow must be able to recover the benzene from the stripping vapor stream passing up the column. The heavier nonaromatics that co-boil with the toluene will require higher column temperatures as compared to an extractive distillation column operating with only benzene in the feed. The higher temperature requires more solvent to recover the benzene since the benzene will have a higher partial pressure.

For a unit that processes benzene, toluene, and xylene in the charge, a conventional unit (that uses an extractor column, a stripper column, and a recovery column) usually makes more economic sense. In a two-column system it may be possible to process such a wide boiling feed if the level of non-aromatics in the xylene fraction of the feed does not need to be reduced significantly, the level of xylene is fairly low, or the non-aromatics in the xylene fraction is low to begin with.

Prima facie, it is logical that if a feed has more aromatics there is a need for more solvent to adsorb the aromatics. This is normally true for extraction. In extractive distillation, the vapor rate of nonaromatics in the extractive distillation column has a significant impact on the required solvent flow. As the amount of nonaromatics in the feed increases, the portion of the feed that must be vaporized increases. Increasing the amount of the feed that must be vaporized increases the vapor traffic and it becomes more difficult to retain the heavier keys. In any stripping operation increasing the stripping rate make it more difficult to retain the heavy key components in the bottom product; in this case it becomes more difficult to retain the aromatics in the solvent. For this reason, it is often the case that as the nonaromatics in the charge increase, the required solvent flow increases as well.

As the aromatics in the feed increase, more is carried down the extractive distillation column with the solvent, reducing the raffinate flow. The flow rate of solvent relative to hydrocarbon phase (at constant S/F) thus increases near the top of the extractive distillation column. This relatively large amount of clean solvent, for a small amount of aromatics (at the solvent introduction point) results in a raffinate with an extremely low aromatic content.

17.4.3.3 Foaming

Foaming is possible in extraction units, and can contribute to not only aromatics losses but costly solvent losses as well. Foaming may make the unit unstable, reduce throughput, or prevent the operation of the unit. Foaming may occur in the various columns of an extraction unit where vapor–liquid separations exist with both solvent and hydrocarbons together. Foaming may also contribute to fouling in the unit. The causes of foaming are varied and not clearly defined or understood. For the purposes of operating the unit, the most reliable method of eliminating this operational issue is to inject antifoam.

Symptoms of foaming vary. During a foaming even it may appear that the column is experiencing flooding: high pressure drop and unstable levels. Foaming near the top of the column may lead to slugs of liquid passing overhead and causing unstable operation of the overhead system. Foam-induced fouling has been identified at very few locations but is usually apparent as poor column performance. In the extractive distillation and recovery columns of a two-column system, foaming is usually evident as difficulty in removing solvent from the overhead products.

Foam or froth is present in most distillation columns initiated by two-phase flow on the trays. It becomes a problem when it is subjected to stabilizing forces that reinforce weak spots in the liquid film separating the vapor bubbles. This hampers the rupture of the film and gives the froth a longer life than normal. The mass transfer between liquid and vapor in a distillation column can generate the stabilizing action required to increase foam life. If the liquid film between vapor bubbles achieves a higher surface tension, the film will then become stronger.

The phenomenon of foaming is very common in extraction processing pyrolysis gasoline and much less common in units processing only reformate from high severity catalytic reforming units. It is also more common in units processing benzene–toluene feeds as opposed to full-range benzene–toluene–xylene feeds.

The extractive distillation column should be operated at the design pressure in order to ensure that vapor rates within it are in line with the design hydraulics. Operating at a lower pressure for any reason will lead to increased vapor traffic, and increase feed flash at constant enthalpy. Increasing feed flash will reduce the actual stripping in the bottom of the extractive distillation column, so it may impact product purity.

The recommended antifoams are 100% polydimethylsiloxane with finely divided silica and have been tested to be effective antifoam compounds. The solids component makes the material a "compound." Poor results have occurred when a material without the solids was substituted. Very little silicon-based antifoam agent is required to be effectively – typically on the order of 0.5–1.0 wt-ppm of active ingredient based on the feed rate to the unit. Required dosage rates may differ depending on the specific agent used. Antifoam agents are very viscous and may require dilution (toluene is typically preferred). Units with downstream processing that may be sensitive to silicon may consider other antifoam products. Normal paraffin products have been used successfully, though their required injection rate is typically 1–2 orders of magnitude higher than what is required for silicon-based antifoams.

17.4.4 Reaction Losses

There are two main reaction-based process units in a typical aromatics complex: a xylene isomerization unit and a transalkylation unit. However, design variations where additional reaction process units exist are not uncommon. This section will discuss losses in these two most common process units.

17.4.4.1 Xylene Isomerization Unit Losses

The xylene isomerization unit typically takes its feed from the raffinate stream of a *para*-xylene adsorptive separation unit, with the intention that it isomerizes the raffinate xylenes back to an equilibrium mixture of xylenes, thereby generating more *para*-xylene. This unit also removes some of the ethylbenzene either by isomerizing it to xylenes via a naphthene intermediate, or by dealkylating it to form benzene and ethane. These catalysts are generally classified by how they convert ethylbenzene, either as EB dealkylation, or EB isomerization. The most common type in the commercial operation is EB dealkylation, so that is what will be discussed in the following paragraphs.

By the second decade of the twenty-first century, catalyst improvements have been made by manufacturers to keep loss of xylenes to 1–2 wt.% or lower as they pass through the reactor during normal operation. For most EB dealkylation-type xylene isomerization catalysts, the vast majority of these losses, generally 80% or more, are due to transalkylation (see Figure 17.4). This means that most xylene loss is by formation of toluene and C_9 aromatics (and to a lesser extent, C_{10} aromatics). If the complex is equipped with a transalkylation unit, this is not a true loss as these molecules are ultimately reconverted mostly back to C_8 aromatics, unless it results in a bottleneck of one of the process units involved. But it would result in a reduced efficiency of the complex due to additional recycle of these molecules.

These catalysts typically have a metal function which serves to hydrogenate the ethyl radical removed during the ethylbenzene dealkylation reaction. If the metal function is insufficient, this ethyl radical may not saturate, and can re-alkylate onto another phenyl ring. If it re-alkylates onto a xylene, it will form a C_{10} aromatic, resulting in another form of xylene loss. The majority of this component would likely find its way back to the transalkylation unit after being processed through several fractionation columns. Once at the transalkylation unit, most (but not all) of the ethyl groups would be dealkylated.

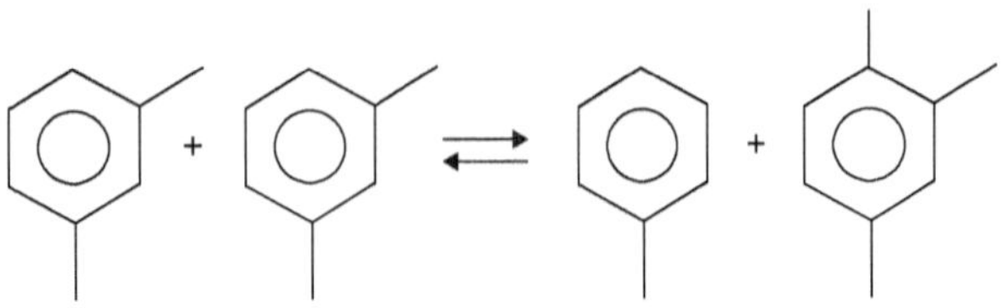

Figure 17.4 Transalkylation side reactions resulting in loss of xylene across xylene isomerization catalyst.

The remaining losses are typically due to aromatics saturation forming a substituted cyclohexane molecule or demethylation. Both of these occur to a very low extent, and are difficult to quantify on a commercial scale. Aromatics saturation may become an issue if the metal function of the catalyst, which typically provides a hydrogenation function, is too strong. Metal function typically does not strengthen over the life of the catalyst. So, if this were to occur, it would likely be apparent upon initial start-up for a load of catalyst. Demethylation is typically a result of overactive catalyst dealkylation function. This function is minimized either during manufacturing or during start-up by an attenuation step such as sulfiding or initial high severity operation.

For the ethylbenzene dealkylation type of xylene isomerization unit, the only operating parameter that is adjusted during normal operation is the reactor inlet temperature. The reactor pressure, recycle gas purity, and hydrogen-to-hydrocarbon ratio (H_2/HC) are typically set to certain values and maintained throughout the catalyst life cycle. The space velocity is set for the reactor based on the catalyst loading and feed rate. Xylene loss is a function of the catalyst activity at a certain temperature and whatever deactivation has occurred thus far for the catalyst. These losses are expected to either stay steady or fall slightly over the life of the catalyst, as the active sites that catalyze these reactions deactivate over time. As the catalyst deactivates and temperature is increased to maintain ethylbenzene conversion, for sake of xylene loss the impact of the higher temperature is not expected to outweigh the impact of catalyst deactivation.

There may be cases where xylene loss is higher than expected or desired, and some adjustment of the other operating variables, which are typically otherwise held constant, may be adjusted to reduce xylene loss. In principal, changing operating conditions such that it becomes less likely for xylenes to contact the catalyst during a given amount of time will reduce the propensity for xylene loss side reactions. Though it also stands to reason that this will limit the xylene isomerization and EB conversion reactions. These will be discussed below, with the assumption that other operating variables are kept constant.

First, ethylbenzene conversion may be reduced. This is perhaps the most obvious option – ethylbenzene conversion is reduced directly by reducing temperature. This reduces the activity for the xylene loss reaction as well. This response is often not desired, however, as reducing the ethylbenzene conversion will increase the ethylbenzene concentration in the *para*-xylene separation unit, which can make the separation more difficult and reduce per-pass *para*-xylene recovery.

Second, if space velocity of the hydrocarbon feed components is increased (synonymous with increasing the feed rate), then the xylene loss side reactions will reduce. This is strictly a residence time effect. This is often not something that can be practically adjusted for sake of reducing xylene loss, as the feed rate is governed by the effluent from the upstream *para*-xylene separation unit.

Third, if the pressure of the reactor circuit can be lowered, this is also expected to reduce xylene losses. This too is similar to the residence time effect, as the frequency of xylenes and catalyst active sites coming into contact is reduced. This is typically a handle that is able to be adjusted to some extent, but is limited by recycle gas compressor limitations and the catalyst licensor's hydrogen partial pressure requirements for sake of catalyst stability and selectivity. Also, if the separator bottoms liquid does not have a pump, the pressure of the reactor circuit may be limited by what is required to send the liquid to downstream fractionation. The potential downside of reducing pressure aside from what has already been listed, is that this may limit the extent of the xylene isomerization and ethylbenzene conversion reactions.

Losses of aromatic rings can occur in a xylene isomerization unit, but these losses are generally very low, to the point that they are difficult to measure on a commercial scale, with the flow meters and analytical test methods used in commercial plants. If these losses were to occur, they generally occur by aromatic ring saturation to naphthenes, potentially followed by nonaromatic ring opening and cracking. Saturation in these units occurs more easily with benzene rings that are formed after ethylbenzene conversion, and this affects benzene product purity, but is not necessarily related to xylene saturation losses.

17.4.4.2 Transalkylation Unit Losses

The transalkylation unit typically takes its feed from an enriched toluene stream (frequently from a benzene–toluene fractionation section) and a C_9 and heavier aromatics stream (frequently as the overhead liquid product from a heavy aromatics column, or side draw product from a column involved in xylenes fractionation). It takes these streams to establish an equilibrium mixture of the phenyl rings and methyl groups, and will dealkylate most alkyl groups 2 carbons and larger.

The primary mechanisms for aromatic losses in a transalkylation unit are saturation to naphthenic nonaromatic compounds, and formation of heavies that end up leaving the complex as a heavies product. There are few handles for limiting these losses during operation. The primary adjustment from start of run to end of run for transalkylation units is increasing temperature as the catalyst deactivates. This increase in temperature allows the conversion reactions to progress to the desired extent. This adjustment may have some impact on the above-mentioned losses. Other operating variables such as reactor circuit pressure, recycle gas hydrogen purity, and H_2/HC are not typically adjusted during the course of a catalyst cycle.

Saturation is favored at lower temperature, so the impact of this may be seen more during start of run while the catalyst is at its most active. This reaction typically does not progress to the extent that it is seen from a production standpoint. However, if the catalyst is expected to achieve high benzene purity without downstream extraction to remove non-aromatics, there may be noticeably higher non-aromatics in the benzene product stream as a result. To address this, more of the benzene produced may be diverted to the extraction unit. Or, if the final benzene product fractionation column has a drag stream from the overhead for removal of light non-aromatics, this drag stream flow rate may need to be increased. Another option would be to run the reactor temperature high enough to reduce the extent of the saturation reaction to a point where benzene purity is acceptable. Each of these options should be discussed with the catalyst supplier if excessive saturation appears to be an issue.

Reducing the hydrogen partial pressure may have some impact on the saturation reaction. However, the typical pressure for transalkylation units is sufficiently high that it would be difficult to make a change significant enough to have any measurable impact on the reaction itself.

The presence of water in the feed or make-up gas to a transalkylation unit can cause issues leading to poor aromatic yield and poor catalyst function. Water is known to reduce the catalyst activity and cause a higher temperature requirement to achieve conversion. The higher temperature requirement, coupled with an inhibited metal function in the presence of water, typically leads to much faster catalyst deactivation compared to operation with a dry feed. Newer transalkylation catalysts will typically contain a metal for metal-catalyzed reactions (typically olefin or radical saturation in the presence of hydrogen). Inhibited metal function will lead to polymerization or indiscriminate re-alkylation, as alkyl radicals find any hydrocarbon species to reattach to if they are not driven to saturate with hydrogen. This re-alkylation can result in heavies formation that can contribute to aromatic molecule losses, though this is just one aspect of the negatives caused by the presence of water in the feed to these units. Keeping water out of the feed of the transalkylation unit is critical to ensure good catalyst performance and long catalyst life.

17.4.5 Methyl Group Losses

Methyl group losses are significant as they limit the complexes' ability to produce *para*-xylene. These losses can occur by very similar mechanisms to aromatic ring losses. Of course, for any aromatic rings that contain methyl groups, the loss of that ring will result in the loss of those attached methyl groups.

17.4.5.1 Fractionation and Separation Losses

Vent losses of methyl groups while attached to aromatic rings occurs to a much lesser extent than benzene rings themselves. This is due to the fact that any additional carbons will decrease the vapor pressure of that particular component. But it can happen when velocities through separators are much higher than design, if the separator temperatures are higher, or pressures are lower.

Losses of methylated aromatics to light streams are likely losses of toluene to a benzene product stream, which is governed by the benzene product specification, and capabilities of the benzene product column. Some toluene loss to an aromatics extraction unit raffinate is also possible. However, losses to heavy waste product streams are by far the most significant source of methyl group fractionation losses. Direct loss of aromatic rings and their associated methyl groups to the heavy aromatics column bottoms will occur by design to protect the stability of the transalkylation unit catalyst. So, it is critical to reduce these losses to the extent allowable by the transalkylation unit fed by the column overhead. Modern transalkylation units are typically able to accept all C_9 and C_{10} aromatics separated in the heavy aromatics column overhead. Some may be able to accept heavier aromatics as well. From an immediate production standpoint, it is beneficial to take all of the $C_{9}+$ aromatic components that the transalkylation unit can handle, as any that are instead sent to the bottoms stream will count directly against potential xylenes production. This, however, may have consequences on catalyst stability, so consult the catalyst supplier if considering changing the heavies component composition of the transalkylation unit feed.

17.4.5.2 Reaction Losses

Most conversion losses of methyl groups are the same as those that would also result in loss of their associated aromatic ring. If an aromatic ring is saturated, it is typically not possible to unsaturate (to any appreciable extent) once the component is downstream of the catalytic reforming unit providing feed. Fortunately, however, for the sake of methyl group retention, saturation tends to occur less on methylated aromatics than to benzene alone.

Demethylation itself is the primary mechanism by which methyl groups are lost without losing the associated aromatic rings. This is typically caused by improper catalyst function, either in the xylene isomerization unit or transalkylation unit. Modern versions of these catalyst have an important metal function, and depending on the metal, may have a propensity to cause demethylation.

However, this demethylation function tends to be attenuated by specific start-up guidelines, such as sulfiding.

17.5 Fouling

Fouling is a serious issue that can both limit production and cause unplanned shutdowns. There are a number of areas to pay particular attention for fouling in an aromatics complex. Fouling may occur due to buildup of corrosion products in a certain part of an operating plant, accumulation of polymerized or heavy aromatics, or deposition of particulates from another part of the plant.

17.5.1 Combined Feed Exchanger Fouling

Combined feed exchangers recover heat from a reactor product to the reactor feed and recycle gas. This reduces the heat input requirement of a feed heater, and reduces the duty required of condensers on the reactor product. These will exist in the transalkylation unit and isomerization unit, and will likely exist on any other catalyst units in the aromatics complex. In very old complexes, these exchangers were large shell and tube exchangers, in some cases several exchangers in series. In newer units, these are most commonly welded plate exchangers (such as the Packinox™ exchanger by Alfa Laval), although twisted tube and spiral-wound exchangers have also been used.

17.5.1.1 Chemical Foulants

The feed and product streams in aromatics plant conversion units are typically very clean and not likely to polymerize as they pass through a combined feed exchanger. Polymerization may occur if large very high amounts of olefins, oxygenates, or acids are present in the feed. This is unlikely for units in an aromatics complex, as the feeds tend to be treated upstream to remove these kinds of compounds. However, this highlights the importance of properly treating any imported feeds that may be sent to the aromatics complex.

17.5.1.2 Particulate Foulants

One potential foulant for these exchangers in an aromatics complex, perhaps surprisingly, is ceramic support material. Depending on the design of the reactor in question, there is a risk of loss of containment through the reactor outlet. While extremely unlikely if the reactor is loaded properly, it is possible for ceramic supports to fracture and exit an outlet collector or basket into the reactor effluent line. Typically, the next piece of equipment from there is the combined feed exchanger. While CFE fouling is typically considered for the cold side (i.e. the fresh feed side), in this case the fouling would occur on the hot side (i.e. the reactor effluent side). This kind of fouling, if it were to occur, would have a high likelihood of causing significant damage to the combined feed exchanger, possibly requiring replacement. Short of shutting down the unit, cleaning the outlet piping, and reloading the reactor (along with any costly repairs required of the exchanger), nothing can be done during operation to fix this issue. The best way to address this is by prevention.

There are two typical reactor designs seen in modern aromatics complexes – downflow and radial flow. The radial flow design incorporates a center pipe, typically lined with profile wire or wire mesh, the openings of which is too small for catalysts to pass through, let alone ceramic balls. So, this issue is less likely with radial flow reactors (Figure 17.5).

Downflow reactors will use either an outlet basket or collector. During catalyst loading, ceramic balls that are larger in diameter than the openings on the basket/collector are loaded to a height that is a certain distance above the top of the basket/collector. This height should be determined by the catalyst supplier and detailed in a loading diagram. Directly above this first layer of large ceramic balls is a smaller layer of ceramic balls, part of a graded bed of ceramic balls. The catalyst itself is loaded above this graded bed. The ceramic balls in the second layer from the outlet collector/basket are typically of a diameter that is smaller than the openings in the collector/basket (Figure 17.6).

The primary protection against losing containment in this way is to ensure that the layer of large ceramic balls between the collector/basket and the smaller ceramic balls is thick enough to avoid migration of the smaller balls downward to the collector/basket. Some technology licensors and catalyst suppliers may use a basket design that incorporates a wire mesh or other features to prevent even these smaller particles from exiting the reactor should this migration occur. However, this will increase the pressure drop through the reactor bed, affecting the circuit hydraulics.

Another possible mechanism for loss of containment is by careless loading of the ceramic supports to save time, or using weaker ceramic supports than are required by the reactor design. Catalyst suppliers and licensors may require that ceramic balls be installed first by bucket, rather than by loading sock. This is to prevent fracturing or weakening of the ceramic balls upon impact with the reactor bottom head. If the ceramic balls are fractured or weakened upon installation, thermal cycling and stress during operation may cause the ceramic balls to crush to a smaller size, allowing them to migrate through the holes in the outlet collector/basket. If this were to happen, these ceramics would have an unobstructed path to the hot side of the combined feed exchanger, which would quickly foul and likely be damaged. Once enough of the ceramics migrate from the reactor, it is likely that smaller ceramics above and catalyst will have a pathway to migrate from

Figure 17.5 Example of a radial flow reactor.

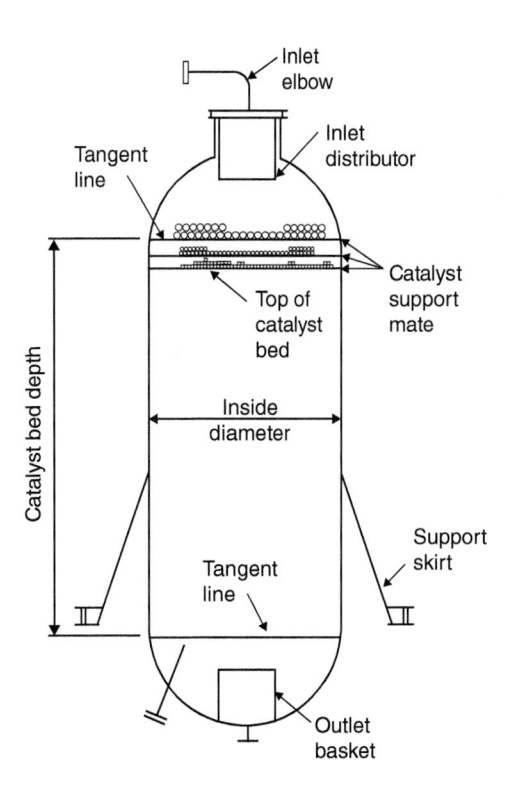

Figure 17.6 Example of a downflow reactor.

the reactor as well. A strainer may be installed on the reactor outlet pipe to catch any particles that leave the reactor, should there be a loss of containment event. This is done purely to protect the combined feed exchanger and verify that a loss of containment has occurred. Still, though, the only way to fix the problem is to shut down, clean the strainer and outlet piping, and dump and reload the reactor. Depending on the catalyst and unload method, it may be possible to reuse the catalyst. Likely an inert dump, screen, and reload would be required.

Feed-side fouling of these exchangers by particulates is also possible. This is easily addressed by including strainers upstream of the feed-side inlet. The pressure drop and condition of these strainer should be periodically checked to ensure their proper function to protect the CFE.

17.5.2 Process Heat Exchanger Fouling

Process exchangers may foul in an aromatics complex, but this again is not expected during normal operation. Typically, this will occur as a result of bringing contaminated feeds into the complex, via lines outside of the main feed route. The contaminants to be most concerned about are those with high amounts of olefins (typically as measured by bromine index) and oxygenates. Highly olefinic feeds are prone to polymerization and fouling the first heat source they see upon introduction to the complex. These feeds should be sent first to an olefin removal unit such as a clay treater or a selective hydrogenation unit such as the Honeywell UOP Olefin Removal Process. It should be noted, however, that as this feed is passed through the process exchanger intended to bring it up to the appropriate temperature for conversion across the clay or catalyst beds, this process exchanger may foul. It is a good practice to have this exchanger spared for the purpose of cleaning off-line while the unit remains operational. Processing this feed as a part of an upstream naphtha hydrotreating unit may be preferred, if the hydrotreating unit is designed to handle the additional olefins. If outside feeds must be processed in the complex, it may be best to consider a feed with low olefin content. Discuss with the complex licensor to determine if the complex can handle a particular outside feed, and if so, where the best location may be to bring the feed into the complex.

Oxygenates present in an aromatic hydrocarbon stream can also cause fouling, as they tend to be reactive and prone to gum formation. Additionally, oxygenates can create issues in the solvent extraction and selective adsorption units (see Sections 17.6 and 17.9.1). Oxygenates can often form in an aromatics feed as it is being shipped, particularly if the shipment is not nitrogen blanketed. Similarly, oxygenates may form in aromatics while in storage at a complex. For this reason, storage tanks should be nitrogen blanket – or better yet, stored in a nitrogen blanketed tank with an internal floating roof. To account for the possible presence of oxygenates in an outside feed stream, this stream should be first sent to an oxygen stripper (otherwise referred to as an oxygen "cooker") to remove any oxygenates by heat (Figure 17.7).

It is not sufficient to merely strip out dissolved oxygen by bubbling nitrogen through, as this will not remove the reactive combined oxygen species. The oxygenated feed typically enters as a feed stream to the top tray of an oxygen stripper, or enters with the reflux stream. Oxygen is thermally removed from the aromatics stream, and stripped aromatics leave the column as a bottoms product.

17.5.3 Heater Fouling

Heater tube fouling is similarly rare, though it may occur in some instances if the process side temperature becomes exceedingly high, particularly at high residence time. The high residence time is most likely in rerun column services where small amounts of relatively high boiling point material is collected in the bottom of a column. A very high temperature in the column bottoms can occur in cases where the composition in the bottom of a column becomes too heavy, resulting in a higher temperature requirement to vaporize the material in the associated reboiler. This material may increase in heavies content particularly when design changes are made to the complex or different feeds are processed that result in more heavies generation.

Consider the example of a xylene rerun column, which would produce a xylene distillate product, and a C_9+ aromatics bottoms product. An example of a design change that would result in a heavier bottoms liquid is the addition of a side draw to this column, for sake of sending this side draw material to a transalkylation unit. A xylene rerun column is typically a relatively high pressure column, and making withdrawing a $C_9A/C_{10}A$ product stream from a lower side draw would result in the bottoms stream being a $C_{11}+$ stream of a much lower flow rate than originally designed. This would result in the bottoms temperature being much higher, and the flow rate much lower, causing the heavy bottoms liquid to recirculate through the column reboiler. A design change such as this may cause coking and fouling of the reboiler heater tubes, as the very heavy bottoms liquid is exposed to much higher temperature for a relatively long time.

17.5.4 Specialty Reboiler Tube Fouling

Modern aromatics complexes are using tighter and tighter temperature approaches in heat exchangers design to minimize utilities consumption. One way to

Figure 17.7 Simple diagram of an example oxygen stripper column.

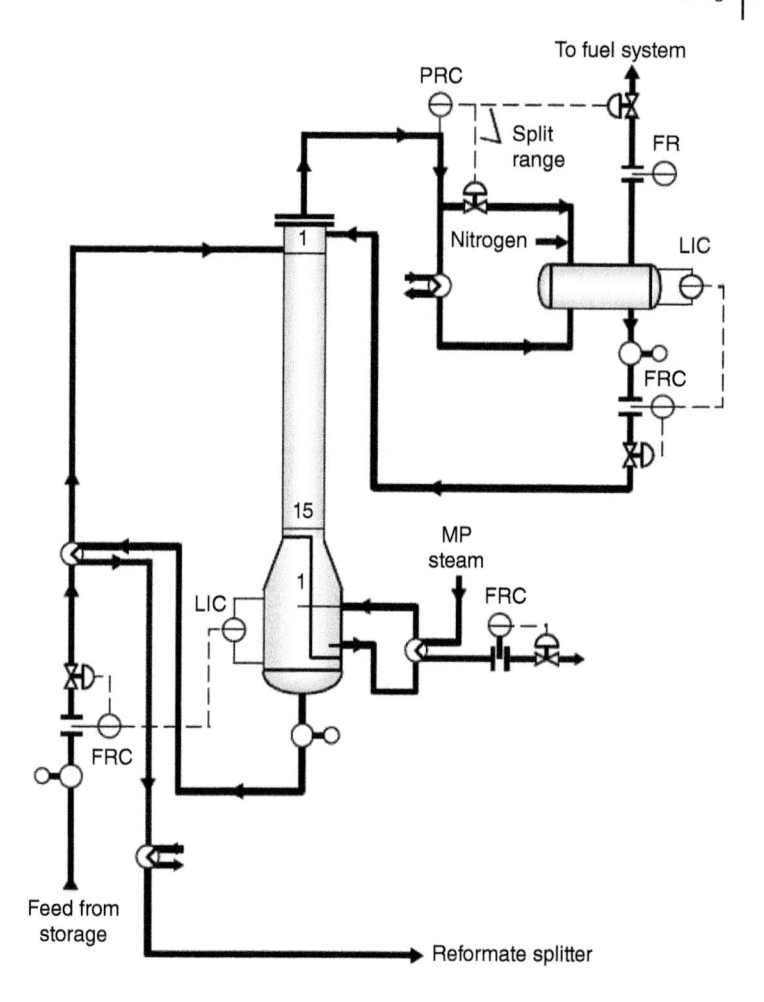

address this has been to install specialty reboiler tubes (such as UOP High Flux™ tubing) designed to achieve very high heat transfer flux through a reasonably sized heat exchanger shell. These tubes typically have modified internal and external surfaces to achieve higher heat transfer compared to bare tubes. These surfaces are applied through specific manufacturing techniques, and can be damaged if proper care is not taken during turnaround activities (Figure 17.8).

One example of this is when a column is steamed out during a turnaround to remove hydrocarbon for subsequent maintenance, the reboiler tubes will also be exposed to the steam-out conditions. If the steam is not removed quickly and purged dry with nitrogen, they can rust very quickly once air is introduced. Rust will both hurt the heat transfer characteristics of the exchanger tubes and impact the hydraulics of the reboiler. This is particularly bad in the case of vertical thermosiphon reboilers, where the reboiler function is closely tied to the pressure drop through the exchanger tubes. So careful consideration must be given to the steam-out and purge conditions of a column that has reboiler tubes such as these. Fortunately, these tubes typically have chemical cleaning procedures that may be carried out to remove corrosion products and restore the factory finish, should rusting occur. Consult with the tube vendor to confirm what procedure may be required to address corrosion, should it occur.

17.5.5 Line Fouling

Line fouling is also uncommon in an aromatics complex, but when it does occur, it is frequently related to inadequate freeze protection, particularly of purified

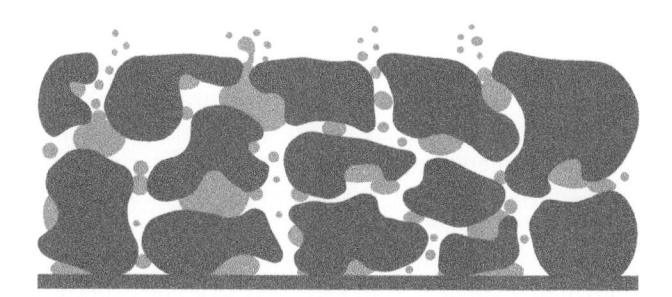

Figure 17.8 Internal surface of UOP High Flux tubing with many nucleation sites.

Table 17.1 Freezing points (at atmospheric pressure) for common pure component streams in an aromatics complex.

Name	Freezing point (°C/°F)
Benzene	5.5/41.9
Toluene	−95.0/−139.0
para-Xylene	13.3/55.7
para-Diethylbenzene	−42.9/−45.1
Sulfolane™ solvent	27.5/81.5

component streams. Shown in Table 17.1 are some of the pure component streams that typically exist in a modern aromatics complex, and their freezing points at atmospheric pressure.

From Table 17.1, it is obvious that benzene, para-xylene, and Sulfolane solvent (for aromatics extraction units that use Sulfolane) are the most problematic for sake of preventing freezing in lines during operation. Their freezing points are well above the typical low ambient temperatures seen in most locations throughout the world. As such, the lines containing these pure components, particularly those that are normally-no-flow (NNF), should be heat traced to a temperature above the freeze point. Some examples of these lines are impulse lines on flow meters or pressure differential instruments, and lines connecting spared pumps, such a spare benzene or para-xylene product pump. If a stream is set-up due to freezing, this is likely because heat tracing on one of these pure streams was missed during plant design or construction, or the existing heat tracing is not functioning properly.

As Sulfolane solvent is imported into the unit from tankage or by drum, the lines that carry this make-up solvent will typically be idle in between make-up events. The make-up lines should be heat traced appropriately. To keep this solvent from setting up during shipment, it is typically shipped with enough dissolved water (several weight percent) to depress the freezing point by around 20 °C. Tetraethylene glycol is another common aromatics extraction solvent (used in UOP Udex units), and also has a relatively high freezing point, though lower than that of Sulfolane solvent, at −9.4 °C (15.1 °F).

17.5.6 Extraction Unit Column Fouling

Aromatics extraction units are notorious for corrosion and fouling. The primary cause of this is degradation of the solvent into organic acids in the presence of oxygen and high temperature. As such, it is critical to keep oxygen and oxygenates out of the extraction unit feed and make-up solvent, and carefully control the temperature of solvent streams. Oxygen can find its way into the unit by other means as well, in particular the vacuum section of the extraction unit itself. Most aromatics extraction units operate with certain equipment under a vacuum, to depress the vaporization temperature of certain components. The purpose of this is to keep the solvent from reaching a temperature where it would start to decompose into organic acids.

Should solvent degradation occur, regardless of the mechanism, it is likely that there will be corrosion in certain parts of the extraction unit. Some pieces of equipment might experience fouling as a result of corrosion products being carried over with circulating solvent flow. In particular, fouling may occur on column trays, heat exchanger tubes, and pump strainers. The primary troubleshooting response is to locate and eliminate oxygen ingress, and to keep the solvent below its degradation temperature. In some cases, it may be helpful to install a filter on the lean circulating solvent to collect any particulates for units prone to fouling. See the following section for more information.

17.6 Aromatics Extraction Unit Solvent Degradation

Sulfolane, tetraethylene glycol, and other glycols are very popular in commercial operating plants for removal of nonaromatics from aromatic streams. In particular, these solvents are used to extract purified benzene streams, and in some cases toluene and xylenes. While these solvents are highly effective at separating aromatics and nonaromatics at reasonable operating conditions, they are prone to degradation in the presence of oxygen and possibly chlorides, and at temperatures just above the normal operating temperature (Noe 2014). If the solvent degrades, it forms organic acids that will corrode equipment and piping components, possibly resulting in unplanned downtime for maintenance, or limiting throughput until the issue can be resolved. Fortunately, these units can be operated effectively without solvent degradation when the typical sources of degradation are mitigated.

If solvent degradation does occur, it will be easily identifiable by two changes in the solvent. First, the solvent pH (as measured in a mixture with water) will drop as the organic acids increase the acidity of the solvent. As pH is a function of the hydronium ion concentration in an aqueous solution, to measure the pH of solvent it must be mixed with water to get a meaningful pH measurement. A 50 : 50 mixture of solvent is chosen to set a consistent standard. Lean solvent pH can be expected to be slightly acidic, normally with a pH of around 5.5–6.0. Another means of measuring acidity is by the acid

Figure 17.9 Photo showing varying quality of solvent condition, from very good on the left, to very poor on the right.

number. This measurement is a titration of acid species using potassium hydroxide (KOH). Healthy solvent should have a value of less than 0.05 mg of KOH per gram of solvent.

The second noticeable change in the solvent is that its color will darken, as the degradation products are not clear and colorless like pure solvent. Color is measured on the Pt/Co scale, and typically ranges from clear and colorless (color < 10) for solvent in good condition, to brownish yellow, to dark brown for solvent in poor condition (Figure 17.9).

The primary response to solvent degradation is to add a neutralizing agent in order to get the solvent pH back under control. This will reduce the effects of corrosion by organic acids, but will not prevent the solvent from degrading further if the degradation mechanism has not been eliminated. So while the acids have been neutralized, the solvent color will still be high as new degradation products continue to form. But with the corrosion under control, the root cause of solvent degradation should be investigated and addressed.

Monoethanolamine (MEA) is a typical neutralizing agent. The benefit of MEA is that it distributes well in both the upper and lower sections of columns, so it will help neutralize acids in both the circulating solvent loops, as well as the piping and equipment with hydrocarbon and water sections in column overheads. In extreme cases of solvent degradation, MEA is not strong enough for neutralization. Caustic NaOH has been used in the past and some operators have been using it more recently as well. The concern about using caustic is that caustic embrittlement may occur particularly on carbon steel surfaces that exist in the unit. But some operators have had good success using caustic in units where corrosion cannot be solved by all other measures. The amount of caustic required is likely in the wt-ppm range of the liquid streams, so true embrittlement of the carbon steel components is unlikely. However, if this measure is being considered, consult a metallurgist as well as the unit licensor.

17.6.1 Oxygen and Oxygenates

Most aromatics extraction units have a certain part of the unit operating under a vacuum, whether it be a recovery column like in a Shell Sulfolane™ unit or in a vacuum regenerator like in a UOP Udex™ unit. This vacuum section is often the culprit when there is oxygen ingress into the extraction unit. The vacuum section should be able to operate under a vacuum without the vacuum-producing equipment operating. The vacuum equipment is typically required at the very least during unit start-up, after which point, if the system is tight enough, it may be shut off. This does not necessarily ensure that the system is leak free, however. If the amount of air leaking into the system is small enough to be dissolved in liquid product streams and exported, this may still allow the vacuum to persist on its own. But if air is leaking in even at this low amount, it may still contain enough oxygen to cause a problem.

If an air leak is suspected, it becomes a matter of diligence to identify the leak location(s) and fix them. There are a number of ways to identify where they may be. This first begins before start-up during pressure tests and vacuum leak tests at ambient temperature. It is difficult to do these leak tests during operation as the equipment will be hot, possibly too hot to conduct safely. Other options are to perform ultrasonic leak testing (which detects the ultrasonic sound of a leak) or by helium leak testing (which uses a mass spectrometer placed downstream of a location where helium is sprayed near a potential leak). Helium leak testing has become an increasingly popular method. One location to definitely focus on is the cover plate header of any air-cooled exchanger in the vacuum section, as there tend to be a large number of bolts to tighten in order to seal properly. In general, hot bolting tends to help reduce leaks.

Another potential area of leakage in the vacuum section is through the relief valves. A licensor will typically specify a specific relief valve with specific seating materials. When these plants are constructed, or maintenance is performed, valves may be replaced or rebuilt with deviations from the intended design. As an example, one relief valve vendor may use a thin piece of Teflon™ and the valve will seal with a relatively low seating pressure, while another vendor uses a thicker piece and requires a much higher seating pressure to seal. Deviations like this may result in leaking past the seats in vacuum service.

This has led some operators to use nonreturn devices (i.e. check valves) to prevent backflow through the relief valve, but these are not always reliable for this type of service. A rupture disk may be a better option to prevent leakage in vacuum service. This is a redundancy with the relief valve, so to address the root cause, make sure to consult with the unit licensor and relief valve vendor to make sure the valve is properly designed and rebuilt for this type of service.

Oxygen may also find its way into the unit is with the feed, or with make-up solvent. Feed can be problematic particularly if it is stored in tanks. Aromatics will oxidize to form reactive oxygenates, and these will contribute to solvent degradation if they find their way into the extraction unit. If the unit is fed by a tank, it should be a nitrogen blanketed tank with an internal floating roof. Otherwise, the feed should be oxygen stripped in an oxygen stripper (a "cooker," rather than with a nitrogen bubbler) prior to entering the unit, see Figure 17.7. Any outside feeds should be oxygen stripped before entering the unit. Make-up solvent is typically brought into the plant by sealed drum or ISO container. A plant inventory and or solvent make-up tank typically exist to plant inventory during a shutdown and store fresh solvent, respectively. Both of these should be nitrogen blanketed internal floating roof tanks. If fresh solvent shows signs of degradation, the nitrogen blanketing system should be the first thing checked.

Because solvent is so sensitive to oxygen, it is critical to keep air out of solvent samples to avoid false pH indications. Many older extraction units use open sample stations in the field to collect samples for lab analysis. With this design, it would be easy to contaminate the sample with atmospheric oxygen. Open solvent sample stations should be switched to close sample stations. Once the sample is taken, it should be kept sealed until it is analyzed, and this analysis should be done quickly. Properly handling a sample from the time it is taken to the time it is analyzed will greatly reduce scatter in the data, and give a much better reading on the actual solvent condition. Solvent left in an open container at room temperature in a laboratory will noticeably darken and become more acidic in a matter of days.

17.6.2 Temperature

The solvents used in most extraction units have a tendency to degrade around the same temperature, around 180 °C (350 °F). So reboilers in this unit are generally designed to serve their function with a return temperature and skin temperature (where tubes may contact the solvent) below this temperature. There is typically some margin added, so operational upsets can be caught before this temperature is reached. Troubleshooting high temperature is quite easy with a functioning data historian and proper maintenance on temperature indicators. Because these temperature limits are well known and modern plants are designed to avoid them, it is rare that temperature is the root cause of degradation. But as it is easy to check the operating data, it is worth doing this quickly at the outset of any solvent condition troubleshooting activity.

17.6.3 Chloride

The impact of chlorides on extraction solvent is not perfectly understood. Some commercial units with relatively high chloride content in the feed do not report corrosion, but certainly some do. Some units with low chlorides in the feed report corrosion, while others do not. Organic chlorides can enter the unit as feed direct from an upstream reforming unit. Reformers will typically have olefins and hydrochloric acid (HCl) in the product stream. Chloride guard beds typically remove this, but if they are not removed, the HCl and olefins can react to form an organic chloride. These chlorides will boil with the hydrocarbons, so they will be present in the extraction unit feed (Figure 17.10).

Once in the extraction unit, these organic chlorides may form back into HCl and an olefin. This reaction proceeds in the presence of another acid and heat. So the leading theory is that if a unit already has some acids present due to solvent degradation, the corrosion problem can be made worse by the presence of organic chlorides (Figure 17.11).

Historically, chlorides have been difficult to quantify at very low levels where they may still cause a problem. New analytical methods have allowed for quantitation down to less than 1 wt-ppm, including ASTM D7359 (ASTM 2014), ASTM D7457 (ASTM 2012), UOP 779 (UOP 2008a), and UOP 991 (UOP 2017).

Most chloride treaters function by converting organic chloride to HCl, and then adsorbing the HCl. A resultant olefin may alkylate across the chloride guard bed and create a heavy that would cause fouling in certain types of extraction units (ones for which it is difficult to remove heavies). So if a chloride treater is installed, it should be placed upstream of a distillation column that will remove the heavy alkylate. A column like a typical reformate splitter will suit this function. Some newer chloride adsorbents may be able to process chlorides without producing this heavy alkylate.

17.6.4 Other Measurements

Corrosion probes are a possibility, and a number of manufacturers provide ones that would be suitable for this service, including the Honeywell SmartCET™ corrosion monitors. These can be used for continuous monitoring

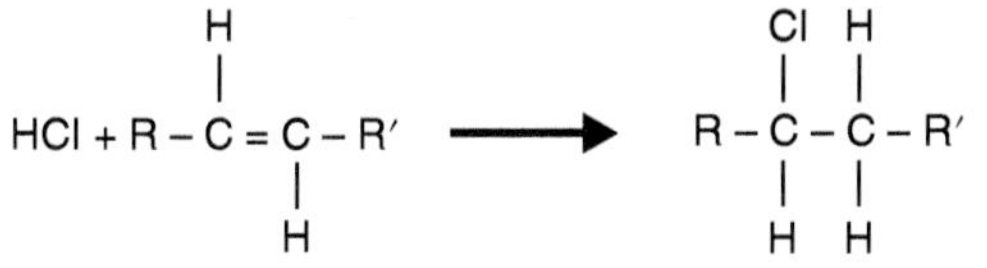

Figure 17.10 Formation of organic chlorides from hydrochloric acid and olefinic hydrocarbon.

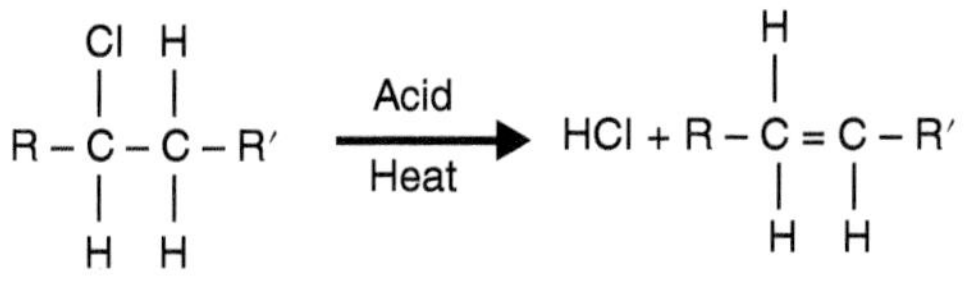

Figure 17.11 Formation of HCl from an organic chloride, in the presence of an acid and heat.

of areas where corrosion has been known to be an issue for a particular plant, and is helpful for scheduling maintenance. This does not necessarily address root causes, unless a correlation can be made to determine what process changes may have occurred around the time a probe indicates an increase in corrosion.

Solvent degradation products are not measured directly with sample analysis. The main difficulty in doing this is the unpredictability of the degradation product's specific composition, and choosing an appropriate laboratory method to detect and quantify these components. Nor are polymers that form as a result of solvent degradation analyzed, again due to the unpredictability in what forms and having a suitable method to use. Though, it hardly matters what degradation products are present, as it is the acidity of these products that cause corrosion, and it is the acidity and degradation themselves that must be addressed.

17.7 Selective Adsorption of *para*-Xylene by Simulated Moving Bed

The process unit where *para*-xylene is finally separated from a mixture of C_8 aromatics is typically done by a simulated moving bed technology, such as the UOP Parex™ Unit. Fractional crystallization is the other main technology used to accomplish this separation. However, since the late 1980s, selective adsorption by simulated moving bed technology has had an ever-increasing market share. So this section will address operational issues with this type of *para*-xylene separation. This type of technology is relatively unique as a separation technique in refining and petrochemical complexes. As such its intricacies are not as well understood throughout the industry as are more common separation technologies such as distillation (Figure 17.12).

In this simulated moving bed technology, there are typically 2 adsorbent chambers with 12 beds of adsorbent in each, for a total of 24 adsorbent beds. A liquid composition profile is established which flows down through one chamber, through a circulation line to the top of the second chamber, and down through that chamber to a circulation line that brings completes the cycle back to the top of the first chamber. The composition profile is established by sending net streams to and from the adsorbent chambers, at successive locations moving down the chambers to keep pace with the composition profile of the downward chamber flow. This coordination is accomplished by equipment such as the UOP Rotary Valve™ in the UOP Parex technology, or a very large number of on–off valves for other selective adsorption licensors. In this composition profile, four discrete zones exist: the adsorption zone (Zone 1) which is defined as the beds between the feed inlet and the raffinate outlet, the purification zone (Zone 2) which is defined as the beds between the extract outlet and feed inlet, the desorption zone (Zone 3) which is defined as the beds between the desorbent inlet and extract outlet, and the buffer zone (Zone 4) which is defined as the beds between the raffinate outlet and desorbent inlet (Jeanneret 1996b). See the diagrams in Figure 17.13.

In Zone 1, feed enters at the top of the zone and as bulk liquid from the zone above carries this liquid down through subsequent beds, *para*-xylene is continually adsorbed. At the end of Zone 1, the bulk chamber liquid should be depleted in *para*-xylene, and contains primarily desorbent and raffinate C_8 aromatics (and any non-aromatic co-boilers). This is the point at which raffinate is removed from the chambers. Increasing the flow rate in Zone 1 will tend to push raffinate components toward the raffinate (and away from the extract point), increasing purity, but will also cause *para*-xylene to be pushed toward the raffinate, reducing recovery. So there is an optimal zone flow that must be achieved for this zone. Its flow rate also needs to be balanced with the rate at which the bedline active with feed is moved from a particular bed to the next bed below. The faster this is done, the more access a given feed has to adsorbent, but this will also require a higher flow rate in this zone to ensure the feed components in the bulk fluid in the chamber keeps pace with the movement of the feed bedline position.

In Zone 2, purified extract is withdrawn at the top of the zone, which bulk fluid from above brings desorbent and *para*-xylene into the zone to reflux the non-desired raffinate components away from the extract point. These components are less strongly adsorbed than *para*-xylene, so they are relatively easy to remove from the adsorbent and refluxed away from the extract point. Similar to Zone 1, when the Zone 2 flow rate is increased, this will push raffinate components further away from the extract

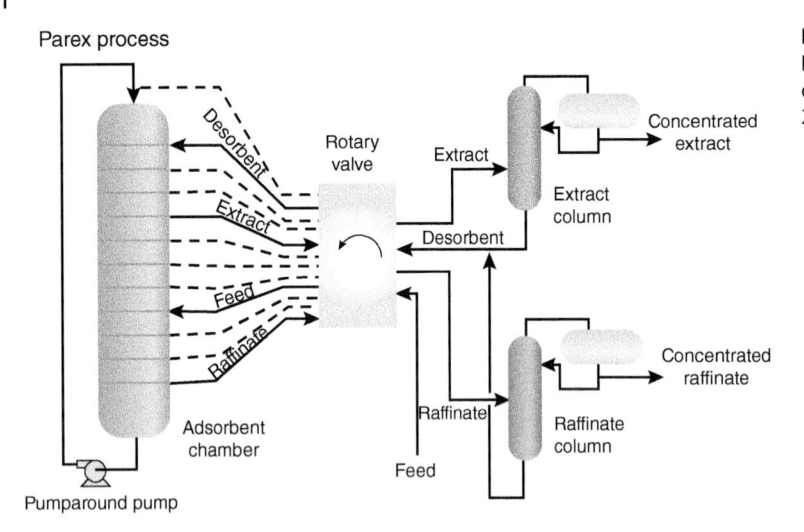

Figure 17.12 Simplified flow diagram for a UOP Parex Unit with a heavy desorbent, a common commercial *para*-xylene separation unit (UOP 2013a).

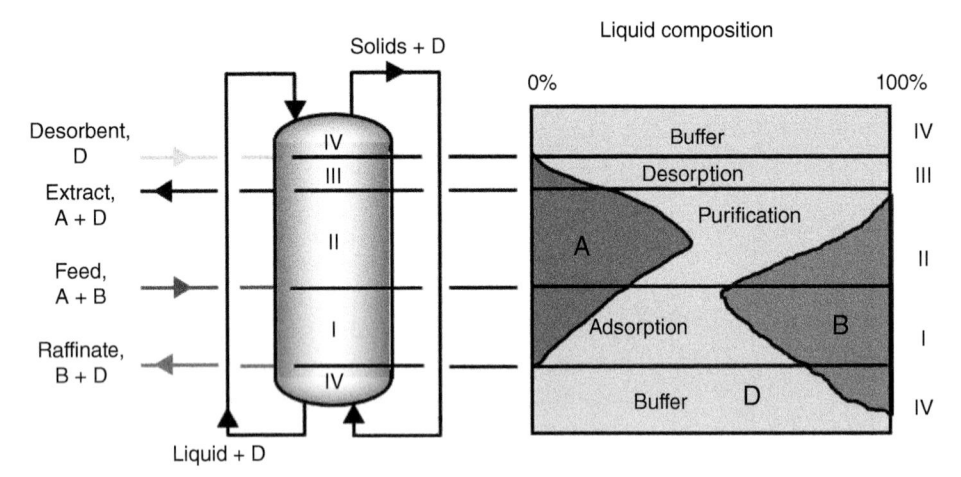

Figure 17.13 Diagram showing the four main zones in a typical simulated moving bed system. The unit simulates a countercurrent upward movement of adsorbent through the system. Component A is the desired component, component B is the undesired raffinate components from the feed, and component D is the desorbent.

point and toward the raffinate point, improving the extract purity. However, this will cause higher amounts of the desired *para*-xylene to be pushed through the zone toward the raffinate point, reducing recovery.

In Zone 3, a desorbent enters the top of the zone and desorbs *para*-xylene from the selective pores of the adsorbent. By the time the adsorbent reaches the top of Zone 3, it should be depleted in *para*-xylene. The bed at the bottom of Zone 3 should contain primarily *para*-xylene and desorbent, with the raffinate components having been refluxed away from this point in Zone 2. It is at this point where the extract is withdrawn. A higher flow rate in Zone 3 will cause *para*-xylene to be desorbed more completely before the desorbent point, leaving less *para*-xylene on the selective pores to be desorbed after the desorbent point. A lower flow rate in Zone 4 will cause more *para*-xylene to remain on the adsorbent once it reaches the desorbent inlet at the top of the zone. When this occurs, the *para*-xylene desorbs in Zone 4,

between the desorbent and raffinate points, and leaves with the raffinate stream. An increase in Zone 3 flow will generally increase *para*-xylene recovery, but have little or no impact on product purity.

In Zone 4, the desorbent enters at the bottom of the zone, with a simulated countercurrent flow pushing up against the non-adsorbed raffinate components toward the raffinate point. This buffer zone is necessary to keep raffinate components from entering Zone 3 past the desorbent point. If that were to occur, they would be carried directly to the extract withdrawal location. Zone 4 can be considered a physical flush of nonselective void volume, where the downward flow is a lower volumetric rate compared to the rate at which the active bedlines are stepping past the adsorbent bed volume. A higher Zone 4 flow rate will cause raffinate components to break through the desorbent point into Zone 3, causing a reduction in purity, primarily in the less strongly adsorbed components such as *meta*-xylene and non-aromatics.

Adjusting the Zone 4 flow rate itself will have little or no impact on *para*-xylene recovery.

In addition to zone flow rates, step time is another important consideration for separation in these units. One step is the amount of time it takes for the active bedlines to remain carrying their current net streams, until the time when the net streams step to the following bedlines below. When the net streams have cycled through all positions and return back to the original position, this is referred to as one cycle. Reducing cycle time (i.e. stepping faster) will generally increase recovery; however, there are practical limitations to how fast the cycle time can be increased.

The above discussion can be visualized by looking at a survey of the chamber composition profile. This survey is created by taking liquid samples of the liquid sent through the chamber circulation lines connecting the bottom of one chamber to the top of the other chamber, taking one sample per step. This survey is commonly called a "Pumparound Survey," as it is taken from liquid circulating through the line "pumped around" from the bottom one chamber and up to the top of the other chamber (Figure 17.14).

17.7.1 Purity and Recovery Relationship

In these process units, purity and recovery of *para*-xylene are closely related. Purity is the primary operating target, and the unit should be optimized at a target purity to achieve as high of a recovery or production rate as is profitable, considering the cost of utilities consumption to do so. When a report of low recovery is made, in nearly all cases it is because a product contaminant shifted, causing the purity to fall. The operator likely responded properly by increasing the Zone 2 flow rate, shifting flow away from the extract and toward the raffinate. This would result in the purity improving, but at lower *para*-xylene recovery.

What should logically follow after this point is to investigate the root cause of the contaminant increase, in particular, which contaminant increased, and what could have caused it specifically to increase. For certain contaminants, this exercise is relatively easy. With *ortho*-xylene, for example, there are few potential root causes that result in an increase in *ortho*-xylene alone (or a stronger response on *ortho*-xylene compared to the other product contaminants). For a heavy desorbent system, if the raffinate fractionation column has inadequate heat input, *ortho*-xylene (the heaviest boiling of the raffinate components) will slump into the column bottoms and contaminate the recycle desorbent. This will ultimately show up in the *para*-xylene product. The other primary possibility for *ortho*-xylene contamination is that the adsorbent is under-hydrated. It should be noted that the selectivity response to hydration is adsorbent specific, though for most modern adsorbents, the *ortho*-xylene selectivity is the most sensitive to water content. For the case of ethylbenzene, this is the second most strongly adsorbed C_8 aromatic after *para*-xylene, and has a relatively low content in the unit feed. The typical causes of increasing ethylbenzene in the product are relatively simple: either the feed ethylbenzene concentration has increased or the Zone 2 flow rate was inadvertently reduced (occasionally by instrumentation error).

If the exact cause of the particularly product contaminate increase can be determined, it can usually be addressed directly. Once this is done, appropriate adjustments can be

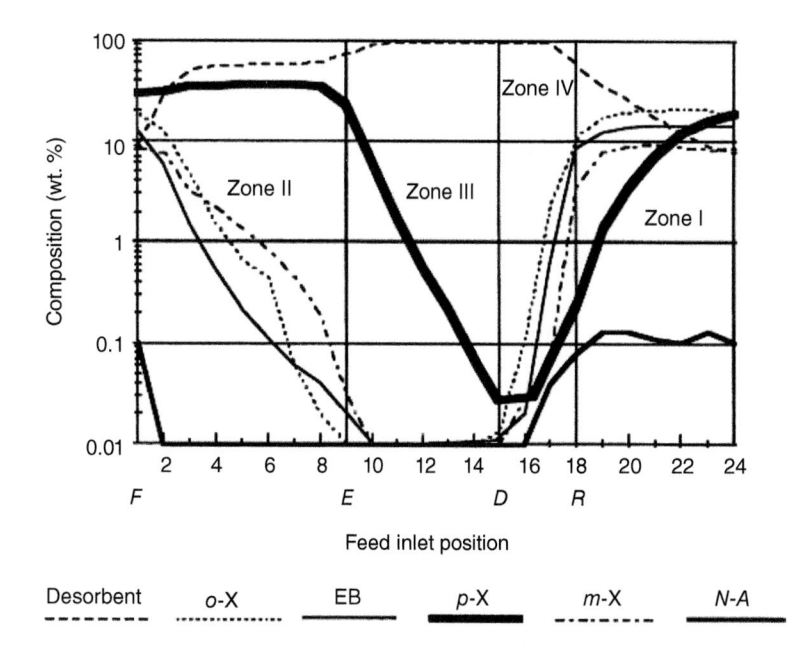

Figure 17.14 A hypothetical Pumparound Survey, showing the typical change in main hydrocarbon components in each of the four main zones of a *para*-xylene selective adsorption unit.

made to return the Zone 2 flow rate to its lower value, and recovery should increase.

17.7.2 *meta*-Xylene Contamination

The feed composition of most *para*-xylene separation units is a near-equilibrium mixture of xylenes, ethylbenzene, and co-boiling nonaromatics. *para*-Xylene is typically in the range of 18–30 wt.%, while *meta*-xylene is typically in the range of 40–60 wt.%. As far as the typical aromatic feed components are concerned, *meta*-xylene is usually the easiest to separate. However, because *meta*-xylene is so prevalent in the feed (and therefore the raffinate as well), if an inefficiency were introduced into the composition profile, it would most likely manifest as an increase in *meta*-xylene in the product over any other contaminant. These inefficiencies should be preventable by design, so it is important to follow the licensor's guidelines on preventing cross contamination of net streams primarily through start-up line isolations, and maintaining proper flow and pressure control in the adsorption section.

17.7.3 Common Poisons

There are several adsorbent poisons to be careful of in these units. Olefins, oxygenates, heavies, and water are the most common. These can each have similar effects on the adsorbent, namely strongly adsorbing in selective pores and resulting in a loss of selective capacity. Units that have been contaminated with these components will often see a loss in selectivity that causes higher *meta*-xylene content in the product, along with a drop in recovery and adsorbent capacity.

17.7.3.1 Olefins

Olefins should be removed upstream, usually to a point that the bromine index is less than 20 mg bromine/100 g. This does not necessarily correlate to a specific wt-ppm of olefinic species, as different olefins contribute differently to the bromine index measurement. The bromine index is a measure of the bromine-reactive species in a sample, and olefins in particular are reactive to bromine. Olefins can find their way into the feed stream if clay treater performance is neglected, so this should be reasonably easy to monitor and control. Feeds from outside the complex should always be clay-treated to remove olefins (and any heavy alkylate rerun from the clay-treated product). These adsorbents have acid sites, similar to those that are found in the clays used in clay treaters, though they have a weaker alkylation function. But over time, the presence of olefins can cause alkylation to occur to such an extent that heavy alkylate molecules can build up on the adsorbent and restrict access to selective pores. This type of poisoning can be considered permanent.

17.7.3.2 Oxygenates

Oxygenates may be present in either the feed or desorbent, though feed should be actively processed to remove oxygenates. Oxygenate content of the feed is frequently measured as a carbonyl number, which determines the amount of ketone or aldehyde carbonyls in a sample. These can easily form when aromatics contact with air, even at ambient conditions. All outside feeds should be oxygen stripped ("cooked") to remove oxygen and oxygenates (see Figure 17.7). Internal feeds will either come directly from a reforming unit, and any reprocessed material should pass through the bottom of a column like a reformate splitter, so they should be devoid of oxygenates. Desorbent, however, is typically stored in atmospheric tanks, and may sit for long periods of time before being introduced to the process. To avoid contamination of this material with atmospheric oxygen, the tanks should be nitrogen blanketed with an internal floating roof.

Oxygenates are polar compounds, and as such will strongly adsorb on the adsorbent. Depending on the particular oxygenate present, and concentration, the impact on performance can vary. In most cases, the difference in selectivity between *para*-xylene and one or more of the raffinate aromatics will drop, causing their concentration in the product to increase. In addition to this, the adsorptive capacity may fall. Poisoning with oxygenates is considered to be mostly permanent. Should the desorbent become contaminated with oxygenates, oxygenate guard beds can be used to process this material remove these oxygenates. If the adsorbent has become contaminated with oxygenates, once the desorbent has been thoroughly cleaned, some of the adsorbed oxygenates may desorb, and some improvement in performance may be realized. But full restoration of selective capacity is unlikely.

17.7.3.3 Heavy Aromatics

Heavy aromatics can also be problematic if they are allowed in the feed or accumulate in the recycle desorbent. The feed to this unit should always be rerun to remove any heavies. This is typically done directly upstream in a fractionation column such as a xylene column, where xylenes are taken as an overhead product and any heavier components are rejected to the bottoms stream. Typically if heavies contamination of the feed occurs, it is with trace C_9 aromatics. So the kinds of heavies that can cause permanent adsorbent damage such as substituted naphthalenes, biphenyls, or larger components are very unlikely to be present. If outside feeds are brought into the complex, direct import to the unit without processing to remove oxygenates, olefins, heavies, and light components is strongly discouraged.

More likely, if heavy aromatics are present, it is a result of not removing heavies as they accumulate in the heavy circulating desorbent. Heavies can be generated, albeit at a very low rate, as aromatic molecules (and possibly with trace levels of olefins and oxygenates) are sent across the adsorbent and through reboiler sections. These heavies are typically oligomers of C_8 and C_{10} aromatic molecules. As the heavy desorbent stream is continuously recycled back to the adsorption section from the extract and raffinate column bottoms, these heavies can continue to build up. If they become present at high-enough levels, they will adsorb permanently to the adsorbent selective pores. For this reason, these units are designed with a rerun column to remove these heavy molecules from the circulating desorbent. These rerun columns typically process a small slipstream of the circulating desorbent, enough to keep the concentration of heavies in the desorbent very low, and are intended to be operated at all times. Units that have shut down the rerun columns for considerable amounts of time (on account of reducing utilities consumption) have suffered permanent losses of adsorptive capacity due to accumulation of heavies on the adsorbent.

17.7.3.4 Water

Water is unique as it is typically a requirement to have a dosed addition to the feed or desorbent stream to maintain proper adsorbent hydration. However, too much water can cause irreversible damage to the adsorbent by hydrothermal damage of the zeolite structure. At slight over-injection rates, the adsorbent selectivity and capacity will fall. Fortunately, typical water injection pumps are not large enough to supply an amount of water that would cause permanent damage to the adsorbent, so performance can recovery if the hydration is returned to normal. For hydrothermal damage to occur, water would likely have to come from another source. In cases where this has occurred, it happens after a turnaround, where a piping segment was inadvertently filled with water, and the pipe was not adequately flushed (or flushed at all) to remove this water prior to lining up the stream to the adsorbent chambers. A pipe segment could fill with water during a turnaround if, for example, the water injection pump, or the seal cooling water of a pump in the circulating desorbent, is left on during a temporary shutdown, and water displaces the hydrocarbon normally in the line. So it is important to make sure these services are accounted for during turnarounds. Most importantly, after a turnaround, it is critical to make sure all lines have been flushed to fractionation for removal of water, prior to lining them up to the adsorbent chambers.

Large amounts of water can also enter the feed or desorbent line if there is a cooling water or steam system that interacts with one of these streams. Some examples of these are a steam generator that may exist in a xylene column overhead, or a water-cooled exchanger that may exist on a circulating desorbent line. The feed and desorbent should be actively monitored for water content, typically by online analyzer, to limit the impact of any water excursion.

17.7.4 Rotary Valve™ Monitoring

Of the commercial operating units that employ selective adsorption by simulated moving bed technology, the UOP Parex Process is by far the most prevalent. At the heart of this process is the Rotary Valve, a multiport valve that directs active net streams to and from specific bedlines in the adsorbent chambers, with valve steps timed to keep pace with the chamber composition profile. While in principle this stepping operation could be duplicated with a large number of separate on–off control valves, the rotary valve configuration simplifies the operation of the unit and improves reliability (Jeanneret 1996b). This piece of equipment is common to the UOP suite of Sorbex™ technologies, which includes Parex. It has gained wide market acceptance due to reliability and capabilities. As it is a valve with moving parts, it is important to perform routine maintenance and operate it as intended to ensure good performance.

17.7.4.1 Dome Pressure

The Rotary Valve is a multiport valve that allows all net to come through a single piece of equipment, and be diverted as required by the process to the appropriate bedlines to and from the adsorbent chambers. The bedline port openings and tracks that carry net streams are all isolated by a compressible polymer seal sheet. This seal sheet is fixed to a plate with crossover lines connecting net stream tracks to appropriate bedline port locations. The seal is compressed by applying a force on the plate, using hydraulic pressure in a dome space above. If this dome pressure is inadequate, then the domes and bedline ports might not be properly isolated. If this is the case, internal back-mixing within the valve may occur, causing contamination of the *para*-xylene product. The dome pressure should be set at the lowest pressure that provides adequate sealing between tracks. This can be determined by lowering the dome pressure until an increase in product contamination occurs – typically *meta*-xylene. Operating at the lowest possible dome pressure will maximize the life of the seal sheet.

Once an optimal dome pressure set point is determined, adjustments are typically not required during operation. Exceptions to this are when there are changes in chamber throughput that change the chamber pressure drop. For example, when the chamber throughput increases, the chamber pressure drop will increase. This will result in an increased upward force on the seal sheet and attached plate, reducing the sealing force. So in order to maintain a consistent downward force on the seal sheet, the dome pressure will need to be increased slightly in this case. A reasonable estimate for a dome pressure at any set of conditions is to set the dome pressure slightly above the highest track pressure. The highest track pressure can be reasonably estimated by adding the average chamber bottom pressure to the average chamber pressure drop (otherwise stated as the average chamber top pressure).

17.7.4.2 Alignment

The rotor plate is attached to a drive shaft that rotates when engaged by a piston and ratchet wheel assembly. As the net downward force on the rotor plate and seal sheet changes with changes in throughput or dome pressure, the torsional load on the drive shaft will also change. With this load taken up by the drive shaft, the rotor plate will rotate a shorter distance at higher seating forces. To account for this, the ratchet will need to be pushed slightly farther with each step when unit operations cause the total downward force on the rotor plate to be higher. Conversely, when unit operations are causing the total downward force to be lower, the ratchet wheel will not need to be pushed as far, as the torsional load on the shaft will be less. The downward force on the rotor plate is directly proportional to the pressure within the dome of the Rotary Valve. So when the dome pressure is increased, for example, the ratchet wheel will need to be pushed slightly farther with each step. Alternatively, if the average chamber pressure increases, then the average pressure within the tracks and bedlines will increase, in turn increasing the upward force on the rotor plate, reducing the net downward force. In this situation, the ratchet wheel will need to be pushed a slightly shorter distance. This adjustment is made by adjusting the distance that the piston which engages the ratchet wheel is extended for each step.

What is conceptually trying to be achieved during these alignment adjustments is to have the seal sheet, which is attached to the rotor plate, perfectly aligned over the bedline port openings that they seal against. This is to limit any excessive overhang of the seal sheet material over the open bedline port during a step.

17.7.4.3 Maintenance

The Rotary Valve has moving parts, which like any valve that is frequently actuated or pump in continuous operation, requires routine maintenance to ensure long-term reliability. Moving parts on the hydraulic system that causes the rotor plate to index require occasional greasing to lubricate and keep wear of these moving parts to a minimum.

The hydraulic system that provides the energy to index the rotor plate into successive positions uses a hydraulic oil. This oil is pumped to a high pressure and stored in accumulator tanks, which will depressure when called upon by the control system. This causes the piston to actuate, rotating the shaft and rotor plate. These tanks store hydraulic energy, as the instantaneous energy demand to step the rotor plate is quite high. If the energy were to be provided by pumps alone, these pumps would have to be very large. Instead, small pumps are used, and the required energy is built up as needed between steps. Occasionally, however, the accumulator tanks may lose their ability to store pressure. This can happen if they leak nitrogen with which they are originally charged, or if the internal bladder breaks. Accumulators are spared, so if one is not functioning properly, the others can store the required pressure. But should the stored pressure become insufficient, the first symptom will be a longer duration for the piston to extend to the appropriate location. Alarms will alert an operator that stepping is taking longer than expected, so this can be investigated and addressed before it causes an impact to the overall process.

Most petrochemical plants turn around on a four or six year schedule. When this is done, it is important to replace or refurbish (as recommended by the licensor) moving parts that have in operation for the past several years. Seal sheets can certainly last longer if properly maintained. Most valves in the hydraulic system are spared, so these can often be replaced during operation. However, a turnaround is a convenient time to perform these while the unit itself is shut down and fully depressured. Other parts such as the piston and ratchet assembly, bearings, and shaft seals should be checked and replaced if necessary.

17.7.5 Flow Meter Monitoring

Adsorption in selective adsorption units is controlled by precisely controlling the net streams and chamber circulation volumetric flowrates. A variation in the expected flowrates will result in a different chamber composition profile, resulting in poor performance. "Extra zone" troubleshooting is a technique that can be used to determine if a net stream flow meter is in error, and if so, by how much.

Typically, in these units, the flow that carries liquid from the bottom of chamber 2 to the top of chamber 1 (referred to as the "pumparound") is flow controlled, with the set point coming from the unit's programmable logic controller (PLC) system. The flow from the bottom of chamber 1 to the top of chamber 2 (referred to as the "pusharound") is typically controlled to maintain the pressure of chamber 1, so as to not over-constrain the system by having too many streams on flow control.

If there is an error in one of the net stream flow meters, when this net stream transitions from chamber 2 to chamber 1, the offset from the expected flow rate will be apparent in the measured pusharound flow rate. If the flows coming into or out of chamber 1 are higher or lower than expected, then the pusharound flow rate must be higher or lower than expected by an equivalent amount, to maintain constant chamber pressure. If this occurs, the shape of the pusharound flow rate chart will no longer match the shape of the pumparound flow rate chart. The differences in these charts can be used to determine which flow meter is in error and by how much.

Because the two-chamber circulation flows are separated by 12 beds from each other, their flow measurement profiles should have the same shape, with the exception of being offset by 12 beds. Note, however, that the pusharound flow is controlled by the chamber 1 pressure controller. Because the pusharound flow does not need to be as accurately measured as the pumparound flow, this flow rate is measured by a pressure differential type of flow meter rather than by turbine meter. Since this flow is set by pressure control, the profile is not as clear due to inherently noisier measurement compared to the direct flow control of the pumparound. In the example shown in Figure 17.15, there are more than four zones as previously discussed. Several of the major zones contain minor zones within them, reflecting the presence of minor net streams required for flushing bedlines clean, carrying other net streams of specific compositions (for example, flushing feed components from bedlines that will need to later carry purified extract).

Consider the example of extra zone troubleshooting for determining a net stream flow meter error. In the figure shown, note that two full cycles are shown in the Figure 17.16.

In the pumparound flow profile for this unit, there are seven zones present in one cycle. However, the pusharound flow profile shows that there are nine zones present in one cycle. These extra zones are due to an error in one or more of the net stream flow meters. In this case, it is the feed flow meter that is in error, an error of magnitude ΔF. The magnitude of the offset in the expected pusharound flow rate is equivalent to the offset of the feed flow meter error.

When the feed transitions through the pumparound from chamber 2 to chamber 1 (see circle #1), there is a step change in the pusharound flow rate where there should actually be no change (see circle #2). Because this unexpected step change occurs when the feed transitions from chamber 2 to chamber 1, it becomes apparent then that the offset is due to an error in the feed flow meter.

Any net stream flow meter error will be reflected in the raffinate flow as well. This is because the raffinate flow is typically used to provide pressure control on the entire adsorption section by way of controlling chamber pressure 2. So in the example shown, there is also an extra zone due to the offset from the expected raffinate flow (circles #3). This occurs when the raffinate flow transitions from the bottom of chamber 2 to the top of chamber 1.

17.7.6 Hydration Monitoring

To maximize the selective capacity of the adsorbent, a specific adsorbed water content must be maintained during operation (Ernst 2014). These adsorbents are very hygroscopic, meaning that they attract water very strongly. If no water is present in the feed streams (meaning feed or desorbent), water that is strongly adsorbed will be gradually desorbed due to the concentration gradient from the adsorbent to the dry bulk liquid surrounding it. A certain amount of water added to one of the feed streams is required in order to maintain the appropriate water content on the adsorbent. This water content is determined by equilibrium principles. Once a steady

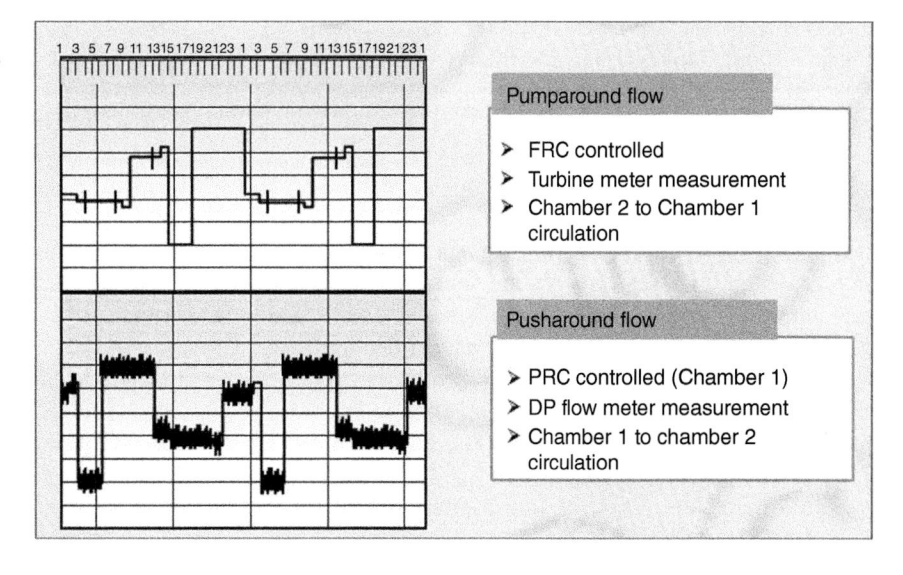

Figure 17.15 Expected flow rate profiles in the pumparound and pusharound lines, the shape of the curves offset by exactly 12 beds.

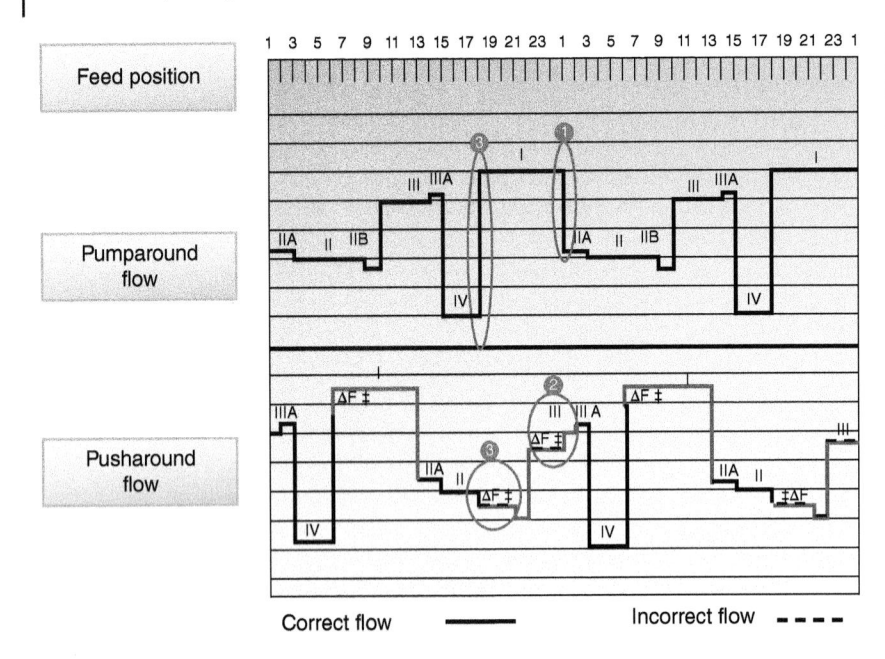

Figure 17.16 Flow rate profiles in the pumparound and pusharound lines for a unit which has an error in the feed flow rate measurement.

Correct flow ———   Incorrect flow - - - -

state is reached at the temperature and pressure within the adsorbent chambers, a certain amount of water in the bulk fluid around the adsorbent will correspond to a certain amount of water on the adsorbent. Maintaining a certain adsorbent hydration will maximize the strength of the interaction between the adsorbent and *para*-xylene, compared to the other hydrocarbons in the feed.

Adsorbent hydration is not straightforward to monitor, however. There is no simple direct way to measure the actual amount of water on the adsorbent during operation. Certainly the adsorbent can be removed by online sampling and analyzed in a laboratory setting. However, in-situ measurement is not trivial. Adsorbent hydration is typically inferred based on calculating the combine moisture content of the total chamber feeds, meaning both the feed itself and desorbent. The optimal concentration of water in the bulk fluid should be determined during the adsorbent supplier's development of the adsorbent. This will become the target for commercial operation. Different commercial adsorbents will have different hydration targets.

Water injection is typically accomplished by a small proportioning pump sending water into one of the chamber feed streams, delivering a very small amount of water compared to the net stream itself. Because it is such a small rate, and it takes time for the system to re-equilibrate once a change is made (on the order of one to two weeks), adjusting and verifying the water injection rate can be overlooked during operation. If unit throughput changes, particularly it is a significant change for more than a few days, the water injection rate should be adjusted in kind.

Additionally, over time the pump internals can wear over time, and deliver less water than expected. If this is not checked for long periods of time, the water injection rate can gradually fall unnoticed, causing an otherwise unexplained shift in adsorbent selectivity. The actual flow rate of water delivered by the pump should be checked periodically.

In these units, the feed streams tend to be dry, typically less than 10 wt-ppm of water. This is because the streams are typically processed through column bottoms prior to reaching the adsorption section. In some units this will not be the case as the streams may come from an overhead system that is expected to knock out water, so some will remain in solution. In any case, these streams may be heat exchanged with water, either in a process cooler such as an overhead water condenser or a steam generator, or come through a column that is reboiled by steam. If any of these heat exchangers have leaks, the pressure differential may send water into the hydrocarbon stream. This will cause a higher water content than expected for these feed streams. So while it is important to monitor the water injection rate, it is also important to monitor the water content of the feed streams themselves independent of what is specifically being injected.

For small amounts of over-injection of water, typically a loss of capacity occurs along with some increase in product contaminants. In such a case, this loss of capacity should be recoverable if the water injection rate is returned to normal. However, for a very large influx of water, the result may be a permanent loss in selective capacity. Hydrothermal damage of the adsorbent may occur, where the crystalline structure of the adsorbent is destroyed.

To avoid improper hydration and potential for hydrothermal damage of the adsorbent, it is important to monitor online water analyzers on the net streams. These give a good indication of any uncontrolled influx of water into the adsorption section. These moisture analyzers do need to be maintained properly to function and be considered reliable, so consult with the vendor periodically to make sure that they are kept in good working order.

17.7.7 Shutdown and Restart Considerations

Occasionally, unit engineers will report a drop in *para*-xylene recovery (or increase in certain contaminants, particularly *meta*-xylene and *ortho*-xylene) following a period where the unit had been shut down. There are a few common potential causes for having poorer performance than expected following a brief shutdown period. The below list is not comprehensive, but highlights where issues following restarts most frequently occur.

17.7.7.1 Severe Start-up or Shutdown Conditions

The start-up and shutdown of these units should be carefully monitored and procedures should limit the severity of conditions that the adsorbent and adsorption equipment experiences. As with most process units, sudden changes in flow rate, pressure, and temperature should be avoided where possible. In particular, conditions that could result in upward flow through the chambers, whether controlled or uncontrolled, should be avoided. If this happens, damage to chamber internals and churning of the upper portion of the adsorbent bed can occur. Once the unit is restarted, preferential flow may occur through the chamber internals and beds themselves, causing internal back-mixing within the adsorbent beds. Additionally, if upward flow causes damage to the chamber internals, this could result in a loss of adsorbent containment. If this were to occur, it typically cannot be permanently fixed without reloading the adsorbent chambers.

17.7.7.2 Oxygenate Ingress

Oftentimes during a temporary shutdown, the adsorbent chambers or associated fractionation columns will be drained to dedicated tanks. If the tanks and associated piping volumes contain oxygen, oxygenates may form in the exported aromatics (both C_8 aromatics and desorbent). When this material is reimported into the unit during restart, oxygenates that may have formed can strongly adsorb on the adsorbent, causing a drop in selective capacity.

To avoid this kind of oxygenate poisoning following restarts, oxygen contamination in tankage and associated piping should be mitigated. The tanks designed to hold this material are typically internal floating roof tanks with nitrogen blankets. If the tank is designed differently, or if the nitrogen blanket is insufficient, formation of oxygenates during the shutdown period becomes more likely. These oxygenates would need to be removed before being reimported into the unit. The piping itself should also be free of air. This piping is not frequently used, so when it is going to be lined up, it may need to be first purged to remove air.

17.7.7.3 Leaking of Adsorption Section Isolation Valves

These units will contain piping sections used only during start-up and shutdown, and these are intended to be isolated during normal operation. Other piping segments may exist to bypass certain pieces of equipment in the adsorption section during normal operation. When shutdowns and subsequent restarts occur, it is critical to make sure these sections that are intended to be re-isolated are in fact closed off, and not leaking. During shutdown and restart, there are quite a few valves to open and close, or blinds to remove or replace, and putting these back in the proper position can be easily missed following restart.

An open start-up line isolation valve can have a huge impact on the product composition. Certain lines connecting piping segments that may contain feed or raffinate components, to those that should contain desorbent or extract components, can cause contamination. This would require operating variables to be adjusted to such an extent to maintain product purity that the ultimate result is much lower recovery of *para*-xylene. Also, valves may be closed but still have leaks, allowing for slow leaks of contaminants into pure streams. This can have a surprisingly high impact on product quality and recovery. Isolated volumes should be periodically checked to make sure that they are properly isolated. If valves are shown to be leaking, they should be marked for repair or replacement during the next turnaround.

17.8 Common Issues with Sampling and Laboratory Analysis

The ability to understand the performance of a unit, and use data for troubleshooting, is only as good as the quality of laboratory data available. Some common issues occur with laboratory analysis that make these activities more difficult.

17.8.1 Bromine Index Analysis for Olefin Measurement

Bromine index is a measure of bromine reactive species in a hydrocarbon sample, in particular, olefins. There are several analytical methods that can be used to determine bromine index. As a side note, bromine number is similarly determined, but is used for higher olefin concentrations more commonly seen in refining services – a bromine number of 1 is equivalent to a bromine index of 1000. Bromine index is measured in terms of milligrams of bromine reacted in 100 g sample, while bromine number is measured in terms of grams of bromine reacted in the same size of sample.

Occasionally, companies will prefer to simply use the PONA (paraffin, olefin, naphthene, aromatics) by carbon number analysis to provide the olefin content of reformate feeding an aromatics complex. After all, the PONA analysis is being done anyway, and provides a value for olefin content. The problem with this approach is that when aromatics olefins are present (for example, styrene), these will be characterized as aromatics, rather than olefins. So, the PONA analysis will underreport the actual amount species that will react as olefins in aromatics processes. The bromine index measurement will characterize the double bonds in aromatic and nonaromatic olefins as bromine reactive.

17.8.2 Atmospheric Contamination of Samples

When taking samples from an operating unit, atmospheric contamination can create uncertainty in the reliability of measurements. There are several places where this is particularly problematic, in particular with air contamination of gas samples, oxygen contamination of reactive liquid samples, and moisture contamination of relatively dry samples.

Gas samples should be taken in closed, airtight sample containers, and thoroughly purged with process gas during sample withdrawal. If not done properly, these samples can show false high values of atmospheric gases, in particular nitrogen, oxygen, and carbon dioxide. In addition, this atmospheric contamination will dilute the actual hydrocarbons present in the sample of the stream to be analyzed. The test method may also have interferences introduced by the presence of these unexpected molecules.

Many liquid samples should be taken in closed sample containers as well, both in consideration of environmental safety and the safety of the people collecting the samples, but also for sake of the sample itself. Aromatic hydrocarbons are reactive to oxygen, as are solvents used for aromatics extraction. Aromatics are frequently tested for their oxygenate content and color, as these will indicate a potential to cause permanent adsorbent poisoning. These compounds form very easily in the presence of oxygen, even in a matter of hours sitting in a laboratory setting if the sample was contaminated during withdrawal. Typically, these samples should have a carbonyl content of less than 2 mg/ml of sample. A false positive can quickly occur if sampling was not done carefully.

In the case of extraction unit solvent, the solvent can quickly react in the presence of oxygen, causing acids and solvent degradation products to form. The sample will physically darken if exposed to air in a matter of days, and give a false high acidity measurement.

Moisture analysis is typically called for on the feed streams to the various process units in an aromatics complex, as adsorbents and catalysts can be sensitive to water. Hot and relatively dry hydrocarbon samples will quickly take on water and give a false high value. Frequently values of 60–100 wt-ppm are seen on streams that, in the process, actually contain 10–50 wt-ppm water. This can make troubleshooting, or targeting a specific hydration amount in a stream, particularly difficult. Taking samples on streams such as these is best done with a closed sample container, which is dried in an oven prior to collecting the sample. Units without this capability may see more spurious results.

17.8.3 Analysis of Unstabilized Liquid Samples

Sampling of a product separator bottoms liquid or the overhead liquid (reflux or distillate product) of a stabilizer column will frequently be required for basic composition analysis. These samples are not straightforward to take and analyze in a laboratory. Portions of these samples will vaporize as the liquid leaves the process and is let down to atmospheric pressure. Often these samples can be seen to boil as they are collected in unpressurized containers. If this occurs, it will give a false low composition of lighter components in the stream, and a false high composition of the heavier components in the stream. To account for this, the sample would need to be taken in a pressurized vessel, and transported to the laboratory in that same vessel, and directly inserted from that vessel into the gas chromatograph equipment. Commercial analytical and sampling methods exist to instruct how this can be done to safely minimize sample losses.

17.8.4 Gas Chromatography

When using gas chromatography (GC) to characterize a sample, it is important to remember that GC is generally used to quantify what is already known to be present in a sample. It is not typically a means of identifying what species are present. For most GC methods, a limited matrix of molecules is considered. If unknown peaks are present in the GC results, then typically another method needs to be used to identify the components causing those peaks. This section will review some common pitfalls in GC composition analysis in aromatics complexes.

17.8.4.1 Nitrogen vs Hydrogen or Helium Carrier Gas

Many GC methods call for hydrogen or helium as the carrier gas, which carries the sample through the GC column. Both provide fast and efficient analysis. However, the global supply of helium is shrinking, resulting in increasing costs. Hydrogen presents an explosion risk for laboratories not already set up to handle hydrogen. Hydrogen can be (and is) used safely throughout the world, but its safety risk occasionally has laboratory managers looking for other options.

Nitrogen has been used as a carrier gas to sidestep the disadvantages of helium and hydrogen. However, nitrogen does not provide as efficient of a separation compared to its lighter molecular weight counterparts. In order for nitrogen to be used effectively, the rate of the carrier gas must be slowed down significantly, and the analysis itself will take longer. If the GC conditions are not appropriately adjusted for the different carrier gas, the results can be unreliable. Nitrogen may also limit the peak resolution for the smallest peaks, reducing the method sensitivity. Most methods are written specifically for helium or hydrogen, so if nitrogen is being considered instead, consult with the GC manufacturer and analytical method publishers to better understand what changes to the equipment and procedure may be required. Unfortunately, as the publisher will generally produce precision statistics (reproducibility and repeatability), these will have been considered only for the method as written. Changing the carrier gas technically changes the method, and the precision statements will no longer be applicable.

17.8.4.2 Resolution of *meta*-Xylene and *para*-Xylene Peaks

In aromatics complexes, it is frequently necessary to know with a high degree of precision both how much *para*-xylene and *meta*-xylene there is in a sample. This is particularly true in purified *para*-xylene or *meta*-xylene product streams. (Although not discussed in this book, *meta*-xylene is commercially purified and produced by a similar method to the selective adsorption method used for *para*-xylene production.)

To resolve the composition of high-purity aromatic streams that have co-boiling aromatic contaminants, a polar GC column is used. Commercial examples of the test method used for this analysis are UOP 720 (UOP 2008b) and ASTM D5917 (ASTM 2015). The resultant *meta*- and *para*-xylene peaks are very close to each other. If one is present in a very high amount (for example, *para*-xylene in a *para*-xylene product stream), the peak of the lesser component is likely located on the shoulder of the large peak. This can make it difficult for integrating the area under the curve of each peak to calculate the composition of the lesser component in particular. See the example chromatogram given in Figure 17.17.

A consistent integration method must be used when discerning between the two components that share peak area. GC software typically makes this integration straightforward, and if the equipment is calibrated using the same method as what is used for the actual sample measurement, then error should be minimized. This can be problematic if different laboratory personnel use different integration methods. Examples of these are to take a vertical drop line from the valley between the two peaks down to the base line of the chromatogram, or to instead follow the slope of the curves either at their lowest point and follow that slope

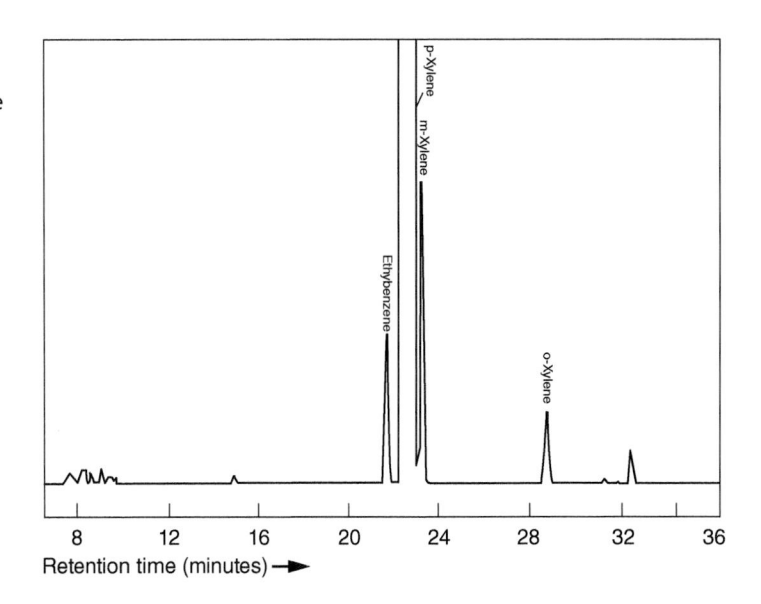

Figure 17.17 Typical chromatogram for a *para*-xylene product stream using lab method UOP 720. Note the close proximity of *meta*-xylene peak on the shoulder of the large *para*-xylene peak.

to the base line (referred to as a tangent skim). The analytical method should prescribe how exactly the peak area is to be determined.

17.8.4.3 Wash Solvent Interference

Analytical methods will call for a rinse of equipment with a solvent between runs. This is to remove any components from the previous sample analyzed, so residue does not lead to a false measurement for future runs. The solvent chosen should be compatible with the equipment, and not result in any interferences of the samples being analyzed. A common solvent specified in certain aromatics composition analyses is carbon disulfide. Laboratory personnel will occasionally switch this with other common solvents, though they can cause interferences and result in very strange measurements on future runs. One example is acetone, which may co-elute with nonaromatics and cause a false high for these components. If the method calls for a specific wash solvent, make sure to not deviate from that specification.

17.8.4.4 Over-Reliance on a Particular Analytical Method

Test methods such as UOP 744 (UOP 2006) for measuring aromatics in hydrocarbons by GC are popular for the wide range of specific components that they will quantify. However, the methods note in their scope statement that certain components that can be measured individually will co-elute if they are both present. For example, heavy nonaromatics (heavier than 10 carbons) can co-elute with benzene and toluene. If specific values for benzene or toluene are required, particularly for samples that are known to contain heavier nonaromatics, additional methods should be used to quantify these components specifically [for example, methods such as UOP 690 (UOP 2013b) and ASTM D5443 (ASTM 2018)]. The particular benefit of a method such as UOP 744 is that it will resolve individual C_9 aromatic components, which other similar test methods will not do, as-written.

17.8.4.5 Impact of Unidentified Components

Different test methods will treat unidentified components differently. Some will lump unidentified components with others in the range where they exist on the chromatogram. Others will treat them as they do not exist, and call for normalization of the identified components. So, for example, some methods will identify species that elute after C_{10} aromatics are characterized as $C_{11}+$. There still may be certain C_9 or C_{10} components that elute after C_{10} aromatics, so they would be mischaracterized as $C_{11}+$. Still other methods, though the chromatogram may show species eluting after C_{10} components, it will treat these as though they do not exist. This can cause confusion as the results may suggest that there are no components heavier than C_{10}. It is important to understand how these components are treated in the analytical method being used. If there is a concern about a specific component that may or may not be present in a sample, it is useful to view the chromatogram itself to see what peaks are actually being considered when reporting a sample's composition.

17.9 Measures of Operating Efficiency in Aromatics Complex Process Units

There are several key performance indicators to consider when optimizing a unit to reduce utilities consumption. Many of these are related to specific pieces of equipment and have been well documented and thoroughly discussed in many publications (for example, heater efficacies, column reflux-to-feed ratios, and heat exchanger U-values). The process units in an aromatics complex similarly have KPIs for energy optimization to be considered. These optimization KPIs should be considered when adjusting unit operating variables to maintain the main KPIs of each unit.

17.9.1 Selective Adsorption *para*-Xylene Separation Unit

The primary KPIs for unit performance are the purity and recovery of *para*-xylene. Both are relatively simple measures to understand. Purity is the weight percent of *para*-xylene in the product stream itself (usually defined by its determination using a particular analytical method such as UOP 720 or ASTM D5917). Recovery is the amount (also measured in weight percent) of *para*-xylene from the feed that makes it into the final product stream. Frequently a "theoretical recovery" is used that calculates recovery without requiring flow meter data, only analytical data. This can be done for systems that have three streams: a feed stream, a product stream, and a raffinate stream (Figure 17.18).

The primary optimization KPI for sake of the complex efficiency itself is the *para*-xylene recovery. As less *para*-xylene from the feed is sent to the extract stream directly, this increases the flow rate of the raffinate stream. This increases the recycle rate of xylenes through the xylene isomerization unit, its downstream fractionation, and the selective adsorption unit itself.

A second set of optimization KPIs directly relate to the utilities consumption within the selective adsorption unit itself, and that is a ratio of either desorbent/feed (D/F) or a desorbent/*p*X product (D/pX). This directly compares the amount of recycle desorbent that is required for the

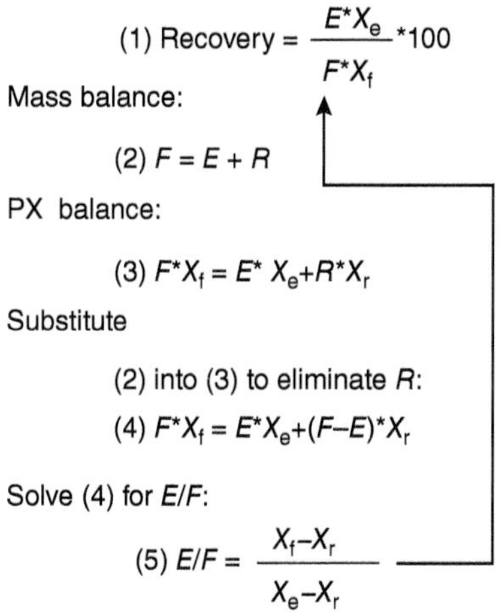

17.9 Measures of Operating Efficiency in Aromatics Complex Process Units

Mass balance:

$$(1)\ \text{Recovery} = \frac{E^* X_e}{F^* X_f} \times 100$$

$$(2)\ F = E + R$$

PX balance:

$$(3)\ F^* X_f = E^* X_e + R^* X_r$$

Substitute (2) into (3) to eliminate R:

$$(4)\ F^* X_f = E^* X_e + (F-E)^* X_r$$

Solve (4) for E/F:

$$(5)\ E/F = \frac{X_f - X_r}{X_e - X_r}$$

Figure 17.18 Derivation of theoretical recovery, starting from classical definition of recovery (total mass of desired components in product divided by total mass of desired components in fresh feed), where F, E, and R are the mass flow rates of feed, extract, and raffinate, respectively, and X_f, X_e, and X_r are the mass fractions of the desired component in the feed, extract, and raffinate, respectively.

current unit throughput, as measured either by feed rate or production rate. As more desorbent for a given throughput is required, more heat input is required to separate this desorbent from xylenes in the downstream extract and raffinate columns. When evaluating whether a process change for sake of unit optimization while maintaining purity and recovery was good or not, if the D/F or D/pX ratio ultimately increased as a result of this change, then the change itself has made the unit's operation less efficient. Higher D/F or D/pX ratios indicate lower operating efficiency, while lower ratios indicate higher operating efficiency.

17.9.2 Xylene Isomerization Unit

The primary KPIs used to evaluate xylene isomerization units are the pX/X ratio and ethylbenzene conversion. pX/X is the ratio of the isomerization unit product *para*-xylene composition relative to the total xylenes content in the product. Note that it is different than a pX/C_8A ratio, as that would also include the ethylbenzene content in the denominator of the ratio. This is not a meaningful measure for most systems, as EB content is not relevant to the extent of the xylene isomerization reactions.

$$pX/X = \left(\frac{pX_{out}}{X_{out}}\right) \times 100 \qquad (17.2)$$

where

- pX_{out} is the mass flow rate of *para*-xylene in the reactor effluent.
- X_{out} is the mass flow rate of all xylene isomers (para + meta + ortho) in the reactor effluent.

Ethylbenzene conversion is the amount of ethylbenzene in the feed that has been converted to other molecules, typically ethane and benzene. This is the typical KPI targeted over the life of the catalyst by adjusting the reactor inlet temperature.

$$\text{Ethylbenzene conversion} = \left(\frac{EB_{in} - EB_{out}}{EB_{in}}\right) \times 100 \qquad (17.3)$$

where

- EB_{in} is the mass flow rate of ethylbenzene in the feed.
- EB_{out} is the mass flow rate of ethylbenzene in the reactor effluent.

Occasionally, engineers will also look at a value for *pX* conversion. This relates the amount of xylenes in the feed excluding the *para*-xylene to the amount of *para*-xylene in the product. This is a valid calculation to perform, but pX/X is convenient as the flow rate terms in the numerator and denominator will cancel out, and will rely only up on lab results, minimizing error. A pX/X approach to equilibrium is another valid calculation, but this can vary based on the equilibrium data or correlation considered. Also, the equilibrium pX/X value is temperature dependent, so the correlation must account for that if it is to be truly accurate.

For sake of optimization, xylene loss is the main KPI. Most xylene loss through these catalyst systems is a result of transalkylation to toluene and C_9 aromatics (and to a much lesser extent C_{10} aromatics). This results in additional recycle within the complex as these transalkylated products will eventually find their way to the transalkylation unit to be ultimately converted back to xylenes. Xylene retention will occasionally be calculated instead of xylene loss - the two performance variables are related as each is 100% less the other.

$$\text{Xylene loss} = 100 - \left(\frac{X_{out}}{X_{in}}\right) \times 100 \qquad (17.4)$$

where

- X_{out} is the mass flow rate of xylenes in the reactor effluent.
- X_{in} is the mass flow rate of xylenes in the feed.

The toluene formed in the isomerization unit will also shift some of the product slate from *para*-xylene toward benzene, reducing *para*-xylene production of the complex. Transalkylation in an isomerization reactor typically increases as the ethylbenzene conversion target is increased. It is not simple enough to say that xylene loss increases as temperature increases, though for individual step change increases in temperature this is true. As the catalyst deactivates, the temperature requirement to maintain ethylbenzene conversion will increase. As the temperature increases over the life of the catalyst, the xylene loss will actually remain fairly stable. This is because catalyst aging attenuates the undesired xylene loss function of the catalyst just as it does the desired ethylbenzene conversion function. Xylene loss is difficult to measure, because it is so strongly impacted by flow meter error. Calculating a transalkylation yield may be more straightforward for evaluation, as it tracks the formation of transalkylation products (toluene, C_9 aromatics, and C_{10} aromatics), which account for most of the xylene loss in a xylene isomerization system.

$$\text{Transalkylation yields} = \left(\frac{(T_{out} + A_{9out} + A_{10out}) - (T_{in} + A_{9in} + A_{10in})}{X_{in}} \right) \times 100 \quad (17.5)$$

where

- T_{in} is the mass flow rate of toluene in the feed.
- A_{9in} is the mass flow rate of C_9 aromatics in the feed.
- A_{10in} is the mass flow rate of C_{10} aromatics in the feed.
- T_{out} is the mass flow rate of toluene in the reactor effluent.
- A_{9out} is the mass flow rate of C_9 aromatics in the reactor effluent.
- A_{10out} is the mass flow rate of C_{10} aromatics in the reactor effluent.
- X_{in} is the mass flow rate of xylenes in the feed.

The H_2/HC ratio may be considered an optimization point, as may the recycle gas hydrogen purity. The values of these operating variables have consequences on the utilities consumption (in particular, steam or electric utility required to drive the compressor) and hydrogen consumption, respectively. For sake of these measures, it is best to minimize each as can be allowed within the unit design constraints, and with consideration of specific catalyst requirements.

17.9.3 Transalkylation Unit

The primary KPI in the transalkylation unit is overall conversion, and secondarily the selectivity (or yield on converted charge). Conversion is simply the amount in weight percent of feed toluene, C_9A, and $C_{10}A$ that converts to other molecules (typically benzene and xylenes). Selectivity calculates how much of this converted feed actually converted to desired benzene and C_8 aromatics. As the catalyst ages, the temperature requirement to achieve a target conversion will increase. This may result in a lower yield due to additional side reactions, so the yield may fall, though slightly.

$$\text{Overall conversion} = \left(\frac{(T_{in} + A_{9in} + A_{10in}) - (T_{out} + A_{9out} + A_{10out})}{(T_{in} + A_{9in} + A_{10in})} \right) \times 100 \quad (17.6)$$

where

- T_{in} is the mass flow rate of toluene in the feed.
- A_{9in} is the mass flow rate of C_9 aromatics in the feed.
- A_{10in} is the mass flow rate of C_{10} aromatics in the feed.
- T_{out} is the mass flow rate of toluene in the reactor effluent.
- A_{9out} is the mass flow rate of C_9 aromatics in the reactor effluent.
- A_{10out} is the mass flow rate of C_{10} aromatics in the reactor effluent.
- X_{in} is the mass flow rate of xylenes in the feed.

Yield can be considered the optimization KPI for the transalkylation unit. As it measures the impact of side reactions, these side reactions will result in side products that will either need to be recovered as a recycle stream through the complex, or result in direct aromatics losses, reducing the production rate per amount of feed.

$$\text{Benzene} + C_8 \text{ Aromatic yield} = \left(\frac{(B_{out} + A_{8out}) - (B_{in} + A_{8in})}{(T_{in} + A_{9in} + A_{10in}) + (T_{out} + A_{9out} + A_{10out})} \right) \times 100 \quad (17.7)$$

where

- B_{out} is the mass rate of benzene in the reactor effluent.
- A_{8out} is the mass rate of C_8 aromatics in the reactor effluent.
- B_{in} is the mass rate of benzene in the feed.
- A_{8in} is the mass rate of C_8 aromatics in the feed.
- T_{in} is the mass flow rate of toluene in the feed.
- A_{9in} is the mass flow rate of C_9 aromatics in the feed.
- A_{10in} is the mass flow rate of C_{10} aromatics in the feed.
- T_{out} is the mass flow rate of toluene in the reactor effluent.
- A_{9out} is the mass flow rate of C_9 aromatics in the reactor effluent.
- A_{10out} is the mass flow rate of C_{10} aromatics in the reactor effluent.

One common limitation that occurs in a transalkylation unit is when the feed contains appreciable amounts of benzene or C_8 aromatics. Since transalkylation units are mainly equilibrium processes, the more of the desired product molecules are already in the feed, the more difficult it becomes to convert other feed components into these desired molecules. When this occurs, operators will frequently push temperature to increase conversion to the design target. A better operating philosophy would be to target an ultimate C_8 content in the product, or another measure of severity such as ethyl group conversion. A value around 90 wt.% is usually a reasonable value for ethyl group conversion (analogous to EB conversion in a xylene isomerization unit). More recent commercially available catalysts are better at dealkylation, so values upward of 95% may be reasonable targets.

Similar to the xylene isomerization unit, the H_2/HC ratio may be considered an optimization point, as may the recycle gas hydrogen purity. The values of these operating variables have consequences on the utilities consumption (in particular, steam or electric utility required to drive the compressor) and hydrogen consumption, respectively. For sake of these measures, it is best to minimize each as can be allowed within the unit design constraints, and with consideration of specific catalyst requirements.

17.9.4 Aromatics Extraction Unit

The primary KPIs for unit performance are the purity and recovery of particular aromatics, typically benzene and toluene, though occasionally C_8 aromatics. Purity is the weight percent of the specific aromatic compound in the extract on a normalized basis, or it may be calculated in downstream fractionation as the final product stream itself. Recovery is the amount (also measured in weight percent) of the particular aromatic from the feed that makes it into the product stream. Similar to that in the selective adsorption process, a "theoretical recovery" is used that calculates recovery without requiring flow meter data, only analytical data.

Aromatics extraction units have many optimization KPIs that can be considered. The most common are the solvent-to-feed ratio, ratios of other internal recycle streams to the solvent or feed, and reflux ratios of the separation columns themselves. The main goal in optimizing is to reduce heat input while maintaining the target product purities and recoveries. As each of the optimization variables are changed, this will typically result in column reflux ratio change. When this is the result of a reboiler heat input reduction, then the unit can generally be considered to be operating more efficiently.

17.10 The Future of Plant Troubleshooting and Optimization

The preceding discussions have highlighted the importance of collecting good quality and comprehensive data for use in troubleshooting and maintaining good unit operations. Various options exist in the way of process data historians and analytical data collection systems. Whichever system or suite of systems is used to effectively use the data, it must be able to be plotted together in historical trends, giving the process engineer freedom to combine dependent and independent variables on the same chart to identify causal relationships. The data can be exported to a spreadsheet such as Microsoft Excel, where it can be manipulated as needed. It is critical to be able to track KPIs in real-time and look at how they have changed from historical baselines.

At the time of publishing this book, cloud-based platforms are becoming available for KPI monitoring and live optimization, with Industrial Internet of Things (IIoT) and software-enabled services to suggest changes based on new capabilities to maintain vast libraries of commercial operating data. As these IIoT systems become more advanced, KPI changes to optimize and troubleshoot are able to be less reactive and move toward being proactive or even predictive. Honeywell's Connected Plant is one such cloud-based platform that has been used successfully in aromatics complexes (as well as refineries). As computing power, modeling capabilities, and data availability improves, these systems are destined to improve as well. The role of the process engineer will change from one of collecting the performance information such as that discussed in this chapter and determining logical courses of action, but one of verifying and validating suggestions made by these IIoT platforms, and evaluating their outcome. New cloud-based and software-enabled systems will enable the process engineer to detect problems earlier than before, prevent them from happening outright, perform real-time rather than reactive optimization, and improve operational reliability and safety (Liebert 2017).

References

ASTM International (2012). ASTM D7457 Standard Test Method for Determining Chloride in Aromatic Hydrocarbons and Related Chemicals by Microcoulometry.

ASTM International (2014). ASTM D7359, Standard Test Method for Total Fluorine, Chlorine and Sulfur in Aromatic Hydrocarbons and Their Mixtures by Oxidative Pyrohydrolytic Combustion followed by Ion

Chromatography Detection (Combustion Ion Chromatography-CIC).

ASTM International (2015). ASTM D5917, Standard Test Method for Trace Impurities in Monocyclic Aromatic Hydrocarbons by Gas Chromatography and External Calibration.

ASTM International (2018). ASTM D5443, Standard Test Method for Paraffin, Naphthene, and Aromatic Hydrocarbon Type Analysis in Petroleum Distillates Through 200 °C by Multi-Dimensional Gas Chromatography.

Ernst, G. (2014). Parex™ Best Practices. UOP Aromatics Technology Users' Conference, Singapore (12–14 November 2014).

Jeanneret, J.J. (1996a). UOP Sulfolane process. In: *Handbook of Petroleum Refining Processes*, 2e (ed. R. Myers). McGraw-Hill.

Jeanneret, J.J. (1996b). UOP Sorbex family of technologies. In: *Handbook of Petroleum Refining Processes*, 2e (ed. R. Myers). McGraw-Hill.

Liebert, R. (2017). Lifecycle management: the new paradigm for service delivery. https://www.linkedin.com/pulse/lifecycle-management-new-paradigm-service-delivery-rebecca-b-liebert (accessed October 2018).

Noe, J. (2014). Zen and the art of Sulfolane™ operation. UOP Aromatics Technology Users' Conference, Singapore (12–14 November 2014).

UOP A Honeywell Company (2006). *UOP Method 744, Aromatics in Hydrocarbons by GC*. Des Plaines, IL: Honeywell UOP.

UOP A Honeywell Company (2008a). *UOP Method 779, Chloride in Petroleum Distillates by Microcoulometry*. Des Plaines, IL: Honeywell UOP.

UOP A Honeywell Company (2008b). *UOP Method 720, Impurities in, and Purity of, High Purity p-Xylene by GC*. Des Plaines, IL: Honeywell UOP.

UOP A Honeywell Company (2013a). UOP Parex™ Process. Honeywell UOP: Des Plaines, IL.

UOP A Honeywell Company (2013b). *UOP Method 690, C_8 and Lower Boiling Paraffins and Naphthenes in Low-Olefin Hydrocarbons by GC*. Des Plaines, IL: Honeywell UOP.

UOP A Honeywell Company (2017). *UOP Method 991, Trace Chloride, Fluoride, and Bromide in Liquid Organics by Combustion Ion Chromatography (CIC)*. Des Plaines, IL: Honeywell UOP.

Yuh, E. (2017). Common operating issues. In: *Hydroprocessing for Clean Energy: Design, Operation, and Optimization* (ed. X.X. Zhu). Wiley.

18

Troubleshooting Case Studies

18.1 Introduction

In this chapter, several situations will be presented that have occurred in actual operating plants. The discussions will include observations and data available to the unit engineer, with the intent that the reader may consider potential root causes and what corrective actions may be taken to address the problem. The actual cause and how it was addressed in the operating plant will be presented for consideration and comparison. These cases are not to be considered in anyway comprehensive, but they should relate to certain aspects discussed in Chapter 17.

18.2 Transalkylation Unit: Low Catalyst Activity During Normal Operation

Current commercial aromatic transalkylation catalysts are known to have long lives compared to what was available only a few decades ago. Initial catalyst activity is quite high, requiring lower temperatures upon start-up to achieve target conversion. These units tend to operate relatively problem free for years, until shortly prior to their end-of-run condition when the catalyst deactivation rate increases. If deactivation occurs within a short time after start-up, this is likely due to a catalyst poisoning event.

18.2.1 Summary of Symptoms

A transalkylation unit had been operating for several months with good performance that would be expected for a newly constructed and loaded unit. About a year into the life of the unit, the catalyst activity rapidly fell as the temperature requirement to maintain overall conversion increased. The unit was expected to deactivate at 5–10 °C per year, but was instead deactivating at a rate of around 50 °C per year. The expected life catalyst life was around 8–10 years, but it appeared that the catalyst would last no longer than 2 years at its current deactivation rate. Looking more closely into the data, the toluene conversion was relatively unchanged, around 35 wt.%, during periods of stable feed composition. But the conversion of C_9 aromatics was falling by around 5 wt.% despite rapidly increasing reactor inlet temperature. See the trends in Figure 18.1.

Common feed poisons such as nitrogen compounds were checked, as well as compositional aspects that may result in faster deactivation such as very high feed heavies, non-aromatics, and olefin content. However, none of these seemed to be any higher than other commercial units that perform well.

A closer look at data highlighted that the feed water content was on the high side, though the online analyzers and laboratory data did not match completely. Historical data however showed that the amount of water in the feed was trending upward, increasing from 20 wt-ppm or lower, to beyond 50 wt-ppm over a short period of time. See the trends in Figure 18.2.

Based on the process flow scheme, the feed water content should have been nearly zero, as the combined feed streams all pass through column bottoms, so water should have been completely stripped. However, the C_7 aromatic fraction of the feed came directly from a toluene column, which was reboiled by condensing steam. This reboiler became a focus of the investigation.

18.2.2 Root Cause and Solution

The unit engineer reported that the toluene column steam reboiler had a leak very shortly after a turnaround, but was soon repaired as the complex was temporarily down for other reasons. It was possible that this same reboiler had developed another leak, or the original leak was not properly repaired and a minor leak remained which had propagated.

The vendor as-built diagram for this reboiler and the exchanger specification from the licensor were compared. The check highlighted that the vendor did not properly groove the tubesheet opening where the tubes

Efficient Petrochemical Processes: Technology, Design and Operation, First Edition.
Frank (Xin X.) Zhu, James A. Johnson, David W. Ablin, and Gregory A. Ernst.
© 2020 John Wiley & Sons, Inc. Published 2020 by John Wiley & Sons, Inc.

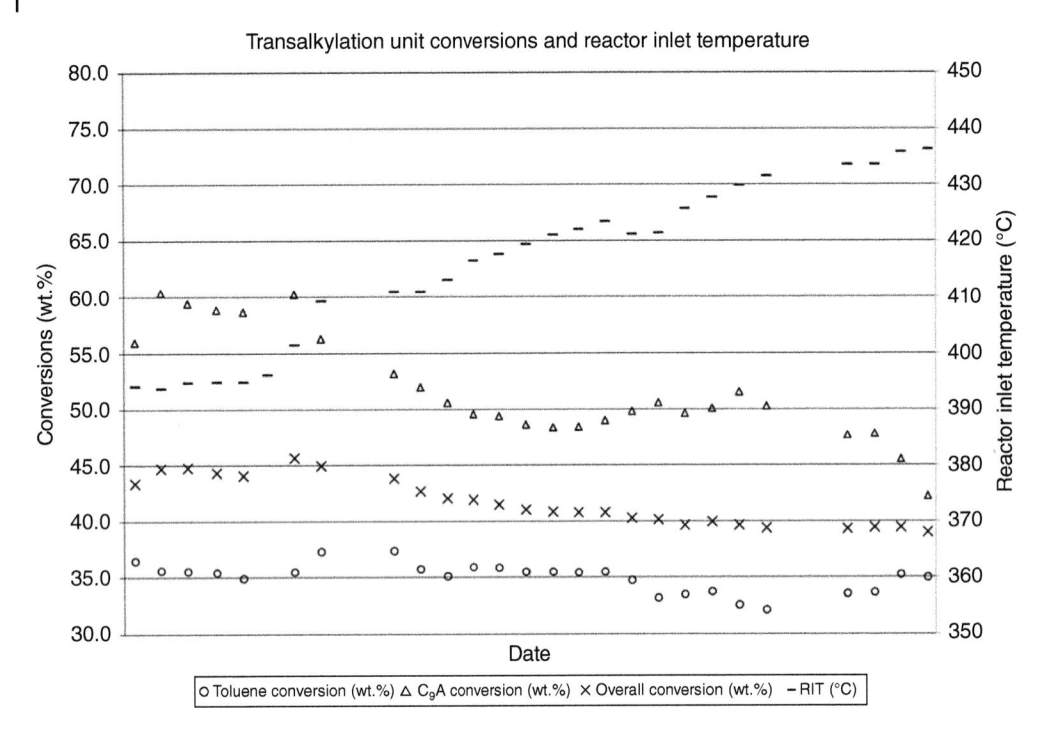

Figure 18.1 Overall, toluene, and C$_9$A conversion, with reactor inlet temperature for the transalkylation unit. The time period trended is approximately 10 months.

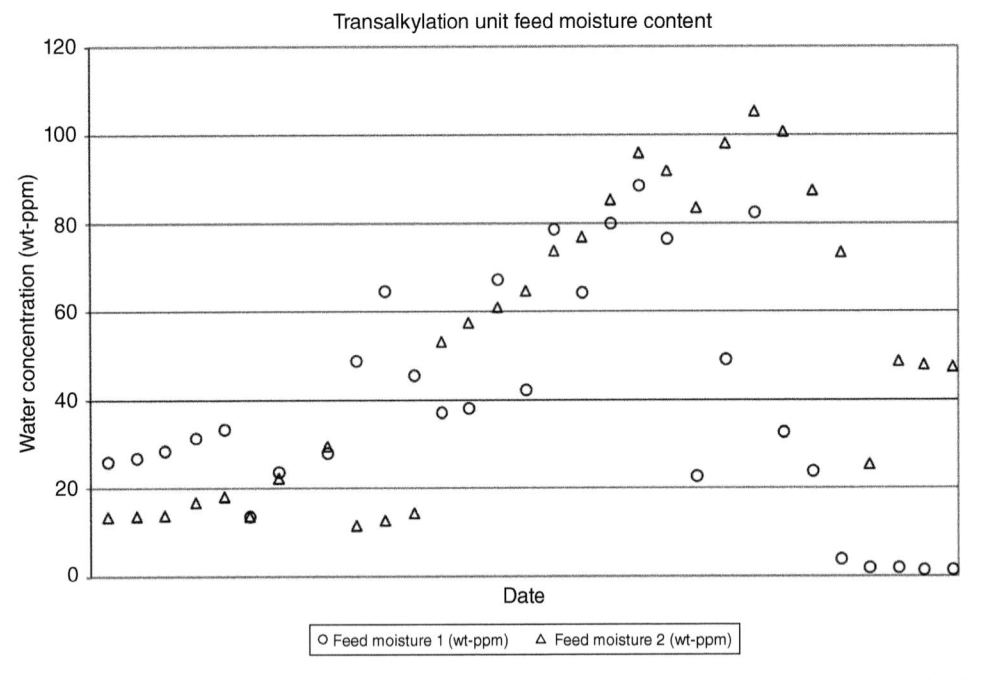

Figure 18.2 Transalkylation unit feed moisture content by separate measurements. The time period is the same as the conversions plot in Figure 18.1, over a range of approximately 10 months.

were inserted. Grooves were specified by the licensor so that when the tube is roll-fitted into the tubesheet, it would extrude into the groove, thus isolating tube-side material from shell-side material. In this case, there was no groove for the tube to grab when rolled, so leakage through the tubesheet would be more likely. Further, the licensor specification called for a 3 mm strength weld at the opening of the tube on the surface of the tubesheet. The vendor drawing instead showed a 1.5 mm weld, so it would have been understandably weaker.

At the next opportunity, the unit was shut down, the toluene column was drained, and the tubesheet welds were strengthened to prevent steam leakage into the process. The unit was restarted shortly thereafter.

Following restart, the water content as measured by online analyzer and lab analysis quickly fell, with all monitors reading between 0 and 50 wt-ppm. Unfortunately, by the time the reboiler had been fixed and significant water was no longer present in the feed, the catalyst had already deactivated through most of the operating temperature window. Though, with water no longer present in the feed, the deactivation rate slowed. The company chose at this point to operate with the installed catalyst for several more months, until which time new catalyst could be supplied and reloaded. Since being reloaded with a new catalyst, the activity has remained strong and the deactivation rate has been low, as the feed has been kept successfully devoid of water.

18.2.3 Lesson Learned

While it is commonly acknowledged that water in transalkylation unit feed can cause poor performance, the actual consequences are not often seen on a commercial scale due to care in design and construction to avoid this scenario. This case serves as a reminder to pay attention to exchanger specifications to avoid steam/water leakage into the process, and to take seriously the online analyzer and laboratory results when they indicate an influx of water. This is important for both feed streams as well as make-up hydrogen.

18.3 Xylene Isomerization Unit: Low Catalyst Activity Following Start-up

Xylene isomerization units convert a raffinate stream from the *para*-xylene separation unit (depleted in *para*-xylene) and convert it to a near-equilibrium mixture of xylenes. This stream will typically contain ethylbenzene that needs to be converted in some way, either by isomerizing to xylenes or by dealkylating to benzene and ethane. Most commercial applications use catalysts that convert ethylbenzene by dealkylation, so that is the style that will be considered in the following discussion.

The catalyst condition is typically judged based on two key performance indicators: its ability to isomerize xylenes, and its ability to dealkylate ethylbenzene. The first should typically be fairly constant over the operating range of the catalyst (although with some shift in the equilibrium compositions as temperature changes), and temperature is increased over the life of the catalyst in order to keep ethylbenzene conversion at its target value. Aside from these two main variables, xylene loss side reactions and in some cases the ability of the catalyst to make a relatively pure benzene stream are also monitored. These catalysts are dual function, with both acid and metal sites, balanced to address each of the expected reaction mechanisms and avoid undesired side reactions.

18.3.1 Summary of Symptoms

A commercial xylene isomerization unit had recently been reloaded with a new load of catalyst, and had not yet been restarted. The complex was standing by, waiting for other units in the complex to complete their recommissioning activities. While this unit was awaiting feed, the preliminary steps of the start-up procedure were begun – the reactor section was oxygen-freed with nitrogen, and nitrogen was subsequently purged with make-up hydrogen so that the compressor and charge heater could be restarted and confirmed to be in good working order. Make-up gas was sourced from the catalytic reforming unit, which had just been reloaded with fresh catalyst as well. This gas was recirculated over the catalyst for a few days, higher than ambient temperature due to heat input from the heater and compressor, though well below normal process temperatures. This catalyst's start-up procedure would call for reduction prior to feed-in, which entails a brief operation in a hydrogen-rich atmosphere at high temperature. This would set the catalyst metal in its active chemical state, reduced rather than oxidized. It would also make the metal atoms less mobile on the catalyst structure.

Once the unit was ready to proceed with start-up, the temperature was increased to complete the reduction procedure, and conditions were adjusted for feed-in. Once final catalyst attenuation was completed, the reactor temperature was adjusted to target the design ethylbenzene conversion.

Performance variables calculated within the first days following start-up indicated immediate problems with the catalyst function, both for xylene isomerization and ethylbenzene conversion. The xylene isomerization function was worse than expected – the pX/X value is expected to be above 23.5 wt.%, though in this case it was instead closer to 23.0 wt.% shortly after start-up, and falling. pX/X is expected to stay fairly consistent throughout the life of the catalyst, except for minor shifts related to shifts in the equilibrium composition as temperature increases. See the trend in Figure 18.3.

The ethylbenzene conversion activity was also low and catalyst stability was poor. The catalyst was expected to require a temperature of around 360 °C to achieve a target ethylbenzene conversion of 75 wt.%. Instead, the temperature shortly after start-up was approximately

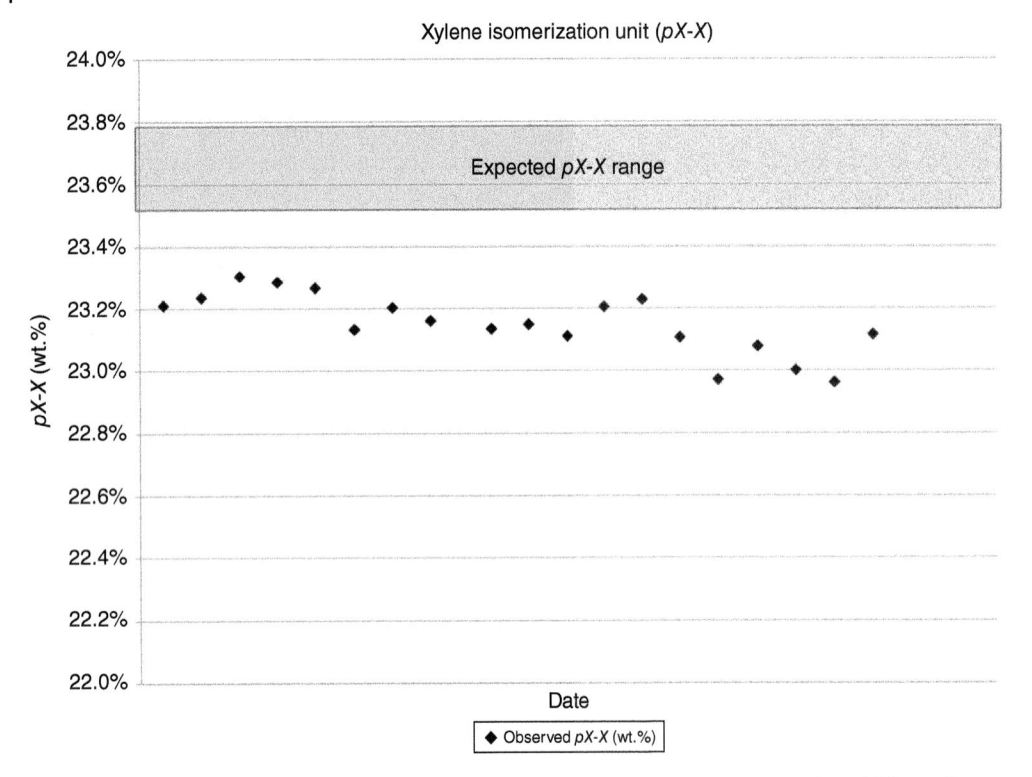

Figure 18.3 Trend of pX/X shortly after start-up. Observed pX/X is lower than expected, and falling. The time period trended is approximately five months.

380 °C, and deactivation was occurring much more quickly than expected. See the trends in Figure 18.4.

In addition to this, the xylene retention was far lower than expected. This catalyst was expected to have a xylene retention (100% − xylene loss) of around 98 wt.% or higher, and was instead around 94–95 wt.% (Figure 18.5).

The catalyst was manufactured to achieve high benzene purity – this was the one parameter that was still okay, creating a benzene product stream with minimal non-aromatic benzene co-boilers.

When a unit is restarted with a new catalyst and immediately does not achieve expectations, the first thought is invariably that the catalyst batch had quality problems related to manufacturing. Catalyst manufacturers typically retain samples when a product is sold, so samples can be retested if there seems to be an issue. In this case, the retained samples were pilot-plant tested and showed that the manufactured catalyst had good performance.

18.3.2 Root Cause and Solution

One of the primary outliers of the start-up of this catalyst compared to others was the prolonged period of recycle gas circulation prior to hydrogen reduction and subsequent feed-in. Another complicating factor was the fact that the make-up gas was sourced by a catalytic reforming unit which had very recently started up with a new catalyst. As such, the make-up gas had higher than normal water and carbon monoxide content. This same gas was used for catalyst reduction once the start-up procedure resumed.

The catalyst was supplied with the impregnated metal in the oxidized state. In this state, the metal is relatively mobile, particularly at elevated temperature, and even more so in the presence of water and carbon monoxide. The metal in this case was platinum. While excellent for certain applications, it can be more prone to mobility in the oxidized state than other metals, so proper care must be taken to avoid metal site agglomeration.

The key performance indicators all suggested an inhibited metal function: lower xylene isomerization, low ethylbenzene conversion activity, high xylene loss, and better benzene purity (less undesired saturation). These led to a review of start-up conditions that may have resulted in metal agglomeration. This review showed that the hydrogen gas circulated over the catalyst had relatively high amounts of water and carbon monoxide. This supported the notion that metal agglomeration occurred while it was still in the relatively mobile oxidized state. Certain metals may be more prone to agglomeration in the presence of water and/or carbon monoxide than others. Unfortunately, this kind of change to the catalyst composition is not easily reversible in situ, and likely requires a reload with a

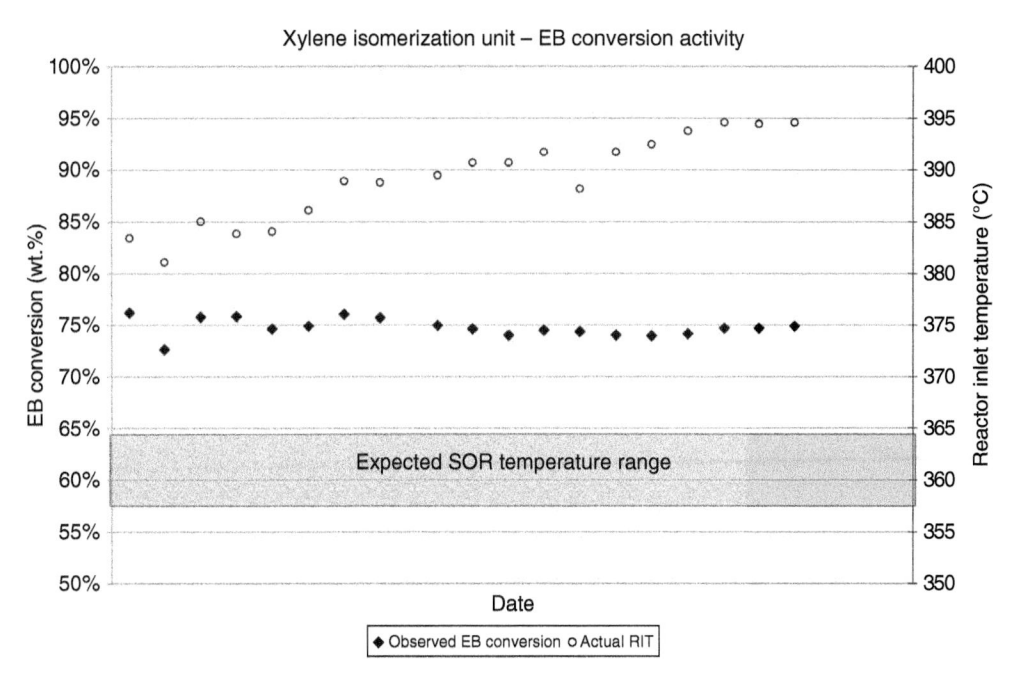

Figure 18.4 Trends of EB conversion and reactor inlet temperature shortly after start-up. Temperature to achieve EB conversion is higher than expected, and rapidly increasing. The time period trended is approximately five months.

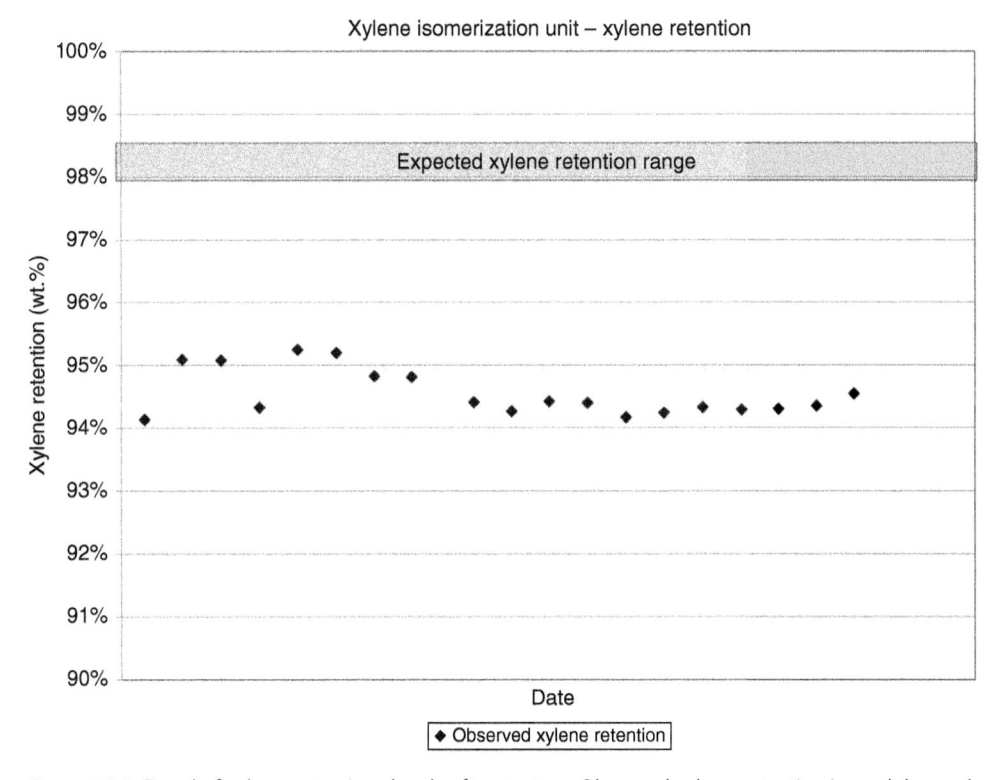

Figure 18.5 Trend of xylene retention shortly after start-up. Observed xylene retention is much lower than expected, and appears to be falling. The time period trended is approximately five months.

new catalyst. Low metal function can hurt the isomerization function due to disruption between the acid and metal balance. The dealkylation function is primarily hurt by the lack of metal function, where ethyl radicals are not properly saturated, and tend to reattach to phenyl rings. In the same way, this would be expected to cause rapid catalyst deactivation as the presence of ethyl and phenyl radicals can contribute strongly to coke formation. The higher temperature requirement and disrupted acid/metal function balance contributed to the very high xylene loss, primarily by promoting transalkylation of methyl groups, and reattachment of ethyl radicals to xylene molecules. The benzene purity improved by the poor metal function further limiting benzene and aromatics saturation. This reduced aromatic saturation activity would have actually been a positive, if the net effect of low metal function was not otherwise such a detriment to performance.

18.3.3 Lesson Learned

For catalysts where metal atoms may be mobile on the catalyst surface prior to completing the procedures called for during start-up, ensure that the make-up and recycle gas composition is carefully checked. Make sure that it does not contain components that contribute to metal atom agglomeration, particularly prior to completing the required start-up steps that activate the metal function and immobilize the metal atoms. The catalyst supplier may also provide a similar catalyst with the metal atoms already conditioned to be in an immobile state (for example, a platinum-based catalyst might be supplied pre-reduced).

18.4 *para*-Xylene Selective Adsorption Unit: Low Recovery After Turnaround

Purified *para*-xylene is typically sold into the market with a minimum purity of 99.7 wt.%, and as such has very little margin for contamination. In a selective adsorption unit that produces *para*-xylene, a shift in adsorbent selectivity, or allowing a purified stream to come into contact with contaminants from the feed or raffinate stream, can quickly cause the *para*-xylene purity to fall to a point that it can no longer be sold. In this case it would have to be stored and slowly reprocessed through the complex. Fortunately, it is very rare that the product purity in a commercial plant falls to the point that it becomes unsalable. Unit operators will closely watch the product quality by online GC and lab analysis each shift, and adjust the adsorption section parameters to increase purity if it appears to be shifting downward. While purity is maintained, these adjustments may result in reduced recovery until the root cause for the product composition change is identified and directly addressed.

18.4.1 Summary of Symptoms

A unit was restarted following a typical complex maintenance turnaround. The unit itself had no major maintenance activities, so following the turnaround, the unit performance was expected to be the same as it was prior to the turnaround. However, upon resuming normal operation, the pX recovery was about 5 wt.% lower than it was prior to the turnaround itself. The product contaminants had shifted such that the *meta*-xylene and *ortho*-xylene were much higher than they were previously. The *meta*-xylene content was about 200 wt-ppm higher than prior to the turnaround, and *ortho*-xylene was about 100 wt-ppm higher than prior to the turnaround.

Unit engineers checked to see if poisons could have come into the unit during restart, but laboratory analysis showed that the feed and recycle desorbent olefins, heavy aromatics, and oxygenates were all very low and within specification. All operating parameters were maintained at the same values as they had been prior to the turnaround, except for those that had been adjusted to maintain purity. The water injection rate and feed stream hydration were similar to what they had been prior to the turnaround. Chamber pressure drop was stable during periods of stable throughput, so the issue appeared unrelated to internal chamber back-mixing.

18.4.2 Root Cause and Solution

Selective adsorptive separation units will have several piping segments that are used to facilitate certain parts of the start-up procedure. Some examples of these are lines that are used for filling of the adsorption chambers, or flushing of bedlines and adsorption section piping to ensure they are clean prior to lining up the chambers for normal operation. Other piping segments may exist in the adsorption section to allow for spare parallel equipment such as critical control valves, or bypasses on flow meters so that they can be serviced without having to shut down the unit.

Since this issue occurred immediately after restart, it was possible that one of these start-up lines or isolation lines was opened during the turnaround to perform maintenance, or facilitate one of the shutdown or start-up procedures. The unit engineers and field operators did a thorough check of the unit to see if any of the isolation valves that were intended to be closed during normal operation happened to be open. It turned out that one of the spare chamber circulation control valves was incompletely isolated – the inlet side block valve of the bypass line was closed, but the outlet side block valve of the bypass line was fully open.

If an unflushed volume such as this is connected to the circulation line, back-mixing will occur at the tee connected to the flowing line. The chamber circulation lines carry the changing composition profile through them. When a composition similar to feed or raffinate passes across the unflushed volume, convective mass transfer and eddy currents will cause some of these feed or raffinate components to remain in the unflushed volume. When the circulation line subsequently carries a desorbent or extract-like composition past this volume, the same effects will cause contamination of this pure stream with the contaminants in the unflushed volume (Figure 18.6).

When this open valve was seen during the field check, it was immediately closed by a field operator. The next day, product contamination had shifted such that the adsorption section operating parameters could be adjusted to maintain purity, and recovery increased around 5 wt.%, back up to the previous baseline.

18.4.3 Lesson Learned

Any time there is an unexpected and otherwise unexplained increase in product contamination, particularly shortly after a restart or maintenance on adsorption section equipment, valves that are expected to be closed during normal operation should be confirmed to be in the closed position. The purity constraints on *para*-xylene are extremely tight, with only 3000 wt-ppm of allowable contaminants before the product becomes unsalable on the open market. An unflushed volume connected to flowing lines in the adsorption section can have a significant impact on product composition, resulting in a very large recovery decline when the operating parameters are readjusted to maintain purity. This will quickly impact unit profitability, and reduce the margin available for future adjustments that may be required to maintain purity if another event occurs.

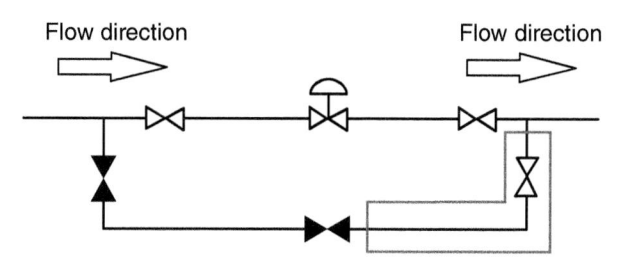

Figure 18.6 Tee in non-flowing line with open isolation valve. Note the piping section outlined in gray, this will contain a composite mixture of the different stream compositions flowing in the normal flowing line to which it is attached.

In addition to checking for open isolation valves, it is also a best practice to periodically ensure that the valves isolating unflushed volumes are actually tightly sealed. If both are leaking slightly, then a slow flow rate of liquid through the volume can cause contaminants to be held up in the volume, eluting on the other end after the main line composition has switched composition to that of a zone which contains predominantly desorbent or extract components. This can similarly cause a measureable impact on product composition and unit operating efficiency. Blinds should be located on start-up lines that are not needed to be opened quickly during operation. These blinds should also be confirmed to be in the proper position during normal operation.

18.5 Aromatics Extraction Unit: Low Extract Purity/Recovery

Aromatics extraction units have very tight specifications on aromatics purity – toluene typically around 99 wt.%, and benzene even tighter at 99.85 or 99.95 wt.% minimum purity. Similar to *para*-xylene production, there is very little margin for contaminants in the aromatics (usually benzene) product. Operation adjustments can frequently be made to improve purity, but this is often done at the expense of aromatics recovery or operating efficiency.

There are numerous handles in each type of extraction unit to adjust both product purity and recovery. Solvent-to-feed ratio, heat input to the various columns, reflux ratios in the various columns, and feed or solvent enthalpies are commonly understood methods, in addition to adjustment of several recycle streams each design may contain. Feed composition can also impact purity and recovery, and when the feed is more or less aromatic, or if the split of aromatics changes (among benzene, toluene, and/or xylenes), operating variables will require adjustment to maintain target aromatics' purity and recovery.

18.5.1 Summary of Symptoms

In extractive distillation type of solvent extraction unit (utilizing an extractive distillation column and recovery column), benzene recovery fell from its recent value of 99.5 to 99.0 wt.%. The non-aromatics content of the recovery column overhead (extract) had increased recently, so operators increased the heat input to the extractive distillation column in order to vaporize the non-aromatics in the extractive distillation column and make sure they did not leave the bottom of the column with the rich solvent. The drop in benzene recovery came after this adjustment.

The process data showed no change to solvent-to-feed ratio. The reflux ratios were stable, until the heat input to the extractive distillation column caused its reflux-to-feed ratio to increase. Column pressures and temperatures were stable, though the extractive distillation pressure drop and bottoms temperature both increased slightly with the heat input increase, as would be expected.

18.5.2 Root Cause and Solution

Due to maintenance on an upstream unit in the complex, the feed to the extraction unit was going to be reduced for one to two days. To keep the unit stable however, material from the normally idle feed tank was added as a supplement to the remaining feed. This feed had been sent to the tank during a period when the reforming unit severity was much lower than normal, and contained significantly more non-aromatics than the typical feed.

The feed non-aromatics content to the extraction unit increased due to adding this tank material to the normal feed stream. This means that the portion of the feed that needs to be vaporized increases, which was done when the operators increased the heat input to the extractive distillation column (Figure 18.7).

However, doing this makes it more difficult to retain the lighter aromatics in the solvent. As such, the proper response at this point should have been to increase the solvent flow rate. This is a somewhat counterintuitive change, as it would normally be thought that the solvent flow rate should increase if there are more aromatics to separate. But increasing the solvent flow when the non-aromatics content increases should result in a lower raffinate aromatics content, improving solvent recovery. This is a result of requiring more solvent to counteract the higher solvent temperature required to drive non-aromatics from the rich solvent (Figure 18.8).

18.5.3 Lesson Learned

When operational changes are required that may change some of the feeds to the aromatics extraction unit, consider what operational changes might be required to maintain purity and recovery. Further than this example showed however, there are practical limitations to how much the heat input to the extractive distillation column can be increased, as well as how much the solvent flow rate can be increased.

18.6 Aromatics Complex: Low *para*-Xylene Production

When *para*-xylene production is lower than expected, it is rarely because *para*-xylene itself is being lost. The precursors to *para*-xylene are most often what is lost, and some of these have a larger impact on potential *para*-xylene production than others. There are many locations where these precursors can be lost, whether it be a reaction loss or separation loss.

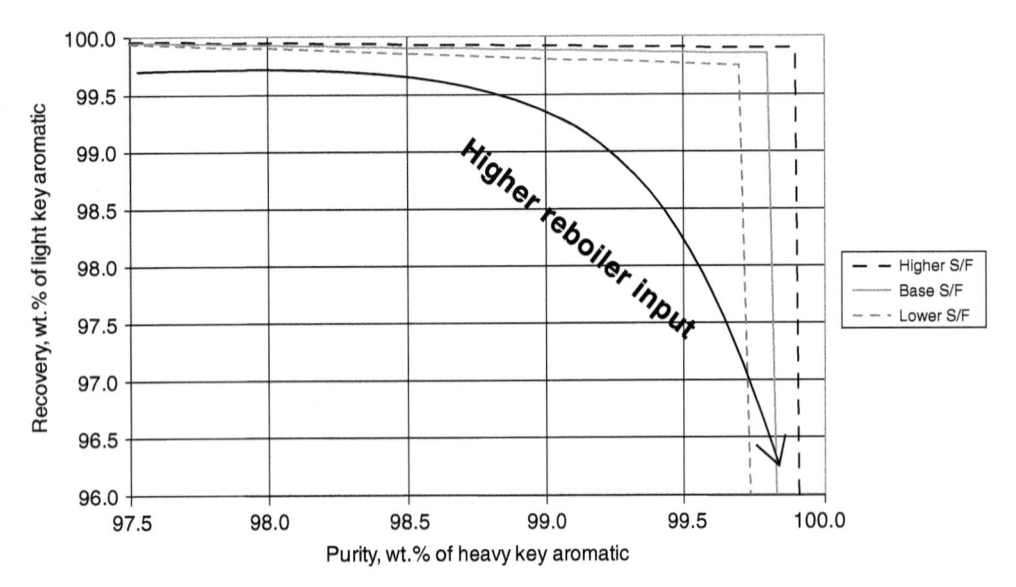

Figure 18.7 Light aromatic key component recovery vs. heavy aromatic key component purity, with changing reboiler heat input. In the case discussed, the light key is benzene, and the heavy key would technically be toluene (most new aromatic complex systems have both benzene and toluene in their purified extract). But the axes here can be simply thought of as bulk aromatics recovery (y-axis) and purity (x-axis). As reboiler input increases, vapor traffic increases and more benzene is liberated from the solvent, and ends up in the raffinate, reducing recovery. The higher reboiler input drives heavier non-aromatics from the solvent as well, ultimately improving the purity of the extract product.

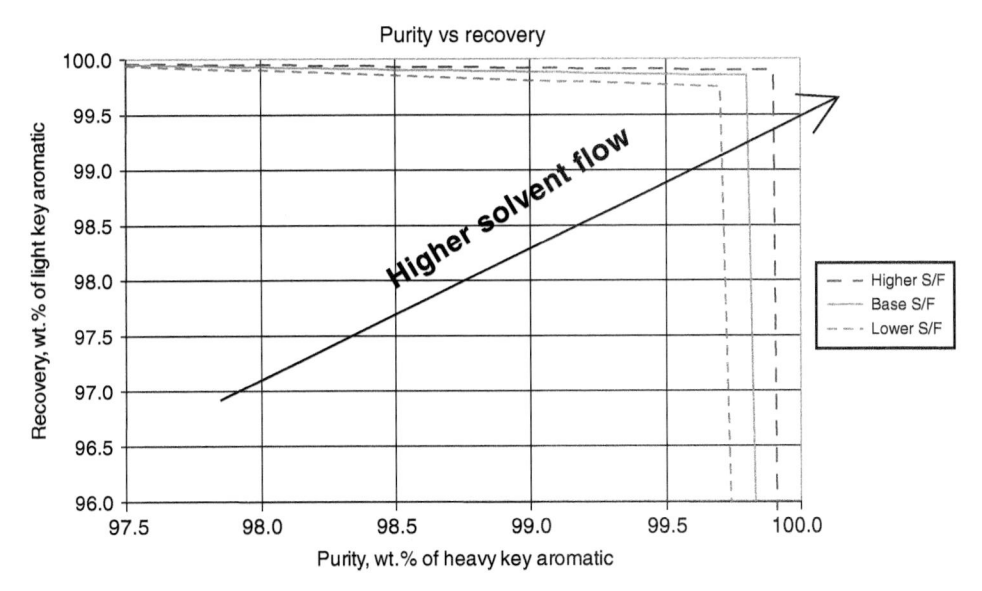

Figure 18.8 Light aromatic key component recovery vs. heavy aromatic key component purity, with changing reboiler heat input. Just as in the trend for Figure 18.7, the light key is benzene, and the heavy key is toluene. Again the axes here can be thought of as bulk aromatics recovery (y-axis) and purity (x-axis). As solvent flow increases, more the capacity for aromatics increases and more benzene can be extracted from the feed by the solvent, improving recovery.

Most complexes will operate such that there is some margin in the ability to produce *para*-xylene, but when that margin erodes for any reason, contracts with *para*-xylene buyers still need to be fulfilled. At that point, either *para*-xylene will be purchased from another source and sold to their buyers (so the profit in this transaction is nearly nonexistent, if at all present), or mixed xylenes or other *para*-xylene precursors will be purchased to supplement the complex feed (at reduced profitability).

18.6.1 Summary of Symptoms

A large facility was losing approximately 10 tons per hour of *para*-xylene production, though the feed rate to the complex was at design rates, and the composition was as expected. The first questions asked in situations like these are typically related to the reaction systems – are aromatics being saturated and cracked to light gases? Or, are they being polymerized to heavies? Engineers did a thorough check of the xylene isomerization unit and transalkylation unit performance, and there was no evidence of unexpected conversion losses. Hydrogen consumption of both units was low as expected, approximately equal to what would be expected by dealkylation reactions. Further, the fractionation columns in these particular units were achieving their expected splits.

The selective adsorption unit was performing well, with reasonable *para*-xylene recovery and the unit was not bottlenecked, though the feed rate was slightly lower than expected considering the complex feed rate was normal. The extractive distillation unit was processing primarily benzene, but even the toluene that it was processing was being separated at better than design recovery.

Engineers also took a look at a few of the component splits in fractionation columns, particularly those that fractionate complex by-product streams from internally recycled streams. Most appeared to be performing as expected. However, the heavy aromatics column had a lighter bottoms product than expected (Figure 18.9).

18.6.2 Root Cause and Solution

Since conversion losses had been ruled out, separation losses seemed the most likely. Many of the overhead and bottoms product streams are internally recycled streams; however, the heavy aromatics column creates a bottoms product that is exported from the complex, typically a $C_{11}+$ aromatics stream to be used as fuel oil. The column overhead stream is typically a C_9–C_{10} aromatics product that is sent to the transalkylation unit to ultimately be reconverted back to xylenes.

At this complex, there was concern about preserving the life of the transalkylation unit as its catalyst was near end-of-life, so the operation of the heavy aromatics column was adjusted to send less heavies to the overhead stream. In doing so however, the bottoms flow rate increased significantly, and a much higher amount of C_{10} aromatics and even C_9 aromatics was sent to the bottoms liquid. The bottoms flow rate had increased by around 15–25 tons per hour, above an expected design rate of

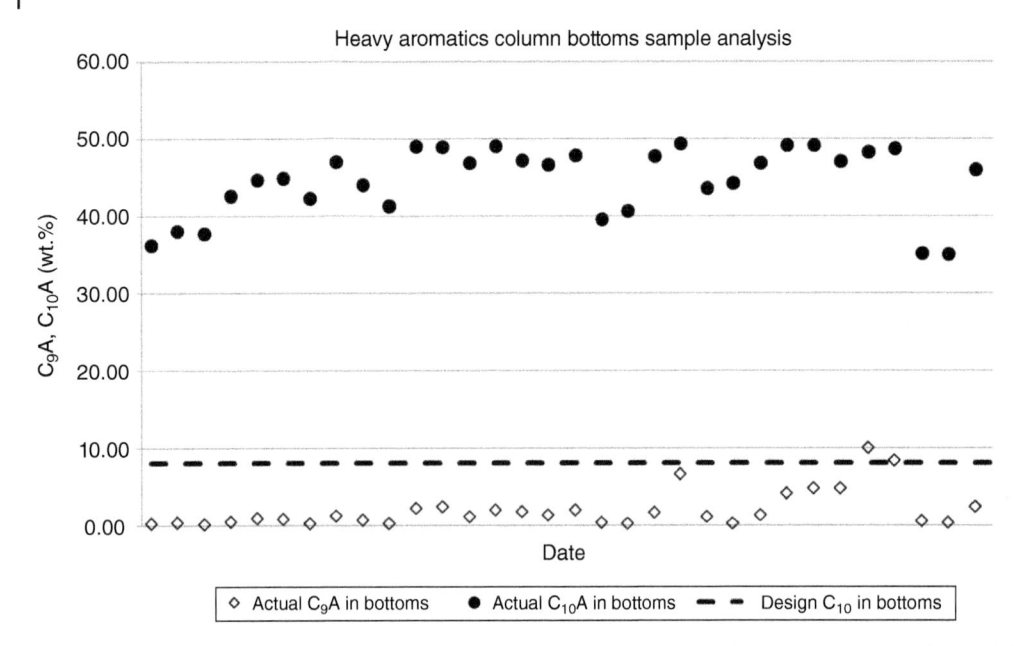

Figure 18.9 Heavy aromatics column bottoms sample analysis showing the bottoms product had a much higher C_{10} aromatic content than the design case. The expected C_{10} aromatic content was around 8 wt.%, but in actuality was around 40–50 wt.%. Analysis also showed elevated C_9 aromatics, which is intended to be nearly zero for this stream, but at times reached as high as 10 wt.%.

around 5 tons per hour. These C_9 and C_{10} aromatics are typically very methyl group rich, and convert at a very high percentage directly to *para*-xylene (Figure 18.10).

As the transalkylation unit was actually designed to process C_{10} aromatics and even some C_{11} aromatics, the composition of the heavy aromatics column overhead was changed back to the expected composition and the bottoms flow rate reduced. This increased the flow rate to the transalkylation unit, and the amount of xylenes generated within the complex. As these xylenes passed through the selective adsorption unit and xylene isomerization unit, the *para*-xylene production rate increased as expected.

18.6.3 Lesson Learned

When xylene precursors are lost through a complex, whether by converting by side reaction in a reactor, or by fractionation to a by-product stream, these losses can very quickly hurt the profitability of a complex. Consider the feed composition that the conversion process can actually tolerate, and take advantage of this for reducing waste streams and increasing complex profitability. Discuss this with the unit licensor and catalyst supplier, particularly when considering catalyst reloads and revamp opportunities.

18.7 Closing Remarks

In each of the preceding examples, the ability to refer to quality operating and laboratory data were critical to understanding the root causes of the problems, and be able to identify corrective actions. Unfortunately, some of the corrective actions were rather severe, such as the requirement to buy a new load catalyst and the cost associated with shutting down to reload. This highlights the importance of ongoing data monitoring to avoid these situations in the first place.

In each case, too, was some misunderstanding about the impact of certain operational changes, or not understanding which process variables or potential poisons to watch at particular times. Much of this comes back to operator and engineer training, or having systems in place to retain institutional knowledge. A common emphasis in twentieth-century businesses is one of employee mobility, keeping engineers moving frequently among different technologies or job descriptions. This is an excellent practice for developing broad knowledge and skills, but often results in detailed knowledge gaps. With this model in mind, it is increasingly difficult for companies to develop true technology specialists. Services exist from technology providers and other resources to help keep engineers and operators current on industry and technology-specific best practices, including software or cloud-enabled service platforms, web-based training, or classroom training. While these services can seem costly on the surface, they can help avoid production losses due to poor unit performance, and unplanned shutdowns due to equipment or catalyst/adsorbent failures (Liebert 2017).

For the process engineer to troubleshooting these events, it is critical to have access to good quality historical data, coupled with a good understanding of the operating plant itself. Even if lacking in experience, other

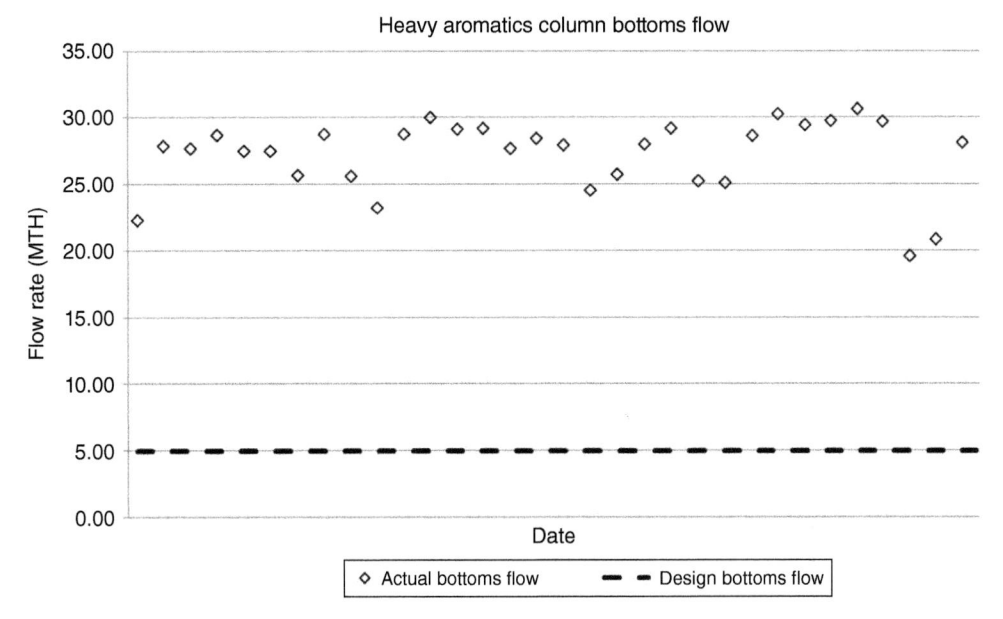

Figure 18.10 The heavy aromatics column bottoms flow rate, which was designed to be around 5 MT/h, was in actuality around 20–30 MT/h.

resources specific to the plant will exist such as operating manuals and design documents. Manuals will typically cover several high-level troubleshooting scenarios, but multivariable troubleshooting may be beyond their scope. Such manuals can be only so comprehensive. Design documents give a good understanding of the operating limits of any piece of equipment in a plant, but do not always explain interactions between different parts of the plants. PFDs, P&ID, and shutdown tables can be instructive on interactions, from these it is not necessarily clear what modes of operation are most appropriate for a given set of conditions. The examples provided here were intended to give the engineer an idea of some potential troubleshooting exercises for the main technologies found in an aromatics complex, and unique circumstances from which trouble may arise. From these, the engineer can consider additional means to supplement their experience and troubleshooting capabilities.

Reference

Liebert, R. (2017). Lifecycle Management: The New Paradigm for Service Delivery, November 2017. https://www.linkedin.com/pulse/lifecycle-management-new-paradigm-service-delivery-rebecca-b-liebert (accessed October 2018).

Index

a

A_{9+} aromatics 15
"accuracy" column 157
acetone 3
acid–base reactions 332
adsorbent bead 65
adsorbent chamber 105
adsorption
   advantage of 316
   defined 316
   separation technology 304
advanced process control (APC) 122, 280, 289, 297
   energy-saving opportunities 287
Alkymax technology 13
alphamethyl styrene 14
alternative feeds, for aromatics 27–28
aluminum chloride catalyst 13
appropriate placement
   for distillation column 246–249
   general principle 244
   for reaction process 245–246
   for utility 245
arbor type heater 98
aromatic building blocks 4
aromatic hydrocarbons 77
aromatic petrochemicals 13
aromatic ring condensation 30
aromatics complex (AC) 252, 386–388
   revamp
      with adsorbent reload 152–153
      transalkylation unit 153
      with xylene isomerization catalyst change 153
aromatics extraction 385–386
   ED
      column sections and functions 93
      operating variables 92–96
      sulfolane aromatics extraction unit flow diagram 93
   liquid-liquid extraction
      hydrocarbon feed 90
      operating variables 91–92
      "rich" absorbing aromatics 89
aromatics process description
   adsorptive separations
      for para-xylene 64–67
      of pure meta-xylene 76–78
   aromatics flow scheme 63–64
   para-selective catalytic technologies, for para-xylene
   para-selective toluene disproportionation 78–79
   para-selective toluene methylation 79–81
   para-xylene purification and recovery, by crystallization 68–71
   transalkylation processes 71–72
   treating feeds, for aromatics production 68
   xylene isomerization 72–76
aromatics process revamp design
   design basis, for revamp projects
      agreement 118–119
      approach, of study 119
      constraints 120
      economic evaluation criteria 121
      feedstock and make-up gas 119
      guarantees 121
      operating data or test run data 120
      processing objectives 119
      product specifications 119
      replacement equipment options 121
      right equipment information 119–120
      utilities 120
   process design, for revamp projects 121–123
   revamp impact, on utilities 123–124
   revamp project approach
      identifying successive bottlenecks 116
      maximize throughput, at minimum cost 115–116
      specified target capacity 115
      target production, with constraints 115
   revamp study methodology and strategies 116–118
   stages and types, of revamp assessment 113–115
aromatics process unit design
   aromatics extraction
      characteristics and tendencies 88–89
      ED 92–96
      liquid-liquid extraction 89–92
   aromatics fractionation
      heavy aromatics fractionation 87–88
      reformate splitter 85–86
      xylene fractionation 86–87
   para-xylene separation 105–106
   process design considerations
      design margin philosophy 106–108
      fractionation optimization 109–110
      operational flexibility 108–109
   safety considerations 110–111
   transalkylation, process flow description 96–100

Index

aromatics ring losses, operating issues
  extraction losses
    feed composition 352–353
    foaming 353
    varibles 352
  feed composition 352–353
  foaming 353
  fractionation and separation losses
    bottoms liquid product 352
    distillate liquid product 351–352
    vent losses 351
  olefin removal 350–351
  reaction losses
    transalkylation unit losses 355
    xylene isomerization unit losses 354–355
aromatic transalkylation 28
Arrhenius activation energy (E_a) 325
Arrhenius pre-exponential factors 340
ASTM D-86 distillation 119
aviation gasoline 3
Axens' HDA process 16
axial flow-type compressors 203

b

backmix reactor 333
back-to-back compressor 206
bad-actor pump 230
Badger cumene process 14
"barrel" compressor 203
bed pressure drop 99–100
benzene 347, 348
  production 16
benzene, toluene, and xylenes (BTX) 321
benzene-toluene (BT) fractionation 85
  DWC 253–254
best efficiency point (BEP) 224–225
beverage packaging 5
biological fouling 177
bi-phenyl, recycle 16
bisphenol-A 14
bridge wall temperature (BWT) 193, 194
bromine index 350

c

C_9 and C_{10} aromatics 28
C_6 and C_7 paraffins 14
capital expense (CAPEX) 63
carbon steel (CS) 126
C_8 aromatic composition, transalkylated product 33
C_8 aromatic isomers 105
C_8 aromatics 18
C_9 aromatics 28
C_{9+} aromatics 15
catalyst deactivation
  defined 341
  principal causes of 342
catalyst geometry 329
catalysts 333
catalytic distillation 13
catalytic reaction, limitation
  external diffusion 331
  important parameters reaction 332–333
  internal pore diffusion 331
  limitation types 332
  mitigating limitations 331–332
  surface reaction 331
catalytic reforming reaction network 23
cavitation 226
C_6–C_{10} hydrocarbons 14
C_5–C_6 nonaromatics 86
C_6–C_7 reformate cut 17
CDTech 13
centrifugal compressors 203, 207
centrifugal pumps 108
chemical kinetics 325
chemical reaction 325
  fouling 177
chemical reactor designs
  considerations 338
  equilibrium constant 335
  general guidelines 338–340
  objective 335
  pressure, reaction conversion, and selectivity 336
  rate-limiting step 337
  reaction time and reactor size 336–337
  temperature constant 335
chemicals and energy demand 8
Chevron Research and Mobil Oil Corp 14
chrome (Cr) alloy tube 126
C_6 hydrocarbons 23
C_6 naphthenes, to aromatics 14
cocurrent flow 172
column flow profile 166
column grand composite curve (CGCC) 246

combined feed exchanger (CFE) 96–98, 277
combustion intensity 189
composite curves 239
  basic pinch concepts 239–240
  cost targeting 240–244
  energy use targeting 240
  pinch design rules 240
compound annual growth rates (CAGR) 13
compressor assessment
  blades, type of 207
  compressors, types of
    multistage beam-type compressor 203–204
    multistage integral geared compressors 204–205
  fundamentals, of centrifugal compressors 208–209
  impeller configurations
    between-bearing configuration 205–206
    integrally geared configuration 206
  inlet throttle valve 212
  partial load control
    inlet guide vane 211–212
    recycle or surge control valve 210
    variable speed control 210–211
  performance curves
    choking 209–210
    design point 209
    surge 209
  process context, for centrifugal compressor 212–213
  radial velocity 207
  relative velocity 207
  selection, of compressor 213
computational fluid dynamic (CFD) flow 99
continuous catalyst regeneration (CCR) platforming 15, 27
continuous flow reactors 333
continuous stirred-tank reactor (CSTR) 333
conventional aromatics complex 251
corrosion fouling 177
countercurrent flow 172
C_6 paraffins, to C_6 naphthenes 14
cross-flow reactions 339
crossflow trays 315–316

cross-pinch (XP) heat transfer   240
crude terephthalic acid (CTA)   51
crystal growth   320
crystallization
   fouling   176–177
   separation technology   305, 320
crystallizer vessel refrigeration
     delivery   70
C_2 splitter, heat and mass
     balances   159
cumene   3, 14
cumene-grade benzene   14
cumene/phenol   21
   production technologies   21
cumene/phenol/bisphenol-A   48–49
cumene service   14
Cyclar catalyst   27
cyclohexane   46
   for nylon   14
cyclo-olefins   15

d

Dacron fabrics   17
dealkylation technology   16
deheptanizer   279
   reboiler   282
   xylene isomerization unit   281
de-hydro-cyclo-dimerization
     (DHCD)   11, 26
deisopentanizers (DIP)   314
   economic improvements   301
   flow scheme   296–297
   objective function   298–299
   optimization benefits   301
   simulation model   298
   variation of   298
depentanizers (DP)   314
design margin philosophy   106–108
detergent purity   14
di-and tri-ethylbenzenes   13
di-ethylene glycol (DEG)   35
diffusion-controlled reactions   331
di-isocyantes   17
2,6-dimethyl naphthalene (2,6-DMN)
     oxidation   20
di-olefins   15
disproportionation   29
distillation   304
   column   305
     column internals   315–316
     difficulty of separation   305–306
     dividing wall column   312–315
     feed preheat optimization *vs.*
        preheat efficiency   310
     heavy oil distillation   311–312
     limitations   316
     operating pressure selection
        306–307
     optimization of design   308–310
     reboiler configurations types
        307–308
     side products   310–311
distillation column, appropriate
     placement
   CGCC   246
   column integration
     against background process
        246–248
     design procedure for   248–249
distillation column assessment
   base case, definition   157–159
   heat balance assessment   164
   material balance assessment
     162–164
   missing and incomplete data
     159–161
   operating profile assessment
     166–168
   process simulation   161–162
   tower efficiency assessment
     164–166
distillation system optimization
   defined   289
   energy optimization
     column bottom temperature
        294
     column reflux ratio   294–295
     economic value function
        293–294
     setting operating pressure
        295–296
   process optimization
     basis   296–297
     current operation assessment
        297–298
     objective function   298–299
     off-line optimization results
        299–300
     online optimization   300–301
     optimization
       implementation   300
     simulation   298
     sustaining benefits   301–302
   tower optimization
     key operating parameters
        289–291
     parameter optimization
        291–292
     parameter relationship   291
     relax soft constraints   292–293
distributed control system (DCS)   297
dividing wall column (DWC)
     252, 253
   application of   313–315
   BT fractionation   253–254
   distillation column   312
   fundamentals   312–313
   guideline technology   313
   reformate splitter   254–255
downstream fractionation   15
dual refrigeration system, for
     crystallization   70
DWC *see* dividing wall column
     (DWC)

e

EB-destruction xylene isomerization
     reactions   33
"EB-isomerization" catalyst   101
EB*One*™ process   13, 17
economic evaluation
   key indicators
     convert time series data   285
     data opetration   284
     patterns opetration   284
     performance operation   285
     variability operation   283
   for revamps
     benefits   149
     costs   147–149
     data requirements   149–150
     economic analyses, types of
        150–152
economic value function   293–294
economizers   277
ED *see* extractive distillation (ED)
elementary reaction rate law   327
end of curve (EOC)   233
end-of-run (EOR)   98
energy benchmarking
   calculations precision for
     272–273
   data extraction, criteria   272
   energy intensity
     calculating   265–267
     definition   263–264
   EPI
     based on industrial peers energy
        performance (PEP)   271–272
     based on the best technology
        energy performance
        (TEP)   272

energy benchmarking (cont'd)
   best-in-operation energy performance (OEP)   271
   FE, for steam and power   264–265, 267–271
energy dashboard, key indicators   285–287
energy intensity, for process   263–264
energy optimization, distillation system
   based on reflux ratio.   293
   column bottom temperature   294
   economic value function   293–294
   reflux ratio   294–295
   setting operating pressure   295–296
energy performance index (EPI)
   based on, best technology energy performance (TEP)   272
   based on, industrial peers energy performance (PEP)   271–272
   best-in-operation energy performance (OEP)   271
equilibrium constant   335
equipment design margins
   air-cooled heat exchangers   108
   compressors   108
   fired heaters   107–108
   fractionation columns   108
   process-process heat exchangers   108
   pumps   108
   reactors   108
   water-cooled heat exchangers   108
ethylbenzene (EB)   13, 44, 349
   isomerization   73
   properties of   317
   yields   13
ethylbenzene/styrene   17, 46–48
   production technologies   17
ex-situ selectivated catalysts   78
external diffusion-controlled reactions   331
extract column, with thermal coupling   252–253
extractive distillation (ED)   15, 321–322
   column sections and functions   93
   operating variables   94–96
   sulfolane aromatics extraction unit flow diagram   93
   technologies, in aromatic separations   35–39
   unit   8

extractive stripper   90
extractor zones, in LLE   37
ExxonMobil TransPlus process   15

f

facilitated transport membrane (FTM)   322, 323
failure mode and effects analysis (FMEA)   110
Faujasite crystal   65
FCC gasoline   3
feed conditioning optimization   248–249
feed exchange energy   279
feed heater   278
feed procedure reaction   339
feed simulation   161
feed stage optimization   248, 250
feedstock sources and technologies   9
feed stream separation   303–304
Fenske equation   306
"final" simulation   121
firebox   193
fired heater assessment
   efficient fired heater operation
      air preheat effects   200
      availability and efficiency   200
      draft effects   199–200
      excess air and reliability   200
      guidelines   200–201
      O_2 analyzer   198
      optimize excess air   198–199
   fired heater design, for high reliability
      burner selection   192–193
      burner to tube clearance   192
      flux rate   189–192
      fuel conditioning system   193
   fired heater revamp   201–202
   operation for high reliability
      bridge wall temperature   194
      draft   194
      excess air or O_2 content   196–197
      flame impingement   196
      flame pattern   197
      tube life   196
      tube wall temperature   194–196
   fired heater evaluation, for revamps
      burners
         flame length   127
         heat release and emission requirements   128

coil pressure drop
   reactor circuit heaters   127
   reboiler pressure drop   127
data required
   data sheets and drawings   124
   operating data   124
detailed analysis   125
heater design limitations   125
metallurgy
   changes, in design practice   126
   tube corrosion rates   126
preliminary analysis   124–125
radiant flux limits
   double-fired heaters   126
   single-fired heaters   125
stack
   revamp basic engineering design   128
   revamp recommendations for stacks   128
   revamp studies   128
tube thickness, using look-up tables   126–127
TWT limits   126
fired heaters   276
fixed bed reactors   334
flame envelope   192
flame impingement   196
flame pattern   197
flow coefficient (FC)   207
fluid catalytic cracking unit (FCCU)   8
fluidized bed reactors   334
flux distribution   191
fouling
   combined feed exchanger fouling
      chemical foulants   356
      particulate foulants   356–358
   extraction unit column fouling   360
   heater fouling   358
   line fouling   359–360
   mitigation, heat exchanger   276
   process heat exchanger fouling   358
   specialty reboiler tube fouling   358–359
fractional crystallization   18, 105
fractionated by-products   119
fractionation optimization   109–110
fractionation section, key indicators   279–280
freezing fouling   177

FTM *see* facilitated transport membrane (FTM)
fuel equivalent (FE) concept
  condensate and water   270–271
  energy intensity based on   265
  fuel factor   264
  power factor   265, 267–268
  steam factor   264–265, 268–270
"fuel gas long" situation   123
fundamental toluene methylation reactions   80
furnace operation   276–277

g

gas chromatography
  meta-xylene and *para*-xylene peaks   373–374
  nitrogen vs hydrogen/helium carrier gas   373
  unidentified components, impact of   374
  wash solvent interference   374
gasoline production   27
global phenol and acetone consumption   22
grand composite curves (GCC)   244
guideline energy performance (GEP)   271

h

hazard and operability (HAZOP)   110–111
heater efficiency energy   279
heater reliability energy   279
heat exchanger assessment
  basic concepts and calculations   171–173
  effects of velocity
    heat exchanger fouling assessment   183–185
    heat exchangers arrangement, in series or parallel   181–183
    rating assessment   179
    suitability assessment   179–181
  fouling
    estimate fouling factor (R_f)   177
    pressure drop due to fouling   177–178
    root causes of   176–177
  identifying performance
    avoiding poor design   186
    fouling resistances   185
    pressure drop monitoring   185–186
    U-value monitoring   185
  performance criterion, U-values
    actual U-value (U_A)   175–176
    clean U-value (U_C)   174–175
    controlling resistance   176
    reaction air cooler   174
    required U-value (U_R)   174
  pressure drop
    shell-side pressure drop   178
    tube-side pressure drop   178
heat exchangers, for revamps
  compressors   143–146
  data required   136
  hydraulics/piping   146–147
  possible recommendations   139–140
  pumps   141–143
  rating procedures   137–139
  special exchanger services   140–141
  thermal rating methods
    constant UA method   136–137
    using key variable relationships   137
  use, of operating data   139
heat-pumped C_3 splitter   158
heat-resistant polyimide resins   13
heat transfer rate   337
Heat Transfer Research, Inc. (HTRI)   137
heavy aromatics fractionation   87–88
"heavy desorbent" Parex   65
heavy-key (LK) component   305, 306
heavy oil distillation   311–312
height equivalent theoretical plate (HETP)   65
heterogeneous reaction   332
heterogeneous reactors   333
high capacity trays   316
high-purity *p*-diethyl benzene   65
Hitachi PTA process   53
HNK octane concentration   168
homogeneous reaction   332
homogeneous reactors   333
"hot vapor bypass" system   86
hybrid reaction   340–341
hydrocarbon
  feed   90
  separation   157
  usage   7
hydro-dealkylation   16
  of C_{2+} alkyl groups   29
hydrogen-to-hydrocarbon ratio (H_2/HC)   100, 122, 279
hydro-processing gas source   123
hydrotreated pygas   15

i

impeller configurations
  between-bearing configuration   205–206
  integrally geared configuration   206
injection molding   4
integrated refining and petrochemicals   57–60
internal pore diffusion limitation   331
internal system battery limit (ISBL)   292
isomerization
  of paraffins   23
  unit deheptanizer column   109–110
isomer type, separations by
  ethylbenzene   45–46
  meta-xylene   44–45
  ortho-xylene   45–46
  para-xylene
    adsorptive separation   41–44
    crystallization   40–41
iso-propyl benzene   14

k

Kern's correlation   178
key energy indicator (KEI)   275, 286
key indicators
  to controllers   287
  defined   277
  economic evaluation   283–285
  energy dashboard   285–287
  fractionation section   279–280
  operation parameter
    fouling mitigation, heat exchanger   276
    furnace operation   276–277
    optimizing energy reaction   275–276
    rotating equipment operation   277
    steam letdown flows   277
    turndown operation   277
  reaction section   278–279
  targets   280–283

key operating parameters
  feed temperature 290
  lower pressure 290
  overflash 290
  overhead temperature 291
  pump around 291
  reflux ratio 290
  stripping steam 290
kinetics 325

l

label claim 7
light cycle oil (LCO) 8
light desorbent 66
light-key (LK) component 305, 306
linear alkyl benzene sulfonate 49–50
liquid-liquid extraction (LLE) 15, 35–38
  hydrocarbon feed 90
  operating variables
    extractor recycle drag 92
    extractor recycle ratio 91
    extractor stages, number of 92
    primary solvent temperature to extractor 91–92
    secondary solvent 92
    stripper receiver temperature and pressure 92
    stripper stages, number of 92
    tertiary solvent 92
  separation technology 304, 320–321
liquid petroleum gas (LPG) 119
liquid-phase alkylation 13
liquid-phase ethylbenzene alkylation plants 13
LLE see liquid-liquid extraction (LLE)
logarithmic mean temperature difference (LMTD) 172

m

management of change (MOC) 301
market information
  alternative feeds, for aromatics 27–28
  aromatics synthesis, technologies 21–27
  benzene 13–14
    production technologies 14–16
    ethylbenzene/styrene 17
    integrated refining 57–60
    molecular weight, separations 39

para-xylene 17–18
  production technologies 18–19
technologies, in aromatic separations
  extractive distillation 35–39
  liquid-liquid extraction 35–38
technologies, in aromatic transformation
  STDP 30–31
  thermal hydro-dealkylation 31–32
  transalkylation 28–30
  xylene isomerization 32–35
toluene 16–17
  production technologies 17
mass transfer 332–333, 337
McCabe–Thiele diagram 164
McDermott 13
MCM-49-based catalysts 13
medium quality terephthalic acid (MTA) 53
mega-plants 11
melt-phase polymerization
  of PET resin 55–57
  of PTA 53–55
membrane, separation technology 305, 322–323
meta-xylene 4, 18, 20, 347, 364
  production technologies 20
meta-xylene diamine (MXDA) 20
methodology, process integration
  column integration, benefit of 256
  column split
    extract column, with thermal coupling 252–253
    xylene column, with thermal coupling 252
  DWC 253–255
  heat pump, for paraxylene column 255–256
  indirect column heat integration 256, 257
  light desorbent, use of 255
  power recovery 257–259
  process-process stream heat integration 256
methyl cyclo-pentane (MCP) 14, 46
metric tons per annum (MTA) 106
microfiltration (MF) membrane 322
minimum continuous stable flow (MCSF) 230
mitigating catalytic reaction 331–332

Mitsubishi HF/BF$_3$ adduct technology 20
mixed xylenes 101
Mobil methanol to gasoline (MTG) process 28
Mobil selective toluene disproportionation (MSTD) process 78
molecular weight (MW) 203
  separations 39
monomers 11
moving bed reactors 334
multistage beam-type compressor 203–204
multistage centrifugal and reciprocating compressors 203
multistage integral geared compressors 204–205
multi-tube reactors 14, 57
MX Sorbex™ process 20

n

nanofiltration (NF) membrane 322
naphtha 3, 11
  components 15
2,6-naphthalene dicarboxylic acid (NDCA) 20
naphtha reforming, of C$_6$ and C$_7$ components 15
naphthene bridge 34, 35
naphthenes 15
  dehydrogenation reactions 23
  ring opening and cracking 30
natural and synthetic fiber demand 5
net positive suction head (NPSH) 226–229
net positive suction head available (NPSH$_A$) 226–227
net positive suction head required (NPSH$_R$) 227
N-formylmorpholine (NFM) 36, 90
nitration-grade benzene 14
nonaromatics 18, 364
nonelementary reaction rate law 327
normal boiling point
  C$_8$ aromatic isomers 66
  C$_9$ aromatics 66
  desorbent components 66
NO$_x$ emission 192
n-paraffins 3, 41

o

O'Connell correlation 165
off-line optimization 299–300
olefinic building blocks 4
Olefin Reduction Process™ (ORP™) 68
olefins 15
once-through reboiler 308
onion diagram 237
online optimization 300–301
operational flexibility 108–109
operation assessment 297–298
operation parameter, key indicators
  fouling mitigation, heat exchanger 276
  furnace operation 276–277
  optimizing energy reaction 275–276
  rotating equipment operation 277
  steam letdown flows 277
  turndown operation 277
order of reaction 326
organic rankine cycle (ORC) 252
ortho-xylene 347
  production technologies 20–21
outside system battery limit (OSBL) 292
overall fouling resistance 171
overflash 290
oxidation
  of *ortho*-xylene 57
  of *para*- and *meta*-xylene 50–53

p

paraffins 15
  cyclization 23
parameter optimization 291–292
para-selective catalytic technologies, for *para*-xylene
  para-selective toluene disproportionation 78–79
  para-selective toluene methylation 79–81
para-STDP technologies 16
para-xylene 4, 347, 348, 384–385
  capacity growth 19
  with cations 65
  production 386–388
  production technologies 18–19
  purification and recovery, by crystallization 68–71
  selective adsorption of water 367
  separation unit 277–278, 281
  xylene isomerization 381
parex adsorption affinities 65
parex-isomar recycle loop 101
parex raffinate stream 77
particulate fouling 177
peers energy performance (PEP) 271–272
pelletized crystalline silicon phosphate 14
Peng–Robinson equation 157
per-capita consumption, of fibers 6
"performance" indicator 173
petrochemical complex, operating issues
  aromatics extraction unit 377
  aromatics losses
    bottoms liquid product 352
    distillate liquid product 351–352
    extraction losses 352
    feed composition 352–353
    foaming 353
    fractionation and separation losses 351–352
    olefin removal 350–351
    reaction losses 353–354
    transalkylation unit losses 355
    variables 352
    vent losses 351
    xylene isomerization unit losses 354–355
  design and maintenance 347
  equipment and instrumentation failures 347
  fouling 356–360
  future of 377
  licensor's experience 347
  loss of production 347
  methyl group losses
    fractionation and separation losses 356
    reaction losses 356
  para-xylene, selective adsorption of
    flow meter monitoring 368–369
    heavy aromatics 366–367
    heavy desorbent 363, 364
    hydration monitoring 369–371
    meta-xylene contamination 366
    olefins 366
    oxygenates 366
    purity and recovery relationship 365–366
    Rotary Valve™ monitoring 367–368
    shutdown and restart considerations 371
    water 367
  sampling and laboratory analysis
    atmospheric contamination 372
    bromine index analysis for olefin measurement 371–372
    gas chromatography 372–374
    unstabilized liquid samples 372
  selective adsorption *para*-xylene separation unit 374–375
  solvent degradation
    caustic NaOH 361
    chloride, impact of 362
    MEA, benefit of 361
    oxygen and oxygenates 361–362
    sample analysis 363
    temperature 362
  start-up considerations
    catalyst attenuation 349
    catalyst reduction 348
    catalyst sulfiding 348–349
    modern transalkylation and isomerization catalysts 348
  suboptimal performance 347
  transalkylation unit 376–377
  xylene isomerization unit 375–376
petrochemical configurations 8–11
petrochemical production
  process designs and operation 11
  refinery integration 10
petrochemical products
  economic significance, of polymers 4–7
  petrochemicals and petroleum utilization 7–8
  phenolic resins 4
  polycarbonate 4
  polyester 4
  polyethylene 3
  polypropylene 3–4
  polystyrene 4
  styrene 4
phase-equilibrium process 320
phenol 3
phenolic resins 4
phosphoric acid 14
physical state reactant 332
piping and instrumentation diagrams (P&IDs) 110

platforming   3
plug flow reactor (PFR)   333, 336
polyalkyl compounds   13
polycarbonate (PC) resin   14
polyester   4
polyethylene   3
polyethylene naphthalate (PEN)   20
polyethylene terephthalate (PET)   3
   value chain and growth rates   19
polymers, economic significance of   4–7
polypropylene   3–4
polystyrene   4
polyurethane applications   17
poly vinyl chloride (PVC)   21
positive displacement compressors   203
pre-exponential factors   340
pre-investment   7
pressure differential indicating controller (PDIC)   86
primary solvent temperature to extractor   91–92
process design, for revamp projects
   adjusting operating conditions   121–122
   design margin   122–123
process hazard analysis (PHA)   110
process integration   259–261
   appropriate placement
      for distillation column   246–249
      general principle   244
      for reaction process   245–246
      for utility   245
   composite curves   239
      basic pinch concepts   239–240
      cost targeting   240–244
      energy use targeting   240
      pinch design rules   240
   definition   237–238
   GCC   244
   methodology   251–259
   systematic approach   249–251
process optimization, distillation
   deisopentanizer   296–297
   model   291
   objective function   298–299
   off-line optimization results   299–300
   online optimization   300–301
   operation assessment   297–298
   optimization implementation   300
   simulation   298
   sustaining benefits   301–302

products condenser   100
profile assessment, operation
   composition profile   167
   flow profile   166
   guidelines   169–170
   LK toluene   168
   pressure drop profile   167
   temperature and pressure profiles   166, 167
   toluene-ethylbenzene separation   168
   tower rating assessment   168–169
propylene   3
   ratios   14
pump arrangement
   parallel arrangement   225–226
   series arrangement   225
pump assessment
   BEP   224–225
   NPSH
      margin   227–228
      measuring pumps   228–229
      $NPSH_A$, calculation of   226–227
      potential causes and mitigation   229
   pump arrangement
      parallel arrangement   225–226
      series arrangement   225
   pump characteristics
      pump curve   222–224
      system curve   221–222
   pump control   230–231
   pump head   215–216
      Bernoulli equation   216–218
      calculation of   218–219
   pump selection and sizing   231–233
   ROE   230
   spillback   229–230
   total head calculation   219–220
pump head   215–216
   Bernoulli equation   216–218
   calculation of   218–219
pump suction   228
purified isophthalic acid (PIA)   19
purified terephthalic acid (PTA)   4, 17
pyrolysis gasoline by-product   8
pyro-mellitic dianhydride   13

q

Q-Max™ process   14, 21

r

radiant flux limit   122
radiant heat flux   189
raffinate stream   90
rate equation based surface kinetics   328–330
rate-limiting step   337
reaction catalyst   325–326
reaction kinetics
   catalyst deactivation   341–343
   catalytic reaction limitation
      external diffusion   331
      important parameters reaction   332–333
      internal pore diffusion   331
      mitigating limitations   331–332
      surface reaction   331
   hybrid reaction and separation   340–341
   modeling basics   326–328
   rate equation based surface kinetics   328–330
   reaction basics   325–326
   reactor design
      considerations   338
      general guidelines   338–340
      objective   335
      pressure, reaction conversion, and selectivity   336
      rate-limiting step   337
      reaction time and reactor size   336–337
      temperature and equilibrium constant   335
   reactor types
      classification   333–334
      practical types of reactors   334–335
reaction mechanism   327–328, 337
reaction rate law   325
reaction section, key indicators
   energy needs   278–279
   process   278
reaction system   237
reactive distillation   342
reactor activity energy   279
reactor design   326
reactors, for revamps
   data required   131
   fractionator evaluation   133–135
   high-capacity trays   135–136
   possible recommendations   133
   process and other modifications   132–133

reactor process evaluation   131
   design pressures and temperatures   132
   flow distribution   131–132
   materials of construction   132
   pressure drop   132
   test run data   133
reboiler configurations   307–308
reboiler pumps   108
recovery column reflux/distillate ratio   95
recycle gas compressor   100, 278–279
recycle gas purity   100
recycle stream
   optimizations   109–110
reducing exposure, to hazardous materials   110
reflux rate optimization   248
reflux ratio (R/D)
   energy optimization   294–295
   key operating parameters   290
reformate C_8 aromatics, composition of   33
reformate product   14
reformate splitter   85–86, 109
Reid vapor pressure (RVP)   297
relative volatility   316
reliability operating envelope (ROE)   230
renewable identification numbers (RINs)   7
research octane number (RONC)   91
"revamp" simulation   121
reverse osmosis (RO) membrane   322
reversible reaction   327
ring saturation   30
rotary valve (RV)   319–320
Rotary Valve™ monitoring
   alignment   368
   dome pressure   367–368
   maintenance   368
rotating equipment operation   277
RZ Platforming   15

s

selective toluene disproportionation (STDP)   16, 30–31
semi-batch reactor   333–334
separation technology
   adsorption   304, 316–317
   crystallization   305, 320
   distillation   304
      column internals   315–316
      difficulty of separation   305–306
      DWC   312–315
      heavy oil distillation   311–312
      operating pressure selection   306–307
      optimization of design   308–310
      reboiler configurations types   307–308
      side products   310–311
   feed stream   303–304
   liquid-liquid extraction   304, 320–321
   membranes   305, 322–323
   method   323–324
   process   304
   simulated moving bed   316–317
      chromatography   305
      concept of moving bed   318–319
      design   319
      rotary valve   319–320
   single-stage separation   304
sequential design approach   238
shale oil sources   7
shell-side pressure drop   178
side condensing/reboiling optimization   249
significant value chains   4
simulated moving bed (SMB)   18
   analogy   318
   chromatography   305
   design   319
   isomers   317
   rotary valve   319–320
simulation fidelity   157
single-stage separation   304
slurry reactors   334–335
SMART™   17
solid catalysts   333
solid phosphoric acid (SPA) catalysts   14, 21
solid state polycondensation (SSP)   55
   crystallinity development   56
   of PET resin   55–57
solid-state polymerization   5
solid zeolitic adsorbent   105
solvent extraction   320–321
spillback   229–230
start-of-run (SOR)   98
steady-state approximation   327
steam letdown flows   277
stirred-tank reactor   333
straight flow-through compressor   206
straight-run naphtha (SRN)   22
   composition   23
stripping steam   290
   ratio   95
styrene   4
sulfolane aromatics extraction   90
sulfolane-based ED process   36
sulfolane-based unit   95
sulfolane solvent selectivity vs. HC type   37
sulfolane solvent system   90
"Sundyne" compressor   204, 205
Sundyne pumps   230
surface reaction limitation   331

t

tatoray process   28
temperature constant   335
tetra-ethylene glycol (TEG)   36
1,2,4,5-tetra-methyl benzene   13
tetramethylene sulfone or 2,3,4,5-tetrahydrothiophene-1,1-dioxide   90
"Tetris-style" diagram   256, 258
theoretical flame temperature (T_{TFT}) 245
thermal hydro-dealkylation   31–32
thermodynamics   325
thermosiphon reboiler   307–308
Thiele modulus (M_T) 329
tire cord   4
toluene   16–17, 347
   production technologies   17
toluene column   109
toluene di-isocyanates (TDI's)   16
toluene disproportionation (TDP)   73
toluene methylation (TM)   11
tower evaluation software   169
tower optimization, distillation
   key operating parameters   289–291
   parameter optimization   291–292
   parameter relationship   291
   relax soft constraints   292–293
traditional design approach   238

transalkylation   71–72, 379–381
   of polyisopropyl benzenes   14
   process flow description
      bed pressure drop   99–100
      catalyst volume   99
      CFE   97–98
      charge heater   98–99
      products condenser   100
      reactor bed dimensions   100
      reactor design   99
      recycle gas compressor   100
      recycle gas purity   100
      separator   100
   yields *vs.* feed C_9 content   15
transmethylation   29
trickle bed reactors   334
tri-ethylene glycol (Tri-EG)   35
tri-mellitic anhydride   13
1,2,4-tri-methyl benzene   13
troubleshooting
   aromatics extraction   385–386
   para-xylene   384–385
      production   386–388
   transalkylation   379–381
   xylene isomerization   381–384
tube metal temperature (TMT)   193
tube-side pressure drop   178
tube thinning   192
tube wall temperature (TWT)   122, 194–196, 276, 292
tubular-flow reactor   333
turndown operation   277

u

ultrafiltration (UF) membrane   322
underwater melt cutting (UMC)   55

UOP cyclar process chemistry   26
UOP isomar process flow diagram   75
UOP RV   319–320
UOP's THDA™ process   16
UOP Tatoray™ process   15
US refining industry   7
utility pinches   245
UZM-8™ zeolites   13

v

vacuum gas oil (VGO)   8
van der Waals force   316
vapor–liquid equilibrium   306
vapor-phase alkylation technology   13
vapor to-liquid ratio *(V/L)*   280
vessels:separators, receivers, and drums
   data required   128
   design pressures and temperatures   129
   materials of construction   129
   possible recommendations   130
   process and other modifications   129–130
   residence time   129
   test run data   130
   vapor liquid separation   128–129

w

weight hourly space velocity (WHSV)   99
wicket-type heaters   98

world hydrocarbon consumption   7
World's toluene consumption   16

x

xylene
   column, with thermal coupling   252
   fractionation   86–87
   impurities   13
   isomers   28
   properties of   317
xylene isomerization   32–35, 72–76, 381–384
   catalyst volume   104
   CFE   102
   charge heater   102–103
   deheptanizer column in   281
   products condenser   104
   radial flow reactor sizing   104
   reactor design   103–104
   recycle gas compressor   105
   recycle gas purity   104–105
   separator   104
   unit flow diagram   101
   UOP aromatics complex   277–278
Xylenes Plus   28

y

yield estimating model(s)   121
Y-zeolite catalysts   13

z

zeolite catalysts   13, 14
zeolitic cumene technologies   14
ZSM-5 zeolite   13